RGANIC AND
IOLOGICAL CHEMISTRY

Table of Atomic Masses and Numbers

Based on the 1989 Report of the Commission on Atomic Weights and Isotopic Abundances of the International Union of Pure and Applied Chemistry and for the elements as they exist naturally on earth. Scaled to the relative atomic mass of carbon-12. The estimated uncertainties in values, between ±1 and ±9 units in the last digit of an atomic mass, are in parentheses after the atomic mass. (From *Journal of Physical and Chemical Reference Data*, Vol. 20 (1991), pp. 1313–1325. Copyright © 1991 IUPAC.)

Element	Symbol	Atomic number	Atomic mass	
Actinium	Ac	89	227.0278	(L)
Aluminum	Al	13	26.981539(5)	
Americium	Am	95	243.0614	(L)
Antimony	Sb	51	121.757(3)	
Argon	Ar	18	39.948(1)	(g, r)
Arsenic	As	33	74.92159(2)	
Astatine	At	85	209.9871	(L)
Barium	Ba	56	137.327(7)	
Berkelium	Bk	97	247.0703	(L)
Beryllium	Be	4	9.012182(3)	
Bismuth	Bi	83	208.98037(3)	
Boron	B	5	10.811(5)	(g, m, r)
Bromine	Br	35	79.904(1)	
Cadmium	Cd	48	112.411(8)	(g)
Calcium	Ca	20	40.078(4)	(g)
Californium	Cf	98	251.0796	(L)
Carbon	C	6	12.011(1)	(r)
Cerium	Ce	58	140.115(4)	(g)
Cesium	Cs	55	132.90543(5)	
Chlorine	Cl	17	35.4527(9)	(m)
Chromium	Cr	24	51.9961(6)	
Cobalt	Co	27	58.93320(1)	
Copper	Cu	29	63.546(3)	(r)
Curium	Cm	96	247.0703	(L)
Dysprosium	Dy	66	162.50(3)	(g)
Einsteinium	Es	99	252.083	(L)
Erbium	Er	68	167.26(3)	(g)
Europium	Eu	63	151.965(9)	(g)
Fermium	Fm	100	257.0951	(L)
Fluorine	F	9	18.9984032(9)	
Francium	Fr	87	223.0197	(L)
Gadolinium	Gd	64	157.25(3)	(g)
Gallium	Ga	31	69.723(4)	
Germanium	Ge	32	72.61(2)	
Gold	Au	79	196.96654(3)	
Hafnium	Hf	72	178.49(2)	
Helium	He	2	4.002602(2)	(g, r)
Holmium	Ho	67	164.93032(3)	
Hydrogen	H	1	1.00794(7)	(g, m, r)
Indium	In	49	114.82(1)	
Iodine	I	53	126.90447(3)	
Iridium	Ir	77	192.22(3)	
Iron	Fe	26	55.847(3)	
Krypton	Kr	36	83.80(1)	(g, m)
Lanthanum	La	57	138.9055(2)	(g)
Lawrencium	Lr	103	262.11	(L)
Lead	Pb	82	207.2(1)	(g, r)
Lithium	Li	3	6.941(2)	(g, m, r)
Lutetium	Lu	71	174.967(1)	(g)
Magnesium	Mg	12	24.3050(6)	
Manganese	Mn	25	54.93805(1)	
Mendelevium	Md	101	258.10	(L)
Mercury	Hg	80	200.59(2)	
Molybdenum	Mo	42	95.94(1)	(g)
Neodymium	Nd	60	144.24(3)	(g)
Neon	Ne	10	20.1797(6)	(g, m)
Neptunium	Np	93	237.0482	(L)
Nickel	Ni	28	58.6934(2)	
Niobium	Nb	41	92.90638(2)	
Nitrogen	N	7	14.00674(7)	(g, r)
Nobelium	No	102	259.1009	(L)
Osmium	Os	76	190.2(1)	(g)
Oxygen	O	8	15.9994(3)	(g, r)
Palladium	Pd	46	106.42(1)	(g)
Phosphorus	P	15	30.973762(4)	
Platinum	Pt	78	195.08(3)	
Plutonium	Pu	94	244.0642	(L)
Polonium	Po	84	208.9824	(L)
Potassium	K	19	39.0983(1)	
Praseodymium	Pr	59	140.90765(3)	
Promethium	Pm	61	144.9127	(L)
Protactinium	Pa	91	231.03588(2)	
Radium	Ra	88	226.0254	(L)
Radon	Rn	86	222.0176	(L)
Rhenium	Re	75	186.207(1)	
Rhodium	Rh	45	102.90550(3)	
Rubidium	Rb	37	85.4678(3)	(g)
Ruthenium	Ru	44	101.07(2)	(g)
Samarium	Sm	62	150.36(3)	(g)
Scandium	Sc	21	44.955910(9)	
Selenium	Se	34	78.96(3)	
Silicon	Si	14	28.0855(3)	(r)
Silver	Ag	47	107.8682(2)	(g)
Sodium	Na	11	22.989768(6)	
Strontium	Sr	38	87.62(1)	(g, r)
Sulfur	S	16	32.066(6)	(g, r)
Tantalum	Ta	73	180.9479(1)	
Technetium	Tc	43	98.9072	(L)
Tellurium	Te	52	127.60(3)	(g)
Terbium	Tb	65	158.92534(3)	
Thallium	Tl	81	204.3833(2)	
Thorium	Th	90	232.0381(1)	(g)
Thulium	Tm	69	168.93421(3)	
Tin	Sn	50	118.710(7)	(g)
Titanium	Ti	22	47.88(3)	
Tungsten	W	74	183.85(3)	
Unnilennium	Une	109	(266)	(L, n, s)
Unnilhexium	Unh	106	263.118	(L, n)
Unniloctium	Uno	108	(265)	(L, n, s)
Unnilpentium	Unp	105	262.114	(L, n)
Unnilquadium	Unq	104	261.11	(L, n)
Unnilseptium	Uns	107	262.12	(L, n)
Uranium	U	92	238.0289(1)	(g, m)
Vanadium	V	23	50.9415(1)	
Xenon	Xe	54	131.29(2)	(g, m)
Ytterbium	Yb	70	173.04(3)	(g)
Yttrium	Y	39	88.90585(2)	
Zinc	Zn	30	65.39(2)	
Zirconium	Zr	40	91.224(2)	(g)

(g) Geologically exceptional specimens of this element are known that have different isotopic compositions. For such samples, the atomic mass given here may not apply as precisely as indicated.

(L) The atomic mass is the relative mass of the isotope of longest half-life. The element has no stable isotopes.

(m) Modified isotopic compositions can occur in commercially available materials that have been processed in undisclosed ways, and the atomic mass given here might be quite different for such samples.

(n) Name and symbol are assigned according to systematic rules developed by the IUPAC.

(r) Ranges in isotopic compositions of normal samples obtained on earth do not permit a more precise atomic mass for the element, but the tabulated value should apply to any normal sample of the element.

(s) Element was not listed in the 1989 report but has been added here.

ORGANIC AND BIOLOGICAL CHEMISTRY

JOHN R. HOLUM

JOHN WILEY & SONS, INC.
New York Chichester Brisbane Toronto Singapore

Acquisitions Editor	Nedah Rose
Marketing Manager	Catherine Faduska
Production Manager	Charlotte Hyland
Designer	Kevin Murphy
Manufacturing Manager	Mark Cirillo
Photo Researcher	Lisa Passmore
Illustration Coordinator	Sigmund Malinowski
Cover Photo	Sidney Moulds/Science Photo Library/Photo Researchers, Inc.
Production Management	Ingrao Associates

This book was set in ITC Souvenir by Progressive Typographers and printed and bound by Von Hoffman Press, Inc. The cover was printed by The Lehigh Press, Inc.

Film preparation was done by Professional Litho.

Recognizing the importance of preserving what has been written, it is a policy of John Wiley & Sons, Inc. to have books of enduring value published in the United States printed on acid-free paper, and we exert our best effort to that end. The paper in this book was manufactured by a mill whose forest management programs include sustained yield harvesting of its timberlands. Sustained yield harvesting principles ensure that the number of trees cut each year does not exceed the amount of new growth.

Library of Congress Cataloging-in-Publication Data

Holum, John R.
[Fundamentals of general, organic, and biological chemistry. Chapters 11–29]
Organic and biological chemistry / John R. Holum.
p. cm.
"This book consists of the last 18 chapters of a longer book, Fundamentals of General, Organic, and Biological Chemistry, 5th ed." —pref.
Includes bibliographical references and index.
ISBN 0-471-12972-0 (cloth : alk. paper)
1. Biochemistry 2. Chemistry, Organic. I. Title.
QP514.2.H643 1996
574.19'2—dc20 95-32942
CIP

Printed in the United States of America

10 9 8 7 6 5 4 3 2 1

PREFACE

This book consists of the last 18 chapters of a longer book, *Fundamentals of General, Organic, and Biological Chemistry,* 5th ed, and is offered as a text for a special course. In many colleges and universities, students intending on careers as allied health care professionals are in a regular general chemistry course for the first term and then are moved into a one-term course that surveys the fields of organic and biological chemistry. This book is for that course.

THE CENTRAL THEME IS THE MOLECULAR BASIS OF LIFE

Two types of students study the material offered here. Those of one type are preparing for careers in the health sciences other than that of a physician. Those of the second type include many who see their futures outside of science but are interested in how human life and parts of the environment work at their molecular levels. This book is suitable for both types of students because of its overarching theme, the molecular basis of life. I have allowed no topics that do not fit the theme or serve as background material for topics that do.

NEARLY ALL CHAPTERS ARE ENRICHED BY SPECIAL TOPICS

Students Crave Relevancy. The question, "why am I studying this?" comes up often. Many teachers anticipate the question and patiently answer it for each chapter and sometimes for each section of a chapter. Another tactic is to sprinkle "asides" into lectures that show the student how a basic topic applies in everyday life or to environmental concerns. Many such "asides" are in this book, both in the main body of the text and in Special Topics. The Special Topics greatly enrich the course but, for the most part, do not constitute core material. Their titles and locations can be found at the end of the Table of Contents. Special Topics include environmental concerns (e.g., "Ozone in Smog"), aspects of personal health (e.g., "Good and Bad Cholesterol"), items of current news interest (e.g., "DNA Typing") and even some organic reaction mechanisms.

Chapters 1–7 Survey Only Those Topics of Organic Chemistry Essential to the Study of Biochemistry The wedding between the theme of the course and the limitation of time results in a very abbreviated survey of organic chemistry. Some of the major topics developed in a one-term course of organic chemistry have had to be excluded, topics such as the theory of resonance, nucleophilic substitution reactions, the Grignard synthesis, and many others.

I have stressed only those functional groups that occur widely among the molecules of life and the reactions of such groups with four kinds of compounds: acids, bases, oxidizing agents, and reducing agents. I have provided some mechanisms because organic reactions otherwise seem too much like magic, and their learning becomes merely rote. Two major reactions that form carbon–carbon bonds, the aldol condensation and the Claisen condensation, are studied in Special Topics only, and these occur not among the organic chemistry chapters but, untraditionally, nearest their potential applications to metabolism.

Chapters 8–18 Constitute One Illustration after Another of the Molecular Basis of Life Carbohydrates, lipids, and proteins begin this section of the book. Because of their importance to all that follows, I next take up enzymes, hormones, neurotransmitters, and the extracellular fluids of the body. A study of nuclei acids completes the survey of the chief kinds of substances in the body and their particular settings.

The citric acid cycle and the respiratory chain serve the metabolism of all types of biochemicals, so these pathways are next studied. Then come treatments of the metabolism of carbohydrates, lipids, and proteins. Finally, after developing a knowledge of the structures, properties, and uses of substances in cells, I close with nutrition and the sources of nutrients.

- Chapters 9 ("Lipids") and 10 ("Proteins") offer an extensive treatment of cell membranes. The lipid components are in Chapter 9 and are used to introduce membrane structure, but the equally important glycoprotein components are in Chapter 10 where cell membranes are revisited. The glycoproteins provide recognition sites for hormones, neurotransmitters, and certain drugs, so Chapter 10 has a special topic of considerable current interest: "Mifepristone (RU 486)—Receptor Binding of a Synthetic Antipregnancy Compound."
- Chapter 10 ("Proteins"), besides the material on cell membranes, also includes information on how cartilage works and how gap junctions between cells enable the synchronized actions by many cells in certain tissues (e.g., the heart).
- Chapter 11 includes recent developments in the area of retrograde chemical messengers such as nitric oxide and carbon monoxide. A special topic—"Molecular Complementarity and Immunity, AIDS, and the ABO Blood Groups"—applies the concepts of Chapter 11 and earlier material on cell membranes to topics of general interest.
- Chapter 12 ("Extracellular Fluids of the Body") includes topics that are *the most important in the entire book for future nurses.* At least the acid–base status of the blood is so regarded by experienced emergency room and cardiac care nurses. This topic is virtually the only one in the course that reaches well beyond background applications for other nursing courses to the lifetime careers of nurses.

  Nearly all medical emergencies involve serious changes in the acid–base status of the blood. The terminology and quantitative values associated with this status are widely encountered both in professional nursing work and in the literature nurses ought to be reading throughout their careers. Where nursing students make up the majority of a class, the acid–base status of the blood must be taught as thoroughly as possible. People going into inhalation therapy or into any aspect of sports medicine also need this material.

  If presented in the right way, liberal arts students with any interest in sports will be fascinated by the subject (and so motivated to learn it). I have given January interim courses on the chemistry of respiration, where all of the students entered the class simply (and reluctantly) to satisfy a general education requirement. But some were athletes and many had enjoyed strenuous outdoor activities, like mountain skiing or wilderness trekking. Grudgingly, many became "believers," and one senior was a bit rueful over missing the opportunity to "go into chemistry." Imagine that!

- Chapter 13 ("Nucleic Acids") includes the factors that stabilize duplex DNA, recent developments in gene therapy, the defective gene in cystic fibrosis, and briefly describes the Human Genome Project.
- Chapter 14 ("Biochemical Energetics") contains a modern treatment of the respiratory chain but in a simplified manner. Exciting details on how ATP forms and is then released from its enzyme are included.

- Chapter 15 ("Metabolism of Carbohydrates") has Special Topics on the aldol condensation and on diabetes.
- Chapter 16 ("Metabolism of Lipids") includes the lipoprotein complexes and "good" and "bad" cholesterol.
- Chapters 17 ("Metabolism of Nitrogen Compounds") and 18 ("Nutrition") conclude the text.

The Design of This Text Enhances Student Learning There are frequent **margin comments** to restate a point, offer data, or simply remind.

Key terms are highlighted in boldface at those places where they are defined and then discussed. A complete **glossary** of these terms plus a few others appears at the end of the book. The ***Study Guide*** also has individual chapter glossaries.

Each section of a chapter begins with a **headline.** This is *not* a one-sentence summary of the section but rather a lead-in to the beginning of the section that tries to state the section's major point.

Each chapter has a **Summary** that uses key terms in a narrative manner. The summaries are not necessarily organized in the same order in which the material occurs in the various chapter sections. The summaries assume that the sections have been studied so that the needed vocabulary is in place.

Many chapters have several **worked examples.** Nearly all are following by **Practice Exercises,** which encourage immediate reinforcements of skills learned in the examples. Answers to all Practice Exercises are in Appendix A. A copious number of **Review Exercises** closes each chapter, including some that are "additional;" they are not identified by topic. Many additional exercises require the use of material from earlier chapters. You will even find a stoichiometry problem here and there.

Three special **icons** are used to draw the attention of a student to places that emphasize various skills that should be mastered or that point out topics of interest from either a health or an environmental view.

This icon draws attention to discussions of special skills, such as predicting structures or naming compounds.

This is the "map sign" icon, and it appears almost exclusively in the organic chemistry chapters. I draw an analogy between the representations of functional groups in organic structures (like an alkene group) and the symbols used by map makers. Thus, we need only a few map symbols to enable us to read almost any map. Similarly, we can see a functional group symbol as representing a relatively short list of properties conferred on the substance. By knowing structural "map signs," we can "read" structural formulas like a map and predict some properties surprisingly well.

This icon, suggesting not only planet earth but also all people on it, draws attention to topics whereby applications of chemical knowledge are made to matters of health or earth-care.

SUPPLEMENTARY MATERIALS FOR STUDENTS AND TEACHERS

The complete package of supplements that are available to help students study and teachers plan includes the following:

***Laboratory Manual for Fundamentals of General, Organic, and Biological Chemistry,* fifth edition** This is a thoroughly revised edition prepared by Dr. Sandra Olmsted, Augsburg College. An *Instructor's Manual* to this laboratory manual is a section of the general Teacher's Manual described below.

***Study Guide for Fundamentals of General, Organic, and Biological Chemistry,* fifth edition** This softcover book contains chapter objectives, chapter glossaries, additional worked examples and exercises, sample examinations, and the answers to all of the Review Exercises. The last 18 chapters of this *Study Guide* match the chapters of this textbook.

Test Bank

Available in both hard copy and software (Macintosh© and IBM© compatible) versions, this test resource contains roughly several hundred questions.

Transparencies

Instructors who adopt this book may obtain from Wiley, without charge, a set of color transparencies that duplicate key illustrations from the text.

ACKNOWLEDGMENTS

My wife Mary has been my strongest supporter, and I am deeply grateful to this wonderful woman. My daughters, Liz, Ann, and Kathryn, now grown, also have been strong champions, and I thank them for what they have meant to Mary and me.

At Augsburg College, I have always enjoyed unstinting support from the Chair of the Chemistry Department, Dr. Earl Alton (now an Assistant Dean); from the Academic Dean, Dr. Ryan LaHurd (now the President of Lenoir Rhyne College); and from the President, Dr. Charles Anderson. Dr. Arlin Gyberg, Dr. Joan Kunz (now Chemistry Department Chair), and Dr. Sandra Olmsted of the Chemistry Department have been important sources of suggestions and corrections.

Extraordinarily nice people are all over the place at John Wiley & Sons. I think particularly of my Chemistry Editor, Nedah Rose, her Administrative Assistant, Marianne Stepanian, and my Supplements' Editor, Joan Kalkut.

The overall design was the responsibility of Kevin Murphy with whom I have worked with immense pleasure on this and other books. Sigmund Malinowski has been skillful, artistic, and faithful in handling the line drawing art work. Lisa Passmore, Associate Photo Editor, produced such a rich supply of outstanding choices for photographs that my choosing became difficult, yet exciting and pleasurable. My copy editor, handled her assignment with grace. Most of the production was supervised by Suzanne Ingrao of Ingrao Associates.

Outstanding proofreaders saved me from innumerable embarrassments—Dr. Melinda Lee (St Cloud State University) checked the answers to all of the Practice and Review Exercises, and she also prevented many glitches. It's hard to imagine that any errors remain, but, based on experience, no doubt some do. They are now entirely my responsibility. Please send a letter to my Chemistry Editor to let me know about them.

The professional critiques of many teachers are part of the process of preparing a manuscript. For their work on the current and previous editions of the longer book, *Fundamentals of General, Organic, and Biological Chemistry,* I am most pleased to acknowledge and to thank the following people again.

Robert Ake
Old Dominion University

Muriel Bishop
Clemson University

Jack Dalton
Boise State University

Donald Harriss
University of Minnesota, Duluth

Herman Knoche
University of Nebraska, Lincoln

Margaret Asirvatham
University of Colorado, Boulder

Lorraine Brewer
University of Arkansas

Arlin Gyberg
Augsburg College

Larry Jackson
Montana State University

John Meisenheimer
Eastern Kentucky University

Sandra Olmsted
Augsburg College

Fred Schell
University of Tennessee, Knoxville

Kent Thomas
Kansas-Newman College

Justine Walhout
Rockford University

Nancy Paisley
Montclair State University

Ram Singhal
Wichita State University

Atilla Tuncay
Indiana University Northwest

Leslie Wynston
California State University,
Long Beach

John R. Holum
Minneapolis, MN

CONTENTS

Index to Special Topics

ORGANIC AND BIOLOGICAL CHEMISTRY

ORGANIC CHEMISTRY. SATURATED HYDROCARBONS

1

Take a moment and marvel. All these people, and a few billion others, are individually unique "packages" of electrons, protons, and atomic nuclei (and actually the nuclei of only a handful of elements). In this chapter we begin the study of another major foundation of the molecular basis of life, the structures and properties of organic compounds.

1.1
ORGANIC AND INORGANIC COMPOUNDS

The major differences between organic and inorganic compounds stem from variations in composition, bond types, and molecular polarities.

Organic compounds are compounds made of carbon atoms covalently bonded to each other and to atoms of other nonmetals, like hydrogen, oxygen, nitrogen, sulfur, or the halogens. All other compounds are called **inorganic compounds** but even they include a few that contain carbon, like the carbonates, bicarbonates, cyanides, and the oxides of carbon.

In the popular press, "organic" (as in "organic foods") has come to mean "produced without the use of pesticides or synthetic fertilizers or hormones." We will use the traditional meaning. **Organic chemistry** is the study of the structures, properties, and syntheses of organic compounds. In this and the next few chapters we will study only those parts of organic chemistry that develop the principles or reactions needed to our study of the molecular basis of life.

Wöhler's Experiment Opened the Doors to the Laboratory Synthesis of Organic Compounds The word *organic* arose from an association with living organisms, because in the early days of organic chemistry all organic compounds were isolated from living systems or their remains. Until 1828, all efforts to synthesize organic from inorganic compounds had failed, and out of such repeated failures the **vital force theory** emerged. It stated that it is actually impossible to make organic compounds in glass vessels, that a special *vital force* said to be found only in living systems was essential.

■ *Vita-* is from a Latin root meaning "life."

In 1828, while trying to make a sample of crystalline ammonium cyanate, NH_4NCO, Friedrich Wöhler (1800–1882) boiled the water from an aqueous solution containing the ammonium ion and the NCO^-, cyanate ion. The white solid he obtained, however, was an unexpected compound, urea. Ammonium cyanate, then as now, is regarded as an inorganic compound, but urea is clearly a product of metabolism. Wöhler had succeeded in making the first organic compound in a glass vessel. The heat used for boiling evidently caused the following reaction.

■ Urea is the chief nitrogen waste in the urine of animals. It is also manufactured from ammonia for use as a commercial fertilizer. Whether made by animals or machinery, urea is urea and is "organic." Plants will accept no fertilizer than what they are used to.

$$\underset{\text{Ammonium cyanate}}{NH_4NCO} \xrightarrow{\text{heat}} \underset{\text{Urea}}{H{-}\underset{\displaystyle H}{\overset{|}{N}}{-}\overset{\displaystyle O}{\overset{\|}{C}}{-}\underset{\displaystyle H}{\overset{|}{N}}{-}H}$$

Following Wöhler's discovery, other organic compounds were made from inorganic substances, and the vital force theory was dead. Today, well over six million organic compounds are known, and all have been or could in principle be made from inorganic substances. The significance to human well-being of the development of organic chemistry as a science is incalculable. Although large numbers of useful organic substances occur in nature and are still obtained from nature, one cannot imagine today's world of synthetic fabrics, dyes, and plastics as well as most pharmaceuticals without the opening to synthetic organic compounds that Wöhler discovered. Among all scientific specialties, there are more organic chemists than any other single kind of chemist. The education of those entering biochemistry, medicinal chemistry, pharmaceutical chemistry, polymer chemistry, molecular biology, and the primary health care fields of nursing and doctoring includes at its core the study of organic chemistry. One specialist at The Johns Hopkins School of Medicine flatly states that we cannot say we *know* what a disease is until we know its *chemistry,* and organic chemical principles are at the heart of this knowledge. The study of these principles begins with the kinds of *bonds* that hold organic molecules together.

Covalent Bonds, Not Ionic Bonds, Dominate Organic Molecules The overwhelming prevalence of nonmetal atoms in organic compounds means that their molecular structures are dominated by *covalent* bonds. In contrast, most inorganic compounds are *ionic.* As we will see, carbon–carbon and carbon–hydrogen bonds are the most prevalent in organic molecules. These bonds are essentially nonpolar, so organic compounds tend to be relatively nonpolar except when atoms of such electronegative elements as oxygen and nitrogen are present. These structural facts are behind several major properties of organic compounds, like melting and boiling points and solubility in water.

Most organic compounds have melting points and boiling points well below 400 °C, whereas most ionic compounds melt or boil far above this temperature. The reason is that the relatively nonpolar molecules in organic substances are unable to attract each other very strongly, in sharp contrast to the oppositely charged ions in ionic compounds. However, the *permanent* polarity of an organic molecule is only one factor that affects a boiling point or melting point. The *size* of the molecule is also a factor because the larger the size, the stronger are the forces of temporary or induced polarity, the *London forces,* between molecules. These depend on the overall size and shape of the large electron cloud of a molecule within which the molecule's nuclei are buried. The larger this cloud is, the more it can be distorted by random collisions with other molecules. Such distortions produce the transient polarities that also result in net forces of attractions between molecules. Large molecules, to repeat, generate larger London forces.

■ Organic *ions* tend to be very soluble in water.

1.2

SOME STRUCTURAL FEATURES OF ORGANIC COMPOUNDS

Organic molecules can have straight or branched chains; they can be open-chained or cyclic, saturated or unsaturated; and ring systems can be carbocyclic or heterocyclic.

The uniqueness of carbon among the elements is that its atoms can bond to each other successively many times and still form equally strong bonds to atoms of other nonmetals. A typical molecule in the familiar plastic polyethylene has hundreds of carbon atoms covalently joined in succession, and each carbon binds enough hydrogen atoms to fill out its full complement of four bonds.

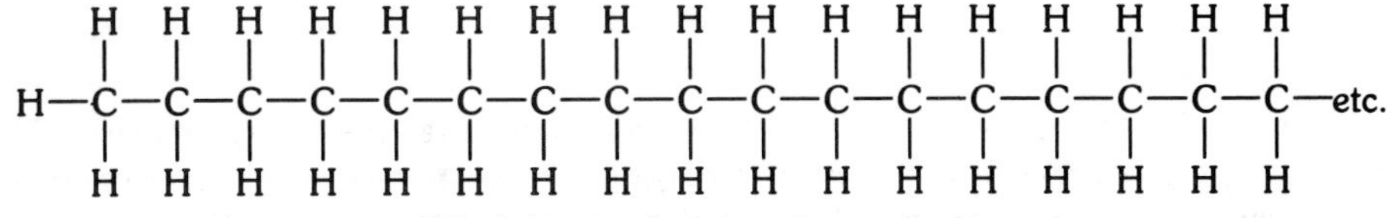

Polyethylene (small segment of one molecule)

■ Only a short segment of a typical molecule of polyethylene is shown here.

The sequence of the heavier atoms, here the carbon atoms, is called the *skeleton* of the molecule, and it holds the hydrogen atoms. Many variations of heavier-atom skeletons occur, and we will look at them next.

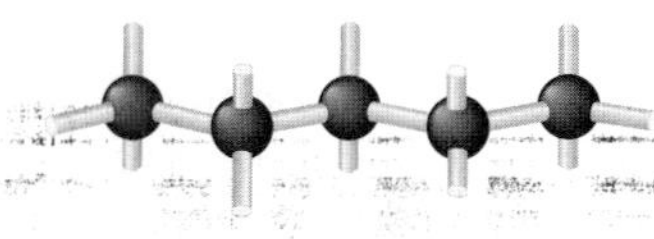

Straight chain

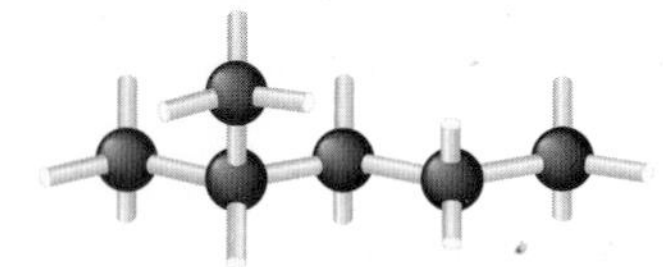

Branched chain

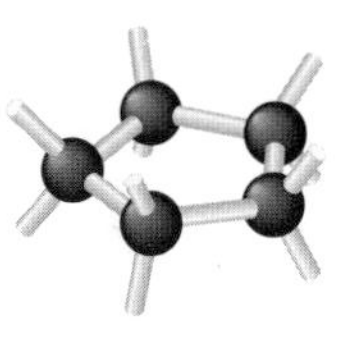

Carbon ring

Carbon Skeletons Can Be in Straight Chains or Branched Chains The carbon skeleton in the polyethylene molecule is described as a **straight chain.** *Straight* has a very limited and technical meaning here: the absence of carbon branches. This means that one carbon follows another, like the pearls in a single-strand necklace, with no additional carbons joined to the skeleton at intermediate points. Pentane illustrates a straight chain of five carbons. The 2-methylpentane molecule has a **branched chain,** a chain with at least one carbon atom joined to the skeleton between the ends of the main chain, like a charm hung on a bracelet.

```
                                      H
                                      |
                                    H-C-H
   H  H  H  H  H                  H   |   H  H  H
   |  |  |  |  |                  |   |   |  |  |
 H-C--C--C--C--C-H              H-C---C---C--C--C-H
   |  |  |  |  |                  |   |   |  |  |
   H  H  H  H  H                  H   H   H  H  H
```

Pentane
(straight chain)

2-Methylpentane
(branched chain)

```
                                      |
                                     -C-
   |  |  |  |  |                  |   |   |  |  |
  -C--C--C--C--C-                -C---C---C--C--C-
   |  |  |  |  |                  |   |   |  |  |
```

Pentane skeleton

2-Methylpentane skeleton

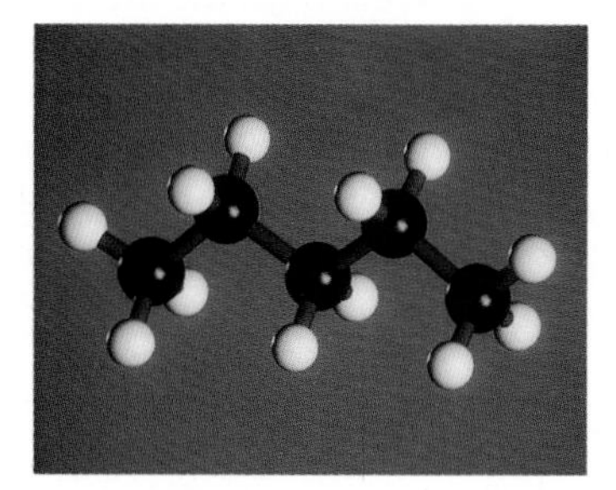

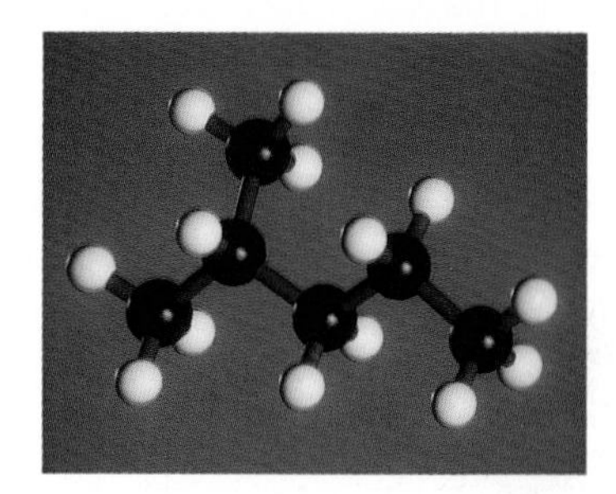

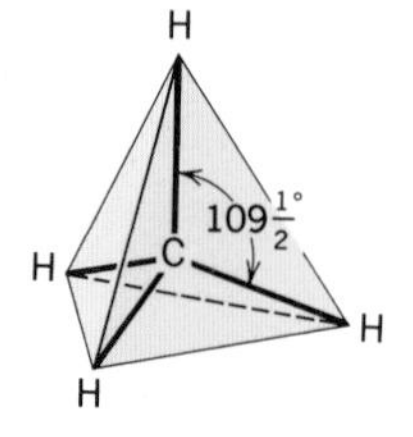

Tetrahedral carbon

Some Features of Molecular Geometry Are Often "Understood" when Writing Structural Formulas When you compare the ball-and-stick models of pentane and 2-methylpentane with their structural formulas, be sure to notice that the written (or printed) structures disregard the correct bond angles at carbon. A carbon that has four single bonds has a tetrahedral geometry with bond angles of 109.5°. The ball-and-stick models faithfully show the correct angles at each carbon, but the printed symbols do not. The point here is that it is perfectly all right to let bond angles be "understood" unless there is some important reason to the contrary.

It is definitely not "all right" to forget about the *geometry* of molecules, however. When we enter the study of biochemistry, particularly the way that enzymes work, we will quickly learn that the geometry of a molecule is just as important as any other feature of its structure. Molecules generally take up whatever shape is permitted by their *bonds* and the *electron clouds* surrounding the individual parts of the molecules. Bonds hold atoms together, and whenever a carbon atom has four *single* bonds, the geometry at that point of the molecule will be tetrahedral. The electron clouds tend to push the parts of molecules away from each other and so influence the *overall* shape of the molecule. This "pushing" cannot succeed in splitting stable molecules apart, but it has a huge influence on how a molecule becomes twisted in shape, at least in molecules that have the flexibility permitted by "free rotation," studied next.

Free Rotation at Single Bonds Is Also "Understood" in Structural Formulas The molecules of pentane and 2-methylpentane are quite flexible, like a necklace. Figure 1.1 shows models of just a few of the many ways the carbon skeleton of the pentane molecule can be flexed. These twistings actually occur as pentane molecules collide with each other in the liquid or gaseous states, and they illustrate an important property of at least large segments of organic molecules, a property called **free rotation** about single bonds. Two clusters of atoms held by a single bond can rotate with respect to each other about the bond.

■ Because of free rotation, we have to be able to interpret zigzags. For example,

$$\begin{array}{l} CH_3 \\ | \\ CH_2CH_2CH_2 \\ \quad\quad\quad\quad | \\ \quad\quad\quad\ CH_3 \end{array}$$

is the same molecule as

$$CH_3CH_2CH_2CH_2CH_3$$

Free rotation is possible because the single bond is a *sigma bond* as described in Figure 1.2. The *strength* of a sigma bond lies in the degree of overlap of the hybrid orbitals that make up the molecular orbital. As long as the degree of such overlap is not affected too much while one group rotates about the bond, the rotation not only is allowed but it happens readily.

The differently twisted forms of pentane in Figure 1.1 are called *conformations,* and samples of liquid or gaseous pentane consisting entirely of molecules in only one conformation cannot be isolated. The physical and chemical properties of pentane, therefore, are the net results that the whole collection of conformations has on whatever physical agent or chemical reactant has been used to observe the property. Generally, however, one conformation is present in a relatively high concentration, the conformation that corresponds to the most stable arrangement of the electron clouds. The top conformation in Figure 1.1, for example, gets the electron clouds of the various parts of the pentane molecule as far apart as they can be within the molecule. With the forces of repulsion thus at a minimum in this conformation, the molecule has its lowest potential energy and so has its greatest stability. It's an example of *nature's preference for the lowest energy, most stable arrangements.* This preference is the basic principle behind all aspects of molecular shape within a given environment.

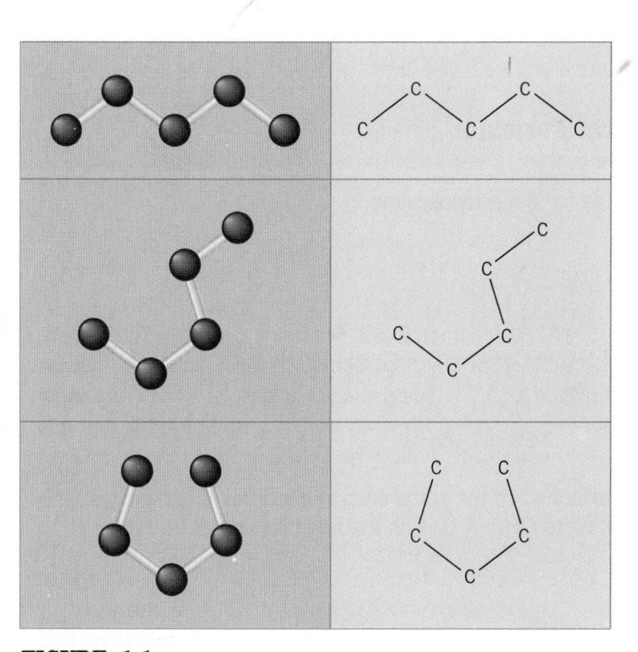

FIGURE 1.1
Free rotation at single bonds. Three of the innumerable conformations of the skeleton of the pentane molecule are shown here. (The hydrogens have been omitted.) Free rotation about single bonds easily converts one conformation into another.

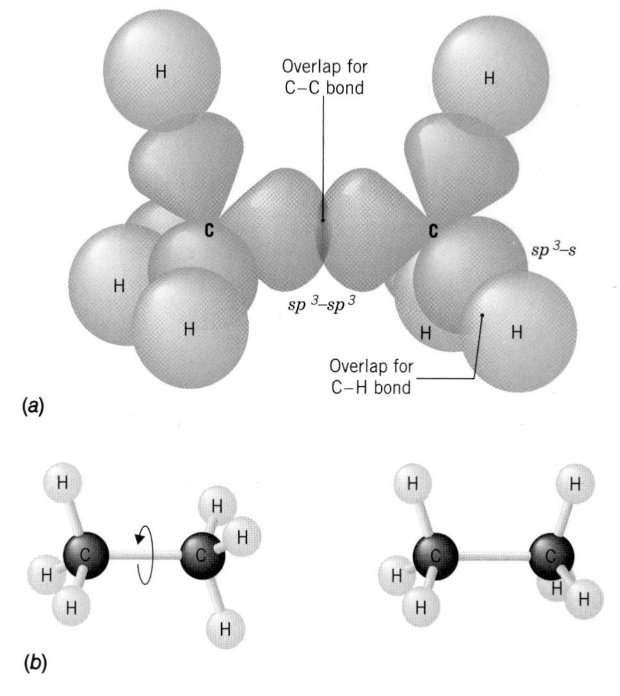

FIGURE 1.2
The sigma bonds in ethane. *(a)* Here are all the orbital overlappings that create the C—C and C—H bonds. *(b)* The sp^3 to sp^3 overlap that forms the C—C single bond, a sigma bond, is affected very little by a rotation of one CH_3 group relative to the other. Because such a rotation costs essentially no energy, it occurs easily.

Condensed Structural Formulas Reduce Clutter with Little Sacrifice in Information Just as we can leave some aspects of molecular geometry to the informed imagination, we can leave most of the bonds in a structural formula to it as well. Remembering that any carbon in a structural formula must have four bonds (or else carry some charge), we can group beside its symbol, C, all the hydrogen atoms held by it. If it holds three hydrogens, we can write CH_3 (or H_3C, but you don't see this as often). Just remember that these three hydrogen atoms are individually joined to the carbon. The structure of ethane illustrates this.

$$\begin{array}{ccccccc} & & H & & H & & \\ & & | & & | & & \\ H & - & C & - & C & - & H \\ & & | & & | & & \\ & & H & & H & & \end{array} \qquad CH_3{-}CH_3 \quad \text{or} \quad H_3C{-}CH_3$$

Ethane (expanded structure)

Ethane (condensed structures)

A carbon holding two hydrogen atoms can be represented as CH_2 or (seen less often) as H_2C. When a carbon holds just one hydrogen, we can write it as CH or HC.

The result of these simplifications is called a **condensed structure,** or simply the **structure.** These kinds of structures will be used almost exclusively in our continuing study.

EXAMPLE 1.1
Condensing a Full Structural Formula

Condense the structural formula for 2-methylpentane.

$$\begin{array}{ccccccccccccc} & & & & H & & & & & & & & \\ & & & & | & & & & & & & & \\ & & H & - & C & - & H & & & & & & \\ & & H & & | & & H & & H & & H & & \\ & & | & & | & & | & & | & & | & & \\ H & - & C & - & C & - & C & - & C & - & C & - & H \\ & & | & & | & & | & & | & & | & & \\ & & H & & H & & H & & H & & H & & \end{array}$$

2-Methylpentane

ANALYSIS Each unit of 3 H attached to the same carbon becomes CH_3. Where 2 H are joined to the same carbon, write CH_2. A carbon holding only one H becomes CH.

SOLUTION

$$\begin{array}{c} \quad\; CH_3 \\ \quad\; | \\ CH_3{-}CH{-}CH_2{-}CH_2{-}CH_3 \end{array}$$

CHECK We formulate here a general check rule for all molecular structures. *Scan all bond connections to verify that the rules of covalence have been obeyed.* In the answer, each C has four bonds and each H one bond. When you find a violation, the answer cannot possibly be correct, so fix it.

PRACTICE EXERCISE 1

Condense the following expanded structural formulas.

```
        H  H  H              H  H  H
        |  |  |              |  |  |
(a) H—C—C—C—H     (b) H—C—C—C—H
        |  |  |              |  |  |
        H  H  H              H  |  H
                                |
                             H—C—H
                                |
                                H
```

```
          H         H         H
          |         |         |
       H—C—H   H—C—H   H—C—H
       H  |         |         |  H
       |  |         |         |  |
(c) H—C—C————————C————————C—C—H
       |  |         |         |  |
       H  |         H         H  H
       H—C—H
          |
          H
```

Even Most Single Bonds Can Be "Understood" When a *single* bond appears on a *horizontal* line, we need not write a straight line to represent it; we can leave such single bonds to the imagination. We do not do this, however, for bonds that are not on a horizontal line. Thus we can write the structure of 2-methylpentane, Example 1.1, as follows. Notice that the vertically oriented bond is shown by a line but that all other single bonds are understood.

```
   CH3
   |
CH3CHCH2CH2CH3
```

2-Methylpentane

PRACTICE EXERCISE 2

Rewrite the condensed structures that you drew for the answers to Practice Exercise 1 and let the appropriate carbon–carbon single bonds be left to the imagination.

PRACTICE EXERCISE 3

Just to be certain that you are comfortable with condensed structures, expand each of the following to make them full, expanded structures with no bonds left to the imagination.

```
                    CH3                 CH3
                    |                   |
(a) CH3CH3  (b) CH3CHCHCH3  (c) CH3CH2CCH2CH2CH3
                       |                |
                       CH3              CH3
```

As indicated in the *check* part of Example 1.1, an important skill in using condensed structures is the ability to recognize errors. The most common is a violation of the rule that every carbon atom in a structure that carries no electrical charge must have exactly four bonds, no more and no fewer. Do the next Practice Exercise to test your skill in recognizing an error in structure.

PRACTICE EXERCISE 4

Which of the following structures cannot represent real compounds?

$$\text{(a)}\ \begin{array}{c} CH_3 \\ | \\ CH_3CCH_3 \\ | \\ CH_3 \end{array} \qquad \text{(b)}\ \begin{array}{c} CH_3 \\ | \\ CH_3CH_2CHCH_3 \end{array} \qquad \text{(c)}\ \begin{array}{c} CH_3 \quad CH_3 \\ | \qquad | \\ CH_3CHCH_2CHCH_2CH_3 \\ | \\ CH_3 \end{array}$$

When atoms other than carbon and hydrogen are present in a molecule, there is no major new problem in writing condensed structures. Remember that every oxygen or sulfur atom carrying no electrical charge must have two bonds, every nitrogen must have three, and every halogen atom must have just one.

Double and Triple Bonds Are Seldom Condensed Another rule about condensed structures is that carbon–carbon double and triple bonds are never left to the imagination. Carbon–oxygen double bonds are sometimes condensed, as some of the following examples illustrate. Study them as illustrations of how to condense structures.

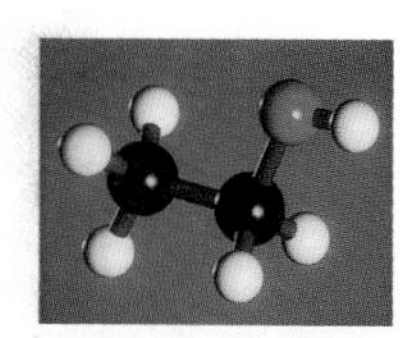

$$\begin{array}{ccccc} & & H & & H & \\ & & | & & | & \\ H & — & C & — & C & —OH \\ & & | & & | & \\ & & H & & H & \end{array} \quad \text{condenses to} \quad CH_3—CH_2—OH \quad \text{or to} \quad CH_3CH_2OH$$

Ethyl alcohol (in alcoholic drinks)

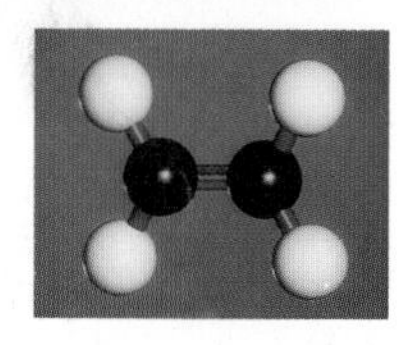

$$\begin{array}{ccccc} & & H & & H & \\ & & | & & | & \\ H & — & C & = & C & —H \end{array} \quad \text{condenses to} \quad CH_2{=}CH_2 \quad \text{or to} \quad H_2C{=}CH_2$$

Ethylene (raw material for making polyethylene)

$$\begin{array}{ccccc} & & H & & O & \\ & & | & & \| & \\ H & — & C & — & C & —OH \\ & & | & & & \\ & & H & & & \end{array} \quad \text{condenses to} \quad \begin{array}{c} O \\ \| \\ CH_3—C—OH \end{array} \quad \text{or to} \quad \begin{array}{c} O \\ \| \\ CH_3COH \end{array}$$

and often to CH_3CO_2H or to CH_3COOH

Acetic acid (in vinegar)

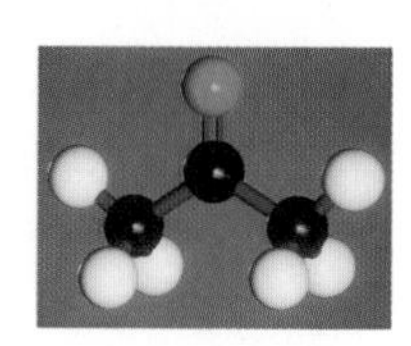

$$\begin{array}{ccccccc} & & H & & O & & H & \\ & & | & & \| & & | & \\ H & — & C & — & C & — & C & —H \\ & & | & & & & | & \\ & & H & & & & H & \end{array} \quad \text{condenses to} \quad \begin{array}{c} O \\ \| \\ CH_3—C—CH_3 \end{array} \quad \text{or to} \quad \begin{array}{c} O \\ \| \\ CH_3CCH_3 \end{array}$$

Acetone (nail polish remover)

$$\begin{array}{ccccc} & & H & & H & \\ & & | & & | & \\ H & — & C & — & N & —H \\ & & | & & & \\ & & H & & & \end{array} \quad \text{condenses to} \quad CH_3—NH_2 \quad \text{or to} \quad CH_3NH_2$$

Methylamine (in decaying fish)

Parentheses Are Sometimes Used to Condense Structures Further Sometimes two or three identical groups that are attached to the same carbon are grouped inside a set of parentheses. For example,

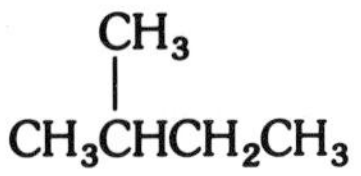

$$\begin{array}{l} \quad CH_3 \\ \quad | \\ CH_3CHCH_2CH_3 \end{array} \quad \text{can be written as} \quad (CH_3)_2CHCH_2CH_3$$

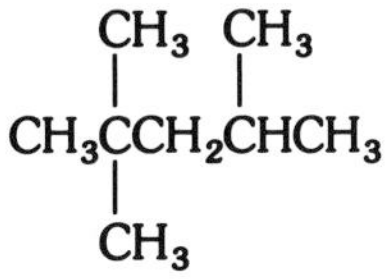

$$\begin{array}{l} \quad CH_3\ CH_3 \\ \quad |\qquad | \\ CH_3CCH_2CHCH_3 \\ \quad | \\ \quad CH_3 \end{array} \quad \text{can be written as} \quad (CH_3)_3CCH_2CH(CH_3)_2$$

We will not do this often, but you will see it in many references and you should be aware of it.

Unsaturated Compounds Have Double or Triple Bonds If its molecules have only single bonds, the compound is called a **saturated compound.** When one or more double or triple bonds are present, the substance is said to be an **unsaturated compound.** Thus ethylene, acetic acid, and acetone, just shown, are all unsaturated compounds, but ethyl alcohol and methylamine are saturated.

■ Molecules of edible oils, like olive oil or corn oil, have many double bonds and are described as polyunsaturated.

Saturated describes any molecule whose atoms are directly holding as many other atoms as they can. Each carbon in ethylene, for example, is holding just three atoms, but in ethyl alcohol each is directly holding four. *Unsaturated* implies that something can be added, and we will see that unsaturated compounds can add certain substances, like hydrogen, to their double or triple bonds.

Many Organic Molecules Contain Rings of Carbon Atoms A carbon **ring** is an arrangement of three or more carbon atoms into a closed cycle. Molecules with this feature are called ring compounds or cyclic compounds. (Sometimes an all-carbon ring is described as *carbocyclic.*) Cyclohexane molecules, for example, have a ring of six carbon atoms. Cyclopropane, once an important anesthetic, is also a cyclic compound.

■ More than 2 billion pounds of cyclohexane are made annually in the United States, with over 90% being used to make nylon.

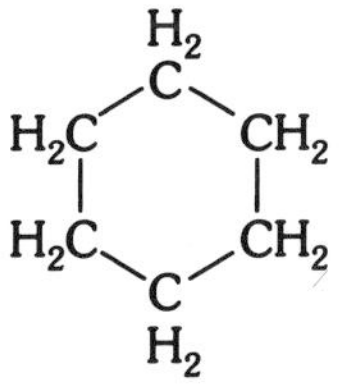

H_2C, CH_2, H_2C, CH_2, C H_2, C H_2

Cyclohexane

H_2 C, H_2C—CH_2

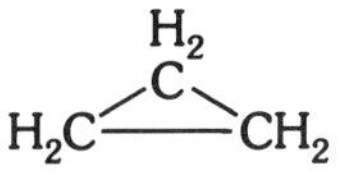

Cyclopropane

Cyclic compounds can have double bonds as in cyclohexene, but always remember that carbon–carbon double bonds are never left to the imagination. Rings, of course, can carry substituents, as in ethylcyclohexane. Not all the ring atoms have to be carbon atoms. They can be O, N, or S, too, and cyclic compounds with ring atoms other than C are called **heterocyclic compounds.** A simple example is tetrahydropyran.

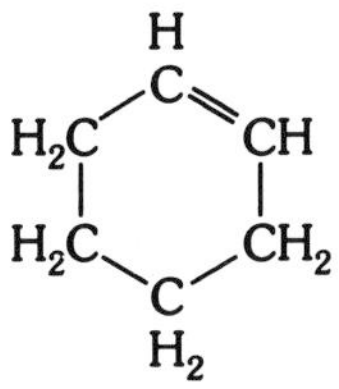

Cyclohexene

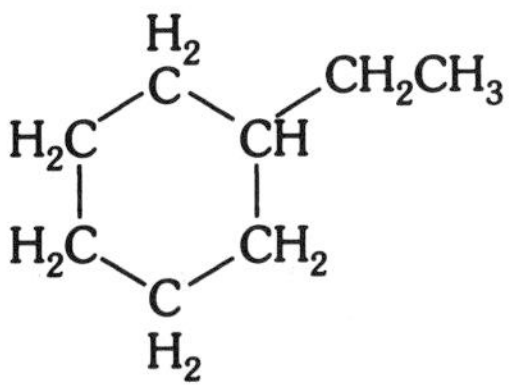

Ethylcyclohexane

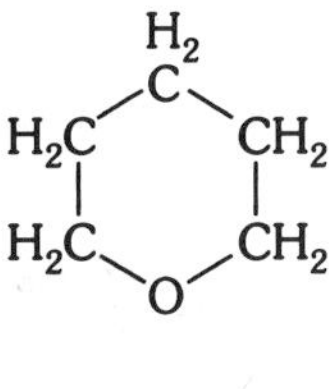

Tetrahydropyran

■ The ring system in tetrahydropyran—5 C plus 1 O in the ring—is widely present among molecules of carbohydrates.

The Rings of Cyclic Compounds Can Be Condensed to Simple Polygons Since we can leave to our imaginations so many structural features, a polygon, a many-sided figure, becomes a handy way to condense rings. A square, for example, can represent cyclobutane. The photograph of the ball-and-stick model of cyclobutane and its progressively more condensed structures show what is left to the imagination when just a square is used. At each corner, we have to understand that there is a CH_2 group. Each line in the square is a carbon–carbon single bond.

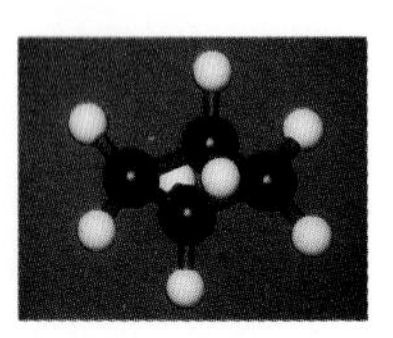

Three ways to represent the structure of cyclobutane

The model of methylcyclopentane and its progressively more condensed structures further show how a polygon can represent a ring.

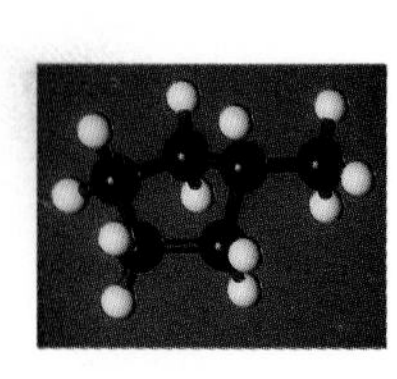

2 H understood

1 H understood

Three ways to represent the structure of methylcyclopentane

By convention, polygons like the hexagon for cyclohexane can be used to represent rings provided that we understand the following rules.

1. C occurs at each corner unless O or N (or another multivalent atom) is explicitly written at a corner.
2. A line connecting two corners is a covalent bond between adjacent ring atoms.
3. Remaining bonds, as required by the covalence of the atom at a corner, are understood to hold H atoms.
4. Double bonds are always explicitly shown.

We can illustrate these rules with the following cyclic compounds.

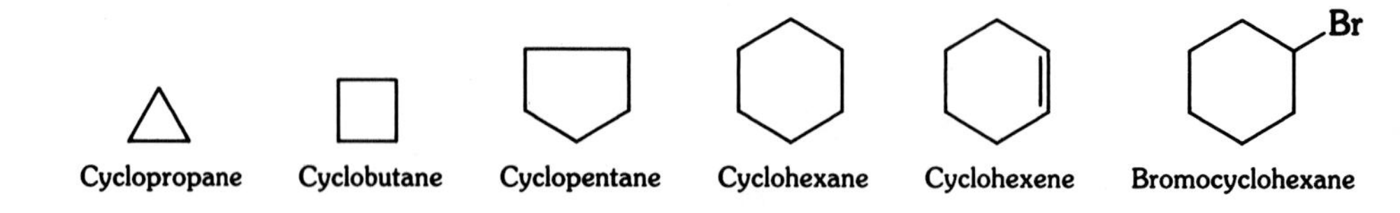

There is no theoretical upper limit to the size of a ring.

EXAMPLE 1.2
Understanding Condensed Structures of Ring Compounds

To make sure that you can read a condensed structure when it includes a polygon for a ring system, expand this structure, including its side chains.

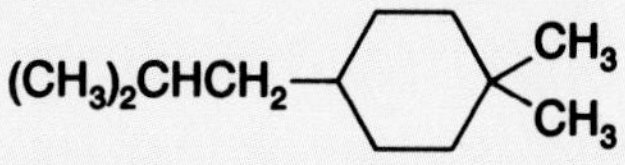

$(CH_3)_2CHCH_2$— (cyclohexane ring) $<^{CH_3}_{CH_3}$

ANALYSIS Every carbon atom, every bond, and every H atom has to be shown explicitly.

SOLUTION

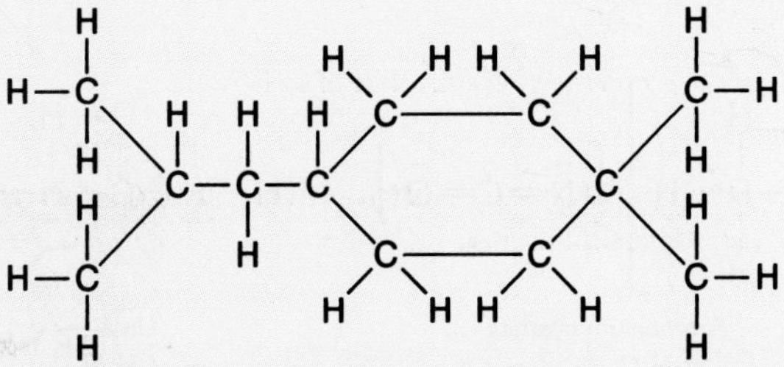

Note especially how we can tell the numbers of hydrogens that must be attached to a ring atom. We need as many as required to fill out a set of four bonds from each carbon. This example also shows a situation (on the right side of the ring) in which no bonding room is left for holding an H atom.

PRACTICE EXERCISE 5

Expand each of these two structures.

(a) (b)

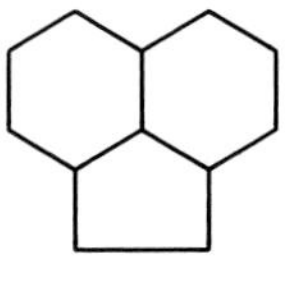

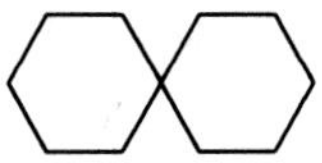

PRACTICE EXERCISE 6

Condense the following structure.

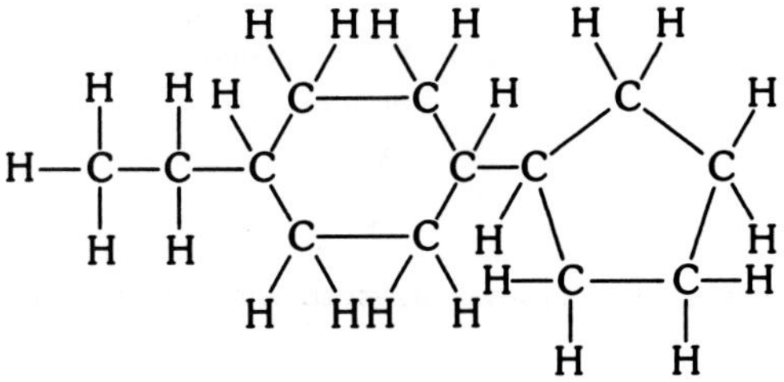

As you might expect, free rotation about the single bonds in a ring is not possible. Although there is a small amount of "flex" in rings, you'd have to break single bonds to get the kind of free rotation we observed with pentane (Fig. 1.1).

When we use polygons to represent saturated rings that have six or more ring atoms, we gloss over one feature of such molecules. The ring atoms that make up the rings of this size do not all lie in the same plane. We will postpone a study of what this fact implies.

1.3

ISOMERISM

Compounds can have identical molecular formulas but different structures. Ammonium cyanate and urea, the chemicals of Wöhler's important experiment, both have the molecular formula CH_4N_2O, but the atoms are organized differently:

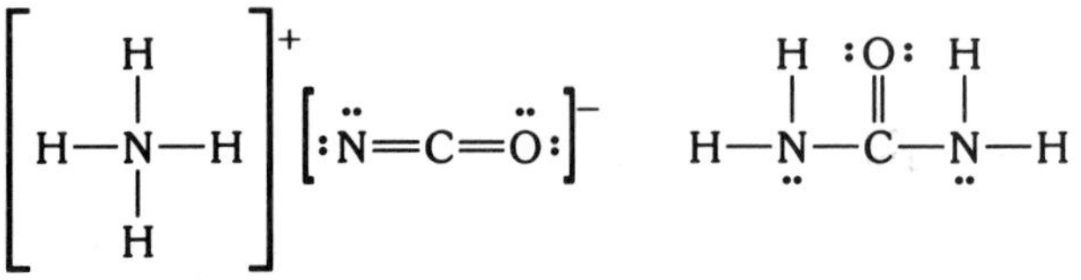

$$\left[\begin{array}{c} \mathrm{H} \\ | \\ \mathrm{H{-}N{-}H} \\ | \\ \mathrm{H} \end{array}\right]^+ \left[:\ddot{\mathrm{N}}{=}\mathrm{C}{=}\ddot{\mathrm{O}}:\right]^- \qquad \mathrm{H{-}\underset{\cdot\cdot}{\overset{H}{N}}{-}\overset{:O:}{\overset{\|}{C}}{-}\underset{\cdot\cdot}{\overset{H}{N}}{-}H}$$

Ammonium cyanate
CH_4N_2O

Urea
CH_4N_2O

"Isomer" has Greek roots — *isos,* the same, and *meros,* parts; that is, "equal parts" (but put together differently).

Compounds that have identical molecular formulas but different structures are called **isomers** of each other, and the existence of isomers is a phenomenon called **isomerism.** Isomerism is one reason why there are so many organic compounds. There are several kinds of isomers, and we will consider just one type here, **constitutional isomers.** Constitutional isomers, once called *structural isomers,*[1] differ in the basic atom-to-atom connectivities. There are three constitutional isomers of C_5H_{12}, for example, pentane, 2-methylbutane, and 2,2-dimethylpropane.

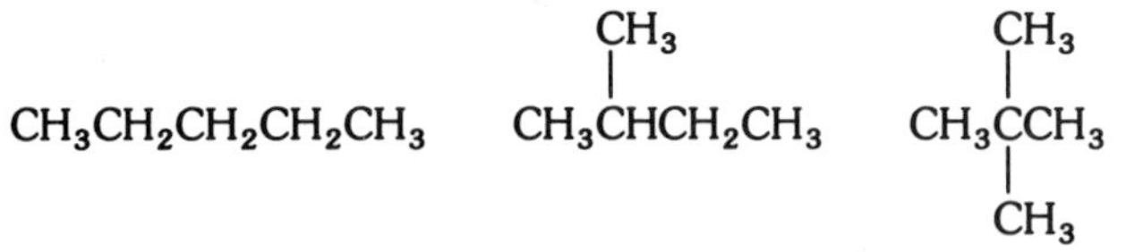

$$CH_3CH_2CH_2CH_2CH_3 \qquad CH_3\overset{\overset{CH_3}{|}}{C}HCH_2CH_3 \qquad CH_3\underset{\underset{CH_3}{|}}{\overset{\overset{CH_3}{|}}{C}}CH_3$$

Pentane (*n*-pentane)

2-Methylbutane (isopentane)

2,2-Dimethylpropane (neopentane)

The names in parentheses are the common names of these compounds. The letter *n* stands for *normal,* meaning the straight-chain isomer. *Neo* signifies *new,* as in a new isomer.

The larger the number of carbon atoms per molecule, the larger the number of isomers. For example, C_8H_{18} has 18 isomers; $C_{10}H_{22}$ has 75, and $C_{20}H_{42}$ has 366,319. Someone has figured out that roughly 6.25×10^{13} isomers are possible for $C_{40}H_{82}$. (Very few have actually been prepared. It would take nearly 200 billion years to make each one at the rate of one per day.)

The isomers of pentane or of $C_{40}H_{82}$ have quite similar chemical properties because their molecules all have only C—C and C—H single bonds. Often, however, isomers have very different properties. There are two ways, for example, to organize the atoms in C_2H_6O into constitutional isomers, as seen near the top of Table 1.1. One isomer is ethyl alcohol and the other is dimethyl ether. They are radically different, as the data in Table 1.1 show. At room temperature, ethyl alcohol is a liquid and dimethyl ether is a gas. Ethyl alcohol reacts with sodium; dimethyl ether does not. It is quite common for isomers to have properties this

[1] The term "structural isomer" has fallen into disfavor because it is regarded as too broad, that it implies not only atom-to-atom connectivities but also geometrical differences. Another kind of isomerism, geometrical isomerism (Section 2.3), deals with the latter.

TABLE 1.1
Properties of Two Isomers: Ethyl Alcohol and Dimethyl Ether

Property	Ethyl Alcohol	Dimethyl Ether
Structure	CH_3CH_2OH	CH_3OCH_3
Boiling point	78.5 °C	−24 °C
Melting point	−117 °C	−138.5 °C
Density (25 °C)	0.79 g/mL (a liquid)	2.0 g/L (a gas)
Solubility in water	Soluble in all proportions	Slightly soluble

different. Therefore, we nearly always use *structural* rather than molecular formulas for organic compounds. Only structures let us see at a glance how the atoms in the molecules are organized. The ability to recognize two structures as identical molecules, or as isomers or something else, is very important.

EXAMPLE 1.3
Recognizing Isomers

Which pair of structures represents a pair of isomers?

1. $CH_3—O—CH_2CH_3$ and $CH_3CH_2—O—CH_3$

2. 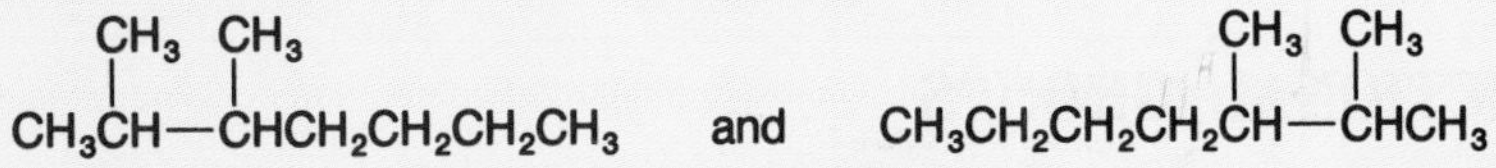

3. 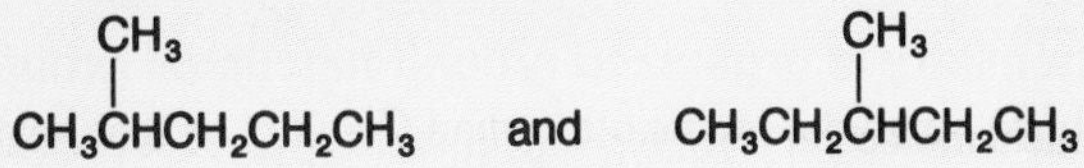

4. 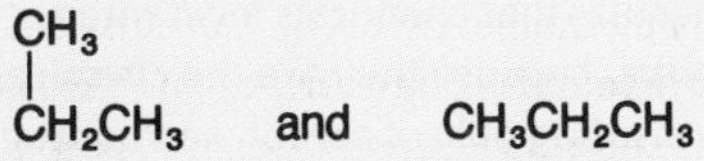

ANALYSIS Unless you spot a difference that rules out isomerism immediately, the *first step* is to see whether the molecular formulas are the same. If they aren't, the two structures are *not* isomers. If the molecular formulas of the two structures are identical, they might be identical or they might be isomers.

In this problem, the members of each pair share the same molecular formula.

Pair 1: C_3H_8O Pair 2: C_9H_{20} Pair 3: C_6H_{14} Pair 4: C_3H_8

Next, to see whether a particular pair represents isomers we try to find at least one structural difference. If we can't, the two structures are identical; they are just oriented differently on the page, or their chains are twisted into different conformations. Don't be fooled by an "east-to-west" versus a "west-to-east" type of difference. The difference must be *internal* within the structure. (Whether you face east or west you're the same person!)

SOLUTION Pair 1 are identical molecules; they're only oriented differently. (Imagine using a pancake turner to flip the one on the left, left to right; it would then be the structure on the right.)

Pair 2 is also an example of an east versus west difference in orientation. These two structures are identical. Their internal sequences—their atom-to-atom connectivities—are the same.

Pair 3 are isomers. In the first, a CH_3 group joins a five-carbon chain at the chain's second carbon, and in the second, this group is attached at the third carbon.

Pair 4 are identical. The two structures differ only in the conformations of their chains. Recall that free rotation about bonds allows us to imagine the straightening out of a continuous, open chain.

PRACTICE EXERCISE 7

Examine each pair to see whether the members are identical, are isomers, or are different in some other way.

(a) $H—O—CH_3$ and $CH_3—O—H$

(b) $CH_3—NH—CH_3$ and $CH_3—CH_2—NH_2$

(c) 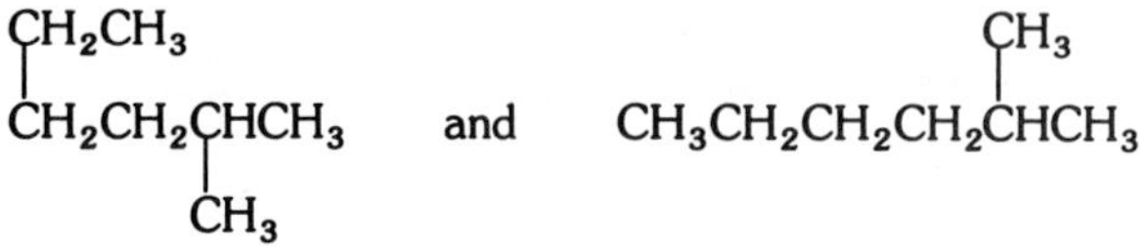 CH_2CH_3 on the first carbon of $CH_2CH_2CHCH_3$ with CH_3 on the CH, and $CH_3CH_2CH_2CH_2CHCH_3$ with CH_3 on the CH

$$\underset{}{\overset{CH_2CH_3}{\underset{|}{\overset{|}{C}}}}H_2CH_2\underset{\underset{CH_3}{|}}{C}HCH_3 \quad \text{and} \quad CH_3CH_2CH_2CH_2\overset{\overset{CH_3}{|}}{C}HCH_3$$

(d) $CH_2{=}CHCH_2CH_3$ and $CH_3CH{=}CHCH_3$

(e)  $CH_3CH_2\overset{\overset{O}{\|}}{C}OH$ and $HO\overset{\overset{O}{\|}}{C}CH_3$

1.4 FUNCTIONAL GROUPS

The study of organic chemistry is organized around functional groups. Regions of molecules that have nonmetal atoms other than C and H or that have double or triple bonds are the specific sites in organic molecules that chemicals most often attack. These small structural units are called **functional groups,** because they are the chemically functioning parts of molecules. Sections of molecules consisting only of carbon and hydrogen and only single bonds are called the **nonfunctional groups.**

Each Functional Group Defines an Organic Family Although over six million organic compounds are known, there are only a handful of functional groups, and each one serves to define a family of organic compounds. Our study of organic chemistry will be organized around just a few of these families, those outlined in Table 1.2. Let's see how the idea of a family will greatly simplify our study.

The *alcohols* are a major organic family of compounds. We have learned, for example, that ethyl alcohol is CH_3CH_2OH. Its molecules have the OH group attached to a chain of two carbons, but *chain length* is not what determines a compound's family. Chain length only bears on the *name* of that specific family member, as we'll soon see. The chain can be any length imaginable, and the substance will be in the alcohol family provided that somewhere on the chain there is an OH group attached to a carbon from which only single bonds extend. This functional group is called the *alcohol group;* it is very common in nature, being present in all carbohydrates and most proteins. Some examples of simple alcohols are

Isopropyl alcohol is commonly used as a rubbing alcohol.

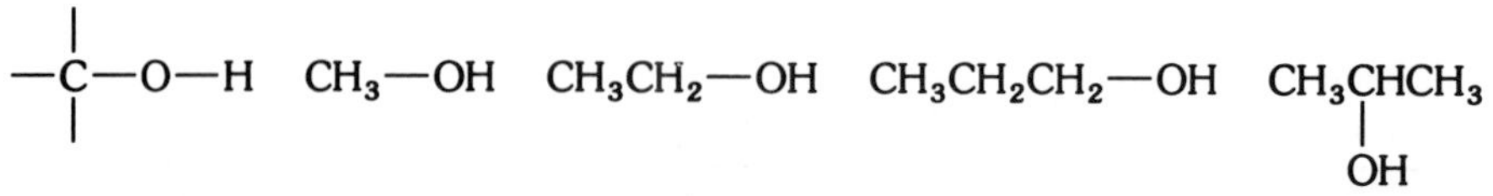

$$\underset{\text{Alcohol group}}{—\overset{|}{\underset{|}{C}}—O—H} \quad \underset{\text{Methyl alcohol}}{CH_3—OH} \quad \underset{\text{Ethyl alcohol}}{CH_3CH_2—OH} \quad \underset{\text{Propyl alcohol}}{CH_3CH_2CH_2—OH} \quad \underset{\text{Isopropyl alcohol}}{CH_3\underset{\underset{OH}{|}}{C}HCH_3}$$

Because all these alcohols have the same functional group, they exhibit the same kinds of chemical reactions. When just one of these reactions is learned, it applies to all members of the family, literally to thousands of compounds. In fact, we will often summarize a particular reaction for an organic family by using a general family symbol. All alcohols, for example, can

TABLE 1.2
Some Important Families of Organic Compounds

Family	Characteristic Structural Feature[a]	Example
Hydrocarbons	Only C and H present	
	Families of Hydrocarbons:	
	Alkanes: only single bonds	CH_3CH_3
	Alkenes: C=C	$CH_2{=}CH_2$
	Alkynes: C≡C	$HC{\equiv}CH$
	Aromatic: benzene ring	 (benzene ring)
Alcohols	ROH	CH_3CH_2OH
Ethers	ROR′	CH_3OCH_3
Thioalcohols	RSH	CH_3SH
Disulfides	RS—SR	$CH_3S{-}SCH_3$
Aldehydes	$R\overset{O}{\overset{\Vert}{C}}H$	$CH_3\overset{O}{\overset{\Vert}{C}}H$
Ketones	$R\overset{O}{\overset{\Vert}{C}}R'$	$CH_3\overset{O}{\overset{\Vert}{C}}CH_3$
Carboxylic acids	$R\overset{O}{\overset{\Vert}{C}}OH$	$CH_3\overset{O}{\overset{\Vert}{C}}OH$
Esters of carboxylic acids	$R\overset{O}{\overset{\Vert}{C}}OR'$	$CH_3\overset{O}{\overset{\Vert}{C}}OCH_3$
Esters of phosphoric acid	$RO\underset{OH}{\underset{\vert}{\overset{O}{\overset{\Vert}{P}}}}OH$	$CH_3O\underset{OH}{\underset{\vert}{\overset{O}{\overset{\Vert}{P}}}}OH$
Esters of diphosphoric acid	$RO\underset{HO}{\underset{\vert}{\overset{O}{\overset{\Vert}{P}}}}O\underset{OH}{\underset{\vert}{\overset{O}{\overset{\Vert}{P}}}}OH$	$CH_3O\underset{HO}{\underset{\vert}{\overset{O}{\overset{\Vert}{P}}}}O\underset{OH}{\underset{\vert}{\overset{O}{\overset{\Vert}{P}}}}OH$
Esters of triphosphoric acid	$RO\underset{HO}{\underset{\vert}{\overset{O}{\overset{\Vert}{P}}}}O\underset{OH}{\underset{\vert}{\overset{O}{\overset{\Vert}{P}}}}O\overset{O}{\overset{\Vert}{P}}(OH)_2$	$CH_3O\underset{HO}{\underset{\vert}{\overset{O}{\overset{\Vert}{P}}}}O\underset{OH}{\underset{\vert}{\overset{O}{\overset{\Vert}{P}}}}O\overset{O}{\overset{\Vert}{P}}(OH)_2$
Amines	RNH_2, $RNHR'$, $RNR'R''$	CH_3NH_2 CH_3NHCH_3 $CH_3\overset{CH_3}{\overset{\vert}{N}}CH_3$
Amides	$R\overset{O}{\overset{\Vert}{C}}{-}\overset{R''(H)}{\overset{\vert}{N}}R'(H)$	$CH_3\overset{O}{\overset{\Vert}{C}}NH_2$

[a] R, R′, and R″ represent hydrocarbon groups—*alkyl groups*—defined in the text. R′(H) or R″(H) signifies that the substituent can be either a hydrocarbon group or hydrogen.

be symbolized by R—OH, where R stands for a carbon chain (or ring), one of whatever length or branchings or rings. All alcohols, for instance, react with sodium metal as follows:

$$2R{-}OH + 2Na \longrightarrow 2R{-}ONa + H_2$$

■ R is from the German word *Radikal*, which we translate here to mean *group* as in a group of atoms.

If we wanted to write the specific example of this reaction that involves, say, ethyl alcohol, all we have to do is replace R by CH_3CH_2.

$$2CH_3CH_2{-}OH + 2Na \longrightarrow 2CH_3CH_2{-}ONa + H_2$$

Notice that this reaction changes only the OH groups of the alcohol. (Dimethyl ether, Table 1.1, which does not have the OH group, cannot give this reaction with sodium, as we learned in the previous section.)

Another organic family is that of the *carboxylic acids*. All their molecules have the *carboxyl group,* and this group makes all its compounds weak acids. We've often illustrated this with acetic acid.

■ The carboxyl group is present in all fatty acids, products of the digestion of the fats and oils in our diets.

$$\underset{\text{Carboxyl group}}{-\overset{\overset{\displaystyle O}{\|}}{C}-O-H} \qquad \underset{\text{Carboxylic acids}}{RCO_2H} \qquad \underset{\text{Acetic acid}}{CH_3\overset{\overset{\displaystyle O}{\|}}{C}OH} \quad \text{or} \quad CH_3CO_2H \quad \text{or} \quad CH_3COOH$$

We know that acetic acid neutralizes the hydroxide ion:

$$CH_3\overset{\overset{\displaystyle O}{\|}}{C}OH + OH^- \longrightarrow \underset{\text{Acetate ion}}{CH_3\overset{\overset{\displaystyle O}{\|}}{C}O^-} + H_2O$$

All molecules with the carboxyl group give the same reaction, so we can represent literally thousands of such reactions by just one simple equation:

$$\underset{\text{Carboxylic acid}}{R\overset{\overset{\displaystyle O}{\|}}{C}OH} + OH^- \longrightarrow \underset{\text{Carboxylate ion}}{R\overset{\overset{\displaystyle O}{\|}}{C}O^-} + H_2O$$

The groups characteristic of both carboxylic acids and the carboxylate ions are present in all proteins and their building blocks, the amino acids.

The amino acids are examples of substances with more than one functional group in the same molecule. They have the carboxyl group as well as the amino group, a group that defines the *amines,* still another simple family.

■ The amino group is a proton acceptor, like ammonia.

$$\underset{\text{Amino group}}{NH_2} \qquad \underset{\text{Methylamine}}{CH_3NH_2} \qquad \underset{\text{Simple amines}}{RNH_2} \qquad \underset{\text{Glycine, the simplest amino acid}}{NH_2CH_2CO_2H}$$

■ This icon will identify structural map signs:

Glycine, like all carboxylic acids, can neutralize hydroxide ion.

You can see how powerful a learning tool the functional group is. In the next few chapters, we will study just a few of the reactions of the most important functional groups found at the molecular level of life. *Learning these reactions will be like mastering a set of map signs.* You can read thousands of maps intelligently once you know their common signs and symbols. Similarly, we'll be able to "read" some of the chemical and physical properties of astonishingly complex molecules with the knowledge of the properties of a few functional groups.

The functional groups of a molecule generally occur in a setting of alkane-like groups. To be able to contrast the properties of functional groups with those of the alkane-like groups we turn our attention next to the alkanes themselves, the least reactive of the organic systems.

1.5

ALKANES AND CYCLOALKANES

Alkanes and cycloalkanes are saturated hydrocarbons.

Petroleum and natural gas are substances that consist almost entirely of a complex mixture of molecular compounds called hydrocarbons. Special Topic 1.1 describes the nature of petroleum and other chemical fuels in greater detail. **Hydrocarbons** are made from the atoms of just two elements, carbon and hydrogen, and the covalent bonds between the carbon atoms can be single, double, or triple. The carbon skeletons can be chains or rings. These possibilities define the various kinds of hydrocarbons, which are outlined in Figure 1.3.

The **alkanes,** whether open-chain or cyclic, are saturated hydrocarbons, those with only single bonds. Table 1.3 gives the first ten straight-chain members of this family. The **alkenes** are hydrocarbons with one or more carbon–carbon double bonds, whether the skeletons are chains or rings. Hydrocarbons with one or more carbon–carbon triple bonds are **alkynes.** We'll study the alkenes (and a little about the alkynes) in the next chapter. It is possible for one molecule to have both double and triple bonds, of course.

■ Because the bond angle at a triple bond is 180°, a ring has to be quite large to have a triple bond, and cycloalkynes are rare.

One important distinction in Figure 1.3 is between aliphatic and aromatic hydrocarbons. **Aliphatic compounds** of whatever family are any that have no benzene ring (or a similar system), and **aromatic compounds** are those with such rings. (We'll postpone the implications of the benzene ring to the next chapter.)

■ The first compounds found with benzene rings had pleasant odors and so were called *aromatic*. *Aliphatic* is from the Greek *aliphatos*, "fat-like."

Hydrocarbons of All Types, Saturated or Not, Have Common Physical Properties Both the carbon-carbon and the carbon-hydrogen bonds are almost entirely nonpolar, so hydrocarbon molecules have very little if any overall polarity. Hydrocarbons of all types, for this reason, are insoluble in water, but they dissolve well in nonpolar solvents (like CCl_4). Indeed, many special mixtures of alkanes are themselves used as nonpolar solvents. Some people, for example, have used gasoline or lighter fluid—both are mixtures of alkanes—to remove tar spots or grease. (If you do, be sure to keep all flames away and work outside, never in a garage or other enclosed room.)

■ In the right proportion in air, hydrocarbon vapors explode when ignited.

Hydrocarbons are not only insoluble in water but are generally less dense than water and so they will float on it. Thus using water to put out a hydrocarbon fire, like flaming gasoline, will only float the flames over a wider area. Nonflammable foams or CO_2 extinguishers must be used instead.

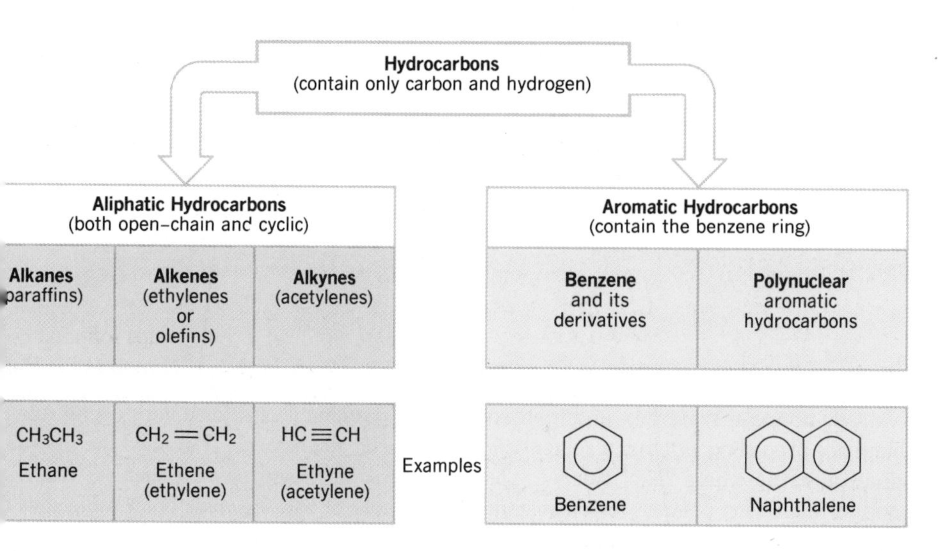

FIGURE 1.3
The several kinds of hydrocarbons. (The circles in the structures for benzene and naphthalene will be explained in the next chapter.)

SPECIAL TOPIC 1.1
ORGANIC FUELS

The Fossil Fuels Long ago, nature used photosynthesis to transform solar energy into the chemical energy of ancient plants and then locked up this energy in the fossilized remains of these plants, the *fossil fuels*—principally petroleum, coal, and natural gas. These fuels are a legacy from the past now being consumed so rapidly that for the first time in history we are concerned about running out of them and about the impact on civilization if their disappearance occurs too suddenly for us to adapt.

Petroleum and Crude Oil The fossil fuels formed over a span of hundreds of thousands of years during the Carboniferous period in geologic history, roughly 280 to 345 million years ago. Vast areas of the continents, little more than monotonous plains, then basked in sunlight near sea level. In the oceans, countless tiny, photosynthesizing plants like the diatoms—tons of them per acre of ocean surface during the early spring—soaked up solar energy to power their fugitive lives. Then they died. According to one theory, the death of each such plant released a tiny droplet of oily material that eventually settled into the bottom muds. The muds grew in thickness and compacted, sometimes into a rock called shale, sometimes into limestone and sandstone deposits. Under the pressure and heat of compacting, the oily matter changed into petroleum, a mixture of crude oil, water, and natural gas. (*Petroleum* is from *petra,* "rock" and *oleum,* "oil.") In some parts of the world, the petroleum managed to move slowly through porous rock and collect into vast underground pools to form the great petroleum reserves of our planet. In other regions, this movement could not occur, and the oily substances remain to this day locked in enormous deposits of oil shales and oil sands.

The story is told of a gentleman in a western state who built a new home and made the fireplace out of a local shale. When he lit the first fire, both the fireplace and his house burned up! His shale was rich in oil; in the Green River region where Utah, Colorado, and Idaho meet, there is a shale formation estimated to contain 2000 billion barrels of *shale oil* (see Figure 1). (The United States uses roughly 5 billion barrels of oil per year.) When the oil shale is crushed and heated to about 260 °C, a substance essentially the same as petroleum is released. Rock qualifies as oil shale if it holds an average of 10 gallons of petroleum per ton of rock. The cost of wringing the petroleum from the rock is presently too high for commercial exploitation, even if the management of the associated environmental problems is excluded from the cost.

FIGURE 1
Oil shale in the Mahogony zone near Parachute, Colorado

Near Lake Athabasca in the province of Alberta, Canada, in a land area about the size of Lake Michigan, there are huge reserves of a material much like crude oil but intermixed with sand, not shale (see Figure 2). Each ton of this *oil sand* holds about two-thirds of a barrel of oil, and the total deposit of oil sand is estimated to contain over 600 billion barrels of oil. This is equivalent to about half of the entire petroleum reserves of the world.

Coal On the marshy lands bordering the ancient oceans, lush vegetation flourished and died in a moist and sunny setting. The rate of decay of the remains of these plants, covered by stagnant, oxygen-poor, often acidic water, was slower than the rate at which the plants died. The slowly rotting mass accumulated to huge depths, became fibrous and turned to *peat,* a woody material used as fuel in some regions of the world.

Where peat layers became thick enough or were compressed by later deposits of sedimentary rock like limestone and sandstone, the peat changed into lignite ("brown coal"). Although lignite is over 40% water, it is

Hydrocarbon Solvents Illustrate the Like-Dissolves-Like Rule Grease and tar ar relatively nonpolar materials, and their solubility in gasoline illustrates a very useful rule c thumb for predicting whether a solvent can dissolve some substance. It's called the **like-dis solves-like rule,** where "like" refers to a likeness in *polarity.* Polar solvents, such as water, ar good for dissolving polar or ionic substances, like sugar or salt, because polar molecules c ions can attract water molecules around themselves, form solvent cages, and in this forı

FIGURE 2
Syncrude's tar sands project in Alberta, Canada

still an important fuel. Many lignite deposits became thick enough for further compaction to occur, and most of the water was squeezed out. Thus bituminous coal ("soft coal") formed, which has less than 5% water but contains considerable quantities of volatile matter. (*Volatile* means "easily evaporated.") When still further compaction took place and nearly all of the volatile matter was forced out, anthracite ("hard coal") formed, which is over 95% carbon.

The energy content in the coal reserves of the world exceeds that of the known petroleum reserves by a wide margin. In estimates that have the available supplies of petroleum lasting less than a century, the coal reserves are considered to have a lifetime of up to three centuries. Coal contains very small quantities of compounds of mercury, sulfur, radioactive elements and other potential pollutants. Their concentrations are very small, as we said, but because of the enormous quantities of coal annually burned, mostly at electric power plants, air and water pollution problems are caused, which often lead to water pollution as well. The mercury pollution in many lakes north of the industrial regions of the United States stems in part from the mercury released by burning coal.

Natural Gas In both soft coal and petroleum, the most volatile hydrocarbons, methane and ethane, also accumulated. Natural gas is largely methane.

Useful Substances from Crude Oil by Refining Crude oil is a complex mixture of organic compounds, but nearly all are hydrocarbons. Small but vexing amounts of sulfur-containing compounds are also present. The object of refining crude oil is to separate this mixture into products of varying uses. Refinery operations yield mixtures of compounds called *fractions* that boil over certain ranges of temperatures (see the accompanying table). Roughly 500 compounds occur among the fractions boiling up to 200 °C; about a third are alkanes, a third are cycloalkanes, and a third are aromatic hydrocarbons. You can see in the table where the chief fuels for transportation—gasoline, diesel oil, and jet fuel—originate.

The gasoline fraction of crude oil does not provide nearly enough of the world's needs for gasoline. One of the strategies at petroleum refineries to make more gasoline is to subject high boiling petroleum fractions to operations called *catalytic cracking* and *reforming.* In the presence of catalysts and when heated, large alkane molecules break up into smaller ones corresponding to lower boiling points that are in the range useful for gasoline engines.

Principal Fractions from Petroleum

Boiling Point Range (in °C)	Molecular Size	Principal Uses
Below 20	C_1-C_4	Natural gas; heating and cooking fuel; raw materials for other chemicals
20–60	C_5-C_6	Petroleum "ether," a nonpolar solvent and cleaning fluid
60–100	C_6-C_7	Ligroin or light naphtha; nonpolar solvents and cleaning fluids
40–200	C_5-C_{10}	Gasoline
175–325	$C_{12}-C_{18}$	Kerosene; jet fuel; tractor fuel
250–400	C_{12} and higher	Gas oil; fuel oil; diesel oil
Nonvolatile liquids	C_{20} and up	Refined mineral oil; lubricating oil; grease (a blend of soap in oil)
Nonvolatile solids	C_{20} and up	Paraffin wax; asphalt; road tar; roofing tar

freely intermingle with water molecules. Nonpolar solvents, like gasoline, do not dissolve sugar or salt, because nonpolar solvent molecules cannot be attracted to polar molecules or ions and form solvent cages.

Because of their low polarity and small size, the hydrocarbons that have one to four carbons per molecule are generally gases at or near room temperature. As the molecular size increases, however, London forces between molecules become stronger. Therefore, the

TABLE 1.3
Straight-Chain Alkanes

IUPAC Name	Carbons	Molecular Formula[a]	Structure	BP (°C)	MP (°C)	Density (g/mL, 20 °C)
Methane	1	CH_4	CH_4	−161.5	−182.5	—
Ethane	2	C_2H_6	CH_3CH_3	−88.6	−183.3	—
Propane	3	C_3H_8	$CH_3CH_2CH_3$	−42.1	−189.7	—
Butane	4	C_4H_{10}	$CH_3(CH_2)_2CH_3$	−0.5	−138.4	—
Pentane	5	C_5H_{12}	$CH_3(CH_2)_3CH_3$	36.1	−129.7	0.626
Hexane	6	C_6H_{14}	$CH_3(CH_2)_4CH_3$	68.7	−95.3	0.659
Heptane	7	C_7H_{16}	$CH_3(CH_2)_5CH_3$	98.4	−90.6	0.684
Octane	8	C_8H_{18}	$CH_3(CH_2)_6CH_3$	125.7	−56.8	0.703
Nonane	9	C_9H_{20}	$CH_3(CH_2)_7CH_3$	150.8	−53.5	0.718
Decane	10	$C_{10}H_{22}$	$CH_3(CH_2)_8CH_3$	174.1	−29.7	0.730

[a] The molecular formulas of the open-chain alkanes fit the general formula C_nH_{2n+2}, where n = the number of carbon atoms per molecule.

boiling points of the straight-chain alkanes increase with chain length. Hydrocarbons that have from 5 to about 16 carbon atoms per molecule are usually liquids at room temperature (see Table 1.3). When alkanes have approximately 18 or more carbon atoms per molecule, the London forces are strong enough to make the alkanes (waxy) solids at room temperature. Paraffin wax, for example, is a mixture of alkanes whose molecules have 20 or more carbon atoms.

■ Most candles are made from paraffin.

Our First Molecular "Map Sign," Hydrocarbon-Like Portions of Molecules Before moving on, let's pause to reflect on what we have done in relating physical properties to structural features. We have introduced the first molecular "map sign" in our study: *substances whose molecules are entirely or even mostly hydrocarbon-like are likely to be insoluble in water but soluble in nonpolar solvents.* When we see an unfamiliar structure, we can tell at a glance whether it is mostly hydrocarbon-like. If it is, we can predict with considerable confidence that the compound is not soluble in water. The structure of cholesterol illustrates our point.

■ The cholesterol structure shown here extends to open-chain systems the formalism of rings. Thus we use a point *where two lines meet* to denote one C and as many Hs as needed to fill out a covalence of 4 for the C atom. A line that terminates with nothing would denote CH_3. Thus, we could represent CH_3CH_2OH simply as

HO

Cholesterol

You probably know that cholesterol can form solid deposits in blood capillaries, and even close them. The heart must then work harder to sustain the flow of blood, and under this stress a heart attack can occur. Notice, now, that virtually the entire cholesterol molecule is hydrocarbon-like. It has only one polar group, the OH or alcohol group, and this takes up too small a portion of the molecule to make cholesterol sufficiently polar to dissolve either in water or in blood (which is mostly water).

What cholesterol illustrates is that by learning one very general fact, one "map sign," we don't have to memorize a long list of separate (but similar) facts about an equally long list of separate compounds that occur at the molecular level of life. With what we have just learned, we can look at the structures of hundreds of complicated compounds and confidently predict particular properties such as the likelihood of their being soluble in water. You have gained a powerful tool, one that immensely reduces what otherwise might have to be memorized.

EXAMPLE 1.4
Predicting Physical Properties from Structures

Study the following two structures and tell which is the structure of the more water-soluble compound.

$$HO{-}CH_2{-}\underset{\displaystyle OH}{\underset{|}{CH}}{-}\underset{\displaystyle OH}{\underset{|}{CH}}{-}\underset{\displaystyle OH}{\underset{|}{CH}}{-}\underset{\displaystyle OH}{\underset{|}{CH}}{-}\overset{\displaystyle O}{\overset{\|}{C}}{-}H$$

$$CH_3CH_2CH_2CH_2CH_2CH_2CH_2CH_2CH_2CH_2CH_2CH_2CH_2CH_2CH_2CH_2CH_2CO_2H$$

ANALYSIS The structure of the first compound has several polar OH groups but the second is almost entirely like an alkane.

SOLUTION The first should be (and is) much more soluble in water. The first compound is glucose (in one of its forms), the chief sugar in the bloodstream. The second is stearic acid, which does not dissolve in water. It forms when we digest the fats and oils (lipids) in the diet.

PRACTICE EXERCISE 8

Which of the following is more soluble in gasoline?

$$HOCH_2{-}\underset{\displaystyle OH}{\underset{|}{CH}}{-}CH_2OH \qquad CH_3CH_2\underset{\displaystyle CH_3}{\underset{|}{C}}HCH_2OH$$

Glycerol — 2-Methyl-1-butanol

1.6

NAMING THE ALKANES AND CYCLOALKANES

An IUPAC name discloses the compound's family, the number of carbons in the parent chain, and the kinds and locations of substituents.

In chemistry, **nomenclature** refers to the rules used to name compounds. The International Union of Pure and Applied Chemistry (IUPAC) is the organization that now develops the rules of chemical nomenclature. All scientific societies in the world accept the rules, known as the **IUPAC rules.** They are so carefully constructed that only one name could be written for each compound, and only one structure could be drawn for each name.

The IUPAC names, unfortunately, are sometimes very long and difficult to write or pronounce. It's much easier to call table sugar *sucrose* than α-D-glucopyranosyl-β-D-fructofuranoside. This illustrates why shorter names, referred to as *common names,* are still widely used. We will want to learn some common names, too, and you will see that even they usually have some system to them.

■ "Nomenclature" is from the Latin *nomen,* "name," and *calare,* "to call." Wealthy Romans had slaves called *nomenclators* whose duty it was to remind their owners of the names of important people who approached them on the street.

IUPAC Rules for Naming the Alkanes

1. The name ending for all alkanes (and cycloalkanes) is *-ane*.
2. The *parent chain* is the longest continuous chain of carbons in the structure. For example, the branched-chain alkane

$$\begin{array}{c} \quad\; CH_3 \\ \quad\; | \\ CH_3CH_2CHCH_2CH_2CH_3 \end{array}$$

is regarded as being "made" from the following parent

$$CH_3CH_2CH_2CH_2CH_2CH_3$$

by replacing a hydrogen atom on the third carbon from the left with CH_3.

$$\begin{array}{c} CH_3 \searrow \nearrow \\ \quad\; H \\ CH_3CH_2CHCH_2CH_2CH_3 \end{array} \longrightarrow \begin{array}{c} \quad\; CH_3 \\ \quad\; | \\ CH_3CH_2CHCH_2CH_2CH_3 \end{array}$$

3. A prefix is attached to the name-ending, *-ane,* that specifies the number of carbon atoms *i* *the parent chain.* The prefixes through parent chain lengths of 10 carbons are as follow *and should be learned.* The names in Table 1.3 show their use.

■ We won't need to know the prefixes for the higher alkanes.

meth-	1 C	hex-	6 C
eth-	2 C	hept-	7 C
prop-	3 C	oct-	8 C
but-	4 C	non-	9 C
pent-	5 C	dec-	10 C

Because the parent chain of our example has six carbons, the parent chain is named hexane—*hex* for six carbons and *ane* for being in the alkane family. The alkane whose name we are devising is regarded as a derivative of this parent, *hexane*.

4. The carbon atoms of the parent chain are numbered starting from whichever end of the chain gives the location of the first branch the lower of two possible numbers. Thus the correct direction for numbering our example is from left to right.

$$\begin{array}{c} \quad\; CH_3 \\ \quad\; | \\ CH_3CH_2CHCH_2CH_2CH_3 \\ 1 \quad 2 \quad 3 \quad 4 \quad 5 \quad 6 \end{array}$$

(correct direction of numbering)

Had we numbered from right to left, the carbon holding the branch would have had a higher number, which is not allowed by the IUPAC rules for alkanes.

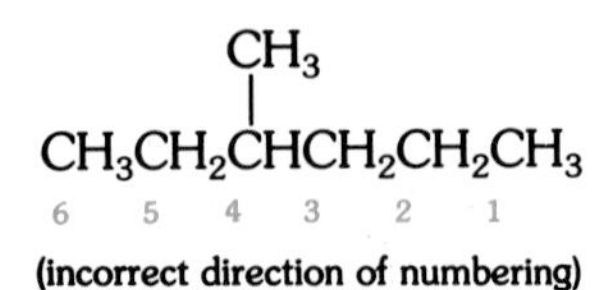

(incorrect direction of numbering)

5. Name each branch attached to the parent chain. We must now pause and learn the names of some of the *alkyl groups,* groups with alkane-like branches.

The Alkyl Groups

Any branch that consists only of carbon and hydrogen and has only single bonds is called an **alkyl group,** and the names of all alkyl groups end in *-yl*. Think of an alkyl group as an alkane minus one H.

$$\text{H—CH}_3 \xrightarrow{\text{remove one H}} \text{H}_3\text{C—} \quad \text{or } CH_3\text{—}$$

Methane → **Methyl**

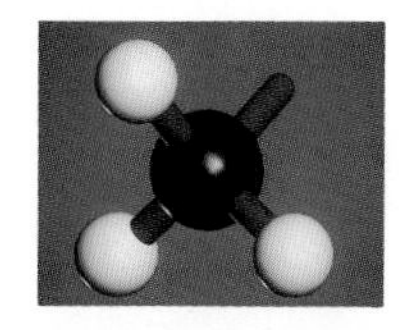

$$\text{H—CH}_2\text{—CH}_2\text{—H} \xrightarrow{\text{remove one H}} \text{H—CH}_2\text{—CH}_2\text{—} \quad \text{or } CH_3CH_2\text{—}$$

Ethane → **Ethyl**

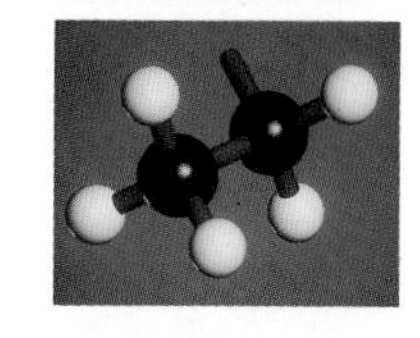

Two alkyl groups can be obtained from propane because the middle position in its chain of three is not equivalent to either of the end positions.

$$\text{H—CH}_2\text{—CH}_2\text{—CH}_2\text{—H} \xrightarrow[\text{(from either end)}]{\text{remove one H}} \text{H—CH}_2\text{—CH}_2\text{—CH}_2\text{—} \quad \text{or } CH_3CH_2CH_2\text{—}$$

Propane → **Propyl**

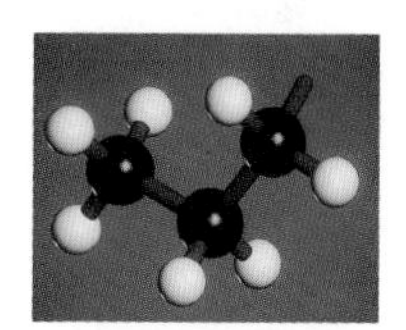

$$\text{H—CH}_2\text{—CH}_2\text{—CH}_2\text{—H} \xrightarrow[\text{(from middle C)}]{\text{remove one H}} \text{H—CH}_2\text{—CH(—)—CH}_2\text{—H} \quad \text{or } CH_3\overset{|}{C}HCH_3$$

Propane → **Isopropyl**

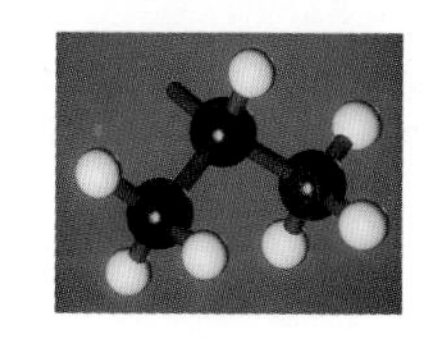

Two alkyl groups can similarly be obtained from butane.

$$\text{H—CH}_2\text{—CH}_2\text{—CH}_2\text{—CH}_2\text{—H} \xrightarrow[\text{(from either end)}]{\text{remove one H}} \text{H—CH}_2\text{—CH}_2\text{—CH}_2\text{—CH}_2\text{—} \quad \text{or } CH_3CH_2CH_2CH_2\text{—}$$

Butane → **Butyl**

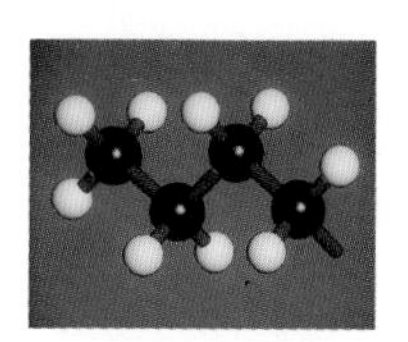

$$\text{H—CH}_2\text{—CH}_2\text{—CH}_2\text{—CH}_2\text{—H} \xrightarrow[\text{(from either of the two interior C atoms)}]{\text{remove one H}} \text{H—CH}_2\text{—CH}_2\text{—CH(—)—CH}_2\text{—H} \quad \text{or } CH_3CH_2\overset{|}{C}HCH_3$$

Butane → ***sec*-Butyl**

The last alkyl group is called the *secondary* butyl group (abbreviated *sec*-butyl) because the open bonding site is at a **secondary carbon,** a carbon that is directly attached to just two other carbons. A **primary carbon** is one to which just one other carbon is directly attached. The open bonding site in the butyl group, for example, is at a primary carbon atom. A **tertiary carbon** is one that holds directly three other carbons. We will encounter a tertiary carbon in a group that we will soon study.

Primary carbons

$$CH_3\text{—}\underset{\text{Tertiary carbon}}{\overset{\overset{\displaystyle CH_3}{|}}{CH}}\text{—}\underset{\text{Secondary carbon}}{CH_2}\text{—}CH_3$$

Tertiary carbon; Secondary carbon

Butane is the smallest alkane to have an isomer. The common name of the isomer is isobutane, and we can derive two more alkyl groups from it.

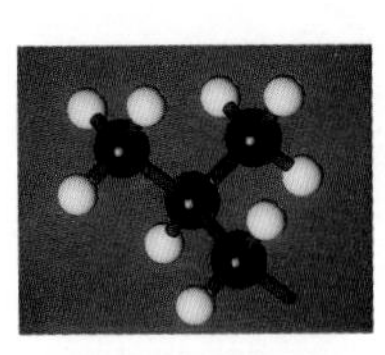

$$\underset{\text{Isobutane}}{\text{H}-\underset{\text{H}}{\overset{\text{H}}{\text{C}}}-\underset{\text{H}}{\overset{\text{H}-\overset{\text{H}}{\text{C}}-\text{H}}{\text{C}}}-\underset{\text{H}}{\overset{\text{H}}{\text{C}}}-\text{H}} \xrightarrow[\text{(from any of the three } CH_3 \text{ groups)}]{\text{remove one H}} \underset{\textbf{Isobutyl}}{\text{H}-\underset{\text{H}}{\overset{\text{H}}{\text{C}}}-\underset{\text{H}}{\overset{\text{H}-\overset{\text{H}}{\text{C}}-\text{H}}{\text{C}}}-\underset{\text{H}}{\overset{\text{H}}{\text{C}}}-} \quad \text{or } CH_3\overset{CH_3}{\overset{|}{C}}HCH_2-$$

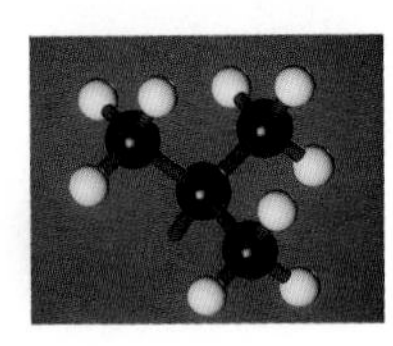

$$\underset{\text{Isobutane}}{\text{H}-\underset{\text{H}}{\overset{\text{H}}{\text{C}}}-\underset{\text{H}}{\overset{\text{H}-\overset{\text{H}}{\text{C}}-\text{H}}{\text{C}}}-\underset{\text{H}}{\overset{\text{H}}{\text{C}}}-\text{H}} \xrightarrow[\text{(from the tertiary C atom)}]{\text{remove one H}} \underset{\textbf{\textit{t}-Butyl}}{\text{H}-\underset{\text{H}}{\overset{\text{H}}{\text{C}}}-\underset{|}{\overset{\text{H}-\overset{\text{H}}{\text{C}}-\text{H}}{\text{C}}}-\underset{\text{H}}{\overset{\text{H}}{\text{C}}}-} \quad \text{or } CH_3\underset{|}{\overset{CH_3}{\overset{|}{C}}}CH_3$$

Notice that the open bonding site in the *tertiary*-butyl group (abbreviated *t*-butyl) occurs at a tertiary carbon.

The names and structures of these alkyl groups must now be learned. If you have access to ball-and-stick models, make models of each of the parent alkanes and then remove hydrogen atoms to generate the open bonding sites and the alkyl groups.

The prefix *iso-* in the name of an alkyl group, such as in isopropyl or isobutyl, has a special meaning. It can be used to name any alkyl group that has the following general features.

■ Here is another way to condense a structure. Thus $CH_3(CH_2)_3CH_3$ represents $CH_3CH_2CH_2CH_2CH_3$.

$$\begin{matrix} CH_3 \diagdown \\ \quad CH(CH_2)_n- \\ CH_3 \diagup \end{matrix} \qquad (n = 0, 1, 2, 3)$$

$n = 0$, isopropyl group
$= 1$, isobutyl group
$= 2$, isopentyl group
$= 3$, isohexyl group

Notice that each of these names has a word fragment (*-prop-*, *-but-*, and so forth) associated with a number of carbon atoms. When these word fragments are attached to *iso*, they specify the *total* number of carbons in the alkyl group. Thus the isopropyl group has three carbons (indicated by *prop*) and the isobutyl group has four carbons (indicated by *but*).

We can now continue with the IUPAC rules for naming alkanes.

6. Attach the name of the alkyl group to the name of the parent as a prefix. Place the location number of the group in front of the resulting name and separate the number from the name by a hyphen. Returning to our original example, its name is 3-methylhexane.

$$CH_3CH_2\overset{CH_3}{\overset{|}{C}}HCH_2CH_2CH_3$$

3-Methylhexane

7. When two or more groups are attached to the parent, name each and locate each with a number. The names of alkyl substituents are assembled in their alphabetical order. Always use *hyphens* to separate numbers from words. Here is an application.

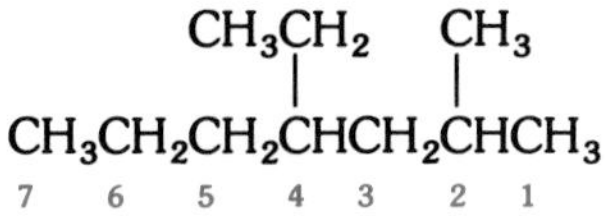

$$
\begin{array}{l}
\qquad\qquad\quad CH_3CH_2 \quad CH_3 \\
\qquad\qquad\qquad\quad | \qquad\quad | \\
CH_3CH_2CH_2CHCH_2CHCH_3 \\
\;7 \quad 6 \quad\; 5 \quad 4 \;\; 3 \quad 2 \;\; 1
\end{array}
$$

4-Ethyl-2-methylheptane

8. When two or more substituents are identical, use such prefixes as di- (for 2), tri- (for 3), tetra- (for 4), and so forth; and specify the location number of every group. Always separate a number from another number in a name by a *comma*. For example,

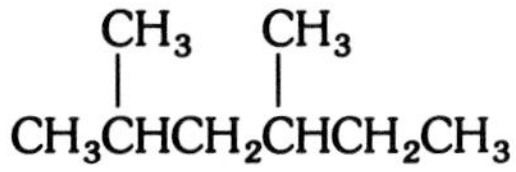

$$
\begin{array}{l}
\quad\; CH_3 \quad CH_3 \\
\quad\;\; | \qquad\;\; | \\
CH_3CHCH_2CHCH_2CH_3
\end{array}
$$

Correct name: 2,4-dimethylhexane
Incorrect names: 2,4-methylhexane
3,5-dimethylhexane
2-methyl-4-methylhexane

9. When identical groups are on the *same* carbon, repeat the number locating this carbon in the name. For example,

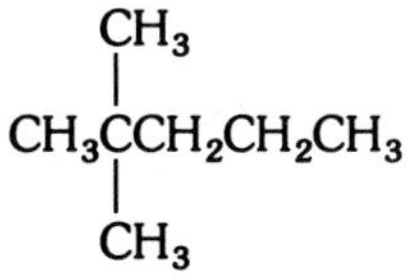

$$
\begin{array}{l}
\quad\;\; CH_3 \\
\qquad | \\
CH_3CCH_2CH_2CH_3 \\
\qquad | \\
\quad\;\; CH_3
\end{array}
$$

Correct name: 2,2-dimethylpentane
Incorrect names: 2-dimethylpentane
2,2-methylpentane
4,4-dimethylpentane

These are not all of the IUPAC rules for alkanes, but they will handle all of our needs. Study the following examples of correctly named compounds. Be sure to notice that in choosing the parent chain we sometimes have to go around a corner as the chain zigzags on the page.

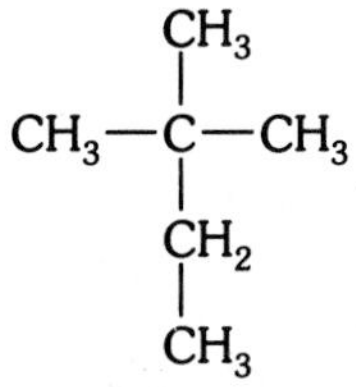

$$
\begin{array}{c}
CH_3 \\
| \\
CH_3-C-CH_3 \\
| \\
CH_2 \\
| \\
CH_3
\end{array}
$$

2,2-Dimethylbutane
not 2-ethyl-2-methylpropane

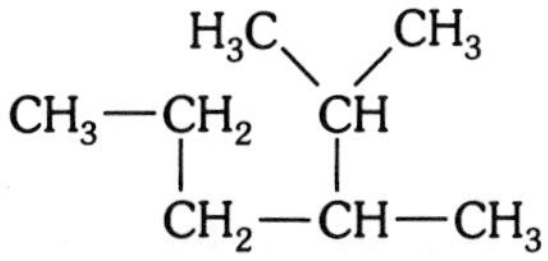

$$
\begin{array}{l}
\qquad\qquad\quad H_3C \diagdown \; \diagup CH_3 \\
CH_3-CH_2 \qquad CH \\
\qquad\quad | \qquad\quad\;\; | \\
\qquad\;\; CH_2-CH-CH_3
\end{array}
$$

2,3-Dimethylhexane
not 2-isopropylpentane

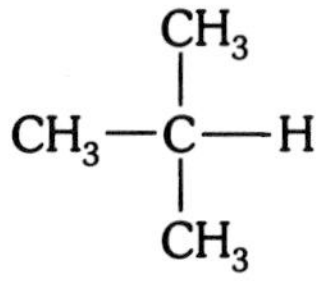

$$
\begin{array}{c}
CH_3 \\
| \\
CH_3-C-H \\
| \\
CH_3
\end{array}
$$

2-Methylpropane
not 1,1-dimethylethane

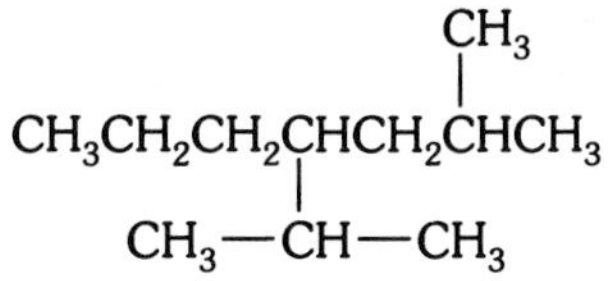

$$
\begin{array}{l}
\qquad\qquad\qquad\qquad\qquad\quad CH_3 \\
\qquad\qquad\qquad\qquad\qquad\quad\; | \\
CH_3CH_2CH_2CHCH_2CHCH_3 \\
\qquad\qquad\quad\;\; | \\
\qquad\quad CH_3-CH-CH_3
\end{array}
$$

4-Isopropyl-2-methylheptane
not 4-isopropyl-6-methylheptane

EXAMPLE 1.5
Using the IUPAC Rules to Name an Alkane

What is the IUPAC name for the following compound?

```
             CH3  CH2CH2CH2CH3
             |    |
     CH3CHCHCHCHCH2CH2CH3
           |    |
          CH3  CH
              /  \
            CH3   CH3
```

ANALYSIS The compound is an alkane because it is a hydrocarbon with only single bonds. We must therefore use the IUPAC rules for alkanes.

```
                   6    7    8    9
             CH3  CH2CH2CH2CH3
      1  2    |    |
     CH3CHCHCHCHCH2CH2CH3
           |  3   |4  5
          CH3    CH
                /  \
              CH3   CH3
```

SOLUTION The ending to the name must be *-ane.* The next step is to find the longest chain even if we have to go around corners. This chain is nine carbons long, so the name of the parent alkane is *nonane.* We have to number the chain from left to right, as follows, in order to reach the first branch with the lower number.
At carbons 2 and 3 there are the one-carbon methyl groups. At carbon 4, there is a three-carbon isopropyl group (not the propyl group, because the bonding site is at the *middle* carbon of the three-carbon chain). At carbon 5, there is a three-carbon propyl group. (It has to be this particular propyl group because the bonding site is the *end* of the three carbon chain in the group.) Alphabetically, *isopropyl* comes before *methyl* which comes before *propyl,* so we must assemble these names as follows to make the final name. (Names of alkyl groups are alphabetized *before* any prefixes such as di- or tri- are affixed.)

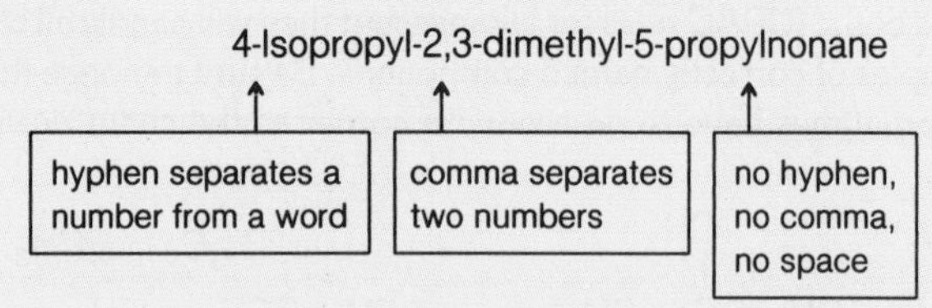

CHECK The most common mistake students make is in the discovery of the longest chain. Check your answer to make sure that you have not erred in this. Then be sure you have numbered from the correct end.

PRACTICE EXERCISE 9

Write the IUPAC names of the following compounds.

```
(a) CH3CH2                (b)  CH3       CH2CH2CH3          (c)      CH3  CH3   CH3
           \                       \     |                             |    |     |
            CHCH3                   CHCHCH2CH3             CH3CH2CHCHCHCH2CHCH3
           /                       /                                  |
    CH2CH2                    CH3CH                                 CH3CH2
    |                              |
    CH3                           CH3
```

IUPAC Rules for Naming Cycloalkanes To name a cycloalkane, place the prefix *cyclo-* before the name of the straight-chain alkane that has the same number of carbon atoms as there are in the ring. This is illustrated in Table 1.4. When necessary, give numbers to the ring atoms by giving location 1 to a ring position that holds a substituent and numbering around the ring in whichever direction reaches the nearest substituent first. For example,

1,2-Dimethylcyclohexane 1,2,4-Trimethylcyclohexane

No number is needed when the ring has only one group. Thus,

CH_3–cyclohexane

is named methylcyclohexane, not 1-methylcyclohexane.

IUPAC Names of Substituents Other than Alkyl Groups When halogen atoms, or nitro or amino groups are joined to a carbon of a chain or ring, the following names are used for them in IUPAC nomenclature.

—F	fluoro	—I	iodo
—Cl	chloro	$—NO_2$	nitro
—Br	bromo	$—NH_2$	amino

For example, $CH_3CH_2CH_2CHCl_2$ is named 1,1-dichlorobutane in the IUPAC system. Nitroethane is $CH_3CH_2NO_2$.

No number is needed in *nitroethane* because the name is unambiguous as it stands.

TABLE 1.4
Some Cycloalkanes

IUPAC Name	Structure	BP (°C)	MP (°C)	Density (20 °C)
Cyclopropane	△	−33	−127	1.809 g/L (0 °C)
Cyclobutane	□	−13.1	−80	0.7038 g/L (0 °C)
Cyclopentane	⬠	49.3	−94.4	0.7460 g/mL
Cyclohexane	⬡	80.7	6.47	0.7781 g/mL
Cycloheptane	heptagon	118.5	−12	0.8098 g/mL

PRACTICE EXERCISE 10

Write the condensed structures of the following compounds.

(a) 1-Bromo-2-nitropentane
(b) 5-Isopropyl-2,2,3,3,4,4-hexamethyloctane
(c) 5-*sec*-Butyl-6-*t*-butyl-2,2-diiodo-4-isopropyl-3-methylnonane
(d) 1-Bromo-1-chloro-2-methylpropane
(e) 5,5-Di-*sec*-butyldecane

PRACTICE EXERCISE 11

Examine the structure of part (e) of Practice Exercise 10. Underline each primary carbon, draw an arrow to each secondary carbon, and circle each tertiary carbon.

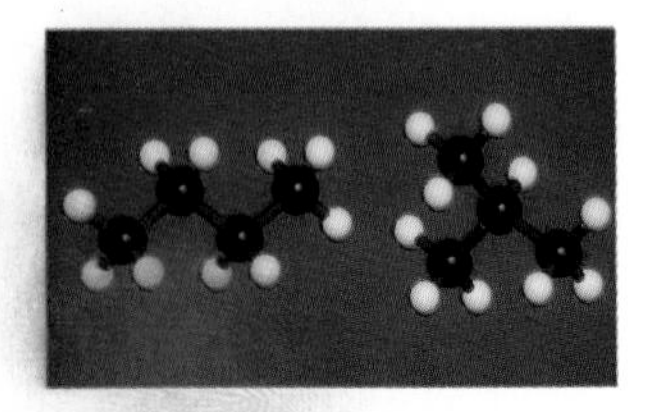
n-Butane and Isobutane.

Common Names In some references you might see the names of straight-chain alkanes with the prefix *n*-, as in *n*-butane, the common name of butane. It stands for *normal*, which is a way of designating that the straight-chain isomer is regarded as the *normal* isomer, as in the common names, *n*-pentane, *n*-hexane, and so forth. It is used only when isomers are possible. (You would never see *n*-propane printed as a name, for example, because there are no isomers of propane.)

Common Names of Alcohols, Amines, and Haloalkanes Employ the Names of the Alkyl Groups The following examples of some halogen derivatives of the alkanes, called *haloalkanes*, illustrate how common names are easily constructed. The IUPAC names are given for comparison.

■ The haloalkanes are examples of *organohalogen compounds*.

Structure	Common Name	IUPAC Name
CH_3Cl	Methyl chloride	Chloromethane
CH_3CH_2Br	Ethyl bromide	Bromoethane
$CH_3CH_2CH_2Br$	Propyl bromide	1-Bromopropane
$CH_3CH(Cl)CH_3$	Isopropyl chloride	2-Chloropropane
$CH_3CH_2CH_2CH_2Cl$	Butyl chloride	1-Chlorobutane
$CH_3CH(Br)CH_2CH_3$	*sec*-Butyl bromide	2-Bromobutane
$CH_3C(CH_3)_2$—Br	*t*-Butyl bromide	2-Bromo-2-methylpropane

PRACTICE EXERCISE 12

Give the common names of the following compounds.

(a) $ClCH_2CH_3$ **(b)** $BrCH_2CH_2CH_2CH_3$ **(c)** $CH_3CH(CH_3)CH_2Cl$ **(d)** $CH_3C(CH_3)(Br)CH_3$

1.7

CHEMICAL PROPERTIES OF ALKANES

Alkanes can burn and can give substitution reactions with the halogens, but they undergo almost no other reaction.

The chemistry of the alkanes and cycloalkanes is quite simple. Very few chemicals react with them. This is why they are nicknamed the *paraffins* or the paraffinic hydrocarbons, after the Latin *parum affinis,* meaning "little affinity" or "little reactivity." The alkanes are not chemically attacked by water, by strong acids such as sulfuric or hydrochloric acid, by strong bases like sodium hydroxide, by active metals such as sodium, by strong oxidizing agents such as the permanganate ion or the dichromate ion, or by any of the reducing agents. Among the few reactions of alkanes are combustion and halogenation — reactions with F_2, Cl_2, Br_2 (but not I_2), the halogens.

■ Mineral oil is a safe laxative (when used with care) because it is a mixture of high-formula-mass alkanes that undergo no chemical reactions in the intestinal tract.

The Combustion of Alkanes or Any Hydrocarbon Gives CO_2 and H_2O Virtually all organic compounds burn, and the hydrocarbons are no exception. We burn mixtures of alkanes, for example, as fuel to obtain energy. Bunsen burner gas is mostly methane. Liquified propane is used as fuel in areas where gas lines have not been built. Gasoline, diesel fuel, jet fuel, and heating oil are all mixtures of hydrocarbons, mostly alkanes.

If enough oxygen is available, the sole products of the complete combustion of *any hydrocarbon,* not just alkanes, are carbon dioxide and water plus heat. The (unbalanced) equation, regardless of the kind of hydrocarbon is

$$\text{Hydrocarbon} + O_2 \longrightarrow CO_2 + H_2O + \text{heat}$$

To illustrate, using propane,

$$CH_3CH_2CH_3 + 5O_2 \longrightarrow 3CO_2 + 4H_2O + 531 \text{ kcal/mol propane}$$

The same products, carbon dioxide and water, are also obtained by the complete combustion of any organic compound that consists only of carbon, hydrogen, and oxygen (for example, the alcohols). If insufficient oxygen is present, some carbon monoxide forms.

The Chlorination of Alkanes Is a *Substitution Reaction* In the presence of ultraviolet radiation or at a high temperature, alkanes react with chlorine to give organochlorine compounds and hydrogen chloride. An atom of hydrogen in the alkane can be replaced by an atom of chlorine, and this kind of replacement of one atom or group by another is called a **substitution reaction.** For example,

$$CH_4 + Cl_2 \xrightarrow[\text{or heat}]{\text{ultraviolet light}} \underset{\text{Methyl chloride}}{CH_3Cl} + HCl$$

The reaction takes place by a chemical chain reaction. We'll use Special Topic 1.2 to explain how this kind of reaction takes place in the chlorination of an alkane, like methane.

The hydrogen atoms in methyl chloride can also be replaced by chlorine, so as methyl chloride starts to form, its molecules compete with those of still unreacted methane for the remaining chlorine. This is how some methylene chloride (dichloromethane) forms when the chlorination of methane is carried out. In fact, when 1 mol of CH_4 and 1 mol of Cl_2 are mixed

SPECIAL TOPIC 1.2

THE CHLORINATION OF METHANE—A FREE RADICAL CHAIN REACTION WITH COUNTERPARTS AT THE MOLECULAR LEVEL OF LIFE

Free Radicals When photons of the proper energy are absorbed by the electron pairs of covalent bonds, the energy of the photons causes the bonds to break. This kind of bond breaking releases not ions but electrically neutral particles in which an atom lacks an octet and has an unpaired electron. Particles with unpaired electrons are called *radicals,* sometimes *free radicals* because they so often form with the freedom to move about. The breaking of the bond in the chlorine molecule, Cl_2, for example, occurs as follows.

$$:\ddot{Cl}:\ddot{Cl}: + \text{UV energy or heat} \longrightarrow :\ddot{Cl}\cdot + \cdot\ddot{Cl}:$$

Two chlorine atoms (two free radicals)

Free Radical Chain Reaction in the Chlorination of Methane The reaction of methane with chlorine is a free radical reaction that begins by the breaking of the bond in Cl_2, as we just described. (It is called the *chain initiating* step.) Chlorine is in group VIIA of the periodic table, so each chlorine atom has but seven valence shell electrons, not an octet. Each Cl atom is thus unstable and is able to launch the first step of a chemical chain reaction. In the end, the Cl atom recovers its outer octet but within a molecule of H—Cl. The first step of the actual *chain* reaction part of the overall reaction is that of a $\cdot$Cl atom with methane to generate $\cdot CH_3$, a neutral particle called the methyl radical. (We'll often explicitly show only the unpaired electron of a radical or atom, not the full population of the valence shell.)

$$\cdot Cl + CH_4 \longrightarrow H—Cl + \cdot CH_3 \quad (1)$$

If we use molecular models we can better imagine what happens when a chlorine atom strikes the H end of an H—C bond in methane hard enough.

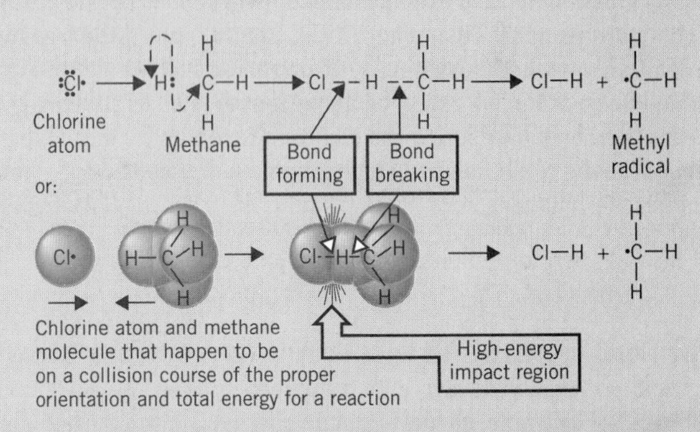

The carbon atom in the methyl radical also lacks an octet; it has seven valence shell electrons, one unshared (and shown by the dot). When it collides hard enough with an unreacted molecule of Cl—Cl, the Cl—Cl bond breaks and the new C—Cl bond in methyl chloride forms.

Methyl radical — Chlorine molecule — Bond forming — Bond breaking — Methyl chloride — Chlorine atom

or:

Orbital with odd electron in methyl radical

This is the second step of the chain itself.

$$\cdot CH_3 + Cl_2 \longrightarrow CH_3Cl + \cdot Cl \quad (2)$$

These two steps are called the *chain propagating* steps.

Notice that one product of reaction (2), $\cdot$Cl, is a necessary reactant for reaction (1). Thus chlorine atoms are generated not only by the action of heat or UV radiation, but also by reaction (2). This is why a relatively small amount of radiation can initiate chains that lead to the formation of a huge number of product molecules. The $\cdot Cl$ forming in reaction (2) appears in the midst of many still unreacted CH_4 molecules, so step (1) can occur again. This only makes another $\cdot CH_3$ radical, so a repeat of reaction (2) is set up. The sequence throughout chain propagation is thus (1) then (2), followed by (1) then (2), and so on.

Two chains are initiated by the breaking of the bond in one molecule of Cl_2 by heat or UV light because the chain-initiating event produces *two* $\cdot Cl$ atoms. Upwards of 8000 molecules of methyl chloride can be generated by one chlorine atom formed from the initiating action of only 1 photon of UV radiation. The chains continue until two radicals happen to find each other and join. Some *chain-termination* reactions are

$$2Cl\cdot \longrightarrow Cl_2$$
$$2CH_3\cdot \longrightarrow CH_3{-}CH_3$$
$$CH_3\cdot + Cl\cdot \longrightarrow CH_3Cl$$

Other chains are launched, however, as some terminate until one or both reactants are used up.

Multiple Chlorinations Because of the nature of the reaction, additional products are inevitable. Methyl chloride, CH_3Cl, forms while some unreacted Cl_2 remains, so a chlorine atom might collide with the H end of a H—C bond in CH_3Cl instead of at a H—C bond in another molecule of CH_4. This launches a new chain of reactions that converts CH_3Cl into CH_2Cl_2. The latter, of course, has its own H—C bonds, so still other chains can be started that convert CH_2Cl_2 into $CHCl_3$. You can see that the latter also has a H—C bond, so further chains can commence that lead to the formation of some CCl_4. These events occur more or less at random as the atoms and molecules whiz about in the gaseous state. Statistical probabilities largely govern what collisions occur but, as we mentioned in the text, a mixture of products is bound to form when chlorination is initiated in a mixture having a 1 : 1 mol ratio of Cl_2 to CH_4.

Cancer, Aging, and Free Radicals Ultraviolet radiation and the generation of free radicals also have roles in the development of skin cancer. Remember that nearly all free radicals are inherently unstable because they lack outer octets. Free radicals are thus rogue species and trigger unwanted events in cells, some leading to cancer.

Atomic radiation is dangerous partly because some of the particles it generates in cells are radicals. The chemical changes that occur during aging, when muscles lose their suppleness and flexibility, involve the formation of free radicals that form not so much from exposure to radiation or sunlight but by natural processes involving peroxides, compounds with such general formulas as R—O—O—H and R—O—O—R′. Their O—O bonds break rather easily to give free radicals of the R—O· type that lead to the crosslinking of protein molecules. Heavy smoking and excessive exposure to sunlight (for example, by overtanning) also contribute, by means of free radical chemistry, to the deep wrinkling of the skin (see Figure 1). Vitamins A and C are known to scavenge and destroy free radicals and so they seem to provide some protection.

FIGURE 1
Prolonged exposure to sunlight contributes to deep wrinkles.

SPECIAL TOPIC 1.3

WHY EQUATIONS FOR ORGANIC REACTIONS CANNOT ALWAYS BE BALANCED

Chemical reactions involve the breaking and reforming of bonds. Because organic molecules have several bonds of similar strengths, several products can sometimes form, even when we are trying to make just one of them. The substances we don't want are called the *by-products,* and the reactions that produce by-products are called *side reactions.* The reaction that produces the largest relative quantity of product is called the *main reaction.* Naturally, the chemist hopes this is the reaction that produces the desired product. In any case, most organic reactions produce a set of products, a mixture that must then be separated into the constituents. This often takes more time and effort than any other aspect of the synthesis.

As an example of a reaction that produces a mixture, we can use the chlorination of ethane. If we were to mix 1 mol of chlorine with 1 mol of ethane, naively trying to prepare 1 mol of ethyl chloride, we would obtain a mixture of mono-, di-, tri-, and possibly still more highly chlorinated molecules (plus HCl).

$$CH_3CH_3 + Cl_2 \xrightarrow{\text{ultraviolet light}} CH_3CH_2Cl + CH_3CHCl_2 + ClCH_2CH_2Cl + CH_3CCl_3 + ClCH_2CHCl_2 + \text{etc.} + HCl$$

It would be foolish to write coefficients in front of any of the products or to change the coefficients in front of the reactants—which, as they stand, say that a 1 : 1 molar ratio was taken—in an effort to balance the equation. If the mixture were to be completely separated, then the molar percentages of the various products could be reported, but seldom is such a thorough separation performed when just one product is sought.

Most organic reactions pose this kind of balancing problem, so the equation written is the equation for the main reaction only. Often it is balanced, but for just one reason—to provide a basis for the selection of the relative numbers of moles of the reactants to be mixed at the start. In nearly all the reactions that we will study, side reactions occur, which we generally will ignore.

and allowed to react, several reactions eventually occur, and a mixture forms of four chlorinated methanes, hydrogen chloride, plus some unreacted methane. It isn't possible to write a balanced equation, but what we can do is represent the reaction by a flow of symbols.

■ Notice how the boiling points of the compounds (in parentheses) increase with formula mass.

$$CH_4 + Cl_2 \xrightarrow[-HCl]{} CH_3Cl \xrightarrow[-HCl]{Cl_2} CH_2Cl_2 \xrightarrow[-HCl]{Cl_2} CHCl_3 \xrightarrow[-HCl]{Cl_2} CCl_4$$

Methane (bp −162 °C); Methyl chloride (bp −24 °C); Methylene chloride (bp 40 °C); Chloroform (bp 61 °C); Carbon tetrachloride (bp 77 °C)

For a discussion of the use of unbalanced reaction sequences to represent organic reactions, see Special Topic 1.3.

■ Chloroform (bp 61 °C) was an anesthetic better suited for use in the tropics than diethyl ether (bp 35 °C) because of its higher boiling point.

Methylene chloride, chloroform, and carbon tetrachloride are examples of *organochlorine compounds,* and all are used as nonpolar solvents. Chloroform has been used as an anesthetic.

Bromine reacts with methane by the same kind of substitution as chlorine. Iodine does not react. Fluorine combines explosively with most organic compounds at room temperature, and complex mixtures form.

The higher alkanes can also be chlorinated. Ethyl chloride, plus more highly chlorinated products, form by the chlorination of ethane, and ethyl chloride (bp 12.5 °C) is used as a local anesthetic. When it is sprayed on the skin, it evaporates very rapidly, and this cools the area enough to prevent the transmission of pain signals during minor surgery at the site.

When propane is chlorinated, both propyl chloride and isopropyl chloride form in roughly equal amounts.

$$\underset{\text{Propane}}{CH_3CH_2CH_3} + Cl_2 \xrightarrow{\text{ultraviolet light}} \underset{\text{Propyl chloride}}{CH_3CH_2CH_2Cl} + \underset{\text{Isopropyl chloride}}{CH_3\overset{\overset{Cl}{|}}{C}HCH_3}$$

Besides these products, some higher chlorinated compounds also form.

PRACTICE EXERCISE 13

(a) How many monochloro compounds of butane are possible? Give both their common and IUPAC names. (Consider only the monochloro compounds with the formula C_4H_9Cl.) (b) How many monochloro derivatives of isobutane are possible? Write both their common and IUPAC names.

SUMMARY

Organic and inorganic compounds Most organic compounds are molecular and the majority of inorganic compounds are ionic. Molecular and ionic compounds differ in composition, in types of bonds, and in several physical properties.

Structural features of organic molecules The ability of carbon atoms to join to each other many times in succession—in straight chains, in branched chains, as well as in rings—accounts in large measure for the existence of several million organic compounds. The skeletons of the rings can be made entirely of carbon atoms or there may be one or more other nonmetal atoms (heterocyclic compounds).

Full structural formulas of organic compounds are usually condensed by grouping the hydrogens attached to a carbon immediately next to this carbon; by letting single bonds on a horizontal line be understood; and by leaving bond angles and conformational possibilities to the informed imagination. Skeletons of rings are usually represented by simple polygons. Free rotation about single bonds is possible in open-chain compounds but not in rings.

Compounds without multiple bonds are saturated. Those with double or triple bonds are unsaturated. Carbon–carbon double or triple bonds are never "understood" in structures.

The families of organic compounds are organized around functional groups, parts of molecules at which most of the chemical reactions occur. Nonfunctional units can sometimes be given the general symbol R, as in ROH, the general symbol for all alcohols. These R groups are hydrocarbon-like groups.

Isomerism Differences in the conformations of carbon chains do not create new compounds, but differences in the organizations of parts do. Isomers are compounds with identical molecular formulas but different structures. Constitutional isomers make up one kind of isomer, those whose molecules have different atom-to-atom connectivities. Sometimes constitutional isomers are in the same family, like butane and isobutane. Often they are not, like ethyl alcohol and dimethyl ether.

Hydrocarbons Hydrocarbons are compounds in which the only elements are carbon and hydrogen. The alkanes are saturated hydrocarbons; the alkenes and alkynes are unsaturated. The alkenes have at least one double bond. The alkynes have at least one triple bond. The aromatic hydrocarbons have a benzene ring, and the aliphatic hydrocarbons do not. Being nonpolar compounds, the hydrocarbons are all insoluble in water, and many mixtures of alkanes are common, nonpolar solvents. The rule like-dissolves-like lets us predict solubilities.

Nomenclature of alkanes In the IUPAC system, a compound's family is always indicated by a name ending, like *ane* for the alkanes. The number of carbons in the parent chain is indicated by a unique prefix for each number, like the prefix *but* to denote four carbons in *but*ane. The locations of side chains or other kinds of atoms or groups are specified in the final name by numbers assigned to the carbons to which they are attached. The numbering of the parent chain is done in the direction that locates the first branch at the lower of two possible numbers.

Alkane-like substituents are called alkyl groups, and the names and formulas of those having from one to four carbon atoms must be learned. Common names are still popular, particularly when the IUPAC names are long and cumbersome.

Chemical properties of alkanes Alkanes and cycloalkanes are generally unreactive at room temperature toward concentrated acids and bases, toward oxidizing and reducing agents, toward even the most reactive metals, and toward water. They burn, giving off carbon dioxide and water, and in the presence of ultraviolet light (or at a high temperature) they undergo useful substitution reactions with chlorine and bromine.

REVIEW EXERCISES

The answers to Review Exercises whose numbers are in color are found in Appendix A. The answers to the other Review Exercises are found in the Study Guide that accompanies this book. The more challenging questions are marked with asterisks.

Organic and Inorganic Compounds

1.1 In terms of *bonding abilities*, what is unique about the element carbon? How does this contribute to the huge *number* of possible organic compounds?

1.2 With respect to the *synthesis* of organic compounds, what specifically was the problem that organic chemists faced prior to 1828? What scientific theory had been devised to meet this problem?

1.3 What was Wöhler's goal when he evaporated an aqueous solution of ammonium cyanate to dryness? What happened instead? With respect to *scientific theory* at the time, what specifically did Wöhler accomplish?

1.4 What kind of bond between atoms predominates among organic compounds?

1.5 How many single bonds are observed in natural molecules at each of the following atoms?
(a) C (b) O (c) N (d) H (e) Cl

1.6 Which of the following compounds are inorganic?
(a) CH_3CH_2OH (b) CO_2 (c) $CHCl_3$ (d) $KHCO_3$
(e) Na_2CO_3

1.7 Are the majority of all compounds that dissolve in water ionic or molecular? Inorganic or organic?

1.8 Explain why very few organic compounds can conduct electricity either in an aqueous solution or as molten materials.

***1.9** Each compound described below is either ionic or molecular. State which it most likely is, and give one reason.
(a) A compound that melts at 281 °C, and burns in air.
(b) A compound that dissolves in water. When hydrochloric acid is added, the solution fizzes and an odorless, colorless gas is released, which can extinguish a burning flame.
(c) A compound that is a colorless gas at room temperature.
(d) A compound that melts at 824 °C and becomes white when heated.
(e) A compound that is a liquid and does not dissolve in water but does burn.
(f) A compound that is a liquid and does dissolve in water as well as burns.

Structural Features of Organic Molecules

1.10 One can write the structure of propane, a common heating gas, as follows.

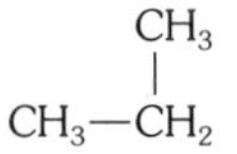

$$CH_3-CH_2(-CH_3)$$

Propane

Are propane molecules properly described as straight chain or as branched chain, in the sense in which we use these terms? Explain.

1.11 Which of the following structures are possible, given the numbers of bonds that various atoms can form?
(a) $CH_3CH_2CH_2OCH_3$
(b) $CH_2CH_2CH_2CH_3$
(c) $CH_3{=}CHCH_2CH_3$
(d) $CH_3CH{=}CHCH_2CH_3$
(e) $NH_2CH_2CH_2CH_3$

***1.12** Write full (expanded) structures for each of the following *molecular* formulas. Remember how many covalent bonds nonmetals have in molecules: C, 4; N, 3; O, 2; H, Cl, and Br, 1 each. In some structures you will have to use double or triple bonds. (*Hint:* A trial-and-error approach will have to be used.)

(a) CH_4O (b) CH_2Cl_2 (c) N_2H_4 (d) C_2H_6
(e) CH_2O (f) CH_2O_2 (g) NH_3O (h) C_2H_2
(i) $CHCl_3$ (j) HCN (k) C_2H_3N (l) CH_5N

1.13 Expand the following structure of nicotinamide, one of the B-vitamins.

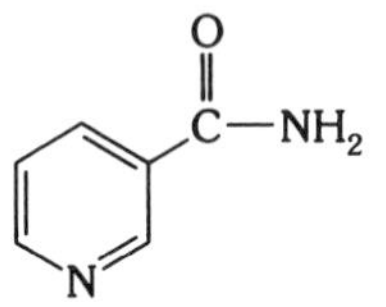

Nicotinamide

1.14 Expand the structure of thiamine, vitamin B_1. Notice that one nitrogen has four bonds so it has a positive charge.

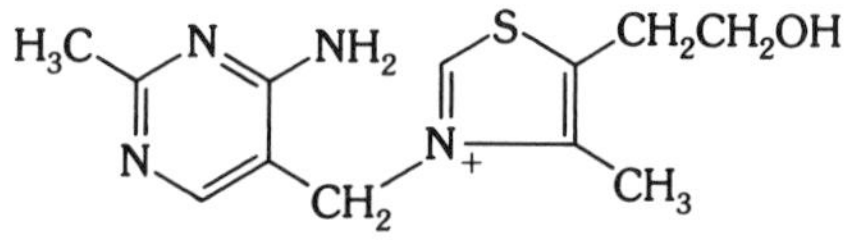

Thiamine

1.15 Write neat, condensed structures of the following.

(a)

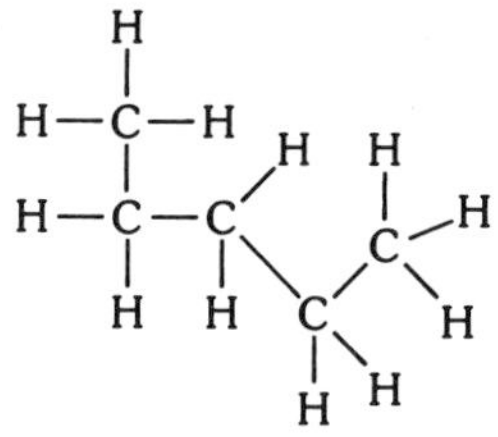

(b)

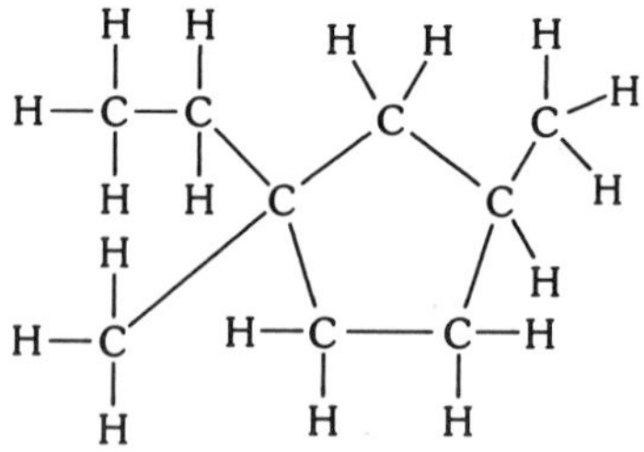

(c)

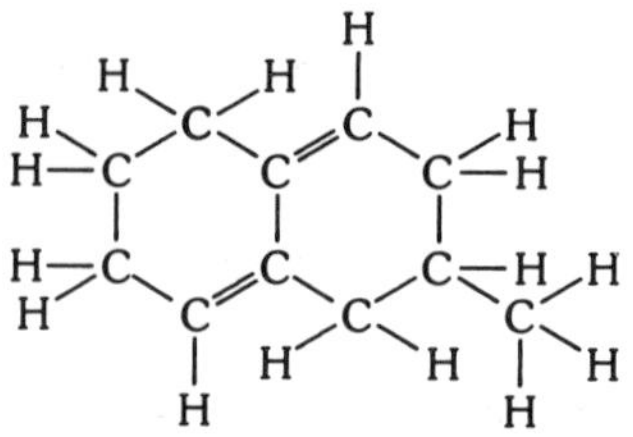

1.16 Why is the topic of molecular shape important at the molecular level of life?

1.17 What kinds of orbitals overlap when the C—C bond forms in ethane?

1.18 Free rotation can occur about single bonds (in open-chain structures) without breaking or weakening the bonds. Why is this possible?

1.19 Which of the following structures represent unsaturated compounds?

(a) 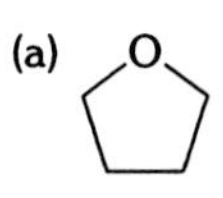

(b) (1,4-dioxene ring: O, O, C=C)

(c) $CH_3\overset{O}{\overset{\|}{C}}CH_2CH_3$ (d) $CH_3C{\equiv}CCH_3$

1.20 Which compounds are saturated?

(a) (cyclopentene ring) (b) $CH_3CH{=}NCH_3$ (c) C_4H_{10}

(d) 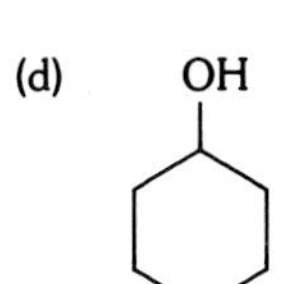

Cyclohexanol

(e) (benzene ring with CO_2H and $O\overset{O}{\overset{\|}{C}}CH_3$)

Aspirin

***1.21** Decide whether the members of each pair are identical, are isomers, or are unrelated.

(a) CH_3—OH (vertical) and $CH_3{-}OH$

(b) CH_3, OH, CH_3 on CH and CH_3, CH_3 on $CH{-}OH$

(c) CH_3CH_2SH and $CH_3CH_2CH_2SH$

(d) $CH_3CH{=}CH_2$ and cyclopropane (CH_2—CH_2—CH_2 ring)

(e) $CH_3\overset{O}{\overset{\|}{C}}CH_2CH_3$ and $CH_3CH_2\overset{O}{\overset{\|}{C}}CH_3$

(f) $CH_3\overset{CH_3}{\overset{|}{C}H}CH_3$ and $CH_3\overset{CH_3}{\overset{|}{C}H}$—$CH_3$ (below)

(g) $CH_3CH_2CH_2NH_2$ and $CH_3NHCH_2CH_3$

(h) $CH_3CH_2\overset{O}{\overset{\|}{C}}OH$ and $HOCH_2\overset{O}{\overset{\|}{C}}CH_3$

(i) $H\overset{O}{\overset{\|}{C}}OCH_2CH_3$ and $CH_3CH_2\overset{O}{\overset{\|}{C}}OH$

(j) $H\overset{O}{\overset{\|}{C}}OCH_2CH_2OH$ and $HOCH_2CH_2\overset{O}{\overset{\|}{C}}OH$

(k) $CH_3OCH_2\overset{O}{\overset{\|}{C}}CH_3$ and $CH_3CH_2\overset{O}{\overset{\|}{C}}OCH_3$

(l) $CH_3{-}CH(-CH_2{-}CH_2{-}CH_2{-}C(CH_3)_2{-}CH(CH_3)_2){-}CH_2{-}CH_3$

and

$CH_3{-}CH(CH_3){-}CH_2{-}CH_2{-}CH_2{-}CH_2{-}C(CH_3)_2{-}CH(CH_3){-}CH_3$

(m) (2-methylcyclohexanone) and (2-methylcyclohexanone, drawn reversed)

(n) $HO{-}O{-}CH_3$ and $HO{-}O{-}CH_2{-}OCH_3$

Families of Organic Compounds

1.22 Name the family to which each compound belongs.

(a) $CH_3CH_2CH_3$ (b) $HOCH_2CH_2CH_3$

(c) (cyclopentane ring with SH) (d) $CH_3C{\equiv}CH$

(e) 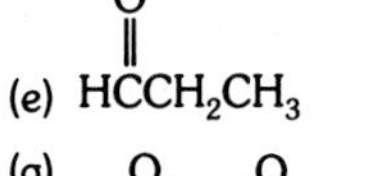 $H\overset{O}{\overset{\|}{C}}CH_2CH_3$ (f) 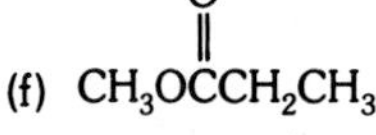 $CH_3O\overset{O}{\overset{\|}{C}}CH_2CH_3$

(g) 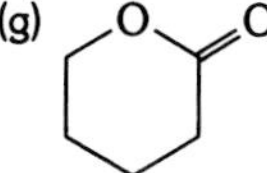 (h) (cyclohexanone ring, =O)

(i) $CH_3CH_2CH_2NH_2$ (j) $CH_3OCH_2CH_3$

***1.23** Name the families to which the compounds in parts (a)–(m) of Practice Exercise 1.21 belong. (A few belong to more than one family.)

Physical Properties and Structure

1.24 Which compound must have the higher boiling point? Explain.

$CH_3CH_2CH_3$ **A** $CH_3CH_2CH_2CH_2CH_3$ **B**

1.25 Which compound must be less soluble in gasoline? Explain.

$ClCH_2CH_2CH_2CH_2CH_2Cl$ **A** $HOCH_2CH_2CH_2CH_2CH_2OH$ **B**

1.26 Suppose that you are handed two test tubes containing colorless liquids, and you are told that one contains pentane and the other holds hydrochloric acid. How can you use just water to tell which tube contains which compound without carrying out any chemical reaction?

1.27 Suppose that you are given two test tubes and are told that one holds methyl alcohol, CH_3OH, and the other hexane. How can water be used to tell these substances apart without carrying out any chemical reaction?

Nomenclature

1.28 There are five isomers of C_6H_{14}. Write their condensed structures and their IUPAC names.

1.29 Which of the isomers of hexane (Review Exercise 1.28) has the common name *n*-hexane? Write its structure.

1.30 Which of the hexane isomers (Review Exercise 1.28) has the common name isohexane? Write its structure.

*1.31 There are nine isomers of C_7H_{16}. Write their condensed structures and their IUPAC names.

1.32 Write the condensed structures of the isomers of heptane (Review Exercise 1.31) that have the following names.
(a) *n*-heptane (b) isoheptane

*1.33 Write the IUPAC names of the following compounds.

```
(a) CH3  CH3                  CH3
      \  /                   /
       CH   CH2CH2CH2CH
       |    |                \
   CH3CCH2CCH2CH3             CH3
       |    |
       CH3  CHCH3
            |
            CH2CH3
```

```
(b)        CH3                 CH3
           |                   |
         CH2CH2              CH2CHCH3
           |                   |
           CH------CH--CH--CH2
           |       |    |
       CH3CCH3    CH2  CH2CHCH3
           |       |       |
           CH3    CH2     CH3
                   |
                  CH3
```

1.34 Write condensed structures for the following compounds.
(a) *n*-Butyl bromide (b) Isohexyl iodide
(c) *sec*-Butyl chloride (d) Isopropylcyclohexane

1.35 Write condensed structures for the following compounds.
(a) Propyl chloride (b) Isobutyl iodide
(c) *t*-Butyl bromide (d) Ethyl bromide

***1.36** The following are incorrect efforts at naming certain compounds. What are the most likely condensed structures and correct IUPAC names?
(a) 1,6-Dimethylcyclohexane (b) 2,4,5-Trimethylhexane
(c) 1-Chloro-*n*-butane (d) Isopropane

1.37 The following names cannot be the correct names, but it is still possible to write structures from them. What are the correct IUPAC names and the condensed structures?
(a) 1-Chloroisobutane (b) 2,4-Dichlorocyclopentane
(c) 2-Ethylbutane (d) 1,3-6-Trimethylcyclohexane

Reactions of Alkanes

1.38 Write the balanced equation for the complete combustion of heptane, a component of gasoline.

1.39 Gasohol is a mixture of ethyl alcohol, CH_3CH_2OH, in gasoline. Write the equation for the complete combustion of ethyl alcohol.

1.40 What are the formulas and common names of all the compounds that can be made from methane and chlorine?

1.41 There are two isomers of $C_2H_4Cl_2$. What are their structures and IUPAC names?

***1.42** When propane reacts with chlorine, in addition to the two isomeric monochloropropanes, some dichloropropanes also form. Write the structures and the IUPAC names for all these possible dichloropropanes.

Fossil Fuels (Special Topic 1.1)

1.43 What are the three principal fossil fuels being used today?

1.44 What is the difference between petroleum, crude oil, and natural gas?

1.45 In general terms, describe what happened to change peat into lignite, lignite into soft coal, and soft coal into hard coal.

1.46 What does *fraction* mean in connection with oil refining?

1.47 What kinds of compounds predominate in the crude oil fractions that boil below 200 °C?

1.48 How do petroleum refineries increase the supply of gasoline?

Free Radical Reactions (Special Topic 1.2)

1.49 Why is the chlorine atom called a free radical but the chlorine molecule is not?

1.50 The bromination of methane proceeds by a series of steps just like those of the chlorination of methane.
(a) Write the equation for the reaction that initiates chains.
(b) Write the two equations that constitute the chain reaction.
(c) Write two equations that illustrate how chains can be broken.

*1.51 The explosive gas-phase reaction of H_2 with Cl_2 that makes HCl is a free radical chain reaction.
(a) Write the balanced equation for the reaction.
(b) It is initiated by the breaking of the bond in Cl_2. Write the equation for this chain-initiating reaction as well as the equations for the chain reaction itself.

On Balancing Organic Reactions (Special Topic 1.3)

1.52 Explain in your own words why it is impossible to write a balanced equation for what actually happens when methane is chlorinated.

Additional Exercises

1.53 When cyclohexane is chlorinated, how many monochloro derivatives are possible? Write the name or names and structure or structures.

1.54 What reaction, if any, will cyclohexane give with each reactant at room temperature?
(a) Aqueous NaOH
(b) Concentrated sulfuric acid
(c) Iodine

1.55 What is the name of the butyl group with nine equivalent hydrogen atoms?

*1.56 If 7.46 g of cyclopentane is chlorinated, what is the maximum number of grams of chlorocyclopentane that can be obtained?

UNSATURATED HYDROCARBONS

2

People who climb mountains, such as this climber near the Half Dome of Yosemite National Park, routinely trust their lives to the strengths of synthetic fibers. The molecules of these synthetics are polymers, and the principles behind their formation are one topic in this chapter.

2.1

OCCURRENCE

The carbon–carbon double bond occurs in nature largely in compounds having other functional groups.

The carbon–carbon double bond is sometimes called the *ene* function.

Unsaturation in hydrocarbons occurs as double bonds, triple bonds, or benzene rings. Hydrocarbons with double bonds are called **alkenes** and the carbon–carbon double bond is the **alkene group.** The double bond is common at the molecular level of life, particularly in the edible fats and oils and in related compounds that make up most of a cell membrane. Special Topic 2.1 briefly describes a few of the large number of compounds with one or more alkene groups. This chapter is mostly about the properties of the alkene group *wherever it occurs,* whether in simple alkenes or in polyfunctional compounds (those having two or more functional groups).

The carbon–carbon triple bond occurs in hydrocarbons called **alkynes** but is uncommon in nature. We will use Special Topic 2.4 later in the chapter to discuss it briefly. The *benzene ring* (Section 2.7) and rings similar to it occur in proteins and nucleic acids, and it is widely present in pharmaceuticals like aspirin (structure on page 65).

Table 2.1 shows the structures and some physical properties of several alkenes. As with the alkanes, the first four alkenes are gases at room temperature, and all are much less dense than water. Like all hydrocarbons, alkenes are insoluble in water and soluble in nonpolar solvents for reasons we discussed in Section 1.1.

TABLE 2.1
Properties of Some 1-Alkenes

Name (IUPAC)	Structure	BP (°C)	MP (°C)	Density (g/mL)[a]
Ethene	$CH_2{=}CH_2$	−104	−169	—
Propene	$CH_2{=}CHCH_3$	−48	−185	—
1-Butene	$CH_2{=}CHCH_2CH_3$	−6	−185	—
1-Pentene	$CH_2{=}CHCH_2CH_2CH_3$	30	−165	0.641
1-Hexene	$CH_2{=}CHCH_2CH_2CH_2CH_3$	64	−140	0.673
1-Heptene	$CH_2{=}CHCH_2CH_2CH_2CH_2CH_3$	94	−119	0.697
1-Octene	$CH_2{=}CHCH_2CH_2CH_2CH_2CH_2CH_3$	121	−102	0.715
1-Nonene	$CH_2{=}CHCH_2CH_2CH_2CH_2CH_2CH_2CH_3$	147	−81	0.729
1-Decene	$CH_2{=}CHCH_2CH_2CH_2CH_2CH_2CH_2CH_2CH_3$	171	−66	0.741
Cyclopentene	(ring structure)	44	−135	0.722
Cyclohexene	(ring structure)	83	−104	0.811

[a] At 10 °C.

SPECIAL TOPIC 2.1
THE ALKENE DOUBLE BOND IN NATURE

Molecules with alkene groups are found among most of the major families of biochemically important substances. Cholecalciferol (vitamin D_3) and retinol (vitamin A_1), for example, are two of the vitamins that have several alkene groups. This group renders both vitamins sensitive to conditions that favor oxidation, such as prolonged standing in air, particularly if it is warm. Food sources of these vitamins, therefore, should not be overcooked. Because the double bond carries two electrons in its pi bond, it has more electron density than a single bond. The defining property of an oxidizing agent is its attraction to electron-rich, electron-donating sites. With increased electron density, the alkene group is thus open to attack by oxidizing agents.

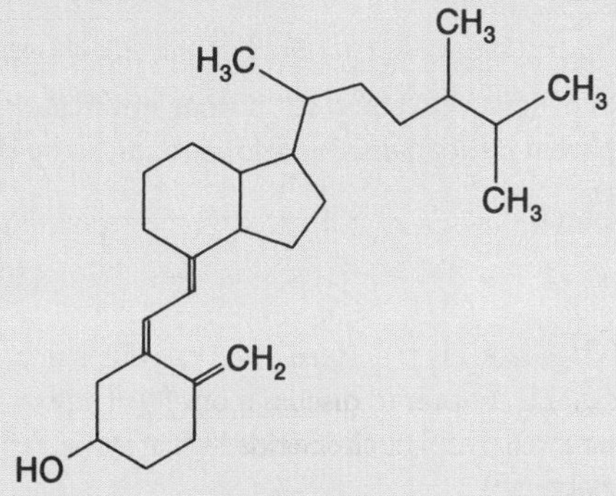

Cholecalciferol (Vitamin D_3)

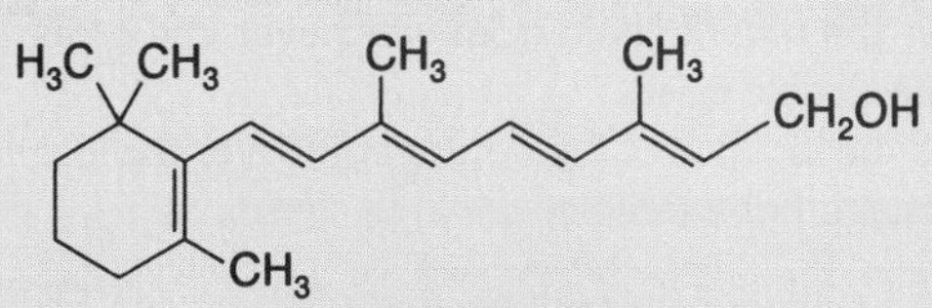

Retinol (Vitamin A_1)

β-Carotene is one of the bright yellow-orange compounds in carrots and is called a *provitamin* because the body is able to convert it into vitamin A. You can see that *β*-carotene has eleven alkene groups and no other functional group. Compounds like this have the general name of *polyene.* By comparing the structures of *β*-carotene and retinol, you can see how the body can obtain the basic retinol system from *β*-carotene.

Several of the human sex hormones have alkene groups, including testosterone, the chief male sex hormone. It is classified as a *steroid* hormone because it has the special four-ring system shown in its structure and called the steroid system.

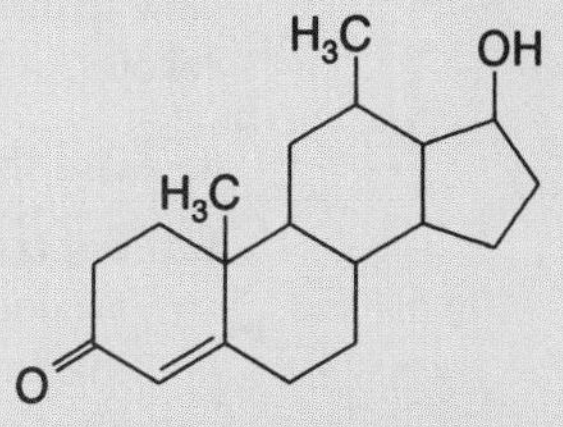

Testosterone (male sex hormone)

The edible fats and oils all have at least one alkene group per molecule, and the vegetable oils generally have two to four. Shown here is the structure of a molecule found in corn oil, a typical vegetable oil and a product widely used to make salad dressing. You can see why vegetable oils are described as *polyunsaturated;* their molecules have several unsaturated sites (specifically meaning alkene groups rather than the carbon–oxygen double bonds). In animal fats like butterfat and tallow (beef fat), the number of double bonds per molecule is smaller so they are more "saturated." Otherwise, animal fats are structurally like the vegetable oils.

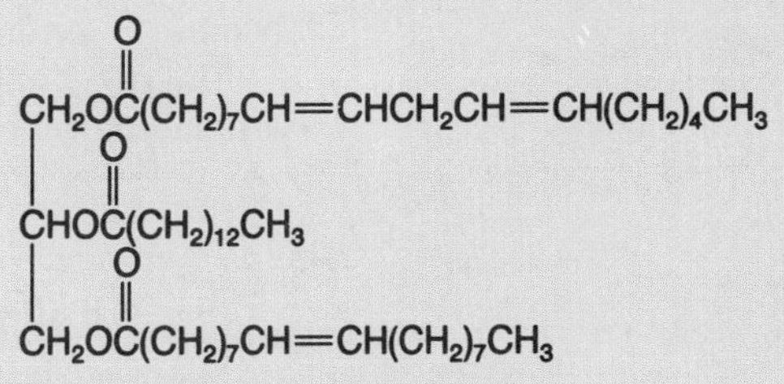

$$
\begin{array}{l}
CH_2O\overset{O}{\overset{\|}{C}}(CH_2)_7CH{=}CHCH_2CH{=}CH(CH_2)_4CH_3 \\
| \\
CHO\overset{O}{\overset{\|}{C}}(CH_2)_{12}CH_3 \\
| \\
CH_2O\overset{O}{\overset{\|}{C}}(CH_2)_7CH{=}CH(CH_2)_7CH_3
\end{array}
$$

A typical molecule in vegetable oils

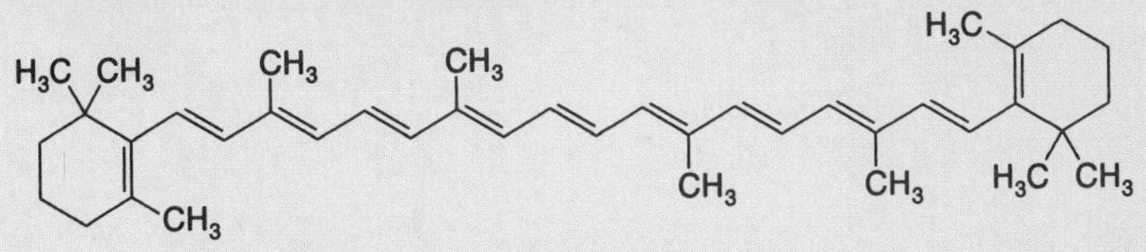

β-Carotene

2.2 NAMING THE ALKENES

The IUPAC names of alkenes end in *-ene,* and the double bond takes precedence over substituents in numbering the parent chain.

Consistent with the IUPAC rules for any family, the rules for the alkenes specify the *name ending,* how to pick out the *parent chain* or *parent ring,* how to *number the chain or ring,* and how to designate substituent groups. For alkenes and cycloalkenes, the IUPAC rules are as follows.

1. Use the ending *-ene* for all alkenes and cycloalkenes.
2. For open-chain alkenes, identify the parent chain as the longest sequence of carbons that *includes the double bond.* Name this chain as if it were that of an alkane and then change the *ane* ending to *ene.* This gives the basic *name* of the parent chain, except that the location of the double bond is yet to be specified. For cyclic alkenes, the ring is the parent in all situations we will encounter.

   For example, the longest chain *that includes the double bond* in the first structure has six carbons. There is a longer chain of seven carbons, but it does not include the double bond. In the second structure, the parent cycloalkane is cyclopentane, so by changing the ending to *ene* we have cyclopentene.

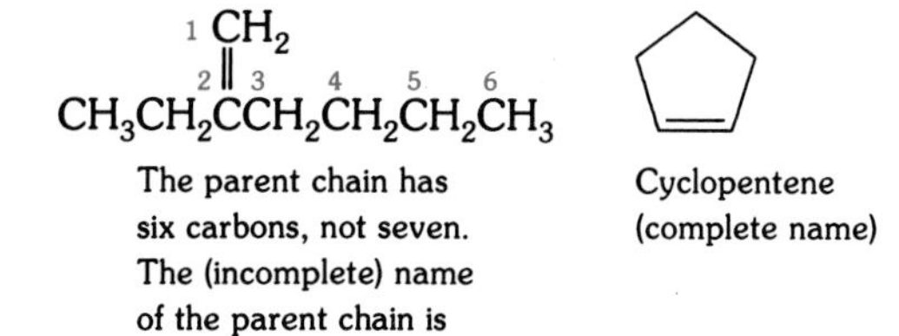

The parent chain has six carbons, not seven. The (incomplete) name of the parent chain is hexene.

Cyclopentene (complete name)

3. For open-chain alkenes, number the parent chain from whichever end gives the lower number to the first carbon of the double bond.

   This rule gives precedence to the location of the double bond over the location of the first substituent on the parent chain. For example,

CH_3 (on C-4)

$CH_3CHCH_2CH{=}CH_2$ The double bond is at position 1, not 4.

5 4 3 2 1

Not 1 2 3 4 5

4-Methyl-1-pentene (complete name)
Not: 2-methyl-4-pentene

4. For cycloalkenes, always give position 1 to one of the two carbons at the double bond. To decide which carbon gets this number, number the ring atoms from carbon 1 *through the double bond* in whichever direction reaches a substituent first.

   For example, the numbers inside the ring represent the correct numbering, not the numbers outside the ring.

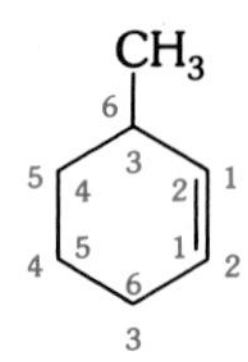

3-Methylcyclohexene (complete name)
Not: 6-methylcyclohexene

5. To the name begun with rules 1 and 2, place the number that locates the first carbon of the double bond as a prefix, and separate this number from the name by a hyphen.
6. If substituents are on the parent chain or ring, complete the name obtained by rule 5 by placing the names and location numbers of the substituents as prefixes.
   Remember to separate numbers from numbers by commas, but use hyphens to connect a number to a word.

Several correctly named alkenes are shown next[1] with the common names of three given in parentheses. The ending *-ylene* characterizes the common names of open-chain alkenes.

■ The common name, ethylene, is also allowed by the IUPAC as the name for ethene.

$CH_2{=}CH_2$ Ethene (ethylene)

$CH_3CH{=}CH_2$ Propene (propylene)

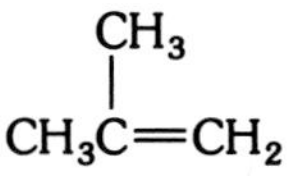

$CH_3C(CH_3){=}CH_2$ 2-Methylpropene (isobutylene)

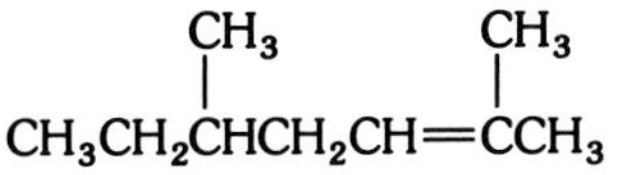

$CH_3CH_2CH(CH_3)CH_2CH{=}C(CH_3)CH_3$ 2,5-Dimethyl-2-heptene

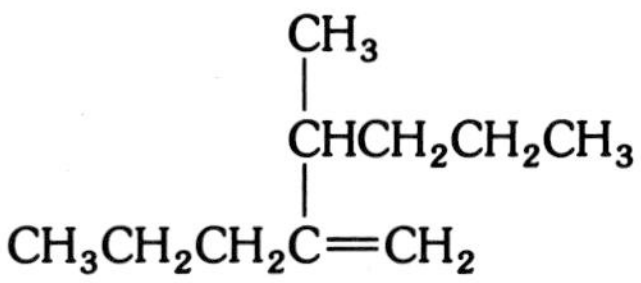

$CH_3CH_2CH_2C(CH(CH_3)CH_2CH_2CH_3){=}CH_2$ 3-Methyl-2-propyl-1-hexene

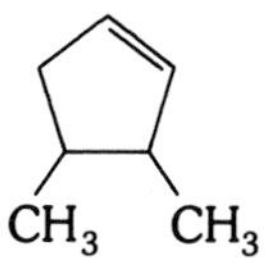

3,4-Dimethylcyclopentene (ring with two CH_3 substituents)
Not 4,5-dimethylcyclopentene

$Cl(CH_2)_6CH{=}CH_2$ 8-Chloro-1-octene

No number is used to locate the double bond in 3,4-dimethylcyclopentene because, by rule 4, the double bond can only be at position 1.

EXAMPLE 2.1
Naming an Alkene

Write the name of the following alkene.

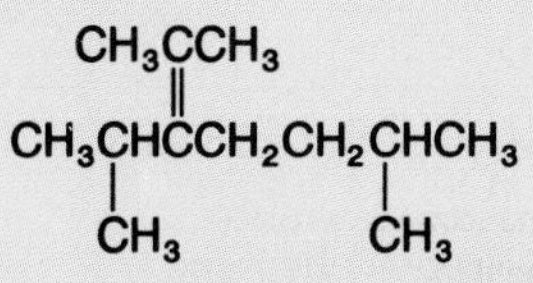

$$\begin{array}{c} CH_3CCH_3 \\ \| \\ CH_3CH(CH_3)CCH_2CH_2CH(CH_3)CH_3 \end{array}$$

ANALYSIS The longest chain that includes the double bond must be identified and numbered from whichever end gives the first carbon of the double bond the lower of two possible numbers. The parent is an *alkene* with a chain of seven carbons, a heptene. The following numbering of the parent chain gives the double bond position 2. (The alternative numbering, right to left, would have given the double bond position 5.)

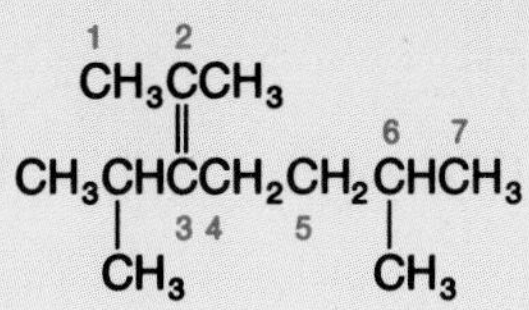

$$\begin{array}{c} \overset{1}{C}H_3\overset{2}{C}CH_3 \\ \| \\ CH_3\overset{3}{C}H(CH_3)\overset{4}{C}\overset{5}{C}H_2\overset{6}{C}H_2\overset{7}{C}H(CH_3)CH_3 \end{array}$$

[1] Propene is not named 1-propene because by rule 3 there is no 2-propene. 2-Methylpropene is not named 2-methyl-1-propene because the 1 is not needed.

The parent alkene is thus 2-heptene. It holds two methyl groups (positions 2 and 6) and one isopropyl group (position 3). The names and location numbers are next assembled into the name.

SOLUTION The correct name is

3-Isopropyl-2,6-dimethyl-2-heptene

A comma separates two numbers

Hyphens separate numbers from names

CHECK The most common error that students make is *to fail to find the longest chain* that includes the double bond. The first check step, then, is to go back over the answer to see if there is a chain holding the alkene group that is longer than 7 carbons. Use a colored pen to draw an enclosure for this chain so that all substituents are outside. Then move in from either end of the chain, counting carbons, to see which starting point yields the lower number for the beginning of the double bond. All the other numbers must then fall into place. Another common error is failure to identify alkyl groups correctly, so double-check these.

PRACTICE EXERCISE 1

Write the IUPAC names for the following compounds.

(a) $(H_3C)(CH_3)C{=}CH_2$

(b) $CH_3CH(CH_3)CH_2C({=}C(CH_3)CH_2CH_3)CH_2CH(CH_3)CH_3$

(c) $CH_3CH{=}CHCl$

(d) $BrCH_2CH{=}CH_2$

(e) $CH_3CH(CH_2CH_3)CH_2CH{=}CH_2$

(f) (cyclohexene ring bearing a CH_3 group)

PRACTICE EXERCISE 2

Write condensed structures for each of the following.

(a) 4-Methyl-2-pentene (b) 3-Propyl-1-heptene

(c) 4-Chloro-3,3-dimethyl-1-butene (d) 2,3-Dimethyl-2-butene

When a compound has two double bonds, it is named as a *diene* with two numbers in the name to specify the locations of the double bonds. For example,

$CH_2{=}C(CH_3)CH{=}CH_2$

2-Methyl-1,3-butadiene

(cyclohexadiene ring, carbons numbered 1–6, double bonds at 1–2 and 4–5)

1,4-Cyclohexadiene

This pattern can be easily extended to *trienes*, *tetraenes*, and so forth.

2.3

GEOMETRIC ISOMERS

The alkenes and cycloalkanes can exhibit geometric isomerism because there is no free rotation at the double bond or in a ring.

The six atoms at a double bond, the two carbons and the four atoms attached to them, all lie in the same plane, as illustrated in Figure 2.1, which shows the simplest alkene. The bond angles are 120°.

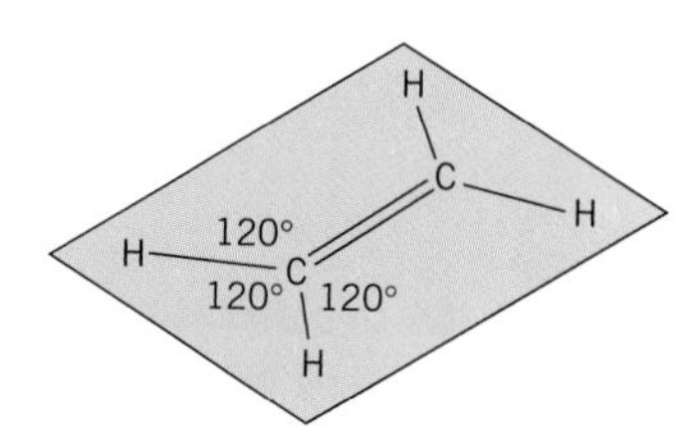

FIGURE 2.1
The geometry at a carbon–carbon double bond.

Some Alkene Isomers Have Identical Constitutions but Different Geometries

Alkenes can exist as isomers in three ways. They can have different carbon skeletons, as shown by 1-butene and 2-methylpropene. These two compounds are *constitutional isomers.* Two alkenes can also have identical carbon skeletons but differ in the locations of their double bonds, as in 1-butene and 2-butene. These two are also constitutional isomers because their H atoms are attached differently to the skeleton.

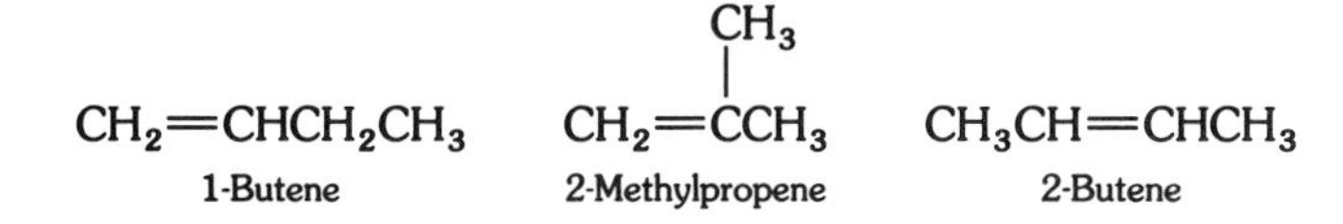

Some alkenes have identical constitutions *including the location of the double bond,* but differ only in the geometry at this bond, as seen in the two isomers with the constitution of 2-butene, $CH_3CH{=}CHCH_3$.

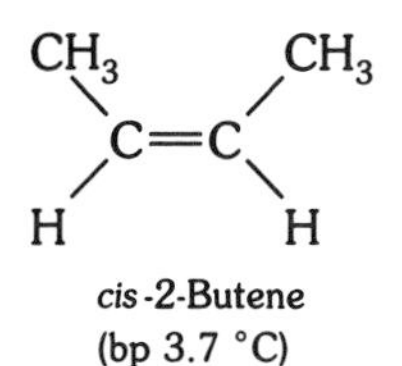

cis-2-Butene
(bp 3.7 °C)

H, CH3, C=C, CH3, H

trans-2-Butene
(bp 0.9 °C)

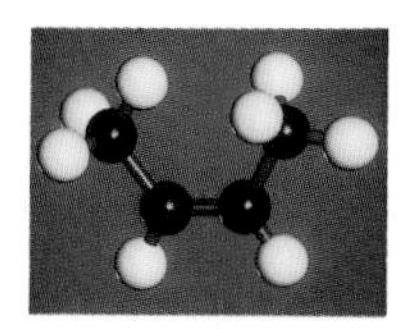

cis-2-Butene and *trans*-2-butene differ only in the *directions* taken by their end-of-chain methyl groups. Isomers that differ in geometry but have identical constitutions are called **geometric isomers,** and the phenomenon is called **geometric isomerism.**

Geometric Isomers Are Possible because There Is No Free Rotation at the Double Bond One of the bonds at a double bond is a pi bond resulting from the side-to-side overlap of two unhybridized 2*p* orbitals, one on each carbon (see Fig. 2.2). The rotation of two groups joined by a double bond could happen only if the pi bond breaks. This ordinarily costs too much energy, making geometric isomers possible.

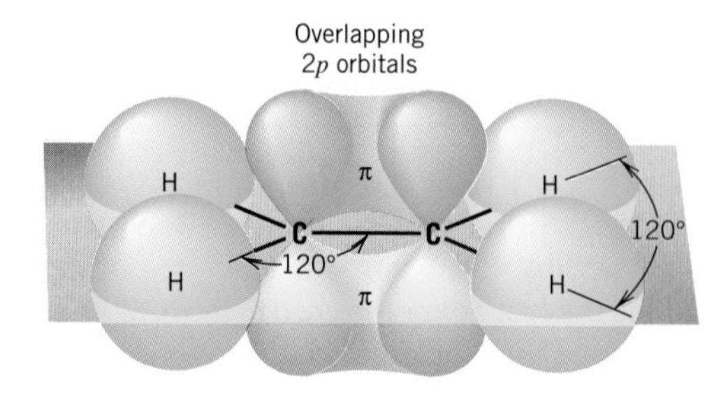

FIGURE 2.2
Overlapping 2*p* orbitals form the pi bond at a double bond.

■ The *side* of a double bond is not the same as the *end* of a double bond.

Same side

C=C } One end

When two designated substituents are on the same side of the double bond, they are said to be *cis* to each other. When they are on opposite sides, they are *trans* to each other. (Sometimes geometric isomerism is called *cis–trans isomerism*.) The designations of cis and trans can be made parts of the names of geometric isomers, as the examples of *cis*- and *trans*-2-butene earlier showed.

When There Are Two Identical Groups at One End of a Double Bond, Geometric Isomers Are Not Possible If one end of a double bond has two *identical* groups, like two Hs or two methyls, then there is nothing for a group at the other end to be *uniquely* cis or trans to. 1-Butene, for example, has two H atoms at one end of its double bond, so the ethyl group at the other end cannot be positioned to give geometric isomers. We can *write* structures that might appear to be isomers:

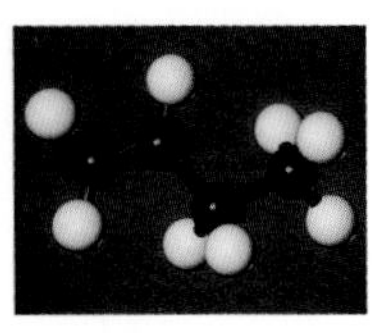

1-Butene

$$\underset{\text{1-Butene}}{(H)(H)C{=}C(H)(CH_2CH_3)} \quad \text{is the same as} \quad \underset{\text{1-Butene}}{(H)(H)C{=}C(CH_2CH_3)(H)}$$

However, they are actually identical. Simply flop the *whole* first structure over, top to bottom, to obtain the second. Whole-molecule flopping, of course, does not reorganize bonds into any new structure or geometry. Thus, there are no geometric isomers of 1-butene.

Geometric isomerism also occurs when the atoms involved at the ends of the double bond are halogen atoms or other groups, for example:

$$\underset{\textit{trans}\text{-1-Chloro-1-propene}}{(CH_3)(H)C{=}C(H)(Cl)} \qquad \underset{\textit{cis}\text{-1-Chloro-1-propene}}{(CH_3)(H)C{=}C(Cl)(H)}$$

EXAMPLE 2.2
Writing the Structures of Cis and Trans Isomers

Write the structures of the cis and trans isomers, if any, of the following alkene.

$$CH_3CH{=}CHCH_2CH_3$$

2-Pentene

ANALYSIS Notice first that geometric isomerism is possible in 2-pentene. At *neither* end of the double bond are the two groups identical. At one end there are H and CH_3; at the other end, H and CH_2CH_3. Therefore, we must draw structures of the geometric isomers.

To show the geometry of each isomer correctly, we start by writing a carbon–carbon double bond without any attached groups *spreading the single bonds at the carbon atoms at angles of about 120°.*

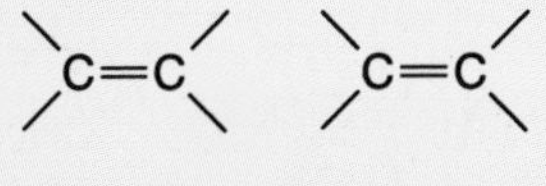

Then we attach the two groups that are at one of the ends of the double bond. We attach them *identically* to make identical partial structures.

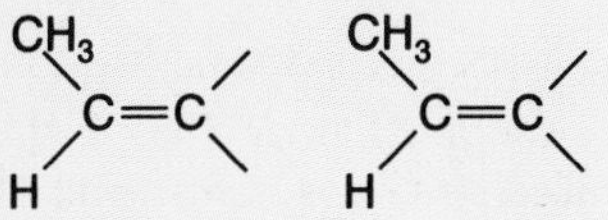

Finally, at the other end of the double bond, we draw the other two groups, only this time be sure that they are switched in their relative positions.

SOLUTION The geometric isomers of 2-pentene are

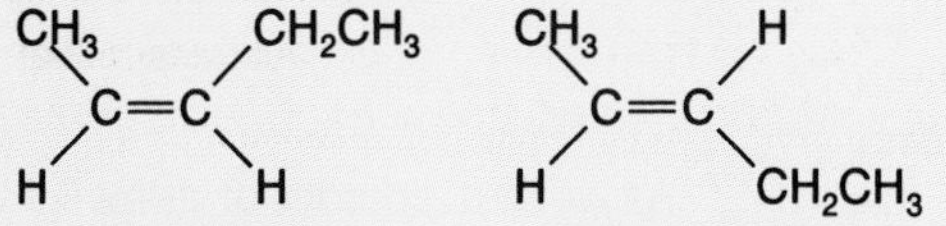

cis-2-Pentene

trans-2-Pentene

CHECK Be sure to check whether the two structures are geometric *isomers* and not two identical structures that are merely flip-flopped on the page.

PRACTICE EXERCISE 3

Write the structures of the cis and trans isomers, if any, of the following compounds.

(a) $CH_3CH_2C(CH_3){=}CHCH_3$ **(b)** $ClCH{=}CHCl$ **(c)** $CH_3C(CH_3){=}CH_2$ **(d)** $ClC(Cl){=}CHBr$

Cyclic Compounds Can Also Have Geometric Isomers The double bond is not the only source of restricted rotation; the ring is another. For example, two geometric isomers of 1,2-dimethylcyclopropane are known, and neither can be twisted into the other without breaking the ring open. This costs too much energy to occur spontaneously even at quite high temperatures.

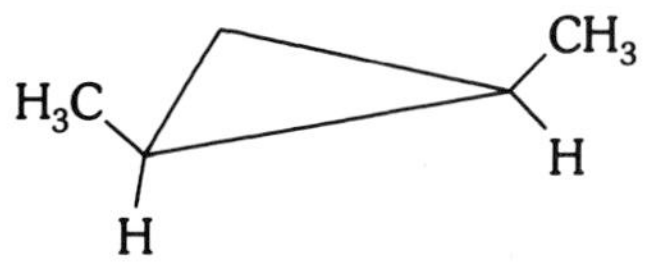

cis-1,2-Dimethylcyclopropane (bp 37 °C)

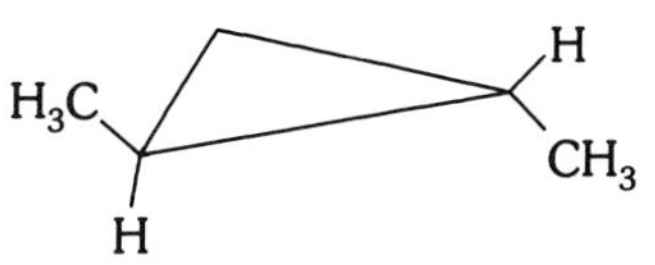

trans-1,2-Dimethylcyclopropane (bp 28 °C)

We'll see this kind of cis–trans isomerism in the many cyclic structures of carbohydrates; their *geometric* differences alone make most carbohydrates unusable in human nutrition. As we have said before, molecular geometry is as important in living processes as functional groups. For an interesting occurrence of cis–trans isomerism at the molecular level of life, see Special Topic 2.2.

SPECIAL TOPIC 2.2

THE PRIMARY CHEMICAL EVENT THAT LETS US SEE

The retina in the eye of a human being has two kinds of cells, rods and cones, that can convert absorbed light into signals which the brain perceives as vision. Rods equip us to see shades of gray; cones give us color vision. We know more about how rods work than cones, so our brief discussion is limited to what happens in the rods when they absorb light.

Light can be described as a stream of tiny energy packets called *photons* having a wide range of energies. Photons of the higher energy regions are in the ultraviolet and have sufficient energy to break sigma (single) bonds. Photons in visible light have less energy, but still enough to break the weaker pi bonds, particularly in molecules with repeating patterns of alternating double and single bonds. Just such a sequence of alternating double double and single bonds occurs in a polyunsaturated compound called 11-*cis*-retinal, the primary absorber of visible light photons in the rods of our eyes. Notice that the chain emerges on the same side or the *cis* side of the double bond at carbons numbered 11 and 12 (according to the numbering used for this system). Otherwise the structures are the same.

If you compare the structures of 11-*cis*- and all-*trans*-retinal to the structure of vitamin A, Special Topic 2.1, you will notice a remarkable similarity. The body needs vitamin A to make the retinals.

Molecules of 11-*cis*-retinal must be joined to a protein called *opsin* in order for the absorption of light photons to be coupled to a nerve signal. The carbon–oxygen double bond in 11-*cis*-retinal makes the joining possible. If we represent everything except the carbon–oxygen double bond of 11-*cis*-retinal by R, then the coupling to opsin can be represented by the structure R—CH=N—{opsin}. This material is called *rhodopsin.*

11-*cis*-Retinal

All-*trans*-retinal

As we said, photons of visible light can break certain pi bonds. When this occurs at a double bond, there is still the sigma bond to hold the molecule together. With the pi bond broken, however, *free rotation now becomes possible.* This is what happens when the vast carbon network of 11-*cis*-retinal in rhodopsin accepts a light photon. One side of the molecule flips to a *trans* configuration at the 11,12 linkage to form all-*trans*-retinal. This change in geometry affects the properties of the opsin portion of rhodopsin, and a series of chemical changes swiftly occurs that sends a signal to the brain. An equally swift set of chemical changes occurs in the cone of the eye to restore rhodopsin to its photon-accepting geometry. Thus cis–trans isomerization is at the heart of vision.

2.4

ADDITION REACTIONS OF THE DOUBLE BOND

The carbon–carbon double bond adds H_2, Cl_2, Br_2, HX, H_2SO_4, and H_2O, and it is attacked by strong oxidizing agents, including ozone.

In an **addition reaction,** pieces of an adding molecule become attached to opposite ends of the double bond which then becomes a single bond. All additions to an alkene double bond thus have the following features, where *X*—*Y* is the adding molecule:

$$\text{>C=C<} + X\text{—}Y \longrightarrow \text{—}\underset{X}{\overset{|}{\underset{|}{C}}}\text{—}\underset{Y}{\overset{|}{\underset{|}{C}}}\text{—}$$

We'll study a few examples and then see how addition reactions take place.

Hydrogen Adds to a Double Bond and Saturates It In the presence of a powdered metal catalyst, like powdered nickel or platinum metal, hydrogen adds to double bonds. The reaction, sometimes called **hydrogenation,** converts an alkene to an alkane as follows.

$$>C=C< + H-H \xrightarrow{\text{Ni catalyst}} -\underset{H}{\overset{|}{C}}-\underset{H}{\overset{|}{C}}-$$

Specific examples are

$$\underset{\text{Ethene}}{CH_2=CH_2} + H-H \xrightarrow{\text{Ni catalyst}} \underset{H}{CH_2}-\underset{H}{CH_2} \text{ or } \underset{\text{Ethane}}{CH_3CH_3}$$

$$\text{3-Methylcyclopentene} + H_2 \xrightarrow{\text{Ni catalyst}} \text{Methylcyclopentane}$$

3-Methylcyclo-pentene

Methylcyclopentane

Notice that each alkane product has the same carbon skeleton as the alkene used to make it.

The Net Effect of Hydrogenation Occurs at the Molecular Level of Life Molecules of H_2 and powdered metal catalysts, of course, are unavailable in the body. Cells, however, have molecular carrier systems that deliver the pieces of H—H to carbon–carbon double bonds. One piece is $H:^-$, donated by a carrier enzyme to one end of the double bond. The other piece is H^+, either donated by the carrier or plucked from a proton donor of the surrounding buffer and given to the carbon that was at the other end of the double bond. The *net* effect is the addition of H_2 because $H:^-$ and H^+ together add up to one H:H molecule.

■ $H:^-$ (the hydride ion) must be donated by the carrier *directly* to the acceptor and not through the solution. When exposed directly to water, $H:^-$ reacts vigorously:

$$H:^- + H-OH \longrightarrow H-H + OH^-$$

EXAMPLE 2.3
Writing the Structure of the Product of the Addition of Hydrogen to a Double Bond

Write the structure of the product of the following reaction.

$$CH_3CH_2\overset{CH_3}{\overset{|}{C}}=\overset{CH_3}{\overset{|}{C}}CH_2CH_2CH_3 + H_2 \xrightarrow{\text{Ni catalyst}} ?$$

ANALYSIS The only change occurs at the double bond, and all the rest of the structure goes through the reaction unchanged. *This is true of all of the addition reactions that we will study.* Therefore copy the structure of the alkene just as it is, except leave only a single bond where the double bond was. Then increase by one the number of hydrogens at each carbon of the original double bond.

■ The addition of hydrogen is often called the *reduction* of the double bond.

SOLUTION The structure of the product can be written as

$$CH_3CH_2\underset{H}{\overset{CH_3}{C}}-\underset{H}{\overset{CH_3}{C}}CH_2CH_2CH_3 \text{ or } CH_3CH_2\overset{CH_3}{CH}-\overset{CH_3}{CH}CH_2CH_2CH_3$$

CHECK Make sure that the product has the identical carbon skeleton as the starting material. Then see that *each* carbon of the original alkene group has one more H.

One major goal in these chapters is to learn some chemical properties of functional groups. We have just learned a chemical "map sign" for the carbon–carbon double bond, one of its important chemical properties. It can be made to add hydrogen, and when it does, it changes to a single bond as each of its carbon atoms picks up one hydrogen atom. This sentence states a chemical fact about the double bond that has to be learned. Learn it, however, by working illustrations involving specific alkenes. There are too many individual reactions to memorize, so use your memory work to learn the *kinds* of reactions and what they do *in general* to change a molecule. As you work each part of the following practice exercise, say to yourself the *general* fact, the chemical map sign about hydrogenation, each time you apply it to a specific case.

PRACTICE EXERCISE 4

Write the structures of the products, if any, of the following.

(a) $CH_3CH{=}CH_2 + H_2 \xrightarrow{\text{Ni catalyst}}$

(b) $CH_3CH_2CH_3 + H_2 \xrightarrow{\text{Ni catalyst}}$

(c) 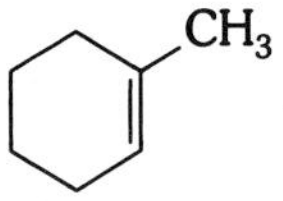 (1-methylcyclohexene, ring with CH_3 on the double-bonded carbon) $+ H_2 \xrightarrow{\text{Ni catalyst}}$

(d) $CH_3(CH_2)_7CH{=}CH(CH_2)_7CO_2H + H_2 \xrightarrow{\text{Ni catalyst}}$

Chlorine and Bromine Also Add to Double Bonds Without any need for a special catalyst, both Cl_2 and Br_2 *rapidly* add to the carbon–carbon double bond. Iodine (I_2) does not add, and fluorine (F_2) reacts explosively with almost any organic compound to give a mixture of products.

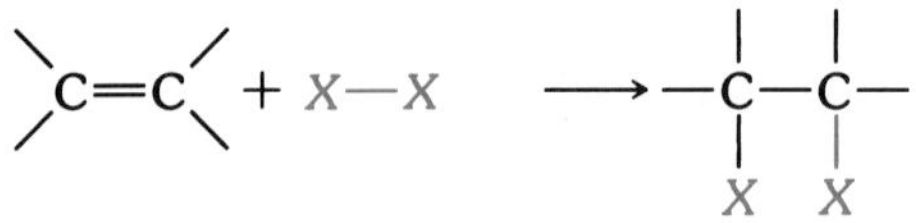

$$\text{>C=C<} + X{-}X \longrightarrow {-}\underset{X}{\overset{|}{\underset{|}{C}}}{-}\underset{X}{\overset{|}{\underset{|}{C}}}{-}$$

X = Cl or Br

Use a good fume hood and protective gloves when dispensing bromine.

Specific examples are

$$CH_3CH{=}CH_2 + Br_2 \longrightarrow CH_3\underset{Br}{\underset{|}{C}H}{-}\underset{Br}{\underset{|}{C}H_2}$$

Propene — 1,2-Dibromopropane

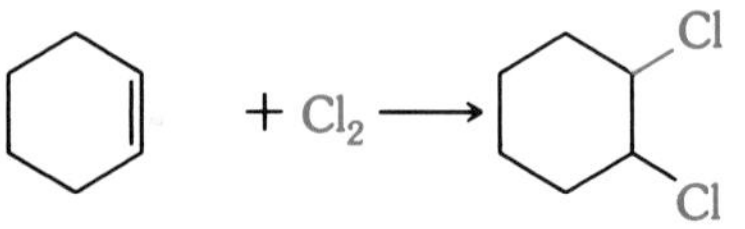

$$\text{Cyclohexene} + Cl_2 \longrightarrow \text{1,2-Dichlorocyclohexane (Cl, Cl on adjacent ring carbons)}$$

Cyclohexene — 1,2-Dichlorocyclohexane

EXAMPLE 2.4

Writing the Structure of the Product of the Addition of Chlorine or Bromine to a Carbon–Carbon Double Bond

What compound forms in the following situation?

$$CH_3CH{=}CHCH_3 + Br_2 \longrightarrow ?$$

ANALYSIS This problem is very similar to that of the addition of hydrogen. We rewrite the alkene except that a single bond is left where the double bond was.

$$CH_3CH{-}CHCH_3 \quad \text{(Incomplete)}$$

Then we attach a bromine atom by a single bond to each carbon of the former double bond.

SOLUTION

$$\underset{\displaystyle Br}{CH_3\overset{}{C}H}{-}\underset{\displaystyle Br}{CHCH_3}$$

2,3-Dibromobutane

CHECK Make sure that the carbon skeleton in the product is identical to that of the original alkene, that the double bond has been replaced by a single bond, and that each carbon of the double bond carries one of the pieces (Br in this example) of the adding molecule.

■ Bromine is dark brown, but the dibromoalkanes are colorless. So an unknown organic compound that rapidly decolorizes bromine quite likely has a double bond.

PRACTICE EXERCISE 5

Complete the following equations by writing the structures of the products. If no reaction occurs under the conditions shown, write "no reaction."

(a) $CH_3\overset{\displaystyle CH_3}{\overset{|}{C}}{=}CH_2 + Br_2 \longrightarrow$

(b) $CH_3CH_2CH_2CH_3 + Cl_2 \longrightarrow$

(c) $CH_2{=}CHCH_2CH_3 + Cl_2 \rightarrow$

(d) $CH_3CH{=}CHCH_3 + H_2 \xrightarrow{\text{Ni catalyst}}$

Hydrogen Chloride, Hydrogen Bromide, and Sulfuric Acid Add Easily to Double Bonds If gaseous hydrogen chloride or hydrogen bromide is bubbled into an alkene or if concentrated sulfuric acid is mixed with it, the following kind of reaction takes place. The pattern is the same in all these additions. We'll let H—*G* represent any of these reactants, where *G* stands for any electron-rich group that we will study, like Cl or OSO_3H.

$$>C{=}C< + H{-}G \longrightarrow -\underset{\displaystyle H}{\overset{|}{\underset{|}{C}}}-\underset{\displaystyle G}{\overset{|}{\underset{|}{C}}}- \qquad (G = Cl, Br, \text{ or } OSO_3H)$$

An example:

$$CH_2{=}CH_2 + H{-}Cl \longrightarrow \underset{\displaystyle H}{\underset{|}{CH_2}}{-}\underset{\displaystyle Cl}{\underset{|}{CH_2}} \quad \text{or} \quad CH_3CH_2Cl$$

Unsymmetrical Reactants Add Selectively to Unsymmetrical Double Bonds We now have a small complication. H—*G* is not a symmetrical molecule, like H—H, Br—Br, or Cl—Cl. When symmetrical molecules add to a double bond, it doesn't matter which end of the double bond gets which half of the adding molecule. But it matters when H—*G* adds, *if the double bond is itself unsymmetrical.* An *unsymmetrical double bond* is one whose two carbon atoms hold unequal numbers of hydrogen atoms. For example, 1-butene, $CH_2{=}CHCH_2CH_3$, and propene, $CH_3CH{=}CH_2$, both have unsymmetrical double bonds. Each molecule has one carbon at the double bond with two Hs, but the other carbon has one H, an unequal number. The double bond in 2-butene, $CH_3CH{=}CHCH_3$, however, is symmetrical; one H is at each carbon.

When H—Cl(*g*) adds to propene we could imagine obtaining 1-chloropropane, 2-chloropropane, or a mixture of the two, perhaps 50 : 50. Let's see what actually happens. We have

$$CH_3CH{=}CH_2 + H{-}Cl \longrightarrow \underset{\substack{|\\H}}{CH_3CH}{-}\underset{\substack{|\\Cl}}{CH_2} \quad \text{or} \quad CH_3CH_2CH_2Cl$$

1-Chloropropane
(Very little forms.)

or

$$CH_3CH{=}CH_2 + H{-}Cl \longrightarrow \underset{\substack{|\\Cl}}{CH_3CH}{-}\underset{\substack{|\\H}}{CH_2} \quad \text{or} \quad \underset{\substack{|\\Cl}}{CH_3CHCH_3}$$

2-Chloropropane
(The major product)

The actual product is largely 2-chloropropane, and very little of its isomer, 1-chloropropane, forms. In other words, the reactant, H—Cl, adds to the unsymmetrical double bond selectively.

Markovnikov's Rule Predicts Directions in Unsymmetrical Additions Vladimer Markovnikov (1838–1904), a Russian chemist, was the first to notice that unsymmetrical alkenes add unsymmetrical reactants in one direction. Which direction can be predicted by **Markovnikov's rule.**[2]

■ "Them that has, gits" applies here, too.

> **Markovnikov's Rule** When an unsymmetrical reactant of the type H—G adds to an unsymmetrical alkene, the carbon with the greater number of hydrogens gets one more H.

The following examples illustrate Markovnikov's rule in action.

$$\underset{\substack{|\\CH_3}}{CH_3C}{=}CH_2 + H{-}Cl \longrightarrow \overset{\substack{Cl\\|}}{\underset{\substack{|\\CH_3}}{CH_3CCH_3}} \quad \left(\textit{Not } \underset{\substack{|\\CH_3}}{CH_3CHCH_2Cl}\right)$$

2-Methylpropene — *t*-Butyl chloride

[2] When hydrogen bromide is used, it is important that no peroxides, compounds of the type R—O—O—H or R—O—O—R, be present. Traces of peroxides commonly form in organic liquids that are stored in contact with air for long periods. When peroxides are present, the addition of H—Br occurs in the direction opposite to that predicted by Markovnikov's rule. Peroxides catalyze the anti-Markovnikov addition *only* of HBr, not of HCl or H_2SO_4.

$$\text{1-Methylcyclohexene} + H{-}Cl \longrightarrow \text{1-Chloro-1-methylcyclohexane} \quad (\textit{Not}\ \text{2-chloro-1-methylcyclohexane})$$

Concentrated Sulfuric Acid Actually Dissolves Alkenes When an alkene is mixed with concentrated sulfuric acid, the hydrocarbon dissolves and heat evolves. How can two substances of such radically different polarities dissolve together? An alkane does not behave this way at all but merely forms a separate layer that floats on the sulfuric acid (see Fig. 2.3).

The alkene dissolves because it reacts by an addition reaction to form an alkyl hydrogen sulfate, which is very polar. For example,

$$\underset{\text{1-Propene}}{CH_3CH{=}CH_2} + \underset{\substack{\text{Sulfuric acid}\\(H_2SO_4)}}{H{-}O{-}S(=O)_2{-}O{-}H} \longrightarrow \underset{\text{Isopropyl hydrogen sulfate}}{CH_3CH(CH_3){-}O{-}S(=O)_2{-}O{-}H}$$

Molecules of the alkyl hydrogen sulfates are very polar, and as soon as they form, they move smoothly into the polar sulfuric acid layer and out of the nonpolar alkene layer. The heat generated by the reaction and the strongly acidic nature of the mixture causes side reactions that generate black by-products.

■ The sodium salts of long-chain alkyl hydrogen sulfates are detergents, for example, $CH_3(CH_2)_{11}OSO_3Na$.

Symmetrical Double Bonds Do Not Add H—*G* Reactants Selectively Markovnikov's rule does not apply when the double bond is symmetrical. Reactants like H—*G* add in both of the two possible directions, and a mixture of isomers can form. For example,

$$\underset{\text{2-Pentene}}{CH_3CH_2CH{=}CHCH_3} + H{-}Br \longrightarrow \underset{\text{3-Bromopentane}}{CH_3CH_2CH(Br)CH_2CH_3} + \underset{\text{2-Bromopentane}}{CH_3CH_2CH_2CH(Br)CH_3}$$

(For a reminder of why we don't try to balance an equation such as this, check back to Special Topic 1.3.)

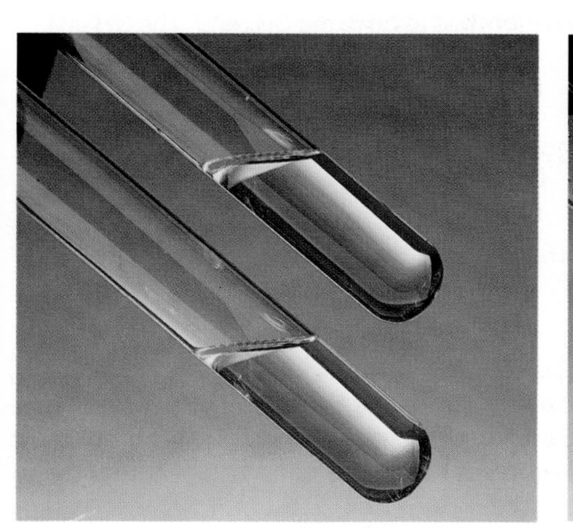

FIGURE 2.3
The effect of concentrated sulfuric acid on an alkane and an alkene. In the photo on the left, the alkane (top tube) and the alkene (bottom tube) are seen as clear, colorless liquids. The photo on the right shows the two systems soon after concentrated sulfuric acid has been added to each. The alkane floats unaffected on the acid (top tube), but the alkene (bottom tube) has already begun to react and form an alkyl hydrogen sulfate (together with darkly colored matter produced by side reactions). The alkyl hydrogen sulfate is soluble in the remaining concentrated sulfuric acid, and so all of the alkene will appear to dissolve. (The alkane is cyclohexane and the alkene is cyclohexene.)

Water Adds to Double Bonds to Give Alcohols Water adds to the carbon–carbon double bond provided that an acid catalyst (or the appropriate enzyme) is present. The product is an alcohol. Water alone, or aqueous bases have no effect on alkenes whatsoever.

$$\text{>C=C<} + \text{H—OH} \xrightarrow[\text{heat}]{\text{H}^+} \text{—}\underset{\text{H}}{\underset{|}{\overset{|}{\text{C}}}}\text{—}\underset{\text{OH}}{\underset{|}{\overset{|}{\text{C}}}}\text{—}$$

Alkene — Alcohol

Specific examples are

$$\underset{\text{Ethene}}{CH_2{=}CH_2} + \text{H—OH} \xrightarrow[240\ ^\circ\text{C (closed vessel)}]{10\%\ H_2SO_4} \underset{\text{Ethyl alcohol}}{CH_3CH_2OH}$$

$$\underset{\text{2-Methylpropene}}{CH_3\underset{CH_3}{\underset{|}{C}}{=}CH_2} + \text{H—OH} \xrightarrow[25\ ^\circ\text{C}]{10\%\ H_2SO_4} \underset{t\text{-Butyl alcohol}}{CH_3\overset{OH}{\overset{|}{\underset{CH_3}{\underset{|}{C}}}}CH_3} \quad \left(\textit{Not}\ CH_3\underset{CH_3}{\underset{|}{C}}HCH_2OH\right)$$

As you can see, Markovnikov's rule applies to this reaction, too. One H in H—OH goes to the carbon with the greater number of hydrogens, and the OH goes to the other carbon of the double bond.

Notice in the last example, and in all previous examples of addition reactions, that the carbon skeleton does not change. Although this is not always true, it will be in all the examples we will use as well as in all the Practice and Review Exercises.

EXAMPLE 2.5
Using Markovnikov's Rule

What product forms in the following situation?

$$CH_3CH{=}CH_2 + \text{H—OH} \xrightarrow[\text{heat}]{\text{H}^+}$$

ANALYSIS As in all of the addition reactions that we are studying, the carbon skeleton of the alkene can be copied over intact, except that a single bond is shown where the double bond was.

$$CH_3CH{—}CH_2 \qquad \text{(Incomplete)}$$

To decide which carbon of the original double bond gets the H atom from the water molecule, we use Markovnikov's rule. The H atom has to go to the CH_2 end. The OH unit from H—OH goes to the other carbon.

SOLUTION The product is

$$CH_3\underset{OH}{\underset{|}{C}}H{—}\underset{H}{\underset{|}{C}}H_2 \quad \text{or} \quad CH_3\underset{OH}{\underset{|}{C}}HCH_3$$

Isopropyl alcohol

CHECK At this stage be sure to see whether each carbon in the product has four bonds. If not, you can be certain that some mistake has been made. This is always a useful way to avoid at least some of the common mistakes made in solving a problem such as this. Then double-check that the carbon of the double bond initially having the greater number of Hs has been given one more.

PRACTICE EXERCISE 6

Write structures for the product(s), if any, that would form under the conditions shown. If no reaction occurs, write "no reaction."

(a) $CH_2{=}CHCH_2CH_3 + HCl \longrightarrow$

(b) $CH_3\overset{\overset{\displaystyle CH_3}{|}}{C}{=}CH_2 + HBr \longrightarrow$

(c) $CH_3CH{=}\overset{\overset{\displaystyle CH_3}{|}}{C}{-}C_6H_{11}$ (cyclohexyl) $+ H_2O \xrightarrow[heat]{H^+}$

(d) H_3C–cyclohexene (methyl on the double-bond carbon) $+ H{-}OH \xrightarrow[heat]{H^+}$

(e) cyclohexane $+ H{-}OH \xrightarrow[heat]{H^+}$

Double Bonds Are Attacked by Many Oxidizing Agents With two pairs of electrons, the double bond is more electron-rich than a single bond, so electron-seeking reagents attack it. These include oxidizing agents. Hot solutions of potassium permanganate ($KMnO_4$) and potassium dichromate ($K_2Cr_2O_7$), for example, vigorously oxidize molecules at carbon–carbon double bonds. The oxidations begin as addition reactions, but continue beyond to the point where alkene molecules are split apart at the double bond. We will not study any of the details or learn how to predict products. We only note the fact that the carbon–carbon double bond makes a molecule susceptible to attack by strong oxidizing agents. This is the fact to remember. The products of alkene oxidations can be ketones, carboxylic acids, carbon dioxide, or mixtures of these. Alkanes, in sharp contrast, are inert toward these oxidizing agents.

$R{-}\overset{\overset{\displaystyle O}{\|}}{C}{-}R'$

Ketones

$R{-}\overset{\overset{\displaystyle O}{\|}}{C}{-}OH$

Carboxylic acid

The permanganate ion is intensely purple in water, and as it oxidizes double bonds it changes to manganese dioxide, MnO_2, a brownish, sludge-like, insoluble solid (see Fig. 2.4). The dichromate ion is bright orange in water, and when it acts as an oxidizing agent, it changes to the bright green, hydrated chromium(III) ion, $Cr^{3+}(aq)$, as seen in Figure 2.5.

(a)

(b)

FIGURE 2.4
Permanganate oxidation. (*a*) Crystals of potassium permanganate are so deeply purple that they appear almost black. The dilute $KMnO_4$ solution has a purple color. (*b*) After an oxidizable compound has been added to the $KMnO_4$ solution and the mixture has been heated to complete the oxidation of the compound, the solution has no color but now contains a precipitate of MnO_2.

FIGURE 2.5
Dichromate oxidation (*a*) Crystals of sodium dichromate are bright orange, and a dilute $Na_2Cr_2O_7$ solution has the orange color of the hydrated dichromate ion. (*b*) After the dichromate solution has been mixed with some alcohol and heated, the solution takes on the green color of the hydrated Cr^{3+} ion.

(*a*)

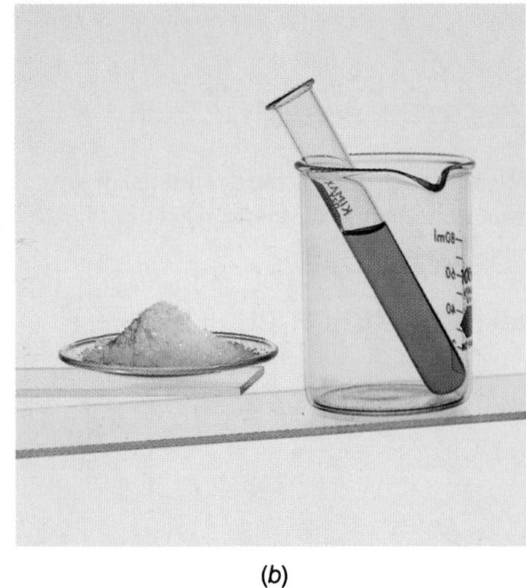

(*b*)

■ Ozone destroys any vegetation that has the green pigment chlorophyll, because chlorophyll contains double bonds.

Ozone Is One of the Most Powerful Oxidizing Agents Ozone, O_3, a pollutant in smog, is dangerous because it is a very powerful oxidizing agent that attacks biochemicals wherever they have carbon–carbon double bonds. Because such bonds occur in the molecules of all cell membranes, you can see that exposure to ozone must be kept very low. Even a concentration in air of only one part per million parts (1 ppm) warrants the declaration of a smog emergency condition. Special Topic 2.3 describes how incompletely burned hydrocarbons released in the exhaust from vehicle fuels contribute to the generation of ozone in smog.

SPECIAL TOPIC 2.3 OZONE IN SMOG

Ozone is dangerous to materials, including plants and lung tissue. The U.S. National Ambient[1] Air Quality Standard for ozone is a daily maximum one-hour average ozone concentration of only 0.12 ppm (120 ppb). Dozens of U.S. urban areas exceed this at least once per year. How does ozone originate in the air where we live?
the air where we live?

The *direct* and only source of ozone in the lower atmosphere is the combination of oxygen atoms with oxygen molecules.

$$O + O_2 + M \longrightarrow O_3 + M \qquad (1)$$

M is any molecule, like N_2 or O_2, that can absorb some of the kinetic energy involved in the collision of O with O_2. The collision that leads to product occurs on the surface of *M*. (Following collisions occurring elsewhere, the ozone molecule breaks up instantly because it carries too much energy.) Thus the earlier question, "How does ozone originate?" becomes a new question, "How are oxygen *atoms* generated?" Once oxygen atoms are generated, there will be ozone. The chief source of oxygen atoms is the breakup of molecules of nitrogen dioxide, an air pollutant, a breakup made possible by solar energy.

$$NO_2 + \text{solar energy} \longrightarrow NO + O \qquad (2)$$

Oxides of Nitrogen The NO_2 for Equation (2) is made from NO, nitrogen monoxide. NO is produced inside vehicle engine cylinders where the direct combination of nitro-

[1] *Ambient* means "all surrounding, all encompassing."

(a)

FIGURE 1 (b)

The effect of smog on visibility. (a) A clear day. (b) A day of heavy smog.

gen and oxygen is possible because of the high temperature and pressure.

$$N_2 + O_2 \xrightarrow[\text{pressure}]{\text{high temperature}} 2NO \qquad (3)$$

As soon as newly made NO, now in the exhaust gas, hits the cooler outside air, it reacts further with oxygen to give nitrogen dioxide, NO_2. NO_2 gives smog its reddish-brown color (see Figure 1).

$$2NO + O_2 \longrightarrow 2NO_2 \qquad (4)$$

There is thus an ample supply of NO_2 in air made smoggy by heavy vehicle traffic.

Ultraviolet Energy Solar energy, which can break down NO_2 (Equation 2), consists of a stream of tiny packages of energy called *photons.* Some have relatively high energies and so are in the ultraviolet region of light. Such photons give us no visual sensation, but their energies cause grave harm, like severe sunburn, eye burn, and even skin cancer.

Virtually all of the dangerous ultraviolet radiation coming from the sun is absorbed high in the stratosphere. However, some ultraviolet energy gets to where we live where it is able to generate ozone in smog-filled air.

An interesting feature of reaction 2 is that it produces nitrogen monoxide, NO, along with oxygen atoms. *Nitrogen monoxide is able to destroy ozone as follows.*

$$NO + O_3 \rightarrow NO_2 + O_2 \qquad (5)$$

Reactions 1, 2, and 5, when added together, give us *no net chemical effect!* (Try it.) How then does ozone develop in our atmosphere at all?

Reactions 1–5 proceed at their own rates under their own kinetic laws. These rates are unequal, so the reactions actually do not exactly cancel each other. The net effect is that a small, steady-state concentration of ozone develops even in clean air. The range of ozone concentration in clean air, however, is very low, between 20 and 50 ppb. In the more polluted urban areas, levels as high as 400 ppb commonly occur for brief periods of time each year.

Unburned Hydrocarbons and the Ozone in Smog Reaction 5 is the reaction that can make most ozone disappear, *but other substances are able to remove NO from reaction 5 before it is used to destroy ozone.* Such removal of NO other than by reaction 5 enables a buildup in the ozone level of the lower atmosphere.

Net destroyers of nitrogen monoxide are themselves present in vehicle exhaust—the unburned or partially oxidized hydrocarbons remaining from the incomplete combustion of the fuel. In sunlight and oxygen, some unburned hydrocarbon molecules are changed into organic derivatives of hydrogen peroxide called *peroxy radicals* usually symbolized as $R{-}O{-}O\cdot$, or simply RO_2. Peroxy radicals originate chiefly by the reaction of unburned hydrocarbons with hydroxyl radicals, HO. These arise by a number of mechanisms in polluted air, but we will not go into the details of how HO radicals form. Suffice it to say, HO radicals and hydrocarbons lead to ROO radicals, and these destroy NO molecules as follows.

$$ROO + NO \longrightarrow RO + NO_2$$

This reaction reduces the supply of ozone-destroying NO.

The development of ozone in smog thus depends on the formation of ROO radicals whose formation, in turn, depends on hydrocarbon emissions. Specialists generally agree that emissions of hydrocarbons must be significantly reduced before the ozone problem will be appreciably solved. Even the loss of hydrocarbons in the vapors from fuel tanks or at gas pumps has to be reduced.
[A reference: J. H. Seinfeld. "Urban Air Pollution: State of the Science." *Science,* February 10, 1989, page 745.]

2.5

HOW ADDITION REACTIONS OCCUR

The carbon–carbon double bond can accept a proton from an acid and change into a carbocation.

Because the double bond is somewhat electron-rich, it functions as a base, a proton acceptor, particularly when the proton donor is strong, like HCl, HBr, or $HOSO_3H$. When alkenes react with water, the *initial* step is the donation of a proton to the double bond from the acid catalyst, not from the water molecule. The water molecule is too weak a proton donor when the acceptor is a double bond.

When an Alkene Accepts a Proton, a Reactive Carbocation Forms Using the symbol H—*G* to represent any proton donor, the reaction we will now explain is the following addition of H*G* to propene.

■ In H—Cl, *G* is Cl. In H_2SO_4, *G* is OSO_3H. In dilute acid, the proton-donating species is H_3O^+, so *G* here is H_2O.

$$CH_3CH{=}CH_2 + H{-}G \longrightarrow CH_3\underset{\displaystyle G}{\underset{|}{C}}HCH_3 \qquad (not\ CH_3CH_2CH_2{-}G)$$

We can write the first step in this addition reaction as follows.

What once were two electrons in a double bond are now the electrons of this bond.

$$CH_3CH{=}CH_2 + H{-}\ddot{\underset{..}{G}}{:} \longrightarrow CH_3\overset{+}{C}H{-}CH_2{-}H + {:}\ddot{\underset{..}{G}}{:}^-$$

Propene Isopropyl carbocation

The curved arrows show how one of the two pairs of electrons of the double bond swings away from one carbon to form a bond to the proton.

One Carbon in a Carbocation Lacks an Octet The isopropyl cation that forms is an example of a **carbocation,** a positive ion in which carbon has a sextet not an octet of electrons. But notice that it is the *isopropyl* carbocation that forms, not the propyl carbocation, $CH_3CH_2CH_2^+$. We'll see why shortly, and when we do we'll understand Markovnikov's rule.

Carbocations are particularly unstable cations. All their reactions are geared to the recovery of an outer octet of electrons for carbon. The instant a carbocation forms, it strongly attracts any electron-rich species, and just such particles are produced when H—Cl, H—Br, or H—OSO_3H donate a proton to the double bond. When the proton leaves any of these acids, their conjugate base is released, which is Cl^-, Br^-, or $^-OSO_3H$ depending on which acid is used. We'll continue to generalize by using the symbol G^- for any of these conjugate bases.

The newly formed carbocation reacts with G^- in the next step of the addition.

$$CH_3\overset{+}{C}HCH_3 + {:}\ddot{\underset{..}{G}}{:}^- \longrightarrow CH_3\overset{\displaystyle {:}\ddot{G}{:}}{\overset{|}{C}}HCH_3$$

Isopropyl carbocation

This restores the octet to carbon at the same time as it gives the product. As we noted earlier, the product is *not* $CH_3CH_2CH_2G$. Let us see why.

The More Stable Carbocation Preferentially Forms The stability of a carbocation, although always very low, varies with the number of neighboring electron clouds surrounding its positively charged center. Alkyl groups provide larger overall electron clouds than H atoms. Therefore packing alkyl groups instead of H atoms around the carbon with the positive charge *helps to stabilize the carbocation.* When the positive charge is on a secondary carbon, as in the isopropyl carbocation, the electron clouds of *two* alkyl groups crowd around it. When the charge is on the end carbon of the propyl carbocation, the electron cloud of only one alkyl group is nearby.

$$CH_3\overset{+}{C}HCH_3 \qquad CH_3CH_2CH_2^+$$

Isopropyl carbocation — Propyl carbocation

As a result, the isopropyl carbocation is more stable than the propyl carbocation. Or, to generalize, a secondary carbocation is more stable than a primary. By extension, a tertiary carbocation is more stable than a secondary. The order of stability of carbocations is thus

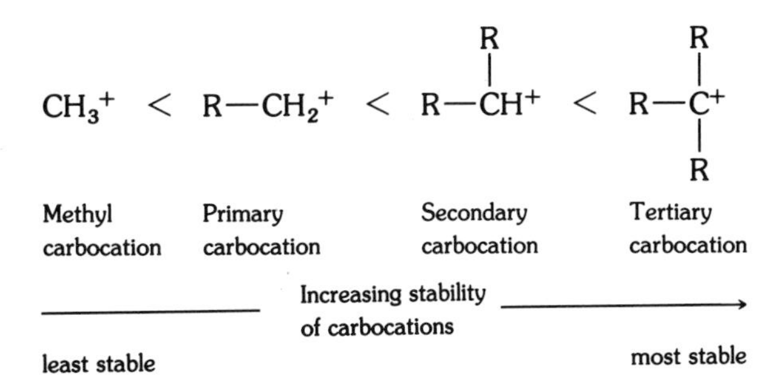

■ The R groups are alkyl groups, and they need not be identical.

This order of stability explains Markovnikov's rule. When the option of two different *kinds* of carbocations exists, as in the addition of H—G to propene, *the most stable carbocation always preferentially forms.* The double bond accepts H^+ from H—G so as to produce the more stable carbocation.

When the two possible carbocations are of the *same* type, both secondary, for example, then there is no preference. Both possible carbocations actually form, both accept G^-, and a mixture of isomeric products is produced. (We'll leave an example to a practice exercise.)

The Water Molecule Is Too Weak a Proton Donor to Make a Carbocation from an Alkene Water does not react with an alkene at all in the absence of an acid catalyst. The H_2O molecule is too weak a donor of H^+, as we said; H_2O holds its protons much too strongly. Thus when water does add to an alkene under acid catalysis, the first step in the mechanism of the reaction is a proton transfer *from the acid catalyst,* not from a water molecule. The water molecule has to wait for this to happen. The acid catalyst converts the double bond temporarily into an electron-poor site, a carbocation.

$$CH_3\overset{CH_3}{\overset{|}{C}}{=}CH_2 + H_2\overset{+}{O}{:}{-}H \xrightarrow[\text{transfer}]{\text{proton}} CH_3\overset{CH_3}{\overset{|}{\underset{+}{C}}}{-}\underset{\underset{H}{|}}{CH_2} + H_2O$$

t-Butyl carbocation (three alkyl groups on C)

The other possible but less stable carbocation, which does *not* form, is

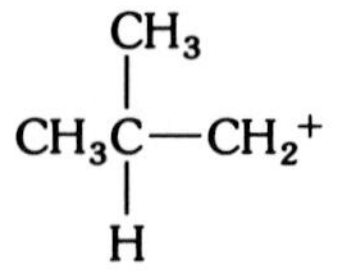

$$CH_3\overset{\displaystyle CH_3}{\underset{\displaystyle H}{\overset{|}{\underset{|}{C}}}}{-}CH_2^+$$

Isobutyl carbocation
(one alkyl group on C)

Once the *t*-butyl carbocation forms, what can it attract? What electron-rich particle is most abundantly available in water containing only a *trace* concentration (a *catalytic* amount) of acid? It is the electron-rich oxygen atom of the water molecule that is most likely to be attracted to the carbocation. The *t*-butyl carbocation, therefore, combines with a water molecule as follows to give what is essentially a hydronium ion in which one H has been replaced by a *t*-butyl group.

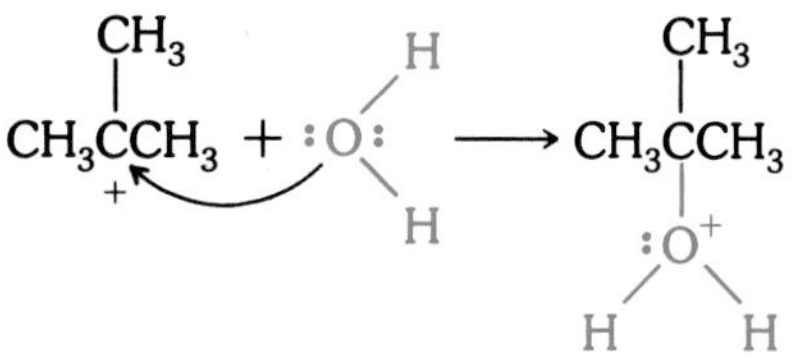

$$CH_3\overset{CH_3}{\underset{+}{C}}CH_3 + :\ddot{O}(H)H \longrightarrow CH_3\overset{CH_3}{C}CH_3 \text{ with } {-}\ddot{O}^+(H)H$$

Now the positive charge is on oxygen, but this is acceptable in this instance because oxygen still has an outer octet.

In the last step, the catalyst, H_3O^+, is recovered by another proton transfer, this time from the *t*-butyl-substituted hydronium ion.

$$CH_3\overset{CH_3}{C}CH_3({-}\ddot{O}^+H_2) + :\ddot{O}H_2 \longrightarrow CH_3\overset{CH_3}{C}CH_3({-}\ddot{O}H) + H{-}\overset{+}{\ddot{O}}H_2$$

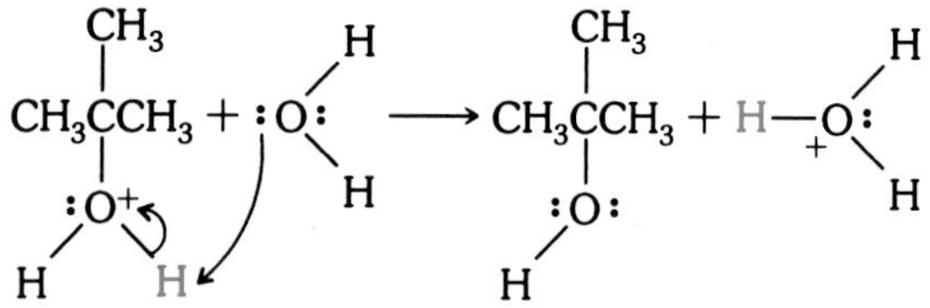

An alkyl substituted hydronium ion; *t*-Butyl alcohol; Recovered catalyst

PRACTICE EXERCISE 7

Write the condensed structures for the two carbocations that could conceivably form if a proton became attached to each of the following alkenes. Circle the carbocation that is preferred. If both are reasonable, state that they are. Then write the structures of the alkyl chlorides that would form by the addition of hydrogen chloride to each alkene.

(a) $CH_3CH_2CH{=}CH_2$ **(b)** $CH_3\overset{CH_3}{C}{=}CH_2$ **(c)** 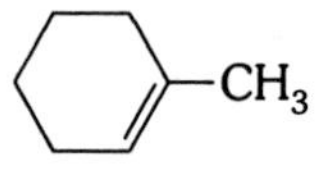 1-methylcyclohexene (ring$-CH_3$)

(d) $CH_3CH{=}CHCH_3$ **(e)** $CH_3CH{=}CHCH_2CH_3$

(f) The addition of water to 2-pentene (part e) gives a mixture of alcohols. What are their structures? Why is the formation of a mixture to be expected here but not when water adds to propene?

13.6

ADDITION POLYMERS

Hundreds to thousands of alkene molecules can join together to make one large molecule of a polymer.

Macromolecules Abound in Nature A **macromolecule** is simply a distinct molecule with a formula mass of several thousand. Out of all of the many known macromolecular substances there are some, the *polymers,* with a unique structural feature. A **polymer** is a substance consisting of macromolecules *all of which have repeating structural units,* up to many thousands. Two of the carbohydrates that we will study—cellulose and one component of starch—are *polysaccharides* characterized by the following system, where Gl stands for a molecular unit made from a glucose molecule (the details of which we'll leave to Chapter 8).

■ *Macro* signifies "huge" or "large scale." *Polymer* has Greek roots: *poly,* "many," and *meros,* "parts."

■ Starch molecules are a storage form for glucose in plants.

etc.—Gl—O—Gl—O—Gl—O—Gl—O—Gl—O—Gl—O—Gl—O—Gl—O—etc.

Section of a polysaccharide

Another component of starch as well as the starch-like polymer glycogen, which we use to store glucose units, have similar sections but with many long branches similarly made of repeating Gl—O units.

■ We store glucose units in the form of glycogen molecules principally in the liver, the kidneys, and in muscles.

Proteins consist almost entirely (and often completely) of polymers called *polypeptides* whose molecules have the following features, showing only the molecular "backbone" and omitting the substituents or *side chains* appended to it. The backbone is like the chain of a charm bracelet, and such chains have *identical* links, but the bracelet acquires its uniqueness in the kinds of its charms (side chains) and the order in which they are hung.

etc.—NHCHC(=O)—NHCHC(=O)—NHCHC(=O)—NHCHC(=O)—NHCHC(=O)—etc.

Section of a polypeptide (The vertical single bonds are sites where *side chains* are attached, about 20 different kinds being the options.)

Like charm bracelets, the backbones (chains) can be of varying lengths.

The molecules of DNA, one of the kinds of nucleic acids, the chemicals of heredity, similarly have identical backbones, but they can be of different lengths and have different side chains (which we again omit).

etc.—OP(=O)(O⁻)O—pentose—OP(=O)(O⁻)O—pentose—OP(=O)(O⁻)O—pentose—OP(=O)(O⁻)O—pentose—etc.

Section of a DNA molecule with the vertical bonds from each pentose unit being sites for the attachment of side chains, four kinds being used. ("Pentose" is a unit made from a sugar called deoxyribose, which gives the D to DNA.)

These examples are intended to illustrate just one feature of all polymers, the fact that they have *repeating structural units.* Thus the study of polymers focuses on the origins of these units and how they are joined together. Let's learn the rudiments of this chemistry by studying easier systems.

Polyethylene Is a Simple but Commercially Important Polymer Under a variety of conditions, many hundreds of ethylene molecules can reorganize their bonds, join together, and change into one large molecule.

■ Certain acids also catalyze this polymerization.

$$n\mathrm{CH_2{=}CH_2} \xrightarrow[\text{trace of } \mathrm{O_2}]{\text{heat, pressure}} \mathrm{+\!CH_2{-}CH_2\!+}_n \qquad (n = \text{a large number})$$

Ethylene → Polyethylene (repeating unit)

The *repeating unit* in polyethylene is $CH_2—CH_2$, and one of these units after another is joined together into an extremely long chain. We see in the structure given a common way to write a polymer structure, by using parentheses to enclose the repeating unit.

Chain-branching reactions also occur during the formation of polyethylene, so the final product includes both straight and branched chain molecules. The molecules in a sample of a commercial polymer like polyethylene are never exact copies of each other. Their chain lengths vary, and the extent of branching varies, but it is still convenient to represent the polymer by showing its most characteristic repeating unit.

The starting material for making a polymer is a compound called a **monomer,** and the reaction in which a monomer changes to a polymer is called **polymerization.** Because alkenes are nicknamed *olefins,* the polymers of alkenes are usually called *polyolefins* (pol-y-**ol**-uh-fins).

Carbocations Are Intermediates in Some Polymerizations The catalysts used industrially to cause polymers to form vary widely. Some, for example, work by generating carbocation intermediates. When the catalyst is a proton donor, like our generalized acid HG, it can convert ethylene into the ethyl carbocation:

■ The negative ions, G^-, are also electron-rich but are present in minute traces compared to molecules of unreactive alkene.

$$G{-}H + \mathrm{CH_2{=}CH_2} \longrightarrow \mathrm{H{-}CH_2{-}CH_2^+} + G^-$$

Acid catalyst + Ethylene → Ethyl carbocation

When the only *abundant* electron-rich species around is unreacted alkene, the new carbocation attracts largely only the electron-rich carrier of an electron pair, *an alkene molecule.* Thus, the new carbocation restores an octet to its positively charged site. For example,

$$\mathrm{CH_3{-}CH_2^+ + CH_2{=}CH_2 \longrightarrow CH_3{-}CH_2{-}CH_2{-}CH_2^+}$$

Ethylene

This reaction, of course, only creates a new and longer carbocation, one still surrounded mostly by unreacted molecules of alkene. So the new carbocation attracts still another molecule of alkene:

$$\mathrm{CH_3{-}CH_2{-}CH_2{-}CH_2^+ + CH_2{=}CH_2 \longrightarrow CH_3{-}CH_2{-}CH_2{-}CH_2{-}CH_2{-}CH_2^+}$$

You can begin to see how this works. Yet another carbocation is produced, and it attracts another molecule of alkene. In this repetitive manner, the chain grows step by step, by one repeating unit of $CH_2—CH_2$ after another, until the positively charged site on a growing chain happens to pick up a stray anion, such as an anion left over from the catalyst. In this way, chains stop growing, some of different lengths than others. Actually, the catalyst should not be called a *catalyst,* because it is finally consumed and not regenerated. The term *promoter* is better than *catalyst* for polymerizations.

■ Polymerization is an example of a *chemical chain reaction* because the product of one step initiates the next step.

The Methyl Side Chains Occur Regularly in Polypropylene When propene (common name, propylene) polymerizes, the methyl groups appear regularly, *on alternate carbons* of the main chain.

$$n\mathrm{CH_2{=}\overset{\overset{\displaystyle CH_3}{|}}{C}H} \xrightarrow{\text{polymerization}}$$

$$\text{etc.}-\mathrm{CH_2-\overset{\overset{\displaystyle CH_3}{|}}{C}H-CH_2-\overset{\overset{\displaystyle CH_3}{|}}{C}H-CH_2-\overset{\overset{\displaystyle CH_3}{|}}{C}H-CH_2-\overset{\overset{\displaystyle CH_3}{|}}{C}H}-\text{etc. or } \mathrm{+CH_2-\overset{\overset{\displaystyle CH_3}{|}}{C}H+_{\mathit{n}}}$$

Polypropylene — Polypropylene (condensed structure)

Polymerizations generally take place in orderly fashions such as this because the reactive intermediates are governed by the same rules of stability that apply to simple carbocations. By the appropriate choice of the polymerization catalyst, however, the methyl groups in the polypropylene product can be made to line up all on the same side of the chain, or to alternate from side to side, or to project at random.

■ Each kind of polypropylene has its own commercial uses.

The Polyolefins Are Chemically Very Stable Because polyolefins, like polypropylene and polyethylene, are fundamentally alkanes, they have all the chemical inertness of alkanes. Polyolefins, therefore, are popular raw materials for making many items including containers that must be inert to food juices and to fluids used in medicines (see Fig. 2.6). Refrigerator boxes and bottles, containers for chemicals, sutures, catheters, various drains, and wrappings for aneurysms are commonly made of polyolefins. Polypropylene fibers are used to make indoor–outdoor carpeting and artificial turf because polypropylene is inert, won't mold, and is wear-resistant.

FIGURE 2.7
Lucite® and Plexiglas® are two trade names for the polymer of methyl methacrylate. The basketball backboard resists shattering when made of this polymer.

Substituted Alkenes Are Monomers for Important Polymers Monomers with carbon–carbon double bonds often carry other functional groups or halogen atoms. The resulting polymers are extremely important commercial substances, which we encounter often in our daily lives. Table 2.2 lists just a few examples (see Fig. 2.7).

Many dienes are used as monomers, too. Natural rubber is a polymer of a diene called isoprene (Table 2.2), and rubber is now made industrially.

FIGURE 2.6
Polypropylene is used to make many items used in clinics and hospitals.

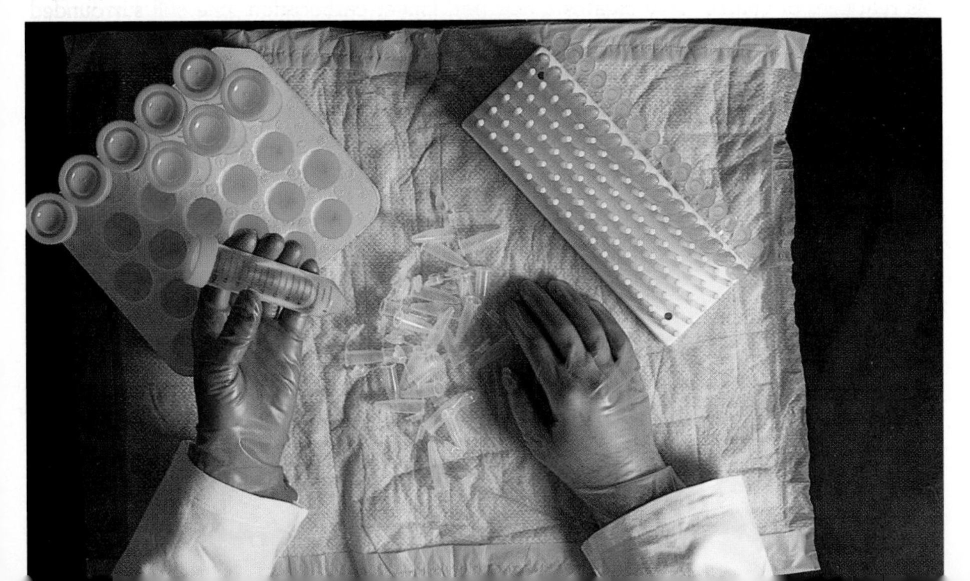

TABLE 2.2
Some Polymers of Substituted Alkenes[a]

Polymer	Monomer	Uses
Polyvinyl chloride (PVC)	$CH_2{=}CHCl$ Vinyl chloride	Insulation, credit cards, bottles, plastic pipe
Saran	$CH_2{=}CCl_2$ Vinylidene chloride and $CH_2{=}CHCl$ Vinyl chloride	Packaging film, fibers, tubing
Teflon	$F_2C{=}CF_2$ Tetrafluoroethylene	Nonstick surfaces, valves
Orlon	$CH_2{=}CH{-}C{\equiv}N$ Acrylonitrile	Fabrics
Polystyrene	$C_6H_5{-}CH{=}CH_2$ Styrene	Foamed items, insulation
Lucite	$CH_2{=}C(CH_3){-}CO_2CH_3$ Methyl methacrylate	Windows, coatings, molded items
Natural Polymer		
Rubber	$CH_2{=}C(CH_3){-}CH{=}CH_2$ Isoprene	Tires, hoses, boots

[a] The common names rather than the IUPAC names of the monomers are given.

2.7

THE BENZENE RING AND AROMATIC PROPERTIES

The benzene ring undergoes substitution reactions instead of addition reactions despite a high degree of unsaturation.

The molecular formula of benzene is C_6H_6, which indicates considerable unsaturation.[3] Its ratio of hydrogen to carbon is much lower than in two simple, saturated hydrocarbons with six carbons, hexane (C_6H_{14}) and cyclohexane (C_6H_{12}). We should expect benzene, therefore, to be some kind of alkene, or alkyne, or a combination. Alkynes give addition reactions very similar to those of alkenes (Special Topic 2.4), and we might expect benzene to give addition reactions just as readily. There is one addition reaction that benzene does give; it adds hydrogen to give cyclohexane. Unlike the addition of hydrogen to an alkene, however, unusually rigorous conditions of pressure and temperature are necessary to make benzene add hydrogen.

$$C_6H_6 + 3H_2 \xrightarrow[\text{high pressure and temperature}]{\text{catalyst}} \text{Cyclohexane } (C_6H_{12})$$

Benzene — Cyclohexane C_6H_{12}

[3] The molecular formulas of all saturated, open-chain alkanes fit the general formula, C_nH_{2n+2}. For cyclic alkanes and open-chain monoalkenes, the general formula is C_nH_{2n}. For open-chain alkynes, it's C_nH_{2n-2}. Thus benzene, C_6H_6, or C_nH_{2n-6}, is highly unsaturated.

SPECIAL TOPIC 2.4
REACTIONS OF ALKYNES

Because the triple bond has pi bonds like the double bond, alkynes give the same kinds of addition reactions as alkenes. Some examples are the following. Notice that the triple bond can add *two* molecules of a reactant. Usually, it is possible to control the reaction so that only one molecule adds.

$$CH_3C{\equiv}CH + H_2 \xrightarrow[\text{pressure}]{\text{Ni, heat,}} CH_3CH{=}CH_2 \xrightarrow{\text{more } H_2} CH_3CH_2CH_3$$

Propyne — Propene — Propane

$$CH_3C{\equiv}CH + HCl \longrightarrow CH_3C(Cl){=}CH_2 \xrightarrow{+HCl} CH_3C(Cl)_2CH_3$$

2-Chloropropene — 2,2-Dichloropropane

Benzene's Typical Reactions Are Substitutions, Not Additions Alkenes (and alkynes) readily *add* chlorine and bromine without a catalyst, but benzene needs a catalyst, and the reaction is not simple addition but substitution. The catalyst is generally an iron halide (or iron itself).

$$C_6H_6 + Cl_2 \xrightarrow[FeCl_3]{\text{Fe or}} C_6H_5Cl + HCl$$

Benzene — Chlorobenzene

$$C_6H_6 + Br_2 \xrightarrow[FeBr_3]{\text{Fe or}} C_6H_5Br + HBr$$

Bromobenzene

■ 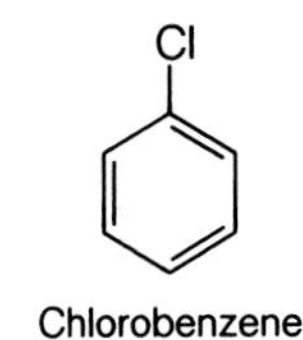

Chlorobenzene

Benzene also reacts, *by substitution,* with sulfur trioxide dissolved in concentrated sulfuric acid. (Recall that alkenes react exothermically with concentrated sulfuric acid by *addition*.)

$$C_6H_6 + SO_3 \xrightarrow[\text{room temperature}]{H_2SO_4\text{ (concd.)}} C_6H_5{-}S(=O)_2{-}O{-}H$$

Benzenesulfonic acid

■ Benzenesulfonic acid is about as strong an acid as hydrochloric acid. It is a raw material for the synthesis of aspirin.

Benzene reacts with warm, concentrated nitric acid when it is dissolved in concentrated sulfuric acid. We will now represent nitric acid as $HO{-}NO_2$, instead of HNO_3, because it loses the HO group during the reaction.

$$C_6H_6 + HO{-}NO_2 \xrightarrow[50-55\ ^\circ C]{H_2SO_4\text{ (concd)}} C_6H_5{-}NO_2 + H_2O$$

Nitric acid — Nitrobenzene

We have learned that alkenes (and alkynes) are readily oxidized by permanganate or dichromate ion, but benzene is utterly unaffected by these strong oxidizing agents even when boiled with them. (Ozone does attack benzene.)

In the light of all these chemical properties, whatever benzene is, it isn't an alkene or alkyne. Yet it surely is unsaturated. The problem of what benzene is wasn't satisfactorily solved in organic chemistry until the early 1930s, roughly a century after its molecular formula was known and half a century after its skeleton structure had been determined. Many of the reactions referred to above helped to establish the structure of benzene. Let's see how this was done.

The Six H Atoms in C_6H_6 Are Chemically Equivalent to Each Other When benzene is used to make chlorobenzene (or any of the other products shown above), only *one* mono-substituted compound forms. Only one $C_6H_5—Cl$ exists. This reminds us of what happens when ethane, CH_3CH_3, is chlorinated. Only one *mono*-substituted compound forms; only one CH_3CH_2Cl exists. It doesn't matter which H in CH_3CH_3 is replaced by Cl. The same is true for benzene. All six H atoms in C_6H_6 are equivalent to each other.

The Benzene Skeleton Has a Six-Membered Ring with a Hydrogen on Each Carbon Cyclohexane forms when benzene is hydrogenated, so the six carbons of a benzene molecule must also form a six-membered ring. Because benzene's six Hs are chemically equivalent, it seems reasonable to put one H on each of the ring carbons. This gives a very symmetrical structure.

■ The older structure, **2**, is still widely used to represent benzene, although it is usually abbreviated further:

1
(Incomplete)

2
(Older structure for benzene)

The trouble with the incomplete structure identified as **1** is that each carbon has only three bonds, not four. To solve this, chemists for several decades simply wrote in three double bonds, as seen in structure **2**. As they well knew, the difficulty with **2** is that it says that benzene is a triene, a substance with three alkene groups per molecule.

■ Open-chain trienes, like 1,3,5-hexatriene, give addition reactions and are easily oxidized, like any ordinary alkene.

$CH_2=CHCH=CHCH=CH_2$
1,3,5-Hexatriene

Because the three double bonds indicated in structure **2** are misleading in a chemical sense, scientists often represent benzene simply by a hexagon with a circle inside, structure **3**.

3
Benzene

The ring system in **3,** the *benzene ring,* is planar. All its atoms lie in the same plane, and all the bond angles are 120°, as seen in the scale model of Figure 2.8.

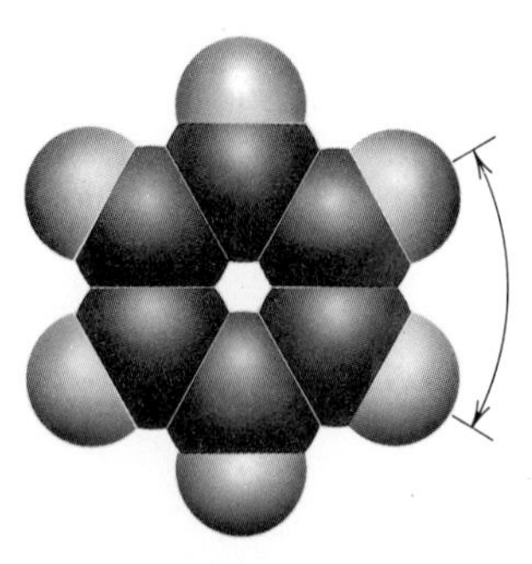

FIGURE 2.8
Scale model of a molecule of benzene. It shows the relative volumes of space occupied by the electron clouds of the atoms.

Three Electron Pairs Are Delocalized in the Benzene Ring The development of the chemical bonding theory of overlapping orbitals finally provided a satisfactory model for the structure of benzene (see Fig. 2.9). Part (*a*) of Figure 2.9 shows the sigma bond network. Each carbon atom holds three other atoms, not four, so each carbon is sp^2-hybridized and each has an unhybridized 2*p* orbital. The ring thus has six 2*p* orbitals, and their axes are coparallel and perpendicular to the plane of the benzene ring (see Fig. 2.9*b*).

The novel and important feature of benzene is that the six 2*p* orbitals overlap side to side *all around the ring*. They don't just pair off as in ethene and form three isolated double bonds. The result is a large, circular, double-doughnut shaped space above and below the plane, as seen in part (*c*) of the figure. Six electrons are in this space, and we can refer to them as the pi electrons of the benzene ring.

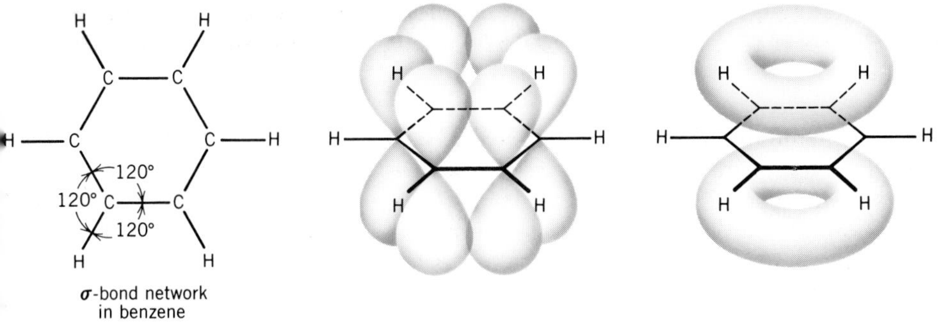

H H C C H C C H 120° 120° C C 120° H H

σ-bond network in benzene

(a) (b) (c)

FIGURE 2.9
The molecular orbital model of benzene. (a) The sigma-bond framework. (b) The six 2*p* orbitals of the ring carbon atoms. (c) The double-doughnut shaped region formed by the side-by-side overlapping of the six 2*p* orbitals.

A molecular orbital, like an atomic orbital, can hold no more than two electrons. The six pi electrons, therefore, are in three molecular orbitals within the double-doughnut shaped space, and they enjoy considerably more room and freedom of movement than if they were in isolated double bonds. The electrons are said to be *delocalized.* When electrons have more room, the system is more stable. The delocalization of the pi electrons, in fact, explains much of the unusual stability of benzene and its resistance to addition reactions. When a hydrogen atom held by a ring carbon is replaced by another group, the closed-circuit pi electron network is not broken up. But if an addition were to occur, the ring system would no longer be that of benzene but, instead, that of a cyclic diene. For example,

■ Electrons repel each other, so they are managed with greater overall stability in systems that allow them more room.

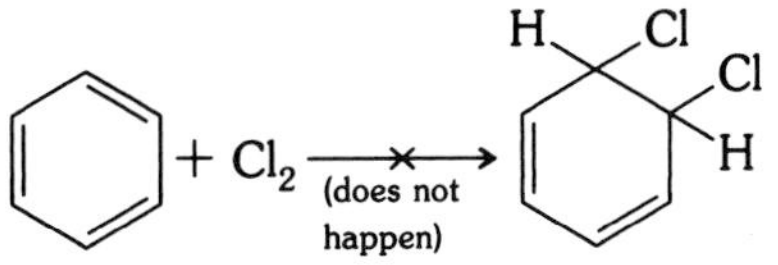

$$+ Cl_2 \xrightarrow[\text{(does not happen)}]{\times}$$ H Cl Cl H

It costs the system far more energy to react this way than to react by a substitution reaction, so the benzene ring strongly resists addition reactions. Just *how* chlorination does occur, however, involves a *temporary* rupture of the closed-circuit pi electron network as explained in Special Topic 2.5.

Aromatic Compounds Have Unsaturated Rings That Give Substitution Reactions, like Benzene Any substances whose molecules have benzene rings and whose rings give substitution reactions instead of addition reactions are called **aromatic compounds.** The term is a holdover from the days when most of the known compounds of benzene actually had aromatic fragrances, but now the term does not refer to odor. Although oil of wintergreen and vanillin do have pleasant fragrances, aspirin does not. Yet all three have the benzene ring and all are classified as aromatic compounds.

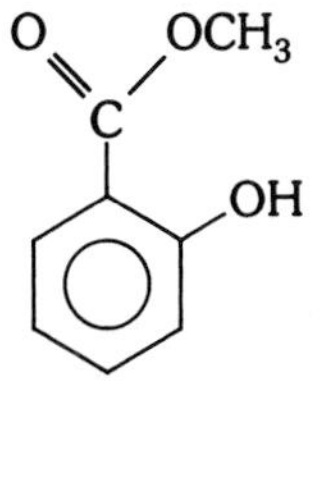

O OCH$_3$ C OH — Oil of wintergreen

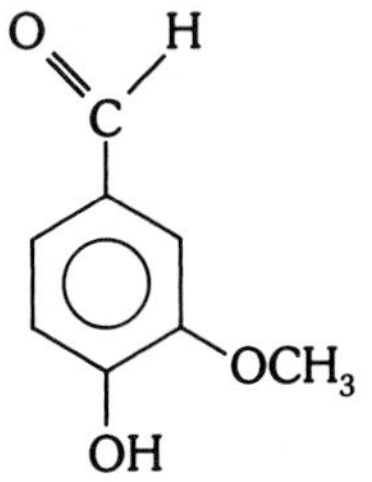

O H C OCH$_3$ OH — Vanillin

O OH C O O—CCH$_3$ — Aspirin

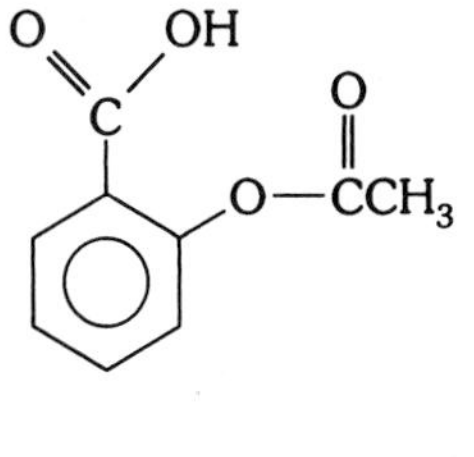

■ = C_6H_5— Phenyl group

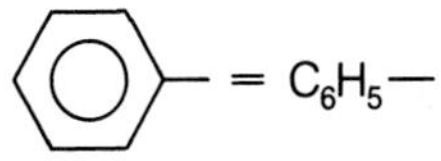

1-Phenylpropane is an example of an aromatic compound with an aliphatic side chain. (In IUPAC nomenclature, the group C_6H_5, derived from benzene by removing one H atom, is called the **phenyl** group and rhyming with "kennel.") So stable is the benzene ring toward

SPECIAL TOPIC 2.5

HOW CHLORINE REACTS WITH BENZENE TO MAKE CHLOROBENZENE

Benzene and chlorine undergo essentially no reaction, even in the presence of ultraviolet radiation. However, when a trace of iron(III) chloride, $FeCl_3$, is present, chlorine changes benzene into chlorobenzene and HCl. The reaction that we'll explain in this Special Topic is

$$Cl_2 + C_6H_6 \xrightarrow{FeCl_3} HCl + C_6H_5Cl$$

To get a chlorine species reactive enough to break into benzene's pi electron network, we have to make the chlorine species temporarily even more reactive than it is in Cl· (Special Topic 1.2). This is the function of the $FeCl_3$ catalyst.

$FeCl_3$ is able to react with Cl_2 as follows to give $Cl^+FeCl_4^-$.

$$:\ddot{Cl}-\ddot{Cl}: + FeCl_3 \longrightarrow :\ddot{Cl}^+ \left[:\ddot{Cl}-FeCl_3\right]^-$$

$FeCl_4^-$

The species $Cl^+FeCl_4^-$ is not very stable, at least not enough to be stored, but it has a chlorine atom in a *positively* charged state, Cl^+. This is like Cl· minus its unpaired electron. In other words, Cl^+ has just *six* electrons in its outside level, even farther from an octet than Cl·. Although Cl^+ is not a freely wandering species, the next step in the chlorination of benzene is envisioned as the attack on Cl^+ by benzene. To show what happens, we have to use one of the Lewis structures of benzene, and we will now show one of the H atoms attached to the benzene ring (leaving other five understood).

[Benzene (H shown) $+ :\ddot{Cl}:^+ \longrightarrow$ cation with Cl and H on the same ring carbon]

This unquestionably is an energy-expensive step, because it disrupts the uniquely stable pi electron network of the ring. However, the chlorine cation, Cl^+, recovers its octet, which is some compensation. Any remaining energy cost is repaid in the next step, because the pi network of the ring is restored and *a species more stable than Cl^+ is released,* namely H^+ *(which becomes joined to chlorine to give HCl).* We'll use the $FeCl_4^-$ species made in the first step to provide the acceptor.

[Cation $+ :\ddot{Cl}-FeCl_3^-$ ($FeCl_4^-$) $\longrightarrow$ Chlorobenzene $+ H-\ddot{Cl}: + FeCl_3$ (Recovered catalyst)]

Thus the equivalent of Cl^- accepts H^+, the pi electron network of the ring is restored, chlorobenzene and HCl form, and *the catalyst is recovered to facilitate further reaction.* The mechanism of the substitution reaction thus involves *ions* and not the free radicals that we've seen in substitution reactions between alkanes and chlorine or bromine.

oxidizing agents that alkylbenzenes like 1-phenylpropane are attacked by hot permanganate at *the side chain* and not at the ring. Benzoic acid can be made by the oxidation of 1-phenylpropane (to use an unbalanced equation).

$$\underset{\text{1-Phenylpropane}}{C_6H_5CH_2CH_2CH_3} \xrightarrow{\text{hot } KMnO_4\ (aq)} \underset{\text{Benzoic acid}}{C_6H_5\overset{O}{\overset{\|}{C}}OH} \quad (+MnO_2)$$

Most of the side chain is destroyed, but the ring is not attacked.

Not all benzene derivatives have rings that resist oxidation as strongly as this. Rings, for example, that hold either the OH or the NH_2 groups are *very* readily oxidized. The unshared electrons on O and N in these groups are also somewhat delocalized into the ring network, which gives the rings more electron density than in benzene. Oxidizing agents, which seek electrons, are therefore able to react with C_6H_5OH and $C_6H_5NH_2$. We won't take this any further because our goal has been to learn *general* properties of the benzene ring. This ring is present in proteins, and benzene-like aromatic rings occur in the heterocyclic rings of all nucleic acids.

■ The liver has enzymes that put OH groups on benzene rings, making the products easier to break down into manageable wastes or to be made into substances needed by the body.

2.8

NAMING COMPOUNDS OF BENZENE

Common names dominate the nomenclature of simple derivatives of benzene. The names of several monosubstituted benzenes are straightforward. The substituent is indicated by a prefix to the word *benzene.* For example,

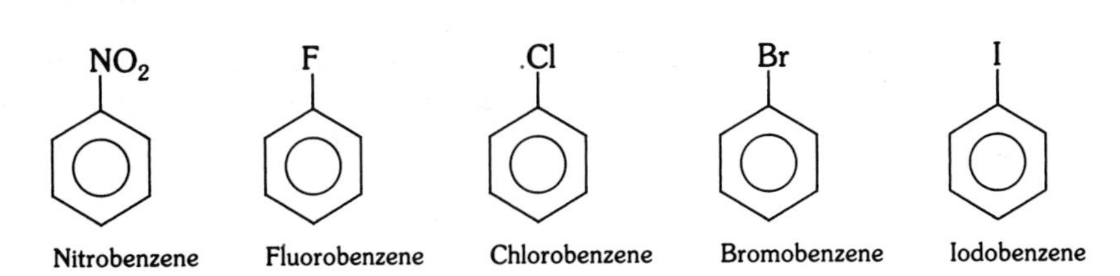

Nitrobenzene Fluorobenzene Chlorobenzene Bromobenzene Iodobenzene

■ All these compounds are oily liquids.

Other derivatives of benzene have common names that are always used.

CH_3 OH NH_2 O OH C O H C SO_3H

Toluene Phenol Aniline Benzoic acid Benzaldehyde Benzene-sulfonic acid

■ Phenol was the first antiseptic used by British surgeon Joseph Lister (1827–1912), the discoverer of antiseptic surgery.

***Ortho*, *Meta*, and *Para* Are Terms for 1,2-, 1,3-, and 1,4-Relationships** When two or more groups are attached to the benzene ring, both what they are and where they are must be specified. One common way to indicate the relative locations of two groups in disubstituted benzenes is by the prefixes *ortho-*, *meta-*, and *para-*, which usually are abbreviated *o-*, *m-*, and *p-*, respectively. Two groups that are in a 1,2-relationship are *ortho* to each other, as in 1,2-dichlorobenzene, commonly called *o*-dichlorobenzene. A 1,3-relationship is designated *meta*, as in *m*-dichlorobenzene. In *p*-dichlorobenzene, the substituents are 1,4- or *para* relative to each other.

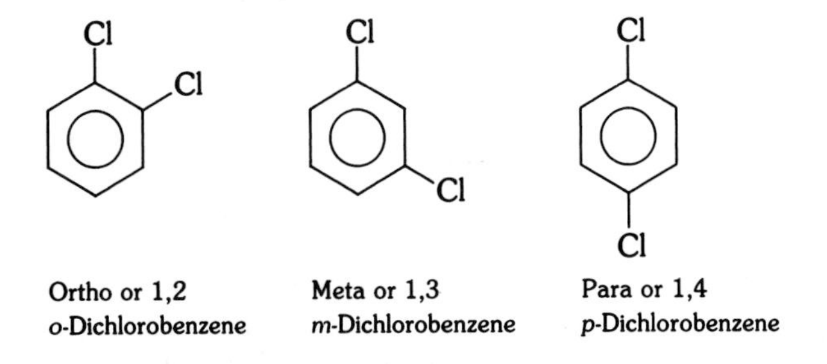

Ortho or 1,2 *o*-Dichlorobenzene Meta or 1,3 *m*-Dichlorobenzene Para or 1,4 *p*-Dichlorobenzene

A disubstituted benzene is usually named as a derivative not of benzene but of a monosubstituted benzene when the latter has a common name, like toluene or aniline. Then the *o-*, *m-*, or *p*- designations are used to specify relative positions of the two groups. For example,

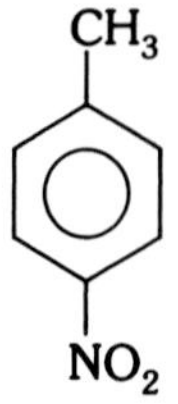

p-Nitrotoluene (*not* 4-nitro-1-methylbenzene)

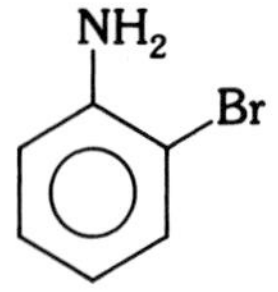

o-Bromoaniline (*not* 2-bromo-1-aminobenzene)

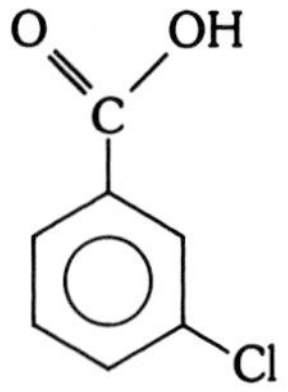

m-Chlorobenzoic acid

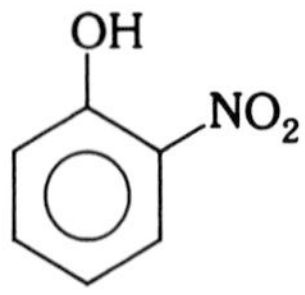

o-Nitrophenol

When we have trisubstituted benzenes (or higher), we cannot use the *ortho, meta,* or *para* designations with sufficient precision to make an unambiguous name. We must now use numbers assigned to ring positions in such a way as to use the lowest numbers possible. For example,

■ TNT is an important explosive. TNB is an even better explosive than TNT, but it is more expensive to make.

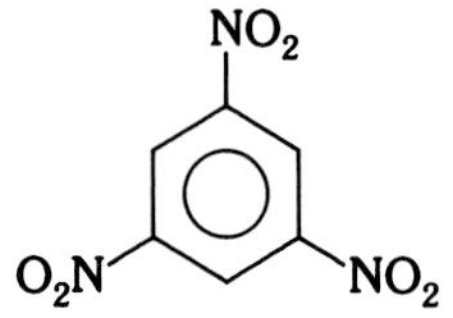

1,3,5-Trinitrobenzene, TNB

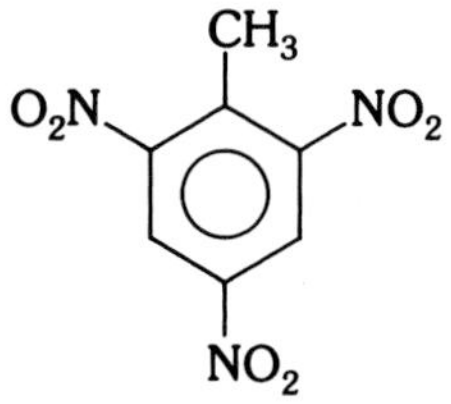

2,4,6-Trinitrotoluene, TNT

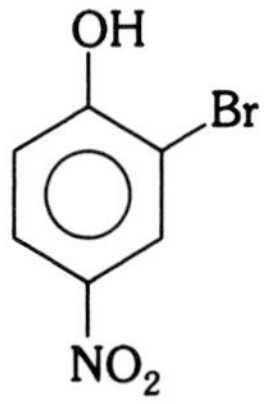

2-Bromo-4-nitrophenol

SUMMARY

Alkenes The lack of free rotation at a double bond makes geometric (cis–trans) isomers possible, but they exist only when the two groups are not identical at *either* end of the double bond. Cyclic compounds also exhibit cis–trans isomerism.

Alkenes and cycloalkenes are given IUPAC names by a set of rules very similar to those used to name their corresponding saturated forms. However, the double bond takes precedence both in selecting and in numbering the main chain (or ring). The first unsaturated carbon encountered in moving down the chain or around the ring through the double bond must have the lower number.

Addition reactions Several compounds add to the carbon–carbon double bond—H_2, Cl_2, Br_2, HCl, HBr, $HOSO_3H$, and H_2O (in the presence of an acid catalyst). The kinds of products that can be made are outlined in the accompanying chart of the reactions of alkenes. When both the alkene and the reactant are unsymmetrical, the addition proceeds according to Markovnikov's rule—the end of the double bond that already has the greater number of hydrogens gets one more. The double bond is vigorously attacked by strong oxidizing agents, like the permanganate ion (MnO_4^-), the dichromate ion ($Cr_2O_7^{2-}$), and ozone (O_3).

How additions occur The carbon–carbon double bond is a proton acceptor toward strong proton donors. The proton becomes attached to one carbon at the double bond, using the pair of electrons in the pi bond to make the sigma bond to this hydrogen. The other carbon of the original double bond becomes positively charged. The result is a carbocation, which then accepts an electron-rich particle to complete the formation of the product.

Polymerization of alkenes The polymerization of an alkene is like an addition reaction. The alkene serves as the monomer, and one alkene molecule adds to another, and so on, until a long chain with a repeating unit forms—the polymer molecule.

Aromatic compounds Aromatic compounds contain rings having alternating double and single bonds, a closed-circuit pi-electron network commonly seen in the benzene ring. This ring is resistant to oxidation except when it holds OH or NH_2 as a substituent. When aromatic compounds undergo reactions at the benzene ring, substitutions rather than additions occur. In this way, the closed-circuit pi-electron network of the ring remains unbroken. This network forms when six unhybridized 2*p* orbitals of the ring carbon atoms overlap side-by-side to form a double-doughnut shaped space found above and below the plane of the ring. There are three sublevels in this space, and the pi electrons enjoy considerable room and freedom of motion.

Chart of chemical properties

An alkene ($>C=C<$):

- $+ H_2$ (catalyst, heat) → Alkanes (—CH—CH—)
- $+ X_2$ (Cl_2 or Br_2) → 1,2-Dihaloalkanes (—CX—CX—)
- $+ HX$ (X = Cl or Br) → Alkyl halides (—CH—CX—)
- $+ H_2SO_4$ (concd) → Alkyl hydrogen sulfates (—CH—C(OSO_3H)—)
- $+ H_2O$ (acid catalyst) → Alcohols (—CH—C(OH)—)
- polymerization of *n* molecules → Polymers ($+C—C+_n$)

Markovnikov's rule applies (to the addition of HX, H_2SO_4, and H_2O)

Benzene:

- $+ HNO_3$, H_2SO_4 catalyst → C_6H_5—NO_2 Nitrobenzene
- $+ X_2$ (Cl_2 or Br_2), Fe or FeX_3 → C_6H_5—X Chloro- or bromobenzene
- $+ SO_3$ (in concd H_2SO_4) → C_6H_5—SO_3H Benzenesulfonic acid

Alkylbenzenes: C_6H_5—R $\xrightarrow{KMnO_4}$ C_6H_5—CO_2H Benzoic acid

REVIEW EXERCISES

The answers to Review Exercises whose numbers are in color are found in Appendix A. The answers to the other Review Exercises are found in the Study Guide that accompanies this book. The more challenging questions are marked with asterisks.

Occurrence and Physical Properties

2.1 Examine the following structural formulas and *use their identifying letters* to answer the questions about them.

A (cyclohexene)   B $CH_3CH{=}CH_2$ C (cyclopentane)   D $CH_3CH_2C{\equiv}CH$

E $CH_2{=}CHCH{=}CHCH_3$

(a) What are the letters of the compounds that are members of the *alkene* family?
(b) Which is a *saturated* compound?
(c) The prefix *pent-* would appear in the names of which compounds?
(d) Which compounds are relatively insoluble in water?
(e) The general formula C_nH_{2n} fits which compounds? Are they all alkenes?
(f) The general formula C_nH_{2n-2} fits which compounds? Are they all alkynes?
(g) In view of your observations regarding parts (e) and (f), how well do the general formulas C_nH_{2n} and C_nH_{2n-2} specify hydrocarbon families?

2.2 In view of the theme of this book, the molecular basis of life, what has justified our including the alkene group in our study?

Nomenclature

2.3 Write the condensed structures of the following compounds. pounds.
(a) Isobutylene
(b) Propylene
(c) *trans*-2-Hexene
(d) 3-Bromo-2-pentene
(e) 1,2-Dimethylcyclohexene
(f) 2,4-Dimethylcyclohexene

2.4 Write the IUPAC names of the following compounds.
(a) $CH_2{=}CH(CH_2)_5CH_3$
(b) $CH_3CH(CH_3)CH{=}CHBr$
(c) $CH_3CH_2CH(CH_3)CH_2C({=}CH_2)CH_2CH_2CH_3$
(d) $CH(CH_3){=}CHC(CH_2CH_3)(CH_3)CH_3$

*2.5 Write the condensed structures and the IUPAC names for all the isomeric pentenes, C_5H_{10}. Include cis and trans isomers.

2.6 Write the condensed structures and the IUPAC names for all the isomeric methylcyclopentenes.

*2.7 Write the condensed structures and IUPAC names for all the isomeric butynes. The IUPAC rules for naming alkynes are identical with the rules for naming alkenes, except that the name ending is *-yne*, not *-ene*.

***2.8** Write the condensed structures and the IUPAC names for all the isomeric dimethylcyclopentenes. Include the cis and trans isomers. Remember that all six atoms directly involved with an alkene system are in the same plane.

*2.9 Write the condensed structures and the names for all the open chain dienes with the molecular formula C_5H_8. (Note that there can be two double bonds from the same carbon.)

Cis–Trans Isomerism

***2.10** What are the *names* of the molecular orbitals in which electron pairs reside between the two carbon atoms of an alkene group?

***2.11** Given the options of the following types of level 2 atomic orbitals—*s*, *p*, *sp*, sp^2, and sp^3—what are the *symbols* of the specific orbitals at a carbon atom that are used to make each of the two bonds to the other carbon of the alkene group?

***2.12** Why is there a lack of free rotation at an alkene group, and why does this matter?

2.13 Which of the following pairs of structures represent identical compounds or isomers?

(a) $CH_3CH{=}CHCH_3$ (both CH_3 on same side) and $CH_3CH{=}CHCH_3$ (CH_3 groups on opposite sides)
(b) $BrCH{=}CHCl$ (Br and Cl on opposite sides) and $ClCH{=}CHBr$ (Cl and Br on opposite sides)
(c) 3-bromocyclopentene (Br on ring carbon adjacent to double bond) and (ring with Br on carbon adjacent to double bond, drawn differently)
(d) $CH_3CH_2C(CH_3){=}CHCH_3$ and $CH_3CH{=}C(CH_3)CH_2CH_3$
(e) $CH_3C(CH_3){=}CHCH(CH_3)CH_3$ and $CH_3C(CH_3){=}CHCH(CH_3)CH_3$ (drawn as $CH(CH_3)CHCH_3$ over CH_3CCH_3 with double bond)

2.14 Study the following structures to discover which are able to exhibit cis–trans isomerism. For those that do, write the structures of the cis and trans isomers.
(a) $CH_3CH{=}CHCH_2CH_3$
(b) $CH_3C(Cl){=}C(Br)CH_2CH_3$
(c) 1,2-dimethylcyclopentane (ring with two CH_3)
(d) $CH_3C(Cl){=}CHCH(CH_3)CH_3$

2.15 Identify which of the following compounds can exist as cis and trans isomers and write the structures of these isomers.

(a) FBrC=CHCl

(b) [structure: cyclopentane with two CH_3 groups on one carbon; labels H_3C, CH_3]

(c) $CH_3CH{=}CHCH{=}CH_2$

Reactions of the Carbon–Carbon Double Bond

2.16 Write equations for the reactions of 2-methylpropene with the following reactants.

(a) Cold, concentrated sulfuric acid
(b) Hydrogen in the presence of a nickel catalyst
(c) Water in the presence of an acid catalyst
(d) Hydrogen chloride
(e) Hydrogen bromide
(f) Bromine

*2.17 The molecules of a hydrocarbon, *A*, with the molecular formula C_7H_{12} have the following structure, which is complete except for the location of one double bond. Compound *A* is *not* capable of existing as cis and trans isomers. However, the *product* of the hydrogenation of compound *A* is capable of existing as cis and trans isomers. Write the structure of *A*. (If more than one structure is possible, write them all.)

[structure: cyclopentane ring with CH_3 on each of two adjacent carbons]

CH_3 CH_3

(Incomplete structure of *A*)

2.18 Write equations for the reactions of 1-methylcyclopentene with the reactants listed in Review Exercise 2.16.

2.19 Write equations for the reactions of 2-methyl-2-butene with the compounds given in Review Exercise 2.16.

2.20 Write equations for the reactions of 1-methylcyclohexene with the reactants listed in Review Exercise 2.16. (Do not attempt to predict whether cis or trans isomers form.)

2.21 Ethane is insoluble in concentrated sulfuric acid, but ethene dissolves readily. Write an equation to show how ethene is changed into a substance polar enough to dissolve in concentrated sulfuric acid, which, of course, is highly polar.

*2.22 Pentene, C_5H_{10}, has several isomers, and most but not all of them add bromine, dissolve in concentrated sulfuric acid, and react with potassium permanganate. Give the structure of at least one isomer of C_5H_{10} that gives *none* of these reactions.

*2.23 One of the raw materials for the synthesis of nylon, adipic acid, can be made from cyclohexene by oxidation using potassium permanganate. The balanced equation for the first step in which $K_2C_6H_8O_4$, the potassium salt of adipic acid, forms is as follows:

How many grams of potassium permanganate are needed for the oxidation of 11.2 g of cyclohexene, assuming that the reaction occurs exactly and entirely as written?

*2.24 Referring to Review Exercise 2.23, how many grams of the potassium salt of adipic acid can be made if 21.0 g of $KMnO_4$ are used in accordance with the equation given?

How Addition Reactions Occur

2.25 When 1-butene reacts with hydrogen chloride, the product is 2-chlorobutane, not 1-chlorobutane. Explain.

2.26 When 4-methylcyclohexene reacts with hydrogen chloride, the product consists of a mixture of roughly equal quantities of 3-chloro-1-methylcyclohexane and 4-chloro-1-methylcyclohexane. Explain why substantial proportions of *both* isomers form.

Polymerization

2.27 Rubber cement can be made by mixing some polymerized 2-methylpropene with a solvent such as toluene. When the solvent evaporates, a very tacky and sticky residue of the polymer (called polyisobutylene) remains, which soon hardens and becomes the glue. The structure is quite regular, like polypropylene. Write the structure of the polymer of 2-methylpropene in two ways.

(a) One that shows four repeating units, one after the other.
(b) The condensed structure.

2.28 Polyvinyl acetate is a soft adhesive that is modified (by reactions that we have yet to study) into a material (Butvar®) that bonds two glass sheets together in safety glass. Thus when safety glass breaks, the broken pieces cannot fly around as easily. Using four vinyl acetate units, write part of the structure of a molecule of polyvinyl acetate. Also write its condensed structure. The structure of vinyl acetate is as follows.

$$CH_2{=}CH{-}O{-}\overset{\overset{\displaystyle O}{\|}}{C}CH_3$$

Vinyl acetate

When it polymerizes, its units line up in a regular way as in the polymerization of propene.

2.29 Gasoline is mostly a mixture of alkanes. However, when a sample of gasoline is shaken with aqueous potassium permanganate, a brown precipitate of MnO_2 appears and the purple color of the permanganate ion disappears. What kind of hydrocarbon was evidently also in this sample?

2.30 If gasoline rests for months in the fuel line of some engine, the line slowly accumulates some sticky material. This can clog the line and also make the carburetor work poorly or not at all. In view of your answer to Review Exercise 2.29, what is likely to be happening, chemically?

Aromatic Properties

2.31 Dipentene has a very pleasant, lemon-like fragrance, but it is not classified as an aromatic compound. Why?

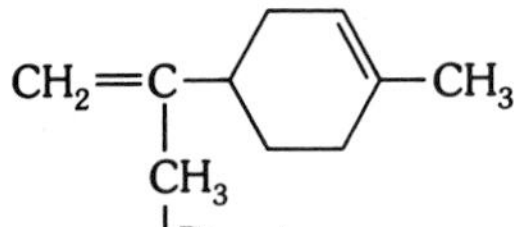

$CH_2{=}C(CH_3)$–(cyclohexene ring)–CH_3

Dipentene

2.32 Sulfanilamide, one of the sulfa drugs, has no odor at all, but it is still classified as an aromatic compound. Explain.

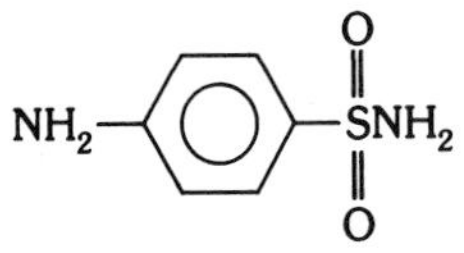

NH_2–(benzene ring)–SO_2NH_2

Sulfanilamide

2.33 Write equations for the reactions, if any, of benzene with the following compounds.

(a) Sulfur trioxide (in concentrated sulfuric acid)
(b) Concentrated nitric acid (in concentrated sulfuric acid)
(c) Hot sodium hydroxide solution
(d) Hydrochloric acid
(e) Chlorine (alone)
(f) Hot potassium permanganate
(g) Bromine in the presence of $FeBr_3$

2.34 Write the structure of any compound that would react with hot potassium permanganate to give benzoic acid.

2.35 Phthalic acid is one of the raw materials for making a polymer that is used in automobile finishes.

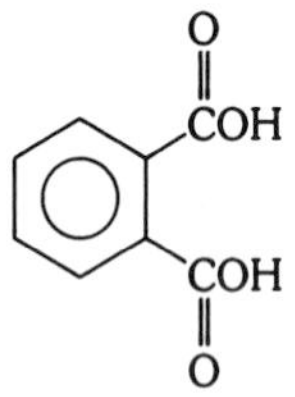

(benzene ring with two adjacent COOH groups)

Phthalic acid

What hydrocarbon with the formula C_8H_{10} could be changed to phthalic acid by the action of hot potassium permanganate? (Write its structure.)

The Bonds in Benzene

2.36 Describe in your own words, making your own drawings, the ways in which the bonds in benzene form from pure or hybrid atomic orbitals.

2.37 What is it about the delocalization of benzene's pi electrons that makes the benzene ring relatively stable?

2.38 Explain why benzene strongly resists addition reactions and gives substitution reactions instead.

Names of Aromatic Compounds

2.39 Write the condensed structure of each compound.

(a) toluene (b) aniline
(c) phenol (d) benzoic acid
(e) benzaldehyde (f) nitrobenzene

2.40 Give the condensed structures of the following compounds.

(a) *p*-nitrobenzoic acid
(b) *m*-bromotoluene
(c) *o*-chloronitrobenzene
(d) 2,4-dinitrophenol

Alkene Group in Nature (Special Topic 2.1)

2.41 What structural feature occurs in vitamins D and A that explains why these vitamins deteriorate somewhat when heated in air?

2.42 What structural difference exists between vegetable oils and animal fats?

Chemistry of Vision (Special Topic 2.2)

2.43 When cis–trans isomerization takes place, do groups at the double bond break off from the chain and exchange places?

2.44 How does a photon of the proper energy enable a cis isomer to change into a trans isomer?

2.45 The retina in the eye of an owl has only rods, not cones. What connection is there between this circumstance and the ability of an owl to see in very dim light? Can an owl see in a situation of total blackness?

Ozone in Smog (Special Topic 2.3)

2.46 What event in a vehicle engine launches the production of ozone in smog? (Write an equation.)

2.47 How is NO_2 formed in smog? (Write an equation.)

2.48 How is NO_2 involved in the production of ozone in smog? (Write equations.)

2.49 Why is ozone dangerous?

2.50 In areas with severe smog problems, air quality authorities seek to reduce emissions of hydrocarbons, even those arising from the use of power lawn mowers and similar machines. What is the connection between these uses of fuels and the ozone in smog?

Reactions of Alkynes (Special Topic 2.4)

2.51 Write the structures of the products that form when one mole of 2-butyne reacts with one mole of each of the following.

(a) H_2 (b) Cl_2 (c) Br_2

2.52 Write the structures of the products that form when one mole of 2-butyne reacts with *two* moles of each of the following.

(a) H_2 (b) Cl_2 (c) Br_2

How Benzene Reacts with Chlorine (Special Topic 2.1)

2.53 Why doesn't the benzene ring simply *add* chlorine molecules, like any alkene system?

2.54 Iron(III) bromide, $FeBr_3$, can be used as a catalyst for *chlorination* with little chance of any bromobenzene forming. Explain.

2.55 When bromine is mixed with benzene in the presence of iron(III) bromide as a catalyst, bromobenzene and HBr form. Write the mechanism; it is just like that of the chlorination of benzene.

Additional Exercises

2.56 Write the structures of the products to be expected in the following situations. If no reaction is to be expected, write "no reaction." To work this kind of exercise, you have to be able to do three things.

(1) *Classify* a specific organic reactant into its proper family. Do this first.

(2) *Recall* the short list of chemical facts about the family. (If there is no matchup between this list and the reactants and conditions specified by a given problem, assume that there is no reaction.)

(3) *Apply* the recalled chemical fact, which might be some "map sign" associated with a functional group, to the specific situation.

Study the next two examples before continuing.

EXAMPLE 2.6 Predicting Reactions

What is the product, if any, of the following?

$$CH_3CH_2CH_2CH_3 + H_2SO_4 \longrightarrow ?$$

Analysis: We note first that the organic reactant is an alkane, so we next turn to the list of chemical properties about all alkanes that we learned. With this family, of course, the list is very short. Except for combustion and halogenation, we have learned no reactions for alkanes, and we assume, therefore, that there aren't any others, not even with sulfuric acid. Hence, the answer is "no reaction."

EXAMPLE 2.7 Predicting Reactions

What is the product, if any, in the following situation?

$$CH_3CH{=}CHCH_3 + H_2O \xrightarrow{\text{acid catalyst}} ?$$

Analysis We first note that the organic reactant is an alkene, so we review our mental "file" of reactions of the carbon–carbon double bond.

1. Alkenes add hydrogen (in the presence of a metal catalyst to form alkanes.
2. They add chlorine and bromine to give 1,2-dihaloalkanes.
3. They add hydrogen chloride and hydrogen bromide to give alkyl halides.
4. They add sulfuric acid to give alkyl hydrogen sulfates.
5. They add water in the presence of an acid catalyst to give alcohols.
6. They are attacked by strong oxidizing agents.
7. They polymerize.

These are the chief chemical facts, the principal "map signs," about the carbon–carbon double bond that we have studied, and we see that the list includes a reaction with water in the presence of an acid catalyst. We remember that in all addition reactions the double bond changes to a single bond and the pieces of the adding molecule end up on the carbons at ends of the double bond. We also have to remember Markovnikov's rule to tell us which pieces of the water molecule go to each carbon. However, in this specific example, Markovnikov's rule does not apply because the alkene is symmetrical.

Solution

$$CH_3CH_2\underset{\displaystyle OH}{\underset{|}{C}}HCH_3$$

Now work the following parts. (Remember that C_6H_6 stands for benzene and that C_6H_5 is the phenyl group.)

(a) $CH_3CH_2CH{=}CHCH_2CH_3 + H_2O \xrightarrow{\text{acid catalyst}}$

(b) $CH_3\overset{\displaystyle CH_3}{\overset{|}{C}}HCH{=}CH_2 + H_2 \xrightarrow{\text{Ni catalyst}}$

(c) $C_6H_6 + Br_2 \xrightarrow{FeBr_3}$

(d) $CH_3CH_2CH_2CH_2CH_2CH_3 + H_2SO_4(\text{concd}) \longrightarrow$

(e) cyclopentene $+ H_2O \xrightarrow{\text{acid catalyst}}$

(f) $CH_2{=}CHCH_2CH_2CH_3 + H_2 \xrightarrow{\text{Ni catalyst}}$

(g) cyclohexane$-CH_2CH_3 + H_2SO_4(\text{concd}) \longrightarrow$

(h) $C_6H_5CH{=}CHC_6H_5 + H_2O \xrightarrow{\text{acid catalyst}}$

(i) cyclopentane$-CH_3 + H_2 \xrightarrow{\text{Ni catalyst}}$

(j) $CH_3CH_2CH(CH_3)_2 + O_2 \xrightarrow[\text{(Balance the equation.)}]{\text{complete combustion}}$

(k) $C_6H_6 + H_2O \xrightarrow{\text{acid catalyst}}$

2.57 Write the structures of the products in the following situations. If no reaction is to be expected, write "no reaction."

(a) $CH_2{=}CHCH_2CH_2CH{=}CH_2 + 2H_2 \xrightarrow{\text{Ni catalyst}}$

(b) 1-methylcyclopentene (CH_3) $+ H_2O \xrightarrow{\text{acid catalyst}}$

(c) $CH_3\overset{H_3C}{\overset{|}{C}}{=}\overset{CH_3}{\overset{|}{C}}CH_3 + HBr \longrightarrow$

(d) cyclohexylcyclopentane $+ H_2SO_4(\text{concd}) \longrightarrow$

(e) methylenecyclopentane ($=CH_2$) $+ Cl_2 \longrightarrow$

(f) 1,2-dimethylbenzene (CH_3, CH_3) $\xrightarrow[\text{(Balance the equation.)}]{\text{complete combustion}}$

(g) $C_6H_6 + Cl_2 \xrightarrow{FeCl_3}$

(h) $C_6H_6 + NaOH(aq) \longrightarrow$

(i) $CH_3CH_2\overset{CH_3}{\overset{|}{C}}{=}CHCH_3 + HCl \longrightarrow$

(j) cyclopentene $+ NaOH(aq) \longrightarrow$

(k) $CH_3CH{=}CHCH_2CH{=}CH_2 + 2Cl_2 \longrightarrow$

(l) $C_6H_5CH{=}CHCH_3 + Br_2 \longrightarrow$

3 ALCOHOLS, PHENOLS, ETHERS, AND THIOALCOHOLS

The sugar in the juice of grapes, as from this vineyard in the Sonoma Valley of California, changes to the ethyl alcohol of wine during fermentation. Substances with alcohol groups occur widely at the molecular level of life and we'll learn several properties of alcohols in this chapter.

3.1

OCCURRENCE, TYPES, AND NAMES OF ALCOHOLS

In the alcohol family, molecules have the OH group attached to a saturated carbon atom.

The alcohol system is one of the most widely occurring in nature. It is present in most of the major kinds of organic substances found in living things such as carbohydrates, proteins, and nucleic acids. The chemistry of carbohydrates, for example, is little more than the chemistry of alcohols (this chapter) and that of aldehydes and ketones (next chapter). Members of the alcohol family also include many common commercial products such as wood alcohol (methanol), rubbing alcohol (2-propanol), beverage alcohol (ethanol), and compounds in antifreezes (see Special Topic 3.1).

■ We'll soon learn about systems with the OH group that are not alcohols.

In **alcohols,** the OH group is covalently held by a *saturated* carbon atom, a carbon atom from which only *single* bonds extend. Only when present in this form is the OH group called the **alcohol group.** Alcohols with one OH group per molecule as the sole functional group are called the *simple alcohols* or **monohydric alcohols** (see Table 3.1). Molecules of **dihydric alcohols,** sometimes called **glycols,** have two OH groups: 1,2-ethanediol (ethylene glycol) is an example (Table 3.1). A **trihydric alcohol** is one whose molecules have three alcohol groups, such as 1,2,3-propanetriol (glycerol).

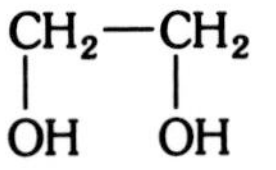

$$\underset{\text{OH}}{\underset{|}{\text{CH}_2}}-\underset{\text{OH}}{\underset{|}{\text{CH}_2}}$$

1,2-Ethanediol (ethylene glycol, a dihydric alcohol)

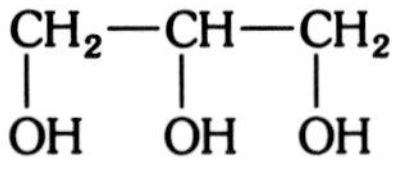

$$\underset{\text{OH}}{\underset{|}{\text{CH}_2}}-\underset{\text{OH}}{\underset{|}{\text{CH}}}-\underset{\text{OH}}{\underset{|}{\text{CH}_2}}$$

1,2,3-Propanetriol (glycerol, a trihydric alcohol)

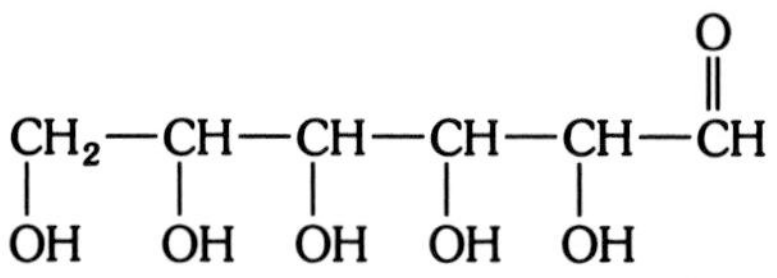

$$\underset{\text{OH}}{\underset{|}{\text{CH}_2}}-\underset{\text{OH}}{\underset{|}{\text{CH}}}-\underset{\text{OH}}{\underset{|}{\text{CH}}}-\underset{\text{OH}}{\underset{|}{\text{CH}}}-\underset{\text{OH}}{\underset{|}{\text{CH}}}-\overset{\text{O}}{\overset{\|}{\text{CH}}}$$

Glucose, a sugar (open form of molecule)

■ The following system, the 1,1-diol, is unstable; only rare examples are known.

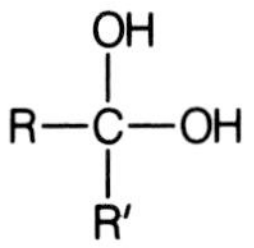

$$\text{R}-\overset{\text{OH}}{\overset{|}{\underset{\text{R}'}{\underset{|}{\text{C}}}}}-\text{OH}$$

1,1-Diols

Many substances, particularly among the carbohydrates, have several OH groups per molecule. One important structural restriction on stable alcohols is that *almost no stable system is known in which one carbon holds two or more OH groups.* If such should form during a reaction, it breaks up. (More will be said about this in Chapter 4.)

The Alcohol Group Occurs as a 1°, 2°, or 3° System An alcohol is classified as primary (1°), secondary (2°), or tertiary (3°) according to the kind of carbon that holds the OH group. When the OH group is held by a 1° carbon, one that has only one carbon atom directly joined to it, the alcohol is a **primary alcohol.** In a **secondary alcohol,** the OH group is held by a 2° carbon. When the OH group is joined to a 3° carbon, the alcohol is a **tertiary alcohol.** We will see the usefulness of these subclasses when we learn that not all alcohols respond in the same way to oxidizing agents.

■ Pronounce 1° as *primary,* 2° as *secondary,* and 3° as *tertiary.*

$$\text{R}-\text{CH}_2-\text{OH}$$

Primary alcohol

$$\text{R}-\overset{\text{R}'}{\overset{|}{\text{CH}}}-\text{OH}$$

Secondary alcohol

$$\text{R}-\overset{\text{R}'}{\overset{|}{\underset{\text{R}''}{\underset{|}{\text{C}}}}}-\text{OH}$$

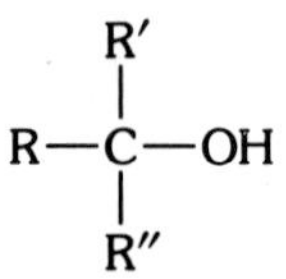

Tertiary alcohol

■ The R— groups in 2° and 3° alcohols don't have to be the same.

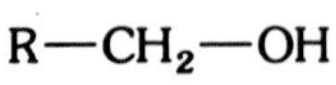

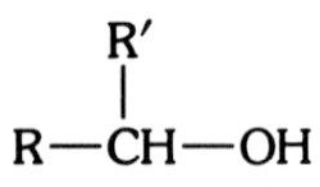

TABLE 3.1
Some Common Alcohols

Name[a]	Structure	BP (°C)
Methanol	CH_3OH	65
Ethanol	CH_3CH_2OH	78.5
1-Propanol	$CH_3CH_2CH_2OH$	97
2-Propanol	$CH_3CH(OH)CH_3$	82
1-Butanol	$CH_3CH_2CH_2CH_2OH$	117
2-Butanol (*sec*-butyl alcohol)	$CH_3CH_2CH(OH)CH_3$	100
2-Methyl-1-propanol (isobutyl alcohol)	$CH_3CH(CH_3)CH_2OH$	108
2-Methyl-2-propanol (*t*-butyl alcohol)	$CH_3C(CH_3)_2OH$	83
1,2-Ethanediol (ethylene glycol)	$CH_2(OH)—CH_2(OH)$	197
1,2-Propandiol (propylene glycol)	$CH_3—CH(OH)—CH_2(OH)$	189
1,2,3-Propanetriol (glycerol)	$CH_2(OH)—CH(OH)—CH_2(OH)$	290

[a] The IUPAC names with the common names in parentheses.

PRACTICE EXERCISE 1

Classify each of the following as monohydric or dihydric. For each found to be monohydric, classify it further as 1°, 2°, or 3°. If the structure is too unstable to exist, state so.

(a) $CH_3CH(OH)CH_3$

(b) cyclohexene ring bearing —OH

(c) $CH_3C(OH)_2CH_3$

(d) cyclohexane ring bearing two adjacent OH groups

(e) $HOCH_2C(CH_3)_2CH_3$

(f) benzene ring bearing —CH_2OH

(g) $CH_3C(CH_3)_2OH$

(h) $CH_3CH_2CH(CH_3)OH$

(i) $CH_3C(OH)_2OH$

SPECIAL TOPIC 3.1
ALCOHOLS IN OUR DAILY LIVES

Methanol (methyl alcohol, wood alcohol) When taken internally in enough quantity, methanol causes either blindness or death. In industry, it is used as the raw material for making formaldehyde (which is used to make polymers), as a solvent, and as a denaturant (poison) for ethanol. It is also used as the fuel in "canned heat" (e.g., Sterno), as well as in burners for fondue pots.

Most methanol is made by the reaction of carbon monoxide with hydrogen under high pressure and temperature:

$$2H_2 + CO \xrightarrow[\text{350-400 °C}]{\text{3000 lb/in.}^2} CH_3OH$$
$$ZnO/Cr_2O_3 \text{ catalyst}$$

Ethanol (ethyl alcohol, grain alcohol) Some ethanol is made by the fermentation of sugars, but most is synthesized by the addition of water to ethene in the presence of a catalyst. A 70% (v/v) solution of ethanol in water is used as a disinfectant.

In industry, ethanol is used as a solvent and to prepare pharmaceuticals, perfumes, lotions, and rubbing compounds. For these purposes, the ethanol is adulterated by poisons that are very difficult to remove so that the alcohol cannot be sold or used as a beverage. (Nearly all countries derive revenue by taxing potable, i.e., drinkable, alcohol.)

Ethanol, a toxic substance responsible for widespread human misery, is absorbed directly into the bloodstream along any part of the intestinal tract. Enzymes in the liver work to detoxify it. One enzyme takes it to the aldehyde stage, producing acetaldehyde (CH_3CHO), which is chiefly responsible for liver fibrosis. In time, the acetaldehyde is oxidized to the acetate ion, $CH_3CO_2^-$, which can be metabolized normally. Fetal alcohol syndrome appears to be caused directly by excessive ethyl alcohol itself, not by its oxidation products. This syndrome leads to babies with impaired mental abilities and many other disorders.

2-Propanol (isopropyl alcohol) 2-Propanol is a common substitute for ethanol for giving back rubs. In solutions with concentrations from 50% to 99% (v/v), 2-propanol is used as a disinfectant. It is twice as toxic as ethanol.

1,2-Ethanediol (ethylene glycol), **and 1,2-Propanediol** (propylene glycol) Ethylene glycol and propylene glycol are the chief components in permanent-type antifreezes. Their great solubility in water and their very high boiling points make them ideal for this purpose. An aqueous solution that is roughly 50% (v/v) in either glycol does not freeze until about −40 °C (−40 °F). Ethylene glycol has a sweet taste, but it is highly toxic. The lethal dose for adults is about 100 mL. 1,2-Propanediol, on the other hand, is far less toxic (possibly because the anions of its oxidation products, pyruvic acid and acetic acid, are normally formed in metabolism).

1,2,3-Propanetriol (glycerol, glycerin) Glycerol, a colorless, syrupy liquid with a sweet taste, is freely soluble in water and insoluble in nonpolar solvents. It is one product of the digestion of the fats and oils in our diets. Because it has three OH groups per molecule, each capable of hydrogen bonding to water molecules, glycerol can draw moisture from humid air. It is sometimes used as a food additive to help keep foods moist.

An oily compound made from glycerol and nitric acid, **nitroglycerin,** is a powerful explosive. When pure, it detonates from concussion. Interestingly, although nitroglycerin is toxic, it has a place in medicine.

$$O_2NOCH_2\underset{\displaystyle ONO_2}{\underset{|}{C}}HCH_2ONO_2$$

Nitroglycerin
(1,2,3-glyceryl trinitrate)

People who have periodic attacks of intense pain (angina pectoris), centered in heart muscle because of vasoconstriction (constriction of the blood vessels), are able to administer to themselves carefully controlled amounts of a vasodilator (a dilator of blood vessels). Nitroglycerin is commonly used for this purpose.

Sugars All carbohydrates consist of polyhydroxy compounds, and we will study them in Chapter 8.

Not All Compounds with the OH Group Are Alcohols As you can see by the following structures, the OH group also occurs in the family of the phenols, where it is attached to a benzene ring, and in the family of the carboxylic acids, where it is attached to a carbon that has a double bond to oxygen.

■ Phenol is only the simplest member of the family of *phenols,* compounds with OH groups attached to benzene rings that may carry any number and kind of other ring substituents.

$$-\overset{|}{\underset{|}{C}}-O-H$$

alcohol group — alcohol system

Phenol — Carboxylic acid ($RCOH$ with C=O) — Enol system (unstable) ($>C=C<$ with OH)

When the OH is attached to an alkene group, the system is called an *enol* (ene + ol), but it is unstable. Most of our attention in this chapter will be given to simple alcohols but with some study of phenols. We will study the carboxylic acids in a later chapter.

PRACTICE EXERCISE 2

Classify the following as alcohols, phenols, or carboxylic acids.

(a) $CH_3-C_6H_4-CH_2OH$ **(b)** $CH_3-C_6H_4-OH$

(c) $CH_3-C_6H_4-\overset{O}{\overset{\|}{C}}OH$ **(d)** $CH_2{=}CHOH$

(e) $CH_3CH_2CH_2CH_2OH$ **(f)** $C_6H_{11}-OH$ (cyclohexane ring)

Common Names of Alcohols Are Popular When Their Alkyl Groups Are Easily Named Simple alcohols have common names devised by writing the word *alcohol* after the name of the alkyl group holding the OH group. For example,

CH_3OH	CH_3CH_2OH	CH_3CHCH_3 (OH on C2)	CH_3CHCH_2OH (CH_3 on C2)	CH_3COH (CH_3 above and below)
Methyl alcohol	Ethyl alcohol	Isopropyl alcohol	Isobutyl alcohol	*t*-Butyl alcohol

IUPAC Names of Alcohols End in *-ol* The IUPAC rules for naming alcohols are similar to those for naming alkanes. The full name of the alcohol is based on the idea of a *parent alcohol* that has substituents on its carbon chain. The rules are as follows.

1. Determine the parent alcohol by selecting the longest chain of carbons *that includes the carbon atom to which the OH group is attached.* Name the parent alcohol by changing the

name ending of the alkane that corresponds to this chain from *-e* to *-ol*. Examples are

CH_3OH	$CH_3CH_2CH_2OH$	$CH_3CH(CH_3)CH_2CH_2OH$
Methanol (complete name; parent alkane is methane)	Propanol (incomplete name; parent alkane is propane)	A substituted butanol (incomplete name; parent alkane is butane)

No number is needed to specify the location of the OH group in the names methanol and ethanol.

2. Number the parent chain from whichever end gives the carbon that holds the OH group the lower number. Be sure to notice this departure from the IUPAC rules for numbering chains for alkanes; in the rules for alkanes the location of the first branch determines the direction of numbering. With alcohols, *the location of the OH group takes precedence over alkyl groups or halogen atoms in deciding the direction of numbering.* Examples are

$\overset{4}{C}H_3\overset{3}{C}H(CH_3)\overset{2}{C}H_2\overset{1}{C}H_2OH$
3-Methyl-1-butanol
Not 2-methyl-4-butanol

$\overset{7}{C}H_3\overset{6}{C}(CH_3)_2\overset{5}{C}H_2\overset{4}{C}H_2\overset{3}{C}H(OH)\overset{2}{C}H_2\overset{1}{C}H_3$
6,6-Dimethyl-3-heptanol
Not 2,2-dimethyl-5-heptanol

3. Write the number that locates the OH group in front of the name of the parent, and separate this number from the name of the parent by a hyphen. Then, as prefixes to what you have just written, assemble the names of the substituents and their location numbers. Use commas and hyphens in the usual way. For illustrations, see the examples above that follow rule 2 for illustrations. For example:

4. When two or more OH groups are present, use name endings such as *-diol* (for two OH groups), *-triol*, and so forth. Immediately in front of the name of the *parent portion* of the alcohol, write the two, three, or more numbers that show the locations of the OH groups. For example,

$CH_3C(CH_3)(OH)CH_2OH$
2-Methyl-1,2-propanediol

$CH_3CH(OH)CH(OH)CH_2OH$
1,2,3-Butanetriol

5. If no parent alcohol name is possible or convenient, then the OH group can be treated as another substituent, and it is named *hydroxy*. An example is in the margin.

$HO-C_6H_4-COOH$
4-Hydroxybenzoic acid

EXAMPLE 3.1
Using the IUPAC Rules to Name an Alcohol

What is the name of the following compound?

$CH_3CH_2CH_2CH(CH_2CH_3)-C(CH_3)(CH_2CH_2CH_3)CH_2OH$

ANALYSIS The compound is in the alcohol family, so the ending to its IUPAC name is *-ol.* The IUPAC rules for alcohols require the parent chain to be the longest carbon chain *that includes the carbon atom to which the OH group is attached.* This is a six-carbon chain in the structure. The IUPAC rules then require us to number the chain from whichever of its ends lets the carbon bearing the OH group have the lower number. Thus

```
              CH3CH2  CH3
                   |   |
CH3CH2CH2CH—CCH2OH
 6   5   4   3   2 | 1
                     CH2CH2CH3
```

The final name must end in -1-hexanol. Carbon 2 holds both a methyl and a propyl group, and carbon 3 has an ethyl group.

SOLUTION The final name, arranging the alkyl groups alphabetically, is

3-Ethyl-2-methyl-2-propyl-1-hexanol

■ The longest carbon chain in the molecule has eight carbon atoms, but it doesn't include the OH group.

PRACTICE EXERCISE 3

Name the following compounds by the IUPAC system.

```
        CH3                          CH3
        |                            |
(a) CH3CHCH2CH2CH2OH    (b) CH3COH
                                     |
                                     CH3

           CH3                    CH3
           |                      |
(c) CH3CH2CCH2OH        (d) HOCH2CHCH2OH
           |
           CH2CH2CH3
```

3.2 PHYSICAL PROPERTIES OF ALCOHOLS

Hydrogen bonding dominates the physical properties of alcohols.

The OH group is quite polar, and it can both donate and accept hydrogen bonds. Hydrogen bonding gives alcohols much higher boiling points and much greater solubilities in water than hydrocarbons of the same formula mass.

The effect of the OH group on water solubility is particularly important at the molecular level of life, because some substances in cells must be in solution and others must not. Moreover, by using special enzymes in the liver, the body attaches OH groups to the molecules of many toxic substances to make them more soluble and thus more easily carried in the blood and eliminated in the urine.

■ The *donor* OH group has the H from which the H-bond (···) extends to the $\delta-$ site on the *acceptor.*

■ H-bond *acceptor*

```
      H
       \δ−
        O—R
        :
        H δ+
       /
   R—O
```

H-bond *donor*

Hydrogen Bonds between Molecules Raise Boiling Points Table 3.2 compares the boiling points of some alcohols to those of alkanes with comparable formula masses. The differences in boiling points are caused by the hydrogen bonding made possible by the considerable difference in the electronegativities of O and H. The hydrogen bond, remember, is a force of attraction between opposite *partial* charges, such as the $\delta+$ charge on H in the OH group and the $\delta-$ charge on the O of another group. No partial charges exist in molecules of alkanes because C and H have nearly identical electronegativities. An individual alcohol mole-

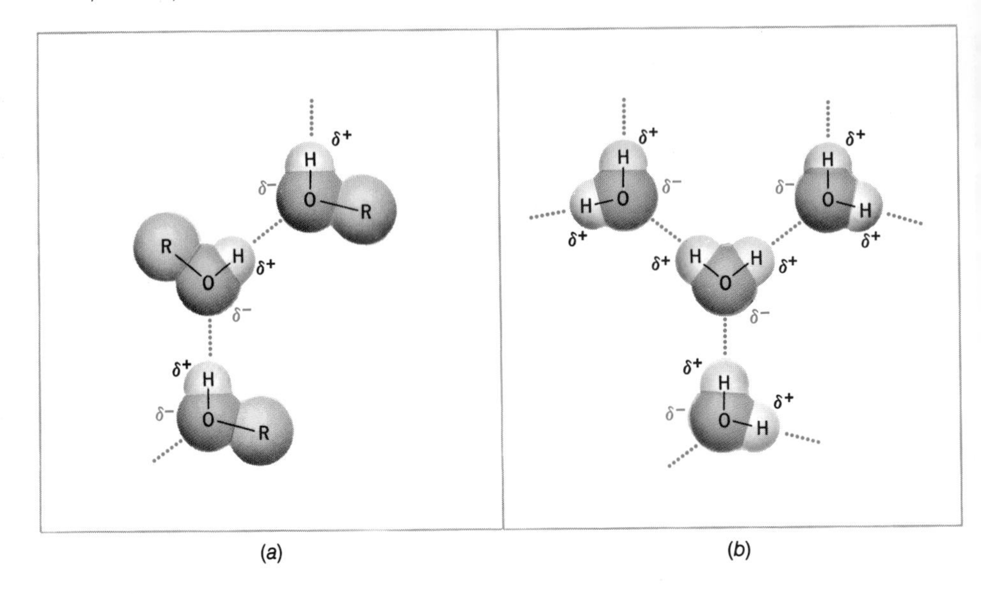

FIGURE 3.1
Hydrogen bonding in alcohols (*a*) and in water (*b*).

cule is attracted to two neighboring molecules by two hydrogen bonds (see Fig. 3.1*a*). Water molecules have three hydrogen bonds between them (see Fig. 3.1*b*). So water, despite having a lower formula mass than methyl alcohol, has a higher boiling point.

Ethylene glycol (Table 3.2), a dihydric alcohol, has four hydrogen bonds between molecules, two from each OH group, and it boils nearly 200 °C higher than butane and 100 °C higher than water. You can see by these data how significantly the OH group provides forces of attraction between molecules.

■ Only three elements, F, O, and N, have atoms that are electronegative enough to participate significantly in hydrogen bonds.

The OH Group Makes Compounds More Soluble in Water Methane, like all hydrocarbons, is insoluble in water.[1] Methyl alcohol, in contrast, dissolves in water in all proportions. The difference is caused by the OH group. Hydrogen bonds can be formed between methyl alcohol molecules and water molecules, and this enables methyl alcohol molecules to slip into the hydrogen-bonding network in water (see Fig. 3.2). The CH_3 group in CH_3OH is too small to interfere.

TABLE 3.2
The Influence of the Alcohol Group on Boiling Points

Name	Structure	Formula Mass	BP (°C)	Difference in BP
Ethane	CH_3CH_3	30	−89	154
Methyl alcohol	CH_3OH	32	65	
Propane	$CH_3CH_2CH_3$	44	−42	120
Ethyl alcohol	CH_3CH_2OH	46	78	
Butane	$CH_3CH_2CH_2CH_3$	58	0	197
Ethylene glycol	$HOCH_2CH_2OH$	62	197	

[1] Nothing is *totally* insoluble in water. The size of Avogadro's number (6.02×10^{23}), the particles in one mole, is so great that even if something can form a solution with a molarity of only, say, 1×10^{-14} mol/L—immeasurably small—there are still roughly a little over a billion (10^9) molecules of solute in each liter of solution!

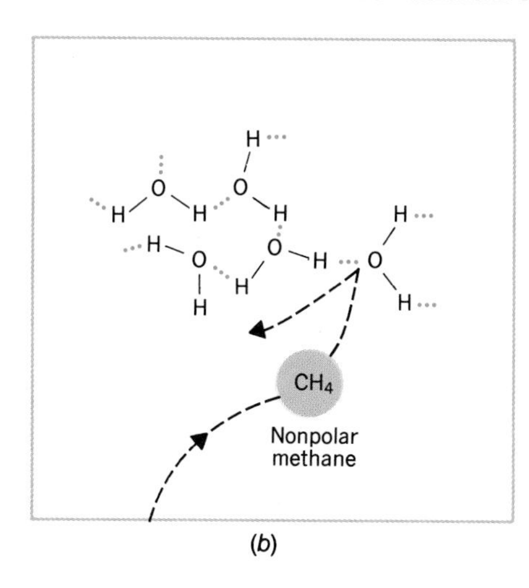

FIGURE 3.2
How a short-chain alcohol dissolves in water. (*a*) The alcohol molecule can take the place of a water molecule in the hydrogen-bonding network of water. (*b*) An alkane molecule cannot break into the hydrogen-bonding network in water, so the alkane cannot dissolve.

As the size of the R group in a monohydric alcohol molecule increases, however, alcohol molecules become more and more alkane-like. A long R group interferes with the alcohol's ability to dissolve in water. In 1-decanol, for example,

$$CH_3CH_2CH_2CH_2CH_2CH_2CH_2CH_2CH_2CH_2OH$$

1-Decanol

the small, water-like OH group is overwhelmed by the long hydrocarbon chain. Water molecules have no attraction for this part of the 1-decanol molecule. The flexings and twistings of its long chain interfere too much with the hydrogen-bonding networks in water, and water molecules will not separate to let 1-decanol molecules into solution. This alcohol, and most with five or more carbons, are thus insoluble in water. However, they do dissolve in such nonpolar solvents as diethyl ether, gasoline, and petroleum ether.

ʼRACTICE EXERCISE 4

1,2-Propanediol (propylene glycol) and 1-butanol have similar formula masses. In which of these two compounds is hydrogen bonding between molecules more extensive and stronger? How do the data in Table 3.1 support the answer? Which would be more soluble in water?

3 3

CHEMICAL PROPERTIES OF ALCOHOLS

The loss of water (dehydration) and the loss of hydrogen (oxidation) are two very important reactions of alcohols.

Alcohols react with both inorganic and organic compounds, but we will study only the inorganic reactants in this chapter. Before continuing, however, we need to point out some properties that the alcohols do *not* have, but which might be assumed because of their OH groups.

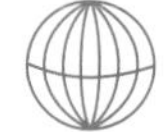

Despite the OH Group, Alcohols Are Not Strong Bases We can think of alcohols as mono-alkyl derivatives of water and, like water, they are extremely weak proton donors (acids) and proton acceptors (bases). Alcohol molecules do not ionize to give either OH^- ions or H^+ ions. Water, as you know, only *very* slightly ionizes; at room temperature, not much more than about 0.0000001% of its molecules are ionized. *The percentage ionization of any*

■ When an alcohol dissolves in water, it doesn't raise or lower the pH.

alcohol group in water is even smaller than that of water. A solution of methyl alcohol (or any alcohol) in water thus has a neutral pH.

Alcohols Can Be Dehydrated to Alkenes In laboratory vessels, the action of heat and a strong acid catalyst causes the dehydration of an alcohol to an alkene. A water molecule splits out and a carbon-carbon double bond emerges. The pieces of the water molecule, one H and one OH, come from *adjacent* carbons.

$$\underset{\text{Alcohol}}{-\overset{|}{\underset{\underset{H}{|}}{C}}-\overset{|}{\underset{\underset{OH}{|}}{C}}-} \xrightarrow[\text{heat}]{H^+ \text{ catalyst}} \underset{\text{Alkene}}{>C=C<} + H-OH$$

Specific examples are as follows.

$$\underset{\text{Ethanol}}{CH_3CH_2OH} \xrightarrow[\text{heat}]{H^+ \text{ catalyst}} \underset{\text{Ethene}}{CH_2=CH_2} + H_2O$$

$$\underset{\text{2-Butanol}}{CH_3CH_2\overset{\overset{OH}{|}}{C}HCH_3} \xrightarrow[\text{heat}]{H^+ \text{ catalyst}} \underset{\substack{\text{2-Butene} \\ \text{(chief product)}}}{CH_3CH=CHCH_3} + \underset{\substack{\text{1-Butene} \\ \text{(minor product)}}}{CH_3CH_2CH=CH_2} + H_2O$$

Special Topic 3.2 explains how the acid works as the catalyst for these reactions.

When It Is Possible for Two Alkenes to Be Made, the More Highly Branched Alkene Forms Water can split out in two ways from 2-butanol (as shown above). Molecules of 2-butanol have H atoms on *two* carbons adjacent to the carbon holding the departing OH group. Two alkenes are therefore possible, 2-butene and 1-butene, and some of each forms. When options like this exist, however, one alkene predominates, generally *the more highly branched alkene*. This is the alkene with the greatest number of alkyl groups attached to the double bond. 2-Butene is more branched than 1-butene because it has two alkyl groups at the double bond (two methyl groups) and 1-butene has only one (an ethyl group).

■ $CH_3-CH=CH-CH_3$
2-Butene
(two alkyl groups)

$CH_2=CH-CH_2CH_3$
1-Butene
(one alkyl group)

In acid-catalyzed dehydrations of alcohols, the more branched alkene predominates because it is the more stable. Information about the relative stabilities of isomeric alkenes comes from combustion experiments. For example, both 1-butene and 2-butene are C_4H_8 and the combustion of either occurs as follows.

$$C_4H_8(g) + 6O_2(g) \longrightarrow 4CO_2(g) + 4H_2O + \text{heat of combustion}$$

However, slightly less heat per mole is released when 2-butene is burned, despite the fact that it has the same number of carbons and hydrogens as 1-butene. The only way by which 2-butene can release less energy than 1-butene is *to have less energy initially*. Having less energy always means being more stable, so 2-butene must be slightly more stable than 1-butene. None of this explains *why* 2-butene is more stable. (The explanation is complex.) The release of less energy by the combustion of 2-butene compared to 1-butene is mentioned only to give one piece of experimental support for the generalization: *the more branched alkene is the more stable.*

The complication of two possible products of dehydration is not a problem when the reaction happens to an alcohol system in the body. Such dehydrations are enzyme-catalyzed, not acid-catalyzed, and enzymes direct reactions in very specific ways. It is even possible for the less stable double bond to be produced when enzymes catalyze the reaction.

■ Enzymes are exceedingly selective in what they do and how they control reactions.

SPECIAL TOPIC 3.2
HOW ACIDS CATALYZE THE DEHYDRATION OF ALCOHOLS

When a strong acid is added to an alcohol, the first chemical event is the ionization of the acid. The reaction is exactly analogous to the ionization of a strong acid when it is added to water; a proton transfers from the acid to the oxygen atom of a solvent molecule. Because sulfuric acid is often used as the catalyst in the dehydration of an alcohol, we will use it to illustrate these reactions. Its ionization in water establishes the following equilibrium, with the products very strongly favored.

$$HO_3SO{-}H + {:}\ddot{O}(H)H \rightleftharpoons HO_3SO^- + H{-}\overset{+}{\ddot{O}}(H)H$$

Sulfuric acid, H_2SO_4 — Water — Hydrogen sulfate ion — Hydronium ion

The same ionization happens in ethyl alcohol. (Be sure to notice the similarity to the previous reaction.)

$$HO_3SO{-}H + {:}\ddot{O}(CH_2CH_3)H \rightleftharpoons HO_3SO^- + H{-}\overset{+}{\ddot{O}}(CH_2CH_3)H$$

Protonated form of ethyl alcohol

All the covalent bonds to the oxygen atoms in either the hydronium ion or in the protonated form of the alcohol are weak, including, in the latter, *the covalent bond to carbon.* This bond is weakened through the action of the catalyst. As ions and molecules bump into one another, some of the ions of the protonated form of the alcohol break up, as we can illustrate using the protonated form of ethyl alcohol.

$$H{-}\overset{+}{\ddot{O}}(CH_2CH_3)H \rightleftharpoons CH_3CH_2^+ + {:}\ddot{O}(H)H$$

Protonated form of ethyl alcohol — Ethyl carbocation

All carbocations are unstable, because a carbon atom does not handle anything less than an octet very well. The octet for carbon in the ethyl carbocation is restored when a proton from the carbon adjacent to the site of the positive charge transfers to a proton acceptor. As this proton transfers, the electron pair of its covalent bond to carbon pivots in to form the second bond of the emerging double bond. It is a smooth, synchronous operation. One acceptor for the proton is the hydrogen sulfate ion, and its acceptance of a proton restores the catalyst, as follows.

$$HO_3SO^- + H{-}CH_2{-}CH_2^+ \longrightarrow CH_2{=}CH_2 + HO_3SO{-}H$$

Ethyl carbocation

Once water starts to appear as another product, its molecules can also accept the proton from the ethyl carbocation (to give H_3O^+), but this is also equivalent to the recovery of the catalyst.

EXAMPLE 3.2
Writing the Structure of the Alkene That Forms When an Alcohol Undergoes Dehydration

What is the product of the dehydration of isobutyl alcohol?

$$CH_3CH(CH_3)CH_2OH$$

Isobutyl alcohol

ANALYSIS Predicting the product of the dehydration of this alcohol involves rewriting the structure of the alcohol but leaving off the OH group and one H from a carbon adjacent to the carbon that holds the OH group. A double bond is then written between these two carbon atoms.

SOLUTION

$$\begin{array}{c} CH_3 \\ | \\ CH_3C{=}CH_2 \end{array}$$

2-Methylpropene

■ When some alcohols undergo dehydration, their carbon skeletons rearrange, but we won't study any examples of this.

CHECK Ask the following questions of the answer. Is it an *alkene?* (Alcohol dehydrations give alkenes.) Is its carbon skeleton the same as in the starting material? (Changes in the carbon skeleton do not occur in all of the examples that we will study.) Does each carbon have four bonds from it? (The rules of covalence must be obeyed.) By answering "yes" to these questions you have made a thorough check.

PRACTICE EXERCISE 5

Write the structures of the alkenes that can be made by the dehydration of the following alcohols.

(a) $CH_3CH_2CH_2OH$ (b) CH_3CHCH_3 with OH on the middle carbon (c) CH_3COH with CH_3 above and CH_3 below the central carbon (d) cyclohexane ring–OH

1° and 2° Alcohols Are Dehydrogenated by Strong Oxidizing Agents In a strictly formal sense, an oxidation number becomes more positive for a species being oxidized. To use this definition to recognize whether something has been oxidized, however, requires the calculation of oxidation numbers. Organic chemists often use a shortcut summarized by the following two rules of thumb. With organic compounds,

1. An *oxidation* is the loss of H or the gain of O by a molecule.
2. A *reduction* is the loss of O or the gain of H by a molecule.

■ In the body, the enzyme *alcohol dehydrogenase* (ADH) oxidizes methyl alcohol (wood alcohol) to formaldehyde, $CH_2{=}O$, which has toxic effects and can cause blindness and death.

■ The term *in vitro* means "in a glass vessel," that is, carried out in laboratory glassware. The term *in vivo* means in a living cell.

The oxidation of an alcohol is an example of the loss of H. Sometimes, therefore, the reaction is called *dehydrogenation*. Enzymes that catalyze such reactions in living systems are called *dehydrogenases*.

In studying dehydrogenation or oxidation, we are particularly interested in what happens to the organic molecule being oxidized. What does it change into? To serve this limited interest, we make two further simplifications. We largely use unbalanced "reaction sequences," not balanced equations, and we'll use the symbol (O) *for any oxidizing agent that can bring about the oxidation given by a reaction sequence.*

One common strong oxidizing agent used *in vitro* is potassium permanganate, $KMnO_4$, in which the actual oxidizing agent is the deeply purple colored permanganate ion, MnO_4^-. Another common strong oxidizing agent used to oxidize alcohols is sodium dichromate, $Na_2Cr_2O_7$, which furnishes the brightly orange colored dichromate ion, $Cr_2O_7^{2-}$.

When an oxidizing agent causes an alcohol molecule to lose hydrogen, molecular hydrogen (H_2) does not itself form either *in vitro* or *in vivo*. Instead, the hydrogen atoms end up in a water molecule whose oxygen atom comes from the oxidizing agent. One H comes from the OH group of the alcohol, and the other H comes from the carbon that has been holding the OH group. Left behind in the organic molecule is a carbon–oxygen double bond, a carbonyl group. Study the following schematic carefully to learn where precisely the two H atoms originate.

$$-\overset{|}{\underset{H}{C}}-O-H + (O) \longrightarrow \;>C{=}O + H-O-H$$

$[H\!:^{-} + H^{+}]$ (O)

What makes the reaction an *oxidation* of the alcohol group is specifically the *loss of the electron pair* of the C—H bond in the alcohol system. In the schematic above, we see this pair first on $H\!:^-$ and then incorporated into the water molecule. The $H\!:^-$ ion (the hydride ion) never becomes free in the aqueous medium. We show it in brackets above only to help you track the oxidation. In the end, a water molecule is exactly where the electron pair of the C—H bond of the alcohol system finally lodges when the alcohol is oxidized *in vivo* by a series of reactions called the *respiratory chain* (to be studied in Chapter 14). When (O) is MnO_4^- or $Cr_2O_7^{2-}$, however, the electrons of the oxidized alcohol group are accepted by these species to change the oxidation states of the central atoms. Thus the oxidation state of Mn is reduced from +7 in MnO_4^- to +4 in MnO_2. When (O) is $Cr_2O_7^{2-}$, the electrons end up in reducing the oxidation state of Cr from +6 in $Cr_2O_7^{2-}$ to +3 in Cr^{3+}. By *accepting* the electrons, MnO_4^- and $Cr_2O_7^{2-}$ are reduced.

■ The powerfully basic hydride ion, if let loose, would react with water to give the hydroxide ion and hydrogen gas.

$$H\!:^- + H_2O \rightarrow H-H + OH^-$$

3° Alcohols Cannot Be Dehydrogenated Only 1° and 2° alcohols can be oxidized by the loss of 2 H atoms. Molecules of 3° alcohols do not have an H atom on the carbon that holds the OH group, so 3° alcohols cannot be oxidized by dehydrogenation.

■ No H here

$$R-\overset{R'}{\underset{R''}{C}}-OH$$

3° Alcohol system

$$R-\overset{O}{\overset{\|}{C}}-H$$

Aldehyde

$$R-\overset{O}{\overset{\|}{C}}-OH$$

Carboxylic acid

$$R-\overset{O}{\overset{\|}{C}}-R'$$

Ketone

In Aqueous Systems, Strong Oxidizing Agents Are Used to Change the 1° Alcohol Group to a Carboxyl Group The organic product of the oxidation of either a 1° or a 2° alcohol has a carbon–oxygen double bond. Each subclass of alcohol, however, is oxidized to a different organic family. A 1° alcohol is oxidized first to an aldehyde, but because *aldehydes are more easily oxidized than 1° alcohols,* particularly in aqueous media, it is seldom practical to use $MnO_4^-(aq)$ or $Cr_2O_7^{2-}(aq)$ to prepare aldehydes. As soon as aldehyde groups appear in the presence of these ions, the aldehyde groups begin to use up oxidizing agent and change to carboxyl groups. In the presence of sufficient oxidizing agent, the successive stages in the oxidation of a 1° alcohol group are

$$\underset{\text{1° Alcohol}}{RCH_2OH} \xrightarrow[MnO_4^-(aq)\text{ or } Cr_2O_7^{2-}(aq)]{(O)} \underset{\text{Aldehyde}}{R\overset{O}{\overset{\|}{C}}H} \xrightarrow{\text{more (O)}} \underset{\text{Carboxylic acid}}{R\overset{O}{\overset{\|}{C}}OH}$$

In the lab, therefore, when either aqueous permanganate or dichromate is used, the oxidation of a 1° alcohol is generally carried out with enough oxidizing agent to take the oxidation of the alcohol all the way to its corresponding carboxylic acid. Carboxylic acids strongly resist further oxidation. The dichromate oxidation of 1-propanol to propanoic acid, for example, occurs by the following equation.

$$\underset{\text{1-Propanol}}{3CH_3CH_2CH_2OH} + 2Cr_2O_7^{2-} + 16H^+ \longrightarrow \underset{\text{Propanoic acid}}{3CH_3CH_2CO_2H} + 4Cr^{3+} + 11H_2O$$

Enzyme-Catalyzed Oxidations Convert a 1° Alcohol Group to an Aldehyde Group The problem of halting the oxidation of a 1° alcohol at the aldehyde stage does not occur in body cells, because *different enzymes are required for each oxidation step.* One

■ Certain vitamins in the diet, like riboflavin and niacin, provide molecular acceptor units for H:⁻ in dehydrogenase enzymes.

enzyme handles the oxidation of a 1° alcohol group to the corresponding aldehyde. A different enzyme is needed to take an aldehyde group to the next oxidation stage, and this enzyme is generally not in the same place where the aldehyde is made. In general, remembering that (O) stands for an oxidizing agent capable of accomplishing the given oxidation, the reaction is

$$\underset{\text{1° Alcohol}}{RCH_2OH} \xrightarrow[\text{(enzyme-catalyzed)}]{(O)} \underset{\text{Aldehyde}}{RCH{=}O} + H_2O$$

■ A dehydrogenase changes (temporarily) to its reduced form when it accepts H:⁻.

Hydrogen is removed by the transfer of the *pieces* of H—H. An oxidizing enzyme, a dehydrogenase, accepts H:⁻ from the CH unit that holds the OH group, and a proton, H^+, slips away from the O atom of the OH group. With some enzymes, H^+ is simply neutralized by the buffer system of the cell fluid. Other enzymes accept both H:⁻ and H^+. We'll leave details to later chapters.

Because our chief interest lies in what can occur in living systems, we will not be further concerned about *in vitro* oxidations using strong, aqueous oxidizing agents except as they may be used in lab tests. What interests us almost entirely is what aldehyde can be made from a 1° alcohol *in vivo* and what carboxylic acid can eventually be formed from a 1° alcohol.

EXAMPLE 3.3
Writing the Structure of the Product of the Oxidation of a Primary Alcohol

What aldehyde and what carboxylic acid could be made by the oxidation of ethyl alcohol?

ANALYSIS The CH_2OH group is changed to CH=O when a 1° alcohol is oxidized to an aldehyde. Anything attached to the carbon of the CH_2OH group, like the CH_3 group in our example, is retained in the final structure of the aldehyde. When the aldehyde is further oxidized to a carboxylic acid, the CH=O group is changed to CO_2H.

SOLUTION The aldehyde corresponding to CH_3CH_2OH is $CH_3CH{=}O$, ethanal. The carboxylic acid is CH_3CO_2H, acetic acid.

CHECK Is the first product an *aldehyde?* (Does it have the CH=O group?) Is the second product a *carboxylic acid?* (Does it have the CO_2H group, sometimes written COOH?) The oxidation of a 1° alcohol first gives an aldehyde and then, with more oxidizing agent, the carboxylic acid. Are the carbon skeletons of the two products identical with that of the 1° alcohol? Does each carbon have four bonds and each oxygen two? By answering "yes" to these questions you have made a thorough check.

PRACTICE EXERCISE 6

Write the structures of the aldehydes and carboxylic acids that can be made by the oxidation of the following alcohols.

(a) $CH_3CH(CH_3)CH_2OH$ (b) C_6H_5—CH_2OH (benzene ring—CH_2OH)

Secondary Alcohols Are Oxidized to Ketones Ketones strongly resist further oxidation, so they are easily made by the oxidation of 2° alcohols using strong oxidizing agents like MnO_4^- or $Cr_2O_7^{2-}$. *In vivo,* dehydrogenases accomplish the identical overall reaction, the dehydrogenation of a 2° alcohol group to a keto group. In general, for 2° alcohols,

$$\underset{\text{2° alcohol}}{\mathrm{R\overset{OH}{\overset{|}{C}}HR'}} + \mathrm{(O)} \longrightarrow \underset{\text{Ketone}}{\mathrm{R\overset{O}{\overset{\|}{C}}R'}} + \mathrm{H_2O}$$

Specific *in vitro* examples are

$$\underset{\text{2-Butanol}}{\mathrm{CH_3\overset{OH}{\overset{|}{C}}HCH_2CH_3}} \xrightarrow{Cr_2O_7^{2-},\ H^+} \underset{\text{2-Butanone}}{\mathrm{CH_3\overset{O}{\overset{\|}{C}}CH_2CH_3}}$$

$$\underset{\text{Cyclohexanol}}{\text{(cyclohexane ring with OH)}} \xrightarrow{Cr_2O_7^{2-},\ H^+} \underset{\text{Cyclohexanone}}{\text{(cyclohexane ring with =O)}}$$

EXAMPLE 3.4
Writing the Structure of the Product of the Oxidation of a Secondary Alcohol

What ketone forms when 2-propanol is oxidized?

ANALYSIS 2-Propanol is a 2° alcohol, and every 2° alcohol has a CHOH group. The oxidation of a 2° alcohol strips the 2 H atoms from CHOH, creates a double bond between C and O, and so changes CHOH to C=O, a keto group. (The two H atoms emerge in a molecule of water.) *The fundamental skeleton of all of the heavy atoms in the 2° alcohol, like C and O, remains intact.*

SOLUTION

$$\underset{\text{2-Propanol}}{\mathrm{CH_3\overset{OH}{\overset{|}{C}}HCH_3}} \text{ is oxidized to } \underset{\text{Propanone}}{\mathrm{CH_3\overset{O}{\overset{\|}{C}}CH_3}} \text{ plus } \mathrm{H_2O}$$

CHECK Is the product a *ketone?* (2° Alcohols are oxidized to ketones.) Does the product have the same carbon skeleton as the starting material? Does each carbon have four bonds and each oxygen atom two?

■ The name chemists commonly use for propanone is *acetone.*

PRACTICE EXERCISE 7

Write the structures of the ketones that can be made by the oxidation of the following alcohols.

(a) $\mathrm{CH_3\overset{OH}{\overset{|}{C}}HCH_2CH_3}$ (b) $\mathrm{C_6H_5\overset{OH}{\overset{|}{C}}HCH_3}$ (benzene ring attached) (c) cyclopentane ring with OH

PRACTICE EXERCISE 8

What are the products of the oxidation of the following alcohols? If the alcohol is a 1° alcohol, show the structures of both the aldehyde and the carboxylic acid that could be made, depending on the conditions. If the alcohol cannot be oxidized, write "no reaction."

(a) $CH_3CH(OH)CH_2CH_2OH$ (b) $HOC(CH_3)_2CH_3$ (c) $CH_3C(CH_3)_2CH_2OH$ (d) $CH_3CH(CH_3)CH(OH)CH_3$

The foregoing discussion of the chemical properties of alcohols leaves us with the following structural "map signs."

1. Alcohols can be dehydrated to alkenes, with the most branched alkene generally forming *in vitro*.
2. 1° Alcohols are oxidized *in vivo* to aldehydes; *in vitro* to carboxylic acids.
3. 2° Alcohols are oxidized both *in vivo* and *in vitro* to ketones.

3.4

PHENOLS

Phenols are weak acids that can neutralize sodium hydroxide, and their benzene rings are easily oxidized.

For a compound to be classified as a *phenol,* its molecules must have at least one OH group directly attached to a benzene ring. The simplest member of this family also carries the name phenol, and it is a raw material for making aspirin. Phenols are widespread in nature and in commerce (see Special Topic 3.3).

■ For phenol itself, $K_a = 1.0 \times 10^{-10}$ (25 °C), which can be compared to $K_a = 1.8 \times 10^{-5}$ for acetic acid.

Phenols Are Weak Acids In sharp contrast to alcohols, phenols are acidic, but they are *weak* acids, as the K_a values in the margin show. Phenol is itself a much weaker acid than acetic acid. Yet phenols in general are strong enough acids to neutralize OH^-. For example,

$$C_6H_5\text{—}OH + OH^-(aq) \longrightarrow C_6H_5\text{—}O^- + H_2O$$

Phenol — Phenoxide ion

Understanding why OH^- can take H^+ from a phenol but not from an alcohol comes down to considering what makes the *anion* of a phenol, a phenoxide ion ($C_6H_5O^-$) more stable than the anion from an alcohol (RO^-). Both anions appear to have one negative charge on oxygen. What the structures do not reveal, however, is that the negative charge on the phenoxide ion is somewhat spread out into the pi-electron network of the benzene ring. *The surest way to stabilize negative charge is to spread it out,* but the negative charge on the RO^- ion cannot spread out as it does in the phenoxide ion. The alcohol's anion, RO^-, therefore, is less stable and it forms to a lesser extent from ROH than does the phenoxide ion from a phenol. Therefore, ROH is an extremely poor proton donor compared to a phenol.

The Ring in a Phenol Is Easily Oxidized Some of the electron density on the oxygen in phenol also spreads out over the pi-electron network of the ring. Oxidizing agents are electron seekers, so this makes the benzene ring in a phenol more easily oxidized than the ring in

SPECIAL TOPIC 3.3
PHENOLS IN OUR DAILY LIVES

Phenol (carbolic acid) Phenol, as we have already noted, was the first antiseptic to be used by Joseph Lister, an English surgeon. It kills bacteria by denaturing their proteins—that is, by destroying the abilities of their proteins to function normally. (Interestingly, phenol does not denature nucleic acids, the chemicals of heredity.) Phenol, however, is dangerous to healthy tissue, because it is a general protoplasmic poison, so other antiseptics have been developed since Lister's time.

Vanillin Vanillin is present in the vanilla bean and is valued as a flavoring agent. Its molecules have three functional groups—phenol, ether, and aldehyde. Most vanillin is manufactured today from another phenol, eugenol.

Eugenol The odor of cloves is caused chiefly by eugenol, which is used in making perfumes and, as we said, for the manufacture of vanillin.

Urushiols The active irritant in poison ivy or poison oak is a mixture of similar compounds called the urushiols. They differ in the nature of the R group shown on the benzene ring. It is a straight 15-carbon chain, but the different members of the urushiol mixture have from zero to three alkene groups in this chain.

BHA and BHT BHA and BHT are widely used as antioxidants in gasoline, lubricating oils, rubber, edible fats and oils, and materials used for packaging foods that might turn rancid. Being phenols, and therefore easily oxidized themselves, they act by competing with the oxidations of the materials they are designed to protect.

OH, OCH_3, $CH{=}O$ — Vanillin

OH, OCH_3, $CH_2CH{=}CH_2$ — Eugenol

OH, OCH_3, R — The urushiols

OH, $C(CH_3)_3$, OCH_3 — BHA (butylated hydroxyanisole)

OH, $(CH_3)_3C$, $C(CH_3)_3$, CH_3 — BHT (butylated hydroxytoluene)

benzene. Oxidations of phenols *in vitro,* however, produce complex mixtures that include some highly colored compounds. Even phenol crystals left exposed to air slowly turn dark because the oxygen in the air can attack phenol. *In vivo,* special hydroxylating enzymes are able to steer the oxidations of the benzene rings in certain phenols to specific products (as we will learn in Section 17.2).

Phenols Are Not Dehydrated Unlike alcohols, phenols cannot be dehydrated. Dehydration would put a *triple* bond into a six-membered ring, and this ring is too small to accommodate the linear geometry at a triple bond.

3.5
ETHERS

The ethers are almost as chemically unreactive as the alkanes.

Ethers are compounds in whose molecules two organic groups are joined to the same oxygen atom, and their family structure is R—O—R′. The carbon joined to the bridging oxygen atom cannot be a carbonyl carbon, the one in C═O. Thus the first three compounds given below are ethers, but methyl acetate is in the ester family, not the ether family, because the bridging oxygen is attached to a carbonyl carbon. (We will study esters in a later chapter.)

$CH_3CH_2—O—CH_2CH_3$ — Diethyl ether

$CH_3—O—C_6H_5$ — Methyl phenyl ether

$C_6H_5—O—C_6H_5$ — Diphenyl ether

$CH_3C({=}O)—O—CH_3$ — Methyl acetate (an *ester*, not an ether)

TABLE 3.3
Some Ethers

Common Name	Structure	BP (°C)
Dimethyl ether	CH_3OCH_3	−23
Methyl ethyl ether	$CH_3OCH_2CH_3$	11
Methyl *t*-butyl ether	$CH_3OC(CH_3)_3$	55.2
Diethyl ether	$CH_3CH_2OCH_2CH_3$	34.5
Dipropyl ether	$CH_3CH_2CH_2OCH_2CH_2CH_3$	91
Methyl phenyl ether	$CH_3OC_6H_5$	155
Diphenyl ether	$C_6H_5OC_6H_5$	259
Divinyl ether	$CH_2{=}CHOCH{=}CH_2$	29

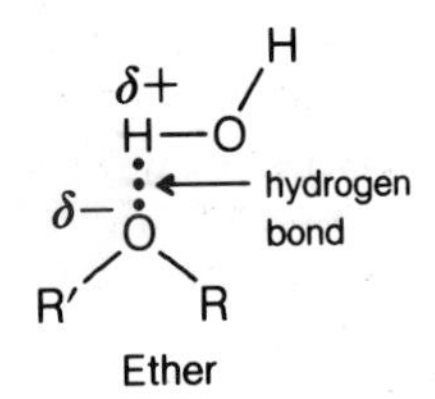

Table 3.3 gives some examples of ethers, and three are described in Special Topic 3.4.

The common names of ethers are made by naming the groups attached to the oxygen and adding the word *ether*, as illustrated in Table 3.3.

Compound	BP (°C)	Solubility in water
Pentane	36	0.036 g/dL[a]
Diethyl ether	35	8.4 g/dL[a]
1-Butanol	118	11 g/dL[b]

[a] At 15 °C [b] At 25 °C

Ethers Cannot Donate Hydrogen Bonds Because the ether group cannot donate hydrogen bonds, the boiling points of simple ethers are more like those of the alkanes of comparable formula masses than those of the alcohols. The oxygen of the ether group can accept hydrogen bonds, however, so ethers are more soluble in water than alkanes. For example, both 1-butanol and its isomer, diethyl ether, dissolve in water to the extent of about 8g/100 mL, but pentane is insoluble in water.

Hydroperoxides are R—O—O—H and peroxides are R—O—O—R′.

Ethers Have Few Chemical Reactions At room temperature, ethers do not react with strong acids, bases, or with strong oxidizing or reducing agents. On long standing in the presence of air, however, liquid ethers slowly react with oxygen, and compounds called *hydroperoxides* and *peroxides* gradually form. These compounds, when concentrated, can explode, so chemists are wary of using aged ether supplies. Like all organic compounds, ethers burn. We will learn no other reactions of ethers, but we must be able to recognize the ether group and to remember that it is not very reactive toward anything, particularly in the environment within the body.

Ethers Can Be Prepared from Alcohols We learned earlier in this chapter that alcohols can be dehydrated by the action of heat and an acid catalyst to give alkenes. The precise temperature that works best has to be discovered experimentally for each alcohol, because if the temperature is not set correctly, a different pathway for dehydration can occur. Water can split out *between* two alcohol molecules rather than from within one alcohol molecule, and the product is an ether. In general,

$$\underset{\text{Two alcohol molecules}}{R{-}O{-}H + H{-}O{-}R} \xrightarrow{\text{acid catalyst}} \underset{\text{Ether}}{R{-}O{-}R} + H_2O$$

A specific example is

Earlier we learned that concentrated H_2SO_4 acts on ethanol to give ethene when the temperature is higher (170 °C).

$$\underset{\text{Ethyl alcohol}}{2CH_3CH_2OH} \xrightarrow[140\ °C]{H_2SO_4} \underset{\text{Diethyl ether}}{CH_3CH_2{-}O{-}CH_2CH_3} + H_2O$$

The dehydration that produces an ether usually requires a lower temperature than that which gives an alkene, but we won't need the details. Our interest is simply in the *possibility* of making an ether from an alcohol as well as the structure of the ether that can be made.

SPECIAL TOPIC 3.4
ETHERS IN MEDICINE

Diethyl Ether ("ether") Diethyl ether is a colorless, volatile liquid with a pungent, somewhat irritating odor; it was once widely used as an anesthetic. It acts as a depressant for the central nervous system and a mild stimulant for the sympathetic system. It exerts an effect on nearly all tissues of the body.

Because mixtures of ether and air in the right proportions can explode, anesthesiologists avoid using it as an anesthetic whenever possible.

Divinyl Ether (vinethene) Divinyl ether is another anesthetic, but it also forms an explosive mixture in air. Its anesthetic action is more rapid than that of diethyl ether.

Methyl *t*-Butyl Ether The standard treatment for stones in the gallbladder or the gallbladder duct—"gallstones"—has been surgical removal of the gallbladder. The chief constituent of gallstones is cholesterol, which is mostly hydrocarbon-like. In the 1980s, methyl *t*-butyl ether, a relatively nontoxic solvent for cholesterol, was discovered by Mayo Clinic scientists to be an effective, nonsurgical agent for removing gallstones. Methyl *t*-butyl ether dissolves the stones without causing serious side effects, and it is less toxic than diethyl ether. It can be inserted into the gallbladder by a catheter, it remains a liquid at body temperature, and it works in just a few hours with no evidence of disagreeable side effects.

Dioxin Dioxin is the name for a family of compounds once present as trace impurities in certain weed killers. One member of the family is TCDD.

Cl Cl O O Cl Cl

TCDD
(2,3,7,8-tetrachlorodibenzo-*p*-dioxin)

The TCDD molecule has two ether groups that bridge two parallel benzene rings each holding two chlorine atoms. There are actually 75 chlorine-containing dioxins, but TCDD is regarded as the most dangerous. In male guinea pigs, TCDD is extrememly toxic, more so than strychnine, the nerve gases, and cyanide. In male or female hamsters, however, the lethal dose is nearly 2000 times greater than for male guinea pigs, demonstrating that toxicity can be very species sensitive. In humans, TCDD appears to be even less toxic. TCDD does cause a form of acne (chloracne) in humans as well as digestive distress, pains in the joints, and psychiatric effects. TCDD is teratogenic (causer of birth defects) and carcinogenic (cause of cancer) in experimental animals. Whether it is carcinogenic or teratogenic in humans is the serious public health issue (and is hotly debated).

TCDD and other dioxins are very stable to heat, and they form in incinerators whenever chlorinated compounds are part of the wastes being burned. The waste gases from such incinerations contain levels of dioxins ranging from several parts per trillion to many parts per million. In the air, the dioxins are degraded by sunlight in a matter of days when moisture is present. TCDD is only slowly broken down in soil, however. None of the dioxins has any known value to humans; they offer nothing but risks. Some believe the risks are so small that municipal incinerators should not be stalled by them; others strongly disagree. [A reference: F. H. Tschirley, "Dioxin," *Scientific American,* February 1986, page 29.]

EXAMPLE 3.5
Writing the Structure of an Ether That Can Form from an Alcohol

If the conditions are right, 1-butanol can be converted to an ether. What is the structure of this ether?

ANALYSIS What this question asks is to complete the following equation:

$$CH_3CH_2CH_2CH_2OH \xrightarrow[\text{heat}]{\text{acid catalyst}} ?$$

As usual, the structure of the starting material gives us most of the answer. We know that the ether must get its organic groups from the alcohol, so to write the structure of the ether we write one O atom with two bonds from it.

—O—

Then we attach the alkyl group of the alcohol, one to each of the bonds.

SOLUTION The structure of the ether is

$$CH_3CH_2CH_2CH_2—O—CH_2CH_2CH_2CH_3$$

Of course, we need *two* molecules of the alcohol to make one molecule of this ether, but always remember, *we balance an equation after we have written the correct formulas for reactants and products.* Remembering that water is the other product, we have as the balanced equation

$$\underset{\text{1-Butanol}}{2CH_3CH_2CH_2CH_2OH} \xrightarrow[\text{heat}]{\text{acid catalyst}} \underset{\text{Dibutyl ether}}{CH_3CH_2CH_2CH_2—O—CH_2CH_2CH_2CH_3} + H_2O$$

CHECK Is the product truly an *ether?* (Is it of the form R—O—R?) Are the R groups identical to the R group of the parent alcohol? (In this example are they *butyl* groups?) Does each carbon atom have four bonds and each oxygen two?

PRACTICE EXERCISE 9

Write the structures of the ethers to which the following alcohols can be converted.

(a) CH_3OH **(b)** $CH_3CH_2CH_2OH$ **(c)** (cyclohexane ring)—OH

3.6

THIOALCOHOLS AND DISULFIDES

Both the SH group, an easily oxidized group, and the S—S system, an easily reduced group, are important groups in proteins.

Alcohols, R—O—H, can be viewed as alkyl derivatives of water, H—O—H. Similar derivatives of hydrogen sulfide, H—S—H, are also known and are in the family called the **thioalcohols** or the **mercaptans.**

■ *Mercaptan* is a contraction of *mercury-capturer.* Compounds with SH groups form precipitates with mercury ions.

R—S—H	R—S—R′	R—S—S—R′
Thioalcohols (mercaptans)	Thioethers	Disulfides

■ Lower-formula-mass thioalcohols are present in and are responsible for the considerable respect usually given to skunks.

Table 3.4 gives the IUPAC names and structures of a few thioalcohols. (We will not study the IUPAC nomenclature as a separate topic because it is straightforward, and because our interest in thioalcohols is limited to just one reaction.) Some very important properties of proteins depend on the presence of the SH group located on one of the building blocks of proteins, the amino acid called cysteine (Table 3.4).

Dialkyl derivatives of water, the ethers (R—O—R′), have their sulfur counterparts, too, the thioethers. (You can see that the prefix *thio-* indicates the replacement of an oxygen atom by a sulfur atom.) Although proteins also have the thioether system, our studies will not require that we learn anything of the chemistry of this group.

Thioalcohols Are Oxidized to Disulfides The one reaction of thioalcohols that will be important in our study of proteins is oxidation. Thioalcohols can be oxidized to **disulfides,** compounds whose molecules have two sulfur atoms joined by a covalent bond, R—S—S—R′. In general,

TABLE 3.4
Some Thioalcohols

Name	Structure	BP (°C)
Methanethiol	CH_3SH	6
Ethanethiol	CH_3CH_2SH	36
1-Propanethiol	$CH_3CH_2CH_2SH$	68
1-Butanethiol	$CH_3CH_2CH_2CH_2SH$	98
Cysteine (a monomer for proteins)	$^{+}NH_3CHCO_2^{-}$ with CH_2SH on the CH	(Solid)

$$R{-}S{-}H + H{-}S{-}R + (O) \longrightarrow R{-}S{-}S{-}R + H_2O$$

Two molecules of a thioalcohol → One molecule of a disulfide

A specific example is

$$2CH_3SH + (O) \longrightarrow CH_3{-}S{-}S{-}CH_3 + H_2O$$

Methanethiol → Dimethyl disulfide

EXAMPLE 3.6
Writing the Product of the Oxidation of a Thioalcohol

What is the product of the oxidation of ethanethiol?

ANALYSIS Because the oxidation of the SH group generates the —S—S— group, begin simply by writing this group down.

—S—S—

Then attach the alkyl group from the thioalcohol, one on each sulfur atom.

SOLUTION Ethanethiol furnishes ethyl groups, so attach one of each of these groups to the S atoms.

$$CH_3CH_2{-}S{-}S{-}CH_2CH_3$$

Diethyl disulfide

If the problem had called for an equation, we would have had to use the coefficient of 2 for the ethanethiol.

$$2CH_3CH_2SH + (O) \longrightarrow CH_3CH_2{-}S{-}S{-}CH_2CH_3 + H_2O$$

However, this coefficient isn't necessary when all we are asked to do is to write the structure of the disulfide that can be made by the oxidation of ethanethiol. Always remember that balancing an equation comes *after* you have written the correct formulas for the reactants and products.

CHECK Is the product a *disulfide?* (Thiols are oxidized to disulfides.) Are the groups attached to the S atoms the same as present in the reactant? Does each carbon have four bonds and each sulfur two?

Disulfides Are Reduced to Thioalcohols The sulfur–sulfur bond in disulfides is very easily reduced, and the products are molecules of the thioalcohols from which the disulfide could be made. We will use the symbol (H) to represent any reducing agent that can do the task, just as we used (O) for an oxidizing agent. In general,

$$R{-}S{-}S{-}R + 2(H) \longrightarrow R{-}S{-}H + H{-}S{-}R$$

One molecule of disulfide — Two molecules of a thioalcohol

A specific example is

$$CH_3CH_2{-}S{-}S{-}CH_2CH_2CH_3 + 2(H) \longrightarrow CH_3CH_2{-}S{-}H + H{-}S{-}CH_2CH_2CH_3$$

PRACTICE EXERCISE 10

Complete the following equations by writing the structures of the products that form. If no reaction occurs, write "no reaction."

(a) $CH_3SSCH_3 + (H) \rightarrow ?$ (b) $CH_3\underset{\displaystyle SH}{\underset{|}{C}}HCH_3 + (O) \rightarrow ?$

(c) (six-membered ring containing S—S) $+ (H) \longrightarrow ?$ (d) (cyclopentane)—$SH + (O) \longrightarrow ?$

SUMMARY

Alcohols The alcohol system has an OH group attached to a saturated carbon. The IUPAC names of simple alcohols end in *-ol*, and their chains are numbered to give precedence to the location of the OH group. The common names have the word *alcohol* following the name of the alkyl group.

Alcohol molecules hydrogen-bond to each other and to water molecules. By the action of heat and an acid catalyst, alcohols can be dehydrated internally to give carbon–carbon double bonds or externally to give ethers. Primary alcohols can be oxidized to aldehydes and ultimately to carboxylic acids. Secondary alcohols can be oxidized to ketones. Tertiary alcohols cannot be oxidized (without breaking up the carbon chain). The OH group in alcohols does not function well as either an acid or a base.

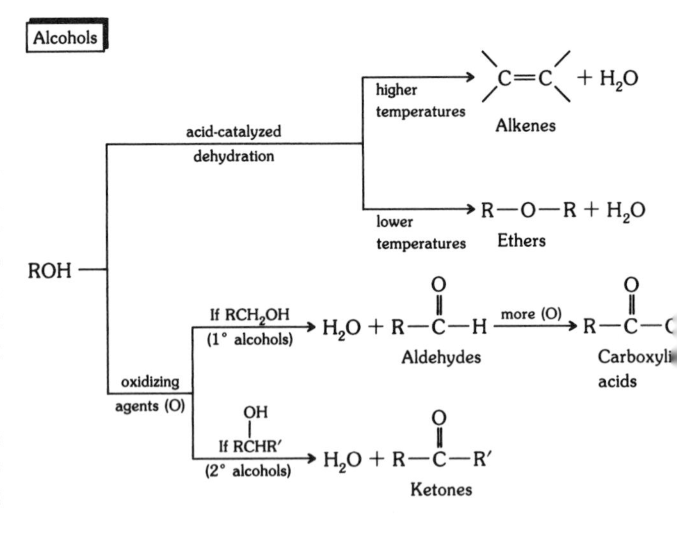

Phenols When the OH group is attached to a benzene ring, the system is the phenol system, and it is now acidic enough to neutralize strong bases. In addition, the ring is vulnerable to oxidizing agents.

Phenols

$$C_6H_5{-}OH \xrightarrow{NaOH(aq)} C_6H_5{-}O^-Na^+ + H_2O$$

Ethers The ether system, R—O—R′, does not react at room temperature or body temperature with strong acids, bases, oxidizing agents, or reducing agents. It can accept hydrogen bonds but cannot donate them.

Thioalcohols The thioalcohols or mercaptans, R—S—H, are easily oxidized to disulfides, R—S—S—R. Disulfides are reduced to the original thioalcohols.

Reactions studied Without attempting to present balanced equations, or even all of the inorganic products, we can summarize the reactions studied in this chapter as follows. (We omit ethers.)

Thioalcohols and Disulfides

$$R{-}S{-}H \underset{\text{(H) reduction}}{\overset{\text{(O) oxidation}}{\rightleftharpoons}} R{-}S{-}S{-}R$$

REVIEW EXERCISES

The answers to Review Exercises whose numbers are in color are found in Appendix A. The answers to the other Review Exercises are found in the Study Guide that accompanies this book. The more challenging questions are marked with asterisks.

Functional Groups

3.1 Name the functional groups identified by the numbers in the structure of cortisone, a drug used in treating certain forms of arthritis. If a group is an alcohol, state if it is a 1°, 2°, or 3° alcohol.

1 CH_2OH, 2 $C{=}O$, 3 HO, H_3C, H_3C, 4 O, 5

Cortisone

3.2 Give the names of the functional groups identified by the numbers in prostaglandin E_1, one of a family of compounds that are smooth muscle stimulants.

1 O, $CH_2CH_2CH_2CH_2CH_2CH_2\overset{O}{\overset{\|}{C}}OH$ (2), $CH{=}CHCHCH_2CH_2CH_2CH_2CH_3$ (4), OH (3), HO (5)

Prostaglandin E_1

Structures and Names

3.3 Write the structure for each compound.

(a) isobutyl alcohol (b) isopropyl alcohol

(c) propyl alcohol (d) glycerol

3.4 Write the structures of the following compounds.

(a) methyl alcohol (b) *t*-butyl alcohol

(c) ethyl alcohol (d) butyl alcohol

3.5 What are the common names of the following compounds?

(a) $CH_3CH_2CH_2OH$ (b) $HOCH_2CH_2CH_2CH_3$

(c) 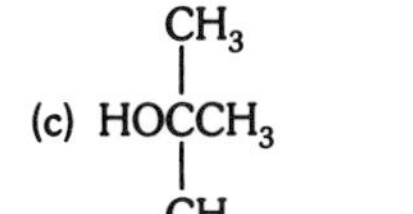

(d) $HOCH_2CH(CH_3)_2$

3.6 What is the structure and the IUPAC name of the simplest, *stable* dihydric alcohol?

3.7 Give the structure and the IUPAC name of the simplest, *stable* trihydric alcohol.

3.8 Give the IUPAC names for the compounds listed in Review Exercise 3.3.

3.9 Give the IUPAC names for the compounds listed in Review Exercise 3.4.

3.10 Give the IUPAC names for the compounds listed in Review Exercise 3.5.

3.11 Write the IUPAC name of the following compound.

$$CH_3CH_2\underset{\displaystyle CH_2OH}{\underset{|}{C}H}CH_2CH_2CH_3$$

Physical Properties

3.12 When ethyl alcohol dissolves in water, its molecules slip into the hydrogen-bonding network in water. Draw a figure that illustrates this. Use dotted lines for hydrogen bonds, and place the $\delta+$ and $\delta-$ symbols where they belong.

***3.13** Arrange the following compounds in their order of increasing boiling points. Place the letter symbol of the compound that has the lowest boiling point on the left end of the series, and arrange the remaining letters in the correct order.

$CH_3CH_2CH_2OH$	CH_3CH_3	$HOCH_2CH_2OH$	$CH_3CH_2OCH_3$
A	B	C	D

___ < ___ < ___ < ___

Lowest bp Highest bp

Chemical Properties of Alcohols

3.14 Write the structures of the alkenes that form when the following alcohols undergo acid-catalyzed dehydration. Where more than one alkene is possible, identify which most likely forms in the greatest relative amount.

(a) $CH_3\underset{\displaystyle CH_3}{\underset{|}{C}H}CH_2OH$ (b) $CH_3\underset{\displaystyle OH}{\underset{|}{C}H}CH_2CH_3$

(c) 1-methylcyclopentanol (HO and CH_3 on the same ring carbon of a cyclopentane ring) (d) $C_6H_5\underset{\displaystyle OH}{\underset{|}{C}H}CH(CH_3)_2$

(e) $CH_3CH_2CH_2\overset{\displaystyle CH_3}{\overset{|}{\underset{\displaystyle OH}{\underset{|}{C}}}}CH_3$ (f) HO–(cyclohexane ring)–CH_3

3.15 Write the structures of the products of the oxidation of the alcohols given in Review Exercise 3.14. If the alcohol is a 1° alcohol, give the structures of both the aldehyde and the carboxylic acid that could be made by varying the quantities of the oxidizing agent.

*3.16 Write the structures of any alcohols that could be dehydrated to give each of the following alkenes. In some instances, more than one alcohol would work.

(a) $CH_2{=}CHCH_3$ (b)  cyclopentene

(c) $CH_3C(=CH_2)CH_3$ (d) 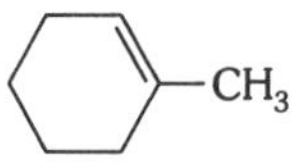 1-methylcyclohexene (ring with CH_3)

*3.17 Write the structure of any alcohol that could be used to prepare each of the following compounds by an oxidation.

(a) $HOCCH_2CH_2CH_3$ (with C=O) (b) $CH_3CH_2CH_2CCH(CH_3)_2$ (with C=O)

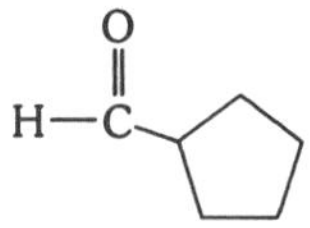

(c) H—C(=O)— cyclopentyl (d) 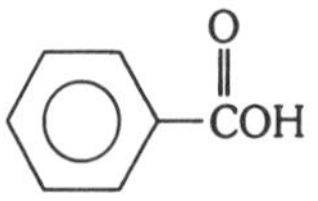 C_6H_5—COH (with C=O)

Phenols

3.18 Write the structures of the three isomeric monochlorophenols and give their names (using the *o*-, *m*-, and *p*- designations).

*3.19 What is one difference in *chemical* properties between **A** and **B**?

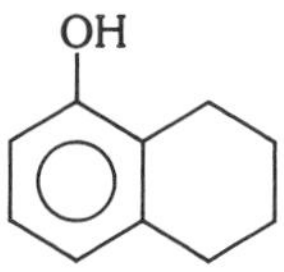

OH OH

A 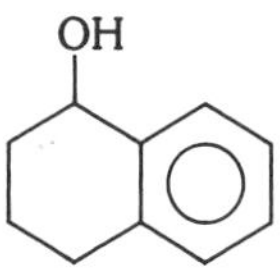 **B**

3.20 A compound was either **A** or **B**.

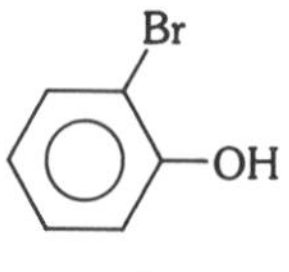

Br

—OH —CH_2OH

A 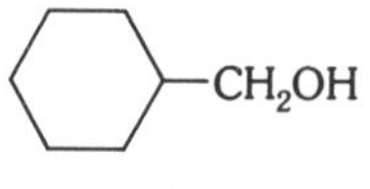 **B**

The compound dissolved in aqueous sodium hydroxide but not in water. Which compound was it? How can you tell? (Write an equation.)

Ethers

3.21 Write the structures of the ethers that can be made from the following alcohols.

(a) $HOCH_2CH_3$ (b) CH_3CHCH_3 (with OH on central C)

(c) $CH_3OCH_2CH_2OH$ (d) 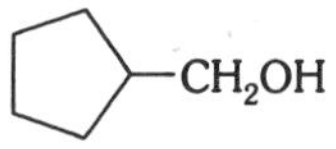 cyclopentyl—CH_2OH

*3.22 Write the structures of the alcohols that could serve as the starting materials to prepare each of the following ethers.

(a) $CH_3CH_2CH_2OCH_2CH_2CH_3$

(b) $CH_3CHCH_2OCH_2CHCH_3$ (with CH_3 on each CH)

(c) cyclohexyl—O—cyclohexyl

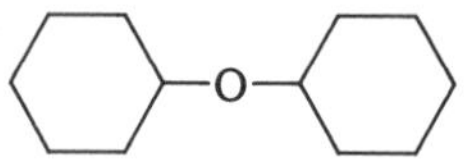

(d) 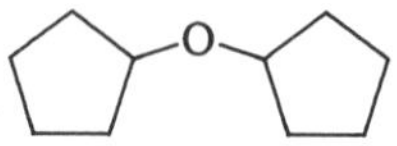 cyclopentyl—O—cyclopentyl

*3.23 Suppose that a mixture of 0.50 mol of ethanol, 0.50 mol of methanol, and a catalytic amount of sulfuric acid is heated under conditions that favor only ether formation. What organic products will be obtained? Write their structures.

3.24 What happens chemically when the following compound is heated with aqueous sodium hydroxide?

$$CH_3CH_2CH_2OCH_2CH_2CH_3$$

Thioalcohols and Disulfides

3.25 We did not study rules for naming thioalcohols, but the patterns of the names in Table 3.4 make these rules obvious. Write the structures of the following compounds.

(a) diethyl disulfide (b) 1,2-propanedithiol
(c) isopropyl mercaptan (d) 1-propanethiol

3.26 Complete the following reaction sequences by writing the structures of the organic products that form.

(a) $CH_3CH_2CH_2SH + (O) \longrightarrow$

(b) $(CH_3)_2CHCH_2{-}S{-}S{-}CH_2CH(CH_3)_2 + (H) \longrightarrow$

(c)  cyclic S—S (five-membered ring) $+ (H) \longrightarrow$

(d) $HSCH(CH_3)_2 + (O) \longrightarrow$

3.27 Ethanol, methanethiol, and propane have nearly the same formula masses, but ethanol boils at 78 °C, methanethiol at 6 °C, and propane at −42 °C. What do the boiling points suggest about the possibility of hydrogen bonding in the thioalcohol family? Does hydrogen bonding occur at all? Are the hydrogen bonds as strong as those in the alcohol family?

Alcohol In Our Daily Lives (Special Topic 3.1)

3.28 Give the name of the alcohol used in each of the following ways.

(a) As the alcohol in beverages
(b) As a rubbing alcohol (Name two.)
(c) As a moisturizer in some food products
(d) As a fuel in "canned heat"
(e) As a permanent antifreeze (Name two.)
(f) To manufacture a vasodilator

3.29 Give the name of the alcohol made by
(a) The digestion of fats or oils in the diet
(b) The fermentation of sugars
(c) The hydrogenation of carbon monoxide

Acid-Catalyzed Dehydration of Alcohols (Special Topic 3.2)

3.30 Write an equation for the equilibrium that forms when sulfuric acid is dissolved in cyclopentanol but before any further steps in the dehydration of this alcohol occur.

3.31 What is the structure of the carbocation that can form following the reaction described in Review Exercise 3.30?

3.32 When the cyclopentyl carbocation changes into cyclopentene, the carbocation must lose something. What is the formula of the species it must lose as the double bond forms? What is the name of the likeliest acceptor of this species when the surrounding medium contains mostly cyclopentanol and catalytic amounts of sulfuric acid?

3.33 Explain (briefly) why carbocations are unstable in water but something like the sodium ion is stable.

*3.34 Cyclopentene can be made to *add* a water molecule to give cyclopentanol when the medium is *dilute* sulfuric acid. The steps in the mechanism are the exact reverse of the kinds of steps for the dehydration of cyclopentanol to cyclopentene. Write the steps in the mechanism of the acid-catalyzed addition of water to cyclopentene.

Phenols in Our Daily Lives (Special Topic 3.3)

3.35 What is the name of a member of the phenol family that is used in or involved in each of the following ways?
(a) A flavoring agent
(b) Lister's original antiseptic
(c) A substitute for cloves
(d) An irritant in poison ivy

3.36 Both BHA and BHT interfere with the air oxidation of food materials. What chemical property do these food additives have that accounts for this?

3.37 Lister's original antiseptic is no longer used. Why?

Ethers in Medicine (Special Topic 3.4)

3.38 What physical property (aside from its anesthetic quality) made diethyl ether workable as an anesthetic?

3.39 What chemical property of diethyl ether makes it undesirable as an anesthetic, a property having nothing to do with its ability to induce the anesthetic state?

3.40 What is it about the structure of methyl *t*-butyl ether that accounts for its ability to dissolve cholesterol (whose structure may be looked up using the index)?

3.41 TCDD is a member of what family of pollutants?

3.42 What is currently an entry for trace quantities of TCDD into the environment?

Additional Exercises

*3.43 Examine each of the following sets of reactants and conditions and decide if a reaction occurs. If one does, write the structures of the organic products. If no reaction occurs, write "no reaction."

Some of the parts involve *alcohols* and their reactions. If the reaction is an oxidation of a 1° alcohol, write the structure of the *aldehyde* that can form, not the carboxylic acid.

When an alcohol is in the presence of an acid catalyst, we have learned that the alcohol might be dehydrated to an *alkene* or an *ether*. To differentiate between these, use the following guide. When the alcohol structure has no coefficient, then write the structure of the alkene that can form. When the alcohol structure has a coefficient of 2, then give the structure of the ether that is possible. (This violates our rule that balancing an equation is the *last* step in writing an equation, but we need a signal here to tell what kind of reaction is intended.)

(a) cyclopentanol (cyclopentyl–OH) $\xrightarrow[\text{heat}]{H_2SO_4}$

(b) $2HOCH_2CH_3 \xrightarrow[\text{heat}]{H_2SO_4}$

(c) $CH_3CH(OH)CH_2CH_3 + (O) \longrightarrow$

(d) 1-methylcyclopentene (cyclopentene ring with CH_3) $+ H_2 \xrightarrow{\text{Ni catalyst}}$

(e) $CH_3CH_2OH + NaOH(aq) \longrightarrow$

(f) $CH_3C(OH)(CH_3)CH_2CH_2CH_3 + (O) \longrightarrow$

(g) $CH_3CH_2CH(CH_3)CH_3 \xrightarrow[\text{heat}]{H_2SO_4}$

(h) $CH_3CH{=}C(CH_3)CH_3 + H_2 \xrightarrow{\text{Ni catalyst}}$

(i) 1-methylcyclohexene (cyclohexene ring–CH_3) $+ H_2O \xrightarrow[\text{heat}]{H_2SO_4}$

(j) HO–cyclopentane–CH_3 $+ (O) \longrightarrow$

***3.44** Write the structure of the principal organic product that would be expected in the following situations. Follow the directions given for Review Exercise 3.43. If no reaction occurs, write "no reaction."

(a) $CH_3CH_2CHCH_3 + NaOH(aq) \longrightarrow$ (with CH_2CH_3 on the CH)

(b) $CH_3CH{=}CCH_3 + HCl(g) \longrightarrow$ (with CH_3 on the C)

(c) $CH_3CHCHCH_3 + (O) \longrightarrow$ (with CH_3 on C-2 and OH on C-3)

(d) $2CH_3CHOH \xrightarrow[\text{heat}]{H_2SO_4}$ (with CH_3 on the CH)

(e) $HOCH_2CH_2CH_3 + (O) \longrightarrow$

(f) 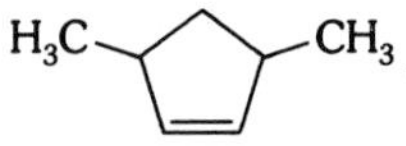 1,3-dimethylcyclopentene (H_3C, CH_3 on the ring) $+ H_2O \longrightarrow$

(g) 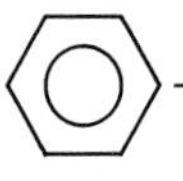 benzene $+ NaOH(aq) \longrightarrow$

(h) CH_3–cyclohexane–$OH + (O) \longrightarrow$

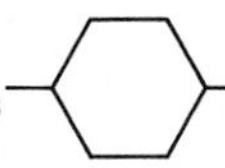

(i) $CH_3CHCH_3 \xrightarrow[\text{heat}]{H_2SO_4}$ (with OH on the CH)

(j) $CH_3CHCH_2OCH_3 + (O) \longrightarrow$ (with HO on the CH)

***3.45** 2-Propanol (C_3H_8O) can be oxidized to acetone (C_3H_6O) by potassium permanganate according to the following equation.

$$3C_3H_8O + 2KMnO_4 \longrightarrow 3C_3H_6O + 2MnO_2 + 2KOH + 2H_2O$$

(a) How many moles of acetone can be prepared from 2.50 mol of 2-propanol?

(b) How many moles of potassium permanganate are needed to oxidize 0.180 mol of 2-propanol?

(c) A student began with 12.6 g of 2-propanol. What is the minimum number of grams of potassium permanganate needed for this oxidation?

(d) Referring to part (c), what is the maximum number of grams of acetone that could be made? How many grams of MnO_2 are also produced?

***3.46** 1-Propanol can be oxidized to propanoic acid.

$$CH_3CH_2CH_2OH + Cr_2O_7^{2-} \longrightarrow CH_3CH_2CO_2H + Cr^{3+}$$

(a) Write the *net ionic equation* using the ion-electron method for balancing redox equations. Assume that the medium is acidic.

(b) Transform the equation of part (a) into a *molecular equation* assuming that the potassium salt of the dichromate ion is used and that the aqueous acid is $HCl(aq)$.

(c) For each mole of 1-propanol used, how many moles (in theory) of potassium dichromate are needed according to the molecular equation?

(d) The stockroom carries potassium dichromate only as its dihydrate. Write the formulas of the dihydrate. How many moles of the dihydrate are needed for each mole of 1-propanol used, according to the equation?

(e) What is the maximum number of grams of propanoic acid that could be obtained if the reaction is carried out starting with 12.4 g of 1-propanol?

(f) What is the minimum number of grams of potassium dichromate dihydrate that would be needed to use up 14.5 g of 1-propanol according to the balanced equation?

***3.47** Write the balanced net ionic equation for the oxidation of cyclopentanol to its corresponding ketone using sodium dichromate in aqueous sulfuric acid. The dichromate ion is changed to the chromium(III) cation. Assume that because of side reactions and losses during the purification of the product ketone, only 50.0% of the theoretically possible ketone can actually be isolated. If you are assigned the task of preparing 10.0 g of the ketone under these limitations, what is the minimum number of grams of cyclopentanol that you must use at the beginning? What is the minimum number of grams of sodium dichromate that you must also use?

ALDEHYDES AND KETONES

4

Architects use mirrored building materials to create spectacular effects. A simple chemical reaction, introduced in this chapter, creates silvered mirrors when one reactant is an aldehyde.

4.1

STRUCTURES AND PHYSICAL PROPERTIES OF ALDEHYDES AND KETONES

Molecules of both aldehydes and ketones contain the carbonyl group.
A knowledge of some of the properties of aldehydes and ketones is essential to understanding carbohydrates. All simple sugars are either polyhydroxy aldehydes or polyhydroxy ketones, as illustrated in the structures of glucose and fructose. (You can see again why the properties of alcohols, which we studied in Chapter 3, are important in carbohydrate chemistry.)

■ We show here only one form of each of the glucose and fructose molecules.

polyhydroxy system; aldehyde group

$$\underset{\mathrm{OH}}{\mathrm{CH_2}}-\underset{\mathrm{OH}}{\mathrm{CH}}-\underset{\mathrm{OH}}{\mathrm{CH}}-\underset{\mathrm{OH}}{\mathrm{CH}}-\underset{\mathrm{OH}}{\mathrm{CH}}-\overset{\mathrm{O}}{\overset{\|}{\mathrm{CH}}}$$

Glucose (open-chain form)

keto group

$$\mathrm{HOCH_2}\underset{\mathrm{OH}}{\mathrm{CH}}-\underset{\mathrm{OH}}{\mathrm{CH}}-\underset{\mathrm{OH}}{\mathrm{CH}}-\overset{\mathrm{O}}{\overset{\|}{\mathrm{C}}}-\mathrm{CH_2OH}$$

Fructose (open-chain form)

Many intermediates in metabolism are also aldehydes or ketones, and Special Topic 4.1 describes just a few.

Aldehydes and Ketones Have the *Carbonyl Group* Both aldehydes and ketones contain the carbon–oxygen double bond, which is called the **carbonyl group** (pronounced car-bon-EEL). The nature of this double bond is discussed in Special Topic 4.2. Like the alkene group, the carbon–oxygen double bond consists of one sigma bond and one pi bond.

■ 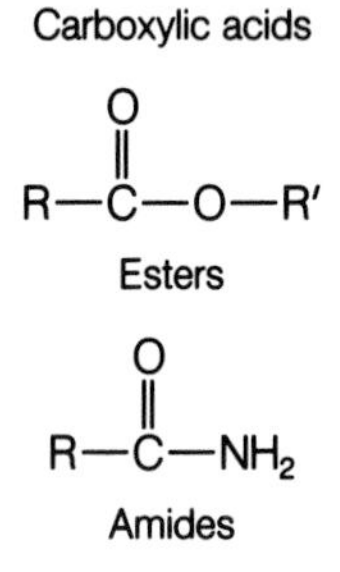

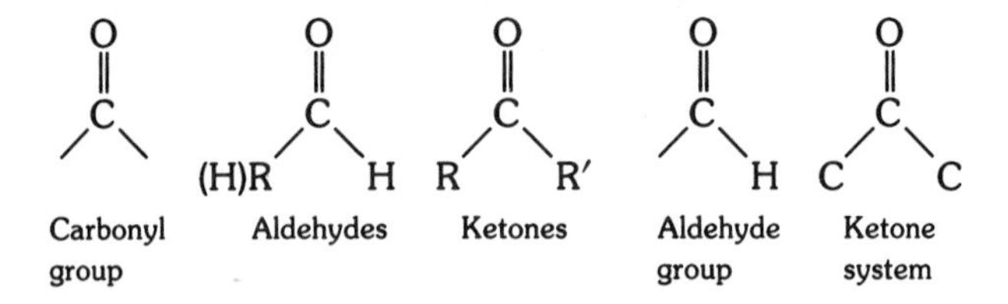

The carbonyl group also occurs in carboxylic acids, and we will study their salts, esters, anhydrides, and amides in the two chapters following this one.

***Aldehydes* Have the CH=O Group** For a compound with a carbonyl group to be called an **aldehyde,** the carbon atom of C=O *must have at least one H atom attached to it.* The other single bond from this carbon *must* be a bond to C (or to a second H), but not a bond to O, N, or S. Thus all aldehydes have the CH=O group, called the **aldehyde group.** (You will often see CH=O written as CHO as in the general formula for aldehydes, RCHO). The simplest aldehyde, HCH=O, methanal (formaldehyde), has two hydrogens on the carbonyl carbon atom. Table 4.1 shows some other relatively simple aldehydes. They have very distinctive, unpleasant odors partly because almost any aldehyde exposed to air also has some of its

TABLE 4.1
Aldehydes

Name	Structure	Formula Mass	BP (°C)	Solubility in Water
Methanal	$CH_2{=}O$	30.0	−21	Very soluble
Ethanal	$CH_3CH{=}O$	44.0	21	Very soluble
Propanal	$CH_3CH_2CH{=}O$	58.1	49	16 g/dL (25 °C)
Butanal	$CH_3CH_2CH_2CH{=}O$	72.1	76	4 g/dL
Benzaldehyde	$C_6H_5CH{=}O$	106.0	178	0.3 g/dL
Vanillin	HO–C₆H₃(OCH₃)–CH=O (HO and CH₃O on ring, CH=O)	152.1	285	1 g/dL

corresponding carboxylic acid. (The aldehyde group slowly oxidizes in air.) The carboxylic acids of lower formula mass, to most people, have some of the most disagreeable odors of all organic compounds.

In *Ketones,* the Carbonyl Carbon Is Joined Directly to Two Carbons For a compound with a carbonyl group to be called a **ketone,** its carbonyl group *must* be joined on *both* sides by bonds to carbon atoms. Only then can the carbonyl group be called the **keto group.** Several ketones are listed in Table 4.2. Sometimes the structure of a ketone is condensed to RCOR′. Ketones also have distinctive but generally not disagreeable odors.

TABLE 4.2
Ketones

Name	Structure	Formula Mass	BP (°C)	Solubility in Water
Acetone	$CH_3C(=O)CH_3$	58.1	56	Very soluble
Butanone	$CH_3C(=O)CH_2CH_3$	72.1	80	33 g/dL (25 °C)
2-Pentanone	$CH_3C(=O)CH_2CH_2CH_3$	86.1	102	6 g/dL
3-Pentanone	$CH_3CH_2C(=O)CH_2CH_3$	86.1	102	5 g/dL
Cyclopentanone	(five-membered ring)=O	84.1	129	Slightly soluble
Cyclohexanone	(six-membered ring)=O	98.1	156	Slightly soluble

SPECIAL TOPIC 4.1
SOME IMPORTANT ALDEHYDES AND KETONES

Formaldehyde Pure formaldehyde is a gas at room temperature, and it has a very irritating and distinctive odor. It is quite soluble in water, so it is commonly marketed as a solution called formalin (37%) to which some methanol has been added. In this and more dilute forms, formaldehyde was once commonly used as a disinfectant and as a preservative for biological specimens. (Concern over formaldehyde's potential hazard to health has caused these uses to decline.) Most formaldehyde today is used to make various plastics such as Bakelite.

Acetone Acetone is valued as a solvent. Not only does it dissolve a wide variety of organic compounds, but it is also miscible with water in all proportions. Nail polish remover is generally acetone. Should you ever use "superglue," it would be a good idea to have some acetone (nail polish remover) handy because superglue can stick your fingers together so tightly that it takes a solvent such as acetone to get them unstuck.

Acetone is a minor by-product of metabolism, but in some situations (e.g., untreated diabetes) enough is produced to give the breath the odor of acetone.

Some Aldehydes and Ketones in Metabolism The aldehyde group and the keto group occur in many compounds at the molecular level of life. The following are just a few examples of substances with the aldehyde group.

$HC(=O)CH(OH)CH_2OPO_3^{2-}$

Glyceraldehyde-3-phosphate, an intermediate in the metabolism of glucose

Pyridoxal, one of the vitamins (B_6)

Just a few of the many substances at the molecular level of life that contain the keto group are the following.

$CH_3C(=O)CO_2^-$

Pyruvate ion, a product of the metabolism of glucose and fructose

$CH_3C(=O)CH_2CO_2^-$

Acetoacetate ion, a product of the metabolism of long-chain carboxylic acids

$HOCH_2C(=O)CH_2OPO_3^{2-}$

Dihydroxyacetone phosphate, an intermediate in the metabolism of glucose and fructose

Estrone, a female sex hormone

Polarization of the carbonyl group

The Carbonyl Group Is Planar, Unsaturated, and Moderately Polar Because the carbonyl group is a *double bond*, it should be no surprise that the group can undergo addition reactions. Whether such reactions produce *stable* products, however, has to be studied on a case-by-case basis. The *direction* of the addition of an *unsymmetrical* reactant to the C=O group is governed by the group's permanent polarity. We'll be studying only one such reactant, ROH. The unsymmetrical reactant's more positive unit (H in ROH) always goes to the O of C=O because oxygen is more electronegative than carbon and so has a permanent $\delta-$ charge. The C of C=O has a permanent $\delta+$ charge and so can accept only the more electron-rich unit of the unsymmetrical reactant (the RO in ROH).

SPECIAL TOPIC 4.2

THE NATURE OF THE CARBON–OXYGEN DOUBLE BOND

We have learned that the carbon–carbon double bond consists of one sigma bond and one pi bond. The carbon–oxygen double bond is exactly like this, except that an oxygen atom has replaced a carbon atom. Both the carbon atom and the oxygen atom of the carbonyl group are sp^2 hybridized, and in methanal the carbon–oxygen sigma bond forms by the overlap of two such hybrid orbitals. The pi bond results from the side-to-side overlap of two unhybridized *p* orbitals (see Figure 1).

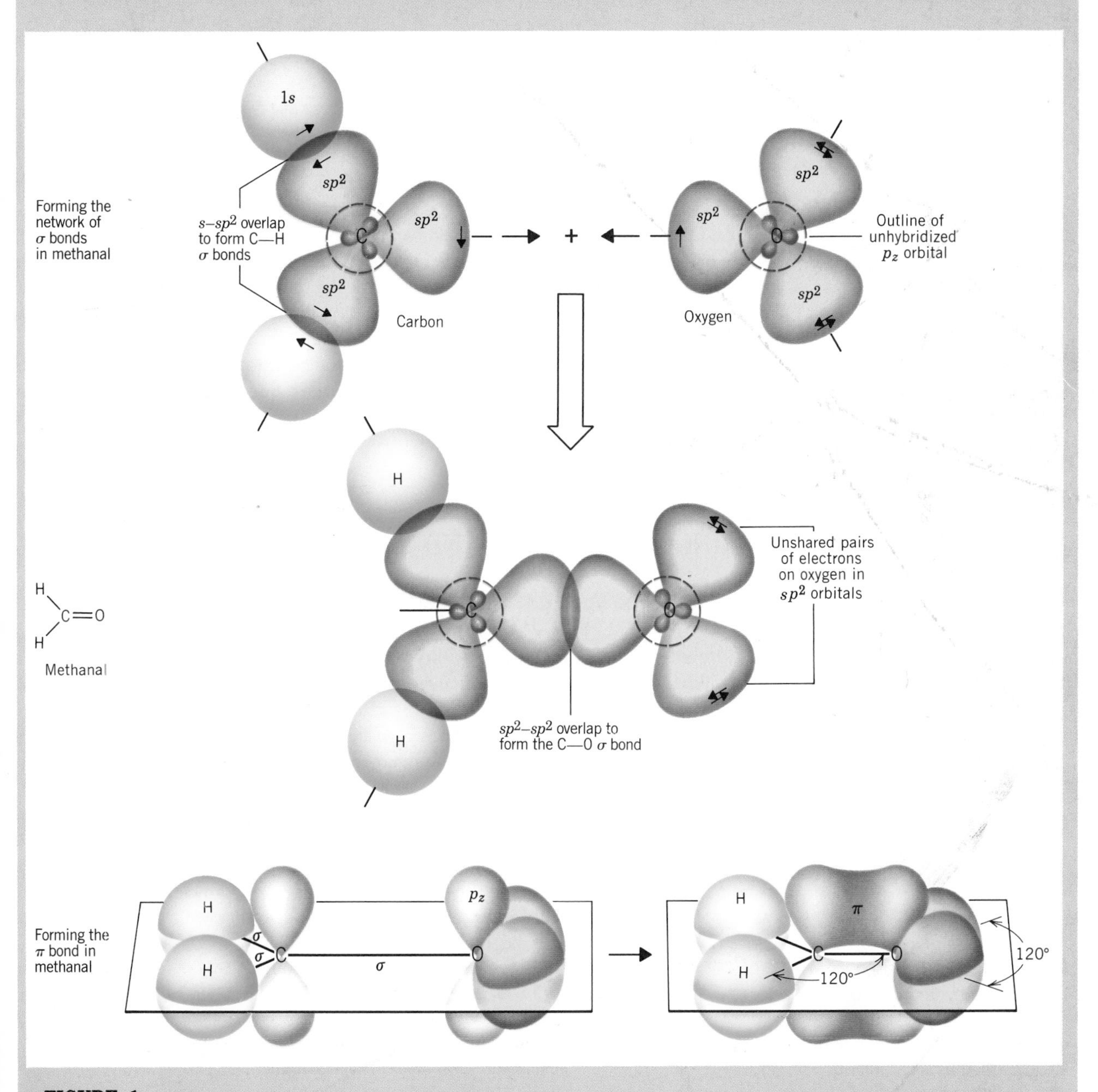

FIGURE 1
The formation of the bonds of the carbonyl group in formaldehyde. The C═O group has one sigma and one pi bond.

TABLE 4.3
Boiling Point Versus Structure

Name	Structure	Formula Mass	BP (°C)
Butane	$CH_3CH_2CH_2CH_3$	58.2	0
Propanal	$CH_3CH_2CH{=}O$	58.1	49
Acetone	$CH_3\overset{O}{\overset{\|}{C}}CH_3$	58.1	56
1-Propanol	$CH_3CH_2CH_2OH$	60.1	98
1,2-Ethanediol (ethylene glycol)	$HOCH_2CH_2OH$	62.1	198

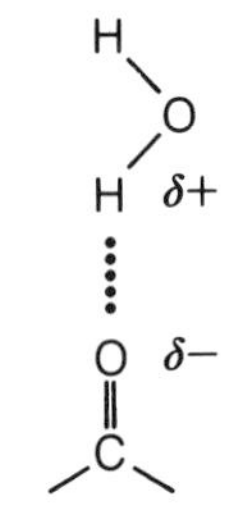

Hydrogen bond (•••••) between a water molecule and a carbonyl group

Because of the polarity of the C=O group, aldehydes and ketones are (moderately) polar compounds. Their molecules are attracted to each other, but not as strongly as they would be if they had OH groups instead and so could participate in hydrogen bonding. Evidence for the polarity of the carbonyl group can be seen in physical properties, like boiling points and solubilities in water. When comparing substances of nearly the same formula masses but from different families, we find that aldehydes and ketones boil higher than alkanes, but lower than alcohols (see Table 4.3). The lack of the OH group means that aldehydes and ketones cannot donate hydrogen bonds, only accept them. This is sufficient to make the low-formula-mass aldehydes and ketones relatively soluble in water, but as the total carbon content increases, their solubility decreases (see Tables 4.1 and 4.2).

4.2

NAMING ALDEHYDES AND KETONES

The IUPAC name ending for aldehydes is *-al* and for ketones is *-one.*

The Common Names of Simple Aldehydes Are Derived from Those of Carboxylic Acids What is easy about the *common* names of aldehydes is that they all (well, nearly all) end in *-aldehyde*. The prefixes to this are the same as found in the common names of the carboxylic acids to which the aldehydes are easily oxidized. We will, therefore, study the common names of these two families here in one place. (Common names are actually more often used than the IUPAC names.)

The simple carboxylic acids have been known for centuries, and their common names are based on natural sources of the acids. Formic acid, for example, is present in the stinging fluid of ants, and the Latin root for ants is *formica*. So this one-carbon acid is called formic acid. The prefix in formic acid is *form-*, so the one-carbon aldehyde is called *formaldehyde*. Here are the four simplest carboxylic acids and their common names together with the structures and names of their corresponding aldehydes.

■ Formic acid also appears to have an aldehyde group, but its second bond from C is to another O, not to H (or C), and it is classified as a carboxylic acid.

■ L. *acetum*, vinegar

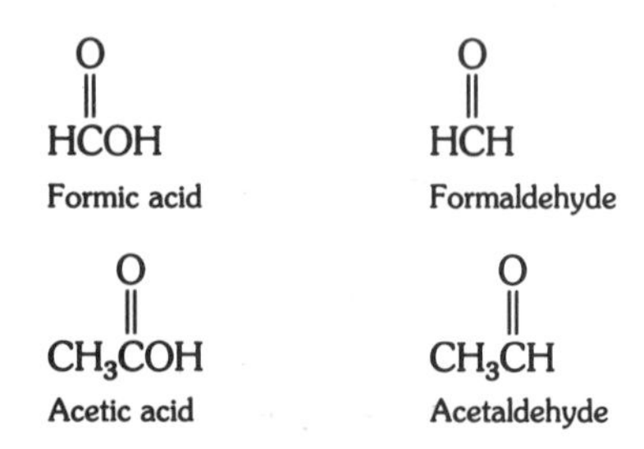

$CH_3CH_2C(=O)OH$ — Propionic acid

$CH_3CH_2C(=O)H$ — Propionaldehyde

■ Gr. *proto,* first, and *pion,* fat

$CH_3CH_2CH_2C(=O)OH$ — Butyric acid

$CH_3CH_2CH_2C(=O)H$ — Butyraldehyde

■ L. *butyrum,* butter

In the aromatic series we have the following examples.

$C_6H_5C(=O)OH$ — Benzoic acid

$C_6H_5C(=O)H$ — Benzaldehyde

The IUPAC Names of Aldehydes End in *-al* As with the IUPAC names of the alcohols, those of the aldehydes are based on the idea of a *parent aldehyde.* Here are the rules.

1. Select as the parent aldehyde the longest chain *that includes the carbon atom of the aldehyde group.*

   The parent aldehyde in the following structure is a five-carbon aldehyde.

$$CH_3CH_2\underset{\underset{CH_3CH_2CH_2}{|}}{C}H\overset{O}{\overset{\|}{C}}H$$

   There is a longer chain in the structure, one of six carbons, but it doesn't include the carbon of the aldehyde group, so this longer chain may not be used in selecting and naming the parent.

2. Name the parent by changing the *-e* ending of the corresponding alkane to *-al.*

   In the example shown with rule 1, the alkane that corresponds to the correct chain is pentane, so the name of the parent aldehyde in this structure is *pentanal.*

3. Number the chain to give the carbon atom of the carbonyl group number 1.

   Precedence is accorded the aldehyde group. Regardless of where other substituents occur on the parent chain, they have to take whatever numbers they receive following the assignment of 1 to the carbonyl carbon atom.

4. Assemble the rest of the name in the same way that was used in naming alcohols, except do not include "1" to specify the location of the aldehyde group.

   The carbonyl carbon cannot have any other number but 1, so we do not include this number. Thus using the example begun under rule 1, we have

$$CH_3CH_2\underset{\underset{\overset{5\ \ 4\ \ 3}{CH_3CH_2CH_2}}{|}}{\overset{2}{C}}H\overset{O}{\overset{\|}{\underset{1}{C}}}H$$

2-Ethylpentanal
Not: 2-ethyl-1-pentanal

EXAMPLE 4.1
Writing the IUPAC Name of an Aldehyde

What is the IUPAC name of the following compound?

$$\begin{array}{c} \quad\quad\quad CH{=}O \\ \quad\quad\quad | \\ BrCH_2CH_2CHCHCH_2CH_3 \\ | \quad\quad \\ CH_3 \quad\quad \end{array}$$

ANALYSIS First, we identify the parent aldehyde. The longest chain that includes the carbon atom of the aldehyde group has five carbons, so the name of the parent aldehyde is *pentanal.* Next, we number the chain beginning with the carbon atom of the aldehyde group.

$$\begin{array}{c} \quad\quad\quad \overset{1}{CH}{=}O \\ \quad\quad\quad | \\ \overset{5}{Br}CH_2\overset{4}{C}H_2\overset{3}{C}H\overset{2}{C}HCH_2CH_3 \\ | \quad\quad \\ CH_3 \quad\quad \end{array}$$

At position 2 there is an ethyl group; at 3, a methyl group; and at 5, a bromo group.

SOLUTION We organize the names of the groups alphabetically. The name is

5-bromo-2-ethyl-3-methylpentanal

CHECK Remember that the most common error is in failing to find the *longest chain that includes the carbonyl group.* Whenever you check your work in writing a name from a structure, test every conceivable way of finding the parent chain.

PRACTICE EXERCISE 1

Write the IUPAC names of the following aldehydes.

(a) $\underset{\displaystyle CH_3}{CH_3\underset{|}{C}HCH{=}O}$ (b) $\underset{\displaystyle Br}{CH_3\underset{|}{C}HCH_2CH{=}O}$ (c) $CH_3\overset{CH_3}{\overset{|}{C}}HCH_2\underset{CH_3CH_2}{\underset{|}{\overset{CH_3}{\overset{|}{C}}}}CH_2\underset{CH_3}{\underset{|}{C}}HCH{=}O$

PRACTICE EXERCISE 2

What is wrong with the name *2-isopropylpropanal?*

■ Pronounce *-one* as *own.*

IUPAC Names of Ketones End in *-one* The IUPAC rules for naming ketones are identical to those for the aldehydes, except for two obvious changes. The name of the parent ketone must end in *-one* (not *-al*) and the keto group must be located by a number. In numbering the chain, the location of the keto group takes precedence, not the locations of substituents.

EXAMPLE 4.2
Writing the IUPAC Name of a Ketone

What is the IUPAC name of the following ketone?

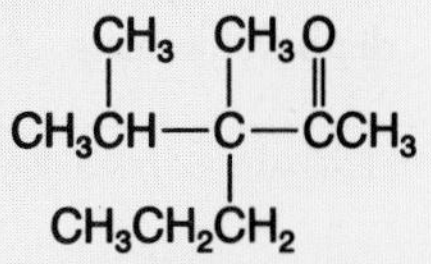

$$\begin{array}{c} \quad CH_3 \;\; CH_3 \; O \\ \quad | \quad\;\; | \quad \| \\ CH_3CH-C-CCH_3 \\ \quad\quad\;\; | \\ CH_3CH_2CH_2 \end{array}$$

ANALYSIS There are two chains that have six carbon atoms, but we have to use the one that has the carbon atom of the carbonyl group. We number this chain to give the location of the carbonyl group the lower number.

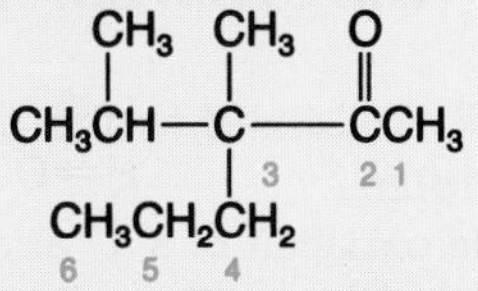

$$\begin{array}{c} \quad CH_3 \;\; CH_3 \quad O \\ \quad | \quad\;\; | \quad\;\; \| \\ CH_3CH-\underset{3}{C}-\underset{2\;1}{CCH_3} \\ \quad\quad\;\; | \\ \underset{6\;\;5\;\;4}{CH_3CH_2CH_2} \end{array}$$

The parent ketone is therefore 2-hexanone. At carbon 3 its chain has a methyl group plus an isopropyl group.

SOLUTION What remains is to assemble these names into the complete name of the ketone:

3-isopropyl-3-methyl-2-hexanone

CHECK Another frequent error is to give incorrect names to alkyl groups. Once you are certain of the parent chain and the direction of numbering, double-check the alkyl groups. If an alkyl group has three carbons *it must be one of the propyl groups* because *prop* goes with three carbons. But which one? When the bonding site is from the middle carbon, it is *isopropyl.*

PRACTICE EXERCISE 3

Write the IUPAC names for the following ketones.

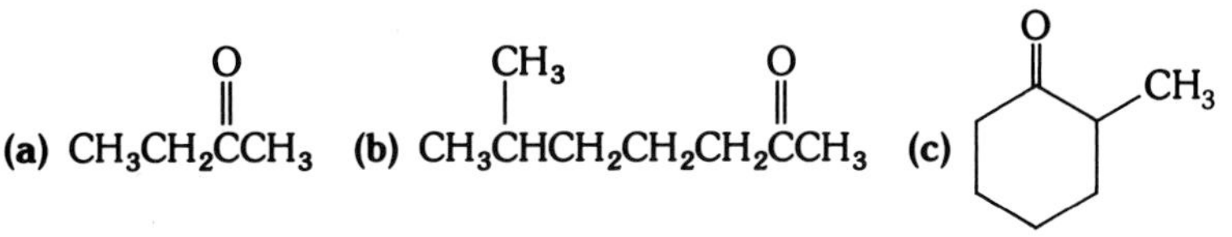

(a) $CH_3CH_2\overset{O}{\overset{\|}{C}}CH_3$ (b) $CH_3\overset{CH_3}{\overset{|}{C}}HCH_2CH_2CH_2\overset{O}{\overset{\|}{C}}CH_3$ (c) 2-methylcyclohexanone (cyclohexane ring with C=O and adjacent CH_3)

The Simplest Ketone Is Usually Called Acetone, Not Propanone Quite often the simpler ketones are given common names that are made by naming the two alkyl groups attached to the carbon atom of the carbonyl group and then following these names by the word *ketone.* For example,

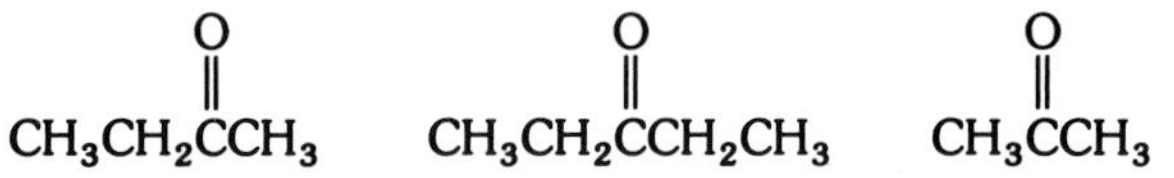

$CH_3CH_2\overset{O}{\overset{\|}{C}}CH_3$ Methyl ethyl ketone

$CH_3CH_2\overset{O}{\overset{\|}{C}}CH_2CH_3$ Diethyl ketone

$CH_3\overset{O}{\overset{\|}{C}}CH_3$ (Dimethyl ketone) Acetone

As we noted, the name *acetone* is almost always used for dimethyl ketone or propanone.

The name *acetone* stems from the fact that this ket*one* can be made by heating the calcium salt of *acet*ic acid.

PRACTICE EXERCISE 4

Write the structures of the following ketones.

(a) ethyl isopropyl ketone (b) methyl phenyl ketone
(c) dipropyl ketone (d) di-*t*-butyl ketone

4.3

THE OXIDATION OF ALDEHYDES AND KETONES

The aldehyde group is easily oxidized to the carboxylic acid group, but the keto group is difficult to oxidize.

We learned in Chapter 3 that the oxidation of a 1° alcohol to an aldehyde requires special reagents, because aldehydes are themselves easily oxidized. We also learned that much less care is needed to oxidize a 2° alcohol to a ketone, because ketones resist further oxidation.

$$\underset{\text{1° Alcohol}}{RCH_2OH} + \underset{\text{Oxidizing agent}}{(O)} \longrightarrow \underset{\text{Aldehyde}}{R\overset{\overset{O}{\|}}{C}H} + H_2O$$

$$\underset{\text{2° Alcohol}}{RCHR'} + (O) \longrightarrow \underset{\text{Ketone}}{R\overset{\overset{O}{\|}}{C}R'} + H_2O$$

The ease with which the aldehyde group is oxidized by even mild reactants has led to some simple test tube tests for aldehydes.

The *Tollens' Test* Produces a Silver Mirror One very mild oxidizing agent, called **Tollens' reagent,** consists of an alkaline solution of the silver ion in combination with two ammonia molecules, $[Ag(NH_3)_2]^+$. This species in Tollens' reagent oxidizes the aldehyde group to a carboxyl group (rather, to its anion form). The silver ion is reduced to metallic silver. In general,

The Tollens' test is sometimes called the *silver mirror test*. Glucose gives this test.

$$RCH{=}O(aq) + 2[Ag(NH_3)_2]^+(aq) + 3OH^-(aq) \longrightarrow RCO_2^-(aq) + 2Ag(s) + 2H_2O + 4NH_3(aq)$$

When this reaction occurs in a thoroughly clean, grease-free test tube, the silver plates to the glass as a beautiful mirror. In fact, this is one technique used to manufacture mirrors. A positive Tollens' test is actually the formation of metallic silver *in any form,* as a mirror in a clean test tube or as a grayish precipitate otherwise. The appearance of silver is dramatic evidence that a reaction occurs, so Tollens' reagent provides a simple test, called **Tollens' test,** to tell whether an unknown compound is an aldehyde or a ketone.

***Benedict's Test* Produces a Brick-Red Precipitate** **Benedict's reagent,** another mild oxidizing agent, consists of a basic solution of the copper(II) ion and the citrate ion, the anion of citric acid, which causes the tart taste of citrus fruits. The medium has to be slightly basic in order for an aldehyde group to be oxidized by the reagent. However, Cu^{2+} normally forms an

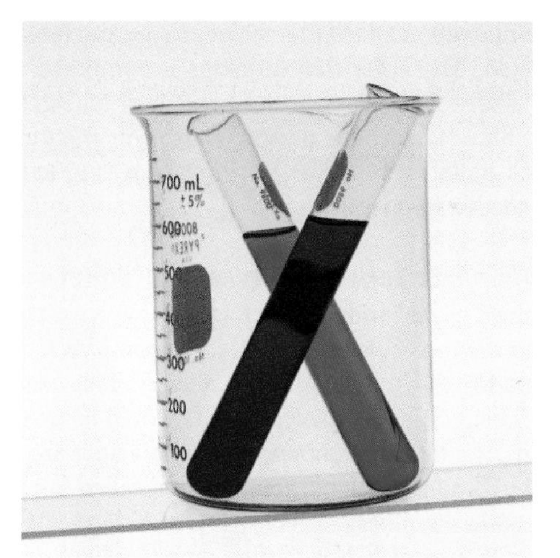

FIGURE 4.1

Benedict's solution is a slightly alkaline solution of citrate ion and Cu^{2+} ion, which makes the solution in the test tube at the rear intensely blue. In a positive Benedict's test, a brick red slurry of Cu_2O forms, seen here in the other test tube before settling.

extremely insoluble precipitate of CuO in a basic environment. The citrate ion prevents this by a mechanism to be studied shortly.

When an easily oxidized compound like glucose is added to a test tube that contains some Benedict's reagent, and the solution is warmed, Cu^{2+} ions are reduced to Cu^{+} ions. The citrate ion is unable to keep the Cu^{+} ion in the dissolved state in the basic medium. The newly formed Cu^{+} ions are instantly changed by the base (OH^-) into a precipitate of copper(I) oxide, Cu_2O. We'll write the equation using the general structure of an aldehyde, RCH=O, but *simple* aldehydes (implied by this structure) give complex results.[1]

■ $CH_2CO_2^-$ — $HOCCO_2^-$ — $CH_2CO_2^-$

Citrate ion

$$RCH{=}O(aq) + 2Cu[citrate]^{2+}(aq) + 2OH^-(aq) \longrightarrow RCO_2^-(aq) + Cu_2O(s) + 3H_2O$$

The Benedict's reagent has a brilliant blue color caused by the copper(II) ion, but Cu_2O has a brick red color (see Fig. 4.1). Therefore the visible evidence of a positive **Benedict's test** is the disappearance of a blue color and the appearance of a reddish precipitate.

Simple aldehydes, as we said, do not give the test as well as aldehydes with neighboring oxygens. Three systems, one not even an aldehyde, and all of which occur among various carbohydrates, give positive Benedict's tests:

RCH(OH)CH=O	RC(=O)—CH=O	RCH(OH)C(=O)R′
α-Hydroxy aldehyde (present in glucose)	α-Keto aldehyde	α-Hydroxy ketone (present in fructose)

■ A carbon atom immediately adjacent to a carbonyl group is often called an alpha (α) carbon:

—C—C(=O)— (Alpha carbon)

Benedict's Test Has Been a Common Method for Detecting Glucose in Urine In certain conditions, like diabetes, the body cannot prevent some of the excess glucose in the blood from being present in the urine, so testing the urine for its glucose concentration has long been used in medical diagnosis. Clinitest® tablets, a convenient solid form of Benedict's reagent, contain all the needed reactants in their solid forms. To test for glucose, a few drops of

[1] Although simple aldehydes give a reaction with the Benedict's reagent that reduces the blue color, the gummy solid that forms is not Cu_2O. Even the equation that we write here for carbohydrates is an oversimplification because in a *warm* alkaline medium, which is involved in the Benedict's test, carbohydrates undergo complex reactions. Yet, Cu_2O does form when carbohydrates are tested with Benedict's reagent. See R. Daniels, C. C. Rush, and L. Bauer, *Journal of Chemical Education,* Vol. 37, page 205 (1960).

urine are mixed with a tablet and, as the tablet dissolves, the heat needed for the test is generated. The color that develops is compared with a color code on a chart provided with the tablets. Specialists in the control of diabetes, however, prefer to monitor the carbohydrate status of a diabetic person by determining the glucose in the *blood* instead of in the urine. Not all patients, however, can or are willing to manage blood tests, particularly when they are needed frequently.

■ Other tests for glucose are based on enzyme-catalyzed reactions, which we will learn more about later.

■ Tollens' reagent must be freshly made, because it deteriorates on standing.

When an aqueous hydroxide is added to aqueous silver nitrate, a tan, mud-like precipitate of silver oxide forms.

***Complex Ions* Are Present in Tollens' and Benedict's Reagents** The reagents for the Tollens' and Benedict's tests involve a species new to our study, one of great importance at the molecular level of life, the *complex ion*. Tollens' reagent is prepared by adding sodium hydroxide to dilute silver nitrate. This causes the very insoluble silver oxide, Ag_2O, to precipitate. Undissolved Ag_2O would be unable to give the Tollen's test. When dilute ammonia is added next, however, its molecules are able to pull silver ions out of the solid silver oxide by forming a soluble complex ion, called the *silver diammine ion*, $[Ag(NH_3)_2]^+$. Thus the silver oxide dissolves.

A **complex ion**—often simply called a **complex**—consists of a metal ion that has strongly attracted a definite number of **ligands,** species that are either negative ions or neutral but electron-rich molecules, like ammonia. Examples of ligands include any of the halide ions (F^-, Cl^-, Br^-, and I^-), the cyanide ion (CN^-), the hydroxide ion, and many anions of organic acids (like the citrate ion). Among the common, electrically neutral ligands are water and ammonia. Other ligands are organic molecules with *amino groups*, NH_2. The most common metal ions that can form complex ions are those of the transition metal elements in the periodic table, like Ag^+ and Cu^{2+}. Two complex ions of Cu^{2+} and neutral ligands are $Cu(H_2O)_4{}^{2+}$ and $Cu(NH_3)_4{}^{2+}$, which are both blue but with strikingly different intensities of color (see Fig. 4.2). The citrate ion is a negatively charged ligand that forms the (blue) complex with Cu^{2+} in Benedict's reagent.

Many Complex Ions Are Important at the Molecular Level of Life Uncomplexed transition metal ions are, in general, insoluble when the pH is greater than 7; they precipitate as their hydroxides or oxides. Thus virtually all of the trace transition metal ions required in nutrition, like Cu^{2+}, Co^{2+}, Fe^{2+}, and several others, can exist in the slightly alkaline fluids of the cell or in blood only as complex ions. Many electron-rich organic ligands, however, are able to form water-soluble complexes with such metal ions and so allow them to be in solution even at pHs greater than 7. The iron(II) ion, for example, is insoluble in base, but in blood (pH 7.35) it occurs in a complex ion called *heme*, the red-colored species in hemoglobin and the oxygen carrier in blood.

■ Hemoglobin exists inside erythrocytes—red blood cells.

Inside cells, the phosphate ion level is sufficiently high to form insoluble phosphates with calcium ion, but this must be prevented for many reasons, not least of which is that cells would mineralize and die. Cells prevent this by forming soluble complex ions between Ca^{2+} and a variety of electron-rich molecular units on protein molecules. Complex ions thus have absolutely vital functions at the molecular level of life.

FIGURE 4.2
The hydrated copper (II) ion, $Cu(H_2O)_4{}^{2+}$ *(left)* gives a bright blue color to its aqueous solution. At the same molar concentration the ammoniated copper (II) ion, $Cu(NH_3)_4{}^{2+}$ *(right)* causes a much deeper blue.

4.4

THE REDUCTION OF ALDEHYDES AND KETONES

Aldehydes and ketones are reduced to alcohols when hydrogen adds to their carbonyl groups.

Aldehydes are reduced to 1° alcohols and ketones are reduced to 2° alcohols by a variety of conditions. We will study two methods, the direct addition of hydrogen and reduction by hydride ion transfer. Either method can be called the *hydrogenation* or the *reduction* of an aldehyde or ketone.

Aldehyde and Keto Groups Can Be Reduced to Alcohol Groups Under heat and pressure and in the presence of a finely divided metal catalyst, the carbonyl groups of aldehydes and ketones add hydrogen.

$$\underset{\text{Aldehyde}}{R\overset{O}{\overset{\|}{C}}H} + H_2 \xrightarrow[\text{heat, pressure}]{\text{Ni}} \underset{1°\text{ Alcohol}}{RCH_2OH}$$

$$\underset{\text{Ketone}}{R\overset{O}{\overset{\|}{C}}R'} + H_2 \xrightarrow[\text{heat, pressure}]{\text{Ni}} \underset{2°\text{ Alcohol}}{R\overset{OH}{\overset{|}{C}}HR'}$$

The experimental *conditions* for these catalytic hydrogenations are impossible in living systems, of course, but they do show the *net effect* of the reductions of aldehyde and keto groups that is also accomplished in living cells. *The aldehyde group is reduced to a 1° alcohol and the keto group to a 2° alcohol.*

The Aldehyde or Keto Group Is Reduced by Acceptance of the Hydride Ion The hydride ion, $H\!:^-$, is a powerful reducing agent. As we have commented before, however, $H\!:^-$ is also an extremely powerful base. *The hydride ion cannot exist as an independent species in the aqueous medium of cells.* In its free form, $H\!:^-$, it reacts vigorously with water to give hydrogen gas, leaving the relatively much weaker base, OH^-.

■ In the lab, $H\!:^-$ can be supplied by $LiAlH_4$ or $NaBH_4$ for these reductions, but water must be rigorously excluded from the reaction mixture.

$$H\!:^- + H{-}OH \longrightarrow H{-}H + OH^-$$

If we add sodium hydride, NaH, to water, for example, the following reaction occurs, and a caustic solution containing sodium hydroxide, lye, forms.

$$NaH(s) + H_2O \longrightarrow H_2(g) + NaOH(aq)$$

The only way hydride ion can be supplied in living systems, therefore, is by a hydride-ion donor that transfers $H\!:^-$ *directly* to the hydride-ion acceptor.

The carbonyl group is an excellent acceptor of hydride ion. We'll represent an organic donor of hydride ion by the symbol *Mtb*:H, where *Mtb* refers to a *metabolite,* a chemical intermediate in metabolism. When an aldehyde or ketone group accepts a hydride ion, the following reaction occurs.

■ *Mtb*:H in the body is often made from a B vitamin unit incorporated into an enzyme.

$$\underset{\substack{\text{Hydride}\\\text{donor}}}{Mtb\!:\!H} + \underset{\substack{\text{Aldehyde}\\\text{or ketone}}}{>C{=}\ddot{O}\!:} \longrightarrow \underset{\substack{\text{Anion of}\\\text{an alcohol}}}{H{-}\overset{|}{\underset{|}{C}}{-}\ddot{\underset{..}{O}}\!:^-}Mtb^+$$

The anion of an alcohol is a stronger proton acceptor than a hydroxide ion. So in the instant when the newly formed anion emerges, it takes a proton either from a water molecule or from some other proton donor in the surrounding buffer system. Thus the final organic product is an alcohol.

$$\text{H—}\overset{|}{\underset{|}{\text{C}}}\text{—}\ddot{\text{O}}\text{:}^- + \text{H—}\ddot{\text{O}}\text{H} \longrightarrow \text{H—}\overset{|}{\underset{|}{\text{C}}}\text{—}\ddot{\text{O}}\text{—H} + \text{:}\ddot{\text{O}}\text{H}^-$$

Anion of an alcohol — Alcohol

■ Remember, a gain of electrons is reduction because it makes oxidation numbers less positive.

As we have already noted, another name for *hydrogenation* is *reduction,* and when a carbonyl carbon atom accepts the pair of electrons carried by the hydride ion, it *gains* this pair and so is reduced.

One of the many examples in cells of reduction by the donation of hydride ion is the reduction of the keto group in the pyruvate ion to form the lactate ion, a step in the metabolism of glucose.

■ NAD+ is a structural unit in several enzymes and is made from a B vitamin. NAD:H, usually written NADH, is the reduced form of NAD+. We explicitly use NAD:H here to emphasize that the species is a donor of H:−. Later, we'll generally write NADH.

$$\underset{\text{Pyruvate ion}}{CH_3\overset{\text{:O:}}{\overset{\|}{C}}CO_2^-} + \underset{\text{Hydride ion donor}}{\text{NAD:H}} \longrightarrow CH_3\overset{\text{:}\ddot{O}\text{:}^-}{\overset{|}{\underset{\underset{H}{|}}{C}}}CO_2^- + \underset{\text{Oxidized form of hydride ion donor}}{NAD^+}$$

$$\xrightarrow[\text{(rapid reaction)}]{\text{HO—H}} \underset{\text{Lactate ion}}{CH_3\overset{OH}{\overset{|}{C}}HCO_2^-} + HO^-$$

In this sequence, NAD^+ stands for *nicotinamide adenine dinucleotide,* a species that we will discuss further in Section 11.1. Right now, all we need to know about NAD^+ is that its reduced form, NAD:H, is a good donor of the hydride ion.

With respect to the reductions of aldehydes and ketones, our needs center on what these reactions produce *in vivo* rather than *in vitro.* We'll study, therefore, only the *net effects* of such reactions and not the specific reagents and conditions commonly used *in vitro.* What we need to learn, therefore, is only how write the products when we know the reactants.

EXAMPLE 4.3
Writing the Structure of the Product of the Reduction of an Aldehyde or Ketone

What is the product of the reduction of propanal?

ANALYSIS All the action is at the carbonyl group. It changes to an alcohol group. Therefore all we have to do is copy over the structure of the given compound, change the double bond to a single bond, and supply the two hydrogen atoms—one to the oxygen atom of the original carbonyl group and one to the carbon atom.

SOLUTION The product of the reduction of propanal is 1-propanol.

$$\underset{\text{Propanal}}{CH_3CH_2\overset{O}{\overset{\|}{C}}H} \xrightarrow{\text{reduction}} \underset{\text{1-Propanol}}{CH_3CH_2\overset{OH}{\overset{|}{C}}H_2}$$

CHECK One common error is to change the structural skeleton, so be sure to check that the *sequence* of all of the atoms heavier than H, like C and O, *has not changed.* Another common error is to write a structure which includes a violation of the covalences of the heavy atoms—4 for C and 2 for O in neutral species. So go down the chain in the answer, atom by atom, to see that each carbon has four bonds and each oxygen has two. If you find an error, fix it.

PRACTICE EXERCISE 5

Write the structures of the products that form when the following aldehydes and ketones are reduced.

(a) $CH_3CH_2\overset{\overset{O}{\|}}{C}CH_3$ (b) $CH_3\underset{\underset{CH_3}{|}}{C}HCH_2\overset{\overset{O}{\|}}{C}H$ (c) cyclohexanone (ring)=O

4.5

THE REACTIONS OF ALDEHYDES AND KETONES WITH ALCOHOLS

1,1-Diethers—acetals or ketals—form when aldehydes or ketones react with alcohols in the presence of an acid or enzyme catalyst.

This section is background to the study of carbohydrates whose molecules have the functional groups introduced here. We will study the simplest examples of these groups now so that carbohydrate structures will be easier to understand.

Alcohols Add to the Carbonyl Groups of Aldehydes and Ketones When a solution of an aldehyde in an alcohol is prepared, molecules of the alcohol add to molecules of the aldehyde and the following equilibrium mixture forms.

$$R'\overset{\overset{:O:}{\|}}{C}H + H\ddot{O}R \rightleftharpoons R'\overset{\overset{:\ddot{O}H}{|}}{C}H\ddot{O}R$$

Aldehyde Alcohol Hemiacetal

The product, a **hemiacetal,** has molecules that always have a carbon atom holding both an OH group and an OR group. When these two groups are this close to each other, they so modify each others' properties that it's useful to place the whole system into its own separate family rather than view the system as only an alcohol–ether combination. For example, hemiacetals, when formed by indirect means, very readily break down to aldehydes and alcohols. Ordinary ethers, we learned, strongly resist reactions that break their molecules. The hemiacetal system occurs among carbohydrates where, however, it is relatively stable. In one form, glucose molecules exist as cyclic hemiacetals (Chapter 8).

The hemiacetal system:

$$C-\overset{\overset{O-H}{|}}{\underset{\underset{O-R}{|}}{C}}-H$$

This originally was the carbon atom of an aldehyde group.

When a ketone is dissolved in an alcohol, a similar reaction occurs to give an equilibrium in which the product is called a **hemiketal** to signify its origin from a ketone.

■ The hemiketal system:

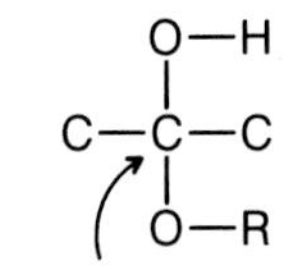

O—H
|
C—C—C
|
O—R

This originally was the carbon atom of an keto group.

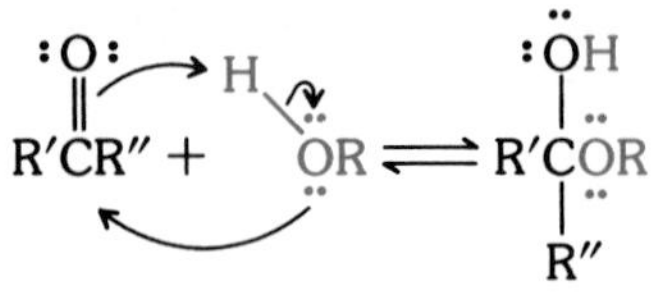

$$R'C(=O)R'' + H\text{—}\ddot{O}R \rightleftharpoons R'C(OH)(OR)R''$$

Ketone Alcohol Hemiketal

The position of equilibrium overwhelmingly favors the reactants, the ketone and alcohol, so hemiketals are even less stable than hemiacetals. The hemiketal system, however, does occur among carbohydrates. One form of fructose, for example, is a cyclic hemiketal (Chapter 8). The continuation of our study of these systems will deal almost entirely with hemiacetals because the extension of the principles to hemiketals is straightforward.

The Polarity of the Carbonyl Group Causes the Alcohol to Add to an Aldehyde in Only One Direction The —OR part of the alcohol molecule has $\delta-$ on O and the carbonyl group of the aldehyde has $\delta+$ on C. Because unlike charges attract, the alcohol's —OR unit *always* ends up attached to the carbon atom of the original carbonyl group, and the H atom of the alcohol always goes to the carbonyl oxygen. Thus the *direction* of this addition reaction is determined by the polarity of the carbonyl group.

Except among carbohydrates, the hemiacetal system is almost always too unstable to exist in a pure compound. If we try to isolate and purify an ordinary hemiacetal, it breaks back down, and only the original aldehyde and alcohol are obtained. Hemiacetals generally exist only in the equilibrium that includes their parent aldehydes and alcohols. Despite this, we're still interested in the hemiacetal system for two reasons. Not only is it relatively stable among carbohydrates, it is an intermediate in the formation of "1,1-diethers" or acetals, which are stable enough to be isolated. The acetal system is also common among carbohydrates.

The relative ease with which hemiacetals break back down means that the hemiacetal system is a site of structural weakness, even among carbohydrates. For this reason, we have to learn how to recognize the system when it occurs in a structure.

EXAMPLE 4.4
Identifying the Hemiacetal System

Which of the following structures has the hemiacetal system? Draw an arrow pointing to any carbon atoms that were initially the carbon atoms of aldehyde groups.

■ The ring system of **3** also occurs in glucose.

CH_3—O—CH_2—CH_2—OH (**1**)

CH_3—O—CH_2—OH (**2**)

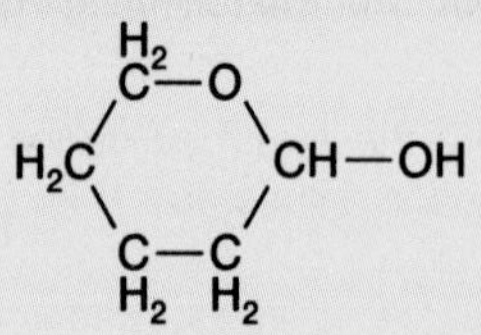

Ring: H_2C—O—CH(OH)—CH_2—CH_2—CH_2— (six-membered ring: H_2C, O, CH—OH, H_2C, H_2C, H_2C) (**3**)

ANALYSIS To have the hemiacetal system, the molecule must have a carbon to which are attached one OH group and one —O—C unit.

SOLUTION In structure **1** there is an OH group and an —O—C unit, but they are not joined to the *same* carbon. Therefore **1** is not a hemiacetal. It has only an ordinary ether group plus an alcohol group.

In structure **2**, the OH and the —O—C are joined to the same carbon, so **2** is a hemiacetal. Similarly, in structure **3**, the carbon on the far right corner of the ring holds both an OH group and —O—C unit, and **3** is also a hemiacetal, a cyclic hemiacetal. Structure **3** shows the way in which the hemiacetal system occurs in many carbohydrates, as a cyclic hemiacetal. The asterisks (*) in the following structures identify carbon atoms that were originally carbonyl carbons.

$$CH_3{-}O{-}\overset{*}{C}H_2{-}OH$$

2

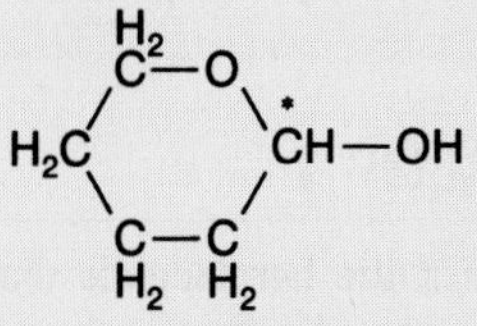

(Cyclic structure: ring of H_2C—O—$\overset{*}{C}H$(—OH)—CH_2—CH_2—CH_2)

3

CHECK Make sure that any structure identified as a hemiacetal has at least one carbon attached to *two* O atoms by single bonds, that *one* of the O atoms is part of the OH group, and that the other is joined by its second single bond to C.

PRACTICE EXERCISE 6

Identify the hemiacetals or hemiketals among the following structures, and draw arrows that point to the carbon atoms that initially were part of the carbonyl groups of parent aldehydes (or ketones).

(a) 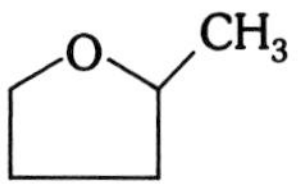 (tetrahydrofuran ring with CH_3 on the carbon adjacent to O)

(b) 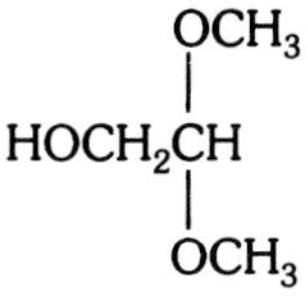 $HOCH_2CH(OCH_3)_2$

(c) $HOCH_2OCH_2CH_3$

(d) 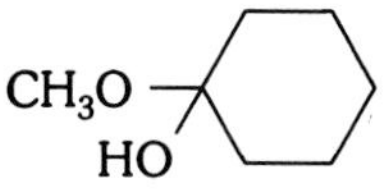 CH_3O and HO on the same carbon of a cyclohexane ring

Another skill that will be useful in our study of carbohydrates is the ability to write the structure of a hemiacetal that could be made from a given aldehyde and alcohol.

EXAMPLE 4.5
Writing the Structure of a Hemiacetal Given Its Parent Aldehyde and Alcohol

Write the structure of the hemiacetal that is present at equilibrium in a solution of propanal in ethanol.

ANALYSIS A hemiacetal must have a carbon atom to which both an OH group and an OR group are attached. This carbon atom *is provided by the aldehyde.* The structure of the alcohol, CH_3CH_2OH, tells us that the R group in OR of the hemiacetal is CH_2CH_3.

SOLUTION The structure of the hemiacetal formed from propanal and ethyl alcohol is

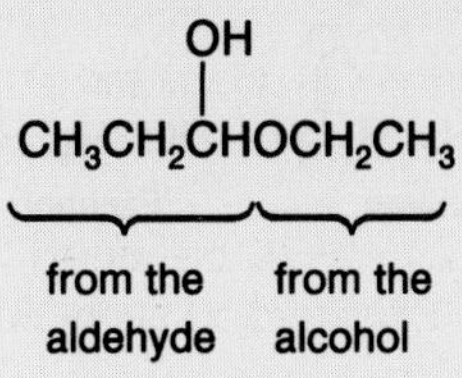

$$\underbrace{CH_3CH_2\overset{OH}{\overset{|}{C}}H}_{\text{from the aldehyde}}\underbrace{OCH_2CH_3}_{\text{from the alcohol}}$$

CHECK Find the carbon holding *two* O atoms and check the other two atoms or groups that it also holds *against the original aldehyde* (or ketone). These two atoms or groups —here, H and CH_3CH_2—must match those of the aldehyde (or ketone). Finally check what else the two O atoms are holding; one must hold H (to make it an OH group) and the other must hold the alkyl group from the original alcohol.

PRACTICE EXERCISE 7

Write the structures of the hemiacetals that are present in the equilibria that involve the following pairs of compounds.

(a) ethanal and methanol **(b)** butanal and ethanol

(c) benzaldehyde and 1-propanol **(d)** methanal and methanol

Still another skill that will be useful in studying carbohydrates is the ability to write the structures of the aldehyde and alcohol that are liberated by the breakdown of a hemiacetal.

EXAMPLE 4.6
Writing the Breakdown Products of a Hemiacetal

What aldehyde and alcohol form when the following hemiacetal breaks down?

$$CH_3CH_2CH_2\overset{\overset{\displaystyle OH}{|}}{C}H{-}O{-}CH_2CH_3$$

ANALYSIS The key step to solving this problem lies in analyzing the given structure. First, pick out the carbon atom of the original carbonyl group; it's the one holding *both* an OH group and an OR unit. Anything else this carbon holds—usually one H and one hydrocarbon group—completes what we need in order to write the structure of the aldehyde. The R group of OR is the hydrocarbon group of the original alcohol. Our analysis thus gives us:

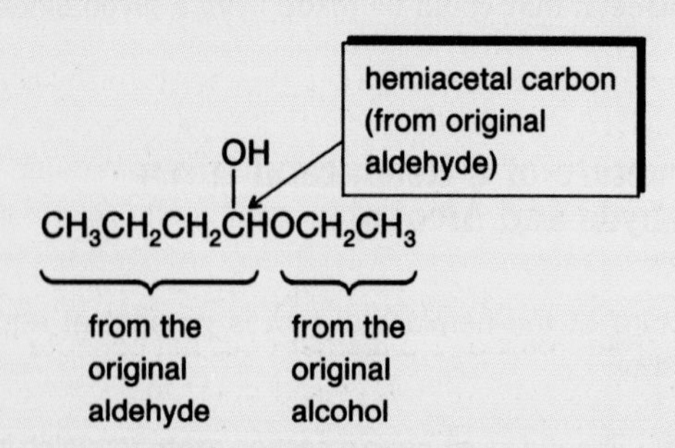

The original aldehyde is thus the unbranched four-carbon aldehyde, butanal, and the original alcohol is seen to be the two-carbon alcohol, ethanol.

SOLUTION The products of the breakdown of the given hemiacetal are

$$\underset{\text{Butanal}}{CH_3CH_2CH_2\overset{\overset{\displaystyle O}{\|}}{C}H} + \underset{\text{Ethanol}}{HOCH_2CH_3}$$

CHECK It is essential to notice that *only one bond* is affected when we disassemble the initial hemiacetal, *the C—O bond of the hemiacetal carbon,* not any other C—O bond, not a C—C bond, and not a C—H bond. Failure to learn this is the most common error that students make in working problems like this. Examine closely, therefore, your answer to see that the original molecule is ruptured *only* at the C—O bond at the hemiacetal carbon. *Break no other bond.*

PRACTICE EXERCISE 8

Write the structures of the breakdown products of the following hemiacetals.

(a) $CH_3CH_2CH(OH)OCH_3$ **(b)** $CH_3CH_2OCH(OH)CH_2CH_3$

Acetals and Ketals Form When Alcohols React Further with Hemiacetals and Hemiketals Hemiacetals and hemiketals are special kinds of alcohols, and they resemble alcohols in one important property. They can undergo a reaction that looks like the formation of an ether. A ordinary ether does not form, however, but a special kind, a 1,1-diether called an **acetal** or a **ketal.** The overall change that leads to an acetal is as follows.

$$\underset{\text{Hemiacetal}}{R'CH(OH)OR} + H{-}OR \xrightarrow{\text{acid catalyst}} \underset{\text{Acetal}}{R'CH(OR)OR} + H_2O$$

Hemiketals give the identical kind of reaction, but the products are called *ketals.* Unlike hemiacetals and hemiketals, both acetals and ketals are stable compounds that can be isolated and stored.

The difference between the formation of an acetal and an ordinary ether is that *acetals form and break more readily.* As a rule, when two functional groups are very close to each other in a molecule, each modifies the properties of the other in some way. Here, the OR group makes the OH group attached to the same carbon much more reactive toward the splitting out of water with an alcohol.

In a structural sense, an acetal is a 1,1-diether, but "1,1" does not refer to the numbering of the chain and the "ether" part of 1,1-diether does not connote "resistance to breaking up." The "1,1" means only that the two OR groups come to the *same* carbon. In this sense, ketals are also 1,1-diethers. The molecules of many carbohydrates, like sucrose (table sugar), lactose (milk sugar), and starch also have the 1,1-diether system, which is why we study acetals and ketals.

Ordinary ethers, R—O—R, do not break up in dilute acid or base, but *acetals and ketals are stable only if they are kept out of contact with aqueous acids.* In water, acids (or enzymes) catalyze the hydrolysis of acetals and ketals to their parent alcohols and aldehydes (or ketones). In aqueous *base,* however, the acetal (ketal) system is stable.

Hydrolysis is the only chemical reaction of acetals and ketals that we need to study; it is the reaction by which carbohydrates are digested. Before we study this reaction further, we must be sure that we can recognize the acetal or ketal system when it occurs in a structure.

■ The acetal system:

C—C(H)(O—R)(O—R)

This originally was the carbon atom of an aldehyde group.

■ The ketal system:

C—C(C)(O—R)(O—R)

This originally was the carbon atom of an keto group.

EXAMPLE 4.7
Recognizing the Acetal or Ketal Systems and Identifying Which Carbon Atoms Came from Parent Carbonyl Carbons

Examine each structure to see whether it is an acetal or a ketal. If it is either, identify the carbon atom furnished by the carbonyl carbon of the parent aldehyde or ketone.

$CH_3OCH_2OCH_3$ (**4**) $CH_3OCH(OCH_3)CH_3$ (**5**) 1,1-dimethoxycyclohexane (cyclohexane ring bearing two OCH_3 groups on one carbon) (**6**)

ANALYSIS We have to find one carbon that holds two OR types of groups. The two *must* join to the *same* carbon.

SOLUTION Structure **4** has such a carbon in its central CH_2 unit. This carbon was initially the carbonyl carbon atom of an aldehyde (methanal), because it also holds at least one H atom. Structure **5** similarly has such a carbon, in the CH unit, a carbon that also came from an aldehyde group (because it has at least one H atom). In structure **6**, we can also find a carbon that holds two O—C bonds. This carbon lacks an H atom, however, so it must have come from the carbonyl group of a ketone system.

(Structure **6**, OCH_3, OCH_3, with arrow to ring carbon:) Initially a ketone carbonyl carbon atom

PRACTICE EXERCISE 9

Which of the following two compounds, if either, is an acetal or a ketal? Identify the carbon atom that came originally from the carbonyl group of a parent aldehyde or ketone.

(a) $CH_3OCH_2CH_2OCH_3$ (b) $CH_3CH_2OC(CH_3)_2OCH_2CH_3$

Because acetals and ketals can be hydrolyzed, we have to be able to examine the structures of such compounds and write the structures of their parent alcohols and aldehydes (or ketones), the hydrolysis products. A worked example shows how this can be done.

EXAMPLE 4.8
Writing the Structures of the Products of the Hydrolysis of Acetals or Ketals

What are the products of the following reaction?

$$CH_3CH(OCH_3)OCH_3 + H_2O \xrightarrow{\text{acid catalyst}} ?$$

ANALYSIS The best way to proceed is to find the carbon atom in the structure that holds *two* oxygen atoms. This carbon is the carbonyl carbon atom of the parent aldehyde (or ketone). Break both of its bonds to these oxygen atoms. *Do not break any other bonds.* Then at the carbon that once held two oxygens, make a carbonyl group. The other groups, those of the OR type (here, OCH_3) become alcohols.

SOLUTION The final products of the hydrolysis of the given acetal are ethanal and methanol.

$$CH_3\overset{\overset{OCH_3}{|}}{C}HOCH_3 + H_2O \xrightarrow{\text{acid catalyst}} CH_3\overset{\overset{O}{\|}}{C}H + 2HOCH_3$$

Initially a carbonyl carbon

PRACTICE EXERCISE 10

Write the structures of the aldehydes (or ketones) and the alcohols that are obtained by hydrolyzing the following compounds. If they do not hydrolyze like acetals or ketals, write "no reaction."

(a) $CH_3OCH_2OCH_3$ (b) $CH_3OCH_2CH_2OCH_2CH_3$ (c) $CH_3\overset{\overset{H_3C}{|}}{C}H\underset{\underset{CH_3}{|}}{\overset{\overset{OCH_3}{|}}{C}}OCH_3$

Other Reactions Involving Aldehydes and Ketones Aldehydes and ketones are able to enter into reactions that create larger molecules from smaller molecules by forming new carbon–carbon bonds. The *aldol condensation* is one such reaction. The simplest illustration is the reaction of ethanal with itself in the presence of dilute sodium hydroxide or an enzyme (called *aldolase*).

$$CH_3\overset{\overset{O}{\|}}{C}H + H{-}CH_2\overset{\overset{O}{\|}}{C}H \underset{\text{enzyme}}{\overset{\text{base or}}{\rightleftharpoons}} CH_3\overset{\overset{OH}{|}}{C}H{-}CH_2\overset{\overset{O}{\|}}{C}H$$

Ethanal; Ethanal, second molecule; 3-Hydroxybutanal "aldol"

New carbon–carbon bond

It is little more than the addition of one aldehyde molecule to the carbonyl double bond of another, but it is a *reversible* reaction, as the equilibrium arrows indicate. The forward reaction makes a new carbon–carbon single bond; the reverse reaction, called the *reverse aldol,* breaks this same bond.

An aldol condensation occurs in those cells of ours where glucose molecules are made from smaller molecules by a series of steps called *gluconeogenesis* (Section 15.4). The breakdown of glucose occurs by a different series of reactions called *glycolysis* (Section 15.3), and one step is a reverse aldol. Because these are relatively complex series, we'll delay further study of the aldol condensation until the place where it most directly is used, Chapter 15.

The mechanism of the aldol condensation is given in Special Topic 15.3. It could easily be brought forward for study here.

SUMMARY

Naming aldehydes and ketones The IUPAC names of aldehydes and ketones are based on a parent compound, one with the longest chain that includes the carbonyl group. The names of aldehydes end in *-al* and of ketones in *-one,* and the chains are numbered so as to give the carbonyl carbons the lower of two possible numbers.

Physical properties of aldehydes and ketones The carbonyl group confers moderate polarity, which gives aldehydes and ketones higher boiling points and solubilities in water than hydrocarbons but lower boiling points and solubilities in water than alcohols (that have comparable formula masses).

Ease of oxidation of aldehydes Aldehydes are easily oxidized to carboxylic acids, but ketones resist oxidation. Aldehydes give a positive Tollens' test and ketones do not. α-Hydroxy aldehydes and ketones give the Benedict's test.

Complex ions The reagents for the Tollens' and Benedict's tests contain complex ions, the silver diammine complex in Tollens' reagent and the copper(II) citrate complex ion in Benedict's test. In the first, Ag^+, the central metal ion, holds two molecules of the ligand, NH_3. The ligand for the Cu^{2+} ion in Benedict's reagent is the negative ion of citric acid. Complex ions help to keep transition metal ions in solution, even in base.

Hemiacetals and hemiketals When an aldehyde or a ketone is dissolved in an alcohol, some of the alcohol adds to the carbonyl group of the aldehyde or ketone. The equilibrium thus formed includes molecules of a hemiacetal (or hemiketal). The chart at the end of this summary outlines the chemical properties of the aldehydes and ketones we have studied.

Hemiacetals and hemiketals are usually unstable compounds that exist only in an equilibrium involving the parent carbonyl compound and the parent alcohol (which generally is the solvent). Hemiacetals and hemiketals readily break back down to their parent carbonyl compounds and alcohols. When an acid catalyst is added to the equilibrium, a hemiacetal or hemiketal reacts with more alcohol to form an acetal or ketal.

Acetals and ketals Acetals and ketals are 1,1-diethers that are stable in aqueous base or in water, but not in aqueous acid. Acids catalyze the hydrolysis of acetals and ketals, and the final products are the parent aldehydes (or ketones) and alcohols.

Summary of reactions (We omit the aldol condensation here.)

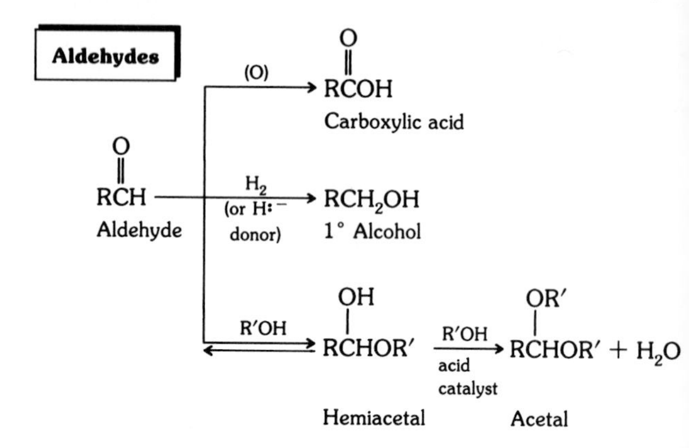

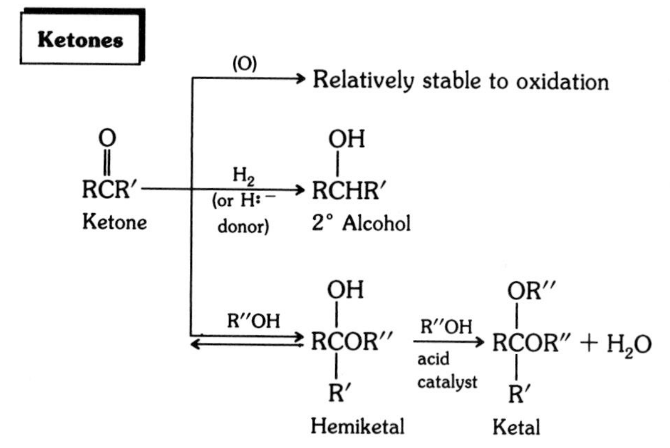

Acetals and Ketals

$$\underset{\text{Acetal or Ketal}}{\mathrm{RC(OR'')(H(R))OR''}} + H_2O \xrightarrow{H^+} \underset{\text{Aldehyde or Ketone}}{\mathrm{RC(=O)H(R)}} + \underset{\text{Alcohol}}{2HOR''}$$

REVIEW EXERCISES

The answers to Review Exercises whose numbers are in color are found in Appendix A. The answers to the other Review Exercises are found in the Study Guide that accompanies this book. The more challenging questions are marked with asterisks.

Names and Structures

4.1 Give the names of the functional groups present in the following structural formulas.

(a) $CH_3CH_2C(=O)CH_2CH_3$ (b) $HC(=O)CH_2CH(CH_3)CH_3$ (c) cyclohexanone (ring with =O)

(d) cyclopentane ring —CO_2H (e) benzene ring —CHO (f) six-membered ring containing O with =O

4.2 To display the *structural* differences among aldehydes, ketones, carboxylic acids, and esters, write the structure of one example of each using three carbons per molecule.

4.3 Write the structure of each of the following compounds.
(a) 2-methylbutanal
(b) 2,3-dichlorocyclohexanone
(c) 1-phenyl-1-ethanone
(d) diisobutyl ketone
(e) hexane-2,5-dione

4.4 What are the structures of the following compounds?
(a) 2-cyclohexylcyclopentanone
(b) 2-methylhexanal
(c) di-*sec*-butyl ketone
(d) 1,3-diphenyl-2-propanone
(e) 1,3,5-cyclohexanetrione

4.5 Although we can write structures that correspond to the following names, when we do, we find that the names aren't proper. How should these compounds be named in the IUPAC system?
(a) 6-methylcyclohexanone
(b) 1-hydroxy-1-propanone (Give common name.)
(c) 1-methylbutanal
(d) 2-methylethanal
(e) 2-propylpropanal

4.6 The following names can be used to write structures, but the names turn out to be improper. What should be their IUPAC names?
(a) 2-*sec*-butylbutanal
(b) 1-phenylethanal
(c) 4,5-dimethylcyclopentanone
(d) 1-hydroxyethanal (Give common name.)
(e) 1-butanone

4.7 If the IUPAC name of **A** is 2-ketopropanal, then what is the IUPAC name of compound **B**?

$CH_3C(=O)-CH(=O)$ **A** $\quad CH_3C(=O)CH_2CH_2CH_2CH(=O)$ **B**

4.8 We can name compound **C** as 3-methylformylcyclopentane. Taking a clue from this, how might compound **D** be named?

C: cyclopentane ring bearing CH_3 and CHO (1,3-positions) $\quad$ **D**: benzene ring bearing CO_2H and CHO (ortho)

4.9 Write the IUPAC names of the following compounds.

(a) $CH_3CH(CH_3)CH_2CHO$

(b) $CH_3CH(CH_2CH_3)CH_2CHO$

(c) $CH_3CH(CH_3)CH_2CH(CH_2CH_3)CH_2CHO$

(d) $CH_3CH(CH_3)CH_2C(=O)CH_3$

(e) cyclopentanone with $CH_3CH_2CH_2$ substituent at C-3

*4.10 What are the IUPAC names of the following compounds?

(a) $CH_3CH_2CH(CH_3)CH_2$ attached to the C-4 of $CH_3CH(CH_3)CH_2CHC(=O)CH_3$

(b) $(CH_3)_3CCH(CH_2CH(CH_3)CH_2CH_3)CH_2CH_2CHO$

(c) $C_6H_5-CH_2CH_2C(=O)CH_3$

(d) cyclohexanone with two H_3C groups (at C-3 and C-4)

(e) cyclooctanone with two CH_3 groups at C-2

4.11 If the common name of $CH_3CH_2CH_2CH_2CO_2H$ is valeric acid, then what is the most likely common name of the following compound?

$CH_3CH_2CH_2CH_2CHO$

4.12 If the common name of **E** is glyceraldehyde, what is the most likely common name of **F**?

$HOCH_2CH(OH)CHO$ **E** $\quad HOCH_2CH(OH)CO_2H$ **F**

Physical Properties of Aldehydes and Ketones

4.13 Arrange the following compounds in their order of increasing boiling points. Do this by placing the letters that identify them in the correct order, starting with the lowest-boiling compound and moving in order to the highest-boiling compound. (They have about the same formula masses.)

CH_3 / O CH_3 / CH_3 OH / OH   OH / CH_3

A   B   C   D

4.14 Arrange the following compounds in their order of increasing boiling points. Do this by placing letters that identify them in the correct order, starting with the lowest-boiling compound on the left in the series and moving to the highest-boiling compound. (They all have about the same formula mass.)

$HOCH_2CH_2CH_2CHOH$ (with CH_3 on the CHOH carbon) **A** $CH_3CH_2CH_2CH_2CCH_3$ (C=O) **B**

$CH_3CH_2CH_2CH_2CHCH_3$ (with CH_3 on CH) **C** $HOCH_2CH_2CH_2CHCH_3$ (with CH_3 on CH) **D**

4.15 Reexamine the compounds of Review Exercise 4.13, and arrange them in their order of increasing solubility in water.

4.16 Arrange the compounds of Review Exercise 4.14 in their order of increasing solubility in water.

4.17 Draw the structure of a water molecule and a molecule of propanal and align them on the page to show how the propanal molecule can accept a hydrogen bond from the water molecule. Use a dotted line to represent this hydrogen bond and place $\delta+$ and $\delta-$ symbols where they are appropriate.

4.18 Draw the structures of molecules of methanol and acetone, and align them on the page to show how a hydrogen bond (which you are to indicate by a dotted line) can exist between the two. Place $\delta+$ and $\delta-$ symbols where they are appropriate.

Oxidation of Alcohols and Aldehydes

4.19 What are the structures and the IUPAC names of the aldehydes and ketones to which the following compounds can be oxidized?

(a) CH_3CH_2CHOH (with CH_3 on CHOH)   (b) $CH_3CHCH_2CH_2CH_2OH$ (with CH_3 on CH)

(c) H_3C–(cyclopentane)–OH   (d) $C_6H_5CH_2CHCH_2OH$ (with CH_3 on CH)

4.20 Examine each of the following compounds to see whether it can be oxidized to an aldehyde or to a ketone. If it can, write the structure of the aldehyde or ketone.

(a) $HOCHCH_2CH_3$ (with CH_3 on HOCH)   (b) $HOCH_2CHCH_3$ (with CH_3 on CH)

(c) (cyclohexane) CH_3, OH   (d) CH_3CH_2COH (C=O)

(e) $CH_3OCH_2CH_2CHOH$ (with CH_3 on CHOH)   (f) $HOCH_2$–(benzene ring)–CO_2H

***4.21** An unknown compound, C_3H_6O, reacted with permanganate ion to give $C_3H_6O_2$, and the same unknown also gave a positive Tollens' test. Write the structures of C_3H_6O and $C_3H_6O_2$

***4.22** An unknown compound, $C_3H_6O_2$, could be oxidized easily by permanganate ion to $C_3H_4O_3$, and it gave a positive Benedict's test. Write structures for $C_3H_6O_2$ and $C_3H_4O_3$.

***4.23** Which of the following compounds can be expected to give a positive Benedict's test? All are intermediates in metabolism.

(a) $HOCH_2CCH_2OH$ (C=O)   (b) $HOCH_2CHCHO$ (with OH on CH)

(c) $CH_3CCH_2CO_2H$ (C=O)   (d) $CH_3CHCH_2CO_2H$ (with OH on CH)

***4.24** Which of the following compounds give a positive Benedict's test? (Most are intermediates in metabolism.)

(a) CH_3CCO_2H (C=O)   (b) CH_3CCHO (C=O)

(c) $HOCH_2CH_2CCH_3$ (C=O)   (d) $HOCH_2CHCHCHO$ (HO and OH on the two CH carbons)

4.25 Concerning complex ions,

(a) The cations of what kinds of elements are usually involved?

(b) What *kinds* of particles commonly are ligands?

(c) Give the *formulas* of three examples of negatively charged ligands that come from the same family in the periodic table.

(d) Give the *formulas* of two electrically neutral ligands that form complex ions with Cu^{2+}. Write the formulas of these complex ions.

4.26 Concerning complex ions in test reagents,

(a) What is the *formula* of the complex ion in Tollens' reagent and why is it important that Ag^+ be so complexed?

(b) What is the *function* of the citrate ion in Benedict's reagent?

4.27 What is the *formula* of the precipitate that forms in a positive Benedict's test?

4.28 Clinitest® tablets are used for what?

4.29 What is one practical commercial application of the Tollens' reagent system?

***4.30** Neither the lactate ion nor the pyruvate ion gives a positive Tollens' test. When the body metabolizes the lactate ion, it oxidizes it to the pyruvate ion, $C_3H_3O_3^-$. Using only these facts, write the structure of the lactate ion.

4.31 One of the steps in the metabolism of fats and oils in the diet is the oxidation of the following compound:

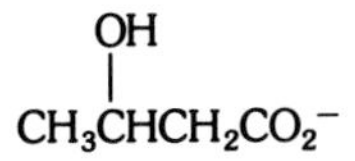

$$CH_3\underset{}{C}H(OH)CH_2CO_2^-$$

Write the structure of the product of this oxidation.

***4.32** One of the important series of reactions in metabolism is called the *citric acid cycle*. Structures **A** and **B** are of compounds (actually, anions) that participate in this cycle. One of them, isocitric acid, is oxidized to a ketone. The other is not. Which one is isocitric acid, **A** or **B**? Write the structure of its corresponding ketone.

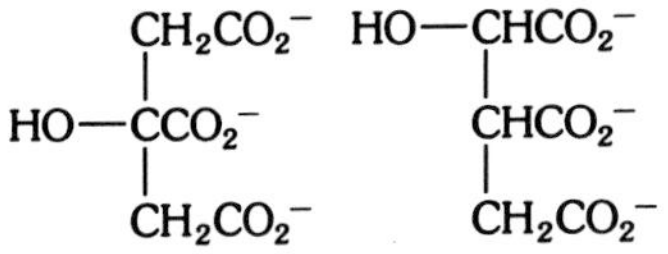

A: $CH_2CO_2^-$ — $HO{-}CCO_2^-$ — $CH_2CO_2^-$

B: $HO{-}CHCO_2^-$ — $CHCO_2^-$ — $CH_2CO_2^-$

Reduction of Aldehydes and Ketones

4.33 The hydride ion, as we learned, reacts as follows with water:

$$H{:}^- + H{-}OH \longrightarrow H{-}H + OH^-$$

The hydride ion reacts in a similar way with CH_3CH_2OH. Write the net ionic equation for this reaction.

4.34 Consider the reaction that occurs when a hydride ion transfers from its donor (which we can write as *Mtb*:H) to ethanal.
(a) Write the structure of the organic anion that forms when the hydride ion is transferred to ethanal.
(b) What is the net ionic equation of the reaction of water with the anion formed in part (a)?
(c) What is the IUPAC name of the organic product of the reaction of part (b)?

4.35 If a donor of a hydride ion (*Mtb*:H) transfers it to a molecule of acetone,
(a) What is the structure of the organic ion that forms?
(b) What happens to the organic ion [formed in part (a)] in the presence of water? (Write a net ionic equation.)
(c) What is the IUPAC name of the organic product of the reaction of part (b)?

***4.36** The metabolism of aspartic acid, an amino acid, occurs by a series of steps. A portion of this series is indicated below, where NAD:H is a reducing agent that becomes NAD^+ as it transfers hydride ion.

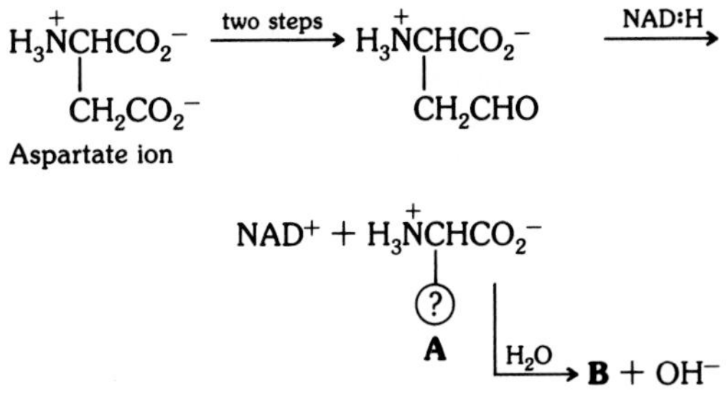

$$H_3\overset{+}{N}CH(CH_2CO_2^-)CO_2^- \xrightarrow{\text{two steps}} H_3\overset{+}{N}CH(CH_2CHO)CO_2^- \xrightarrow{\text{NAD:H}}$$

Aspartate ion

$$NAD^+ + H_3\overset{+}{N}CH(?)CO_2^- \;(\mathbf{A}) \xrightarrow{H_2O} \mathbf{B} + OH^-$$

Complete the structure of **A**, and write the structure of **B**.

***4.37** One of the steps the body uses to make long-chain carboxylic acids is a reaction similar to the following reaction.

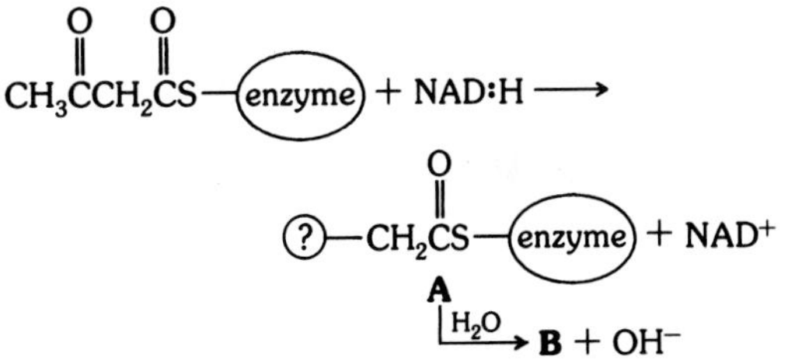

$$CH_3\overset{O}{\overset{\|}{C}}CH_2\overset{O}{\overset{\|}{C}}S{-}(\text{enzyme}) + NAD{:}H \longrightarrow$$

$$(?){-}CH_2\overset{O}{\overset{\|}{C}}S{-}(\text{enzyme}) + NAD^+$$

A $\xrightarrow{H_2O}$ **B** + OH^-

Complete the structure of **A** and write the structure of **B**.

4.38 Write the structures of the aldehydes or ketones that could be used to make the following compounds by reduction (hydrogenation).

(a) 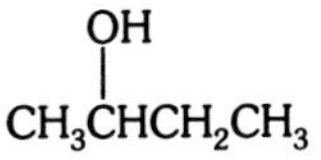 $CH_3CH(OH)CH_2CH_3$

(b) 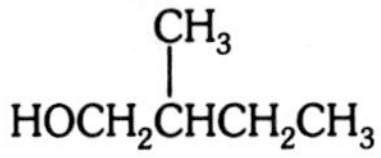 $HOCH_2CH(CH_3)CH_2CH_3$

(c) 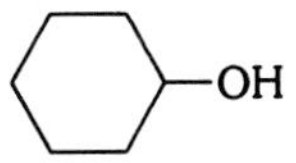 cyclohexyl–OH

(d) 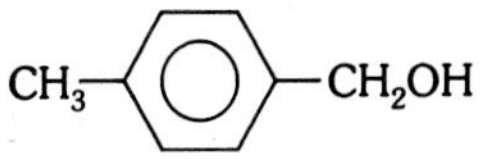 CH_3–(benzene ring)–CH_2OH

4.39 Either an aldehyde or a ketone could be used to make each of the following compounds by hydrogenation. Write the structure of the aldehyde or ketone suitable in each part.

(a) $HOCH_2CH_2OCH_3$

(b) 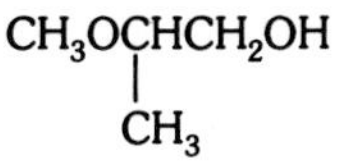 $CH_3OCH(CH_3)CH_2OH$

(c) 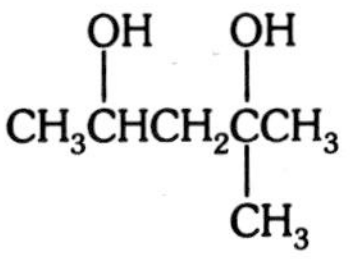 $CH_3CH(OH)CH_2C(OH)(CH_3)CH_3$

(d) 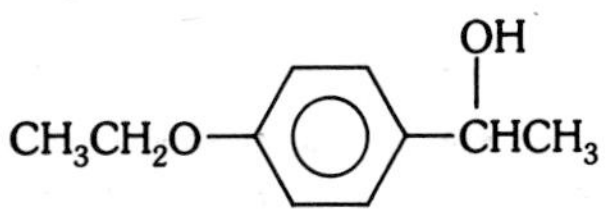 CH_3CH_2O–(benzene ring)–$CH(OH)CH_3$

Hemiacetals and Acetals. Hemiketals and Ketals

4.40 Examine each structure and decide whether it represents a hemiacetal, hemiketal, acetal, ketal, or something else.

(a) $CH_3OCH(CH_3)OH$ (b) $CH_3CH(OCH_3)OCH_3$

(c) $CH_3OCH(CH_3)CH_2OCH_3$ (d) $CH_3OC(CH_3)_2OCH_3$

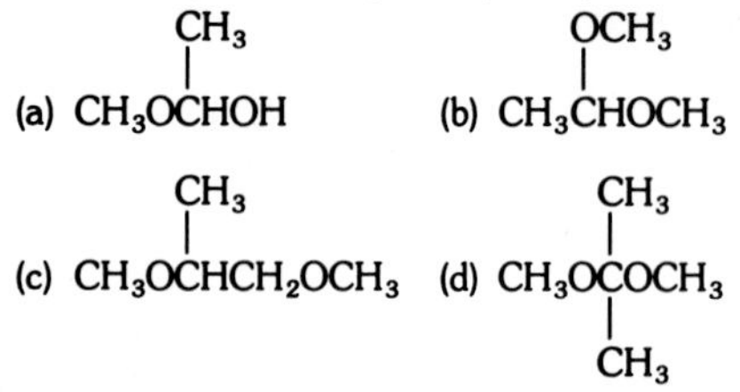

4.41 Examine each structure and decide whether it represents a hemiacetal, hemiketal, acetal, ketal, or something else.

(a) $HOCH_2CH_2CH(OCH_2CH_3)OCH_3$ (b) six-membered ring containing O, with —OH on the ring carbon next to O

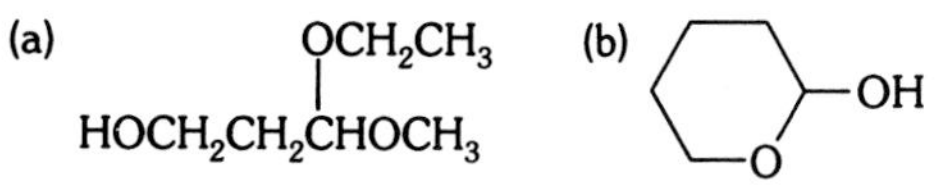

(c) $CH_3CH_2OCH_2OH$

(d) $HOCH_2CH_2OCH_2CH_2CH_3$

4.42 Write the structures of the hemiacetals and the acetals that can form between propanal and the following two alcohols.
(a) methanol (b) ethanol

4.43 What are the structures of the hemiketals and the ketals that can form between acetone and these two alcohols?
(a) methanol (b) ethanol

***4.44** Write the structure of the hydroxyaldehyde (a compound having both the alcohol group and the aldehyde group in the same molecule) from which the following hemiacetal forms in a ring-closing reaction. (You may leave the chain of the open-chain compound somewhat coiled.)

Ring: CH_3—CH—O—CH—OH, H_2C, CH_2—CH_2

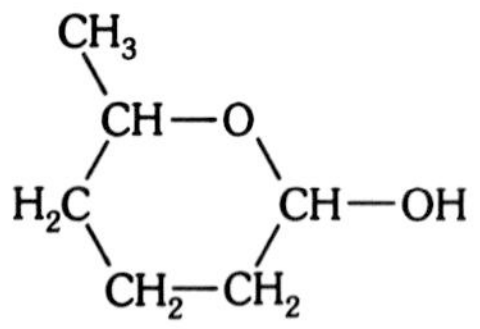

***4.45** One form in which a glucose molecule exists is given by the following structure. (*Note:* The atoms and groups that are attached to the carbon atoms of the six-membered ring must be seen as projecting *above* or *below* the ring.)

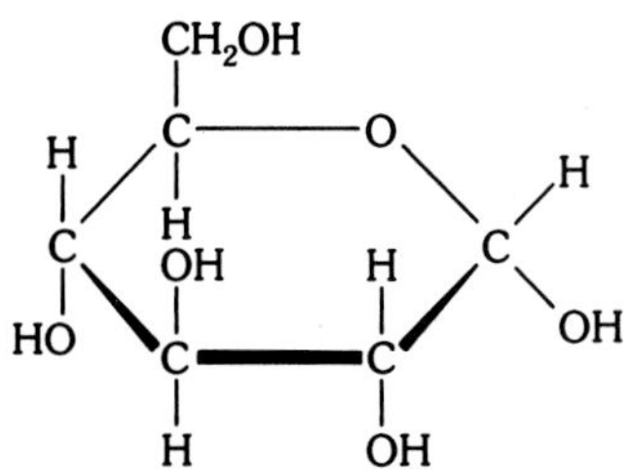

Ring: CH_2OH, C, O, C(H)(OH), C(H)(OH), C(OH)(H), C(H)(OH)

(a) Draw an arrow that points to the hemiacetal carbon.
(b) Write the structure of the open-chain form that has a free aldehyde group. (You may leave the chain coiled.)

***4.46** Write the structure of a hydroxy ketone (a molecule that has both the OH group and the keto group) from which the following hemiketal forms in a ring-closing reaction. (You may leave the chain of the open-chain compound somewhat coiled.)

Ring: H, O, CH_3, H H, H OH, H H

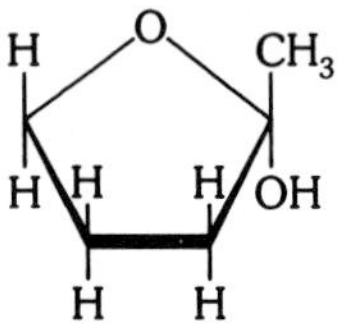

***4.47** Fructose occurs together with glucose in honey, and it is sweeter to the taste than table sugar. One form in which a fructose molecule can exist is given by the following structure.

Ring: $HOCH_2$, O, CH_2OH, H H, HO OH, HO H

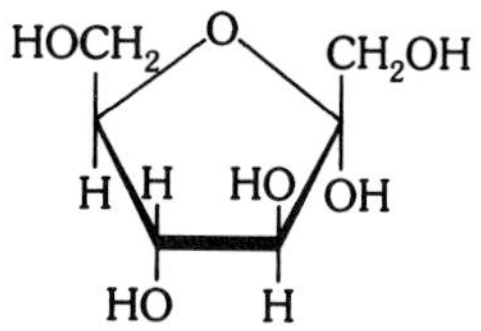

(a) Draw an arrow to the carbon of the hemiketal system that came initially from the carbon atom of a keto group.
(b) In water, fructose exists in an equilibrium with an open-chain form of the given structure. This form has a keto group in the same molecule as five OH groups. Draw the structure of this open-chain form (leaving the chain coiled somewhat as it was in the structure that was given).

4.48 The digestion of some carbohydrates is simply their hydrolysis catalyzed by enzymes. Acids catalyze the same kind of hydrolysis of acetals and ketals. Write the structures of the products, if any, that form by the action of water and an acid catalyst on the following compounds.

(a) $CH_3OCH(CH_2CH_3)OCH_3$ (b) $CH_3OCH_2CH(CH_3)OCH_3$

(c) $CH_3CH_2OC(CH_3)_2OCH_2CH_3$ (d) cyclopentane with two OCH_3 groups on one carbon

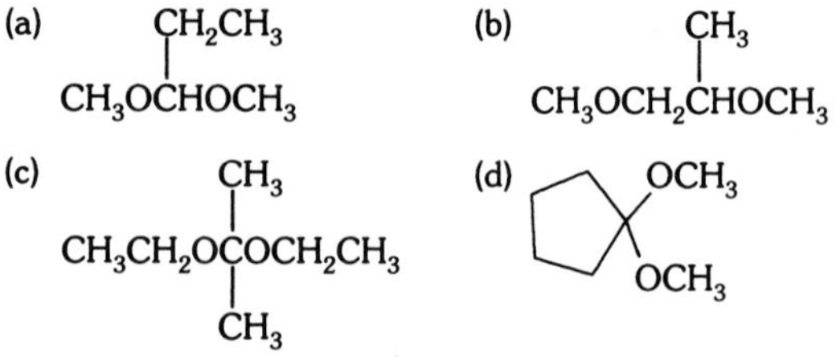

4.49 What are the structures of the products, if any, that form by the acid-catalyzed reaction of water with the following compounds?

(a) $CH_3CH_2CH(OCH(CH_3)_2)OCH_2CH_3$

(b) $CH_3OCH_2CH(CH(CH_3)_2)OCH_2CH_3$

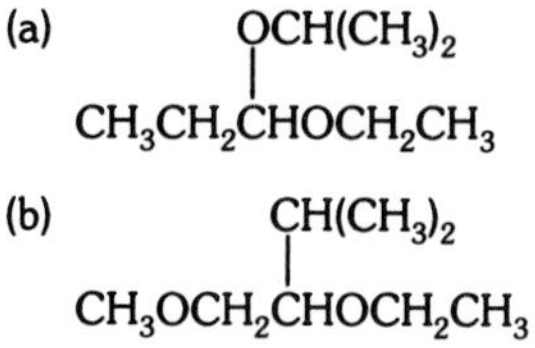

(c) 2,2-dimethyl-1,3-dioxane ring with two CH_3 groups

(d)
$$CH_3OCH_2\underset{|}{\overset{OCH_2CH_3}{C}}HOCH_3$$

Important Aldehydes and Ketones (Special Topic 4.1)

4.50 Give the name of a specific aldehyde or ketone described in Special Topic 4.1 that is
(a) a female sex hormone
(b) a good nail polish remover
(c) a preservative

The Bonds in the Carbonyl Group (Special Topic 4.2)

4.51 What kinds of atomic orbitals (pure or hybridized) overlap to form the following bonds in formaldehyde?
(a) the C—H bonds
(b) the sigma bond in the carbonyl group
(c) the pi bond in the carbonyl group

4.52 What are the bond angles in formaldehyde? Do all its atoms lie in the same plane?

Additional Exercises

*__4.53__ Complete the following reaction sequences by writing the structures of the organic products that form. If no reaction occurs, write "no reaction." (Reviewed here too are some reactions of earlier chapters.)

(a) $CH_3\overset{CH_3}{C}HCHO + H_2 \xrightarrow[\text{heat, pressure}]{\text{Ni catalyst}}$

(b) $(CH_3)_2CH\overset{OH}{C}HCH_3 \xrightarrow[\text{(e.g., } Cr_2O_7^{2-}, H^+)]{(O)}$

(c) cyclopentane bearing OH and CH_2CH_3 on the same carbon $\xrightarrow[\text{(e.g., } MnO_4^-, OH^-)]{(O)}$

(d) $CH_3CH{=}CHCH_2CH_3 + H_2 \xrightarrow{\text{Ni catalyst}}$

(e) $CH_3OH + CH_3CH_2CHO \rightleftharpoons$

(f) $CH_3CHO + Mtb:H \xrightarrow[\text{by } H^+)]{\text{(followed}}$
(where *Mtb* :H is a metabolite able to donate hydride ion)

(g) $CH_3CHO + 2CH_3CH_2OH \xrightarrow[\text{catalyst}]{\text{acid}}$

(h) $CH_3\overset{OCH_3}{C}HOCH_3 + H_2O \xrightarrow[\text{catalyst}]{\text{acid}}$

(i) cyclopentyl–CHO $\xrightarrow[\text{(e.g., } Cr_2O_7^{2-}, H^+)]{(O)}$

(j) cyclohexyl–CH_3 $\xrightarrow[\text{(e.g., } MnO_4^-, OH^-)]{(O)}$

*__4.54__ Write the structures of the organic products that form in each of the following situations. If no reaction occurs, write "no reaction." (Some of the situations constitute a review of reactions in earlier chapters.)

(a) cyclohexanone (cyclohexane ring =O) $\xrightarrow[\text{(e.g., } MnO_4^-, OH^-)]{(O)}$

(b)
$$CH_3CH_2\overset{CH_3}{\underset{CH_3CH_2}{C}}OH \xrightarrow[\text{(e.g., } MnO_4^-, OH^-)]{(O)}$$

(c) $CH_3CH_2CH_2OH + CH_3CHO \rightleftharpoons$

(d) tetrahydropyran ring (O in ring) with $-OCH_3$ on the carbon adjacent to O $+ H_2O \longrightarrow$

(e)
$$CH_3\overset{OH}{\underset{CH_3}{C}}OCH_2CH_3 \rightleftharpoons$$

(f) $CH_3OCH_2CH_2CHO + Mtb:H \xrightarrow[\text{by } H^+)]{\text{(followed}}$
(where *Mtb* :H is a metabolite able to donate hydride ion)

(g)
$$CH_3CH_2\overset{OCH_2CH_3}{\underset{OCH_2CH_3}{C}}CH_3 + H_2O \xrightarrow[\text{catalyst}]{\text{acid}}$$

(h) $CH_3\overset{OH}{C}H$–C_6H_4 (benzene ring) $\xrightarrow[\text{(e.g., } Cr_2O_7^{2-}, H^+)]{(O)}$

(i) cyclopentene ring with OCH_3 $+ H_2 \xrightarrow{\text{Ni catalyst}}$

(j) C_6H_5–$CHO + 2CH_3OH \xrightarrow[\text{catalyst}]{\text{acid}}$

*__4.55__ Catalytic hydrogenation of compound **A** (C_3H_6O) gave **B** (C_3H_8O). When **B** was heated strongly in the presence of sulfuric acid, it changed to compound **C** (C_3H_6). The acid-catalyzed addition of water to **C** gave compound **D**

(C_3H_8O); and when **D** was oxidized, it changed to **E** (C_3H_6O). Compounds **A** and **E** are isomers, and compounds **B** and **D** are isomers. Write the structures of compounds **A** through **E**.

***4.56** When compound **F** ($C_4H_{10}O$) was gently oxidized, it changed to compound **G** (C_4H_8O), but vigorous oxidation changed **F** (or **G**) to compound **H** ($C_4H_8O_2$). Action of hot sulfuric acid on **F** changed it to compound **I** (C_4H_8). The addition of water to **I** (in the presence of an acid catalyst) gave compound **J** ($C_4H_{10}O$), a compound that could not be oxidized. Compounds **F** and **J** are isomers. Write the structures of compounds **F** through **J**.

***4.57** A student was assigned the preparation of the dimethyl acetal of butanal for which the equation is

A solution of 12.5 g of butanal in 50.0 mL of methanol was prepared for accomplishing this reaction. The density of methanol is 0.787 g/mL.

(a) How many moles of butanal were taken?
(b) How many moles of methanol were used?
(c) Was sufficient methanol taken? (Calculate the minimum number of grams of methanol that would be required.)
(d) How many grams of water would be obtained?
(e) Offer a reason for using an excess quantity of methanol.

***4.58** In review exercise 2.28 (page 71), polyvinyl acetate was described as a polymer from which Butvar® is made and that Butvar® is used to make safety glass. You may wish to refer to your answer to 2.28 for the parts of this Review Exercise.

(a) Prepare the structure of a segment of the polyvinyl acetate molecule consisting of *two* repeating units.
(b) Polyvinyl acetate can be converted into the corresponding *polyvinyl alcohol* by the replacement of all of the CH_3CO groups attached to the oxygens of polyvinyl acetate's chain by hydrogen atoms. This leaves a very long alkane chain with OH groups on every other carbon atom. Write the structure of a segment of polyvinyl alcohol consisting of *two* repeating units.
(c) Write the structure of the *monomer* of polyvinyl alcohol. This monomer does not exist (explaining why its polymer must be made indirectly). Is the monomer properly called an *alcohol* by our definition? What kind of an "alcohol" is it? Is this kind of "alcohol" stable? Does polyvinyl alcohol have OH groups that are properly called alcohol groups? Explain.
(d) Butvar® is made by the combination of butanal with polyvinyl alcohol. *Cyclic* acetal systems form in which every other carbon of the main alkane chain is part of the ring system of these cyclic acetals. The generic name of this new polymer is polyvinyl butyral. Write the structure of a segment of this polymer that includes one cyclic acetal system.

CARBOXYLIC ACIDS AND ESTERS

5

Strong, light-weight fabrics made of synthetic polymers have transformed virtually all recreational activities, such as extended backpacking and tenting ventures into remote wilderness areas. Dacron, described in this chapter, is one of many polyesters.

5.1

OCCURRENCE, NAMES, AND PHYSICAL PROPERTIES OF CARBOXYLIC ACIDS

The carboxylic acids are polar compounds whose molecules form strong hydrogen bonds to each other.

The two principal types of organic acids are the carboxylic acids and the sulfonic acids. Sulfonic acids are much less common than carboxylic acids, and we will not study them.

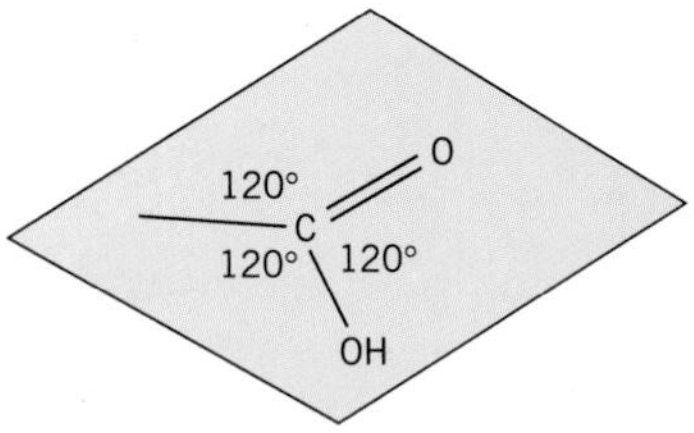

120° 120° 120° C O OH

The carboxyl group has a planar geometry at the carbonyl group.

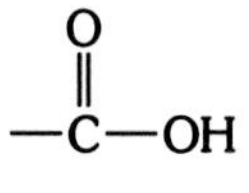

$$-\overset{\displaystyle O}{\overset{\|}{C}}-OH \quad \text{or} \quad -CO_2H \quad \text{or} \quad -COOH$$

Carboxyl group

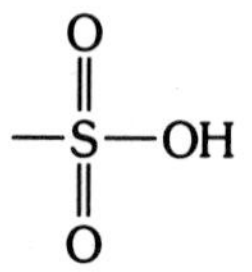

$$-\overset{\displaystyle O}{\overset{\|}{\underset{\underset{\displaystyle O}{\|}}{S}}}-OH \quad \text{or} \quad -SO_3H$$

Sulfonic acid group

■ "Carboxyl" comes from *carb*onyl + hydrox*yl*.

In **carboxylic acids,** the carbonyl carbon holds a hydroxyl group and either another carbon atom or a hydrogen atom. A number of the specific examples are given in Table 5.1, and information in the table shows how widely the carboxyl group occurs in nature. The acids with long, straight, alkane-like chains are often called the **fatty acids,** because they are products of the hydrolysis of butterfat, olive oil, and similar substances in the diet.

Formic acid, the simplest acid, has a sharp, irritating odor and is responsible for the sting of nettle plants and certain ants. The next acid in Table 5.1, acetic acid, gives tartness to vinegar, where its concentration is 4% to 5%. Butyric acid causes the odor of rancid butter. Valeric acid gets its name from the Latin *valerum,* meaning to be strong. What is strong about valeric acid is its odor. Other acids with vile odors are caproic, caprylic, and capric acids, which get their names from the Latin *caper,* meaning "goat," a reference to their odor.

Some acids are *dicarboxylic acids* with two carboxyl groups per molecule. Oxalic acid, the simplest example, gives the sour taste to rhubarb. Citric acid, a *tricarboxylic acid,* causes the tartness of citrus fruits. Lactic acid, which has both a carboxyl group and a 2° alcohol group, gives the tart taste to sour milk.

■ The lactate ion is produced in muscles engaged in strenuous exercise.

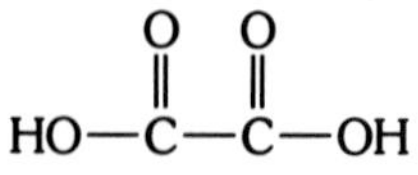

$$HO-\overset{\displaystyle O}{\overset{\|}{C}}-\overset{\displaystyle O}{\overset{\|}{C}}-OH$$

Oxalic acid
(in rhubarb)

$$\begin{array}{c} CH_2CO_2H \\ | \\ HOCCO_2H \\ | \\ CH_2CO_2H \end{array}$$

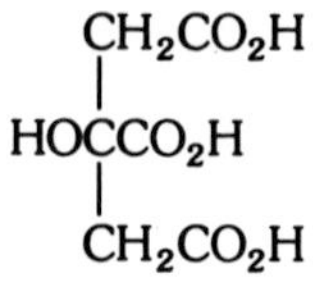

Citric acid
(in lemon juice)

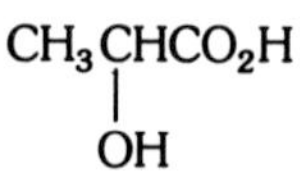

$$\begin{array}{c} CH_3CHCO_2H \\ | \\ OH \end{array}$$

Lactic acid
(in sour milk)

All carboxylic acids are weak Brønsted acids, and they exist in the form of their anions both in basic solutions and in their salts. It is largely as their anions that they occur in living cells and body fluids.

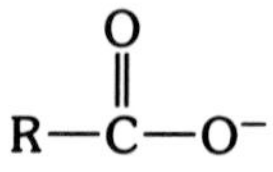

$$R-\overset{\displaystyle O}{\overset{\|}{C}}-O^- \quad \text{or} \quad RCO_2^- \quad \text{or} \quad RCOO^-$$

Symbols of anions of carboxylic acids

TABLE 5.1
Carboxylic Acids

n[a]	Structure	Name[b]	Origin of Name	MP (°C)	BP (°C)	Solubility in Water[c]	K_a (25 °C)
1	HCO_2H	Formic acid (methanoic acid)	L. *formica*, ant	8	101	∞	1.8×10^{-4} (20°C)
2	CH_3CO_2H	Acetic acid (ethanoic acid)	L. *acetum*, vinegar	17	118	∞	1.8×10^{-5}
3	$CH_3CH_2CO_2H$	Propionic acid (propanoic acid)	L. *proto, pion*, first, fat	−21	141	∞	1.3×10^{-5}
4	$CH_3(CH_2)_2CO_2H$	Butyric acid (butanoic acid)	L. *butyrum*, butter	−6	164	∞	1.5×10^{-5}
5	$CH_3(CH_2)_3CO_2H$	Valeric acid (pentanoic acid)	L. *valere*, valerian root	−35	186	4.97	1.5×10^{-5}
6	$CH_3(CH_2)_4CO_2H$	Caproic acid (hexanoic acid)	L. *caper*, goat	−3	205	1.08	1.3×10^{-5}
7	$CH_3(CH_2)_5CO_2H$	Enanthic acid (heptanoic acid)	Gr. *oenanthe*, vine blossom	−9	223	0.26	1.3×10^{-5}
8	$CH_3(CH_2)_6CO_2H$	Caprylic acid (octanoic acid)	L. *caper*, goat	16	238	0.07	1.3×10^{-5}
9	$CH_3(CH_2)_7CO_2H$	Pelargonic acid (nonanoic acid)	Pelargonium, geranium	15	254	0.03	1.1×10^{-5}
10	$CH_3(CH_2)_8CO_2H$	Capric acid (decanoic acid)	L. *caper*, goat	32	270	0.015	1.4×10^{-5}
12	$CH_3(CH_2)_{10}CO_2H$	Lauric acid (dodecanoic acid)	Laurel	44	—	0.006	—
14	$CH_3(CH_2)_{12}CO_2H$	Myristic acid (tetradecanoic acid)	*Myristica*, nutmeg	54	—	0.002	—
16	$CH_3(CH_2)_{14}CO_2H$	Palmitic acid (hexadecanoic acid)	Palm oil	63	—	0.0007	—
18	$CH_3(CH_2)_{16}CO_2H$	Stearic acid (octadecanoic acid)	Gr. *stear*, solid	70	—	0.0003	—
Miscellaneous Carboxylic Acids							
	$C_6H_5CO_2H$	Benzoic acid	Gum benzoin	122	249	0.34 (25 °C)	6.5×10^{-5}
	$C_6H_5CH{=}CHCO_2H$	Cinnamic acid (trans isomer)	Cinnamon	132	—	0.04	3.7×10^{-5}
	$CH_2{=}CHCO_2H$	Acrylic acid	L. *acer*, sharp	13	141	Soluble	5.6×10^{-5}
	OH, —CO_2H (benzene ring)	Salicylic acid	L. *salix*, willow	159	211	0.22 (25 °C)	1.1×10^{-3} (19 °C)

[a] n = total number of carbon atoms per molecule

[b] In parentheses below each common name is the IUPAC name.

[c] In grams of acid per 100 g H_2O at 20 °C except where noted otherwise.

IUPAC Names of Carboxylic Acids end in *-oic Acid* In the IUPAC rules for naming carboxylic acids, the *parent acid* is that of the longest chain that includes the carboxyl group. To name the parent acid, change the ending of the name of the alkane having the same number of carbons (the parent alkane) from *-e* to *-oic acid.* To number the chain of the parent acid, *always* begin with the carbon atom of the carbonyl group, giving it position 1. The names in parentheses in Table 5.1 are IUPAC names.

■ Use the name of the *acid,* not the name of the alkane, to devise the name of the anion of the acid.

To name anions of carboxylic acids, the *carboxylate ions,* change the ending of the name of the parent *acid* from *-ic* to *-ate,* and omit the word *acid.* This rule applies both to the IUPAC names and to the common names. For example,

■ Pronounce "oate" as "oh-ate."

$H-\overset{\overset{O}{\|}}{C}-OH$	$H-\overset{\overset{O}{\|}}{C}-O^-$	$H-\overset{\overset{O}{\|}}{C}-ONa$
Methanoic acid (formic acid)	Methanoate ion (formate ion)	Sodium methanoate (sodium formate)

EXAMPLE 5.1
Naming a Carboxylic Acid and Its Anion

The following carboxylic acid has the common name of isovaleric acid. What is the IUPAC name of this acid and its sodium salt?

$$CH_3\underset{}{\overset{\overset{CH_3}{|}}{C}}HCH_2\overset{\overset{O}{\|}}{C}OH$$

ANALYSIS The longest chain that includes the carboxyl group has four carbon atoms, so the parent acid is named by changing the name *butane,* the parent alkane, to *butanoic acid.* Next, we have to number the chain, starting with the carboxyl group's carbon.

$$\underset{4}{CH_3}\underset{3}{\overset{\overset{CH_3}{|}}{C}H}\underset{2}{CH_2}\underset{1}{\overset{\overset{O}{\|}}{C}}OH$$

SOLUTION The methyl group is at position 3, so the IUPAC name of this acid is 3-methylbutanoic acid.

To name its anion, we drop *-ic acid* from the name of the acid and add *-ate.* Therefore the name of the anion is 3-methylbutanoate, and the name of the sodium salt of this acid is sodium 3-methylbutanoate. (The common name is sodium isovalerate.)

PRACTICE EXERCISE 1

What are the IUPAC names of the following compounds?

(a) $CH_3\underset{\underset{CH_3}{|}}{\overset{\overset{CH_3}{|}}{C}}CO_2H$

(b) $CH_3CH_2CH_2\underset{\underset{CH_3CHCH_3}{|}}{\overset{\overset{CH_3CH_2}{|}}{C}}CH_2\overset{\overset{CH_3}{|}}{C}HCH_2CO_2H$

(c) CH_3CO_2Na

(d) $CH_3CH_2\underset{\underset{Cl}{|}}{C}HCH_2\overset{\overset{CH_3}{|}}{C}HCH_2CO_2H$

PRACTICE EXERCISE 2

If the IUPAC name of $HO_2CCH_2CO_2H$ is propanedioic acid (and not 1,3-propanedioic acid), what must be the IUPAC name for $HO_2CCH_2CH_2CH_2CO_2H$?

PRACTICE EXERCISE 3

If the IUPAC name of $CH_3CH{=}CHCH_2CH_2CO_2H$ is 4-hexenoic acid, what must be the IUPAC name of the following acid? Its common name is *oleic acid,* and it is one of the products of the hydrolysis of almost any edible vegetable oil or animal fat. (The name of the straight-chain alkane with 18 carbon atoms is octadecane.)

$$CH_3(CH_2)_7CH{=}CH(CH_2)_7CO_2H$$

Oleic acid (common name)

Oleic acid is actually the *cis* isomer. The name of the *trans* isomer is elaidic acid.

Carboxylic Acid Molecules Hydrogen-Bond to Each Other Carboxylic acids have higher boiling points than alcohols of comparable formula masses, because molecules of carboxylic acids form hydrogen-bonded pairs:

Hydrogen bonds (•••) hold two molecules of a carboxylic acid together.

This makes the effective formula mass of a carboxylic acid much higher than its calculated formula mass, and therefore the boiling point is higher.

The lower-formula-mass carboxylic acids (C_1-C_4) are soluble in water largely because the carboxyl group has *two* oxygen atoms that can accept hydrogen bonds from water molecules. In addition, the carboxyl group has the OH group that can donate hydrogen bonds.

Remember, we're interested in how structure affects solubility in water because water is the fluid medium in the body.

5.2
THE ACIDITY OF CARBOXYLIC ACIDS

The carboxylic acids are weak Brønsted acids toward water but strong Brønsted acids toward the hydroxide ion.

Aqueous solutions of carboxylic acids contain the following species in equilibrium:

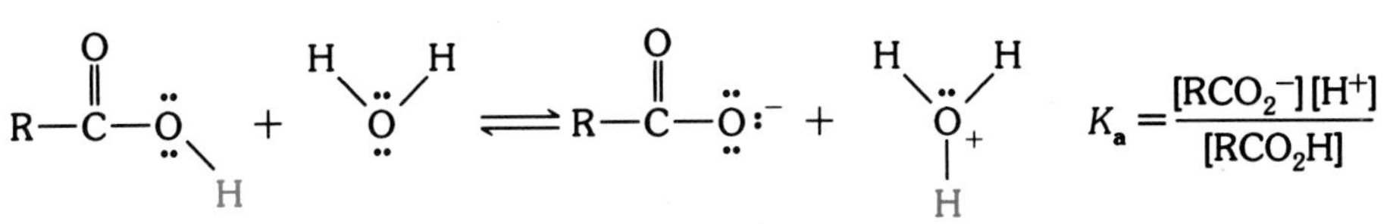

$$K_a = \frac{[RCO_2^-][H^+]}{[RCO_2H]}$$

The K_a values of several carboxylic acids are given in Table 16.1, and you can see that most are on the order of 10^{-5}. Thus the carboxylic acids are weak acids toward water, and their percentage ionizations are low. For example, in a 1 *M* solution at room temperature, acetic acid is ionized only to about 0.5%.

Carboxylic Acids Are Stronger Acids than Phenols or Alcohols Carboxylic acids ($K_a \approx 10^{-5}$) are several billion times stronger acids than alcohols ($K_a \approx 10^{-16}$), and roughly 100,000 times stronger acids than phenols ($K_a \approx 10^{-10}$). The greater acidity of carboxylic acids reflects the greater relative stability of carboxylate anions compared to the anions of alcohols or phenols. In the carboxylate ion, the negative charge is adjacent to the strongly electronegative carbonyl group. This helps to stabilizes the anion of the acid by helping to spread the negative charge over a wider area than is possible in the anions of alcohols or phenols. A more stable ion means one easier to form.

■ Phenols are strong enough acids to neutralize OH^-, a strong base, but are not strong enough acids to neutralize HCO_3^-, a weak base.

Carboxylic Acids Are Neutralized by Strong Bases The hydroxide ion, the carbonate ion, and the bicarbonate ion are bases strong enough to neutralize carboxylic acids. This reaction is important at the molecular level of life, because the carboxylic acids we normally produce by metabolism must be neutralized. Otherwise, the pH of body fluids, such as the blood, would fall too low to sustain life.

With hydroxide ion, the reaction is as follows.

$$\underset{\text{Stronger acid}}{RCO_2H} + \underset{\text{Stronger base}}{OH^-} \longrightarrow \underset{\text{Weaker base}}{RCO_2^-} + \underset{\text{Weaker acid}}{H{-}OH}$$

■ The negative charges on the anions are balanced by the presence of some cation, the Na^+ or K^+ ion, for example. However, we'll work largely with net ionic equations.

With bicarbonate ion, the chief base in the buffer systems of the blood, the following reaction occurs.

$$\underset{\text{Stronger acid}}{RCO_2H} + \underset{\text{Stronger base}}{HCO_3^-} \longrightarrow \underset{\text{Weaker base}}{RCO_2^-} + \underset{\text{Weaker acid}}{H_2O} + CO_2$$

Some specific examples are as follows.

$$\underset{\text{Acetic acid}}{CH_3CO_2H} + OH^- \longrightarrow \underset{\text{Acetate ion}}{CH_3CO_2^-} + H_2O$$

$$\underset{\text{Stearic acid (insoluble in water)}}{CH_3(CH_2)_{16}CO_2H} + OH^- \longrightarrow \underset{\text{Stearate ion (soluble in water)}}{CH_3(CH_2)_{16}CO_2^-} + H_2O$$

$$\underset{\text{Benzoic acid (insoluble in water)}}{C_6H_5CO_2H} + HCO_3^- \longrightarrow \underset{\text{Benzoate ion (soluble in water)}}{C_6H_5CO_2^-} + H_2O + CO_2$$

■ The stearate ion is one of several organic ions in soap. The ion is so large that it might be better to say that it is "dispersed" in water rather than "soluble."

PRACTICE EXERCISE 4

Write the structures of the carboxylate anions that form when the following carboxylic acids are neutralized.

(a) $CH_3CH_2CO_2H$ **(b)** $CH_3O{-}C_6H_4{-}CO_2H$ (para-substituted benzene ring) **(c)** $CH_3CH{=}CHCO_2H$

Carboxylate Ions Are More Soluble in Water than Their Parent Acids The purified salts that combine carboxylate ions and metal ions are genuine salts, assemblies of oppositely charged ions, so all are solids at room temperature. Table 5.2 gives a few examples of sodium salts. Like all sodium salts, they are soluble in water but completely insoluble in such nonpolar solvents as ether or gasoline. Several are used as decay inhibitors in foods, as described in Special Topic 5.1.

SPECIAL TOPIC 5.1
SOME IMPORTANT CARBOXYLIC ACIDS AND SALTS

Acetic Acid Most people experience acetic acid directly in the form of its dilute solution in water, which is called vinegar. Because blood is slightly alkaline, acetic acid circulates as the acetate ion, and we will meet this species many times when we study metabolic pathways. The acetate ion, in fact, is one of the major intermediates in the metabolism of carbohydrates, lipids, and proteins.

Acetic acid is also an important industrial chemical, and more than 3 billion pounds (23 billion moles) are manufactured each year in the United States. Acetate rayon is just one consumer product made using it (see Figure 1).

Propanoic Acid and Its Salts Propanoic acid occurs naturally in Swiss cheese in a concentration that can be as high as 1%. Its sodium and calcium salts are food additives used in baked goods and processed cheese to retard the formation of molds or the growth of bacteria. (On ingredient labels, these salts are listed under their common names, sodium or calcium propionate.)

FIGURE 1
Acetate rayon is a lustrous fabric made from cellulose and acetic acid.

Sorbic Acid and the Sorbates Sorbic acid, or 2,4-hexanedienoic acid, $CH_3CH{=}CHCH{=}CHCO_2H$, and its sodium or potassium salts are added in trace concentrations to a variety of foods to inhibit the growth of molds and yeasts. The sorbates often appear on ingredient lists for fruit juices, fresh fruits, wines, soft drinks, sauerkraut and other pickled products, and some meat and fish products. For food products that usually are wrapped, such as cheese and dried fruits, solutions of sorbate salts are sometimes sprayed onto the wrappers.

Sodium Benzoate Traces of sodium benzoate inhibit molds and yeasts in products that normally have pH values below 4.5 or 4.0. (The sorbates work better at slightly higher pH values—up to 6.5.) You'll see sodium benzoate on ingredient lists for beverages, syrups, jams and jellies, pickles, salted margarine, fruit salads, and pie fillings. Its concentration is low—0.05 to 0.10%—and neither benzoic acid nor the benzoate ion accumulates in the body.

TABLE 5.2
Some Sodium Salts of Carboxylic Acids

Common Name[a]	Structure	MP (°C)	Solubility[a] Water	Ether
Sodium formate (sodium methanoate)	HCO_2Na	253	Soluble	Insoluble
Sodium acetate (sodium ethanoate)	CH_3CO_2Na	323	Soluble	Insoluble
Sodium propionate (sodium propanoate)	$CH_3CH_2CO_2Na$	—	Soluble	Insoluble
Sodium benzoate	$C_6H_5CO_2Na$	—	66 g/100 mL	Insoluble
Sodium salicylate	o-$HOC_6H_4CO_2Na$ (benzene ring with OH and CO_2Na)	—	111 g/100 mL	Insoluble

[a] At 20 °C.

Carboxylate Ions Are Good Proton Acceptors or Bases Because carboxylate ions are the anions of *weak* acids, they themselves must be relatively good bases, especially toward the hydronium ion, a strong proton donor. At room temperature, the following neutralization of a strong acid by a carboxylate ion occurs virtually instantaneously. It is the most important reaction of the carboxylate ion that we will study, because it makes the carboxylate ion group a neutralizer of excess acid at the molecular level of life.
For example,

$$R{-}\overset{\overset{\large O}{\|}}{C}{-}\ddot{\underset{\cdot\cdot}{O}}{:}^{-} + H_3\ddot{O}^{+} \longrightarrow R{-}\overset{\overset{\large O}{\|}}{C}{-}\ddot{\underset{\cdot\cdot}{O}}{-}H + H{-}\ddot{\underset{\cdot\cdot}{O}}{-}H$$

Carboxyate ion (stronger base) | Hydronium ion (stronger acid) | Carboxylic acid (weaker acid) | Water (weaker base)

$$C_6H_5CO_2^- + H_3O^+ \longrightarrow C_6H_5CO_2H + H_2O$$

Benzoate ion (soluble in water) → Benzoic acid (insoluble in water)

$$CH_3(CH_2)_{16}CO_2^- + H_3O^+ \longrightarrow CH_3(CH_2)_{16}CO_2H + H_2O$$

Stearate ion (soluble in water) → Stearic acid (insoluble in water)

The Carboxylic Acid Group Is a Solubility "Switch" In the reactions just studied, we noted the solubilities in water of several species to draw attention to a very important property given to a molecule by the carboxylic acid group. This group can be used to switch on or off the solubility in water of any substance that contains it. When we try to increase the pH of a solution—by adding a strong base—a water-insoluble carboxylic acid almost instantly dissolves, because it changes to its carboxylate ion. Similarly, when we try to decrease the pH of a solution—by adding a strong acid—a water-soluble carboxylate anion instantly changes to its much less soluble, free carboxylic acid form. In other words, by suitably adjusting the pH of an aqueous solution, we can make a substance with a carboxyl group more soluble or less soluble in water.

PRACTICE EXERCISE 5

Write the structures of the organic products of the reactions of the following compounds with dilute hydrochloric acid at room temperature.

(a) $CH_3O{-}C_6H_4{-}CO_2^-K^+$ (para-substituted benzene ring) **(b)** $CH_3CH_2CO_2^-Li^+$

(c) $(CH_3CH{=}CHCO_2^-)_2Ca^{2+}$

Carboxylic Acids Strongly Resist Oxidation Simple carboxylic acids (those with no other functional groups) or their anions are the stable end products of the oxidations of 1° alcohols and aldehydes, as we have learned. Not even hot solutions of permanganate or dichromate ion break down the carboxyl group. (The carboxylic acids will burn, of course, to give carbon dioxide and water.) The same end products, CO_2 and H_2O, form when carboxylic acids are completely metabolized in the body, but an elaborate, multistep, oxidative process is required.

5.3

THE CONVERSION OF CARBOXYLIC ACIDS TO ESTERS

Carboxylic acids can be used directly or indirectly to make esters from alcohols. The carboxylic acids are the parent compounds for several families that collectively are called **acid derivatives.** These include the **acid chlorides,** the **anhydrides** (both common and mixed anhydrides), the **esters,** and the **amides.** They are called acid *derivatives* because they can be made from the acids and they can be hydrolyzed back to the acids.

■ We'll learn how to name esters soon, but we will not develop the rules for naming acid chlorides or acid anhydrides.

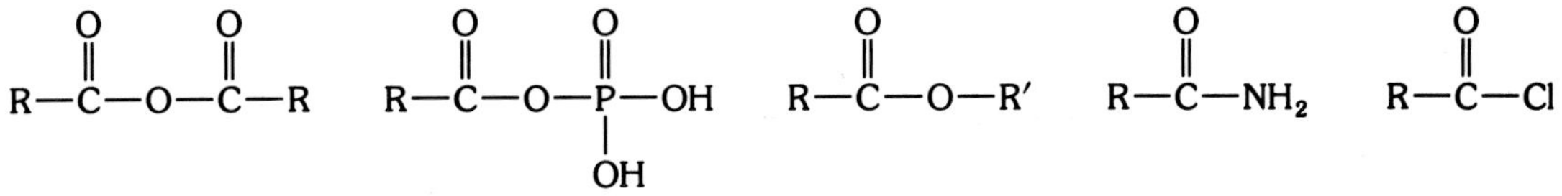

$$\underset{\text{Acid anhydrides}}{R{-}\overset{O}{\overset{\|}{C}}{-}O{-}\overset{O}{\overset{\|}{C}}{-}R} \qquad \underset{\text{Mixed anhydrides with phosphoric acid}}{R{-}\overset{O}{\overset{\|}{C}}{-}O{-}\underset{OH}{\underset{|}{\overset{O}{\overset{\|}{P}}}}{-}OH} \qquad \underset{\text{Esters}}{R{-}\overset{O}{\overset{\|}{C}}{-}O{-}R'} \qquad \underset{\text{Amides}}{R{-}\overset{O}{\overset{\|}{C}}{-}NH_2} \qquad \underset{\text{Acid chlorides}}{R{-}\overset{O}{\overset{\|}{C}}{-}Cl}$$

The molecules of all of these compounds possess the **acyl group,** which is simply a carboxylic acid minus its OH group.

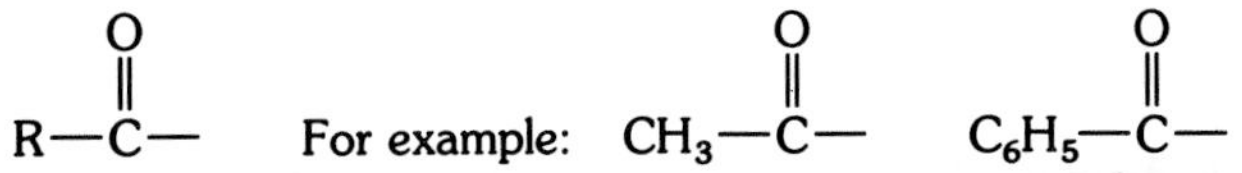

$$\underset{\text{Acyl group}}{R{-}\overset{O}{\overset{\|}{C}}{-}} \qquad \text{For example:} \quad \underset{\text{Acetyl group}}{CH_3{-}\overset{O}{\overset{\|}{C}}{-}} \qquad \underset{\text{Benzoyl group}}{C_6H_5{-}\overset{O}{\overset{\|}{C}}{-}}$$

The reactions by which the acid derivatives are made as well as the reactions of the derivatives themselves generally occur as **acyl group transfer reactions.** The ability to transfer an acyl group, however, varies widely among the acid derivatives, as we will see.

Acid Chlorides Are the Most Reactive of Acid Derivatives Acid chlorides react readily and exothermically with water to give the parent acids and hydrogen chloride (which, because of the excess water, forms as hydrochloric acid). In this reaction an acyl group transfers from Cl to OH. We call the chloride ion a *leaving group*.

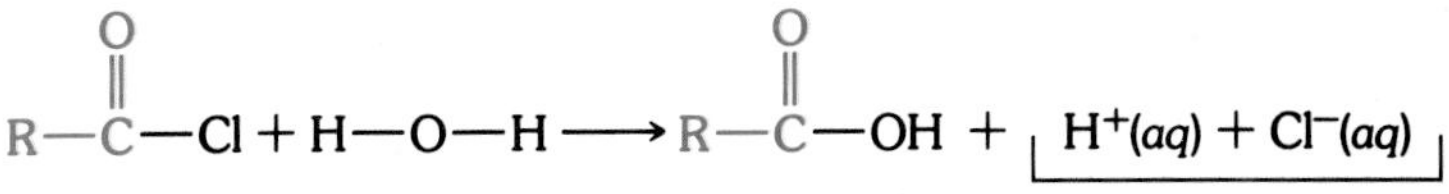

$$\underset{\text{Acid chloride}}{R{-}\overset{O}{\overset{\|}{C}}{-}Cl} + H{-}O{-}H \longrightarrow \underset{\text{Carboxylic acid}}{R{-}\overset{O}{\overset{\|}{C}}{-}OH} + \underset{\text{Hydrochloric acid}}{H^+(aq) + Cl^-(aq)}$$

Acid chlorides also react vigorously with alcohols to give esters. In this reaction an acyl group transfers to an alcohol unit.

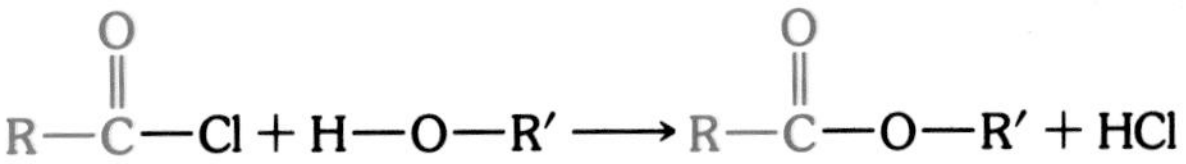

$$\underset{\text{Acid chloride}}{R{-}\overset{O}{\overset{\|}{C}}{-}Cl} + \underset{\text{Alcohol}}{H{-}O{-}R'} \longrightarrow \underset{\text{Ester}}{R{-}\overset{O}{\overset{\|}{C}}{-}O{-}R'} + HCl$$

For example,

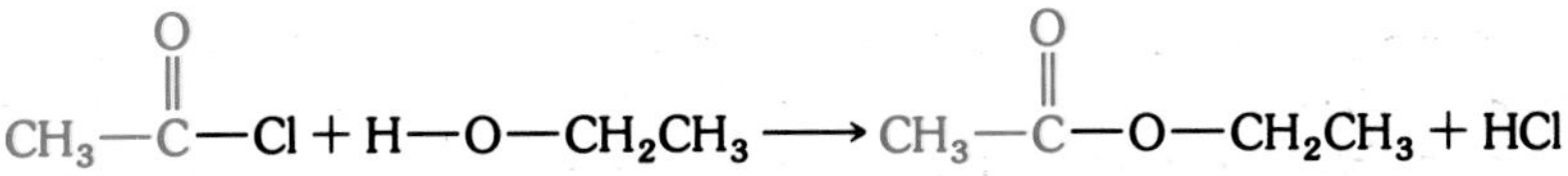

$$\underset{\text{Acetyl chloride}}{CH_3{-}\overset{O}{\overset{\|}{C}}{-}Cl} + \underset{\text{Ethyl alcohol}}{H{-}O{-}CH_2CH_3} \longrightarrow \underset{\text{Ethyl acetate}}{CH_3{-}\overset{O}{\overset{\|}{C}}{-}O{-}CH_2CH_3} + HCl$$

This is an example of an **esterification** reaction, the synthesis of an ester. We say that both the alcohol and carboxylic acid are *esterified*.

Acid Anhydrides Are Good Acyl Transfer Reactants When a carboxylic acid anhydride reacts with an alcohol, an acyl group transfers from a carboxylate group to OR. This reaction occurs with roughly the same ease as the reaction of an acid chloride with an alcohol. Acid anhydrides react with alcohols as follows:

■ Either of the two acyl groups of the acid anhydride could be transferred. We've picked one.

$$\underset{\text{Acid anhydride}}{R-\overset{O}{\overset{\|}{C}}-O-\overset{O}{\overset{\|}{C}}-R} + \underset{\text{Alcohol}}{H-O-R'} \xrightarrow[\text{heat}]{H^+} \underset{\text{Ester}}{R-\overset{O}{\overset{\|}{C}}-O-R'} + \underset{\text{Carboxylic acid}}{H-O-\overset{O}{\overset{\|}{C}}-R}$$

Phenols, like alcohols, can be also esterified by acid anhydrides. When the acid anhydride is acetic anhydride and the phenol group is in salicylic acid, one product is aspirin.

■ Both salicylic acid and acetic anhydride are common, readily available organic chemicals.

$$\underset{\text{Salicylic acid}}{o\text{-HOC}_6H_4\text{COOH}} + \underset{\text{Acetic anhydride}}{CH_3\overset{O}{\overset{\|}{C}}O\overset{O}{\overset{\|}{C}}CH_3} \xrightarrow[\text{heat}]{H^+} \underset{\text{Acetylsalicylic acid (aspirin)}}{o\text{-}CH_3\overset{O}{\overset{\|}{C}}OC_6H_4\text{COOH}} + \underset{\text{Acetic acid}}{HO\overset{O}{\overset{\|}{C}}CH_3}$$

Direct Esterification of Acids Is Another Synthesis of Esters When a solution of a carboxylic acid in an alcohol is heated in the presence of a strong acid catalyst, the following species become involved in an equilibrium.

$$\underset{\text{Carboxylic acid}}{R-\overset{O}{\overset{\|}{C}}-O-H} + \underset{\text{Alcohol}}{H-O-R'} \overset{H^+}{\rightleftharpoons} \underset{\text{Ester}}{R-\overset{O}{\overset{\|}{C}}-O-R'} + H-O-H$$

When the alcohol is in *excess*, the equilibrium shifts so much to the right (in accordance with Le Châtelier's principle) that this reaction is a good method for making an ester. This synthesis of an ester is called *direct esterification*. Some specific examples are as follows.

$$\underset{\text{Acetic acid}}{CH_3-\overset{O}{\overset{\|}{C}}-O-H} + \underset{\text{Ethyl alcohol (large excess)}}{H-O-CH_2CH_3} \xrightarrow{H^+} \underset{\text{Ethyl acetate}}{CH_3-\overset{O}{\overset{\|}{C}}-O-CH_2CH_3} + H_2O$$

$$\underset{\text{Salicylic acid}}{o\text{-HOC}_6H_4\overset{O}{\overset{\|}{C}}-O-H} + \underset{\text{Methyl alcohol (large excess)}}{H-O-CH_3} \xrightarrow[\text{heat}]{H^+} \underset{\text{Methyl salicylate (oil of wintergreen)}}{o\text{-HOC}_6H_4\overset{O}{\overset{\|}{C}}-O-CH_3} + H_2O$$

Notice that salicylic acid has *two* groups that can form an ester. Esterfication of its phenolic OH group by acetic anhydride gives aspirin; esterification of its carboxyl group by methyl alcohol gives methyl salicylate.

EXAMPLE 5.2
Writing the Structure of a Product of Direct Esterification

What is the ester that can be made from benzoic acid and methyl alcohol?

ANALYSIS We need the *structures* of the starting materials, and in an esterification it is sometimes helpful to let the OH groups "face" each other.

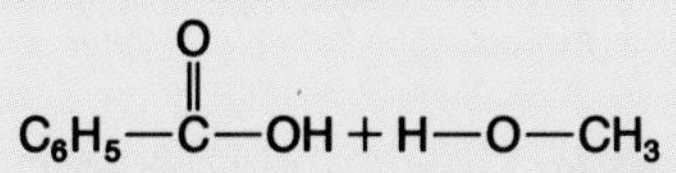

$$\underset{\text{Benzoic acid}}{C_6H_5-\overset{\overset{\displaystyle O}{\|}}{C}-OH} + \underset{\text{Methyl alcohol}}{H-O-CH_3}$$

We know that a molecule of water splits out between the acid and the alcohol during esterification. For the pieces of H_2O, we take the OH from the acid and the H from the alcohol. This operation leaves the fragments:

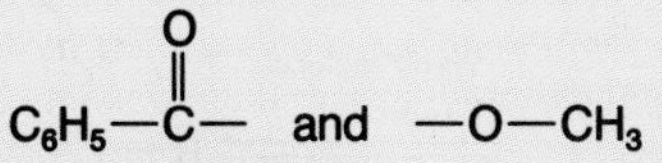

$$C_6H_5-\overset{\overset{\displaystyle O}{\|}}{C}- \quad \text{and} \quad -O-CH_3$$

All that remains is to join these fragments.

SOLUTION The ester that forms is methyl benzoate.

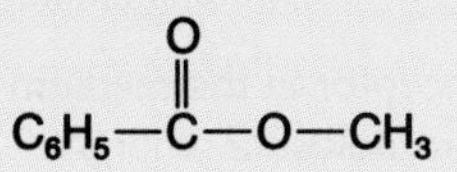

$$C_6H_5-\overset{\overset{\displaystyle O}{\|}}{C}-O-CH_3$$

■ Although it is not important in predicting correct structures of products, always erase the OH group from the carboxylic acid, not the alcohol. This will make it easier to learn a reaction coming up in the next chapter.

PRACTICE EXERCISE 6

Write the structures of the esters that form by the direct esterification of acetic acid by the following alcohols.
(a) methyl alcohol **(b)** propyl alcohol **(c)** isopropyl alcohol

PRACTICE EXERCISE 7

Write the structures of the esters that can be made by the direct esterification of ethyl alcohol by the following acids.
(a) formic acid **(b)** propionic acid **(c)** benzoic acid

The Acid Catalyst Helps Direct Esterification Without an acid catalyst, direct esterification would proceed very slowly. The acid catalyst, by donating H^+ to the O atom of the carbonyl group of the carboxylic acid, makes the C atom of the carbonyl group much more positive in charge.

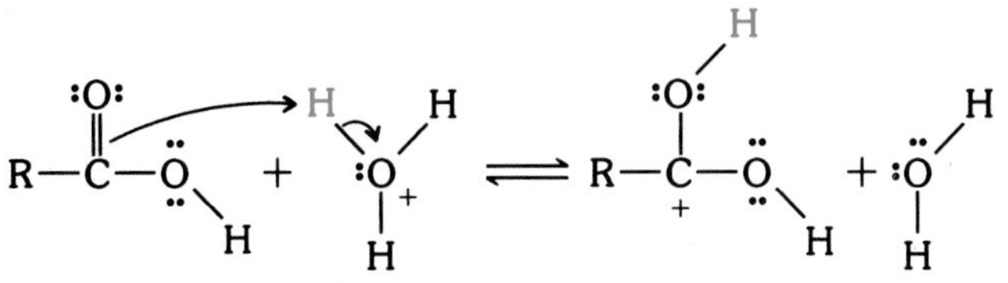

$$R-\overset{\overset{\displaystyle :O:}{\|}}{C}-\ddot{O}-H + \underset{\text{Acid catalyst}}{H-\overset{+}{\ddot{O}}(H)-H} \rightleftharpoons \underset{\text{Protonated form of the carboxylic acid}}{R-\overset{\overset{\displaystyle :O-H}{|}}{\overset{+}{C}}-\ddot{O}-H} + :\ddot{O}(H)-H$$

Acid catalyst

Protonated form of the carboxylic acid

In the protonated form of the carboxyl group, the original carbonyl carbon atom is now much more attractive to the O atom of the alcohol. The alcohol molecule now attacks, and there follows a shift of a proton from one O atom to another.

$$\text{R—C}^{+}(\text{—}\ddot{\text{O}}\text{H})(\text{—}\ddot{\text{O}}\text{H}) + \text{R}'\ddot{\text{O}}\text{H} \ \underset{\text{the carboxylic acid}}{\overset{\text{Alcohol attacks the protonated form of}}{\rightleftharpoons}} \ \text{R—C}(\text{OH})(\text{OH})(\text{—O}^{+}\text{HR}') \ \underset{\text{atom to another}}{\overset{\text{Transfer of proton occurs from one O}}{\rightleftharpoons}} \ \text{R—C}(\text{OH})(\text{—O}^{+}\text{H}_2)(\text{—}\ddot{\text{O}}\text{R}')$$

Alcohol

After the proton transfer occurs, the system now has a very stable leaving group built into it, the water molecule, which leaves.

$$\text{R—C}(\text{OH})(\text{—O}^{+}\text{H}_2)(\text{—}\ddot{\text{O}}\text{R}') \ \overset{\text{Water molecule leaves}}{\rightleftharpoons} \ \text{R—C}^{+}(\text{OH})(\text{—}\ddot{\text{O}}\text{R}') + :\ddot{\text{O}}\text{H}_2$$

Finally, a proton transfers to an acceptor in the medium (thus freeing the proton catalyst for more chemical work) and the ester molecule emerges.

$$\text{R—C}^{+}(\text{OH})(\text{—}\ddot{\text{O}}\text{R}') \ \underset{\text{(The catalyst is regenerated.)}}{\overset{\text{Proton transfers to an acceptor}}{\rightleftharpoons}} \ \text{R—}\overset{\text{:O:}}{\overset{\|}{\text{C}}}\text{—}\ddot{\text{O}}\text{—R}' + \text{H}^{+}$$

Ester

Every step involves an equilibrium. We have gone into detail about the mechanism of esterification to show what might seem to be very mysterious is not mysterious at all. Each step in direct esterification involves simple principles of proton transfers and the attractions of unlike charges. What can drive all of the equilibria to the right in favor of the ester *in vitro* is simply an excess of the alcohol. *All* of the equilibria shift to the right in accordance with Le Châtelier's principle.

■ The molecules of fats and oils — triacylglycerols — have three ester groups per molecule.

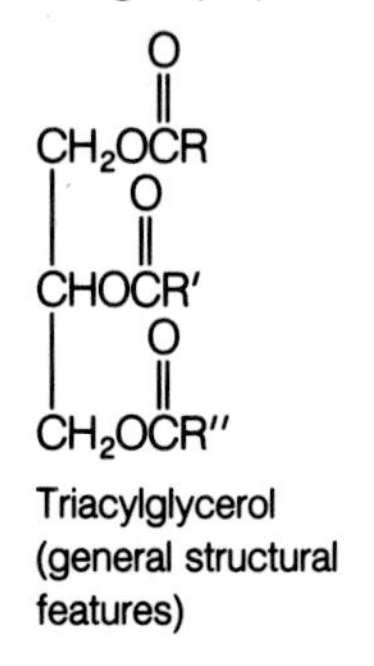

Triacylglycerol (general structural features)

5.4

OCCURRENCE, NAMES, AND PHYSICAL PROPERTIES OF ESTERS

Esters can be broken apart by reactions with water and aqueous base.
The functional group of an ester is the central structural feature of all the edible fats and oils as well as a number of constituents of body cells. Be sure that you can recognize this group and can pick out what we will call the *ester linkage,* the single bond between the carbonyl carbon atom and the oxygen atom that holds the ester's alkyl group. This linkage is where an ester breaks apart when it reacts with water.

$$\underset{\text{Two general formulas for esters}}{(H)R-\overset{\overset{\displaystyle O}{\|}}{C}-O-R' \quad \text{or} \quad (H)RCO_2R'}$$

$$-\overset{\overset{\displaystyle O}{\|}}{C}-O-\overset{|}{\underset{|}{C}}-$$

Ester linkage

Ester group (carbonyl–oxygen–carbon system)

Table 5.3 lists several common esters.

TABLE 5.3
Esters of Carboxylic Acids

Name[a]	Structure	MP (°C)	BP (°C)	Solubility in Water[b]
Ethyl Esters of Straight-Chain Carboxylic Acids, $RCO_2C_2H_5$				
Ethyl formate (ethyl methanoate)	$HCO_2C_2H_5$	−79	54	Soluble
Ethyl acetate (ethyl ethanoate)	$CH_3CO_2C_2H_5$	−82	77	7.35[c]
Ethyl propionate (ethyl propanoate)	$CH_3CH_2CO_2C_2H_5$	−73	99	1.75
Ethyl butyrate (ethyl butanoate)	$CH_3(CH_2)_2CO_2C_2H_5$	−93	120	0.51
Ethyl valerate (ethyl pentanoate)	$CH_3(CH_2)_3CO_2C_2H_5$	−91	145	0.22
Ethyl caproate (ethyl hexanoate)	$CH_3(CH_2)_4CO_2C_2H_5$	−68	168	0.063
Ethyl enanthate (ethyl heptanoate)	$CH_3(CH_2)_5CO_2C_2H_5$	−66	189	0.030
Ethyl caprylate (ethyl octanoate)	$CH_3(CH_2)_6CO_2C_2H_5$	−43	208	0.007
Ethyl pelargonate (ethyl nonanoate)	$CH_3(CH_2)_7CO_2C_2H_5$	−45	222	0.003
Ethyl caproate (ethyl decanoate)	$CH_3(CH_2)_8CO_2C_2H_5$	−20	245	0.0015
Esters of Acetic Acid, CH_3CO_2R				
Methyl acetate	$CH_3CO_2CH_3$	−99	57	24.4
Ethyl acetate	$CH_3CO_2CH_2CH_3$	−82	77	7.39[c]
Propyl acetate	$CH_3CO_2CH_2CH_2CH_3$	−93	102	1.89
Butyl acetate	$CH_3CO_2CH_2CH_2CH_2CH_3$	−78	125	1.0[d]
Miscellaneous Esters				
Methyl acrylate	$CH_2{=}CHCO_2CH_3$	—	80	5.2
Methyl benzoate	$C_6H_5{-}CO_2CH_3$	−12	199	Insoluble
Natural waxes	$CH_3(CH_2)_nCO_2(CH_2)_nCH_3$	$n = 23–33$, carnauba wax $n = 25–27$, beeswax $n = 14–15$, spermaceti		

[a] Common names; IUPAC names are in parentheses.

[b] In grams of ester per 100 g H_2O at 20 °C (unless otherwise specified).

[c] At 25 °C.

[d] At 22 °C.

SPECIAL TOPIC 5.2
SOME IMPORTANT ESTERS

FIGURE 2
The knitted tubing for this aortic heart valve is made of Dacron fibers.

Esters of *p*-Hydroxybenzoic Acid — the Parabens Several alkyl esters of *p*-hydroxybenzoic acid — referred to as *parabens* on ingredient labels — are used to inhibit molds and yeasts in cosmetics, pharmaceuticals, and food.

Salicylates Certain esters and salts of salicylic acid are analgesics, the pain suppressants; and antipyretics, the fever reducers. The parent acid, salicylic acid, is itself too irritating to the stomach for these uses, but sodium salicylate and acetylsalicylic acid (aspirin) are commonly used. Methyl salicylate, a pleasant-smelling oil, is used in liniments, for it readily migrates through the skin.

[Structures: benzene ring with CO_2Na and OH — Sodium salicylate; benzene ring with CO_2H and $OCCH_3$ (C=O) — Acetylsalicylic acid (aspirin); benzene ring with CO_2CH_3 and OH — Methyl salicylate (oil of wintergreen)]

Sodium salicylate | Acetylsalicylic acid (aspirin) | Methyl salicylate (oil of wintergreen)

Dacron Dacron, a polyester of exceptional strength, is widely used to make fabrics and film backing for recording tapes. (Actually, the name *Dacron* applies just to the fiber form of this polyester. When it is cast as a thin film, its name is *Mylar*.) Dacron fabrics have been used to repair or replace segments of blood vessels (see Figure 1).

The formation of Dacron and many other polyfunctional polymers starts with two difunctional monomers, *aAa* and *bBb*. Their functional groups are able to react with each other to split out a small molecule, *ab*. The monomer fragments, *A* and *B*, join end to end to make a very long, polymer molecule. In principle, the polymerization can be represented as follows:

$$aAa + bBb + aAa + bBb + aAa + bBb + \ldots \text{etc.} \longrightarrow -A-B-A-B-A-B-\ldots \text{etc.} + n(ab)$$

A copolymer

Because two monomers are used, the reaction is called *copolymerization*.

One monomer used to make Dacron is ethylene glycol, which has two alcohol OH groups. The other monomer is dimethyl terephthalate, which has two methyl ester groups. The copolymerization of these two monomers depends on a reaction of esters with alcohols that we will not study; the *ab* molecule that splits out is methyl alcohol. The copolymerization proceeds as follows.

$$\text{etc.}-\overset{O}{\overset{\|}{C}}-OCH_3 + H-OCH_2CH_2O-H + CH_3O-\overset{O}{\overset{\|}{C}}-C_6H_4-\overset{O}{\overset{\|}{C}}-OCH_3 + H-OCH_2-\text{etc.}$$

Ethylene glycol | Dimethyl terephthalate

$$\downarrow$$

$$\text{etc.}-\overset{O}{\overset{\|}{C}}\left(-OCH_2CH_2O-\overset{O}{\overset{\|}{C}}-C_6H_4-\overset{O}{\overset{\|}{C}}\right)_n-OCH_2-\text{etc.} + nCH_3OH$$

(Repeating unit)
Dacron/Mylar

TABLE 5.4
Fragrances of Some Esters

Name	Structure	Fragrance
Ethyl formate	$HCO_2CH_2CH_3$	Rum
Isobutyl formate	$HCO_2CH_2CH(CH_3)_2$	Raspberries
Pentyl acetate	$CH_3CO_2CH_2CH_2CH_2CH_2CH_3$	Bananas
Isopentyl acetate	$CH_3CO_2CH_2CH_2CH(CH_3)_2$	Pears
Octyl acetate	$CH_3CO_2(CH_2)_7CH_3$	Oranges
Ethyl butyrate	$CH_3CH_2CH_2CO_2CH_2CH_3$	Pineapples
Pentyl butyrate	$CH_3CH_2CH_2CO_2(CH_2)_4CH_3$	Apricots
Methyl salicylate	OH, $-CO_2CH_3$ (on benzene ring)	Oil of wintergreen

One interesting feature about acids and their esters is that the low-formula-mass acids have vile odors, but their esters have some of the most pleasant fragrances in all of nature (see Table 5.4). Special Topic 5.2 describes some important esters in more detail.

The Acid Portions of Esters and Carboxylate Ions Have Identical Names Common and IUPAC names of esters are devised in the same way. For the moment, simply ignore the R′ group of an ester, the part supplied by the alcohol, and focus on the acid portion. Pretend you are naming the *anion* of the acid. Remember that in both the common and the IUPAC names for this anion, the *-ic* ending of the name of the parent acid is changed to *-ate*. Thus salts of acetic acid (ethanoic acid) are called acetate salts (common name) or ethanoate salts (IUPAC). Similarly, esters of this acid are called acetate esters (common) or ethanoate esters (IUPAC).

Acid portion (the acyl group)

$$R-\overset{O}{\overset{\|}{C}}-O-R'$$

Alcohol portion of the ester

Once you have the name of the acid portion of the ester, simply write the name of the alkyl group in the ester's alcohol portion in front of this name (as a separate word). Here are some examples that show the pattern. (The IUPAC names are in parentheses.)

Ester	Name of Parent Acid	Name of Acid Portion of Ester	Alkyl Group in Ester	Name of Ester
$CH_3\overset{O}{\overset{\|}{C}}OCH_3$	Acetic acid (ethanoic acid)	Acetate (ethanoate)	Methyl	Methyl acetate (methyl ethanoate)
$CH_3CH_2O\overset{O}{\overset{\|}{C}}H$	Formic acid (methanoic acid)	Formate (methanoate)	Ethyl	Ethyl formate (ethyl methanoate)

EXAMPLE 5.3
Writing IUPAC Names for Esters

What is the IUPAC name for the following ester?

$$CH_3CH_2CH_2CH_2CH_2\overset{O}{\overset{\|}{C}}-O-\overset{CH_3}{\overset{|}{C}}HCH_3$$

ANALYSIS To devise the name of any ester we start with naming its parent carboxylic acid and then change *-ic acid* to *-ate.* The first step toward the ester's name, therefore, is to determine which part of the ester is contributed by a carboxylic acid. It is the *acyl group* in the molecule, that section of the molecule with the carbonyl group and anything joined to it by a *carbon–carbon bond,* not a carbon–oxygen bond.[1] Thus, the given ester has a straight-chain, six-carbon acyl group, so the parent acid is hexanoic acid.

The non-acyl part of the ester came from a parent alcohol. In the given ester, the non-acyl group is isopropyl.

SOLUTION We change the *-ic acid* ending of hexanoic acid to *-ate,* so the ester is a hexanoate ester. We add the name *isopropyl* as a separate word. The ester's name is

Isopropyl hexanoate

CHECK Compare the parts of the name with the original structure. The R group attached to O is a three-carbon, branched group named *isopropyl.* The rest of the ester unit has six carbon atoms and so the prefix *hex-* must be in its name.

PRACTICE EXERCISE 8

Write the IUPAC names of the following esters.

(a) $CH_3CH_2\overset{\overset{\Large O}{\|}}{C}OCH_3$ (b) $CH_3CH_2\overset{\overset{\Large CH_3}{|}}{C}HCH_2\overset{\overset{\Large O}{\|}}{C}OCH_2CH_2CH_3$

PRACTICE EXERCISE 9

Using the patterns developed, write the common names of the following esters.

(a) $CH_3CO_2\underset{\underset{\Large CH_3}{|}}{\overset{\overset{\Large CH_3}{|}}{C}}CH_3$ (b) $CH_3CH_2CH_2CO_2CH_2CH_3$

The Ester Group Is Polar but It Cannot Donate Hydrogen Bonds The inability of the ester group to donate hydrogen bonds affects the boiling points of esters. Thus esters of the lower-formula-mass alcohols, like methyl and ethyl alcohol, have lower boiling points than their parent acids. For example, although methyl acetate has a higher formula mass than acetic acid, it boils at 57 °C, whereas acetic acid boils at 118 °C.

The ester group, because it has oxygen atoms, can *accept* hydrogen bonds, however. This allows the lower-formula-mass esters to be relatively soluble in water.

Although the ester group is itself polar, the *overall* polarity of ester molecules is relatively low. Esters, therefore, are generally soluble in nonpolar solvents. In Chapter 9, we'll learn how molecules with three ester groups—the lipids (e.g., butterfat or olive oil)—are relatively only weakly polar and so are insoluble in water but soluble in solvents like ether and carbon tetrachloride.

[1] A bond from the carbonyl carbon to *hydrogen* occurs in formate esters. Esters of formic acid have no carbon attached to the carbonyl carbon.

5.5

SOME REACTIONS OF ESTERS

Ester molecules are broken apart by water in the presence of either acids or bases.

The reaction of esters with water is very slow unless some catalyst or promoter is present. Strong acids as well as special enzymes are good catalysts, and strong bases promote the reaction while becoming neutralized. Ester hydrolysis, catalyzed by enzymes, is the chemistry of the digestion of fats and oils.

Esters Hydrolyze to Their Parent Acids and Alcohols An ester reacts with water to give the carboxylic acid and the alcohol from which the ester could be made. This reaction is called the *hydrolysis of an ester,* and a strong acid catalyst is generally used. (In the body, an enzyme acts as the catalyst.)

In general,

$$\underset{\text{Ester}}{R-\overset{\overset{\Large O}{\|}}{C}-O-R'} + H-O-H \xrightarrow[\text{heat}]{H^+} \underset{\text{Carboxylic acid}}{R-\overset{\overset{\Large O}{\|}}{C}-O-H} + \underset{\text{Alcohol}}{H-O-R'}$$

Specific examples are

$$\underset{\text{Ethyl acetate}}{CH_3-\overset{\overset{\Large O}{\|}}{C}-O-CH_2CH_3} + H_2O \xrightarrow{H^+} \underset{\text{Acetic acid}}{CH_3-\overset{\overset{\Large O}{\|}}{C}-O-H} + \underset{\text{Ethyl alcohol}}{H-O-CH_2CH_3}$$

$$\underset{\text{Methyl benzoate}}{CH_3-O-\overset{\overset{\Large O}{\|}}{C}-C_6H_5} + H_2O \xrightarrow[\text{heat}]{H^+} \underset{\text{Methyl alcohol}}{CH_3O-H} + \underset{\text{Benzoic acid}}{H-O-\overset{\overset{\Large O}{\|}}{C}-C_6H_5}$$

To avoid a mistake that students often make, notice that the *only* bond to break in ester hydrolysis is the one that joins the carbonyl group to the oxygen atom, the "ester bond." Notice also that the products are always the "parents" of the ester and that the names of these parents are strongly implied in the name of the ester itself. Thus methyl benzoate hydrolyzes to *methyl* alcohol and *benzoic* acid. Let's now work an example.

EXAMPLE 5.4
Predicting the Products of an Ester Hydrolysis

What are the products of the hydrolysis of the following ester?

$$CH_3\overset{\overset{\Large O}{\|}}{C}-O-CH_2CH_2CH_3$$

ANALYSIS Finding the ester bond, the carbonyl-to-oxygen bond, is the crucial step because this is the bond that is broken when an ester hydrolyzes. It doesn't matter in which direction this bond happens to point on the page, it is the *only* bond that breaks.

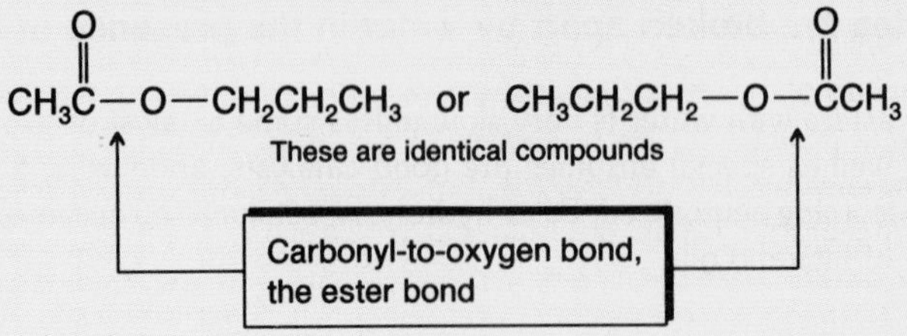

$$CH_3\overset{\overset{\displaystyle O}{\|}}{C}-O-CH_2CH_2CH_3 \quad \text{or} \quad CH_3CH_2CH_2-O-\overset{\overset{\displaystyle O}{\|}}{C}CH_3$$

These are identical compounds

Carbonyl-to-oxygen bond, the ester bond

Break the carbonyl–oxygen bond. Erase it and separate the fragments. If the ester were written as follows:

$$CH_3\overset{\overset{\displaystyle O}{\|}}{C}-O-CH_2CH_2CH_3 \dashrightarrow CH_3\overset{\overset{\displaystyle O}{\|}}{C} + O-CH_2CH_2CH_3$$

On the other hand, if the ester's structure were written in the opposite direction:

$$CH_3CH_2CH_2-O-\overset{\overset{\displaystyle O}{\|}}{C}CH_3 \dashrightarrow CH_3CH_2CH_2-O + \overset{\overset{\displaystyle O}{\|}}{C}CH_3$$

Either way gives the same results. Next we attach the pieces of the water molecule to make the "parents" of the ester. We attach OH to the carbonyl carbon and we put H on the oxygen atom of the other fragment. The products therefore are propyl alcohol and acetic acid.

$$CH_3CH_2CH_2OH + HO\overset{\overset{\displaystyle O}{\|}}{C}CH_3$$

CHECK Reexamine the structure of the ester. Its alcohol portion, the R group on O, has three carbons in a straight chain, so the alcohol we have written is correct. The acid portion has two carbons, so the acid we wrote is correct. (Double-check each C and O for the correct number of bonds.)

PRACTICE EXERCISE 10

Write the structures of the products of the hydrolysis of the following esters.

(a) $CH_3O\overset{\overset{\displaystyle O}{\|}}{C}CH_3$ (b) $CH_3CH_2\overset{\overset{\displaystyle O}{\|}}{C}-O-\overset{\overset{\displaystyle CH_3}{|}}{C}HCH_3$ (c) $CH_3\overset{\overset{\displaystyle CH_3}{|}}{C}H-\overset{\overset{\displaystyle O}{\|}}{C}-OCH_2CH_2CH_3$

Ester hydrolysis is the reverse of direct ester formation. Both involve the identical species in a chemical equilibrium, the equilibrium that we studied in the previous section for direct esterification. So if we were to take an ester and water in a 1 : 1 mole ratio, not all of the ester molecules would change into molecules of the parent acid and alcohol. Some of the ester molecules and some of the water molecules would still be unchanged. When we want to make sure that all of the ester is hydrolyzed, we use a large excess of water. In accordance with Le Châtelier's principle, this excess of one reactant shifts the equilibrium in favor of making the products of hydrolysis.

Esters Are Saponified by Bases If an aqueous base instead of a strong acid is used to promote the breakup of an ester, the products are the salt of the parent acid and the parent alcohol. The reaction is called **saponification,** and it requires a full mole of base (not just a catalytic trace) for each mole of ester bonds. The base *promotes* the reaction but, unlike a true catalyst, it is permanently changed (neutralized). No equilibrium forms, because one product, the *anion* of the parent acid, cannot be converted into an ester by a direct reaction with alcohols. In the lab, the base often used is sodium hydroxide. We'll write net ionic equations, but remember that wherever you see an anion, there is always some cation also around (the Na^+ ion when NaOH is used). In general:

■ L. *sapo*, soap, and *onis*, to make. Ordinary soap is made by the saponification of the ester groups in fats and oils.

$$R{-}\overset{\overset{\displaystyle O}{\|}}{C}{-}O{-}R' + OH^-(aq) \xrightarrow[\text{heat}]{} R{-}\overset{\overset{\displaystyle O}{\|}}{C}{-}O^- + H{-}O{-}R'$$

Ester — Carboxylate anion — Alcohol

Specific examples are as follows. (Assume that OH^- comes from an aqueous solution of NaOH or KOH.)

$$CH_3{-}\overset{\overset{\displaystyle O}{\|}}{C}{-}O{-}CH_2CH_3 + OH^-(aq) \xrightarrow[\text{heat}]{} CH_3{-}\overset{\overset{\displaystyle O}{\|}}{C}{-}O^- + H{-}O{-}CH_2CH_3$$

Ethyl acetate — Acetate ion — Ethyl alcohol

$$C_6H_5{-}\overset{\overset{\displaystyle O}{\|}}{C}{-}O{-}CH_3 + OH^-(aq) \xrightarrow[\text{heat}]{} C_6H_5{-}\overset{\overset{\displaystyle O}{\|}}{C}{-}O^- + H{-}O{-}CH_3$$

Methyl benzoate — Benzoate ion — Methyl alcohol

Saponification occurs approximately as follows.

■ The steps occur much more smoothly than implied by the way the mechanism has been written.

Step (1), Hydroxide ion attacks δ+ site:

$$R{-}\overset{\overset{\displaystyle :O:}{\|}}{\underset{\delta+}{C}}{-}\ddot{O}{-}R' + {}^-:\ddot{O}{-}H \xrightarrow{(1)} R{-}\overset{\overset{\displaystyle :\ddot{O}:^-}{|}}{\underset{\underset{\displaystyle :\ddot{O}{-}R'}{|}}{C}}{-}\ddot{O}{-}H$$

Step (2), New CO double bond forms; ^-OR breaks off:

$$\xrightarrow{(2)} R{-}\overset{\overset{\displaystyle :O:}{\|}}{C}{-}\ddot{O}{-}H + {}^-:\ddot{O}{-}R'$$

Step (3), Proton transfers; anion of acid and alcohol form:

$$\xrightarrow{(3)} R{-}\overset{\overset{\displaystyle :O:}{\|}}{C}{-}\ddot{O}:^- + H{-}\ddot{O}{-}R'$$

Carboxylate ion — Parent alcohol

In step (1), the hydroxide ion is attracted to the $\delta+$ site on the carbonyl group's carbon atom, and the double bond breaks open to a single bond. In the step (2), the double bond re-forms as $^-OR'$ is expelled. This species is an even stronger base than OH^-, and in step (3) it takes a proton where indicated. The end products are the anion of the parent acid and the parent alcohol. The steps are not reversible because the $\delta-$ site on the O atom of the alcohol has no attraction for the carboxylate ion; like charges repel.

EXAMPLE 5.5
Writing the Structures of the Products of Saponification

What are the products of the saponification of the following ester?

$$CH_3CH_2\overset{\overset{\displaystyle O}{\|}}{C}-O-CH_3$$

ANALYSIS Saponification is very similar to ester hydrolysis. The ester bond is broken. *Break only this bond.* Separate the fragments:

$$CH_3CH_2\overset{\overset{\displaystyle O}{\|}}{C}-O-CH_3 \dashrightarrow CH_3CH_2\overset{\overset{\displaystyle O}{\|}}{C} + -O-CH_3$$

Now change the fragment that has the carbonyl group into the *anion* of a carboxylic acid. Do this by attaching O^- to the carbonyl carbon atom. Then attach an H atom to the oxygen atom of the other fragment to make the alcohol molecule.

SOLUTION The products are

$$CH_3CH_2\overset{\overset{\displaystyle O}{\|}}{C}-O^- + H-O-CH_3$$

PRACTICE EXERCISE 11

Write the structures of the products of the saponification of the following esters.

(a) 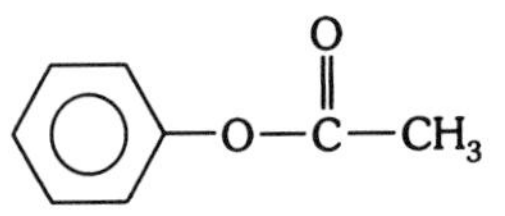 $C_6H_5-O-\overset{\overset{\displaystyle O}{\|}}{C}-CH_3$ (b) 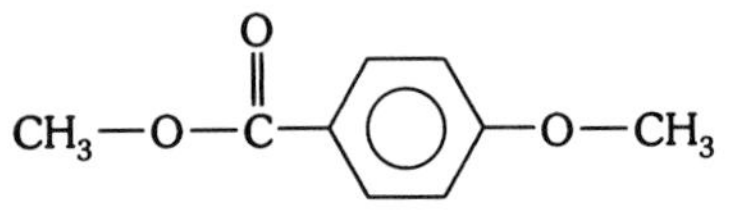 $CH_3-O-\overset{\overset{\displaystyle O}{\|}}{C}-C_6H_4-O-CH_3$

Other Reactions of Esters Esters participate in a reaction called the *Claisen ester condensation* that generates considerably larger molecules from smaller ones. We'll take this somewhat complicated reaction up in Chapter 14, closer to where it is needed in an application involving metabolism.

5.6
ORGANOPHOSPHATE ESTERS AND ANHYDRIDES

Some of the most widely distributed kinds of esters and anhydrides in living organisms are the anions of esters of phosphoric acid, diphosphoric acid, and triphosphoric acid.

Phosphoric acid appears in several forms and anions in the body, but the three fundamental parents of all these forms are phosphoric acid, diphosphoric acid, and triphosphoric acid.

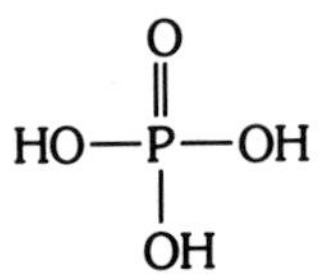

$$HO-\overset{\overset{\displaystyle O}{\|}}{\underset{\underset{\displaystyle OH}{|}}{P}}-OH \qquad HO-\overset{\overset{\displaystyle O}{\|}}{\underset{\underset{\displaystyle OH}{|}}{P}}-O-\overset{\overset{\displaystyle O}{\|}}{\underset{\underset{\displaystyle OH}{|}}{P}}-OH \qquad HO-\overset{\overset{\displaystyle O}{\|}}{\underset{\underset{\displaystyle OH}{|}}{P}}-O-\overset{\overset{\displaystyle O}{\|}}{\underset{\underset{\displaystyle OH}{|}}{P}}-O-\overset{\overset{\displaystyle O}{\|}}{\underset{\underset{\displaystyle OH}{|}}{P}}-OH$$

Phosphoric acid 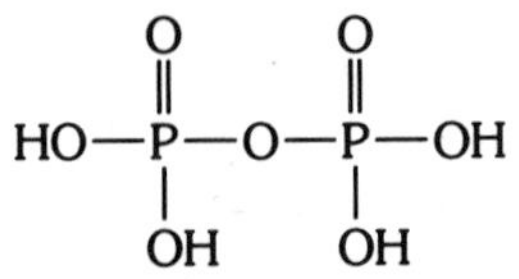 — Diphosphoric acid 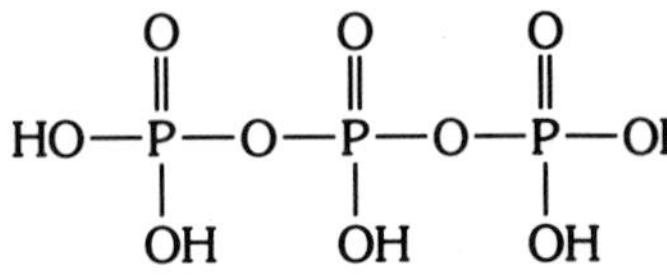 — Triphosphoric acid

These are all polyprotic acids, but at the slightly alkaline pHs of body fluids, they cannot exist as free acids. They occur, instead, as a mixture of negative ions. The net charge on each ion and the relative amounts of the ions are functions of the pH of the medium.

Esters of Alcohols and Phosphoric Acid Are Monophosphate Esters If you look closely at the structure of phosphoric acid, you can see that part of it resembles a carboxyl group.

HO—P(=O)—	HO—C(=O)—
Part of a phosphoric acid molecule	Part of a carboxylic acid molecule

It isn't surprising therefore that esters of phosphoric acid exist and that they are structurally similar to esters of carboxylic acids.

RO—P(=O)—	RO—C(=O)—
Part of a phosphate ester	Part of a carboxylate ester

One large difference between a phosphate ester and a carboxylate ester is that a phosphate ester is still a diprotic acid. Its molecules still carry two proton-donating OH groups. Therefore, depending on the pH of the medium, a phosphate ester can exist in any one of three forms, and usually there is an equilibrium mixture of all three.

RO—P(=O)(O—H)—O—H	RO—P(=O)(O—H)—O^-	RO—P(=O)(O^-)—O^-
Phosphate ester (as a diprotic acid)	Phosphate ester (as a singly ionized species)	Phosphate ester (as a doubly ionized species)
Favored at low pH	Favored at pH values just below 7	Favored at pH values above 7

At the pH of most body fluids (just slightly more than 7), phosphate esters exist mostly as the doubly ionized species—as the di-negative ion. All forms, however, are generally soluble in water, and one reason that the body converts so many substances into their phosphate esters may be to improve their solubilities in water.

■ Each of the two OH groups in a phosphate ester can be converted into an ester. Nucleic acid molecules, for example, have the phosphate diester system.

RO—P(=O)(O—H)—OR′

A phosphate diester

Alcohols and Diphosphoric Acid Form Diphosphate Esters A diphosphate ester actually has three functional groups: the phosphate ester group, the proton-donating OH groups, and the phosphoric anhydride system.

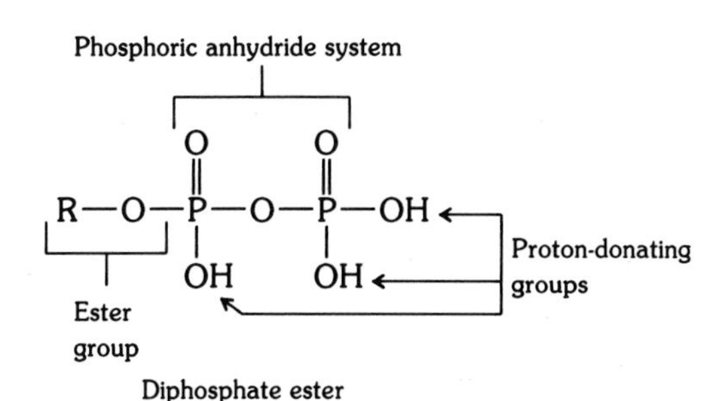

Diphosphate ester

Notice the similarity of part of the structure of this diphosphate ester to that of an anhydride of a carboxylic acid:

$$-\overset{\overset{\displaystyle O}{\|}}{\underset{|}{P}}-O-\overset{\overset{\displaystyle O}{\|}}{\underset{|}{P}}- \qquad -\overset{\overset{\displaystyle O}{\|}}{C}-O-\overset{\overset{\displaystyle O}{\|}}{C}-$$

Part of the diphosphate system
The phosphoric anhydride group

Part of a carboxylic acid anhydride system
The carboxylic anhydride group

One of the many diphosphate esters in the body is called adenosine diphosphate, or ADP. We will show its structure as its triply charged anion, because it exists largely in this fully ionized form at the pH of most body fluids.

Adenosine diphosphate, ADP
(fully ionized form)

The Phosphoric Anhydride Group Is a Major Storehouse of Chemical Energy in Living Systems ADP can be hydrolyzed to adenosine and two phosphate ions, or to adenosine monophosphate and one phosphate ion. Although either hydrolysis is *very* slow in the absence of an enzyme, the breaking up of the phosphoric anhydride group generates considerable energy.

ADP can also react with alcohols. This reaction resembles hydrolysis because it breaks up the phosphoric anhydride group. This group in ADP and similar compounds (like ATP, below) turns out to be the chief means for storing chemical energy in cells. The phosphoric anhydride system is so important in this way that we should learn how it holds its chemical energy.

The source of the internal energy in the triply charged anion of ADP is the tension up and down the anhydride chain. The central chain bears oxygen atoms with full negative charges, and these charges repel each other. The internal repulsion primes the phosphoric anhydride system for breaking apart exothermically when it is attacked by a suitable reactant.

Molecules with alcohol groups are examples of such reactants in the body, but these reactions require enzymes for catalysis. Without a catalyst, the negatively charged ADP anion actually *repels* electron-rich species. Yet if an alcohol group is to attack the phosphoric anhydride system and break it apart, the oxygen atom of the alcohol group must be able to strike a phosphorus atom. We can visualize this attack as follows.

$$RO-\overset{\overset{\displaystyle O}{\|}}{\underset{\underset{\displaystyle O^-}{|}}{P}}-O-\overset{\overset{\displaystyle O}{\|}}{\underset{\underset{\displaystyle O^-}{|}}{P}}-O^- \longrightarrow RO-\overset{\overset{\displaystyle O}{\|}}{\underset{\underset{\displaystyle O^-}{|}}{P}}-O^- + R'-O-\overset{\overset{\displaystyle O}{\|}}{\underset{\underset{\displaystyle O^-}{|}}{P}}-O^-$$

$R'-\ddot{O}-H$ ---> (proton to be buffered)

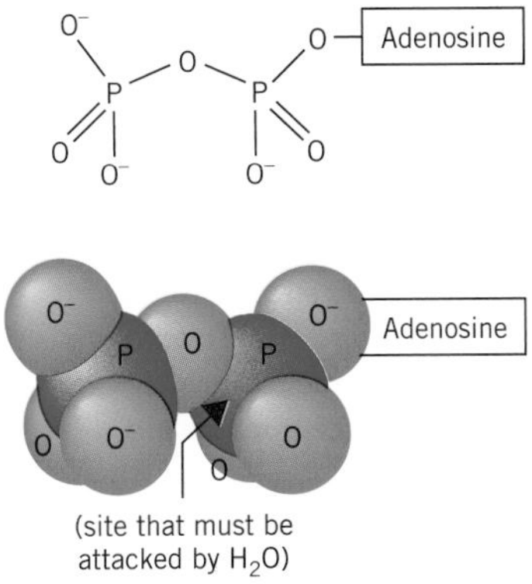

FIGURE 5.1
The oxygen atoms in the phosphoric anhydride system of ADP screen the phosphorus atoms. The negative charges on these oxygen atoms deflect incoming, electron-rich particles such as molecules of an alcohol or of water. Therefore this kind of anhydride system reacts very slowly with these reactants, unless a special catalyst such as an enzyme is also present.

(site that must be attacked by H_2O)

As seen in Figure 5.1, however, the phosphorus atoms in the chain are buried within a clutch of negatively charged oxygen atoms that repel the alcohol molecule. Thus the internal tension cannot be relieved by this kind of reaction *unless an enzyme for the reaction is present.* You have probably already realized that *the body exerts control over energy-releasing reactions of diphosphates by its control of the enzymes for these reactions.*

Water could make the same kind of exothermic attack on a diphosphate ester as an alcohol, but *the body has no enzymes inside cells that catalyze this reaction.* Hence, energy-rich diphosphates can exist in cells despite the abundance of water.

Alcohols and Triphosphoric Acid Form Triphosphate Esters Adenosine triphosphate or ATP is the most common and widely occurring member of a small family of energy-rich triphosphate esters. Because the triphosphates have two phosphoric anhydride systems in each molecule, on a mole-for-mole basis the triphosphates are among the most energy-rich substances in the body.

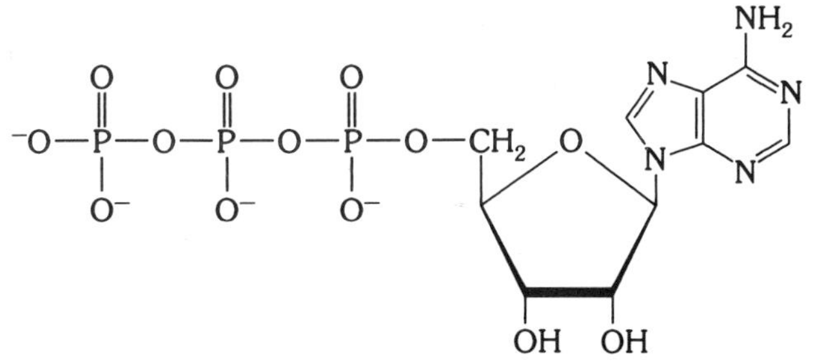

Adenosine triphosphate, ATP
(fully ionized form)

Triphosphates are much more widely used in cells as sources of energy than the diphosphates. The overall reaction for the contraction of a muscle, for example, can be written as follows. We now introduce the symbol P_i to stand for the set of inorganic phosphate ions, mostly $H_2PO_4^-$ and HPO_4^{2-}, produced in the breakup of ATP and present at equilibrium at body pH.

$$\text{Relaxed muscle} + \text{ATP} \xrightarrow{\text{enzyme}} \text{contracted muscle} + \text{ADP} + P_i$$

Muscular work requires ATP, and if the body's supply of ATP were used up *with no way to remake it,* we'd soon be helpless.

The resynthesis of ATP from ADP and P_i is one of the major uses of the chemical energy in the food we eat. What we have learned here is *how* these tri- and diphosphates can be energy-rich and, at the same time, not be destroyed by uncatalyzed reactions in the body. Later, we will study more details of how a cell makes and uses these phosphate systems.

SUMMARY

Acids and their salts The carboxyl group, CO_2H, is a polar group that confers moderate water solubility to a molecule without preventing its solubility in nonpolar solvents. This group is very resistant to oxidation and reduction. Carboxylic acids are strong proton donors toward hydroxide ions, whereas alcohols are not. Toward water, carboxylic acids are weak acids. Therefore their conjugate bases, the carboxylate anions, are good proton acceptors toward the hydronium ions of strong acids.

Salts of carboxylic acids are ionic compounds, and the potassium or sodium salts are very soluble in water. Hence, the carboxyl group is one of nature's important "solubility switches." An insoluble acid becomes soluble in base, but it is thrown out of solution again by the addition of acid.

The derivatives of acids studied in this chapter—acid chlorides, anhydrides, and esters—can be made from the acids and are converted back to the acids by reacting with water. We can organize the reactions we have studied for the carboxylic acids and esters as follows.

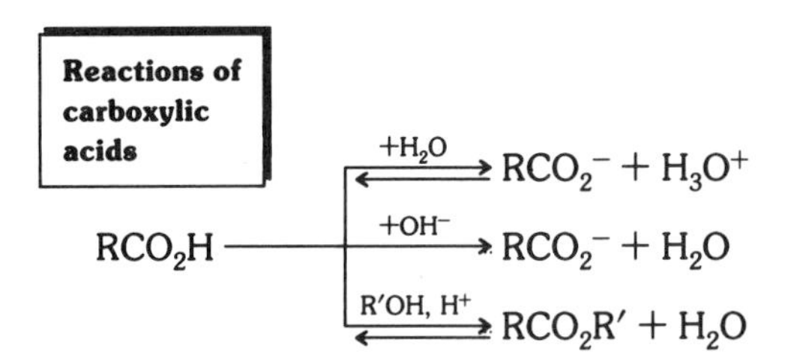

Formation of esters

$R-C(=O)-Cl + HOR' \rightarrow R-C(=O)-O-R' + HCl$

$R-C(=O)-O-C(=O)-R + HOR' \rightarrow R-C(=O)-O-R' + RCO_2H$

$R-C(=O)-OH + HOR' \underset{}{\overset{H^+}{\rightleftharpoons}} R-C(=O)-O-R' + H_2O$

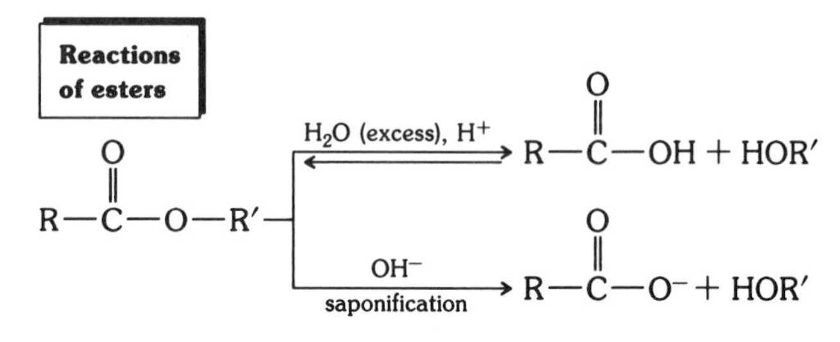

Esters and anhydrides of the phosphoric acid system Esters of phosphoric acid, diphosphoric acid, and triphosphoric acid occur in living systems largely as anions, because these esters are also polyprotic acids. In addition, esters of diphosphoric and triphosphoric acid are phosphoric anhydrides. These anhydrides are energy-rich compounds. Their reactions with water or alcohols are very exothermic, but the reactions are also very slow unless a catalyst (an enzyme) is present.

REVIEW EXERCISES

The answers to Review Exercises whose numbers are in color are found in Appendix A. The answers to the other Review Exercises are found in the Study Guide that accompanies this book. The more challenging questions are marked with asterisks.

Structures and Names of Carboxylic Acids and Their Salts

5.1 Of the following structures, which has the following functional group?
(a) alcohol (b) carboxyl group (c) ketone (d) enol

$CH_3CH_2C(=O)OH$ **A** $\quad CH_3C(=O)CH_2OH$ **B** $\quad CH_3CH_2C(=CH_2)OH$ **C**

5.2 The carboxylic acids obtained by the hydrolysis of fats and oils in the diet have what general name?

5.3 What is the common name of the acid in vinegar? In sour milk?

5.4 Write the structures of the following substances.
(a) propionic acid (b) benzoic acid
(c) acetic acid (d) formic acid

5.5 What are the structures of the following?
(a) 2,2-dimethylbutanoate ion
(b) 4-chloro-2-methylheptanoic acid
(c) butanedioic acid
(d) benzoate ion

5.6 Write the IUPAC names of the following compounds.

(a) $CH_3CH(CH_3)CO_2H$ (b) $HO_2CC(CH_2CH_3)(CH_3)CH_3$

(c) $CH_3CH_2CH(Cl)CH_2CO_2Na$ (d) C_6H_5—CO_2K

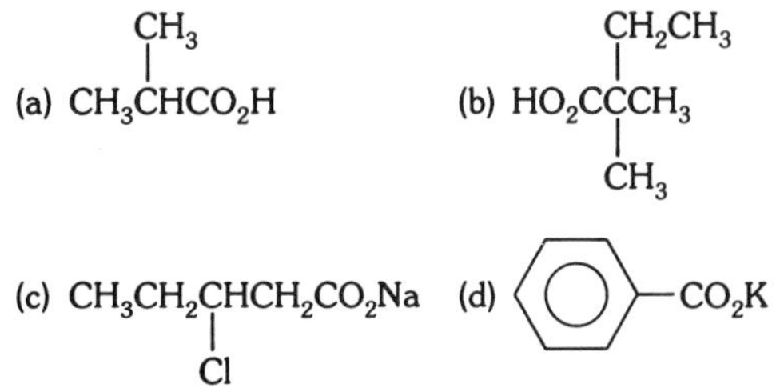

5.7 What are the IUPAC names of the following compounds?

(a) $CH_3CH(CH_3)CH_2CH(CH_3)CO_2H$ (b) $HO_2CCH_2CHCH_3$ with CH_3CHCH_3 on the CH

(c) $CH_3(CH_2)_8CO_2Na$ (d) CH_3CHBr with CO_2K on the CH

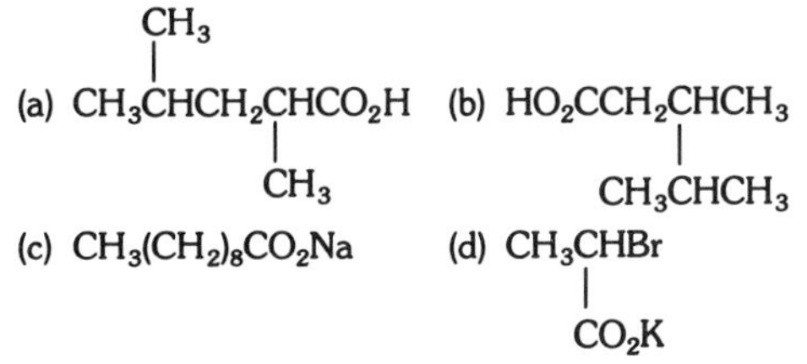

5.8 One of the compounds whose concentration in blood increases in unchecked diabetes has the following structure.

$CH_3CH(OH)CH_2CO_2H$

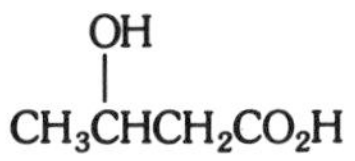

If its IUPAC name is 3-hydroxybutanoic acid and its common name is β-hydroxybutyric acid, what are the IUPAC and common names for its sodium salt?

***5.9** The citric acid cycle is one of the major metabolic sequences of reactions in the body. One of the acids in this series of reactions is commonly called fumaric acid, which has the following structure.

$HO_2C(H)C{=}C(H)CO_2H$ (trans: HO_2C and H on the left carbon, H and CO_2H on the right carbon)

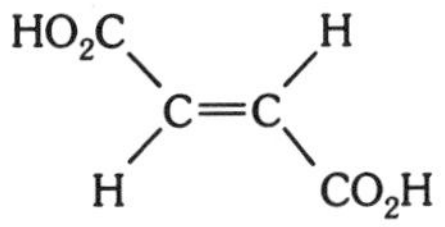

Which of the following names is its correct IUPAC name?
(a) *trans*-dibutenoic acid
(b) *trans*-butenedioic acid
(c) *trans*-dibutenedioic acid
(d) *cis*-ethenedicarboxylic acid

Physical Properties of Carboxylic Acids

5.10 Draw a figure that shows how two acetic acid molecules can pair in a hydrogen-bonded form.

***5.11** The hydrogen bond system in formic acid includes an array of molecules, one after the other, each carbonyl oxygen of one molecule attracted to the HO group of the next molecule in line. Represent this linear array of hydrogen-bonded molecules of formic acid by a drawing.

5.12 Give the following compounds in their order of increasing solubility in water. Do this by arranging their identifying letters in a row in the correct order, placing the letter of the least soluble on the left.

$CH_3CH_2CH_2CH_2CO_2H$ (**A**) HCO_2H (**B**)

$CH_3CH_2CH{=}CHCH_2CH_3$ (**C**)

***5.13** Give the following compounds in their order of increasing boiling points by arranging their identifying letters in a row in the correct order. Place the letter of the lowest-boiling compound on the left.

$HO_2CCH_2CH_2CO_2H$ (**A**) $CH_3CH_2CH_2CH_3$ (**B**)

CH_3CH_2OH (**C**) CH_3CO_2H (**D**)

Carboxylic Acids as Weak Acids

5.14 Concerning an aqueous solution of acetic acid:
(a) Write the equation for the equilibrium that is present.
(b) In what direction will this equilibrium shift, toward acetic acid or toward the acetate ion, if hydrochloric acid is added? Explain.
(c) Write the K_a equation for acetic acid.
(d) Using data in Table 5.1, is acetic acid a stronger or a weaker acid than salicylic acid?

5.15 Consider a solution of formic acid in water.
(a) Write the equation for the equilibrium that is present.
(b) In what direction will this equilibrium shift, toward formic acid or toward the formate ion, if sodium hydroxide is added? Explain.
(c) Write the K_a equation for formic acid.
(d) Using data in Table 5.1, is formic acid a stronger or a weaker acid than benzoic acid?

5.16 Arrange the following compounds in their order of increasing acidity by placing their identifying letter in a row in the correct sequence. (Place the letter of the least acidic compound on the left.)

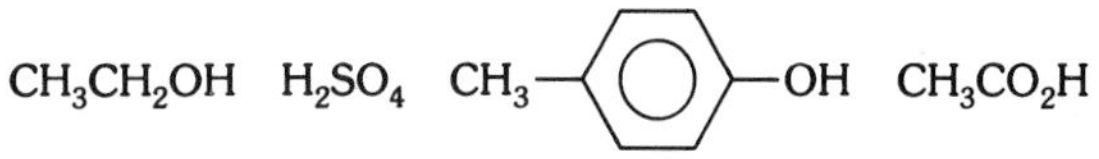

CH_3CH_2OH (**A**) H_2SO_4 (**B**) CH_3—C_6H_4—OH (**C**) CH_3CO_2H (**D**)

***5.17** Give the order of increasing acidity of the following compounds. Arrange their identifying letter in the order that corresponds to their acidity, with the letter of the least acidic compound on the left.

CH_3—C_6H_4—OH (**A**) CH_3—C_6H_4—CH_2OH (**B**)

CH_3—C_6H_4—CO_2H (**C**) HNO_3 (**D**)

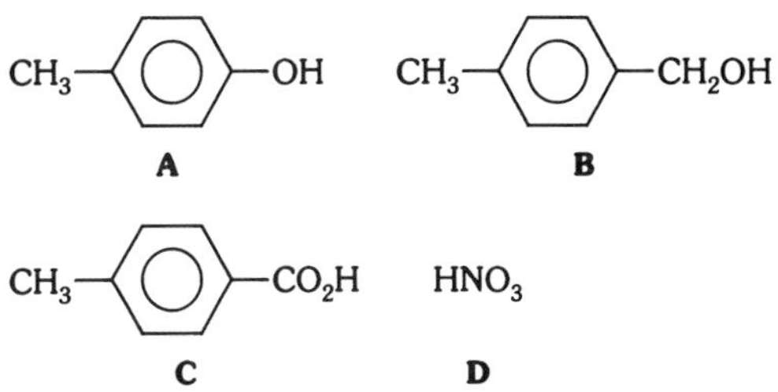

5.18 Write the net ionic equation for the complete reaction, *if any*, of aqueous sodium hydroxide with each of the compounds in Review Exercise 5.17 at room temperature.

5.19 What are the net ionic equations for the reactions of the following compounds with aqueous potassium hydroxide at room temperature?

(a) $HO_2CCH_2CH_2CO_2H$ (b) $HOCH_2CH_2CH_2CO_2H$

(c) $HCCH_2CH_2CH_2CO_2H$ (with C=O on the first carbon) (d) O= ⬡ $-CO_2H$

Salts of Carboxylic Acids

5.20 Which compound, **A** or **B,** is more soluble in ether? Explain.

$CH_3(CH_2)_6CO_2Na$ (**A**) $CH_3(CH_2)_6CO_2H$ (**B**)

5.21 Which compound, **A** or **B,** is more soluble in water? Explain.

$CH_3CH_2-C_6H_4-ONa$ (**A**) $CH_3CH_2-C_6H_4-OH$ (**B**)

***5.22** Suppose that you add 0.1 mol of hydrochloric acid to an aqueous solution that contains 0.1 mol of the compound given in each of the following parts. If any reaction occurs rapidly at room temperature, write its net ionic equation.
(a) $CH_3CH_2CO_2^-$
(b) $^-O_2CCH_2CH_2CH_2CO_2^-$
(c) NH_3

***5.23** Suppose that you have each of the following compounds in a solution in water. What reaction, if any, will occur rapidly at room temperature if an equimolar quantity of hydrochloric acid is added? Write net ionic equations.

(a) $HOCH_2CH_2CO_2^-$
(b) $HOCH_2CH_2CO_2H$
(c) $C_6H_5-O^-$

Esterification and Reactivity

5.24 What are the structures of the reactants that are needed to make ethyl propanoate from ethanol and each of the following kinds of starting materials?
(a) an acid chloride
(b) a carboxylic acid anhydride
(c) by direct esterification

5.25 In order to prepare methyl benzoate, what are the structures of the reactants needed for each kind of approach?
(a) by direct esterification
(b) from an acid chloride
(c) from an acid anhydride

***5.26** The reaction of ethyl alcohol with acetyl chloride is rapid.
(a) What is the structure of each organic product?
(b) How might the relatively high speed of the reaction be explained?

***5.27** Methyl alcohol reacts rapidly with acetic anhydride to give methyl acetate and acetic acid. How might the very rapid rate of reaction be explained?

5.28 What are the structures of the products of the esterification by ethyl alcohol of each compound?
(a) propionic acid
(b) 2-methylpropanoic acid
(c) *p*-nitrobenzoic acid
(d) terephthalic acid (Show the esterification of both of the carboxyl groups.)

$HO_2C-C_6H_4-CO_2H$

Terephthalic acid

5.29 When ethanoic acid is esterified by each of the following compounds, what are the structures of the esters that form?
(a) methanol
(b) 2-methyl-1-propanol
(c) phenol
(d) $HOCH_2CH_2OH$ (1,2-ethanediol) (Show the esterification of both alcohol groups.)

***5.30** Explain by means of equations how H^+ works as a catalyst in the direct esterification of acetic acid by ethyl alcohol.

***5.31** Water does not react readily with methyl acetate to form methyl alcohol and acetic acid. The hydroxide ion, on the other hand, more readily attacks methyl acetate (to form methyl alcohol and the acetate ion). What might be the reason for the higher reactivity of the OH^- ion over H_2O toward methyl acetate?

5.32 Suppose that a way could be found to remove H_2O as rapidly as it is produced in direct esterification. What would this do to the equilibrium in this reaction, shift it to the right (favoring the ester) or to the left (favoring the carboxylic acid and the alcohol)? Explain.

***5.33** How do we explain the fact that esters react much more slowly with water than acid chlorides do?

Structures and Physical Properties of Esters

5.34 Write the structures of the following compounds.
(a) ethyl formate (b) ethyl *p*-chlorobenzoate

5.35 What are the structures of the following compounds?
(a) *t*-butyl propanoate (b) isopropyl 2-methylbutanoate

***5.36** Arrange the following compounds in their order of increasing boiling points. Do this by placing their identifying letters in a row, starting with the lowest-boiling compound on the left.

$CH_3CO_2CH_2CH_2CH_3$ (**A**) $CH_3CH_2CH_2CH_2CO_2H$ (**B**)

$CH_3CH_2CH_2CH_2CH_3$ (**C**) $HO_2CCH_2CH_2CH_2CH_2CH_2CH_3$ (**D**)

***5.37** Arrange the following compounds in their order of increasing solubilities in water by placing their identifying letters in the correct sequence, beginning with the least soluble on the left.

$CH_3CH_2CH_2CH_2CO_2Na$ (**A**) $CH_3CH_2CH_2CH_2CO_2CH_2CH_3$ (**B**)

$CH_3CH_2CH_2CH_2CO_2H$ (**C**) $CH_3CH_2CH_2CH_2CH_3$ (**D**)

Reactions of Esters

5.38 Write the equation for the acid-catalyzed hydrolysis of each compound. If no reaction occurs, write "no reaction."

(a) $CH_3\overset{O}{\overset{\|}{C}}OCH_2\overset{CH_3}{\overset{|}{C}H}CH_3$ (b) $CH_3CH_2O\overset{O}{\overset{\|}{C}}$–(cyclohexyl)

(c) $CH_3\overset{O}{\overset{\|}{C}}CH_2OCH_3$ (d) $CH_3CH_2OCH_2CH_2CH_3$

5.39 What are the equations for the acid-catalyzed hydrolyses of the following compounds? If no reaction occurs, write "no reaction."

(a) C_6H_5–$O\overset{O}{\overset{\|}{C}}\underset{CH_3}{\underset{|}{C}H}CH_3$

(b) C_6H_5–$\overset{O}{\overset{\|}{C}}O\underset{CH_3}{\underset{|}{C}H}CH_3$

(c) C_6H_5–$\overset{O}{\overset{\|}{C}}CH_2OCH_3$

(d) $CH_3CH_2O\overset{O}{\overset{\|}{C}}CH_2CH_2\overset{O}{\overset{\|}{C}}OCH_2CH_3$

5.40 The metabolism of fats and oils involves the complete hydrolysis of molecules such as the following. What are the structures of its hydrolysis products?

$$CH_3(CH_2)_{12}\overset{O}{\overset{\|}{C}}OCH_2\underset{O\underset{\underset{O}{\|}}{C}(CH_2)_{10}CH_3}{\underset{|}{C}H}CH_2O\overset{O}{\overset{\|}{C}}(CH_2)_{14}CH_3$$

***5.41** Cyclic esters are known compounds. What is the structure of the product when the following compound is hydrolyzed?

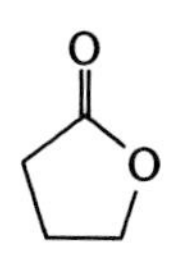

5.42 What are the structures of the organic products of the saponification of the compounds in Review Exercise 5.38 by aqueous NaOH?

5.43 What forms, if anything, when the compounds of Review Exercise 5.39 are subjected to saponification by aqueous KOH? Write the structures of the organic products.

5.44 What are the products of the saponification of the compound given in Review Exercise 5.40? (Assume that aqueous NaOH is used.)

5.45 Write the structure of the organic ion that forms when the compound of Review Exercise 5.41 is saponified.

***5.46** A pharmaceutical chemist needed to prepare the ethyl ester of an extremely expensive and rare carboxylic acid in order to test this form of the drug for its side effects. Direct esterification had to be used. How could the conversion of all the acid to its ethyl ester be maximized? Use RCO_2H as a symbol for the acid in any equations you write.

***5.47** Write the steps in the mechanism of the acid-catalyzed hydrolysis of methyl acetate. (Remember, this is the exact reverse of the acid-catalyzed, direct esterification of acetic acid by methyl alcohol.)

Phosphate Esters and Anhydrides

5.48 Write the structures of the following compounds.
(a) monomethyl phosphate
(b) monoethyl diphosphate
(c) monopropyl triphosphate

5.49 State one apparent advantage to the body of its converting many compounds into phosphate esters.

5.50 What part of the structure of ATP is particularly responsible for its being described as an *energy-rich* compound? Explain.

5.51 Why is ATP more difficult to hydrolyze than acetyl chloride?

Common Acids and Salts (Special Topic 5.1)

5.52 Name a compound that is
(a) Used to manufacture a kind of rayon.
(b) Present in vinegar.
(c) A food additive put into wrappers of cheese.

Common Esters (Special Topic 5.2)

5.53 Esters of *p*-hydroxybenzoic acid are referred to by what common name? How are these esters used in commerce?

5.54 What is meant by a *copolymer?*

5.55 What copolymer has been used in surgical grafts?

5.56 Salicylates are described as analgesics and antipyretics. What do these terms mean?

5.57 Why is salicylic acid, the parent of the salicylates, structurally modified for medicinal uses?

5.58 Concerning salicylic acid,

(a) What two functional groups does it have?

(b) Which functional group is esterified in acetylsalicylic acid?

(c) Which group is esterified in methyl salicylate?

Additional Exercises

5.59 Consider compounds **A** and **B**.

[benzene ring]$-CO_2H$ (**A**)   [benzene ring]$-OH$ (**B**)

(a) Which is the stronger acid?

(b) Write the structure of the conjugate base of each compound.

(c) Which is the stronger conjugate base of the two you wrote for part (b)?

***5.60** To prepare a sample of oil of Niobe, methyl benzoate, a student heated a solution of 5.64 g of benzoic acid in 25.0 mL of methyl alcohol in the presence of a small amount of sulfuric acid as a catalyst.

(a) What is the maximum number of grams of methyl benzoate obtainable from the mass of benzoic acid used?

(b) What is the minimum number of grams and milliliters of methyl alcohol needed for the complete conversion of benzoic acid to methyl benzoate? (The density of methyl alcohol is 0.787 g/mL.)

(c) What advantage is there in using an excess of methyl alcohol? (What principle is involved?)

***5.61** Complete the following reaction sequences by writing the structures of the organic products. If no reaction occurs, state so. (These constitute a review of this and earlier chapters on organic chemistry.)

(a) $CH_3CHCO_2H + NaOH(aq) \longrightarrow$
with CH_3O on the CH carbon

(b) $CH_3CH_2\overset{O}{\overset{\|}{C}}OCH_3 + H_2O \xrightleftharpoons[\text{catalyst}]{\text{acid}}$

(c) $CH_3CHO \xrightarrow{K_2Cr_2O_7(aq)}$

(d) [cyclohexane ring]$-OH \xrightarrow[\text{heat}]{H_2SO_4}$

(e) $CH_3\overset{CH_3}{\overset{|}{C}}HCH_2\overset{O}{\overset{\|}{C}}OCH_3 + NaOH(aq) \longrightarrow$

(f) $CH_3\overset{H_3C}{\overset{|}{C}}H\overset{OH}{\overset{|}{C}}HCH_3 \xrightarrow{KMnO_4(aq)}$

(g) $CH_3CH_2\overset{O}{\overset{\|}{C}}Cl + CH_3CH_2OH \longrightarrow$

(h) $CH_3CH_2\overset{OCH_3}{\overset{|}{C}}HOCH_3 + H_2O \xrightarrow[\text{catalyst}]{\text{acid}}$

(i) [cyclopentane ring] $+ H_2SO_4 \longrightarrow$

(j) $CH_3CH_2OH +$ [benzene ring]$-CO_2H \xrightleftharpoons[\text{catalyst}]{\text{acid}}$

(k) $H_2C{=}CHCH_2CH_3 + HCl(g) \longrightarrow$

(l) $CH_3CH_2OCH_2CH_2\overset{O}{\overset{\|}{C}}CH_3 + H_2O \longrightarrow$

***5.62** Write the structures of the organic products, if any, that form in the following situations. If no reaction occurs, state so. (Some of these constitute a review of the reactions of earlier chapters.)

(a) [1-methylcyclopentene ring, CH_3 on the double-bonded carbon] $+ H_2 \xrightarrow{\text{Ni or Pt}}$

(b) $CH_3\overset{O}{\overset{\|}{C}}O\overset{O}{\overset{\|}{C}}CH_3 + HOCH_2CH_2CH_3 \longrightarrow$

(c) $CH_3\overset{CH_3}{\overset{|}{C}}HCH_2CO_2^- + HCl(aq) \longrightarrow$

(d) [cyclopentane ring with H_3C and OH on the same carbon] $\xrightarrow{KMnO_4(aq)}$

(e) $CH_3CH_2O\overset{O}{\overset{\|}{C}}CH_2CH_2\overset{O}{\overset{\|}{C}}OCH_3 + \underset{\text{(excess)}}{NaOH(aq)} \longrightarrow$

(f) [cyclohexane ring]$-CO_2H + CH_3OH \xrightleftharpoons[\text{catalyst}]{\text{acid}}$

(g) $CH_3CH_2O\overset{O}{\overset{\|}{C}}CH_2CH_2O\overset{O}{\overset{\|}{C}}CH_2CH_3 + \underset{\text{(excess)}}{NaOH(aq)} \longrightarrow$

(h) [benzene ring]$-\overset{OCH_3}{\overset{|}{C}}HOCH_3 + H_2O \xrightarrow[\text{catalyst}]{\text{acid}}$

(i) $HO_2CCH_2CH_2CH_2CH_3 + NaOH(aq) \longrightarrow$

(j) $CH_3CH_2CH_2O\overset{O}{\overset{\|}{C}}CH_2CH_2O\overset{O}{\overset{\|}{C}}CH_3 + \underset{\text{(excess)}}{H_2O} \xrightleftharpoons[\text{catalyst}]{\text{acid}}$

(k) HO_2C-[benzene ring]$-CO_2H + \underset{\text{(excess)}}{CH_3OH} \xrightleftharpoons[\text{catalyst}]{\text{acid}}$

(l) $CH_3OCH_2CH_2CO_2^- + HCl(aq) \longrightarrow$

AMINES AND AMIDES

6

One of the very first uses of nylon was to make parachutes for the U. S. Army Airforce in World War II, when silk became unavailable. Now sky diving is a sport enjoyed by many who trust their lives to nylon, a polyamide described in this chapter.

6.1
OCCURRENCE, NAMES, AND PHYSICAL PROPERTIES OF AMINES
6.2
CHEMICAL PROPERTIES OF AMINES
6.3
AMIDES OF CARBOXYLIC ACIDS

6.1

OCCURRENCE, NAMES, AND PHYSICAL PROPERTIES OF AMINES

The amino group, NH_2, has some of the properties of ammonia, including the ability to be involved in hydrogen bonding.

Both the amino group and its protonated form occur in living things in proteins, enzymes, and nucleic acids (the chemicals that carry our genes). When a carbonyl group is attached to nitrogen, the properties change sufficiently to make it convenient to create a different chemical family, the amides. The amide system also occurs widely in living things.

$$-NH_2 \qquad -NH_3^+ \qquad -\overset{\overset{\displaystyle O}{\|}}{C}-\overset{|}{N}-$$

Amino group — Protonated amino group — Amide system

Amines Are Ammonia-like Compounds The **amines** are organic relatives of ammonia in which one, two, or all three of the hydrogen atoms on an ammonia molecule have been replaced by a hydrocarbon group. Some examples are

These are the common names, not the IUPAC names.

$$CH_3NH_2 \qquad CH_3NHCH_3 \qquad CH_3\overset{\overset{\displaystyle CH_3}{|}}{N}CH_3 \qquad CH_3NHCH_2CH_3$$

Methylamine — Dimethylamine — Trimethylamine — Methylethylamine

Several amines are listed in Table 6.1. All these are classified as *amines,* and all are basic, like ammonia.

It's quite important to realize that for a compound to be an amine, not only must its molecules have a nitrogen with three bonds, but also none of these bonds can be to a carbonyl group. If such a system is present—a carbonyl–nitrogen bond—the substance is an **amide.** Thus the structure given by **1** is an amide, not an amine. However, the structure given by **2** is not an amide, because there is no carbonyl–nitrogen bond. Instead, **2** has two functional groups, a keto group and an amino group. The chemical difference is that amines are basic and amides are not. Another difference is that the carbon–nitrogen bond in amines can't be broken by water, but the carbonyl–nitrogen bond in amides can, provided an appropriate catalyst is present.

The carbonyl–nitrogen bond occurs in protein molecules, where it is called the **peptide bond.**

$$\underset{\mathbf{1}}{R-\overset{\overset{\displaystyle O}{\|}}{C}-NH_2} \qquad \underset{\mathbf{2}}{R-\overset{\overset{\displaystyle O}{\|}}{C}-CH_2-NH_2}$$

If one or more of the groups attached directly to nitrogen in an amine is a benzene ring, then the amine is an *aromatic* amine. Otherwise, it is classified as an *aliphatic* amine. Thus benzylamine is an aliphatic amine and aniline, *N*-methylaniline, and *N,N*-dimethylaniline are all aromatic amines.

C_6H_5—NH_2 Aniline C_6H_5—$NHCH_3$ *N*-Methylaniline C_6H_5—$N(CH_3)_2$ *N,N*-Dimethylaniline C_6H_5—CH_2NH_2 Benzylamine

The Common Names of Amines Usually End in *-amine* The common names of the simple, aliphatic amines are made by writing the names of the alkyl groups attached to nitrogen in front of the word *amine* (and leaving no space). We have already seen how this works. Here are three more examples.

$CH_3CH(CH_3)CH_2NH_2$ Isobutylamine $CH_3CH(CH_3)NHCH_2CH_3$ Ethylisopropylamine $CH_3CH_2N(CH_3)CH_2CH_2CH_3$ Methylethylpropylamine

In complex systems, the names of some amines use the name *amino* as a substituent name for the NH_2 group. Thus isobutylamine can be named 1-amino-2-methylpropane. We will not develop IUPAC names for amines.

PRACTICE EXERCISE 1

Give common names for the following compounds.

(a) $(CH_3)_2NCH(CH_3)_2$ **(b)** 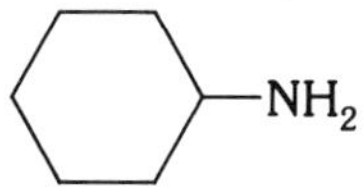 cyclohexyl—NH_2 **(c)** $(CH_3)_2CHCH_2NHC(CH_3)_3$

PRACTICE EXERCISE 2

Write the structures of the following compounds.

(a) *t*-butyl-*sec*-butylamine **(b)** *p*-nitroaniline

(c) *p*-aminobenzoic acid (the PABA of sun-screening lotions)

Heterocyclic Amines Have N as a Ring Atom Both proteins and nucleic acids are rich in nitrogen-containing heterocyclic rings. For example, the amino acid proline has a saturated ring that includes a nitrogen atom. Unsaturated rings are present in other amino acids such as tryptophan.

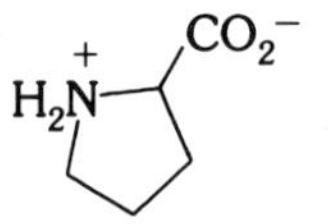

Proline (H_2N^+ ring, CO_2^-)

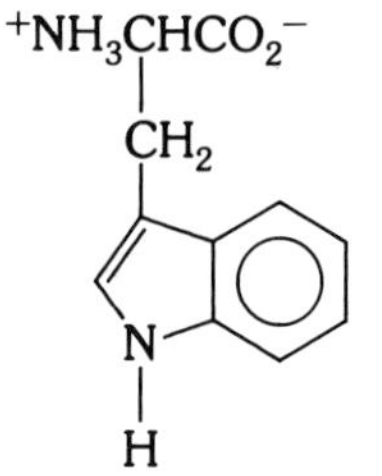

Tryptophan ($^+NH_3CHCO_2^-$, CH_2, indole ring with N—H)

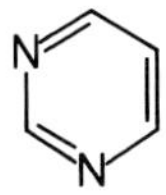

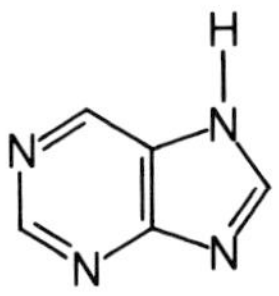

In molecules of nucleic acids, heterocyclic amine systems generally involve either the pyrimidine ring or the purine ring system.

Pyrimidine

Purine

FIGURE 6.1
Hydrogen bonds in *(a)* amines and in *(b)* aqueous solutions of amines.

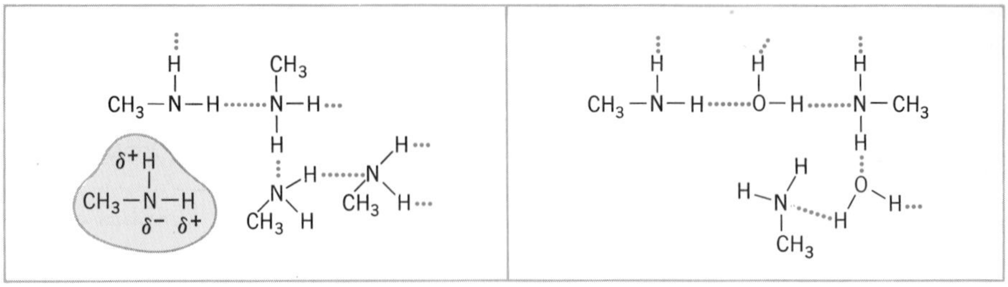

(a) *(b)*

Compound	Formula Mass	BP (°C)
CH_3CH_3	30	−89
CH_3NH_2	31	−6
CH_3OH	32	65

N—H Groups in Amines Are Involved in Hydrogen Bonding We have sometimes used boiling point data to tell us something about forces between molecules. The data in the margin tell us, for example, that when compounds of similar formula mass are compared, the boiling points of amines are higher than those of alkanes but lower than those of alcohols. This suggests that the forces of attraction between molecules are stronger in amines than in alkanes, but they are weaker in amines than in alcohols.

We can understand these trends in terms of hydrogen bonds. When a hydrogen atom is bound to oxygen or nitrogen but not to carbon, the system can donate and accept hydrogen bonds. Nitrogen, however, has a lower electronegativity than oxygen, so the polarity of the N—H bond in amines is weaker than the polarity of the O—H bond in alcohols. *The N—H system, therefore, develops weaker hydrogen bonds than the O—H system.* As a result, amine molecules can't attract each other as strongly as alcohol molecules, so amines boil lower than alcohols (of comparable formula masses). But amine molecules, nevertheless, do develop some hydrogen bonds (Fig. 6.1), which alkane molecules cannot do, so amines boil higher than alkanes (of comparable formula masses). Hydrogen bonding also helps amines to be much more soluble in water than alkanes, as Figure 6.1 also depicts.

Although hydrogen bonding between molecules of amines is weaker than hydrogen bonding between molecules of alcohols, it has a very important function at the molecular level of life. Among the molecules of proteins and nucleic acids, for example, the hydrogen bond stabilizes their special shapes, without which the substances lose their abilities to carry out their biological functions.

■ The processes in the body that lead to the sensations of odor or taste begin with *chemical* reactions.

Odor isn't actually a *physical* property, but we should note that the amines with lower formula masses smell very much like ammonia. The odors of amines become very "fishy" at slightly higher formula masses.

6.2

CHEMICAL PROPERTIES OF AMINES

The amino group is a proton acceptor, and the protonated amino group is a proton donor.

We will examine two chemical properties of amines that will be particularly important to our study of biochemicals: the basicity of amines, in this section, and their conversion to amides, in the next.

Aliphatic Amines Are About as Basic as Ammonia When ammonia dissolves in water, the following equilibrium becomes established and a small concentration of hydroxide ion forms together with the ammonium ion.

$$\underset{\text{Ammonia}}{NH_3} + H_2O \rightleftharpoons \underset{\text{Ammonium ion}}{NH_4^+} + OH^-$$

TABLE 6.1
Amines

Common Name	Structure	BP (°C)	Solubility in Water (20 °C)	K_b (25 °C)
Methylamine	CH_3NH_2	−6	very soluble	4.4×10^{-4}
Dimethylamine	$(CH_3)_2NH$	8	very soluble	5.3×10^{-4}
Trimethylamine	$(CH_3)_3N$	3	very soluble	0.5×10^{-4}
Ethylamine	$CH_3CH_2NH_2$	17	very soluble	5.6×10^{-4}
Diethylamine	$(CH_3CH_2)_2NH$	55	very soluble	9.6×10^{-4}
Triethylamine	$(CH_3CH_2)_3N$	89	14 g/dL	5.7×10^{-4}
Propylamine	$CH_3CH_2CH_2NH_2$	49	very soluble	4.7×10^{-4}
Aniline	$C_6H_5NH_2$	184	4 g/dL	3.8×10^{-10}

A very similar equilibrium forms when an amine dissolves in water. All that is different is that an alkyl group has replaced one of the H atoms of NH_3 (or NH_4^+).

$$R{-}NH_2 + H_2O \rightleftharpoons R{-}NH_3^+ + OH^-$$

Amine — Protonated amine

How much the products are favored in this equilibrium is expressed by the value of the **base ionization constant, K_b.**

$$K_b = \frac{[RNH_3^+][OH^-]}{[RNH_2]}$$

■ Remember, the brackets [ ] denote the moles-per-liter concentration of the compound or ion that the brackets enclose.

Table 6.1 gives the K_b values for several amines. The K_b of ammonia is 1.8×10^{-5}, so you can see that most amines have K_b values slightly greater than that of ammonia. Remember, the larger the K_b value, the *stronger* is the base, because a larger value means that the terms in the numerator, including $[OH^-]$, have to be larger. Thus, the aliphatic amines are generally slightly stronger bases than ammonia. Like ammonia, water-soluble amines cause the hydroxide ion concentration of the medium to become greater than the hydrogen ion concentration, and the aqueous solutions of amines are basic (and turn red litmus blue). Compounds with aliphatic amino groups, in other words, tend to increase the pH of an aqueous solution.

Like ammonia, compounds with amino groups can also neutralize hydronium ions. The following acid–base neutralization occurs rapidly and essentially completely at room temperature.

$$R{-}\ddot{N}H_2 + H{-}\overset{+}{O}H_2 \longrightarrow R{-}\overset{+}{N}H_3 + H_2\ddot{O}$$

Amine (or ammonia, when R = H) — Hydronium ion — Protonated amine (or the ammonium ion when R = H)

For example, hydrochloric acid is neutralized by methylamine as follows.

$$CH_3NH_2 + H_3O^+ + Cl^- \longrightarrow CH_3NH_3^+Cl^- + H_2O$$

Methylamine — Hydrochloric acid — Methylammonium chloride

■ Tetraalkylammonium ions:

$$R-\overset{R}{\underset{R}{\overset{|}{\underset{|}{N^+}}}}-R$$

are also known species, but such cations can't be basic because they have no unshared pair of electrons on nitrogen.

It doesn't matter if the nitrogen atom in an amine bears one, two, or three hydrocarbon groups. The amine can still neutralize strong acids, because the reaction involves just the unshared pair of electrons on the nitrogen, not any of the bonds to the other groups.

$$(CH_3)_2NH: + H-\overset{+}{O}H_2 \longrightarrow CH_3-\overset{+}{N}H_2-CH_3 + :\ddot{O}H_2$$

The previously unshared electron pair on N now holds the H atom to N.

Dimethylamine | Hydronium ion | Dimethylammonium ion

EXAMPLE 6.1
Writing the Structure of the Product when an Amine Is Neutralized by a Strong Acid

What organic cation forms when hydrochloric acid (or any strong acid) neutralizes each of the following amines?

(a) $CH_3CH_2NH_2$ (b) $CH_3CH_2N(CH_3)CH_3$ (c) N-methylpiperidine (ring N—CH_3)

ANALYSIS The nitrogen atom of any amine group can accept H^+, so all that we do is increase the number of H atoms attached to the nitrogen atom by one, and then write a positive sign to show the charge.

SOLUTION The protonated forms of the given amines are

(a) $CH_3CH_2NH_3^+$ (b) $CH_3CH_2\overset{+}{N}HCH_3$ with CH_3 on N (c) ring N^+ bearing H and CH_3

PRACTICE EXERCISE 3

What are the structures of the cations that form when the following amines react completely with hydrochloric acid?

(a) aniline **(b)** trimethylamine **(c)** $NH_2CH_2CH_2NH_2$

Protonated Amines Can Neutralize Strong Bases A combination of a protonated amine and an anion make up an organic salt called a **amine salt.** Table 6.2 gives some examples and, like all salts, amine salts are crystalline solids at room temperature. In addition,

TABLE 6.2
Amine Salts

Name	Structure	MP (°C)
Methylammonium chloride	$CH_3NH_3^+Cl^-$	232
Dimethylammonium chloride	$(CH_3)_2NH_2^+Cl^-$	171
Dimethylammonium bromide	$(CH_3)_2NH_2^+Br^-$	134
Dimethylammonium iodide	$(CH_3)_2NH_2^+I^-$	155
Tetramethylammonium hydroxide[a]	$(CH_3)_4N^+OH^-$	130–135 (decomposes)

[a] This compound is as strong a base as NaOH because its OH^- ion fully dissociates in water.

like the salts of the ammonium ion, nearly all amine salts of strong acids are soluble in water *even when the parent amine is not*. Amine salts are much more soluble in water than amines because the *full* charges carried by the ions of an amine salt can be much better hydrated by water molecules than the amine itself, where only the small, partial charges of polar bonds occur.

Some amine salts are internal salts, like all of the amino acids, the building blocks of proteins.

$$^+NH_3—\underset{\underset{R}{|}}{CH}—CO_2^-$$

General formula of all amino acids. R = an organic group but not always a simple alkyl group

Protonated amine cations also neutralize the hydroxide ion and revert to amines in the following manner (where we show only skeletal structures).

$$—\overset{+}{N}—H + :\ddot{O}^-—H \longrightarrow —\ddot{N}< + H—\ddot{O}—H$$

Protonated amine · Hydroxide ion · Amine

For example,

$$CH_3NH_3^+ + OH^- \longrightarrow CH_3NH_2 + H_2O$$

$$CH_3CH_2\overset{+}{N}H_2CH_3 + OH^- \longrightarrow CH_3CH_2NHCH_3 + H_2O$$

$$C_5H_{10}\overset{+}{N}(H)CH_3 + OH^- \longrightarrow C_5H_{10}N—CH_3 + H_2O$$

The Amino Group Is a Solubility Switch We have just learned that putting a proton on an amino group and taking it off are easily done at room temperature simply by adding an acid and then a base. We have also learned that the protonated amine is more soluble in water than the amine. This makes the amino group an excellent "solubility switch."

The solubility of an amine can be switched on simply by adding enough strong acid to protonate it. Triethylamine, for example, is insoluble in water, but we can switch on its solubility by adding a strong acid, like hydrochloric acid. The amine dissolves as its protonated ionic form is produced.

The other important solubility switch that we have studied involves the carboxylic acid group:

$$RCO_2H \text{ (less soluble)} \xrightarrow{OH^-} RCO_2^- \text{ (more soluble)} \xrightarrow{H^+} RCO_2H$$

$$(CH_3CH_2)_3N: + HCl(aq) \longrightarrow (CH_3CH_2)_3\overset{+}{N}HCl^-(aq)$$

Triethylamine (water insoluble) · Hydrochloric acid · Triethylammonium chloride (water soluble)

We can just as quickly and easily bring the amine back out of solution by adding a strong base, like the hydroxide ion. It takes the proton off the protonated amine and gives the less soluble form.

$$(CH_3CH_2)_3\overset{+}{N}HCl^-(aq) + OH^- \longrightarrow (CH_3CH_2)_3N\!: + H_2O$$

Triethylammonium chloride (water soluble) — OH^- (supplied by NaOH, for example) — Triethylamine (water insoluble)

The significance of this "switching" relationship is that the solubilities of complex compounds that have the amine function can be changed almost instantly simply by adjusting the pH of the medium.

One application of this property involves medicinals. A number of amines obtained from the bark, roots, leaves, flowers, or fruits of various plants are useful drugs. These naturally occurring, acid-neutralizing, physiologically active amines are called **alkaloids,** and morphine, codeine, and quinine are just three examples.

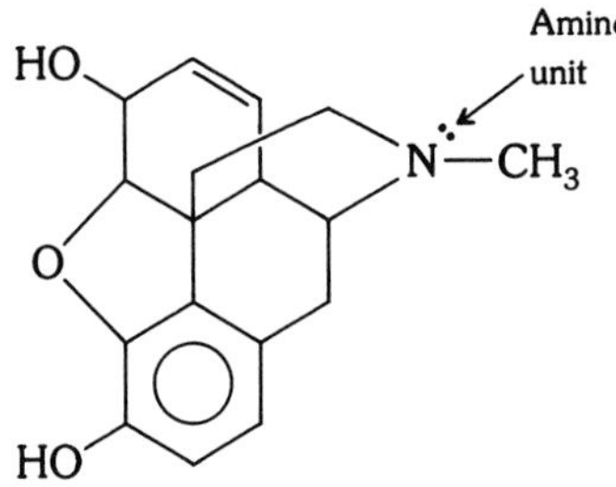

Morphine

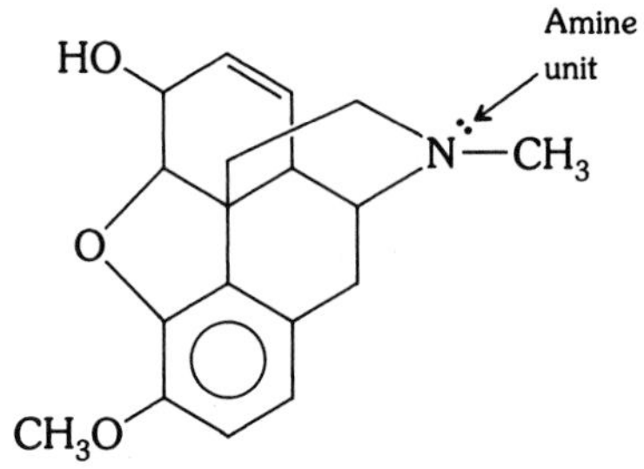

Codeine

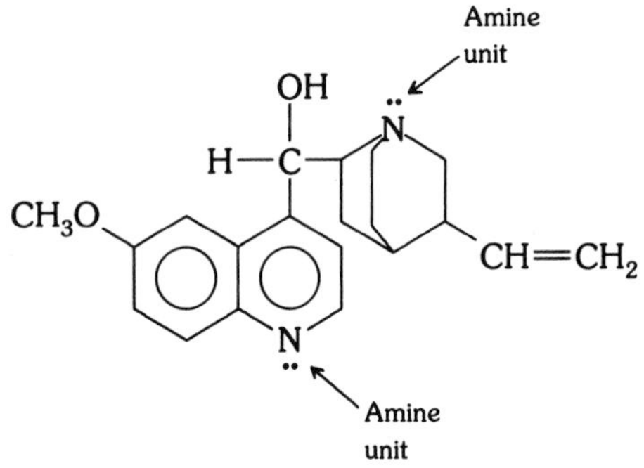

Quinine

To make it easier to administer alkaloidal drugs in the dissolved state, we often prepare them as their water-soluble amine salts. Morphine, for example, a potent sedative and painkiller, is often given as morphine sulfate, the salt of morphine and sulfuric acid. Quinine, an antimalarial drug, is available as quinine sulfate. Codeine, sometimes used in cough medicines, is often present as codeine phosphate. Special Topic 6.1 tells about a few other physiologically active amines, most of which are also prepared as their amine salts.

EXAMPLE 6.2
Writing the Structure of the Product of Deprotonating the Cation of an Amine Salt

The protonated form of amphetamine is shown below. What is the structure of the product of its reaction with OH^-?

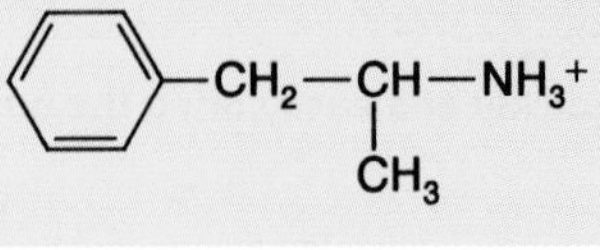

$$C_6H_5-CH_2-\underset{\displaystyle CH_3}{\underset{|}{CH}}-NH_3^+$$

SPECIAL TOPIC 6.1
SOME PHYSIOLOGICALLY ACTIVE AMINES

Epinephrine and norepinephrine are two of the many hormones in our bodies. We will study the nature of hormones in a later chapter, but we can use a definition here. **Hormones** are compounds the body makes in special glands to serve as chemical messengers. In response to a stimulus somewhat unique for each hormone, such as fright, food odor, sugar ingestion, and others, the gland secretes its hormone into circulation in the bloodstream. The hormone then moves to some organ or tissue where it activates a particular metabolic series of reactions that constitute the biochemical response to the initial stimulus. Maybe you have heard the expression, "I need to get my adrenalin flowing." Adrenalin—or epinephrine, its technical name—is made by the adrenal gland. If you ever experience a sudden fright, a trace amount of epinephrine immediately flows and the results include a strengthened heartbeat, an increase in blood pressure, and a release of glucose into circulation from storage—all of which get the body ready to respond to the threat.

Norepinephrine has similar effects, and because these two hormones are secreted by the adrenal gland, they are called **adrenergic agents.**

HO—C_6H_3(HO)—CH(OH)CH_2NHCH_3

Epinephrine

HO—C_6H_3(HO)—CH(OH)CH_2NH_2

Norepinephrine

C_6H_5—CH(OH)CH_2NH_2

β-Phenylethanolamine

Several useful drugs mimic epinephrine and norepinephrine, and all are classified as *adrenergic drugs.* Most of them, like epinephrine and norepinephrine, are related structurally to β-phenylethanolamine (which is not a formal name, obviously). In nearly all their uses, these drugs are prepared as dilute solutions of their amine acid salts. Several of the β-phenylethanolamine drugs are even more structurally like epinephrine and norepinephrine, because they have the structural features of 1,2-dihydroxybenzene, commonly called *catechol.* The catechol-like adrenergic drugs are called the **catecholamines.** Synthetic epinephrine (an agent in Primatene Mist), ethylnorepinephrine, and isoproterenol are examples.

HO—C_6H_3(HO)—CH(OH)CH(CH_2CH_3)NH_2

Ethylnorepinephrine
(Used against asthma
in children)

HO—C_6H_3(HO)—CH(OH)$CH_2NHCH(CH_3)_2$

Isoproterenol (Used in treating
emphysema and asthma)

The β-phenylethylamines are another family of physiologically active amines. For example, dopamine (which is also a catecholamine) is the compound the body uses to make norepinephrine. Its synthetic form is used to treat shock associated with severe congestive heart failure.

The amphetamines are a family of β-phenylethylamines that include Dexedrine ("speed") and Methedrine ("crystal," "meth"). The amphetamines can be legally prescribed as stimulants and antidepressants, and sometimes they are prescribed for weight-control programs. However, millions of these "pep pills" or "uppers" are sold illegally, and this use of amphetamines constitutes a serious drug abuse problem. The dangers of overuse include suicide, belligerence and hostility, paranoia, and hallucinations.

HO—C_6H_3(HO)—$CH_2CH_2NH_2$

Dopamine

C_6H_5—$CH_2CH(CH_3)NH_2$

Dexedrine

C_6H_5—$CH_2CH(CH_3)NHCH_3$

Methedrine

In a later chapter we will discuss the mechanisms by which these drugs and the naturally occurring hormones work.

Amphetamine sulfate (also known as benzedrine sulfate) is the form in which this drug is administered. It can be prescribed as an anorexigenic agent—one that reduces the appetite.

ANALYSIS Because OH^- removes just one H^+ from the nitrogen atom of a protonated amine's cation, all we have to do is reduce the number of H atoms on the nitrogen by one and cancel the positive charge.

SOLUTION Amphetamine is

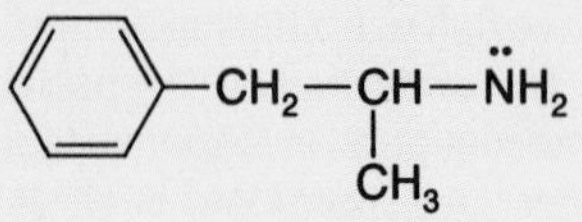

$$C_6H_5-CH_2-\underset{\displaystyle CH_3}{\underset{|}{CH}}-\ddot{N}H_2$$

Amphetamine

(The structure of amphetamine, given here, appears to be identical with that of Dexedrine shown in Special Topic 6.1. However, there is an important difference that we will explore in the next chapter.)

PRACTICE EXERCISE 4

Write the structures of the products after the following protonated amines have reacted with OH^- in a 1 : 1 mole ratio.

(a)

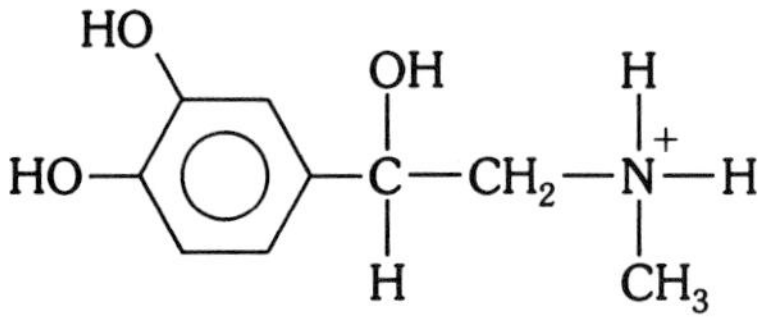

$$(HO)_2C_6H_3-\overset{\displaystyle OH}{\overset{|}{\underset{\displaystyle H}{\underset{|}{C}}}}-CH_2-\overset{\displaystyle H}{\overset{|}{\underset{\displaystyle CH_3}{\underset{|}{N^+}}}}-H$$

Epinephrine (adrenaline), a hormone given here in its protonated form. As the chloride salt in a 0.1% solution, it is injected in some cardiac failure emergencies. (See also Special Topic 6.1.)

(b)

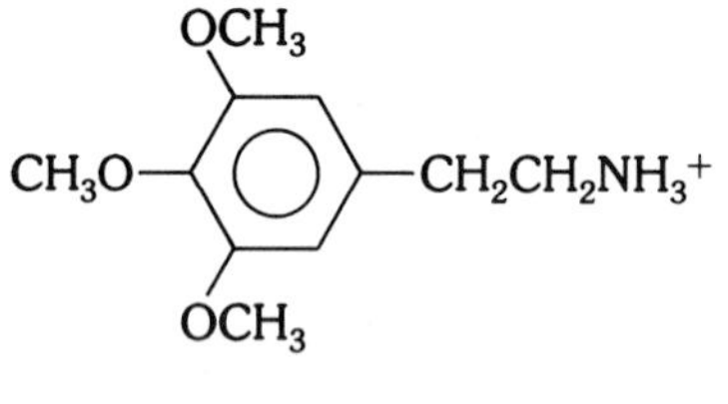

$$(CH_3O)_3C_6H_2-CH_2CH_2NH_3^+$$

Mescaline, a mind-altering hallucinogen shown here in its protonated form. It is isolated from the mescal button, a growth on top of the peyote cactus. Indians in the southwestern United States have used it in religious ceremonies.

Hallucinogens are drugs that cause illusions of time and place, make unreal experiences or things seem real, and distort the qualities of things.

6.3

AMIDES OF CARBOXYLIC ACIDS

Amides are neutral nitrogen compounds that can be hydrolyzed to carboxylic acids and ammonia (or amines).

The carbonyl–nitrogen bond is sometimes called the **amide bond,** because it is the bond that forms when amides are made, and it is the bond that breaks when amides are hydrolyzed. As the following general structures show, an amide can be derived either from ammonia or from amines. Those derived from ammonia itself are often referred to as *simple* amides.

$$\underset{\text{Amides of ammonia (simple amides)}}{R{-}\overset{\overset{\displaystyle O}{\|}}{C}{-}NH_2} \qquad \underset{\text{Amides of amines}}{R{-}\overset{\overset{\displaystyle O}{\|}}{C}{-}NHR' \qquad R{-}\overset{\overset{\displaystyle O}{\|}}{C}{-}\overset{\overset{\displaystyle R''}{|}}{N}R'} \qquad \underset{\text{Amide group}}{{-}\overset{\overset{\displaystyle O}{\|}}{C}{-}\overset{|}{N}{-}}$$

(Amide bond: the C—N bond of the amide group)

We study the amide system because all proteins are essentially polyamides, polymers whose molecules have regularly spaced amide bonds. Nylon is a synthetic polymer with repeating amide groups (see Special Topic 6.2).

Table 6.3 lists several low-formula-mass amides. Their molecules are quite polar, and when they have an H atom bonded to N, they can both donate and accept hydrogen bonds. These forces add up so much in simple amides that all except methanamide are solids at room temperature. Simple amides have considerably higher boiling points than alkanes, alcohols, or even carboxylic acids of comparable formula mass, as the data in the margin show. When we study proteins, we'll see how hydrogen bonding is involved in stabilizing the shapes of protein molecules, shapes that are as important to the functions of proteins as anything else about their structures.

Compound	Formula mass	BP (°C)
$CH_3CH_2CH_2CH_3$	58	−42
$CH_3CH_2CH_2OH$	60	97
CH_3CO_2H	60	118
CH_3CONH_2	59	222

The Names of Simple Amides End in *-amide* The common names of simple amides are made by replacing *-ic acid* by *-amide,* and their IUPAC names are devised by replacing *-oic acid* by *-amide.* In the following examples, notice how we can condense the structure of the amide group.

$$CH_3\overset{\overset{\displaystyle O}{\|}}{C}NH_2 \text{ or } CH_3CONH_2 \qquad CH_3CH_2CH_2\overset{\overset{\displaystyle O}{\|}}{C}NH_2 \text{ or } CH_3CH_2CH_2CONH_2$$

Acetamide (common name)
Ethanamide (IUPAC name)

Butyramide (common name)
Butanamide (IUPAC name)

The simplest aromatic amide is called benzamide, $C_6H_5CONH_2$, where C_6H_5 signifies the phenyl group. We'll not need to know the rules for naming other kinds of amides.

$C_6H_5 =$ (phenyl ring)—

TABLE 6.3
Amides of Carboxylic Acids

IUPAC Name	Structure	MP (°C)
Methanamide	$HCONH_2$	3
N-Methylmethanamide	$HCONHCH_3$	−5
N,N-Dimethylmethanamide	$HCON(CH_3)_2$	−61
Ethanamide	CH_3CONH_2	82
N-Methylethanamide	$CH_3CONHCH_3$	28
N,N-Dimethylethanamide	$CH_3CON(CH_3)_2$	−20
Propanamide	$CH_3CH_2CONH_2$	79
Butanamide	$CH_3CH_2CH_2CONH_2$	115
Pentanamide	$CH_3CH_2CH_2CH_2CONH_2$	106
Hexanamide	$CH_3CH_2CH_2CH_2CH_2CONH_2$	100
Benzamide	$C_6H_5CONH_2$	133

SPECIAL TOPIC 6.2
NYLON, A POLYAMIDE

The term *nylon* is a coined name that applies to any synthetic, long-chain, fiber-forming polymer with repeating amide linkages. One of the most common members of the nylon family, nylon-66, is made from 1,6-hexanediamine and hexanedioic acid.

$$NH_2CH_2CH_2CH_2CH_2CH_2CH_2NH_2$$

1,6-Hexanediamine

$$HO\overset{\overset{\large O}{\|}}{C}CH_2CH_2CH_2CH_2\overset{\overset{\large O}{\|}}{C}OH$$

Hexanedioic acid

$$\text{etc.}\!-\!\!\!\!+\overset{\overset{\large O}{\|}}{C}(CH_2)_4\overset{\overset{\large O}{\|}}{C}NH(CH_2)_6NH\!+\!\!\!\!-_n \text{ etc.}$$

Repeating unit in nylon-66

(The "66" means that each monomer has six carbon atoms.) To be useful as a fiber-forming polymer, each nylon-66 molecule should contain from 50 to 90 of each of the monomer units. Shorter molecules form weak or brittle fibers.

When molten nylon resin is being drawn into fibers, newly emerging strands are caught up on drums and stretched as they cool. Under this tension, the long polymer molecules within the fiber line up side by side, overlapping each other, to give a finished fiber of unusual strength and beauty (see Figure 1). Part of nylon's strength comes from the innumerable hydrogen bonds that extend between the polymer molecules and that involve their many regularly spaced amide groups.

Nylon is more resistant to combustion than wool, rayon, cotton, or silk, and it is as immune to insect attack as fiberglass. Molds and fungi do not attack nylon molecules either. In medicine, nylon is used in specialized tubing, and as velour for blood contact surfaces. Nylon sutures were the first synthetic sutures and are still commonly used.

FIGURE 1
This woman's life depends on the strength of nylon when she is parasailing.

PRACTICE EXERCISE 5

Write the IUPAC names of the following amides.

(a) $CH_3CH_2\underset{\underset{\large CH_3}{|}}{C}HCH_2CH_2\overset{\overset{\large O}{\|}}{C}NH_2$ (b) $CH_3CH_2\underset{\underset{\large CH_3CH_2}{|}}{C}H\overset{\overset{\large O}{\|}}{C}NH_2$

Unlike the Amines, Amides Are Not Proton Acceptors One reason for creating a separate family for the amides apart from the amines is that, unlike amines, amides are not proton acceptors or bases. They're not proton donors or acids either. *Amides are neutral in an*

acid–base sense. The amide group, in other words, does not affect the pH of an aqueous system.

The electronegative carbonyl group on the nitrogen atom causes the acid–base neutrality of amides. Although both an amide and an amine have an unshared pair of electrons on nitrogen, in the amide this pair is drawn back so tightly by the electron-withdrawing ability of the carbonyl group that the electron pair shown on the nitrogen atom of the amide cannot actually accept and hold a proton.

■ The oxygen atom of the carbonyl group is what makes the whole group electronegative.

Amides Are Made from Amines by Acyl Group Transfer Reactions Amides can be made from amines just as esters can be made from alcohols. Either acid chlorides or acid anhydrides react smoothly with ammonia or amines to give amides. (The amine, of course, must have at least one hydrogen atom on nitrogen, because one hydrogen has to be replaced as the amide forms.) We can illustrate these reactions using ammonia.

$$\underset{\text{Acid chloride}}{R{-}\overset{\overset{\displaystyle O}{\|}}{C}{-}Cl} + 2NH_3 \longrightarrow \underset{\text{Amide}}{R{-}\overset{\overset{\displaystyle O}{\|}}{C}{-}NH_2} + \underset{\text{Ammonium chloride}}{NH_4^+Cl^-}$$

$$\underset{\text{Acid anhydride}}{R{-}\overset{\overset{\displaystyle O}{\|}}{C}{-}O{-}\overset{\overset{\displaystyle O}{\|}}{C}{-}R} + 2NH_3 \longrightarrow \underset{\text{Amide}}{R{-}\overset{\overset{\displaystyle O}{\|}}{C}{-}NH_2} + \underset{\text{Ammonium salt of the carboxylic acid}}{NH_4^{+}\,{}^{-}O{-}\overset{\overset{\displaystyle O}{\|}}{C}{-}R}$$

■ When heated strongly, the ammonium salts of carboxylic acids change to the corresponding simple amides as water is expelled.

These reactions are further examples of *acyl group transfer reactions*. The acyl group in the acid chloride, for example, transfers from the Cl atom to the N atom of the amine (or ammonia). An acyl group can transfer from an acid anhydride to N, also. In both of these reactions, a stable, weakly basic leaving group, Cl^- or RCO_2^-, is released by the transferring acyl group. Direct acyl transfer from a carboxylic acid, however, is more difficult because the leaving group is the less stable strong base, OH^-.

In the body, other kinds of acyl carrier molecules serve instead of ordinary acid chlorides and anhydrides as sources of the acyl group. When proteins are made from amino acids, for example, the acyl portions of amino acids—they are called *aminoacyl units*—are held by carrier molecules.

$$\overbrace{NH_2{-}\underset{\underset{\displaystyle R}{|}}{CH}{-}\overset{\overset{\displaystyle O}{\|}}{C}}^{\text{Aminoacyl unit}}{-}\boxed{\text{Carrier molecule}}$$

■ R is some organic group, but not necessarily an alkyl group.

When a cell makes an amide bond, it transfers an aminoacyl group from its carrier molecule to the nitrogen atom of the amino group. The carrier molecule is released to be reused.

$$NH_2{-}\underset{\underset{\displaystyle R}{|}}{CH}{-}\overset{\overset{\displaystyle O}{\|}}{C}{-}\boxed{\text{Carrier molecule}} + NH_2{-}R' \xrightarrow[\text{group transfer}]{\text{Aminoacyl}} NH_2{-}\underset{\underset{\displaystyle R}{|}}{CH}{-}\overset{\overset{\displaystyle O}{\|}}{C}{-}NH{-}R' + \boxed{\text{Carrier molecule}} + H^+$$

■ The molecule given here as NH_2—R can be another aminoacyl group that is bound to another carrier molecule.

This is the aspect of making amides—aminoacyl transfers—that is of greatest interest as we prepare for our upcoming study of biochemistry. The skill we'll need is to figure out the structure of the amide that can be made from ammonia (or some amine) and a carboxylic acid regardless of the exact nature of the acyl transfer agent. We'll practice this in the next worked example.

EXAMPLE 6.3
Writing the Structure of an Amide That Can Be Made from the Acyl Group of an Acid and an Amine

What amide can be made from the following two substances, assuming that a suitable acyl group transfer process is available?

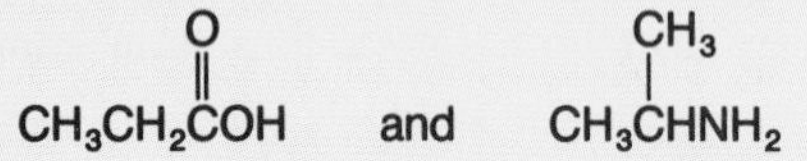

$$CH_3CH_2\overset{O}{\overset{\|}{C}}OH \quad \text{and} \quad CH_3\overset{CH_3}{\overset{|}{C}}HNH_2$$

ANALYSIS The amide system must be part of the structure we seek, so the best way to proceed is to write the skeleton of the amide system and then build on it. It doesn't matter how we orient this skeleton, left-to-right or right-to-left, as we'll demonstrate by showing both approaches.

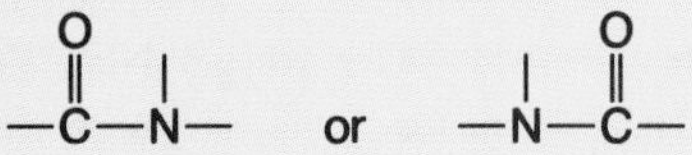

$$-\overset{O}{\overset{\|}{C}}-\overset{|}{N}- \quad \text{or} \quad -\overset{|}{N}-\overset{O}{\overset{\|}{C}}- \quad \text{(Incomplete)}$$

Then we look at the acid to see what else must be on the carbon atom of this skeleton. It's an ethyl group, so we write it in:

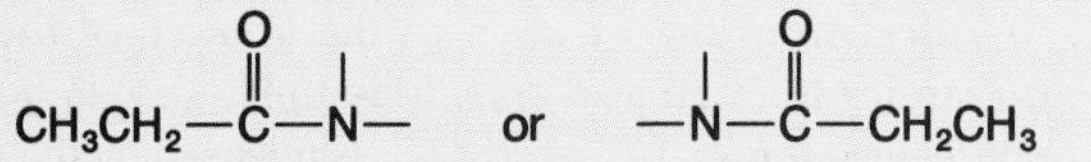

$$CH_3CH_2-\overset{O}{\overset{\|}{C}}-\overset{|}{N}- \quad \text{or} \quad -\overset{|}{N}-\overset{O}{\overset{\|}{C}}-CH_2CH_3 \quad \text{(Incomplete)}$$

Then we look at the amine to see what group(s) it carries. It has an isopropyl group, so we attach it to the N atom. (If there *had been two* organic groups on N in the amine, we would attach both, of course.)

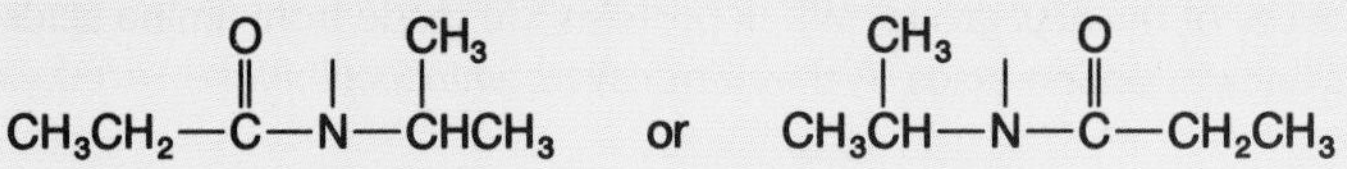

$$CH_3CH_2-\overset{O}{\overset{\|}{C}}-\overset{|}{N}-\overset{CH_3}{\overset{|}{C}}HCH_3 \quad \text{or} \quad CH_3\overset{CH_3}{\overset{|}{C}}H-\overset{|}{N}-\overset{O}{\overset{\|}{C}}-CH_2CH_3 \quad \text{(Incomplete)}$$

Finally, of the two H atoms on N in the amine, one survives, and our last step is to write it in. (Recall that N needs three bonds in a neutral species.)

SOLUTION The final answer is

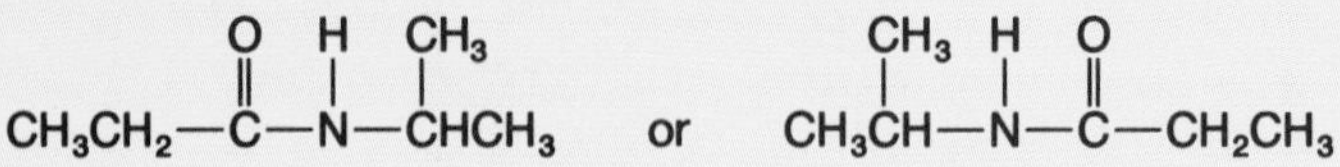

$$CH_3CH_2-\overset{O}{\overset{\|}{C}}-\overset{H}{\overset{|}{N}}-\overset{CH_3}{\overset{|}{C}}HCH_3 \quad \text{or} \quad CH_3\overset{CH_3}{\overset{|}{C}}H-\overset{H}{\overset{|}{N}}-\overset{O}{\overset{\|}{C}}-CH_2CH_3$$

These structures, of course, are identical.

PRACTICE EXERCISE 6

What amides, if any, could be made by suitable acyl group transfer reactions from the following pairs of compounds?

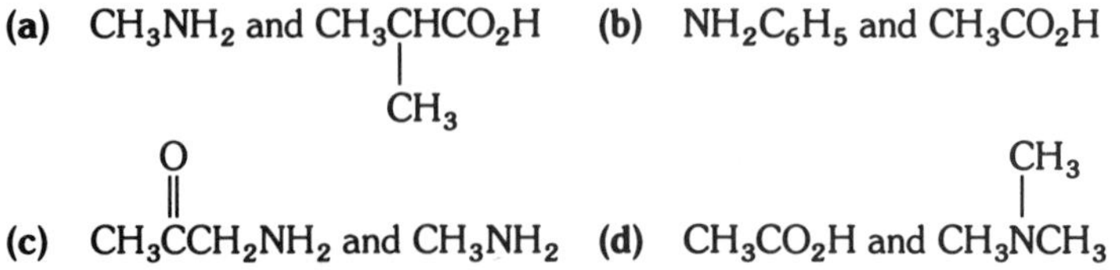

(a) CH_3NH_2 and $CH_3\underset{CH_3}{\underset{|}{C}}HCO_2H$ (b) $NH_2C_6H_5$ and CH_3CO_2H

(c) $CH_3\overset{O}{\overset{\|}{C}}CH_2NH_2$ and CH_3NH_2 (d) CH_3CO_2H and $CH_3\overset{CH_3}{\overset{|}{N}}CH_3$

Amides Are Hydrolyzed to Their Parent Amines and Acids The only reaction of amides that we will study is their hydrolysis, a reaction in which the amide bond breaks and we obtain the amide's parent acid and amine (or ammonia). The hydrolysis of an amide does not occur easily. *In vitro,* either acids or bases are needed to promote the reaction. *In vivo,* enzymes catalyze amide hydrolysis, and this reaction is all that is involved in the overall chemistry of the digestion of proteins.

■ The digestive tract provides several protein-digesting enzymes called *proteases.*

In vitro, when an acid promotes the hydrolysis of an amide, one of the products, the amine, neutralizes the acid. (This is why we don't say that the acid *catalyzes* the hydrolysis. Catalysts, by definition, are reaction promoters that are not used up.) Thus instead of obtaining the amine itself, we get the salt of the amine. For example,

$$R-\overset{\overset{\large O}{\|}}{C}-NH-CH_3 + H-OH + HCl(aq) \longrightarrow R-\overset{\overset{\large O}{\|}}{C}-OH + H-\overset{+}{\underset{H}{\overset{H}{N}}}-CH_3\ Cl^-$$

On the other hand, if we use a base to promote amide hydrolysis, then the carboxylic acid that forms neutralizes the base, and we get the salt of the carboxylic acid. For example,

$$R-\overset{\overset{\large O}{\|}}{C}-NH-CH_3 + NaOH(aq) \longrightarrow R-\overset{\overset{\large O}{\|}}{C}-O^-Na^+ + H-\overset{\overset{\large H}{|}}{N}-CH_3$$

When enzymes catalyze this hydrolysis, they are not used up by the reaction. Because our applications of this reaction concern *in vivo* situations at the molecular level of life, we'll write amide hydrolysis as a simple reaction with water to give the free carboxylic acid and the free amine. Here are some examples. How the acid and the amine actually emerge depends on the pH of the medium and the buffers present.

$$R-\overset{\overset{\large O}{\|}}{C}-NH_2 + H_2O \xrightarrow{\text{enzyme}} R-\overset{\overset{\large O}{\|}}{C}-OH + NH_3$$

$$R-\overset{\overset{\large O}{\|}}{C}-NHR' + H_2O \xrightarrow{\text{enzyme}} R-\overset{\overset{\large O}{\|}}{C}-OH + NH_2-R'$$

$$R-\overset{\overset{\large O}{\|}}{C}-\overset{\overset{\large R''}{|}}{N}R' + H_2O \xrightarrow{\text{enzyme}} R-\overset{\overset{\large O}{\|}}{C}-OH + H-\overset{\overset{\large R''}{|}}{N}-R'$$

EXAMPLE 6.4
Writing the Products of the Hydrolysis of an Amide

Acetophenetidin (phenacetin) was once used in some brands of headache remedies.

$$CH_3CH_2-O-C_6H_4-NH-\overset{\overset{\large O}{\|}}{C}-CH_3$$

Acetophenetidin

If this compound is an amide, what are the products of its hydrolysis?

ANALYSIS Acetophenetidin does have the amide bond, NH to carbonyl, so it can be hydrolyzed. (The functional group on the left side of this structure is an *ether*, and *ethers do not react with water.*) Because the amide bond breaks when an amide is hydrolyzed, simply erase this bond from the structure and separate the parts. *Do not break any other bond.*

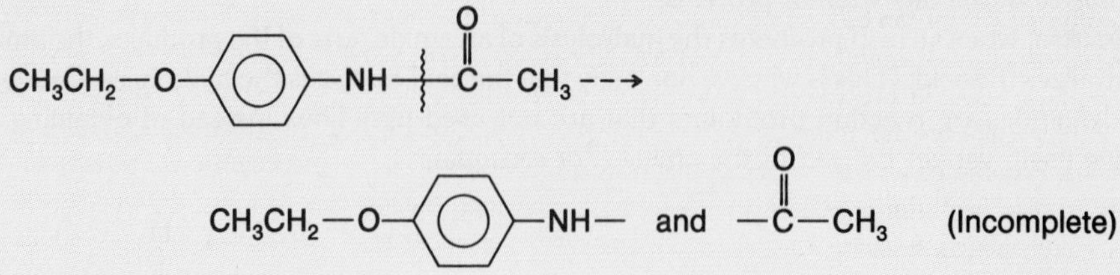

$$CH_3CH_2{-}O{-}C_6H_4{-}NH \,\vert\, {-}\overset{O}{\overset{\|}{C}}{-}CH_3 \dashrightarrow$$

$$CH_3CH_2{-}O{-}C_6H_4{-}NH{-} \quad \text{and} \quad {-}\overset{O}{\overset{\|}{C}}{-}CH_3 \qquad \text{(Incomplete)}$$

We know that the hydrolysis uses HO—H to give a *carboxylic acid* and an *amine,* so we put a HO group on the carbonyl group we put H on the nitrogen of the other fragment.

SOLUTION The products of the hydrolysis of acetophenetidin are

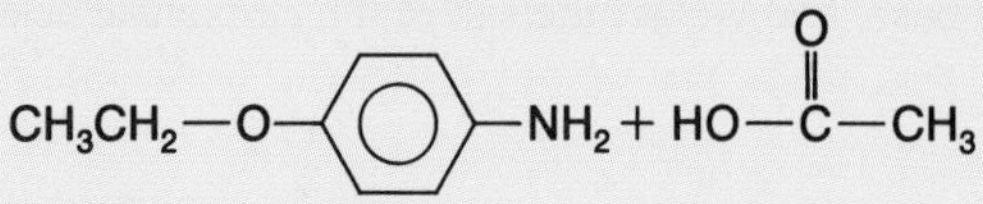

$$CH_3CH_2{-}O{-}C_6H_4{-}NH_2 + HO{-}\overset{O}{\overset{\|}{C}}{-}CH_3$$

PRACTICE EXERCISE 7

For all compounds in the following list that are amides, write the products of their hydrolysis.

(a) 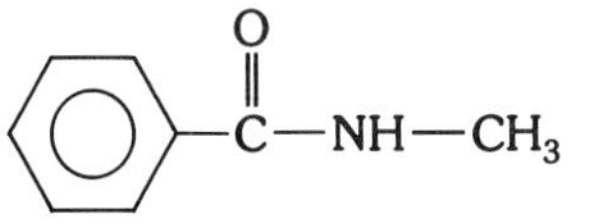 $C_6H_5{-}\overset{O}{\overset{\|}{C}}{-}NH{-}CH_3$ (b) $C_6H_5{-}\overset{O}{\overset{\|}{C}}{-}CH_2{-}NH_2$

(c) $C_6H_5{-}NH{-}\overset{O}{\overset{\|}{C}}{-}CH_3$ (d) $CH_3{-}\overset{O}{\overset{\|}{C}}{-}NH{-}CH_2CH_2{-}NH{-}\overset{O}{\overset{\|}{C}}{-}CH_3$

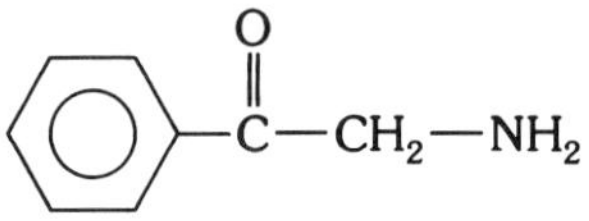

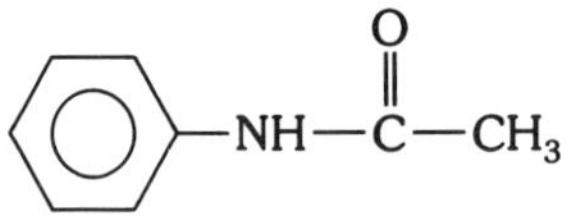

PRACTICE EXERCISE 8

The following structure illustrates some of the features of protein molecules. What are the products of the complete, enzyme-catalyzed hydrolysis (the digestion) of this substance? (A typical protein would hydrolyze to give several hundred and up to several thousand of the kinds of small molecules produced by hydrolysis in this example.)

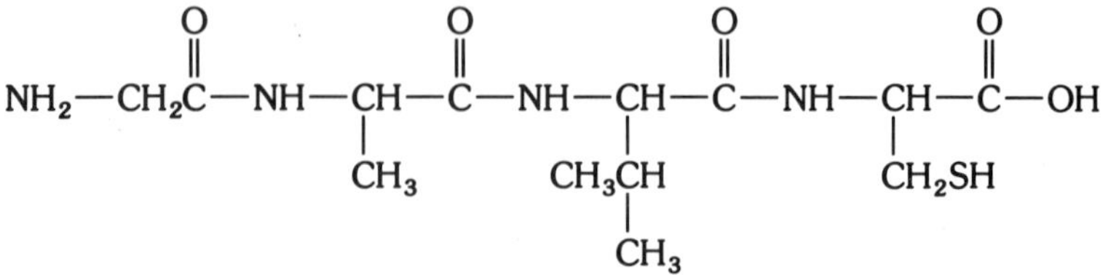

$$NH_2{-}CH_2\overset{O}{\overset{\|}{C}}{-}NH{-}\underset{CH_3}{\underset{|}{CH}}{-}\overset{O}{\overset{\|}{C}}{-}NH{-}\underset{CH_3\underset{CH_3}{\underset{|}{CH}}}{\underset{|}{CH}}{-}\overset{O}{\overset{\|}{C}}{-}NH{-}\underset{CH_2SH}{\underset{|}{CH}}{-}\overset{O}{\overset{\|}{C}}{-}OH$$

SUMMARY

Amines and protonated amines When one, two, or three of the hydrogen atoms in ammonia are replaced by an organic group (other than a carbonyl group), the resulting compound is an amine. The nitrogen atom can be part of a ring, as in heterocyclic amines. Like ammonia, the amines are weak bases, and all can form salts with strong acids. The cations in these salts are protonated amines.

Amine salts are far more soluble in water than their parent amines. Protonated amines are easily deprotonated by any strong base to give back the original and usually far less soluble amine. Thus any compound with the amine function has a "solubility switch," because its solubility in an aqueous system can be turned on by adding acid (to form the amine salt) and turned off again by adding base (to recover the amine).

Amides The carbonyl–nitrogen bond, the amide bond, can be formed by letting an amine or ammonia react with anything that can transfer an acyl group (e.g., an acid chloride or an acid anhydride). Amides are neither basic nor acidic, but are neutral compounds. Amides can be made to react with water to give back their parent acids and amines. The accompanying chart summarizes the reactions studied in this chapter.

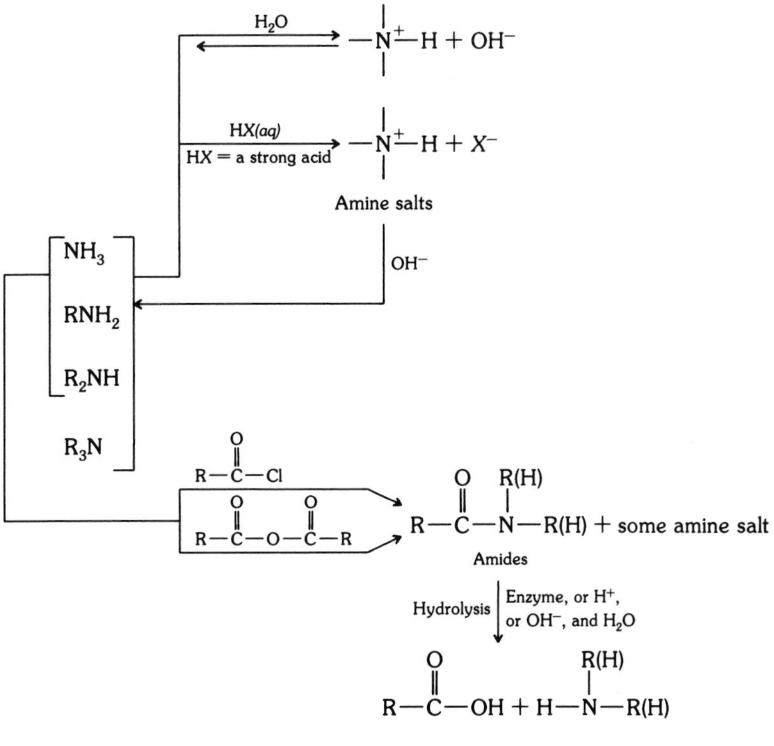

NH_3, RNH_2, R_2NH, R_3N

$$\xrightleftharpoons{H_2O} -\overset{|}{\underset{|}{N}}{}^{+}\!-H + OH^-$$

$$\xrightarrow[HX\,=\,\text{a strong acid}]{HX(aq)} -\overset{|}{\underset{|}{N}}{}^{+}\!-H + X^-$$

Amine salts

$\xrightarrow{OH^-}$ (back to RNH_2, etc.)

$$R-\overset{O}{\overset{\|}{C}}-Cl \quad \text{or} \quad R-\overset{O}{\overset{\|}{C}}-O-\overset{O}{\overset{\|}{C}}-R \longrightarrow R-\overset{O}{\overset{\|}{C}}-\overset{R(H)}{\overset{|}{N}}-R(H) + \text{some amine salt}$$

Amides

Hydrolysis ↓ Enzyme, or H^+, or OH^-, and H_2O

$$R-\overset{O}{\overset{\|}{C}}-OH + H-\overset{R(H)}{\overset{|}{N}}-R(H)$$

REVIEW EXERCISES

The answers to Review Exercises whose numbers are in color are found in Appendix A. The answers to the other Review Exercises are found in the Study Guide that accompanies this book. The more challenging questions are marked with asterisks.

Structures of Amines and Amides—Review of Functional Groups

6.1 Classify the following as aliphatic, aromatic, or heterocyclic amines or amides, and name any other functional groups, too.

(a) $CH_3OCH_2\overset{O}{\overset{\|}{C}}NH_2$ (b) $CH_3O\overset{O}{\overset{\|}{C}}CH_2NH_2$

(c) 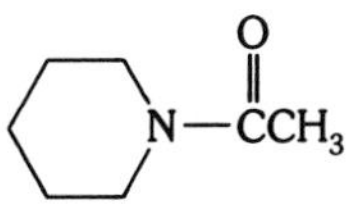 (piperidine ring)$N-\overset{O}{\overset{\|}{C}}CH_3$ (d) 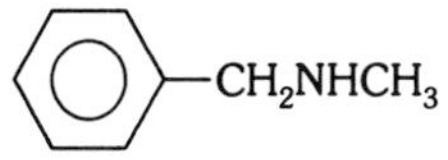 (benzene ring)$-CH_2NHCH_3$

6.2 Classify each of the following as aliphatic, aromatic, or heterocyclic amines or amides. Name any other functional groups that are present.

(a) 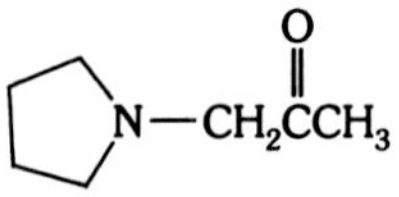 N—CH_2CCH_3 (b) 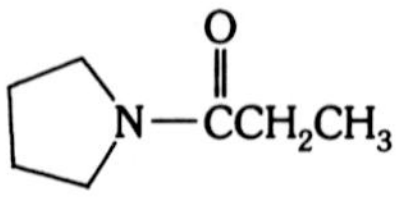 N—CCH_2CH_3

(c) 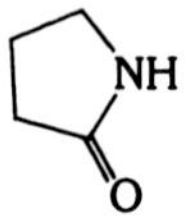 NH (d) 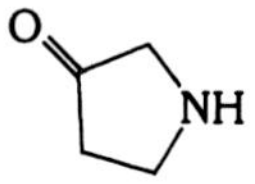 NH

*6.3 The following compounds are all very active physiological agents. Name the numbered functional groups that are present in each.

(a) 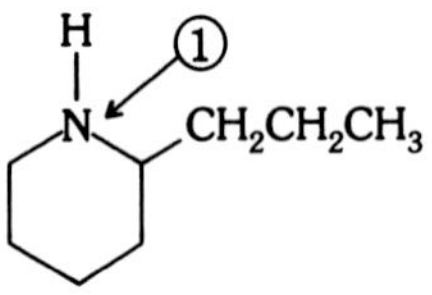 H, N, $CH_2CH_2CH_3$

Coniine, the poison in the extract of hemlock that was used to execute the Greek philosopher Socrates

(b) 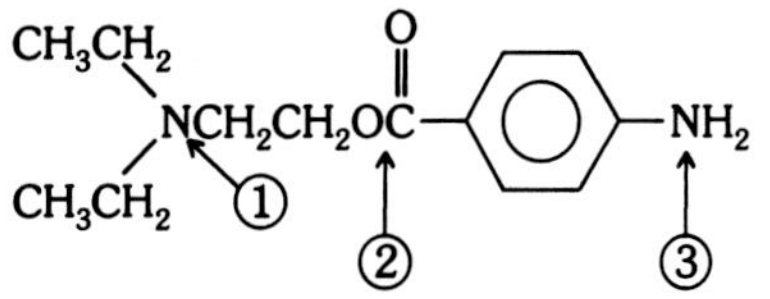 CH_3CH_2, NCH_2CH_2OC, CH_3CH_2, NH_2

Novocaine, a local anesthetic

(c) 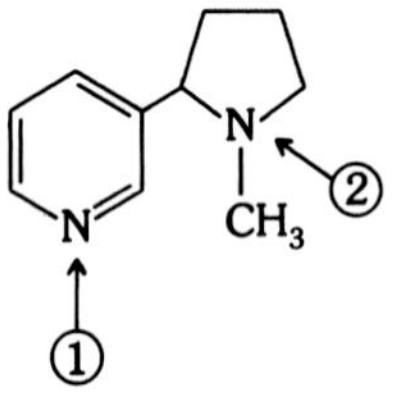 N, N, CH_3

Nicotine, a poison in tobacco leaves

(d) 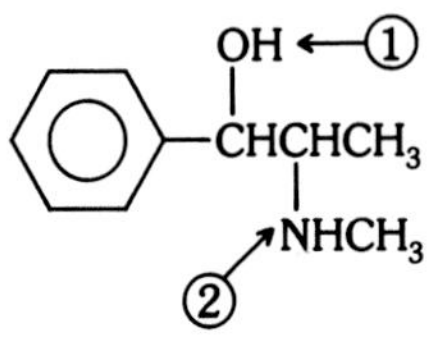 OH, $CHCHCH_3$, $NHCH_3$

Ephedrine, a bronchodilator

*6.4 Some extremely potent, physiologically active compounds are in the following list. Name the functional groups that they have.

(a) 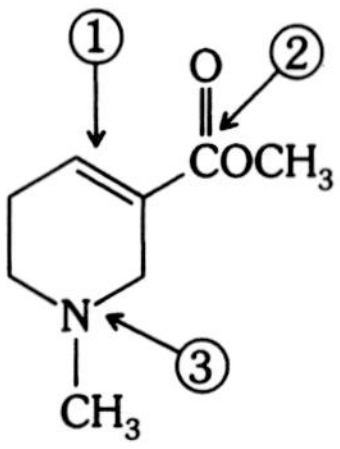 O, $COCH_3$, N, CH_3

Arecoline, the most active component in the nut of the betel palm. This nut is chewed daily as a narcotic by millions of inhabitants of parts of Asia and the Pacific islands.

(b) 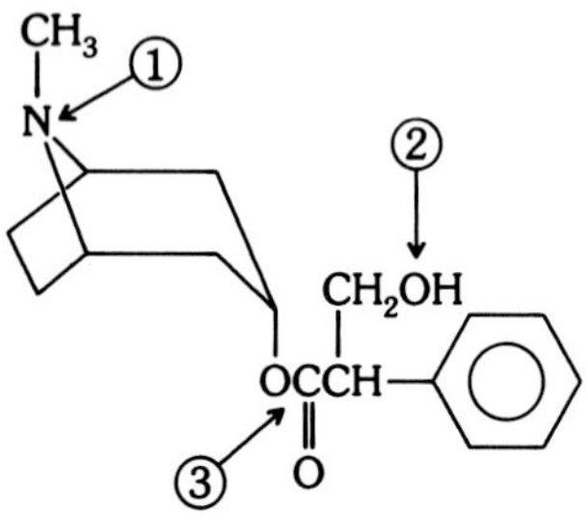 CH_3, N, CH_2OH, OCCH, O

Hyoscyamine, a constituent of the seeds and leaves of henbane, and a smooth muscle relaxant. (A similar form is called atropine, a drug used to counteract nerve poisons.)

(c) 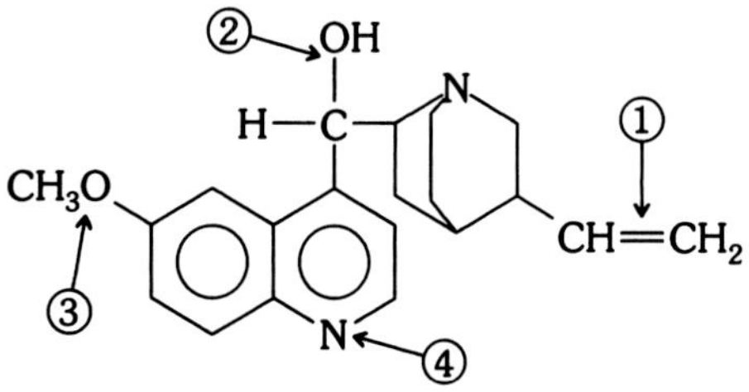 OH, H—C, N, CH_3O, $CH{=}CH_2$, N

Quinine, a constituent of the bark of the chinchona tree in South America and used to treat malaria

(d) 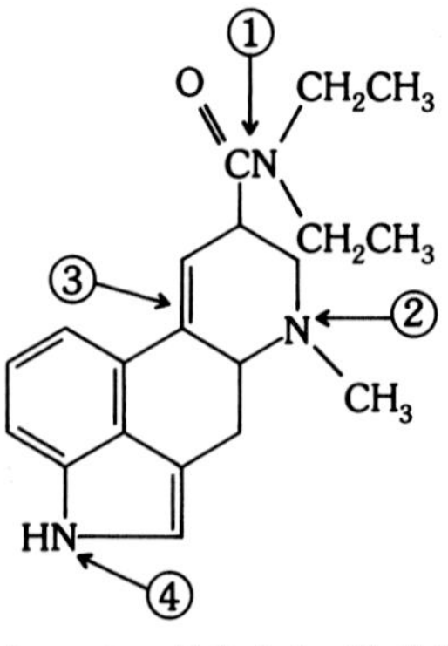 O, CN, CH_2CH_3, CH_2CH_3, N, CH_3, HN

Lysergic acid diethylamide (LSD), a constituent of diseased rye grain and a notorious hallucinogen

6.5 Give the common names of the following compounds or ions.

(a) $CH_3CH_2CH_2NHCHCH_3$ (with CH_3 on the CH)

(b) $CH_3CH_2CH_2N(CH_3)CH_2CH_3$

(c) NH_2–C_6H_4–Br

(d) $CH_3CH_2CH_2NHCH_2CH_2CH_3$

6.6 What are the common names of the following compounds?

(a) $(CH_3)_3NH^+Cl^-$ (b) C_6H_{11}–$NHCH_3$

(c) 3,5-dichloro-C_6H_3–NH_2 (d) $[(CH_3)_2CH]_3N$

Chemical Properties of Amines and Amine Salts

6.7 Complete the following reaction sequences by writing the structures of the organic products. If no reaction occurs, write "no reaction."

(a) $CH_3CH_2CH_2NH_2 + HCl(aq) \longrightarrow$
(b) $CH_3CH_2CH_2NH_3^+Cl^- + NaOH(aq) \longrightarrow$
(c) $CH_3CH_2CH_2NH_3^+Cl^- + HCl(aq) \longrightarrow$
(d) $CH_3CH_2CH_2NH_2 + NaOH(aq) \longrightarrow$

6.8 Write the structures of the organic products that form in each situation. Assume that all reactions occur at room temperature. (Some of the named compounds are described in Review Exercises 6.3 and 6.4.) If no reaction occurs, write "no reaction."

(a) Piperidine (ring NH) $+ HCl(aq) \longrightarrow$

(b) Ring with $COCH_3$ (ester, C=O) and $\overset{+}{N}$ bearing H and CH_3 $+ OH^-(aq) \longrightarrow$

Protonated form of arecoline

(c) Pyridine ring attached to pyrrolidine ring with $\overset{+}{N}$ bearing H and CH_3 $+ OH^-(aq) \longrightarrow$

Protonated form of nicotine

(d) C_6H_5–$CH(OH)CH(NHCH_3)CH_3 + HCl(aq) \longrightarrow$

Ephedrine

(e) Bicyclic amine (N–CH_3) with ester group $OC(=O)CH(CH_2OH)$–C_6H_5 $+ HCl(aq) \xrightarrow[\text{temperature}]{\text{at room}}$

Hyoscyamine

6.9 Which is the stronger base, **A** or **B?** Explain.

A: $NH_2CH_2CCH_3$ (with C=O) **B**: $CH_3CNHCH_2CH_3$ (with C=O)

6.10 Which is the stronger proton acceptor, **A** or **B?** Explain.

A: Piperidinium ring with $\overset{+}{N}$ bearing two CH_3 groups **B**: Cyclohexyl–$N(CH_3)_2$

Names and Structures of Amides

6.11 What are the IUPAC names of the following compounds?

(a) $CH_3CH_2CH_2CNH_2$ (with C=O)

(b) $CH_3CH(CH_3)CH_2CNH_2$ (with C=O)

6.12 If the common name of hexanoic acid is caproic acid, what is the common name of its simple amide?

6.13 If $C_6H_5CONHCH_3$ is the structure of *N*-methylbenzamide, what is the structure of *N,N*-dimethylbenzamide?

6.14 If ethanediamide has the structure shown, what is the structure of butanediamide?

$NH_2C(=O)$–$C(=O)NH_2$

Ethanediamide

***6.15** What is the structure of lysergic acid? The structure of its *N,N*-diethylamide was given in Review Exercise 6.4, part (d).

6.16 What is the structure of the amide that can form between acetic acid and ephedrine? (The structure of ephedrine is given in part (d) of Review Exercise 6.3).

Synthesis of Amides

6.17 Write the equations for two ways to make acetamide using ammonia as one reactant.

6.18 What are two different ways to make *N*-methylacetamide if methylamine is one reactant? Write the equations.

***6.19** Examine the following acyl group transfer reaction.

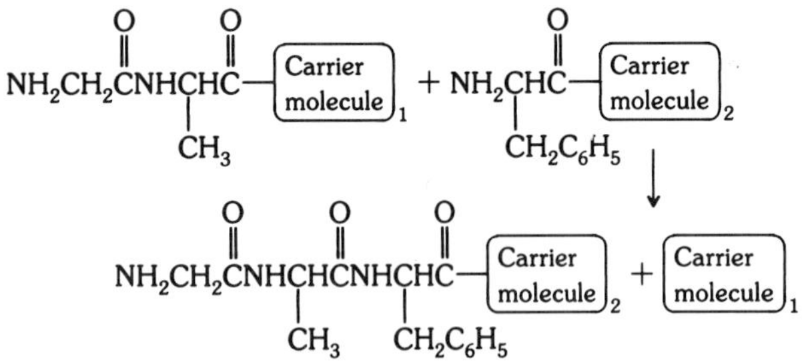

$$NH_2CH_2\overset{O}{\overset{\|}{C}}NH\underset{CH_3}{\underset{|}{C}}H\overset{O}{\overset{\|}{C}}-\boxed{\text{Carrier molecule}}_1 + NH_2\underset{CH_2C_6H_5}{\underset{|}{C}}H\overset{O}{\overset{\|}{C}}-\boxed{\text{Carrier molecule}}_2$$

$$\downarrow$$

$$NH_2CH_2\overset{O}{\overset{\|}{C}}NH\underset{CH_3}{\underset{|}{C}}H\overset{O}{\overset{\|}{C}}NH\underset{CH_2C_6H_5}{\underset{|}{C}}H\overset{O}{\overset{\|}{C}}-\boxed{\text{Carrier molecule}}_2 + \boxed{\text{Carrier molecule}}_1$$

(a) Which specific acyl group transferred? (Write its structure.)
(b) How many amide bonds are showing (or implied) in the product?

***6.20** If the following anhydride is mixed with ammonia, what possible organic products that are not salts can form? Write their structures.

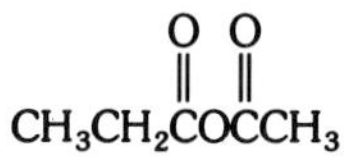

$$CH_3CH_2\overset{O}{\overset{\|}{C}}O\overset{O}{\overset{\|}{C}}CH_3$$

Reactions of Amides

6.21 What are the products of the hydrolysis of the following compounds? (If no hydrolysis occurs, state so.)

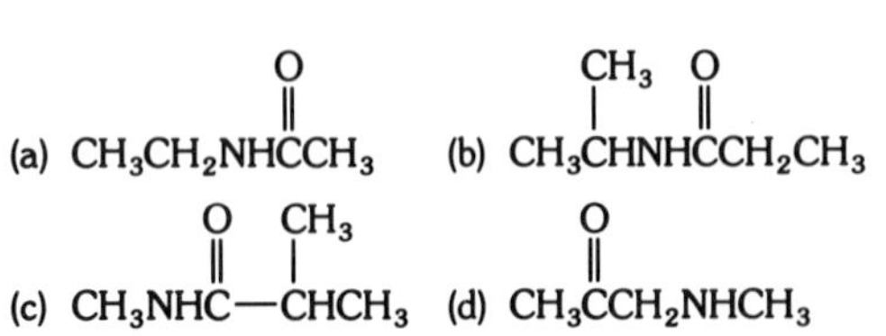

(a) $CH_3CH_2NH\overset{O}{\overset{\|}{C}}CH_3$ (b) $CH_3\overset{CH_3}{\overset{|}{C}}HNH\overset{O}{\overset{\|}{C}}CH_2CH_3$

(c) $CH_3NH\overset{O}{\overset{\|}{C}}-\overset{CH_3}{\overset{|}{C}}HCH_3$ (d) $CH_3\overset{O}{\overset{\|}{C}}CH_2NHCH_3$

6.22 Write the structures of the products of the hydrolysis of the following compounds. If no reaction occurs, state so. If more than one bond is subject to hydrolysis, be sure to hydrolyze all of them.

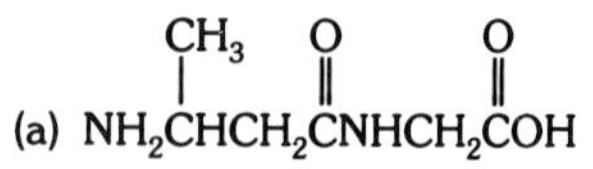

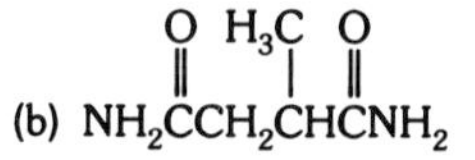

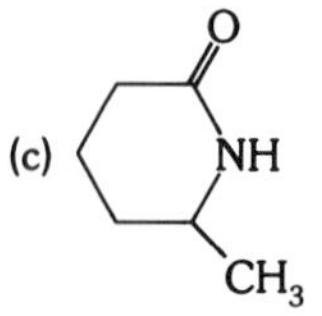

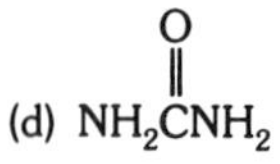

(a) $NH_2\overset{CH_3}{\overset{|}{C}}HCH_2\overset{O}{\overset{\|}{C}}NHCH_2\overset{O}{\overset{\|}{C}}OH$

(b) $NH_2\overset{O}{\overset{\|}{C}}CH_2\overset{H_3C}{\overset{|}{C}}H\overset{O}{\overset{\|}{C}}NH_2$

(c) six-membered ring lactam: C=O adjacent to NH, with CH_3 on the ring carbon adjacent to NH (ring atoms: O, NH, CH_3)

(d) $NH_2\overset{O}{\overset{\|}{C}}NH_2$

Physiologically Active Amines (Special Topic 6.1)

6.23 What are hormones and, in very broad terms, what is th[e] function?

6.24 Hormones secreted by the adrenal gland are called what kin[d] of agents?

6.25 Name two hormones secreted by the adrenal gland.

6.26 Drugs that tend to mimic the two hormones secreted by t[he] adrenal gland are called what kinds of drugs?

6.27 To be a *catecholamine* as well as a β-phenylethanolamine, [a] compound must have what structural features?

6.28 Is dopamine, a β-phenylethylamine, also a catecholamine?

6.29 In what general family of the physiologically active amines a[re] the amphetamines found?

6.30 What is the chemical name of each?
(a) "Speed" (b) "Uppers"

Nylon (Special Topic 6.2)

6.31 What functional group is present in nylon-66?

6.32 The strength of a nylon fiber is attributed in part to what rel[a]tively weak bond?

Additional Exercises (A Review of Organic Reactions)

***6.33** What are all the functional groups we have studied that can b[e] changed by each of the following reactants? Write the equ[a]tions for the reactions, using general symbols such as ROH [or] RCO_2H and so forth to illustrate these reactions, and nam[e] the organic families to which the reactants and products b[e]long.
(a) Water, either with an acid or an enzyme catalyst.
(b) Hydrogen (or a hydride ion donor) and any needed cat[a]lysts and special conditions.
(c) An oxidizing agent represented by (O), such as $Cr_2O_7^{2-}$ [or] MnO_4^-, but not ozone and not oxygen as used in combu[s]tion.

6.34 We have described three functional groups that typify thos[e] involved in the chemistry of the digestion of carbohydrate[s,] fats and oils, and proteins. What are the names of thes[e] groups and to which type of food does each belong?

***6.35** A student performed an experiment that hydrolyzed 1.64 g [of] benzamide.
(a) What is the maximum number of grams of benzoic ac[id] that could be obtained?

(b) How many milliliters of 0.482 *M* HCl would be needed to convert all of the ammonia that can form from the hydrolysis of the sample of benzamide into ammonium chloride?

*6.36 Write the structures of the organic products that would form in the following situations. If no reaction occurs, state so. These constitute a review of nearly all the organic reactions we have studied, beginning with Chapter 1.

(a) $CH_3C(=O)OCH_3 + H_2O \xrightarrow[\text{catalyst}]{\text{acid}}$

(b) $CH_3C(=O)CH_2CH_3 + H_2 \xrightarrow[\text{pressure}]{\text{Ni or Pt}}$

(c) $CH_3CH_2CH_2CH_2CH_3 + MnO_4^-(aq) \longrightarrow$

(d) $CH_3CH_2C(=O)Cl + NH_3\ (\text{excess}) \longrightarrow$

(e) $C_6H_{11}{-}CH_2CH_3 + Cr_2O_7^{2-}(aq) \longrightarrow$ (cyclohexyl)

(f) $CH_3CH(CH_3)CH_2CH_3 + NaOH(aq) \longrightarrow$

(g) $CH_3CH(OCH_2CH_3)OCH_2CH_3 + H_2O \xrightarrow[\text{catalyst}]{\text{acid}}$

(h) $CH_3CH_2C(=O)H + Cr_2O_7^{2-}(aq) \longrightarrow$

(i) $CH_3CH_2C(=O)CH_3 + Cr_2O_7^{2-}(aq) \longrightarrow$

(j) $CH_3OH + CH_3CH_2C(=O)OH \overset{H^+}{\rightleftharpoons}$

(k) $CH_3CH_2C(=O)NH_2 + NaOH(aq) \xrightarrow[\text{heat}]{}$

(l) $CH_3CH_2C(=O)H + 2CH_3OH \xrightarrow[\text{catalyst}]{\text{acid}}$

(m) $CH_3CH_2SH + (O) \longrightarrow$

(n) $C_6H_5C(=O)OCH_2CH(CH_3)CH_3 + NaOH(aq) \longrightarrow$

(o) $NH_2CH_2CH_2CH(CH_3)CH_3 + HCl(aq) \longrightarrow$

(p) $CH_3CH{=}CHCH_2OCH_3 + H_2 \xrightarrow{\text{Ni or Pt}}$

*6.37 What are the structures of the organic products that form in the following situations? (If there is no reaction, state so.) These reactions review most of the chemical properties of functional groups we have studied, beginning with Chapter 1.

(a) $HOC(=O)CH_2CH_2CH_3 + NaOH(aq) \longrightarrow$

(b) (tetrahydrofuran, cyclic ether C_4H_8O) $+ NaOH(aq) \longrightarrow$

(c) $CH_3CH_2OCH_2CH_2C(=O)H + MnO_4^-(aq) \longrightarrow$

(d) $NH_2CH_2CH_2NH_2 + HCl(aq) \longrightarrow$ (excess)

(e) $C_6H_{11}{-}C(=O)CH_2CH_3 + H_2 \xrightarrow[\text{pressure}]{\text{Ni or Pt}}$ (cyclohexyl)

(f) $CH_3(CH_2)_5CH_3 + Cr_2O_7^{2-}(aq) \longrightarrow$

(g) $CH_3{-}C_6H_4{-}C(=O)H + 2CH_3OH \xrightarrow[\text{catalyst}]{\text{acid}}$ (para-substituted benzene)

(h) $CH_3{-}C_6H_4{-}C(=O)OCH_2CH(CH_3)CH_3 + NaOH(aq) \longrightarrow$ (para-substituted benzene)

(i) $CH_3CH(OH)CH_2CH_3 + MnO_4^-(aq) \longrightarrow$

(j) $CH_3C(OCH_2CH_3)(CH_3)OCH_2CH_3 + H_2O \xrightarrow[\text{catalyst}]{\text{acid}}$

(k) $CH_3CH_2OC(=O)(CH_2)_3C(=O)OCH_2CH_3 + H_2O \xrightarrow[\text{catalyst}]{\text{acid}}$

(l) $CH_3OCH(CH_2CH_3)OCH_3 + H_2O \xrightarrow[\text{catalyst}]{\text{acid}}$

(m) $CH_3C(=O)OC(=O)CH_3 + NH_3\ (\text{excess}) \longrightarrow$

(n) $CH_3SSCH_3 \xrightarrow{\text{reduction (2H)}}$

(o) $CH_3(CH_2)_5CO_2H + CH_3OH \overset{H^+}{\rightleftharpoons}$

(p) $CH_3NHC(=O)CH_2CH_2NHC(=O)CH_2CH_3 + H_2O \xrightarrow{\text{enzyme}}$ (excess)

6.38 Write the IUPAC names of the following compounds.

(a) $CH_3CH_2CH_2CO_2CH_3$
(b) $(CH_3)_2CHCH_2Br$
(c) $O{=}CHCH_2CH(CH_3)_2$
(d) $CH_3CH_2CH{=}CHCH_3$
(e) $(CH_3)_3CCH_2CH(CH_3)_2$
(f) $CH_3CH_2CH_2COCH(CH_3)_2$
(g) $(CH_3)_3COH$
(h) $HO_2C(CH_2)_3CH_3$
(i) CH_3SH
(j) CH_3CO_2Na

6.39 Write the common names of the following compounds.

(a) $C_6H_5CO_2Na$
(b) $NH_2CH_2CH_2CH_3$
(c) $C_6H_5NH_2$
(d) $HO_2CCH_2CH_3$
(e) $O{=}CHCH_2CH_2CH_3$
(f) $HOCH_2CH(CH_3)_2$
(g) HOC_6H_5
(h) $CH_3CH_2OCH_2CH_3$
(i) $CH_3CH_2CH_2CO_2CH_2CH_3$
(j) CH_3CONH_2
(k) CH_3COCH_3

6.40 Identify by letter which of the following compounds would be more soluble in water at a pH of 12 than at a pH of 7. Explain.

$CH_3(CH_2)_6CO_2CH_3$ **A** $CH_3(CH_2)_6CO_2H$ **B**

$CH_3(CH_2)_6CH_2NH_2$ **C**

6.41 Identify by letter which of the following compounds would be more soluble in water at a pH of 2 than at a pH of 7. Explain.

$C_6H_5CH_2CH_2CONH_2$ **A** $C_6H_5CH_2COCH_2NH_2$ **B**

$C_6H_5CH_2CH_2CO_2H$ **C**

STEREOISOMERISM

7

Light that reflects from a flat surface is rich in polarized light, which causes glare. A polarizing lens, however, removes polarized light and reduces glare, making this photo of a stream near Sedona, Arizona, very pleasant.

7.1

TYPES OF ISOMERISM

Constitutional isomers and stereoisomers are the two broad classes of isomers. Compounds that have identical molecular formulas can be different in two general ways, as *constitutional isomers* or as *stereoisomers*.

■ Constitutional isomers are often called *structural isomers*, an older name now being replaced.

Constitutional Isomers Differ in Molecular Frameworks Butane and isobutane, both C_4H_{10}, have different chains. Ethanol and dimethyl ether, both C_2H_6O, have different functional groups. Each pair illustrates **constitutional isomerism** because their members differ in the ways by which the atoms are joined to each other. The molecules of **constitutional isomers** have different atom-to-atom sequences.

$CH_3CH_2CH_2CH_3$ Butane

$CH_3CH(CH_3)CH_3$ 2-Methylpropane (isobutane)

CH_3CH_2OH Ethanol

CH_3OCH_3 Dimethyl ether

Stereoisomers Differ Only in Geometry *cis*-2-Butene and *trans*-2-butene illustrate **stereoisomerism** because they have the same constitutions—identical molecular formulas, functional groups, and heavy-atom skeletons—but display different geometries.

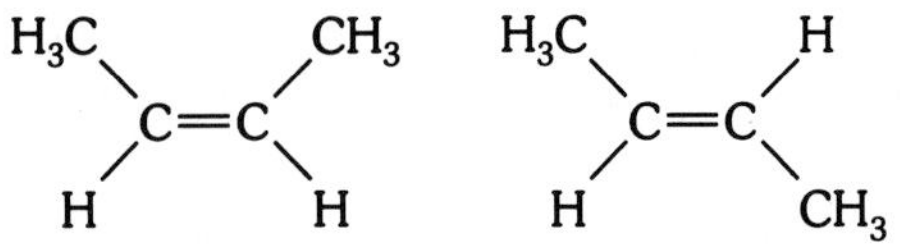

$(H_3C)(H)C{=}C(CH_3)(H)$ *cis*-2-Butene

$(H_3C)(H)C{=}C(H)(CH_3)$ *trans*-2-Butene

■ STER-ee-oh-EYE-som-ers

Stereoisomers have identical constitutions but different geometries. Thus *cis*- and *trans*-2-butene are identical in sharing the same constitution, $CH_3CH{=}CHCH_3$, but they differ in geometry. The lack of free rotation at the double bond ensures that one isomer cannot easily switch to the other.

■ Recall that the four atoms attached directly to the carbon atoms of the double bond all lie in the same plane.

■ dye-a-STER-ee-o-mers en-AN-tee-o-mers

There are two broad families of stereoisomers, *diastereomers* (illustrated by *cis*- and *trans*-2-butene) and *enantiomers*. This chapter is mostly about the latter, but to define them requires more background. Enantiomers come about as close to being identical as your left and right hands, yet they display dramatic differences in chemistry at the molecular level of life. One such difference is illustrated by the bittersweet story of asparagine.

Asparagine is a white solid that was first isolated in 1806 from the juice of asparagus. The asparagine obtained from this source has a bitter taste. Its molecular formula is $C_4H_8N_2O_3$, and its structure is given by structure **1**.

■ Asparagine is a building block for making proteins in the body.

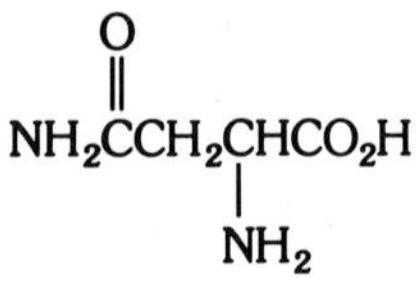

$NH_2C(=O)CH_2CH(NH_2)CO_2H$

1 Asparagine

In 1886, a chemist isolated from sprouting vetch a white substance with the same molecular formula *and constitution,* but it had a sweet taste. To have names for these, the one isolated from asparagus is now called L-asparagine, and the one from vetch sprouts is called D-asparagine.

■ Vetch is a member of a genus of herbs, some of which are useful as fodder for cattle.

Here were two substances seemingly answering to the same structure despite what is almost a dogma in chemistry, the principle of *one-substance – one-structure*. If two samples of matter have identical physical and chemical properties, then they must be identical at the level of their individual formula units. If two samples differ in even one way in their fundamental properties, then there has to be at least one difference in the way that the atoms are put together into their molecules.

Taste is a chemical sense, so the two samples of asparagine do have one difference in chemical property. Under the one-substance – one-structure rule, therefore, the molecules of D- and L-asparagine must be structurally different in at least one way. This difference arises from a peculiar lack of symmetry in their molecules that makes possible two different relative configurations of their molecular parts.

■ The letters D and L will acquire more specific meaning in the next chapter. Consider them to be only labels now.

7.2 MOLECULAR CHIRALITY

The molecules of many substances have a handedness like that of the left and the right hands.

Two partial ball-and-stick molecular models of asparagine are shown in Figure 7.1*a*. Examine each model to make sure that it faithfully represents asparagine, structure **1**. To make the study of these molecules easier, we have simplified them as shown in Figure 7.1*b*. Note again how alike the two molecular structures are. Notice particularly that the same four groups are attached to a central carbon atom, and that there is no cis–trans kind of difference between them. Yet the fact remains that one represents a bitter-tasting compound, and the other is of a sweet-tasting compound. In some way these two similar structures have to be different, and the difference isn't something that can be removed by rotating groups around single bonds.

Two Materials Whose Molecules Can Be Superimposed Are Samples of the Same Compound To understand how the two asparagine structures are different, we first have to learn the ultimate test for deciding whether two structures are the same. *For two structures to be identical, it must be possible to superimpose them.* To superimpose two structures means to do a manipulation that you can carry to completion only in your mind, but the use of molecular models of the structures is a great help. **Superimposition** is the mental blending of one molecular model with another so that the two would coincide in every atom and bond simultaneously if the operation could actually be completed. (When working with molecular models, it's fair to twist parts about single bonds to find conformations that can be superimposed, but it's not legal to break any bonds.) Superimposition is illustrated in the lower left-hand side of Figure 7.1*c*, where two *identical* models of *one* of the asparagines are used.

■ In some references you'll see the word *superposition* used for *superimposition*.

As we said, the fundamental criterion of identity of two structures is that they pass the test of superimposition, a test that we will see is failed by the two *different* asparagines of Figure 7.1*b*. In Figure 7.1*c*, the model on the left in Figure 7.1*b* has been turned counterclockwise by 120° around the vertical bond from C to the group **g** and then placed as the object in front of the mirror. Now look at the reflection of this model in the mirror. If you made an exact molecular model of its reflection, the new model would be identical to that of the *other* asparagine, the one on the right in Figure 7.1*b*. This is how nearly identical the two asparagines are. In the lower right of part Figure 7.1*c*, you can see that these two models, the one in front of the mirror as the object and the model of its reflection, do not superimpose.

FIGURE 7.1
The two stereoisomers of asparagine. (*a*) Two ways of joining the four groups to the carbon marked by the asterisk. (*b*) Simplified representations of the models in part *a*. (*c*) What is in front of the mirror is identical with the model on the left in part *b*. What is seen in the mirror as the image is identical with the model on the right in part *b*. You'll mentally have to spin the mirror image 120° counterclockwise about the bond from C to **g** to see that they are the same. The object and its mirror image do not superimpose, so they can't be identical. Instead, they are enantiomers.

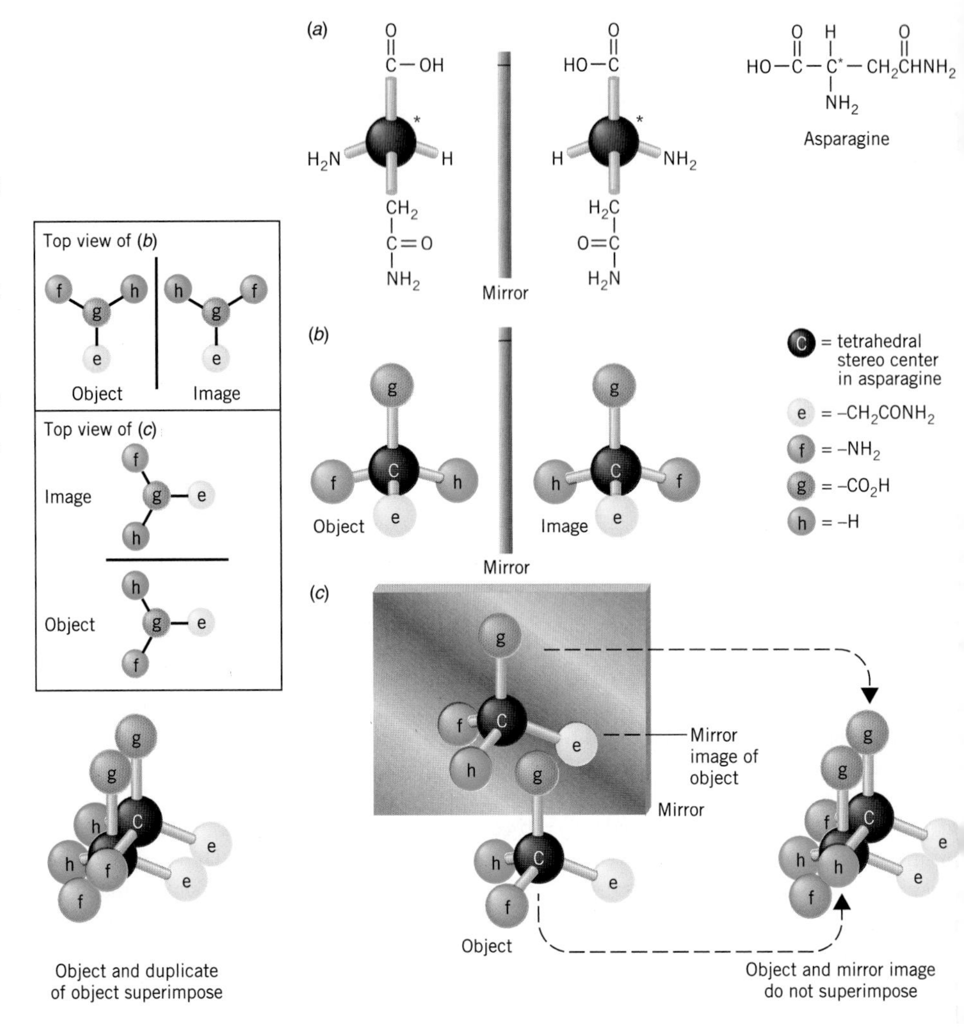

■ **Constitutional isomers:** Identical molecular formulas but different arrangements of atoms.
Stereoisomers: Identical *constitutions* but different geometries.
Enantiomers: Pairs of *stereoisomers* whose molecules are related as object to mirror image but cannot be superimposed.
Diastereomers: *Stereoisomers* that are not enantiomers.

The Molecules of Enantiomers Are Related as an Object to a Mirror Image but Cannot Be Superimposed As we said, the two different asparagine molecules are so alike that the molecule of one is the mirror image of the other. Yet the model and its mirror image do not superimpose. Pairs of stereoisomers whose molecules are related as object to mirror image that cannot be superimposed are called **enantiomers.**

Always remember two general facts about isomers of any kind. They are truly *different substances,* different compounds, but they share the same molecular formula while differing in the arrangements of their atoms. Enantiomers are just special kinds of stereoisomers. All other isomers that qualify as *stereoisomers* are called **diastereomers.** Diastereomers are stereoisomers that are not enantiomers. Thus all purely geometric (cis–trans) isomers are diastereomers but other examples do exist, as we'll see in Special Topic 7.1 later in this chapter. Figure 7.2 sorts out the kinds of isomers we have studied.

Molecules of a Pair of Enantiomers Have Opposite Configurations Lack of free rotation is not the cause of the asparagine enantiomers. Their molecules, instead, are described as having *opposite configurations.* To show what this means, we have repositioned their abbreviated molecular models in Figure 7.3. (Imagine a mirror standing between the two and perpendicular to the page to see that they are related as object to mirror image.)

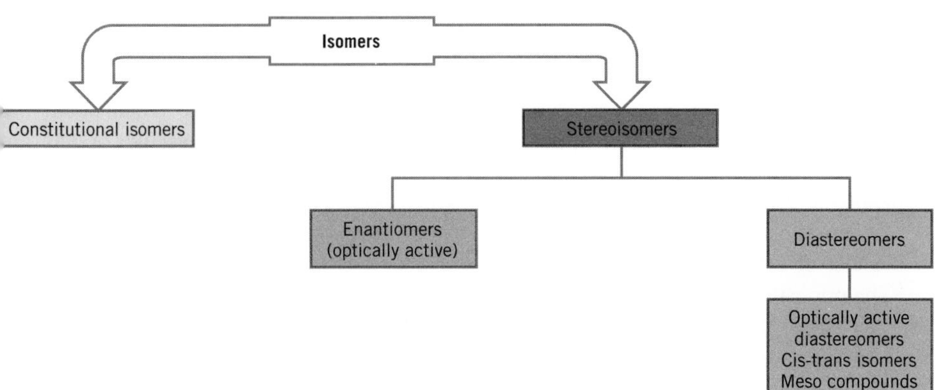

FIGURE 7.2
The relationships of various kinds of isomers.

You're now going to let your eyes make a special scanning trip around each molecule. Imagine that the bond from C to H in each model is the steering column of the steering wheel of a car. Then imagine that the remaining three groups—**e, f,** and **g**—are distributed around the steering wheel itself. Now move your eyes from **g** to **e** and then to **f.** When you do this with the asparagine model on the left in Figure 7.3, your eyes move clockwise. But to make the identical trip—**g** to **e** to **f**—in the model on the right, your eyes move counterclockwise. These clockwise versus counterclockwise arrangements of identical parts around the same central axis are what having *opposite configuration* means. The four groups on the central carbon in one asparagine are the same four groups as in the other, but they are configured oppositely in space.

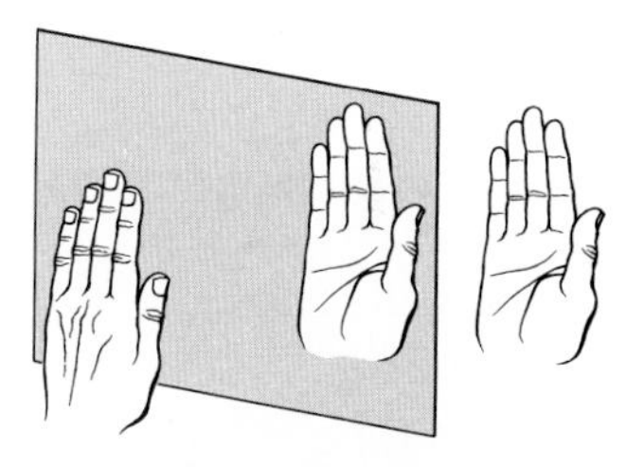

The image of the left hand is just like the right hand in the relative orientations of the fingers, thumb, and palm.

Molecules of a Pair of Enantiomers Have Opposite Chirality We have to remind ourselves now that any object has a mirror image. Spheres, cubes, broom handles, water glasses, and so on, can all be reflected in a mirror. It's only when the model of the object and the model of the mirror image cannot be superimposed that we call the two enantiomers.

Your two hands are like a pair of enantiomers, if you disregard small differences such as wrinkles, scars, rings, and fingerprints. Place your left hand in front of a mirror near its edge. Place your other hand just off the edge of the mirror and notice that the reflection of your left hand in the mirror is the same as your right hand (disregarding, as we said, the small differences). Your right hand is the mirror image reflection of your left hand.

Next, try to superimpose the two hands. Because the mirror image of your left hand is your real right hand, use your two hands to see whether they superimpose. If you put them palm to palm it seems as though all the fingers do superimpose. But remember, you have to carry this blending through to completion (in your mind), and when you do, the palm of one hand comes out on the back side of the other. The palms won't superimpose when the fingers seem to. And if you try to get the palms to come out right, then the fingers come out all wrong. Left and right hands, although related as object to mirror image, don't superimpose. They are related as enantiomers.

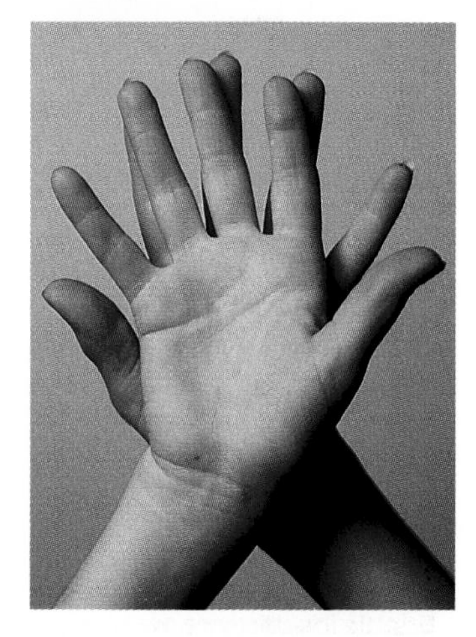

The two hands are related as an object to its mirror image, but they can't be superimposed.

Notice, now, how the two hands have opposite configurations. Look at them down the same axis, as we did with the asparagine models in Figure 7.3, say, down the axis from palm to backside. This means that both palms will face you. To make the trip from thumb to little

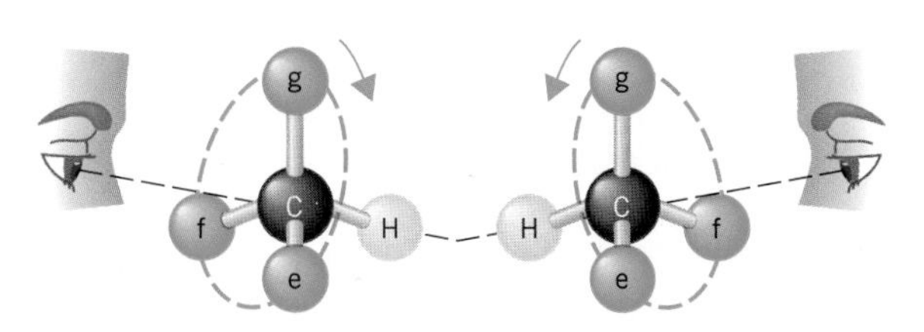

FIGURE 7.3
When the two stereoisomers of asparagine are viewed down the same axis, the C—H bond axis, the remaining three groups at the central carbon atom have opposite configurations.

finger, touching every other finger on the way, you have to scan in one direction for one hand and in exactly the opposite direction for the other. Thus the two hands have opposite configurations.

The little experiment with the hands has been used for decades to teach about the configurational differences of enantiomers, and this is why we say that molecules of different enantiomers have different *handedness*. The technical term for this configurational property is **chirality** (from the Greek word for "hand," which is *cheir*). We say that *the molecules of enantiomers have opposite chirality,* meaning opposite handedness. We also say that the molecules of any given enantiomer are **chiral**—they possess handedness or chirality.

The opposite of chiral is **achiral.** An achiral molecule is one that is symmetrical enough so that its model and the model of its mirror image do superimpose. The methane molecule is achiral, for example. Some examples of larger achiral objects include a cube, a sphere, a broom handle, and a water glass.

Chirality Can Make Enormous Differences at the Molecular Level of Life One asparagine enantiomer tastes sweet and the other tastes bitter—a large (although a somewhat trivial) difference that chirality can make. The details are not fully known, but the taste mechanism probably begins with a chemical reaction that is catalyzed by an enzyme. The enzymes involved, *like all enzymes,* themselves consist of very large, chiral molecules with molecular surfaces that are different for each enzyme.

A left-hand glove does not fit the right hand.

The substance whose reaction an enzyme catalyzes is called the **substrate** for that enzyme. An enzyme works by letting a molecule of the substrate come and temporarily fit into the contours on the enzyme's surface. This idea of fitting can be illustrated by a return to our hands, only now we'll add gloves. Gloves, like hands, are chiral, and a glove fits well only to its matching hand. The left hand fits well into the left glove, not the right glove. Now suppose that an enzyme responsible ultimately for the sensation of a sweet taste is like a glove for the right hand. This means that only the substrate molecules that have the matching handedness can interact with this enzyme. Substrate molecules of the opposite handedness can't fit to this enzyme.

We can now shift from this analogy of hands and gloves to chiral molecules with the aid of Figure 7.4. To make it easier, we have used simple geometric forms to create two enantiomers and indentations that match these forms are part of the enzyme surface. One enantiomer can fit to the enzyme surface, but no matter how you turn the model of the other enantiomer you can't get it to fit to the same enzyme surface.

There is, evidently, a different enzyme whose surface chirality matches the other asparagine enantiomer that lets us know that this other enantiomer has a different taste. The

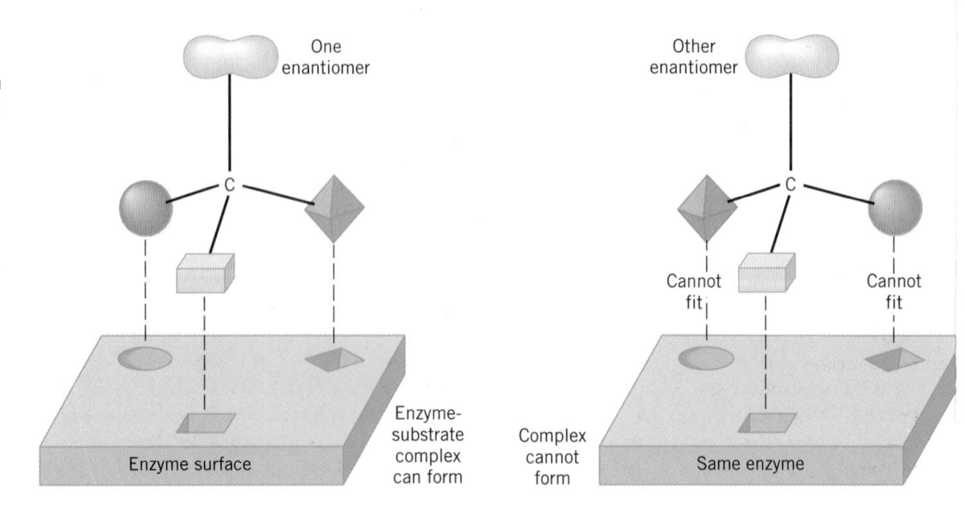

FIGURE 7.4
Because an enzyme is chiral, it can accept substrate molecules of only one of a pair of enantiomers. To illustrate this difference, we have used simple geometric forms. On the left, the enzyme can accept as a substrate the molecule of one enantiomer. On the right, the same enzyme can't accept a molecule of the other enantiomer, because the shapes don't match.

phenomenon of chirality is absolutely central to this difference. And the different chemical properties that relate to the taste of asparagine are illustrated in countless ways at the molecular level of life. We'll see time after time that *differences in geometry and configuration are as important as functional groups to the chemical reactions of life.*

Chirality Does Not Affect Reactions with Achiral Compounds If the enzyme molecule were not chiral, it could not discriminate between enantiomers. In fact, *enantiomers have identical chemical properties toward all reactants whose molecules are achiral*—reactants such as H_2O, NaOH, HCl, Cl_2, NH_3, and H_2. A broom handle, which isn't chiral, is an analogy. It fits just as easily to the left hand as to the right. It can't discriminate between the hands. In like manner, *the molecules of an achiral reactant cannot tell the difference between molecules of enantiomers.* To summarize, enantiomers react differently toward reactants whose molecules are chiral but identically toward achiral reactants.

Two Enantiomers Have Identical Physical Properties, with One Exception Two enantiomers have identical melting points, boiling points, densities, and solubilities in common (i.e., achiral) solvents like water, diethyl ether, or ethanol. This *must* be so, because the molecules of two enantiomers *must* have identical molecular polarities. They have, after all, identical intranuclear distances and bond angles.

Returning once again to our analogy with the hands, you can see that the distance, say, from the tip of the thumb to the tip of the little finger is the same in both hands. You can pick any other such intrahand distance that you please, and it is the same in both hands. Similarly, the angle between any two fingers is the same in both hands when the hands are spread identically. In like manner, if we pick any distance between atoms or any bond angle in one asparagine enantiomer, we will find it to be the same in the other. When all these distances and angles are identical in the two enantiomers, their molecules as a whole must have identical polarities. This is responsible for the identical physical properties that we mentioned. There is one difference in physical property, however, which we will study in Section 7.3.

Molecules with One Carbon Holding Four Different Groups Are Chiral It is important that we be able to recognize when a potential substrate is chiral, because molecules of opposite chirality have such different chemical properties at the molecular level of life. How, then, can we look at a structure and tell whether its molecules are chiral without making molecular models of both object and mirror image?

In all examples of chiral molecules that we'll encounter, their molecules always have at least one carbon to which *four different groups* are attached. A carbon that holds four different atoms or groups is called a **tetrahedral stereocenter.** A *stereocenter* is an atom in a molecule whose attached groups can be arranged in different configurations to give different stereoisomers. A *tetrahedral* stereocenter is simply a stereocenter having four single bonds arranged tetrahedrally.[1]

■ An older (but still widely used) name for *tetrahedral stereocenter* is **chiral atom.** When the atom is C, the older term was *chiral carbon.*

The asparagine molecule has one (and just one) tetrahedral stereocenter, and when a molecule has only one such center, we can be absolutely certain that the molecule as a whole is itself chiral. Here, then, is one way to predict if a substance consists of chiral molecules; we look for a tetrahedral stereocenter. If we find *one,* we can be certain that the molecules are chiral.[2]

[1] Alkenes that are cis and trans isomers of each other (e.g., the geometric isomers of 2-butene) have two stereocenters, two carbon atoms each holding *three* atoms or groups that, when differently arranged, give the two stereoisomers—the cis and the trans. The two carbon atoms at the double bond are thus not tetrahedral stereocenters but *trigonal* stereocenters.

[2] If we find more than one tetrahedral stereocenter, we have to be careful. The complication is treated in Special Topic 7.1. Probably 99.99% of all examples of substances that have two or more tetrahedral stereocenters also have chiral molecules. A few exceptions, however, exist where the molecule is achiral despite having tetrahedral stereocenters. These are the *meso compounds* discussed in Special Topic 7.1.

SPECIAL TOPIC 7.1
MESO COMPOUNDS AND DIASTEREOMERS

Tartaric acid is a normal constituent of grapes. During the fermentation of grape juices into wine, the monopotassium salt separates as an insoluble substance called "tartar."

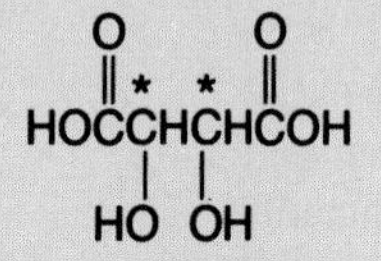

```
 O       O
 ‖ *   * ‖
HOCCHCHCOH
    |  |
   HO  OH
```

Tartaric acid

Tartaric acid molecules have two tetrahedral stereocenters. They are *identical* because the sets of four atoms or groups attached to each are identical; both centers hold HO, H, CO_2H, and $CH(OH)CO_2H$. When we prepare perspective drawings of the structures of all of the possible stereoisomers that have the same constitutions as tartaric acid, we obtain those shown below, where solid wedges denote bonds coming forward and dashed-line wedges mean bonds going rearward. You can see that the first two structures, those of D- and L-tartaric acid, are related as an object to its mirror image and that by no allowed manipulation can the two be superimposed. The two are thus *enantiomers*. Notice that they have identical melting points and identical degrees but opposite signs of specific rotation.

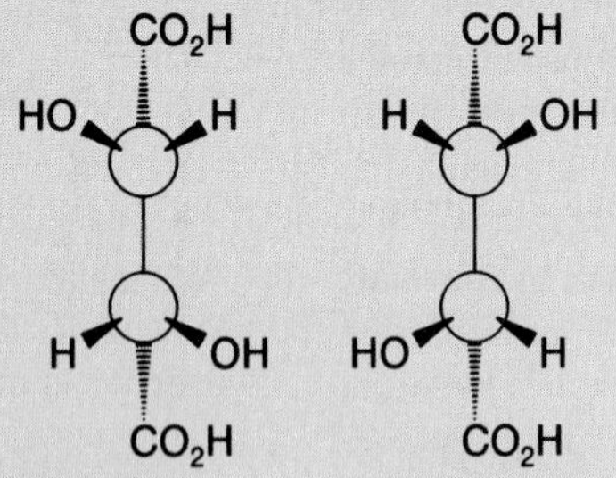

D-Tartaric acid
$[\alpha]_D^{20}$ −11.98°
mp 170°C

L-Tartaric acid
$[\alpha]_D^{20}$ +11.98°
mp 170°C

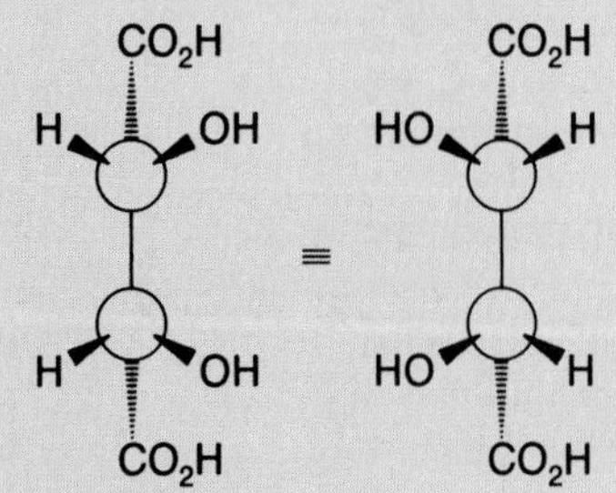

meso-Tartaric acid
$[\alpha]_D^{20}$ 0°
mp 140°C

Only one of the remaining two perspective structures has been given a name—*meso*-tartaric acid. Although its mirror image structure has been drawn to make a point, the object and mirror image are actually superimposable. If you rotate the mirror image of *meso*-tartaric acid by 180° in the plane of the paper and around an axis perpendicular to the plane (piercing the paper at the bond *between* the two carbons), you soon realize that the two structures are identical (and so superimposable). Despite having two tetrahedral stereocenters, the *meso*-tartaric acid molecule is achiral. Therefore *meso*-tartaric acid is optically inactive; it has zero specific rotation. Optical activity *requires* chirality, and *meso*-tartaric acid doesn't have it. According to our definitions, because *meso*-tartaric acid is a nonenantiomeric isomer of a pair of enantiomers, it is a *diastereomer* of each enantiomer.

meso-Tartaric is one example of a common occurrence and, historically, it gave part of its name to the kind of diastereomer it exemplifies. An achiral diastereomer of a set of stereoisomers that includes some that are chiral is called the **meso** isomer. There are thus only three stereoisomers of tartaric acid. The equation, 2^n = number of stereoisomers, cannot be applied to tartaric acid because it works only when the *n* tetrahedral stereocenters are *different,* when the sets of four atoms or groups at the center differ in at least one way. No general equation exists for calculating the number of stereoisomers when two (or more) stereocenters are identical.

In Example 7.2, the structures of the four stereoisomers of threonine were shown. (Two bear the name "threonine" and two have the name "allothreonine.") Their molecules also have two tetrahedral stereocenters, but the centers are not identical. Hence the correct number of stereoisomers is predicted by the equation we have used. Each of the threonine enantiomers is a diastereomer of each of the allothreonine enantiomers.

Diastereomers do not have identical physical or chemical properties (although the chemical properties will be very similar to all achiral reactants). You can see in the threonine—allothreonine system how melting points and specific rotations for diastereomers are different. The set of intramolecular distances and bond angles in one diastereomer isn't exactly duplicated in any other diastereomer. Hence, molecules of one diastereomer should be expected to have at least slightly different polarities than those of any other diastereomer in the set. Such differences cannot help but cause differences in physical properties.

EXAMPLE 7.1
Identifying Tetrahedral Stereocenters

Amphetamine exists as a pair of enantiomers. One of them has its own name—Dexedrine. Find the tetrahedral stereocenter in amphetamine, and list the four groups attached to it.

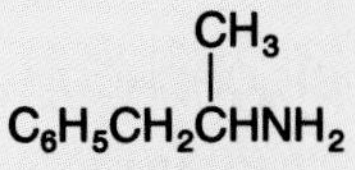

$$C_6H_5CH_2\underset{}{C}(CH_3)HNH_2$$

Amphetamine

ANALYSIS A tetrahedral stereocenter has *four different* attached atoms or groups.

SOLUTION Amphetamine has one tetrahedral stereocenter labeled with an asterisk:

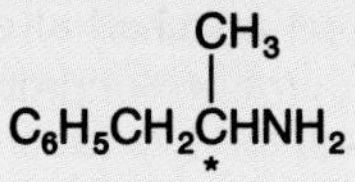

$$C_6H_5CH_2\overset{CH_3}{\underset{*}{C}}HNH_2$$

The four groups are $C_6H_5CH_2$, H, CH_3, and NH_2.

■ Amphetamine, a stimulant, is a controlled substance in the United States.

PRACTICE EXERCISE 1

Place an asterisk next to each tetrahedral stereocenter in the following structures.

(a) 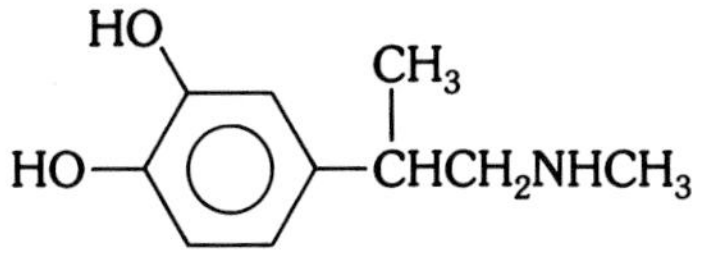 $(HO)_2C_6H_3\text{—}\overset{CH_3}{C}HCH_2NHCH_3$ Epinephrine, a hormone (See Special Topic 6.1.)

(b) 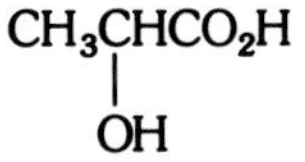 $CH_3\underset{OH}{C}HCO_2H$ Lactic acid, the sour constituent in sour milk

(c) 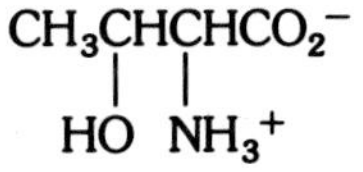 $CH_3\underset{HO}{C}H\underset{NH_3^+}{C}HCO_2^-$ Threonine, one of the amino acid building blocks of proteins

(d) 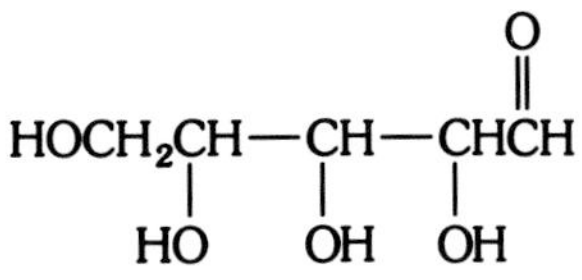 $HOCH_2\underset{HO}{C}H\text{—}\underset{OH}{C}H\text{—}\underset{OH}{C}H\overset{O}{C}H$ Ribose, a sugar unit in one of the two kinds of nucleic acids (ribonucleic acid or RNA)

A Molecule with *n* Different Tetrahedral Stereocenters Has 2^n Stereoisomers

When a molecule has two or more tetrahedral stereocenters, as in parts (c) and (d) of Practice Exercise 1, then it becomes useful to judge whether these centers are *different*. When used in this context, *different* means that the sets of four atoms or groups at the various tetrahedral stereocenters have at least one difference. Two tetrahedral stereocenters are said to be *different* if the set of four groups at one is not duplicated by the set at the other. Whenever the tetrahedral stereocenters in a molecule are different in this sense—as they were in parts (c) and (d) of Practice Exercise 1—then the substance can exist in the forms of 2^n stereoisomers, where *n* is the number of different tetrahedral stereocenters. These 2^n stereoisomers occur as half as many *pairs* of enantiomers. We'll see this illustrated in the next example.

EXAMPLE 7.2
Judging Whether Tetrahedral Stereocenters Are Different and Calculating the Number of Stereoisomers

The threonine molecule, part (c) of Practice Exercise 1, has two tetrahedral stereocenters, labeled here by asterisks.

$$CH_3\overset{*}{C}H(OH)\overset{*}{C}H(NH_3^+)CO_2^-$$

Threonine

Are these tetrahedral stereocenters different? If so, how many stereoisomers of threonine are there?

ANALYSIS To compare the groups attached at each tetrahedral stereocenter, we should make a list of the sets of four different groups and compare them. If the lists aren't identical in every respect, then the two tetrahedral stereocenters are different.

At one tetrahedral stereocenter: CH_3, H, HO, $CHCO_2^-$ (with NH_3^+ on the CH)

At the other tetrahedral stereocenter: CH_3CH (with HO on the CH), H, CO_2^-, NH_3^+

SOLUTION The sets are obviously different, so $n = 2$ is the number of different tetrahedral stereocenters in a threonine molecule. Therefore, $2^n = 2^2 = 4$, the number of stereoisomers of threonine. These occur as half of 4 or 2 pairs of enantiomers. The complete set has the following structures. One pair of enantiomers is on the left. Just imagine that the mirror is between them and is perpendicular to the page. The other pair of enantiomers is on the right.

■ Only L-threonine works as a building block for making proteins in the body. There is no enzyme that can accept any of the other optical isomers as substrates.

D-Threonine: front carbon CO_2^-, H, $\overset{+}{N}H_3$; rear carbon HO, H, CH_3. $[\alpha]_D^{26}$ +28.3°

L-Threonine: front carbon CO_2^-, $H_3\overset{+}{N}$, H; rear carbon H, OH, CH_3. $[\alpha]_D^{26}$ −28.3°

D-Allothreonine: front carbon CO_2^-, H, $\overset{+}{N}H_3$; rear carbon H, OH, CH_3. $[\alpha]_D^{26}$ −9.6°

L-Allothreonine: front carbon CO_2^-, $H_3\overset{+}{N}$, H; rear carbon HO, H, CH_3. $[\alpha]_D^{26}$ +9.6°

■ The labels D and L will be explained in the next chapter, as we said. The meaning of the experimental values given for the symbol $[\alpha]_D^{26}$ is explained in the next section.

PRACTICE EXERCISE 2

Examine the structure of ribose that was given in part (d) of Practice Exercise 1. **(a)** How many different tetrahedral stereocenters does it have? **(b)** How many stereoisomers are there of this structure? (Only one is actually the ribose that can be used by the body.) **(c)** How many pairs of enantiomers correspond to this structure?

PRACTICE EXERCISE 3

Write the structure of 2,3-butanediol and place an asterisk by each carbon that is a tetrahedral stereocenter. Are they *different* tetrahedral stereocenters?

7.3 OPTICAL ACTIVITY

The members of a pair of enantiomers affect polarized light in equal and opposite ways when compared under identical conditions.

We mentioned earlier that the two members of a pair of enantiomers differ in one physical property, and to describe it we first have to learn something about polarized light.

The Electromagnetic Oscillations of Polarized Light Are All in the Same Plane Light is electromagnetic radiation in which the intensities of the electric and magnetic fields set up by the light source oscillate in a regular way. In ordinary light, these oscillations occur equally in all directions about the line that defines the path of the light ray.

Certain materials, such as the polarizing film in the lenses of Polaroid sunglasses, affect ordinary light in a special way. Polarizing film interacts with the oscillating electrical field of any light passing through it to make this field oscillate *in just one plane.* The light that emerges is now **plane-polarized light** (see Figures 7.5 and 7.6*a*).

If we look at some object through polarizing film and then place a second film in front of the first, we can rotate one film until the object can no longer be seen (see Figures 7.6*b* and 7.7). If we now rotate one film by 90°, we'll see the object at maximum brightness again. The first film seems to act as a lattice fence, forcing any light that goes through it to vibrate only in the direction allowed by the long spaces between the slats. This light then moves on to the molecular slats of the second film. If the second film's slats are perpendicular to those of the first, the light has no freedom to oscillate, and it cannot get through the second film. At intermediate angles, fractional amounts of light can go through the second film. Only when the slats of both films are *parallel* to each other can the light leaving the first film slip easily through the second film with its maximum intensity.

■ You can try this out using two Polaroid sunglass lenses. The lenses of these glasses reduce glare by cutting out the plane-polarized light produced when sunlight reflects from a plane surface such as a road, a snowfield, or a lake.

An Enantiomer Can Rotate the Plane of Plane-Polarized Light When a solution of D-asparagine in water is placed in the path of plane-polarized light, the *plane* of polarization is

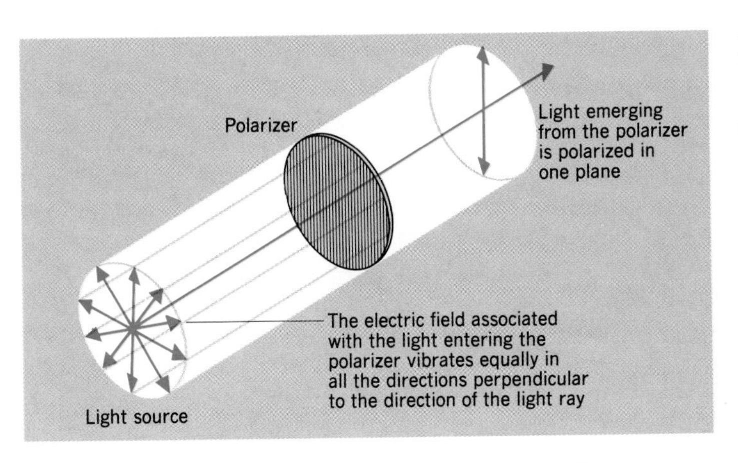
Polarizer
Light emerging from the polarizer is polarized in one plane
The electric field associated with the light entering the polarizer vibrates equally in all the directions perpendicular to the direction of the light ray
Light source

FIGURE 7.5
When light passes through polarizing film, it becomes polarized light.

FIGURE 7.6
The principal working parts of a polarimeter and how optical rotation can be measured.

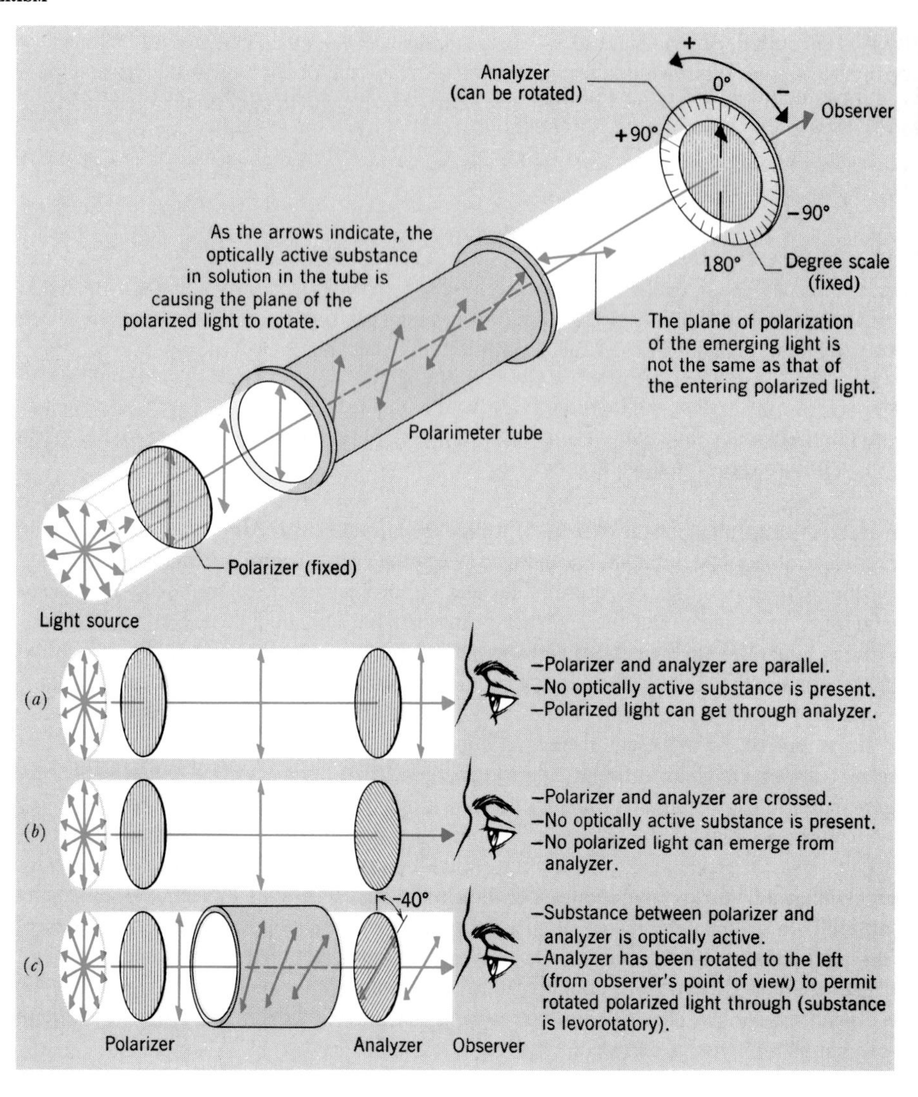
Analyzer
(can be rotated)
+
0°
−
Observer
+90°
−90°
180°
Degree scale
(fixed)
As the arrows indicate, the optically active substance in solution in the tube is causing the plane of the polarized light to rotate.
The plane of polarization of the emerging light is not the same as that of the entering polarized light.
Polarimeter tube
Polarizer (fixed)
Light source
(a)
–Polarizer and analyzer are parallel.
–No optically active substance is present.
–Polarized light can get through analyzer.
(b)
–Polarizer and analyzer are crossed.
–No optically active substance is present.
–No polarized light can emerge from analyzer.
-40°
(c)
–Substance between polarizer and analyzer is optically active.
–Analyzer has been rotated to the left (from observer's point of view) to permit rotated polarized light through (substance is levorotatory).
Polarizer
Analyzer
Observer

FIGURE 7.7
When two polarizing films are "crossed," no light can get through to the observer.

POLAROID
POLAROID

twisted or rotated. Any substance that can rotate the plane of plane-polarized light is said to be **optically active.** We have not actually explained *how* this phenomenon happens—we are unable to do so. We have only reported that it does take place. Quite often the members of a set of stereoisomers in which any or all are optically active are called **optical isomers.** Thus the two asparagines are optical isomers. All of the threonine stereoisomers of Example 7.2 are all likewise optical isomers of each other.

The *Polarimeter* Measures the Degree of Optical Activity The instrument used to detect and measure optical activity is called a **polarimeter** (see Fig. 7.6). Its principal working parts consist of a *polarizer* for the light beam, a tube for holding solutions in the path of the polarized light, an *analyzer* (actually, just another polarizing device), and a circular scale for measuring the number of degrees of rotation. When the "slats" of the polarizer and the analyzer are parallel and the tube contains no optically active material, the polarized light emerges from the analyzer with maximum intensity. Let's assume that we start with this parallel orientation of the polarizer and analyzer.

When a solution of one pure enantiomer is placed in the light path, the plane-polarized light encounters molecules of just one chirality. They cause the plane of oscillation of the polarized light to be rotated. The plane of oscillation of the polarized light that leaves the solution is now no longer parallel with the analyzer (see Fig. 7.6c). Consequently, not as much light gets through the analyzer, so the observed light intensity is now less than the original maximum. To restore the original intensity, the operator can rotate the analyzer to the right or to the left a definite number of degrees until the analyzer once again is parallel *with the light that emerges from the tube.*[3]

The operator, looking *toward* the light source, might find that rotating the analyzer to the right (clockwise) restores the original light intensity with fewer degrees of rotation than rotating the analyzer to the left (counterclockwise). When such a rightward rotation works, the degrees are recorded as positive, and the optically active substance is said to be **dextrorotatory.** If the fewer degrees of rotation are found by a leftward rotation, then the degrees are recorded as negative and the substance is said to be **levorotatory.** In Figure 7.6c, the reading is $\alpha = -40°$, where α (including the plus or minus sign) stands for the observed **optical rotation,** the observed number of degrees of rotation caused by the solution.

■ Latin, *dextro*, right, and *levo*, left.

The value of α varies both with the temperature of the solution and the frequency of the light used, but not in any simple direct way. Consequently, when α is recorded, both the temperature and the light frequency must also be recorded.

The *Specific Rotation* of an Optically Active Compound Is One of Its Physical Constants, Like Its Density, Boiling Point, or Melting Point The observed rotation is related in simple ways to the concentration of the solution and the length of the tube.

$$\alpha \propto \text{concentration}$$
$$\alpha \propto \text{path length}$$

What both of these proportionalities are really saying is that the degree of rotation of the plane-polarized light is a function of the *population* of the chiral molecules. Either by increasing the concentration or by making the tube longer, we can force the polarized light to be in greater contact with chiral molecules.

Because α is *directly* proportional both to concentration (c) and to path length (l), α is proportional to the products of these two.

$$\alpha \propto cl \tag{7.1}$$

The unit used for length, l, is the decimeter.

■ This is a rare use of the decimeter unit in chemistry.

1 decimeter (dm) = 10 centimeters (cm)

[3] This description of the measurement emphasizes only the essential principle involved. The operator has other options that yield identical results.

We can convert expression 7.1 to an equation by inserting a constant of proportionality, which is given the symbol $[\alpha]$ and the name **specific rotation.**

$$\alpha = [\alpha]cl \tag{7.2}$$

■ Be sure to notice that the units of c are g/mL, not g/100 mL. If an optically active pure liquid is in the polarimeter tube, then the units of its concentration equal those of its density.

The unit traditionally used for concentration when reporting specific rotations is g/mL. As we said earlier, both the temperature (t) and the light frequency (λ) must be reported with any value of specific rotation. By rearranging Equation 7.2 and placing symbols for t and λ by the closing bracket, we obtain the usual form of the definition of specific rotation.

$$[\alpha]_{\lambda}^{t} = \frac{\alpha}{cl} \tag{7.3}$$

■ L-Asparagine is the more common form.

Quite often polarimeters are used with the intense yellow light of a sodium vapor lamp like those that illuminate the streets in some cities. The symbol for this light is D, so when we write that $[\alpha]_{D}^{20} = +5.42°$ for D-asparagine, we mean that the temperature of the solution was 20 °C and that a sodium vapor lamp was used. For L-asparagine, $[\alpha]_{D}^{20} = -5.42°$. Thus we see that the *only* physical difference between the two enantiomers of asparagine is the *direction* of the rotation of the plane of plane polarized light. The numbers of degrees are identical; only the signs of rotation are opposite. Any pair of enantiomers is like this. All physical properties of a pair of enantiomers are the same—densities, boiling points, and melting points, for example—but the *signs* of the numerically identical degrees of rotation are opposite.

Specific Rotation Provides an Analytical Tool The specific rotation of an optically active compound is an important physical constant, comparable to its melting point, boiling point, or density. If we know the value of $[\alpha]_{\lambda}^{t}$ for a compound, it's easy to see from Equation 7.3 that we have a way to determine the concentration of a solution. We measure the observed rotation, α, for the solution when it is in a tube of known path length, l, and use these data together with the specific rotation to calculate the concentration, c.

Sometimes the measurement of optical activity is used to identify a substance. A measurement of the observed rotation, α, of a solution of known concentration, c, in a tube of known path length, l, is made and the specific rotation is calculated. The calculated value is then compared to a table of specific rotations to see what matches.

A 50 : 50 Mixture of Enantiomers Is Optically Inactive It is important to realize that *optical activity* and *optical isomerism* are not the same. *Optical activity* refers to a phenomenon observable with a special instrument, and what we see is the number of degrees of rotation of the solution. *Optical isomerism* is part of our explanation of optical activity. We *infer* optical isomerism from the observation of optical activity.

What happens if we mix two enantiomers together in a 1 : 1 ratio? Now as the plane-polarized light travels through the tube it encounters some molecules forcing its plane to twist to the right but it meets an identical number forcing its plane to twist just as much to the left. *The result is no net change to the plane of vibration of the plane-polarized light.* The operator of the polarimeter would be bound by the definitions that we have introduced to report that the substance in the tube is *optically inactive.* Any 50 : 50 mixture of enantiomers is optically inactive and is called a **racemic mixture.** Thus a substance, like a racemic mixture, can be made entirely of chiral molecules and yet be optically inactive.

■

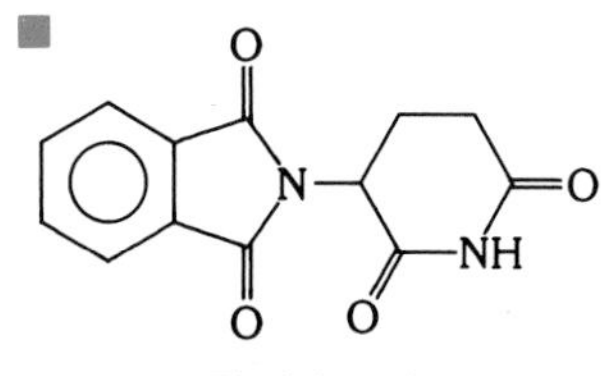

Thalidomide

Thalidomide was never approved for use in the United States because Dr. Frances Oldham Kelsey of the U.S. Food and Drug Administration insisted *before* birth defects appeared in Europe that the testing of thalidomide had not been thorough enough.

Enantiomers Normally Have Large Differences in Biological Properties It's both interesting and significant that if a racemic mixture of asparagine enantiomers is given in the diet, the body can use only one. This is generally true about pairs of enantiomers. If the body uses one enantiomer, it cannot use the other, which may even be a very dangerous substance. Thalidomide, for example, can exist as a pair of enantiomers. (Can you spot the tetrahedral stereocenter?) It was once widely prescribed in Europe as a sedative-tranquilizer, particularly for pregnant women. Tragically, the prescribed drug was the racemic mixture and only one enantiomer gives the desired effect. The other enantiomer disrupts fetal development during the first 12 weeks of pregnancy causing phocomelia—seal or flipperlike arms—and often abnormalities of the digestive tract, the eyes, and the ears. Between 1959 and 1962, from 2000 to 3000 babies were born in Germany with such thalidomide-caused problems. The drug was withdrawn from the market but not before many tragic births had occurred. The episode furnishes a dramatic example of how great the differences can be between enantiomers, even though their molecules are so nearly identical as to be related as object and mirror image.

■ In 1993, thalidomide was found to reduce the rate of activation of HIV-1, the virus that causes AIDS.

SUMMARY

Stereoisomerism Stereoisomers are isomers whose molecules have identical constitutions but different geometries. The two kinds of stereoisomers are enantiomers and diastereomers. Enantiomers are pairs of stereoisomers whose molecules are mirror images but they do not superimpose. Diastereomers are stereoisomers that are not related as enantiomers. Almost always, an enantiomer molecule has a tetrahedral stereocenter, which is usually a carbon atom holding four different atoms or groups. If a molecule has *n different* tetrahedral stereocenters, then the number of stereoisomers is 2^n, and these occur as half as many pairs of enantiomers. A 50 : 50 mixture of enantiomers, a racemic mixture, is optically inactive.

Optical activity Optical activity is a natural phenomenon detected by means of a polarimeter. A substance is optically active if polarized light that passes through a substance or its solution undergoes a rotation in its plane of polarization.

Specific rotation The specific rotation of an optically active substance is what its observed rotation is at one unit of concentration (1 g/mL) in one unit of path length (1 dm). It varies, but not in a simple way, with the temperature and the wavelength of the light used. Values of specific rotation can be used to determine concentrations of optically active substances.

Properties of enantiomers Enantiomers are identical in every physical property except the signs of their specific rotation. They are also identical in every chemical respect provided that the molecules or ions of the reactant are achiral. When the reactant particles are chiral, then one enantiomer reacts differently with the reactant than the other enantiomer, a phenomenon always observed when an enzyme is acting as a (temporary) reactant.

REVIEW EXERCISES

The answers to Review Exercises whose numbers are in color are found in Appendix A. The answers to the other Review Exercises are found in the Study Guide that accompanies this book. The more challenging questions are marked with asterisks.

Structural Isomers and Stereoisomers

7.1 What specifically must be true about two compounds before they can be called constitutional isomers?

7.2 What are the structures of the *simplest* alcohols that can exhibit constitutional isomerism?

7.3 What must be true about two compounds before they can be called stereoisomers?

7.4 There are two kinds of stereoisomers. What are their names and how is each kind defined?

7.5 Classify the following pairs of structures as constitutional isomers or as stereoisomers.

(a)

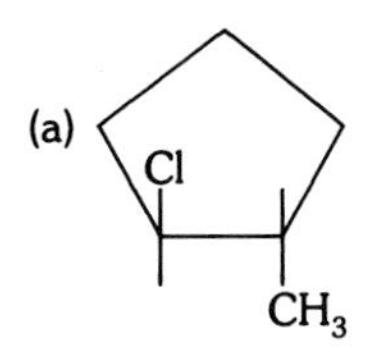

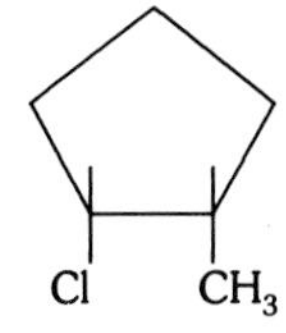

(b)

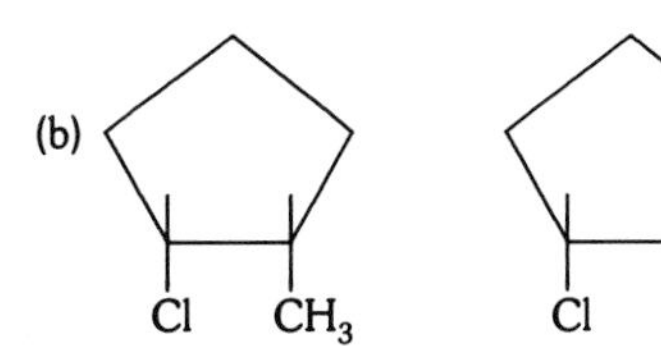

7.6 What is the "one-substance–one-structure" principle?

Optical Isomers

7.7 The general structure of several stereoisomers that include glucose is that of 2,3,4,5,6-pentahydroxyhexanal:

$$HOCH_2-\underset{OH}{CH}-\underset{OH}{CH}-\underset{OH}{CH}-\underset{OH}{CH}-CH=O$$

(a) Place an asterisk by each tetrahedral stereocenter.
(b) How many of these tetrahedral stereocenters qualify as *different* tetrahedral stereocenters?
(c) How many stereoisomers of this compound are possible?
(d) These optical isomers occur as how many *pairs* of enantiomers?

7.8 One of the approximately 20 building blocks of protein molecules is glycine: $^{+}NH_3CH_2CO_2^{-}$. Does glycine have stereoisomers? How can you tell?

7.9 One of the important intermediate substances in the body's energy-producing metabolism is an anion of the following acid, citric acid.

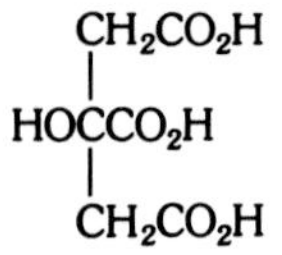

$$\begin{array}{c} CH_2CO_2H \\ | \\ HOCCO_2H \\ | \\ CH_2CO_2H \end{array}$$

(a) Does citric acid have stereoisomers? How can you tell?
(b) One of the possible monomethyl esters of citric acid has chiral molecules. Write its structure. Place an asterisk by its tetrahedral stereocenter.

Properties of Enantiomers

7.10 The melting point of (−)-cholesterol is 148.5 °C. What is the melting point of (+)-cholesterol? How can we know what this melting point *must* be without actually making the measurement?

7.11 Explain why enantiomers should have identical physical properties (except for the sign of specific rotation).

7.12 Explain why enantiomers cannot have different chemical properties toward reactants whose molecules or ions are achiral. (Use an analogy if you wish.)

7.13 Explain why enantiomers have different chemical properties toward reactants whose molecules or ions are chiral. (Use an analogy if you wish.)

Specific Rotation

***7.14** Pantothenic acid was once called vitamin B_3.

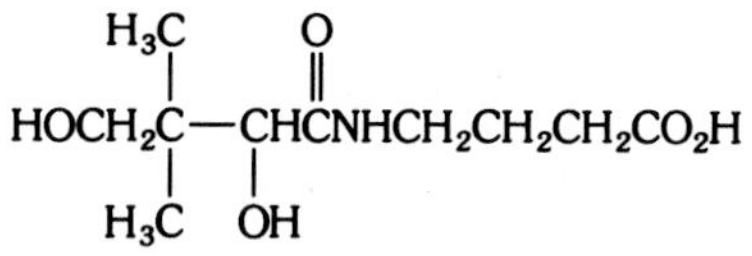

$$\begin{array}{c} H_3C \quad\quad O \\ | \quad\quad \| \\ HOCH_2C{-}CHCNHCH_2CH_2CH_2CO_2H \\ | \quad\quad | \\ H_3C \quad OH \end{array}$$

Panthothenic acid

Only its dextrorotatory form can be used by the body, and its specific rotation is $[\alpha]_D^{25} = +37.5°$. In a tube 1.00 dm long and at a concentration of 1.00 g/100 mL, what is the *observed* optical rotation of the levorotatory enantiomer?

7.15 Ascorbic acid is also known as vitamin C. Only its dextrorotatory enantiomer can be used by the body.

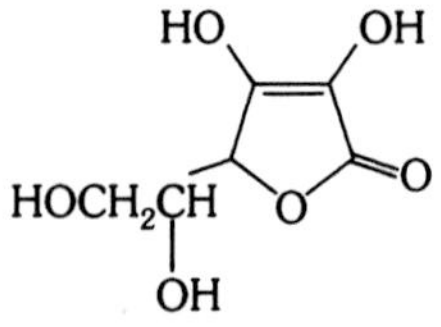

[Structure: five-membered lactone ring with HO and OH on the double bond, =O, ring O, and side chain $HOCH_2CH$ bearing OH]

Ascorbic acid
(vitamin C)

Its specific rotation is $[\alpha]_D^{25} = +21°$. What is the specific rotation of (−)-ascorbic acid?

7.16 Explain why the observed rotation is proportional to the concentration of the optically active compound and to the length of the tube through which the polarized light travels.

7.17 What becomes of the optical activity of a substance if it cor sists of a mixture of equal numbers of moles of its two enar tiomers? Explain.

***7.18** A solution of sucrose (table sugar) in water at 25 °C in a tub that is 10.0 cm long gives an observed rotation of +2.00 The specific rotation of sucrose in water at this temperatur and the same wavelength of light is +66.4°. What is the cor centration of sucrose in g/100 mL?

7.19 Quinine sulfate, an antimalarial drug, has a specific rotation i water at 17 °C of −214°. If a solution of quinine sulfate in th solvent in a 1.00-dm tube and under the same conditions c temperature and wavelength has an observed rotation c −10.4°, what is its concentration in g/100 mL?

***7.20** Strychnine and brucine are structurally similar compound that have extremely bitter tastes, and both are very poisonous The specific rotation in chloroform at 20 °C of brucine i −127° and that of strychnine under identical conditions i −139°. If a solution of one of these in chloroform at a cor centration of 1.68 g/100 mL in a tube 1.00 dm long gave a observed rotation of −2.34°, which of the two compound was it? Do the calculation.

7.21 Corticosterone and cortisone are two substances used to trea arthritis. Under identical conditions of solvent and tempera ture, the specific rotation of corticosterone is +223° and tha of cortisone is +209°. If a solution of one of these at a cor centration of 1.48 g/100 mL and in a tube 1 dm long has a observed rotation of +3.10°, which compound was in th solution? Do the calculation.

Other Optical Isomers (Special Topic 7.1)

7.22 Examine the structure of D-threonine and D-allothreonine a given in Example 7.2.
(a) Why don't these qualify as enantiomers?
(b) Why are they described as being members of the *same se* of optical isomers?
(c) Why aren't they called geometric isomers?
(d) What kind of stereoisomers are they?

7.23 Two models of methane, related as object to mirror image superimpose. Why isn't methane called a meso compound?

7.24 Although *meso*-tartaric acid is optically inactive, it is describe as an *optical isomer* of D- or L-tartaric acid. Why?

Additional Exercises

7.25 Explain how a substance could consist entirely of chiral mole cules and yet be optically inactive.

7.26 What is the IUPAC name of each of the following?
(a) The monohydric alcohol of lowest formula mass whose molecules are chiral.
(b) The alkane of lowest formula mass whose molecules are chiral.

***7.27** There are *four* stereoisomers of the following compound. Ex plain.

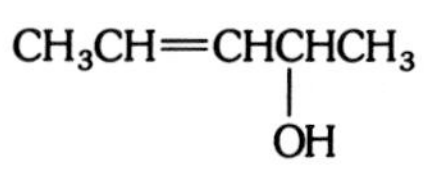

$$\begin{array}{c} CH_3CH{=}CHCHCH_3 \\ | \\ OH \end{array}$$

CARBOHYDRATES

8

In what a wonderful, roundabout way do we tap into solar energy! The chemical energy in the milk we drink comes through cows that feed on grass made possible by photosynthesis, one of the Special Topics in this chapter.

8.1

BIOCHEMISTRY—AN OVERVIEW

Building materials, information, and energy and are basic essentials for life. **Biochemistry** is the systematic study of the chemicals of living systems, their organization, and the principles of their participation in the processes of life.

The Cell Is the Smallest Unit That Lives The molecules of living systems are lifeless, yet life has a molecular basis. Whether studied in cells or when isolated from them, the chemicals at the foundation of life obey all of the known laws of chemistry and physics. Yet, in isolation, not one compound of a cell has life. The intricate *organization* of compounds in a cell is as important to life as the chemicals themselves. Thus the cell is the smallest unit of matter that lives and that, in the proper environment, can make a new cell like itself.

The Life of a Cell Requires Materials, Information, and Energy Our purpose in the remainder of this book is to study the molecular basis of meeting the three basic needs of a living system, its needs for materials, information, and energy. Without the daily satisfaction of these, life at any of the many loftier levels, like creativity, relationships, and love, would be severely constrained. Most of our focus will be on the molecular basis of life in the human body.

■ Cornstarch, potato starch, table sugar, and cotton are all carbohydrates.

■ Butter, lard, margarine, and corn oil are all examples of lipids known as fats and oils.

We will begin in the next three chapters to study the organic materials of life, starting with the three main classes of foodstuffs: carbohydrates, lipids, and proteins. We use their molecules to build and run our bodies and to try to stay in some state of repair. Plants rely heavily on carbohydrates for cell walls, and animals obtain considerable energy from carbohydrates made by plants. Lipids (fats and oils) serve many purposes. They are used as materials for cell membranes, and as sources of chemical energy. Proteins are particularly important in both the structures and functions of cells, whether of plants or of animals. Because of the central catalytic role of proteins in regulating chemical events in cells, we will immediately follow our study of proteins with an examination of a particular family of proteins, the enzymes. We've mentioned them often as the special catalysts in living systems.

The Circulatory System Delivers Needed Compounds and Carries Wastes Away Few of the substances in the diet are in forms directly usable by our bodies. Carbohydrates, lipids, and proteins must be broken down (hydrolyzed) to much smaller molecules. This work is done by the enzymes of our digestive juices, and the chemical reactions of digestion will be a part of our study. The small molecular products of digestion are delivered into circulation via the bloodstream, the vital conduit on which all tissues depend for raw materials and oxygen, for chemical signals such as hormones, for disease-fighting agents, and for the removal of wastes. One of the special emphases, already begun in our study, is the molecular basis for using oxygen and releasing carbon dioxide during metabolism.

■ The reactions of *digestion* process food molecules into (usually) smaller molecules which then enter circulation and later participate in *metabolism.*

Every Cell Has an Information System Enzymes, hormones, and neurotransmitters are components of the intricate information system in an organism. Without information—plans or blueprints—materials and energy could combine to produce only rubble and rubbish.

■ Neurotransmitters carry chemical signals from one nerve cell to the next.

Monkeys swinging hammers would only reduce a stack of lumber to splinters. Carpenters, using the same materials and expending no more raw energy, can build a building, because they possess information in the form of plans and experience.

■ The field of *molecular biology* deals with the work of nucleic acids, their synthesis, and uses.

Although enzymes are elements in the cell's information system, enzymes do not *originate* the blueprints. They only help to carry them out. The blueprint for any one member of a species is encoded in the molecular structures of its nucleic acids. These compounds are able to direct the synthesis of a cell's enzymes. Hormones and neurotransmitters, other elements of cellular information, depend on the presence of the right enzymes for their own existence. Thus a study of the enzyme-makers must be included in any study of the molecular basis of life. The study of nucleic acids will help us to see how different species can take essentially the same raw materials and energy, synthesize their enzymes, and thus lay the basis for making everything else needed.

There is both contentment and chemical energy in a peanut butter sandwich.

Some Materials Are Used Mainly for Their Chemical Energy The molecular basis of energy for life is another broad topic of our study. One of the kinds of questions that we will address is "How can one get the energy for running, skipping, and laughing out of a sandwich?" As we study biochemical energetics and its enzymes and metabolic pathways, we'll have numerous occasions to peer deeply into the molecular basis of some disorders and diseases.

To supply materials for any use—parts, information, or energy—each organism has basic nutritional needs. These include not just organic materials, but also minerals, water, and oxygen. Thus, after learning about the materials of life and how they are processed and used, we will close our study of the molecular basis of life with a broad survey of nutritional needs.

We Launch a New Beginning with the Study of Carbohydrates We have an exciting trip ahead. In the preceding chapters we have slowly and carefully built a solid foundation of chemical principles. It's been like a mountain-climbing trip where the route for a large part of the trek is through country with few grand vistas, and yet with a beauty of its own. Now we're moving to elevations where the vistas begin to open. It's like a new beginning, and we start it with a study of the first of the three chief classes of food materials, the carbohydrates.

■ When people *understand* how much health depends on the timely coming together of all of the proper substances, they become more interested in good nutrition.

8.2

INTRODUCTION TO THE MONOSACCHARIDES

The monosaccharides—polyhydroxyaldehydes and ketones—are carbohydrates that are not converted into substances with smaller molecules by hydrolysis.

Carbohydrates are aldehydes and ketones with many OH groups, or substances that form these when hydrolyzed. They include the simple sugars, like glucose, as well as table sugar, starch, and cellulose. Carbohydrates are the primary products of **photosynthesis,** the complex series of reactions in plants by which CO_2, H_2O, and minerals are converted to plant chemicals and oxygen using the solar energy absorbed by the green pigment, chlorophyll. Special Topic 8.1 discusses the photosynthesis further.

■ The oxidized and reduced forms of polyhydroxy aldehydes and ketones as well as certain amino derivatives are also in the family of carbohydrates.

The Simple Sugars Do Not React with Water The carbohydrates that cannot be hydrolyzed are called the **monosaccharides** or **simple sugars.** Their empirical formula is $(CH_2O)_n$. Those with aldehyde groups are called **aldoses** and those with keto groups are **ketoses.** Like these terms, the names of virtually all carbohydrates end in *-ose*.

Whether they are aldoses or ketoses, monosaccharides with three carbons are trioses, those with four are tetroses, and this pattern continues with pentoses, hexoses, and higher

SPECIAL TOPIC 8.1
PHOTOSYNTHESIS

The energy released when a piece of wood burns came originally from the sun. The wood, of course, isn't just bottled sunlight. It's a complex, highly organized mixture of compounds, mostly organic. The solar energy needed to make these compounds is temporarily stored in wood in the form of distinctive arrangements of electrons and nuclei that characterize energy-rich molecules.

They are made from very simple, energy-poor substances such as carbon dioxide, water, and soil minerals. In the living world only plants have the ability to use solar energy to convert energy-poor substances into complex, energy-rich, organic compounds. The overall process by which plants do this is called **photosynthesis.**

The simplest statement of photosynthesis in equation form is

$$nCO_2 + nH_2O + \begin{matrix}\text{solar}\\ \text{energy}\end{matrix} \xrightarrow[\text{plant enzymes}]{\text{chlorophyll}} (CH_2O)_n + nO_2$$

To make glucose, a hexose, n must equal 6. The symbol (CH_2O) stands for a molecular unit in carbohydrates, but plants can use the energy of carbohydrates (which came from the sun) to make other substances as well—proteins, lipids, and many others. In the final analysis, the synthesis of all the materials in our bodies consumes solar energy, and all our activities that use energy ultimately depend on a steady flow of solar energy through plants to the plant materials we eat. The meat and dairy products in our diets also depend on the consumption of plants by animals.

Chlorophyll is the green pigment in the solar-absorbing systems of plants, usually their leaves. Chlorophyll molecules absorb solar energy and, in their energized states, trigger the subsequent reactions leading to carbohydrates. A large number of steps and several enzymes are involved. The rate of photosynthesis increases as the air temperature increases and as the concentration of CO_2 in air increases.

Notice that another product of photosynthesis is oxygen, and this process continuously regenerates the world's oxygen supply. Roughly 400 billion tons of oxygen are set free by photosynthesis each year, and about 200 billion tons of carbon (as CO_2) is converted into compounds in plants. Of all of this activity, only about 10 to 20% occurs in land plants. The rest is done by tiny phytoplankton and algae in the earth's oceans. In principle it would be possible to dump so much poison into the oceans that the cycle of photosynthesis would be gravely affected. It is quite clear that the nations of the world must see that this does not happen.

When plants die and decay, their carbon atoms end up eventually in carbon dioxide again, and the reactions of decay consume oxygen. The combustion of fuels such as petroleum, coal, and wood also uses oxygen. And animals consume oxygen during respiration. Thus there exists a grand cycle in nature in which atoms of carbon, hydrogen, and oxygen move from CO_2 and H_2O into complex forms plus molecular O_2. The latter then interact in various ways to regenerate CO_2 and H_2O.

Someone has estimated that all the oxygen in the earth's atmosphere is renewed by this cycle once in about 20 centuries, and that all the CO_2 in the atmosphere and the earth's waters goes through this cycle every three centuries.

sugars as well. Two trioses, glyceraldehyde and dihydroxyacetone, occur in metabolism, and two pentoses, ribose and 2-deoxyribose, are essential to the nucleic acids.

■ *Deoxy* means lacking an oxygen where one normally is.

$$HOCH_2CH(OH)CH{=}O \qquad HOCH_2C({=}O)CH_2OH \qquad HOCH_2CH(OH){-}CH(OH){-}CH(OH)CH{=}O \qquad HOCH_2CH(OH){-}CH(OH)CH_2CH{=}O$$

Glyceraldehyde — Dihydroxyacetone — Ribose — 2-Deoxyribose

Glucose is a hexose, $(CH_2O)_6$ or $C_6H_{12}O_6$, with an aldehyde group, so it is also called an **aldohexose.** (We're interested only in terms here; structures will come soon.) Galactose is also an aldohexose, a stereoisomer of glucose. Fructose, also $(CH_2O)_6$ or $C_6H_{12}O_6$, has a keto group, so it's a **ketohexose.** You can see how parts of words can be combined into one very descriptive term. Glucose, galactose, and fructose are the nutritionally important monosaccharides.

Disaccharides and Polysaccharides Make up the Other Families of Carbohydrates The monosaccharides are the monomer units of di- and polysaccharides. **Disaccharides** are carbohydrates that can be hydrolyzed to two monosaccharides. Sucrose, maltose, and lactose are common examples of disaccharides. Starch and cellulose are called **polysaccharides,** because when one of their molecules reacts with water, it gives hundreds of monosaccharide molecules.

■ *Oligosaccharide* molecules yield from three to a few dozen monosaccharide molecules when they are hydrolyzed.

All Monosaccharides Are Reducing Carbohydrates Carbohydrates are sometimes described by their abilities to react with Tollens' and Benedict's reagents (pages 110 and 111). Something is reduced in these tests (e.g., Ag^+ or Cu^{2+}), so carbohydrates that give these tests are called **reducing carbohydrates.** All monosaccharides and nearly all disaccharides are reducing carbohydrates. Sucrose (table sugar) is not, and neither are the polysaccharides. We'll see why shortly.

■ Glucose is also called corn sugar, because it can be made by the hydrolysis of cornstarch.

Glucose Is Nature's Most Widely Used Organic Monomer If we count all its combined forms, (+)-glucose is perhaps the most abundant organic species on earth. It's the building block for molecules of cellulose, a polysaccharide that makes up about 10% of all the tree leaves of the world (on a dry mass basis), about 50% of the woody parts of plants, and nearly 100% of cotton. Glucose is also the monomer for starch, a polysaccharide in many of our foods, particularly grains and tubers. Glucose and fructose are the major components of honey. Glucose is also commonly found in plant juices. Because it is by far the most common carbohydrate in blood, glucose is often called **blood sugar,** although this term strictly applies to the mixture of all the carbohydrates in blood.

■ Massachusetts General Hospital regards a concentration of glucose in the blood of 70–100 mg/100 mL (3.9–6.1 mmol/L) to be the "normal" range for a healthy adult who has not eaten for a few hours.

One Form of Glucose Is a Pentahydroxy Aldehyde Simple alcohols, ROH, can form acetate esters, CH_3CO_2R. Glucose forms a pentaacetate, so five of the six oxygens in $C_6H_{12}O_6$ are in alcohol groups. The sixth oxygen is in an aldehyde group because glucose is easily oxidized to a C_6 monocarboxylic acid by reagents, like Tollens' reagent, that do not oxidize alcohol groups.

Under strong, forcing conditions, glucose can be reduced to a straight-chain derivative of hexane, so the six carbons in glucose must be in a straight chain. The five OH groups must be strung out, one on each of five carbons, because 1,1-diols are not stable. These data support the conclusion that glucose is 2,3,4,5,6-pentahydroxyhexanal. In fact, all aldohexoses are optical isomers of each other and so all have the same basic skeleton:

■ A 1,1-diol consists of the following system:

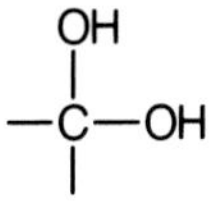

$$-\overset{\overset{\displaystyle OH}{|}}{\underset{|}{C}}-OH$$

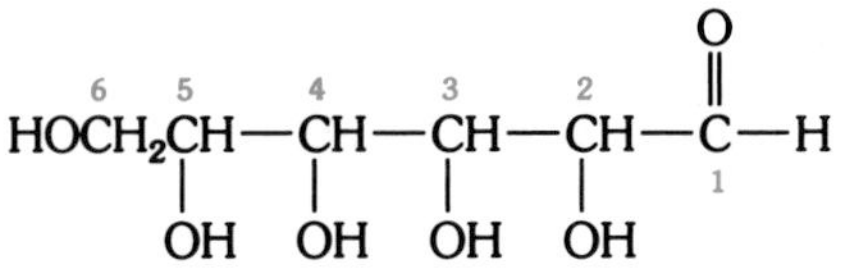

$$\overset{6}{HOCH_2}\overset{5}{CH}(OH)-\overset{4}{CH}(OH)-\overset{3}{CH}(OH)-\overset{2}{CH}(OH)-\overset{1}{C}(=O)-H$$

Basic structure of all aldohexoses, including glucose

Carbons 2, 3, 4, and 5 in the glucose chain are all tetrahedral stereocenters. Each center has a *unique* set of four different groups, so the centers are all different. We learned in the previous chapter that the number of stereoisomers of a compound whose molecules have n different tetrahedral stereocenters is 2^n. In 2,3,4,5,6-pentahydroxyhexanal, $n = 4$, so there must be $2^4 = 16$ stereoisomers, or eight pairs of enantiomers. (+)-Glucose is one of these 16; galactose is another. None of the remaining 14 stereoisomers is nutritionally important. Clearly, to know what glucose really is, we must look more closely at the stereoisomers of glucose.

8.3

D- AND L-FAMILIES OF MONOSACCHARIDES

The tetrahedral stereocenter farthest from the carbonyl group of all naturally occurring monosaccharides has the same configuration as (+)-glyceraldehyde, which puts these monosaccharides in the D-family.

In this section we will study the question of the actual orientations or configurations of the tetrahedral stereocenters in the monosaccharides.

All Naturally Occurring Monosaccharides Belong to the Same Optical Family To simplify this study, we'll retreat from the complexities of (+)-glucose and go back to the simplest aldose, glyceraldehyde. The structures of the two enantiomers of glyceraldehyde are shown in Figure 8.1. Both are known, and the enantiomer labeled D-(+)-glyceraldehyde actually has the absolute configuration shown. **Absolute configuration** refers to the actual arrangement in space about each stereocenter in a molecule. When we know the absolute configuration of (+)-glyceraldehyde, we also know that of (−)-glyceraldehyde, because its molecules *must* be the mirror image of the molecules of (+)-glyceraldehyde. (See also Figure 8.1.)

■ When we use the term "stereocenter" we'll always mean "tetrahedral stereocenter." (The older term for such a center at a C atom is **chiral carbon.**)

Chemists have used the absolute configurations of the enantiomers of glyceraldehyde to devise configurational or optical families for the rest of the carbohydrates. Any compound that has a configuration like that of (+)-glyceraldehyde and can be related to it by known reactions is said to be in the **D-family.** For example, (−)-glyceric acid is in the D-family because it can be made from D-(+)-glyceraldehyde by an oxidation that doesn't disturb any of the four bonds to the stereocenter, as the following equation shows.

■ As long as no bond to the stereocenter is disturbed in the reaction, no change in configuration can possibly occur.

O=C—H, H, OH, CH_2OH (D-(+)-Glyceraldehyde) —— oxidation of CH=O group (configuration at the stereocenter does not change) ——→ O=C—OH, H, OH, CH_2OH (D-(−)-Glyceric acid)

When the molecules of a compound are the mirror images of an enantiomer in the D-family, the compound is in the **L-family.**

The letters D and L are only family names. *They have nothing to do with actual signs of the values of their specific rotations, [α].* No way exists, in fact, to tell from the *sign* of specific rotation whether a compound is in the D- or L-family. These letters signify something about absolute configuration only. Later we'll see that D-glucose is dextrorotatory but D-fructose is levorotatory.

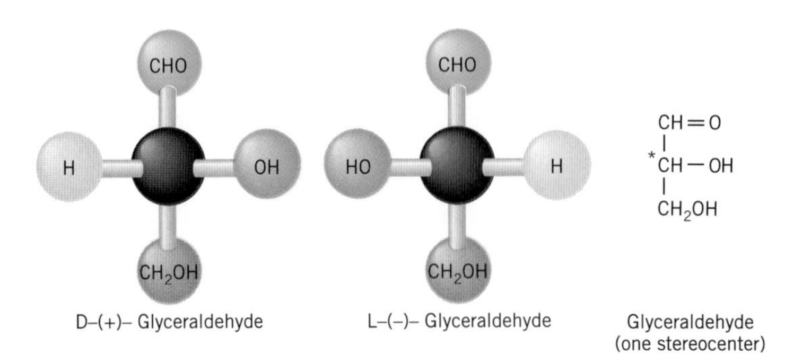

FIGURE 8.1
The absolute configurations of the enantiomers of glyceraldehyde.

Fischer Projection Formulas Simplify Absolute Configurations When a molecule has several stereocenters, it becomes quite difficult for most people to make a perspective, three-dimensional drawing of an absolute configuration. Emil Fischer, a chemist who unraveled most of the carbohydrate structures, devised a way around this, and his structural representations are called *Fischer projection formulas.* To make them, we follow a set of rules that let us project onto a plane surface the three-dimensional configuration of each stereocenter in a molecule.

■ Emil Fischer (1852–1919), a German chemist, won the second Nobel prize in chemistry in 1902.

Rules for Writing Fischer Projection Formulas

1. Visualize the molecule with its main carbon chain vertical and with the bonds that hold the chain together projecting to the rear at each stereocenter. *Carbon-1 is at the top.*
2. Mentally flatten the structure, stereocenter by stereocenter, onto a plane surface. See Figures 8.2 and 8.3.
3. In the projected structure, represent each stereocenter either as the intersection of two lines or conventionally as C.
4. The horizontal lines at a stereocenter actually represent bonds that project *forward,* out of the plane of the paper.
5. The vertical lines at a stereocenter actually represent bonds that project *rearward,* behind the plane.

A Fischer projection formula can have more than one intersection of lines, each representing a stereocenter, as seen in Figure 8.3. Always remember that at each stereocenter, a horizontal line is a bond coming toward you and a vertical line is a bond going away from you.

Once we have one plane projection structure, it's easy to draw the mirror image, as we saw in Figures 8.2 and 8.3. We can easily test for superimposition, too, provided we strictly heed one important additional rule. We may never (mentally) lift a Fischer projection formula out of the plane of the paper. We may only slide it and rotate it within the plane and a rotation must be by 180°, not 90°. This rule is necessary because if we turn a Fischer projection formula out of the plane and over or rotate it by only 90°, we actually make groups that project in one direction project oppositely, but the operation that we do on the paper will not show this reversal.

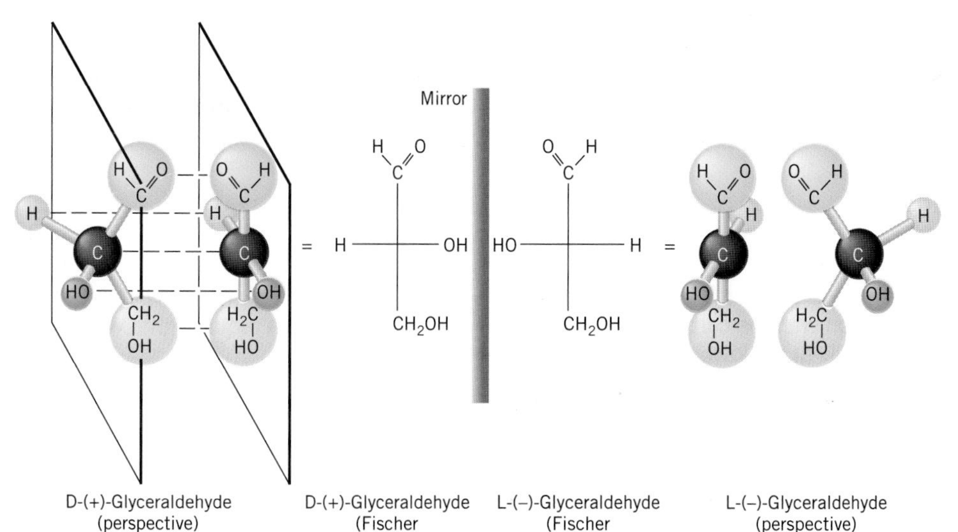

FIGURE 8.2
The relationships of the perspective (three-dimensional) drawings of D-(+)-glyceraldehyde and L-(−)-glyceraldehyde to their corresponding Fischer projection formulas.

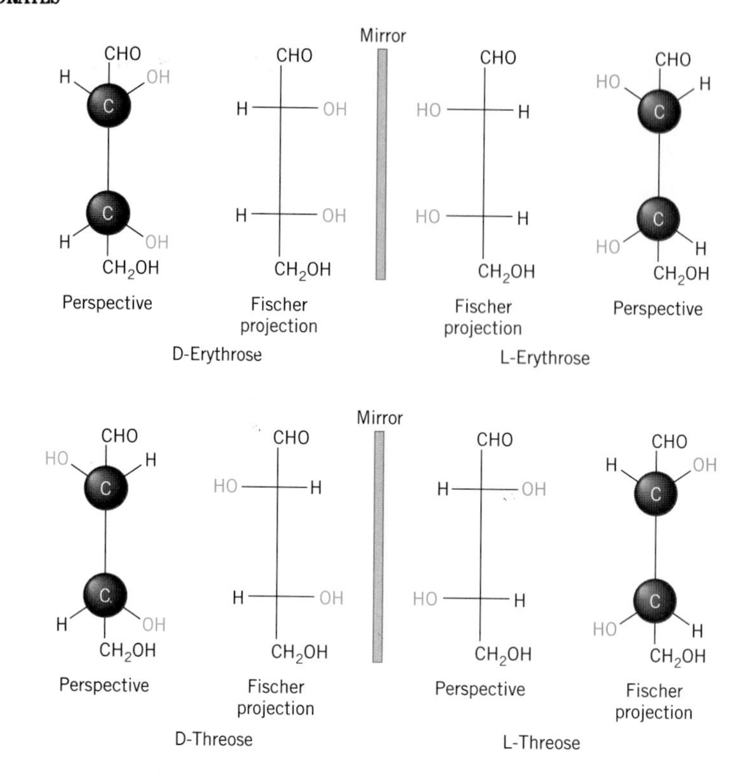

FIGURE 8.3
The four aldotetroses in their perspective and Fischer projection formulas. There are two different tetrahedral stereocenters, so there are $2^2 = 4$ stereoisomers that occur as two pairs of enantiomers, those of D- and L-erythrose and those of D- and L-threose.

EXAMPLE 8.1
Writing Fischer Projection Formulas

Write the Fischer projection formulas for the stereoisomers of glyceric acid.

$$\mathrm{HOCH_2CHCO_2H}$$
$$\mathrm{|}$$
$$\mathrm{OH}$$

Glyceric acid

ANALYSIS There are three carbons in the chain, but only one is joined to four different atoms or groups. Only the center carbon is a stereocenter. Therefore in our equation for calculating the number of stereoisomers, $n = 1$, so $2^n = 2$ and glyceric acid has two enantiomers. We represent its lone stereocenter by the intersection of two perpendicular lines, and we make two of these, one for each enantiomer:

(Incomplete)

Then we attach the other two carbons. According to the rules, we have to put C-1, the carbon with the carbonyl group in CO_2H, at the top. This then requires that we place CH_2OH at the lower end of the vertical line.

CO_2H CO_2H

CH_2OH CH_2OH (Incomplete)

SOLUTION We know that the OH group at C-2 can be either on the right or the left, so we finish the Fischer projection formulas:

$$\begin{array}{ccc} CO_2H & & CO_2H \\ H-\!\!\!+\!\!\!-OH & & HO-\!\!\!+\!\!\!-H \\ CH_2OH & & CH_2OH \end{array} \quad \text{(Complete)}$$

These are the two enantiomers of glyceric acid. The one on the left is D-glyceric acid, and the other is L-glyceric acid.

D-Glyceric acid is levorotatory but its salts are dextrorotatory, which illustrates again that the sign of a specific rotation cannot be deduced from the D- or L-family membership.

PRACTICE EXERCISE 1

Fischer projection formulas that correspond to the various optical isomers of tartaric acid are given below. (a) Which are identical? (b) Which are related as enantiomers? (c) One is a *meso* compound (Special Topic 7.1). Which is it?

$$\begin{array}{ccccc} CO_2H & CO_2H & CO_2H & CO_2H & CO_2H \\ H-\!\!+\!\!-OH & H-\!\!+\!\!-OH & H-\!\!+\!\!-OH & HO-\!\!+\!\!-H & HO-\!\!+\!\!-H \\ HO-\!\!+\!\!-H & HO-\!\!+\!\!-H & H-\!\!+\!\!-OH & H-\!\!+\!\!-OH & HO-\!\!+\!\!-H \\ CO_2H & CO_2H & CO_2H & CO_2H & CO_2H \\ \textbf{(a)} & \textbf{(b)} & \textbf{(c)} & \textbf{(d)} & \textbf{(e)} \end{array}$$

PRACTICE EXERCISE 2

Write Fischer projection formulas for the stereoisomers of the following compound.

$$HOCH_2\underset{HO}{\underset{|}{C}}H\underset{OH}{\underset{|}{C}}HCO_2H$$

All Naturally Occurring Carbohydrates Are in the D-Family Monosaccharides are assigned to the D-family or the L-family according to the projection of the OH group *at the stereocenter farthest from the carbonyl group.* The compound is in the D-family when this OH group projects to the right *in a Fischer projection formula* oriented so that the carbonyl group is at or near the top. When this OH group projects to the left, the substance is in the L-family. We have already illustrated these rules by D- and L-glyceraldehyde (Fig. 8.2), and by the enantiomers of threose and erythrose (Fig. 8.3). It doesn't matter how the other OH groups at the other stereocenters project. Membership in the D- or L-family is determined solely by the projection of the OH on the stereocenter farthest from the carbonyl carbon in a properly drawn Fischer projection structure.

Because the nutritionally important carbohydrates are all in the D-family, throughout the rest of this book we'll assume that the D-family is meant whenever the family membership of a carbohydrate isn't given.

Figure 8.4 gives the Fischer projection formulas of all the aldoses in the D-family from the aldotriose through the aldohexoses. There are eight D-aldohexoses. The enantiomers of these constitute eight L-aldohexoses (not shown). In all, therefore, there are 16 optical isomers of the aldohexoses, as we calculated earlier. Figure 8.5 gives the Fischer projection formulas of several important ketoses. The ketotriose dihydroxyacetone, the two ketopentoses ribulose and xylulose, and the ketohexose fructose are the biologically important ketoses.

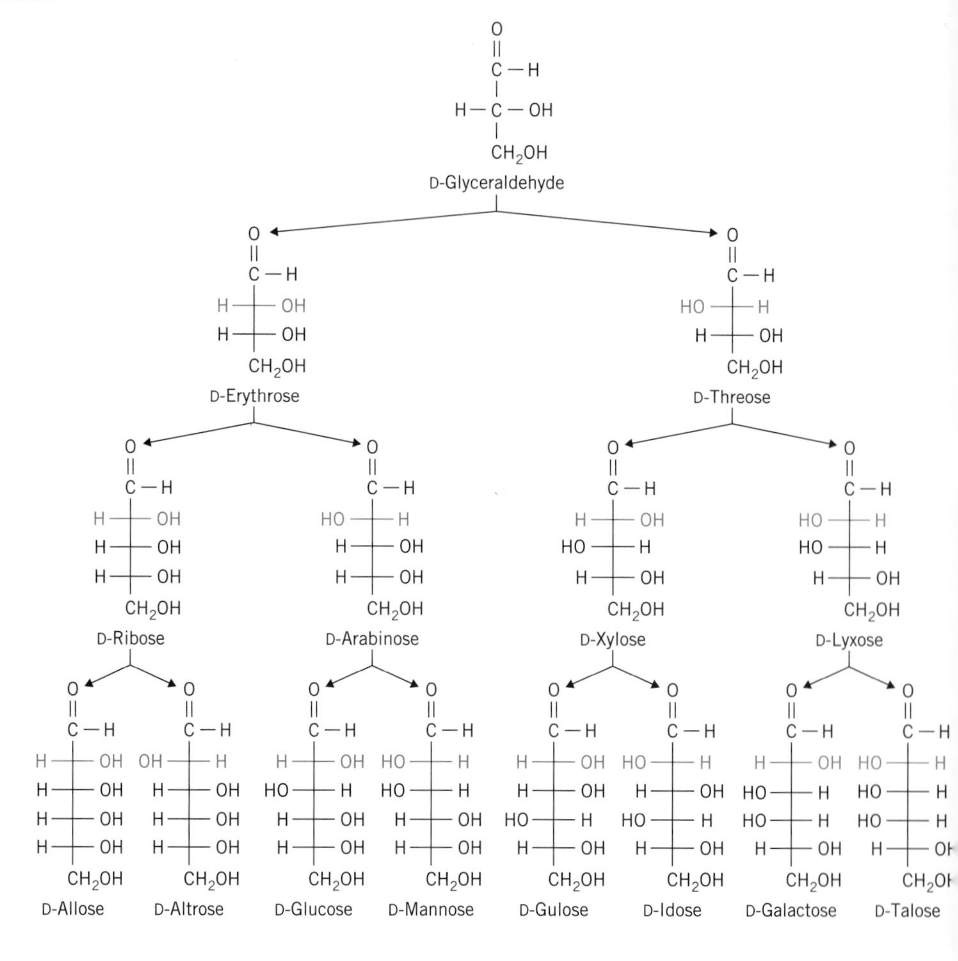

FIGURE 8.4
The D-family of the aldoses through the aldohexoses. Notice that in all of them, the OH group on the stereocenter that is farthest from the carbonyl group projects to the right. Each pair of arrows points to a pair of aldoses whose configurations are identical except at C-2.

8.4

CYCLIC FORMS OF THE MONOSACCHARIDES

The internal addition of an OH group to the double bond of the aldehyde or keto group makes possible two cyclic forms of monosaccharides.

■ The "aging" of a glucose solution occurs almost instantly when hydroxide ion is present.

Fresh Glucose Solutions Gradually Change in Optical Rotation Although glucose is optically active, when we try to measure its optical activity, it behaves in a very strange way. A *freshly prepared* solution of (+)-glucose has a specific rotation of $[\alpha]_D^{20} = +113°$. As this solution ages, however, its specific rotation slowly changes until it stabilizes at a value of +52°. We'll call a glucose solution with this specific rotation an "aged glucose solution."

By a special method of recovery (which we'll not discuss), we can recover crystalline (+)-glucose from the aged solution, the same glucose in every respect as before. A *freshly prepared* solution of this recovered glucose again shows a specific rotation of +113°. This new solution ages in the identical way until its specific rotation stabilizes at +52°. This cycle can be repeated as often as we please.

It is possible by another method of recovering glucose from an aged solution to obtain a slightly different crystalline compound. Its freshly prepared solution has a specific rotation of +19°, but it also changes with time to +52°, the same value that was observed for the other aged solution. Using this second method to recover glucose, we can repeat this cycle as often as we please, too.

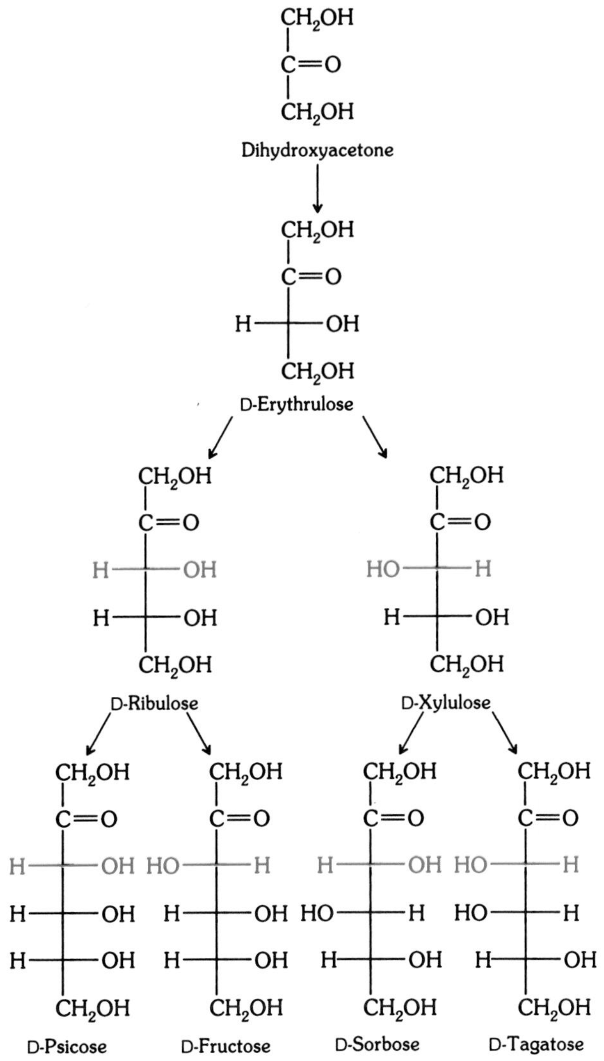

CH_2OH
$C{=}O$
CH_2OH
Dihydroxyacetone

CH_2OH
$C{=}O$
H—OH
CH_2OH
D-Erythrulose

CH_2OH
$C{=}O$
H—OH
H—OH
CH_2OH
D-Ribulose

CH_2OH
$C{=}O$
HO—H
H—OH
CH_2OH
D-Xylulose

CH_2OH
$C{=}O$
H—OH
H—OH
H—OH
CH_2OH
D-Psicose

CH_2OH
$C{=}O$
HO—H
H—OH
H—OH
CH_2OH
D-Fructose

CH_2OH
$C{=}O$
H—OH
HO—H
H—OH
CH_2OH
D-Sorbose

CH_2OH
$C{=}O$
HO—H
HO—H
H—OH
CH_2OH
D-Tagatose

FIGURE 8.5
The D-ketoses having three to six carbon atoms. Notice that in all of them, the OH group on the stereocenter that is farthest from the carbonyl group projects to the right. Each pair of arrows points to a pair of ketoses whose configurations are identical except at C-3.

To summarize, from the original aged solution we can use one method to recover the solute and get back glucose with a specific rotation of $+113°$. With the second recovery technique, we get a glucose with a specific rotation of $+19°$. We can interconvert these forms through the aged solution as often as we please. This change over time of the optical rotation of an optically active substance, one that can be recovered from an aged solution without any other apparent change, is called **mutarotation.** All the hexoses and most of the disaccharides mutarotate. Let us now see what is behind it.

Glucose Molecules Exist Mostly in Cyclic Forms Built into the *same* molecule of 2,3,4,5,6-pentahydroxyhexanal are the two functional groups needed to make a hemiacetal, the OH group and the CH=O group. Recall that hemiacetal formation is represented as follows:

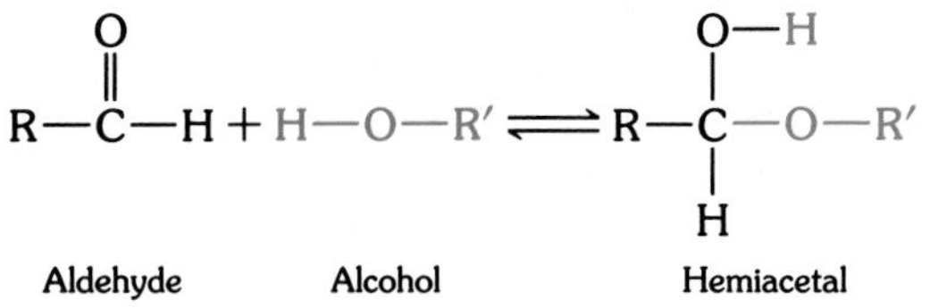

$$\underset{\text{Aldehyde}}{R-\overset{\overset{\displaystyle O}{\|}}{C}-H} + \underset{\text{Alcohol}}{H-O-R'} \rightleftharpoons \underset{\text{Hemiacetal}}{R-\overset{\overset{\displaystyle O-H}{|}}{\underset{\underset{\displaystyle H}{|}}{C}}-O-R'}$$

Suppose now that the OH group is on the *same chain* as the CH=O group. We would have something like

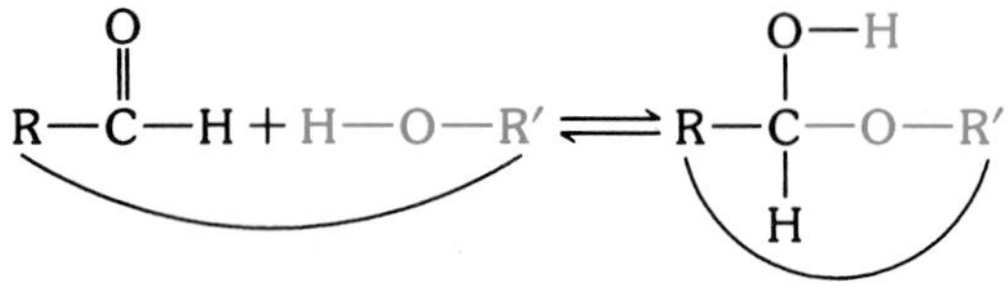

$$R-\overset{O}{\overset{\|}{C}}-H + H-O-R' \rightleftharpoons R-\underset{H}{\underset{|}{\overset{O-H}{\overset{|}{C}}}}-O-R'$$

R and R′ would be joined if CH=O and HO were in the *same* molecule

A cyclic hemiacetal

This is what happens to the open form of glucose. The C-5 OH group adds to the aldehyde group. The open structure has to coil for this to happen, as shown in structure **2**, below. It undergoes ring closure, and a new OH group, the hemiacetal OH, appears at C-1. This C-1 OH, however, can emerge on one side of the ring or the other. It depends on the way the O atom of the C=O group points just before ring closure.

If, at the moment of ring closure, the C-1 OH comes out on the side of the ring opposite to the CH_2OH unit (involving C-6), one cyclic form of glucose emerges. It is the alpha form, **1**, called α-glucose. If the new C-1 OH group comes out on the same side of the ring as the CH_2OH unit, the beta form of glucose, called β-glucose, **3**, forms.

The H at C-2 and C-5 and the OH at C-3 do not stick *inside* the ring. They stick *above or below the plane* of the ring.

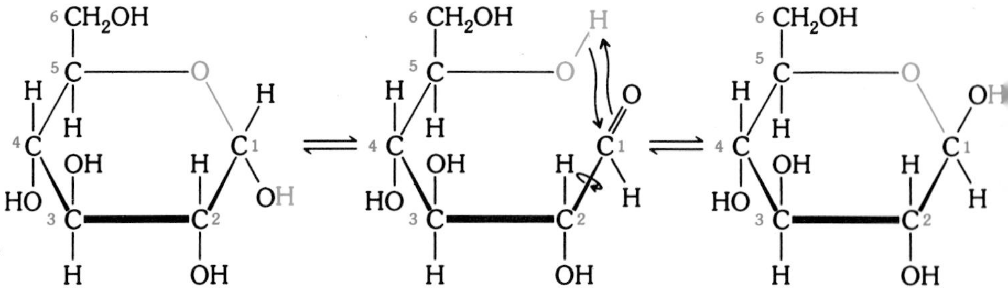

1
α-Glucose

2
Open form of glucose

3
β-Glucose

The Six-Membered Rings of Glucose Are Actually Not Flat The carbon atoms in glucose are all tetrahedral, so the bonds from them normally are at angles of 109.5°. The two bonds from the O atom in the ring are also close to this. A *flat* hexagon ring, however, would have internal angles of 120°. A *saturated* six-membered ring, therefore, cannot be flat as indicated by structures **1** or **3**. Instead, such rings are nonplanar so that normal bond angles are possible. We show next the three forms of glucose as they are known to exist in nonplanar conformations called *chair forms*.

The isomeric, cyclic forms of any given carbohydrate that differ *only* in the configuration of the hemiacetal (or hemiketal) carbon are called **anomers.** Thus **4** (=**1**) and **6** (=**3**) are anomers.

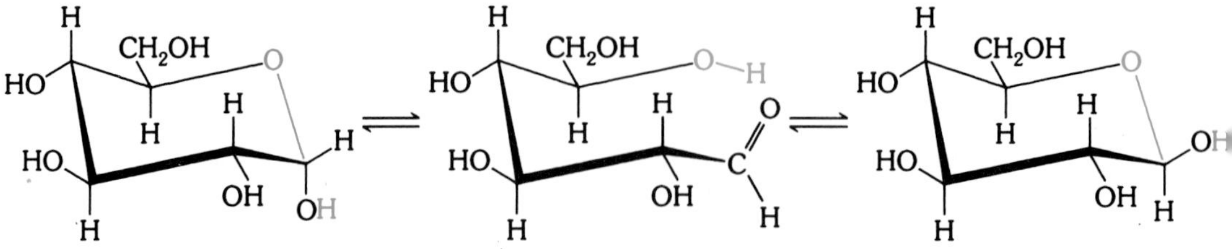

4 (= **1**)
α-Glucose

5 (= **2**)
Open form of glucose

6 (= **3**)
β-Glucose

The nonplanar forms of glucose

Special Topic 8.2 discusses these conformations in more detail and indicates why they are preferred. Having called this to your attention, we will often use the planar designations anyway, because major references in biochemistry do so. Many times we will show both forms, however.

n Aged Glucose Solutions, All Three Forms of Glucose Exist in Equilibrium

ɑ-Glucose is the form of the glucose molecules when a freshly prepared aqueous solution has a pecific rotation of +113°. Hemiacetals are unstable, however, and glucose in its cyclic forms is hemiacetal. First one molecule of **1** and then another opens up to give form **2**. As soon as nolecules of **2** appear, they can and do reclose. Figure 8.6 shows how free rotation about the C-1 to C-2 bond can reposition the aldehyde group, so that either one side or the other side of he carbonyl system faces the C-5 OH group at the moment of ring closure.

To keep the two glucose forms straight, use the CH_2OH group and the ring O atom as oints of reference. Notice that in both of the cyclic forms of glucose the CH_2OH unit sticks upward from the plane *when the ring is drawn with its oxygen in the upper right-hand corner.* When we use these specific orientations, we are certain to be drawing a member of the D-family of the aldohexoses. Now notice in structure **3** that the OH group at C-1 projects *upward* in β-glucose and is on the same side of the ring as the CH_2OH group to the left of the ing O atom. In α-glucose, structure **1**, you can see that the OH at C-1 projects *downward* on he opposite side of the ring from the CH_2OH group. We should now use the names β-D-glucose and α-D-glucose for **3** and **1** but, as we have said, we'll always mean the D-family unless something else is stated). The projections of all the other OH groups in both structures and **3** are identical. If we changed any of their orientations, we would have the structure of a nolecule that isn't any form of glucose, but one of its stereoisomers instead.

■ This arrangement of CH_2OH relative to the ring O atom also ensures that the optical configurations at all stereocenters of natural glucose are correctly displayed in our structures.

Ring Opening and Ring Closing Occur during Mutarotation The two cyclic forms of glucose differ only in the orientation of the OH group at C-1. With this in mind, let's review what happens during mutarotation. As we said, when α-glucose is freshly dissolved in water, it has a specific rotation of +113°. But its molecules open and close, because the hemiacetal ystem easily breaks apart and reforms. Some of the newly formed open-chain molecules eclose as α-glucose, and some as β-glucose. These events take place during mutarotation, whether we start with α- or β-glucose, until one grand, dynamic equilibrium involving all three

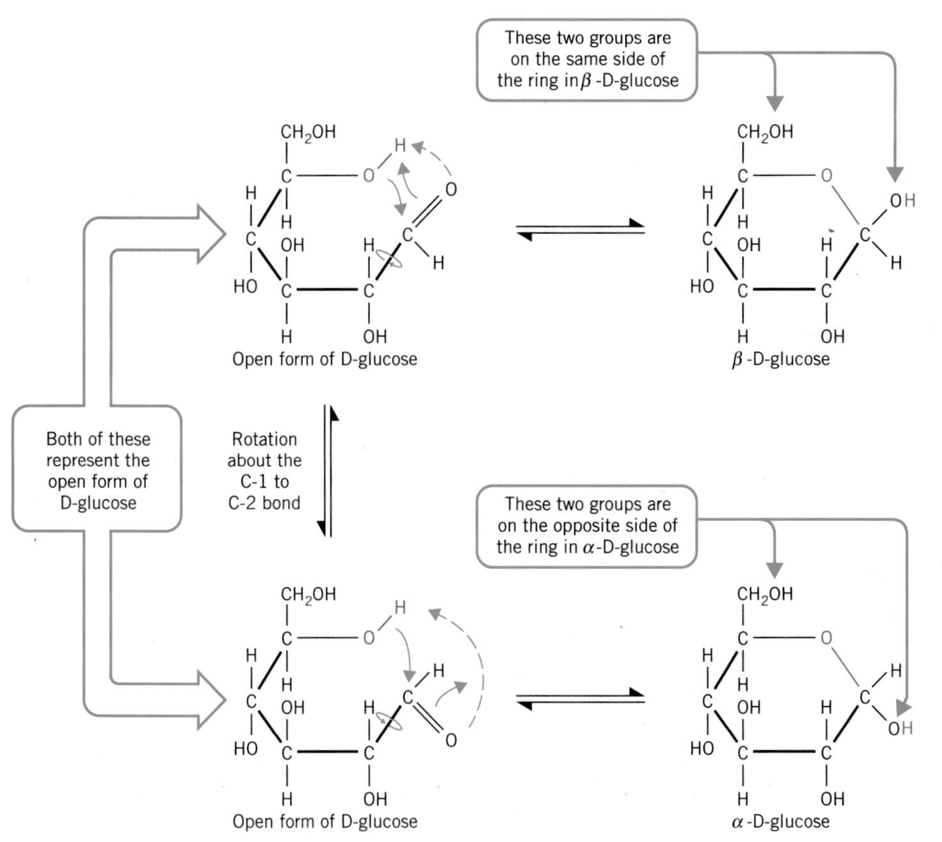
These two groups are on the same side of the ring in β-D-glucose
Open form of D-glucose
β-D-glucose
Both of these represent the open form of D-glucose
Rotation about the C-1 to C-2 bond
These two groups are on the opposite side of the ring in α-D-glucose
Open form of D-glucose
α-D-glucose

FIGURE 8.6 The α- and β-forms of D-glucose arise from the same intermediate, the open-chain form. Depending on how the aldehyde group, CH=O, is pointing when the ring closes, one ring form or the other results.

SPECIAL TOPIC 8.2
THE BOAT AND CHAIR FORMS OF SATURATED, SIX-MEMBERED RINGS

The inside angle of a regular hexagon is 120°, so the six atoms of a saturated six-membered ring cannot lie in the same plane and also have the tetrahedral bond angles of 109.5°. The cyclohexane ring resolves this by twisting into a nonplanar shape called the **chair form.**

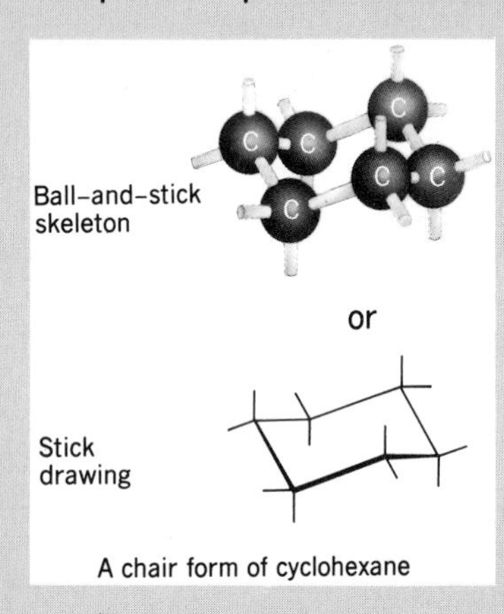

A chair form of cyclohexane

Even when one CH_2 unit of this ring is replaced by an oxygen atom, as in the rings of the glucose forms, the same kind of chair conformation predominates.

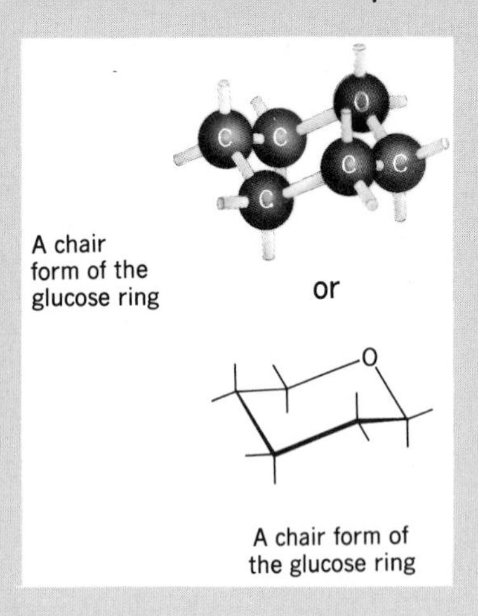

A chair form of the glucose ring

The **boat form** is another conformation of the ring that permits normal bond angles, and the ring has enough flexibility to be able to twist from the chair to the boat form. As the drawings show, if you twist one end of the chair form upward, you get the boat, and if you twist the opposite end downward you get an alternative chair form. Thus there are *two* chair forms and one boat. All three have normal bond angles.

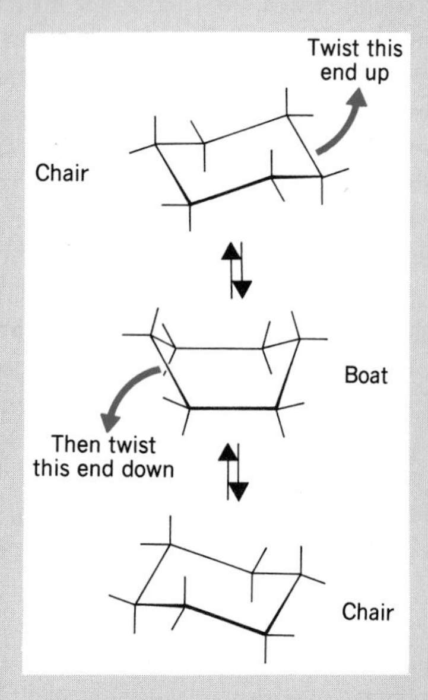

The two chair forms of cyclohexane are equally stable, but the boat form is less stable than a chair. In the boat form, as you can see in the scale model, the electron clouds by the hydrogen atoms are closer to one another, particularly the two hydrogens at opposite ends, marked *a*, one at the "prow" and one at the "stern." They nudge each other in the boat form, but are as far from each other as possible in the chair form.

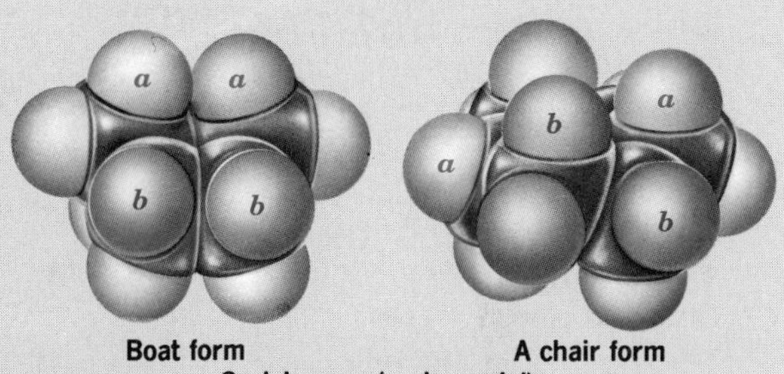

Boat form A chair form
Cyclohexane (scale model)

Similarly, the electron clouds marked *b* are closer to each other in the boat than in the chair form. Electron clouds repel each other, so the boat form is less stable than the chair form.

When the six-membered ring holds substituents, as it does among the carbohydrates, the alternative chair forms are no longer equivalent. We must, therefore, have labels to distinguish the two ways that bonds and substituents can be oriented. Positions around the perimeter are called *equatorial positions* because they are located roughly on the ring's equator. They are indicated by the black bonds and their attached H atoms in structure **A**. The positions that project above or below the average plane parallel to the axis through this plane are called *axial substituents,* indicated in color in **A**.

A

⇌

B

When chair form **A** twists into its alternative chair form, **B**, every equatorial position changes to axial and every axial to equatorial.

In an actual sample of cyclohexane, the two chair forms exist in equilibrium, and they constantly flip-flop back and forth. In a sense, the flat hexagonal structure that we usually draw for the ring of cyclohexane is an average of these two forms, and in most situations we can ignore the true bond angles of the six-membered ring.

The equivalency of the two chair forms vanishes, as we said, when the ring bears substituents. The electron clouds of axial substituents nudge one another more than do those of equatorial substituents. Thus equatorial orientations are more stable. As a rule, therefore, saturated six-membered rings take up whichever chair form puts the maximum number of bulky substituents in equatorial positions. This important fact dominates the conformations of the ring forms of the aldohexoses. The beta form of glucose, for example, is able to have every ring substituent oriented equatorially, the only aldohexose in which this is possible. In its alternative chair form, however, all substituents would be oriented axially, and this form does not occur.

β-Glucose
(more stable;
all substituents
are equatorial)

β-Glucose
(less stable;
all substituents
are axial)

Since most people find it easier to draw flat rings for the aldohexoses, and most references in biochemistry use them widely, we will generally use them also. They correctly show *relative* projections of groups on a ring—the up or down orientations—but their bond angles are not correct.

FIGURE 8.7
How to draw the cyclic forms of D-glucose in a highly condensed way.

1. First write a six-membered ring with an oxygen in the upper right-hand corner.
2. Next "anchor" the terminal CH_2OH unit on the carbon to the left of the oxygen. (Let all the Hs attached to ring carbons be "understood.") — or condense to
3. Continue in a *counterclockwise* way around the ring, placing the OHs first down, then up, then down. — or condense to
4. Finally, at the last site on the trip, how the last OH is positioned depends on whether the alpha or the beta form is to be written. The alpha is "down," the beta "up." — β-D-Glucose or condense to β-D-Glucose or α-D-Glucose

If this detail is immaterial, or if the equilibrium mixture is intended, the structure may be written as — or condense to — α or β

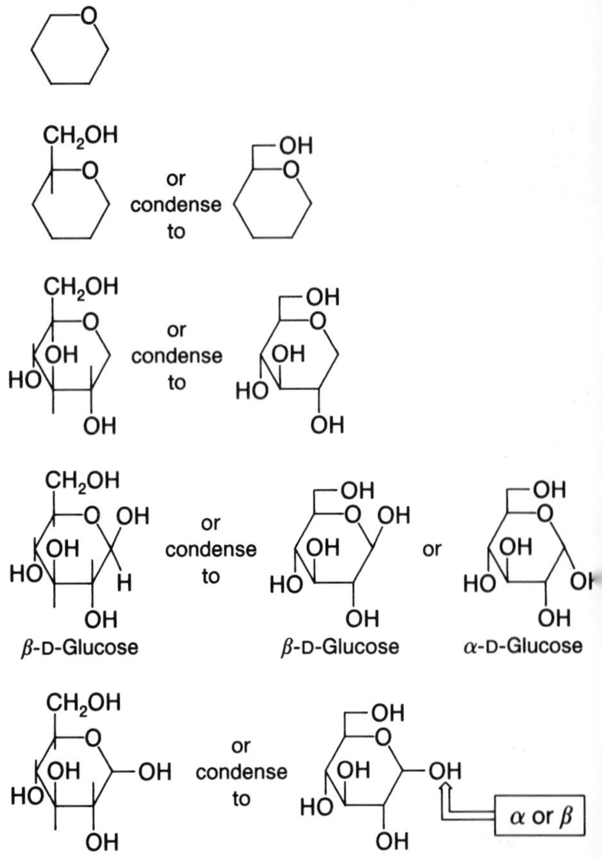

forms of glucose is established. The identical equilibrium develops from either cyclic form of glucose, because we obtain an aged solution that has the identical specific rotation, +52°. We can express the equilibrium in words as follows.

■ All forms of glucose are used by living systems.

$$\alpha\text{-Glucose} \rightleftharpoons \text{open-chain form of glucose} \rightleftharpoons \beta\text{-glucose}$$

At equilibrium, the solute molecules are 36% α-glucose, 64% β-glucose, and scarcely a trace (<0.05%) of the open-chain form.

The method of recovering solid glucose from an aged solution determines the form of glucose in the crystals. One recovery method succeeds in getting just α-glucose molecules to start the formation of crystals. The molecules of β-glucose can't fit to this crystal and help it grow, so they remain in solution. But the loss of molecules of α-glucose from the equilibrium puts a stress on it, and the equilibrium shifts (in accordance with Le Châtelier's principle) to replace the lost molecules. In this way, all the β-glucose molecules eventually get changed to α-glucose molecules and nestle into the growing crystals.

The other method of crystallization succeeds in getting crystals started from just the β-glucose molecules, so eventually all the α-glucose molecules get switched over to the β-form and join the growing crystals.

■ This is another example of Le Châtelier's principle in action.

Molecules of the open-chain form never crystallize. They occur only in the solution. Of course, they are the ones attacked by Tollens' or Benedict's reagents and, as they are removed, they are replaced by a steady shifting of the equilibrium from closed forms to the open form. This is why glucose gives the chemical properties of a pentahydroxy aldehyde despite the fact that glucose is in either one cyclic form or the other—entirely in the solid state and almost entirely in solution. This is also why it is acceptable to define a monosaccharide as an aldehyde (or ketone) with multiple OH groups rather than as a cyclic hemiacetal.

Before continuing, you should now pause to learn how to write the cyclic forms of glucose (using flat rings). Figure 8.7 outlines the steps in mastering this that have worked well for many students. Although six-membered rings are not *flat*, as we said, the projection of the groups on the ring, relative to the CH_2OH unit, are faithfully shown by these kinds of drawings.

Galactose Is a Stereoisomer of Glucose Galactose is an aldohexose that occurs in nature mostly as a structural unit in larger molecules such as the disaccharide lactose. It is also a sugar in peas. Galactose differs from glucose only in the orientation of the C-4 OH group. Like glucose, it is a reducing sugar, it mutarotates, and it exists in solution in three forms, α, β, and open.

■ Carbohydrates that differ *only* in the orientation of the OH at *only one* stereocenter other than the hemiacetal or hemiketal carbon are called **epimers** of each other.

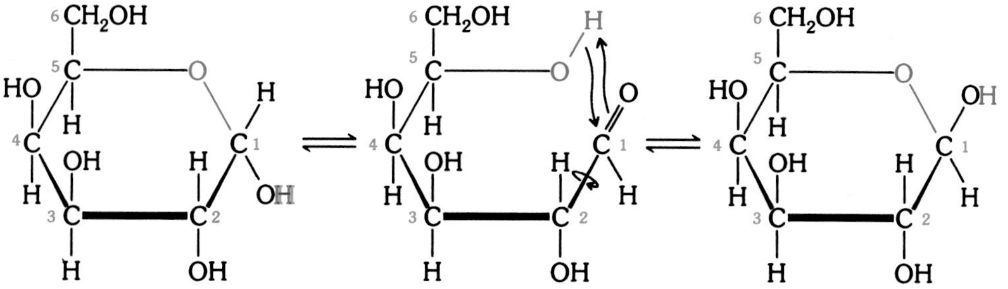

7 α-Galactose

8 Open form of galactose

9 β-Galactose

■ D-Glucose and D-galactose are *epimers* at C-4.

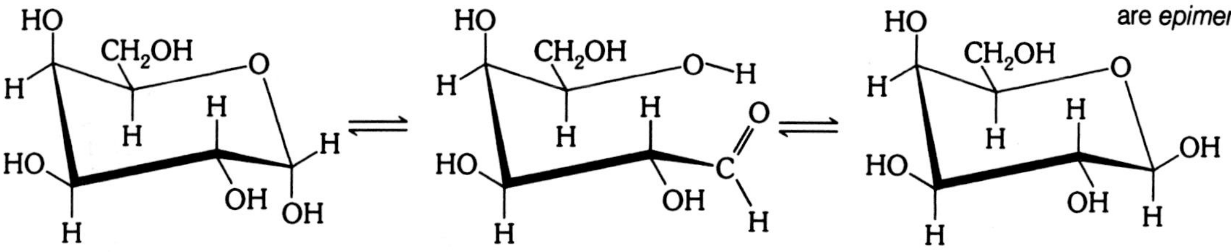

10 (= **7**) α-Galactose

11 (= **8**) Open form of galactose

12 (= **9**) β-Galactose

Fructose Occurs in a Five-Membered, Cyclic Hemiketal Form Fructose, the most important ketohexose, is found with glucose and sucrose in honey and in fruit juices. Fructose exists in more than one form, including cyclic hemiketals. The hemiketal carbon is C-2.

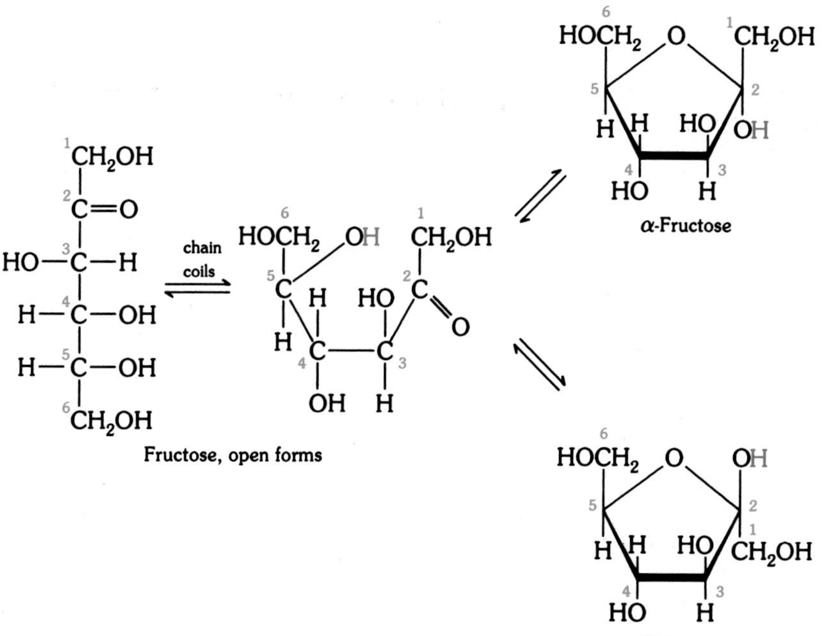

chain coils

Fructose, open forms

α-Fructose

β-Fructose

■ The internal angle of a pentagon (108°) is very close to the tetrahedral angle, so the five-membered ring is nearly flat.

■ An old name for fructose is *levulose*, after its levorotatory power.

■ Mono- and diphosphoric acid esters of fructose and glucose are important compounds in metabolism.

D-Fructose is strongly levorotatory with a specific rotation of $[\alpha]_D^{20} = -92.4°$. Fructose is a reducing sugar, not because it's a ketone (simple ketones do not easily oxidize) but because its molecules have the α-hydroxyketone system mentioned on page 111 as a system that gives the Benedict's test.

Ribose and 2-Deoxyribose Are Aldopentoses Important to Nucleic Acid Structure Both ribose and 2-deoxyribose, as we noted earlier, are building blocks of nucleic acids, and each of these aldopentoses can exist in three forms, two cyclic hemiacetals and an open form. We show just one form of each. A *deoxycarbohydrate* is one with a molecule that lacks an OH group where normally such a group is expected. Thus 2-deoxyribose is the same as ribose except that there is no OH group at C-2, just two Hs instead.

■ Ribose gives the R to RNA, a nucleic acid. Deoxyribose gives the D to DNA, another nucleic acid and the chemical of genes.

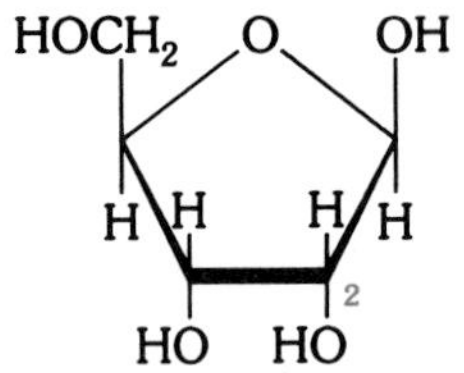

HOCH₂ O OH / H H H H / HO HO — β-Ribose

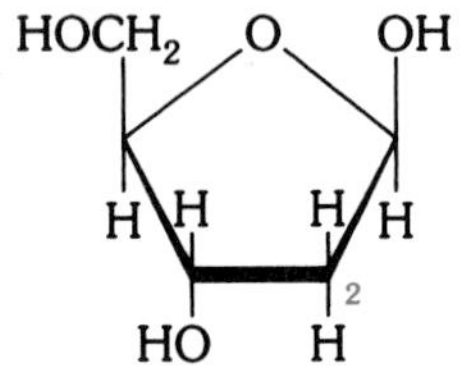

HOCH₂ O OH / H H H H / HO H — β-2-Deoxyribose

Ribose furnishes the sugar units to molecules of ribonucleic acid or RNA. Deoxyribose provides sugar units to the molecules of deoxyribonucleic acid, DNA.

8.5

DISACCHARIDES

The disaccharides are glycosides (sugar acetals) that can be hydrolyzed to monosaccharides.

The aldoses, as we have just seen, are hemiacetals. Like all hemiacetals, they react with alcohols in the presence of a catalyst to give acetals.

$$\underset{\text{Hemiacetal}}{\text{R—CH(O—R')—OH}} + \underset{\text{Alcohol}}{\text{H—O—R'}} \xrightarrow[\text{heat}]{H^+} \underset{\text{Acetal}}{\text{R—CH(O—R')—O—R'}} + \text{H—OH}$$

The Sugar Acetals Are Called Glycosides Like all hemiacetals, those of the cyclic forms of the monosaccharides can also form acetals. Methanol, for example, can be made to react with the cyclic hemiacetal system in glucose to give either an α- or a β-acetal, depending on how the new OCH_3 group becomes oriented at the ring. If it's on the same side of the ring as our reference CH_2OH group, then we have the β-form. If it is on the opposite side, then we have the α-form.

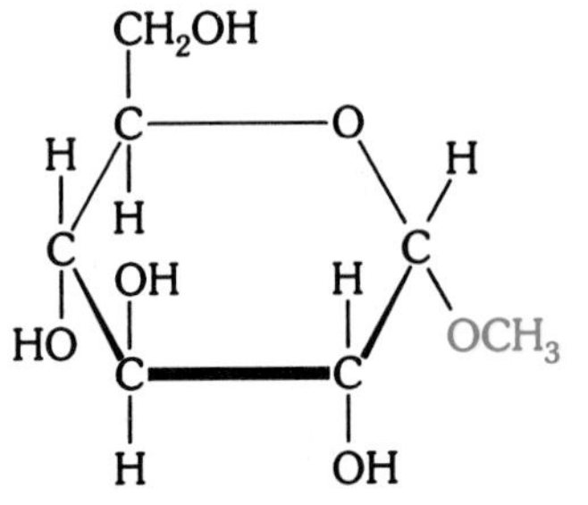

An alpha glucoside

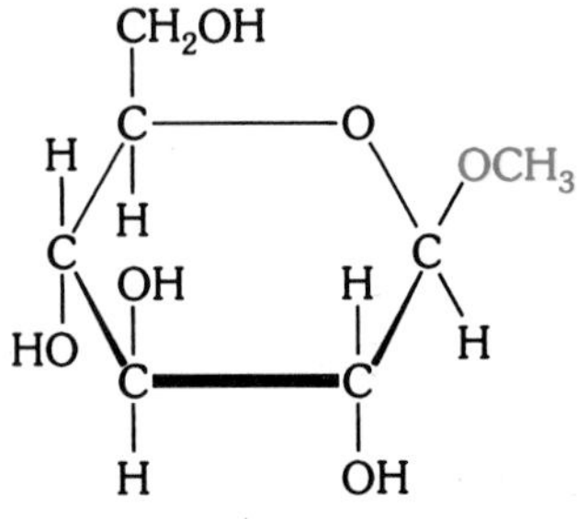

A beta glucoside

Alternatively,

Methyl α-D-glucoside
$[\alpha]_D^{25} = +158°$

Methyl β-D-glucoside
$[\alpha]_D^{25} = -33°$

All sugar acetals have the general name of **glycosides.** To name the glycoside of a specific sugar, the *-ose* in the name of the sugar is replaced by *-oside*. Thus a glycoside made from glucose is called a *glucoside*. One made from galactose is a *galactoside*.

The sugar acetals or glycosides made from simple alcohols are stable enough to isolate, but they are readily hydrolyzed when an acid catalyst (or the appropriate enzyme) is present. The glycosides shown above do not mutarotate and do not give positive Benedict's tests because their rings cannot open to expose an aldehyde group.

The Disaccharides Are Glycosides That Use a Second Sugar as the Alcohol All the disaccharides are glycosides made from the cyclic hemiacetal unit of one sugar and one of the alcohol groups of another. *An acetal oxygen "bridge" thus links two monosaccharide units in disaccharides.* This acetal unit, like all acetals, reacts readily with water in the presence of an acid or enzyme catalyst, and this hydrolysis frees the original monosaccharide molecules.

The three nutritionally important disaccharides are maltose, lactose, and sucrose. All are in the D-family. We'll first show their relationships to monosaccharides by word equations, and then we'll look more closely, but briefly, at their structures.

$$\text{Maltose} + H_2O \xrightarrow[\text{enzyme (maltase)}]{H^+ \text{ or}} \text{glucose} + \text{glucose}$$

$$\text{Lactose} + H_2O \xrightarrow[\text{enzyme (lactase)}]{H^+ \text{ or}} \text{glucose} + \text{galactose}$$

$$\text{Sucrose} + H_2O \xrightarrow[\text{enzyme (sucrase)}]{H^+ \text{ or}} \text{glucose} + \text{fructose}$$

■ You can see why glucose is of such central interest to carbohydrate chemists.

Maltose Is Made from Two Glucose Units Maltose or malt sugar does not occur widely as such in nature, although it is present in germinating grain. It occurs in corn syrup, which is made from cornstarch, and it forms from the partial hydrolysis of starch. As the first equation indicated, maltose is made from two glucose units. They are joined by an acetal oxygen bridge that in carbohydrate chemistry is called a **glycosidic link.**

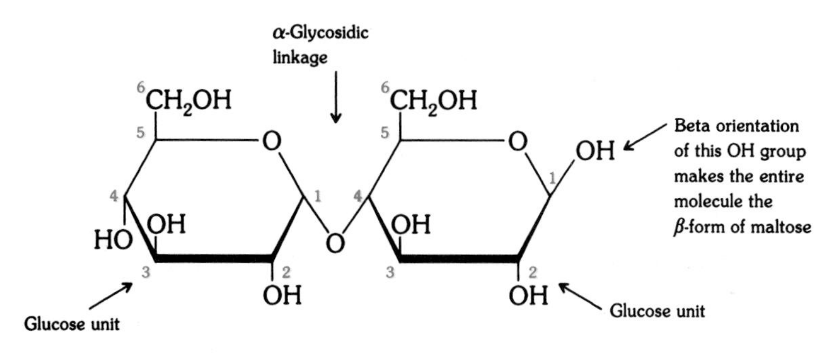

β-Maltose

■ If the OH group at C-1 on the far-right glucose unit projected downward instead of upward, the structure would be that of α-D-maltose instead of β-D-maltose.

Alternatively,

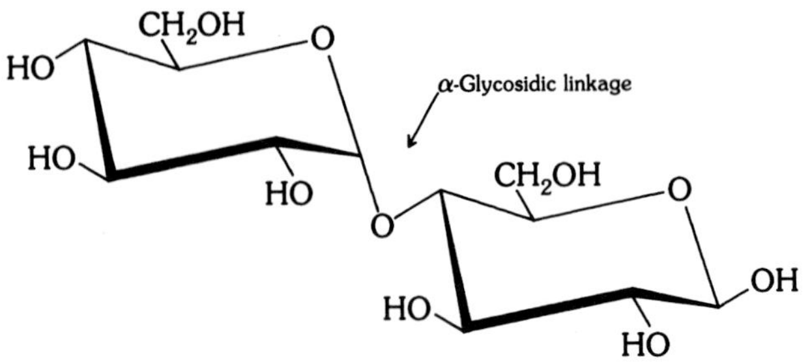

α-Glycosidic linkage

β-Maltose

In maltose, the bridging oxygen of the glycosidic link joins the C-1 position of the glucose unit that serves as the hemiacetal partner, the unit on the left, to the C-4 of the glucose unit that is the alcohol partner, on the right in the structure. Such a glycosidic link is designated as $(1 \rightarrow 4)$.

The bond to the bridging oxygen from C-1 (on the left) points in the *alpha* direction, so the glycosidic link is more fully described as $\alpha(1 \rightarrow 4)$. Had this link pointed in the beta direction, it would have been described as $\beta(1 \rightarrow 4)$. But then the disaccharide would not have been maltose but a different disaccharide, cellobiose.

The purely geometric difference between $\alpha(1 \rightarrow 4)$ and $\beta(1 \rightarrow 4)$ that marks the difference between maltose and cellobiose may seem to be a trifle, but the difference to us is that we can digest maltose but not cellobiose. Just this difference in *geometry* bars humans from an enormous potential food source, for nature could supply much of it from the polysaccharide cellulose. We have an enzyme, maltase, that catalyzes the digestion (the hydrolysis) of maltose. We have no enzyme for cellobiose (or, for that matter, cellulose), although some organisms do.

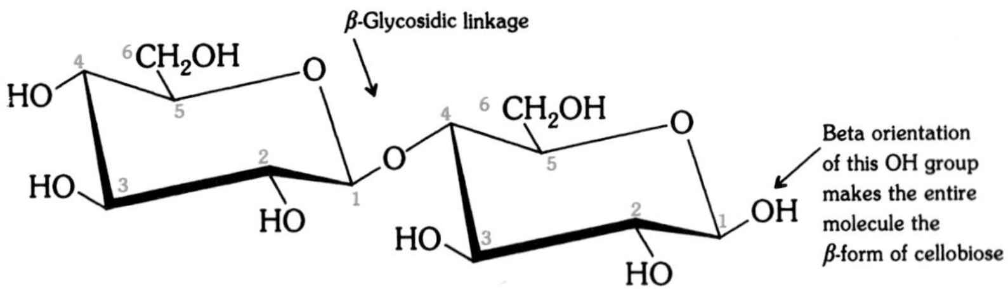

β-Glycosidic linkage

Beta orientation of this OH group makes the entire molecule the β-form of cellobiose

β-Cellobiose

Maltose Retains a Hemiacetal Unit, so It Is a Reducing Sugar and Mutarotates

The glucose unit on the 4-side of an $\alpha(1 \rightarrow 4)$ glycosidic link in maltose still has a hemiacetal group. This part of maltose, therefore, can open and close, so maltose can exist in three forms, α-, β-, and the open form. Maltose therefore mutarotates and is a reducing sugar. The ring opening action occurs only at the hemiacetal part, not at the oxygen bridge.

Lactose Links Galactose by a $\beta(1 \rightarrow 4)$ Bridge to Glucose Lactose or milk sugar occurs in the milk of mammals—4 to 6% in cow's milk and 5 to 8% in human milk. It is also a by-product in the manufacture of cheese.

Lactose is a galactoside. From C-1 of its galactose unit there is a $\beta(1 \rightarrow 4)$ glycosidic link to C-4 of a glucose unit. The glucose unit therefore still has a free hemiacetal system, so lactose mutarotates and is a reducing sugar.

Oxygen bridge

Galactose unit

Glucose unit

β-Lactose

Alternatively,

β-Glycosidic linkage

Beta orientation of this OH group makes the entire molecule the β-form of lactose

β-Lactose

Sucrose Links a Glucose Unit to a Fructose Unit Sucrose, our familiar table sugar, is obtained from sugar cane or from sugar beets. Its structure links a glucose to a fructose unit by an oxygen bridge in such a way that *no hemiacetal or hemiketal group remains.* Neither ring in sucrose, therefore, can open and close spontaneously in water. Hence, sucrose neither mutarotates nor gives positive tests with Tollens' or Benedict's reagents. It's our only common nonreducing disaccharide.

■ Beet sugar and cane sugar are identical compounds, sucrose.

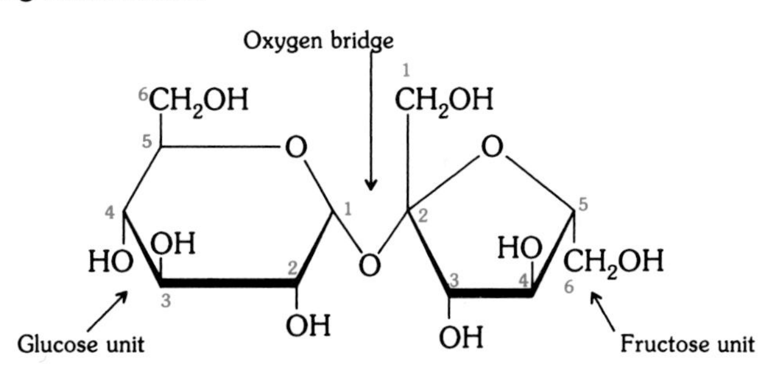

Sucrose

The 50 : 50 mixture of glucose and fructose that forms when sucrose is hydrolyzed is called *invert sugar,* and it makes up the bulk of the carbohydrate in honey. (The sign of specific rotation inverts from + to − when sucrose, $[\alpha]_D^{20}$ +66.5°, changes to invert sugar, $[\alpha]_D^{20}$ −19.9°, and this inversion of the sign is the origin of the term *invert* sugar.)

8.6

POLYSACCHARIDES

Starch, glycogen, and cellulose are all polyglucosides.

In this section we will study the structures and some of the properties of three polymers of glucose—starch, glycogen, and cellulose.

■ Up to a million glucose units per molecule have been found in some amylose samples, making amylose one of nature's largest molecules.

Plants Store the Chemical Energy of Glucose in Starch Molecules When glucose is made by photosynthesis, solar energy becomes stored as chemical energy, which the plants can use for chemical work. Free glucose, however, is very soluble in water. A plant, therefore, would have to retain considerable water if its cells had to hold free glucose molecules in solution. Otherwise, the concentration of the cell fluid would be too high, thereby causing osmotic pressure problems and upsetting proper movement of plant fluids. This problem of storing glucose without too much water is avoided by the conversion of glucose to its much less soluble polymer, starch. It is particularly abundant in plant seeds and tubers, where its energy is used for sprouting and growth. Animals that include plants in their diets also take advantage of the chemical energy in starch.

Starch is a mixture of two kinds of polymers of α-glucose, *amylose* and *amylopectin*. In amylose, the glucose units are joined by a linear succession of $\alpha(1 \rightarrow 4)$ glycosidic links, as seen in Figure 8.8. The lengths of the amylose "chains" vary within the same sample, but over 1000 glucose units occur per amylose molecule. Formula masses ranging from 150,000 to 600,000 have been measured. The long amylose molecules coil into spiral-like helices, which tuck a significant fraction of the OH groups inside and away from contact with water. Thus amylose is only slightly soluble in water.

Amylopectin molecules have both $\alpha(1 \rightarrow 4)$ and $\alpha(1 \rightarrow 6)$ glycosidic links, as seen in Figure 8.9. The $\alpha(1 \rightarrow 6)$ bridges link the C-1 ends of linear amylose-type units to C-6 positions of glycose units in other long amylose chains, as seen in Figure 8.9. There are hundreds of such links per molecule, so amylopectin is heavily branched, and the branches prevent any coiling of the polymer. This leaves many more OH groups exposed to water than in amylose, so amylopectin tends to be somewhat more soluble in water than is amylose. However, neither dissolves well. The "solution" is actually a colloidal dispersion, because it gives the Tyndall effect.
gives the Tyndall effect.

Natural starches are about 10 to 20% amylose and 80 to 90% amylopectin. Neither is a reducing carbohydrate and neither gives a positive Tollens' or Benedict's test. One unique test that starch does give is called the **iodine test,** and it can detect extremely minute traces of starch.[1] When a drop of iodine reagent is added to starch, an intensely purple color develops as the iodine molecules become trapped within the vast network of starch molecules. In a starch sample undergoing hydrolysis, this network gradually breaks up so the system slowly loses its ability to give the iodine test.

(a)

(b)

(c)

The iodine test for starch. (*a*) The starch dispersion is so dilute it does not appear cloudy. (*b*) A few drops of iodine reagent cause a purple color to develop, which becomes quite intense. (*c*) The development of a purple color when a dilute iodine solution is added to a starch dispersion is a positive starch iodine test.

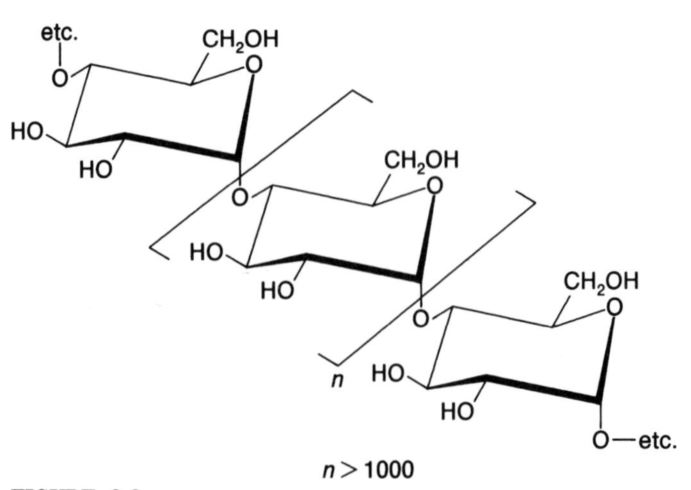

FIGURE 8.8
Amylose—partial structure.

[1] The starch-iodine reagent is made by dissolving iodine, I_2, in aqueous potassium iodide, KI. Iodine by itself is very insoluble in water, but iodine molecules combine with iodide ions to form the triiodide ion, I_3^-. Molecular iodine is readily available from this ion if some reactant is able to react with it.

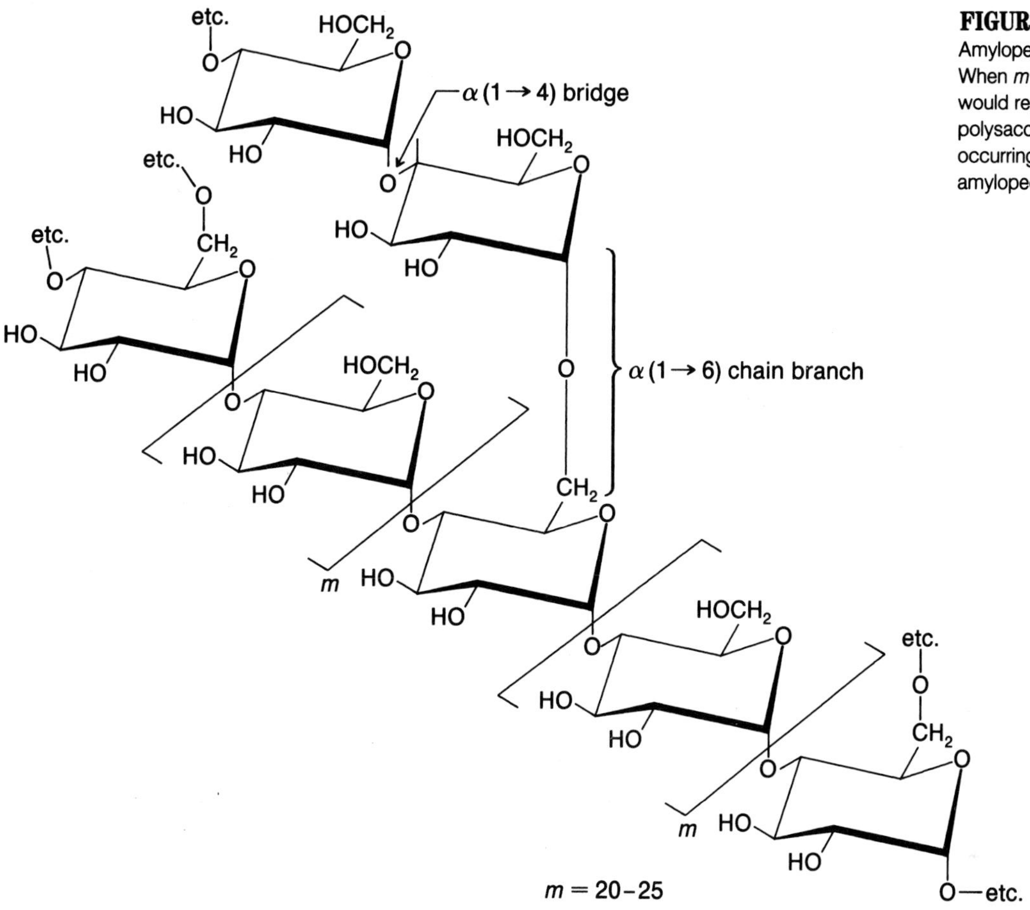

FIGURE 8.9
Amylopectin—partial structure. When $m = 6-12$, the structure would represent glycogen, a polysaccharide with branches occurring more frequently than in amylopectin.

The glycosidic links in starch are easily hydrolyzed in the presence of acids or the appropriate enzymes, which humans have. Thus the complete digestion of starch gives us only glucose. The partial hydrolysis of starch produces smaller polymer molecules that make up a substance known as *dextrin,* which has been used to manufacture mucilage and paste.

■ So-called *soluble* starch is partially hydrolyzed starch, and its smaller molecules more easily dissolve in water.

Glycogen Is the Storage Form of Glucose in Animals We and many animals use plant starch for food. Digestion hydrolyzes starch, and what glucose our bodies cannot use right away is changed into an amylopectin-like polymer called *glycogen.* In this form, we can store the chemical energy of glucose units. Normally we don't excrete any excess glucose. If we eat enough to replenish glycogen reserves in various tissues, any additional glucose is converted to fat (to the satisfaction of a huge weight-watcher industry).

■ Glycogen is sometimes called animal starch.

Glycogen molecules are essential like those of amylopectin, perhaps even more branched. When the values of m in the structure of amylopectin (Fig. 8.9) are in the range of 6 to 12, the structure would be that of glycogen. The formula masses of various samples of glycogen have been reported in the range of 300,000 to 100,000,000, which correspond roughly to 1700 to 600,000 glucose units per molecule. We store glucose as glycogen principally in the liver and in muscle tissue.

Cellulose Is a Polymer of β-Glucose Much of the glucose a plant makes by photosynthesis goes to make cellulose and other substances that it needs to build its cell walls and its rigid fibers. Cellulose is thus a major component of the food fiber in our diets. Cellulose, unlike starch or glycogen, has a geometry that allows its molecules to line up side by side, overlap each other, and twist into fibers.

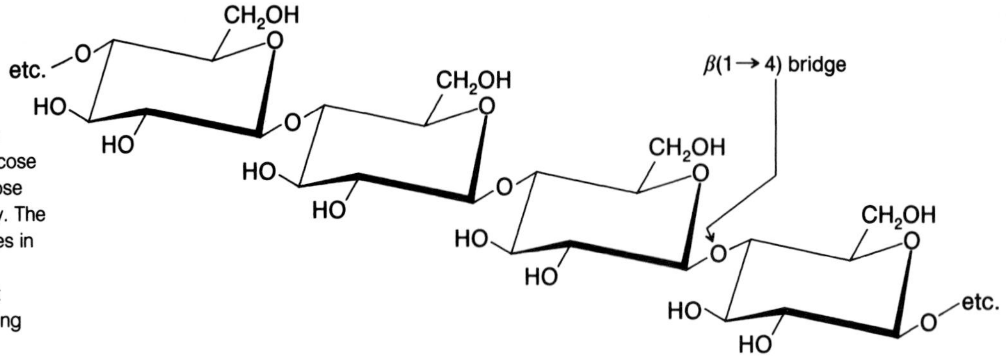

FIGURE 8.10
Cellulose—partial structure. In cotton, this polymer of β-D-glucose has from 2000 to 26,000 glucose units, depending on the variety. The strength of a cotton fiber comes in part from the thousands of hydrogen bonds that can exist between parallel and overlapping cellulose molecules.

■ Cellobiose is to cellulose what maltose is to starch.

The huge geometric difference that allows cellulose to form fibers but not amylose is the orientation of the oxygen bridge. Cellulose is a polymer of the beta form of glucose. All the oxygen bridges are $\beta(1 \rightarrow 4)$. See Figure 8.10. Cellulose molecules, moreover, have no branches corresponding to the $\alpha(1 \rightarrow 6)$ branches in amylopectin. All the substituents in the rings in cellulose project in the most stable directions (the equatorial directions as discussed in Special Topic 8.2). The cellulose molecule is thus quite ribbon-like, so it's easy for neighboring molecules to nestle to each other where hydrogen bonds between molecules stabilize the aggregations. With twisting of these collections, cellulose fibers of considerable strength are possible.

■ The adults in some ethnic human groups lack the digestive enzyme lactase that catalyzes hydrolysis of the β-glycosidic link between galactose and glucose units in lactose.

As we have noted, humans have no enzyme that can catalyze the hydrolysis of a beta-glycosidic link in cellulose, so none of the huge supply of cellulose in the world, or the cellobiose that could be made from it, is nutritionally useful to us. Many bacteria have this enzyme, however, and some strains dwell in the stomachs of cattle and other animals. Bacterial action converts cellulose in hay and other animal feed into small molecules that the larger animals can then use. Fungi and termites also can hydrolyze cellulose, enabling them to cause the decay of woody debris.

The oxygen bridges in the cellulose of cotton fabrics, being acetal systems, are hydrolyzed when a trace of acid catalyst is present. Perhaps you have discovered this the morning after you spilled some acid on your jeans in lab. (If you know that you have spilled a small amount of dilute acid on jeans, put a small spatulaful of sodium bicarbonate or sodium carbonate on a towel, make a paste, and daub the spot with it. A towel moistened with dilute ammonia also works, but watch out for the ammonia odor. You might be able to save the fabric if you act quickly.)

■ If you spill *concentrated* acid on your clothes or skin, flush the area with water immediately.

SUMMARY

Carbohydrates Carbohydrates are aldehydes or ketones with multiple OH groups or are glycosides of these. Those that can't be hydrolyzed are the monosaccharides, which in pure forms exist as cyclic hemiacetals or cyclic hemiketals that can mutarotate and that are reducing sugars.

D- and L families of carbohydrates Fischer projection structures of open-chain forms of monosaccharides are made according to a set of rules. The carbon chain is positioned vertically with any carbonyl group as close to the top as possible. At each stereocenter, this chain projects toward the back. Any groups on bonds that appear horizontal project forward. If the OH group of a carbohydrate that is farthest from the carbonyl group projects to the right in a Fischer projection structure, the carbohydrate is in the D-family. If this OH group projects to the left, the substance is in the L-family.

Monosaccharides The three nutritionally important monosaccharides are glucose, galactose, and fructose—all in the D-family. Glucose is the chief carbohydrate in blood. Galactose, which differs from glucose only in the orientation of the OH at C-4, is obtained (together with glucose) from the hydrolysis of lactose. Fructose, a reducing ketohexose, differs from glucose only in the location of the carbonyl group. It's at C-2 in fructose and at C-1 in glucose.

Disaccharides The disaccharides are glycosides whose molecules hydrolyze into two monosaccharide molecules when they react with water. Maltose is made of two glucose units joined by an $\alpha(1 \rightarrow 4)$ glycosidic link. In a molecule of lactose (milk sugar)—a galactoside—a galactose unit joins a glucose unit by a $\beta(1 \rightarrow 4)$ oxygen bridge. In sucrose (cane or beet sugar), there is an oxygen bridge from C-1 of a glucose unit to C-2 of a fructose unit. Both maltose and lactose retain hemiacetal systems, so both mutarotate and are reducing sugars. They also exist in α- and β-forms. Sucrose is a nonreducing disaccharide. The digestion of these disaccharides gives their monosaccharide units.

Polysaccharides Three important polysaccharides of glucose are starch (a plant product), glycogen (an animal product), and cellulose (a plant fiber). In molecules of each, $(1 \rightarrow 4)$ glycosidic links occur. They're alpha bridges in starch and glycogen and beta bridges in cellulose. In the molecules of the amylopectin portion of starch as well as in glycogen, numerous $\alpha(1 \rightarrow 6)$ bridges also occur. No polysaccharide gives a positive test with Tollens' or Benedict's reagents. Starch gives the iodine test. As starch is hydrolyzed its molecules successively break down to dextrins, maltose, and finally glucose. Humans have enzymes that catalyze the hydrolysis of $\alpha(1 \rightarrow 4)$ and $\alpha(1 \rightarrow 6)$ glycosidic links, but not the $\beta(1 \rightarrow 4)$ glycosidic links of cellulose.

REVIEW EXERCISES

The answers to Review Exercises whose numbers are in color are found in Appendix A. The answers to the other Review Exercises are found in the Study Guide that accompanies this book. The more challenging questions are marked with asterisks.

Biochemistry

8.1 Substances in the diet must provide raw materials for what three essentials for life?

8.2 What are the three broad classes of foods?

8.3 What kind of compound carries the genetic "blueprints" of a cell?

8.4 What is as important to the life of a cell as the chemicals that make it up or that it receives?

Carbohydrate Terminology

8.5 Examine the following structures and identify by letter(s) which structure(s) fit each of the labels. If a particular label is not illustrated by any structure, state so.

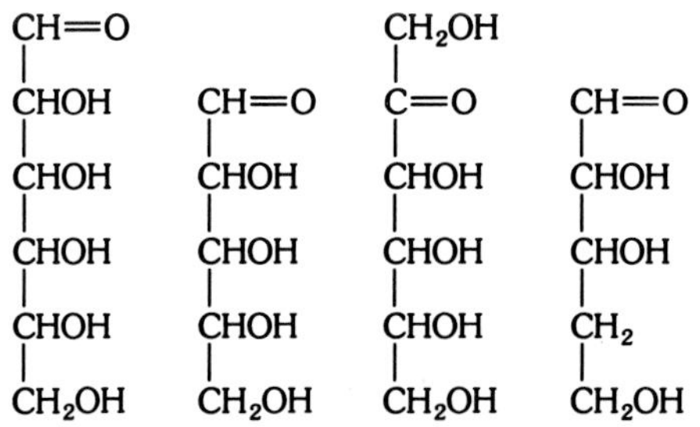

A	B	C	D
CH=O		CH_2OH	
CHOH	CH=O	C=O	CH=O
CHOH	CHOH	CHOH	CHOH
CHOH	CHOH	CHOH	CHOH
CHOH	CHOH	CHOH	CH_2
CH_2OH	CH_2OH	CH_2OH	CH_2OH

(a) ketose (b) deoxy sugar
(c) aldohexose (d) aldopentose

8.6 Write the structure (open-chain form) that illustrates
(a) any ketopentose
(b) any aldotetrose

8.7 What is the structure and the common name of the simplest aldose?

8.8 What is the structure and the common name of the simplest ketose?

8.9 A sample of 0.0001 mol of a carbohydrate reacted with water in the presence of a catalyst and 1 mol of glucose was produced. Classify this carbohydrate as a mono-, di-, or polysaccharide.

8.10 An unknown carbohydrate gives a positive Benedict's test. Classify it as a reducing or nonreducing carbohydrate.

8.11 An unknown carbohydrate, **A**, gives the following reaction.

$$A + H_2O \xrightarrow{H^+ \text{ catalyst}} \text{galactose} + \text{glucose} + \text{xylose}$$

(a) What should be the coefficient of H_2O to balance this equation?
(b) How is **A** classified, as a mono-, di-, or trisaccharide?

***8.12** A student in an advanced lab was assigned the task of determining the structure of a carbohydrate. The empirical formula was found to be CH_2O and the compound had a molecular mass of 150. When it was allowed to react with as much acetic anhydride as it could, it was changed to $C_{13}H_{18}O_9$. Very gentle oxidation (Tollens' reagent) changed the carbohydrate into a monocarboxylic acid, $C_5H_{10}O_6$. Vigorous reduction yielded pentane. Write a structure (open-chain) for the carbohydrate that is consistent with these observations.

***8.13** An unknown carbohydrate could be reduced by a series of steps to butane. It gave a positive Benedict's test, and its molecules had just one stereocenter. What is a structure consistent with these facts?

8.14 A student proposed the following two structures as the likeliest candidate structures for a carbohydrate being studied.

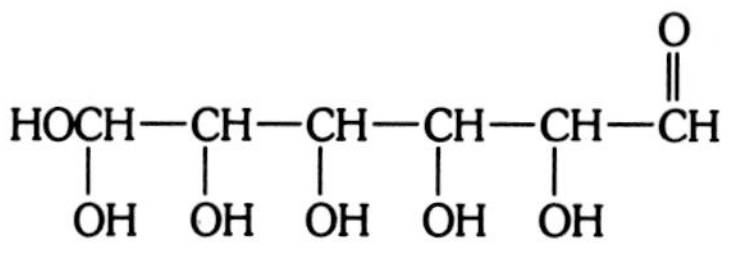

```
                            O
                            ‖
HOCH—CH—CH—CH—CH—CH
  |     |     |     |     |
 OH   OH   OH   OH   OH
              A
```

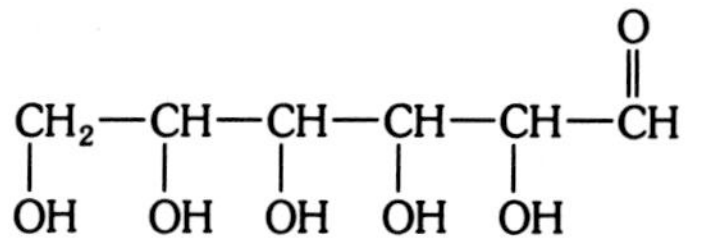

```
                              O
                              ‖
CH2—CH—CH—CH—CH—CH
 |       |     |     |     |
OH     OH   OH   OH   OH
              B
```

One of these structures is highly unlikely. Which one, and why?

8.15 What is the name of the most abundant carbohydrate in blood?

Absolute Configurations

8.16 Consider the monomethyl ether of glyceraldehyde

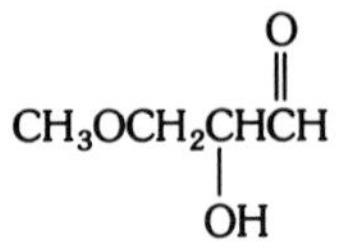

$$CH_3OCH_2CH(OH)CHO$$

(a) How many stereoisomers are possible for this compound?
(b) Draw the Fischer projection structures of the stereoisomers according to the conventions used for carbohydrates.
(c) Correctly label each structure as D or L according to the conventions used for carbohydrates.

8.17 The simplest ketotriose is never drawn in the form of a Fischer projection structure. Why not?

8.18 Write the Fischer projection formula of L-glucose. (Refer to Figure 8.4.)

8.19 What is the Fischer projection formula of D-2-deoxyribose?

8.20 Suppose that the aldehyde group of D-glyceraldehyde is oxidized to a carboxylic acid group, and that this is then converted to a methyl ester under conditions that do not touch any of the four bonds to C-2. What is the Fischer projection formula of this methyl ester? To what family, D or L, does it belong? Explain.

8.21 Sorbose has the structure given below. It is made by the fermentation of sorbitol, and hundreds of tons of sorbose are used each year to make vitamin C.

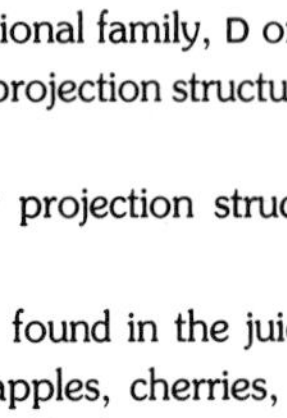

CH_2OH
C=O
HO—H
H—HO
HO—H
CH_2OH

(a) In what configurational family, D or L, is sorbose?
(b) Write the Fischer projection structure of the enantiomer of sorbose.
(c) Write the Fischer projection structure of D-3-deoxysorbose.

8.22 Sorbitol, $C_6H_{14}O_6$, is found in the juices of many fruits and berries (e.g., pears, apples, cherries, and plums). It can be made by the addition of hydrogen to D-glucose:

$$\underset{\text{D-Glucose}}{C_6H_{12}O_6} + H_2 \xrightarrow[\text{heat and pressure}]{\text{Ni}} \underset{\text{D-Sorbitol}}{C_6H_{14}O_6}$$

Write the Fischer projection structure of D-sorbitol.

8.23 The magnesium salt of D-gluconic acid (Glucomag®) is used as an antispasmodic and to treat dysmenorrhea. D-Gluconic acid, $C_6H_{12}O_7$, forms by the mild oxidation of D-glucose. What is the Fischer projection formula of D-gluconic acid?

Cyclic Forms of Carbohydrates

***8.24** Consider the following cyclic hemiacetal.

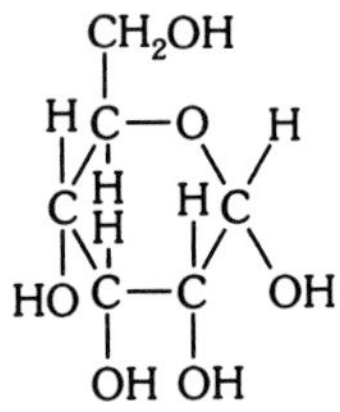

(a) What is the structure of its open form? (Write the open form with its chain coiled in the same way it is coiled in the closed form, above.)
(b) At which specific carbon (by number) does this compound differ from naturally occurring glucose? (The hemiacetal carbon has position 1 in the ring, and the ring is numbered clockwise from it.)
(c) Is the compound in the D or the L-family? (How can you tell without writing a Fischer projection structure?)
(d) With the aid of your answer to part (b) and Figure 8.4, what is the name of this compound?
(e) Is this compound an *anomer* or an *epimer* of D-glucose?

8.25 Examine the following structure. If you judge that it is either a cyclic hemiketal or a cyclic hemicetal, write the structure of the open-chain form (coiled in like manner as the chain of the ring).

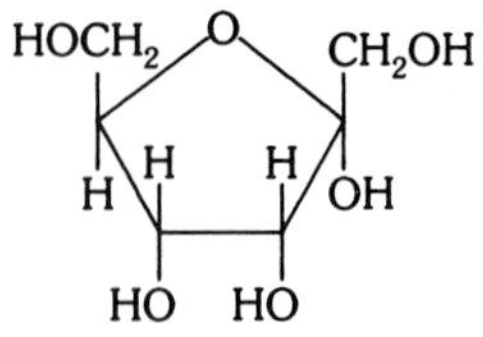

***8.26** Mannose mutarotates like glucose. Mannose is identical with glucose except that in the cyclic structures the OH at C-2 in mannose projects on the same side of the ring as the CH_2OH group. Write the structures of the three forms of mannose that are in equilibrium after mutarotation gives a steady value of specific rotation. Identify which corresponds to α-manose and which to β-mannose.

8.27 Allose is identical with glucose except that in its cyclic form the OH group at C-3 projects on the opposite side of the ring from the CH_2OH group. Allose mutarotates like glucose. Write the structures of the three forms of allose that are in equilibrium after mutarotation gives a final value of specific rotation. Which structures are α- and β-allose?

8.28 If less than 0.05% of all galactose molecules are in their open-chain form at equilibrium in water, how can galactose give a strong, positive Tollens' test, a test good for the aldehyde group?

8.29 At equilibrium, after mutarotation, a glucose solution consists of 36% α-glucose and 64% β-glucose (and just a trace of the open form). Suppose that in some enzyme-catalyzed process the beta form is removed from this equilibrium. What becomes of the other forms of glucose?

8.30 Study the cyclic form of β-ribose on page 212 again. If its designation as β-ribose signifies a particular relationship between the CH_2OH group at C-4 and the OH group at C-1, what must the cyclic formula of α-ribose be?

8.31 Is the following structure that of α-fructose, β-fructose, or something else? Explain.

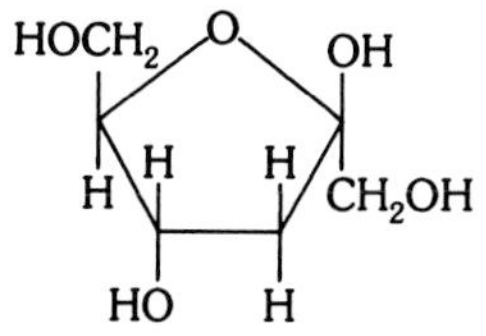

8.32 Write the cyclic structure of α-3-deoxyribose and draw an arrow that points to its hemiacetal carbon.

8.33 Could 4-deoxyribose exist as a cyclic hemiacetal with a five-membered ring (one of whose atoms is O)? Explain.

***8.34** With the aid of Figure 8.5 and the cyclic forms of D-fructose given in this chapter, write the structures of the cyclic forms of D-sorbose and correctly label your structures as α- or β-forms.

Glycosides

8.35 Using cyclic structures, write the structures of ethyl α-glucoside and ethyl β-glucoside. Are there two enantiomers or are they some other kind of stereoisomers?

8.36 What are the structures of methyl α-galactoside and methyl β-galactoside? Could these be described as cis–trans isomers? Explain.

Disaccharides

8.37 What are the names of the three nutritionally important disaccharides?

8.38 What is invert sugar?

8.39 Why isn't sucrose a reducing sugar?

***8.40** Examine the following structure and answer the questions about it.

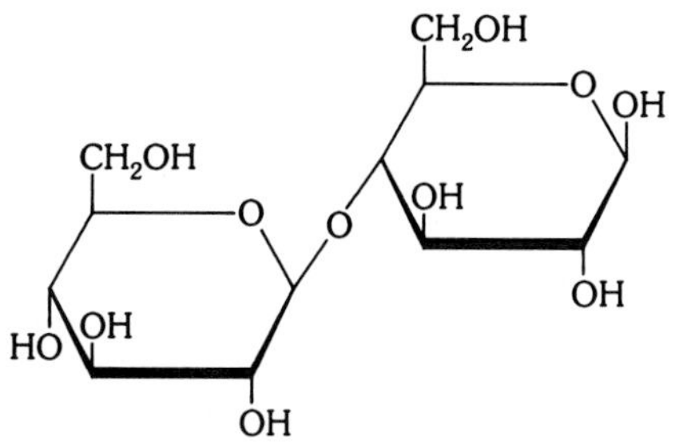

(a) Does it have a hemiacetal system? Where? (Draw an arrow to it or circle it.)
(b) Does it have an acetal system? Where? (Circle it.)
(c) By what specific symbolism would the oxygen bridge be described? As an example of the kind of symbolism meant, recall that a bridge might be described as $\alpha(1 \rightarrow 6)$.
(d) Does this substance give a positive Benedict's test? Explain.
(e) In what specific structural way does it differ from maltose?
(f) What are the names of the products of the acid-catalyzed hydrolysis of this compound?

8.41 Trehalose is a disaccharide found in young mushrooms and yeast, and it is the chief carbohydrate in the hemolymph of certain insects. On the basis of its structural features, answer the following questions.

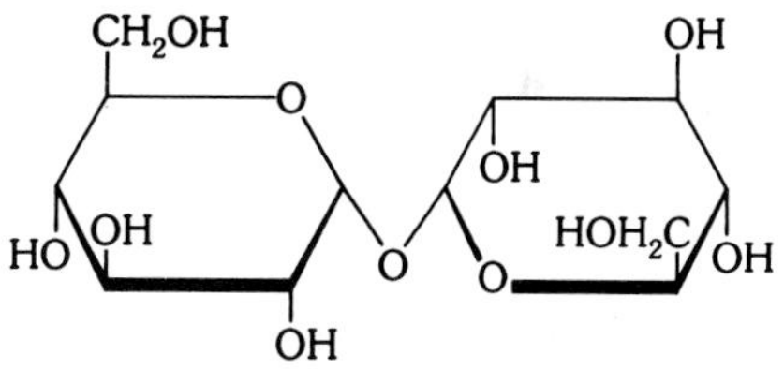

(a) Is trehalose a reducing sugar? Explain.
(b) Can trehalose mutarotate? Explain.
(c) Identify, by name only, the products of the hydrolysis of trehalose.

***8.42** Maltose has a hemiacetal system. Write the structure of maltose in which this group has changed to the open form.

***8.43** When lactose undergoes mutarotation, one of its rings opens up. Write this open form of lactose.

Polysaccharides

8.44 Name the polysaccharides that give only D-glucose when they are completely hydrolyzed.

8.45 What is the main structural difference between amylose and cellulose?

8.46 How are amylose and amylopectin alike structurally?

8.47 How are amylose and amylopectin different structurally?

8.48 Why can't humans digest cellulose?

8.49 What is the iodine test? Describe the reagent and state what it is used to test for and what is seen in a positive test.

8.50 How do amylopectin and glycogen compare structurally?

8.51 How does the body use glycogen?

Photosynthesis (Special Topic 8.1)

8.52 The energy available in glucose originated in the sun. Explain in general terms how this happened.

8.53 Write the simple, overall equation for photosynthesis.

8.54 What is the name and color of the energy-absorbing pigment in plants?

8.55 In what general region of the planet earth is most of the photosynthesis carried out? By what organisms?

8.56 Describe in general terms the oxygen cycle of our planet including the function of photosynthesis in it.

Boat and Chair Forms of Saturated Six-Membered Rings (Special Topic 8.2)

Start by simply tracing the structures:, then practice drawing the skeletons of the two chair forms of six-membered rings.

8.57 Why are chair forms of saturated, six-membered rings more stable than boat forms.

8.58 Why are equatorial positions for groups attached to a six-membered ring more stable than axial?

*8.59 Draw the structures of the following.

(a) A chair form of cyclohexane. Label the bonds that can hold substituents as being axial (*a*) or equatorial (*e*).

(b) The least stable structure of *trans*-1,2-dimethylcyclohexane.

(c) The most stable structure of *trans*-1,2-dimethylcyclohexane.

(d) The most stable form of D-glucose.

(e) The most stable form of D-allose. (Hint: Refer to th Fischer projection structure of D-allose in Figure 8.4 ar note where it is the same as the Fischer projection of glucose and where it differs.)

Additional Exercises

***8.60** A freshly prepared aqueous solution of sucrose gives a neg tive Tollens' test (as expected), but when the solution has stoc at room temperature for about a week it gives this test. E plain.

*8.61 A freshly prepared solution (actually a dispersion) of starch water gives a positive iodine test. If this solution is warme with a trace of human saliva, however, the ability of the sol tion to give this test gradually disappears. How might this o servation be explained?

***8.62** How many atoms (C and O) are there in the largest size rin possible for a cyclic hemiacetal form of an aldotetrose?

*8.63 Glyceraldehyde does not form a cyclic hemiacetal. Offer a explanation for this fact.

9

LIPIDS

Peanuts, olives, and corn. Humble fare, yet delightful to eat. That oils can be pressed from them was discovered ages ago, and such oils, members of the lipid family studied in this chapter, are the polyunsaturated oils widely recommended over saturated fats for heart-friendly diets.

9.1
WHAT LIPIDS ARE

The lipids include the edible fats and oils whose molecules consist of esters of long-chain fatty acids and glycerol.

When undecomposed plant or animal material is crushed and ground with a nonpolar solvent such as ether, whatever dissolves is classified as a **lipid.** This operation catches a large variety of relatively nonpolar substances, and all are lipids. Thus this broad family is defined not by one common structure but by the technique used to isolate its members, *solvent extraction.* Among the many substances that won't dissolve in nonpolar solvents are carbohydrates, proteins, other very polar organic substances, inorganic salts, and water. The chart in Figure 9.1 outlines the many kinds of lipids.

■ "Lipid" is from the Greek *lipos,* fat.

■ Extraction means to shake or stir a mixture with a solvent that dissolves just part of the mixture.

Lipids Are Broadly Subdivided According to the Presence of Hydrolyzable Groups One of the major classes of lipids, the **hydrolyzable lipids,** consists of compounds with one or more groups that can be hydrolyzed. In nearly all examples, these are *ester groups.* A number of families are in the lipid group, including the neutral fats, the waxes, the phospholipids, and the glycolipids. The *neutral fats* include such familiar food products as butterfat, lard (pork fat), tallow (beef fat), olive oil, corn oil, and peanut oil. Thus some neutral fats are solids and others are liquids at room temperature. The solid neutral fats are generally from animals and so are called the *animal fats.* The liquid neutral fats are from plants and are called the *vegetable oils.* The fats and oils are the high-calorie components of the diet. The conversion factor used by the National Academy of Sciences is 9.0 kcal/g for food fat, which can be compared to 4.0 kcal/g for proteins and carbohydrates.

■ What is *neutral* about the neutral fats is the absence of electrical charges on their molecules.

The **nonhydrolyzable lipids** lack groups that can be hydrolyzed. These include the steroids such as cholesterol and many sex hormones (Section 9.4). Many plants produce another group of nonhydrolyzable lipids called the *terpenes,* which often are responsible for the very pleasant odors of plant oils. Oil of rose, for example, is 40–60% geraniol, a terpene alcohol with a sizeable hydrocarbon unit.

■ 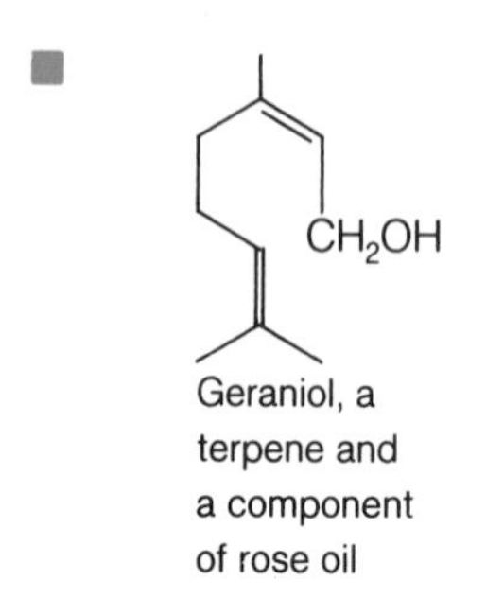

Geraniol, a terpene and a component of rose oil

The *Fatty Acids* Are Mostly Long-Chain, Unbranched Monocarboxylic Acids When hydrolyzable lipids react with water, among the products are carboxylic acids (or their anions) that are dubbed the **fatty acids.** The fatty acids obtained from the lipids of most plants and animals share the following features.

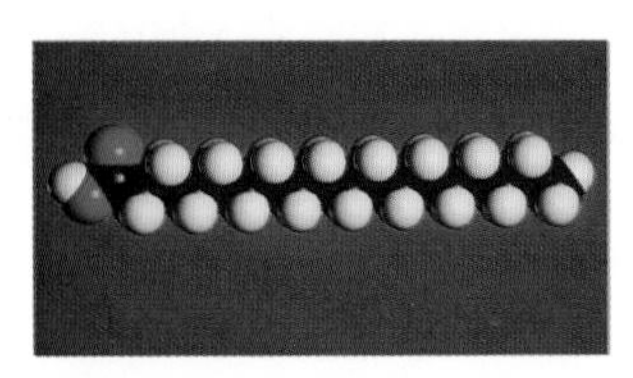
Model of stearic acid, a saturated fatty acid.

Structural Features of the Common Fatty Acids

1. They are usually *mono*carboxylic acids, RCO_2H.
2. The R group is usually a long *unbranched* chain.
3. The number of carbon atoms is almost always *even.*
4. The R group can be saturated, or it can have one or more double bonds, which are cis.

LIPIDS

Hydrolyzable lipids

(Only C, H, O)

Waxes

Triacylglycerols

Animal fats

Vegetable oils

(C, H, O, P, N)

Esters of glycerol

Phosphatides

Plasmalogens

Esters of sphingosine

Sphingomyelins

Cerebrosides

Nonhydrolyzable lipids

Steroids

Terpenes

Others

FIGURE 9.1
Lipid families.

Thus just two functional groups are present in the fatty acids, both often in the same molecule—the alkene double bond and the carboxyl group.

The most abundant *saturated* fatty acids are palmitic acid, $CH_3(CH_2)_{14}CO_2H$, and stearic acid, $CH_3(CH_2)_{16}CO_2H$, which have 16 and 18 carbons, respectively. Refer back to Table 5.1 for the other saturated fatty acids obtainable from lipids—the acids with more carbon atoms than acetic acid but with *even* numbers of carbons, like butanoic, hexanoic, octanoic, and decanoic acids. Fatty acids with fewer than 16 carbons, however, are relatively rare in nature.

The ***unsaturated*** fatty acids most commonly obtained from lipids are listed in Table 9.1 and include palmitoleic acid (C_{16}) and the C_{18} acids, oleic, linoleic, and linolenic acids. The double bonds in the unsaturated fatty acids of Table 9.1 are cis. Oleic acid is the most abundant and most widely distributed fatty acid in nature.

The presence of cis alkene groups causes kinks in the long hydrocarbon groups of the unsaturated fatty acids, which affects their melting points, as you can see in Table 9.1. As more alkene groups per molecule are present, the melting points of the fatty acids decrease. The structural kinks inhibit the closeness of the packing of molecules in crystals, and such closeness is required for stronger forces of attraction between molecules that cause higher melting points. The relationship between double bonds per molecule and melting point carries over to the neutral fats. The animal fats happen to have fewer alkene groups per molecule than the vegetable oils and so the animal fats are likely to be solids at room temperature and the vegetable oils tend to be liquids.

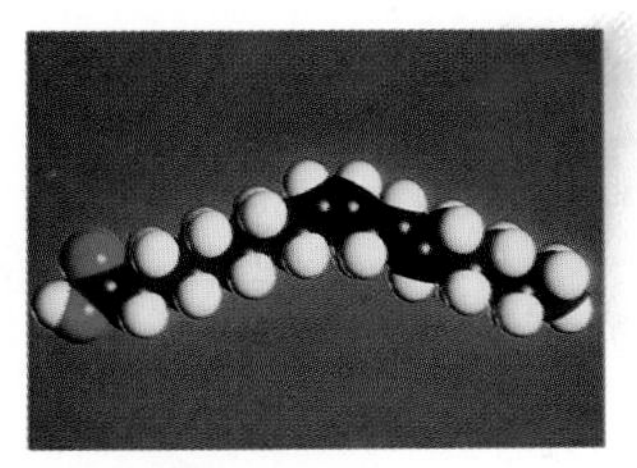

Model of linoleic acid.

PRACTICE EXERCISE 1

To visualize how a cis double bond introduces a kink into a molecule, write the structure of oleic acid in a way that correctly shows the cis geometry of the alkene group. (Without the double bond, the entire sidechain can stretch out into a perfect zigzag conformation, as in stearic acid. This makes it easy for two sidechains to nestle very close to each other.)

TABLE 9.1
Common Unsaturated Fatty Acids

Name	Double Bonds	Total Carbons	Structure	MP (°C)
Palmitoleic acid	1	16	$CH_3(CH_2)_5CH{=}CH(CH_2)_7CO_2H$	32
Oleic acid	1	18	$CH_3(CH_2)_7CH{=}CH(CH_2)_7CO_2H$	4
Linoleic acid	2	18	$CH_3(CH_2)_4CH{=}CHCH_2CH{=}CH(CH_2)_7CO_2H$	−5
Linolenic acid	3	18	$CH_3CH_2CH{=}CHCH_2CH{=}CHCH_2CH{=}CH(CH_2)_7CO_2H$	−11
Arachidonic acid	4	20	$CH_3(CH_2)_4CH{=}CHCH_2CH{=}CHCH_2CH{=}CHCH_2CH{=}CH(CH_2)_3CO_2H$	−50

SPECIAL TOPIC 9.1 THE PROSTAGLANDINS 

The prostaglandins were discovered in the mid-1930s by a Swedish scientist, Ulf von Euler (Nobel prize, 1970), but they didn't arouse much interest in medical circles until the late 1960s, largely through the work of Sune Bergstrom. It became apparent that these compounds, which occur widely in the body, affect a large number of processes. Their general name comes from an organ, the prostate gland, from which they were first obtained. About 20 are known, and they occur in four major subclasses designated as PGA, PGB, PGE, and PGF. (A subscript is generally placed after the third letter to designate the number of alkene double bonds that occur outside of the five-membered ring.) The structures of some typical examples are shown here.

PGA_1

PGB_1

PGE_1

Prostaglandins are made from twenty-carbon fatty acids, like arachidonic acid. By coiling a molecule of this acid, as shown, you can see how its structure needs only a ring closure (suggested by the dashed arrow) and three more oxygen atoms to become PGF_2. The oxygen atoms are all provided by molecular oxygen itself.

Arachidonic acid

Several steps (inhibited by aspirin)

$PGF_{2\alpha}$

Prostaglandins as Chemical Messengers The prostaglandins are like hormones in many ways, except that they do not act globally, that is, over the entire body. They do their work within the cells where they are made or in nearby cells, so they are sometimes called *local hormones*. This is perhaps why the prostaglandins have such varied functions; they occur and express their roles in such varied tissues. They work together with hormones to modify the chemical messages that hormones bring to cells. In some cells, the prostaglandins inhibit enzymes and in others they activate them. In some organs, the prostaglandins help to regulate the flow of blood within them. In others, they affect the transmission of nerve impulses.

Some prostaglandins enhance inflammation in a tissue, and it is interesting that aspirin, an inflammation reducer, does exactly the opposite. This effect is caused by aspirin's ability to inhibit the work of an enzyme needed in the synthesis of prostaglandins.

Prostaglandins as Pharmaceuticals In experiments that use prostaglandins as pharmaceuticals, they have been found to have an astonishing variety of effects. One prostaglandin induces labor at the end of a pregnancy. Another stops the flow of gastric juice while the body heals an ulcer. Other possible uses are to treat high blood pressure, rheumatoid arthritis, asthma, nasal congestion, and certain viral diseases.

SPECIAL TOPIC 9.2
THE OMEGA-3 FATTY ACIDS AND HEART DISEASE

Omega-3 refers to the location of a double bond third in from the far end of a long-chain fatty acid, particularly those with 18, 20, and 22 carbons. Just as omega, ω, is the last letter in the Greek alphabet, so the omega position in a fatty acid is the one farthest from the carboxyl group. Arachidonic acid (Table 9.1) is an omega-6 C-20 fatty acid because it has a double bond at the sixth carbon from the omega end. Linolenic acid (Table 9.1) is an omega-3 fatty acid. Two other omega-3 acids are considered by some to be important in metabolism:

$$CH_3CH_2CH{=}CHCH_2(CH{=}CHCH_2)_4(CH_2)_2CO_2H$$

ω-3-Eicosapentaenoic acid

$$CH_3CH_2CH{=}CHCH_2(CH{=}CHCH_2)_5CH_2CO_2H$$

ω-3-Docosahexaenoic acid

The basis of the interest in these is that Eskimos have low incidences of heart disease despite relatively high cholesterol levels in their diets (from fish oils and fish liver). Some scientists believe that the high level of the omega-3 fatty acids in marine oils provides protection against disease.

The properties of the fatty acids are those to be expected of compounds with carboxyl groups, double bonds (where present), and long hydrocarbon chains. Thus they are insoluble in water and soluble in nonpolar solvents. The fatty acids are neutralized by bases to form salts, and they can be esterified by reacting with alcohols (see Section 5.5). Fatty acids with alkene groups react with bromine or hydrogen in the presence of a catalyst.

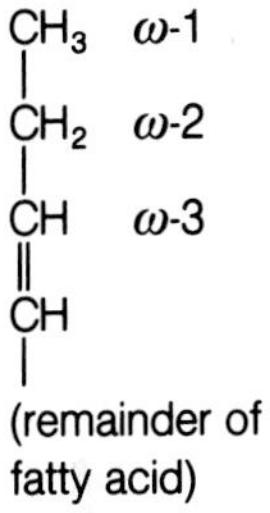

The *prostaglandins* are an unusual family of fatty acids with 20 carbons, five-membered rings, and a wide variety of effects in the body. See Special Topic 9.1.

In the late 1980s, one small group of fatty acids, the omega-3 fatty acids, appeared in scientific debates about the value of fish or marine oils in the diet. Special Topic 9.2 describes them further.

Omega (ω) designation

CH_3 ω-1
|
CH_2 ω-2
|
CH ω-3
||
CH
|
(remainder of fatty acid)

The Triacylglycerols (or Triglycerides) Are Triesters of Glycerol and Fatty Acids The molecules of the most abundant lipids are the **triacylglycerols** or the **triglycerides.** They are triesters between glycerol and three fatty acids.

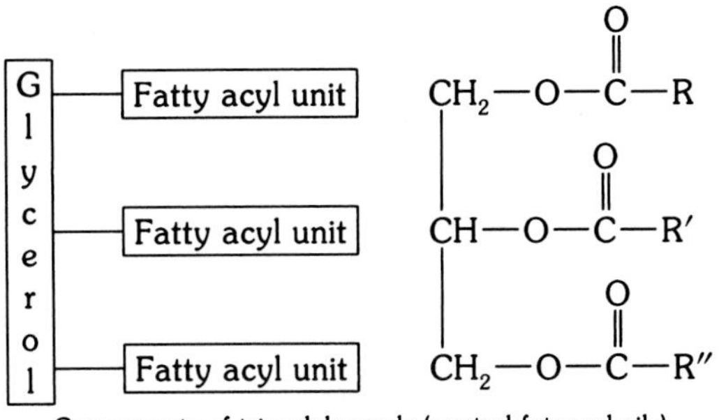

Glycerol		Structure
G l y	Fatty acyl unit	$CH_2-O-\overset{O}{\overset{\|\|}{C}}-R$
c e r	Fatty acyl unit	$CH-O-\overset{O}{\overset{\|\|}{C}}-R'$
o l	Fatty acyl unit	$CH_2-O-\overset{O}{\overset{\|\|}{C}}-R''$

Components of triacylglycerols (neutral fats and oils)

As you can see, there are no (+) or (−) charges on triacylglycerol molecules and so they are unlike the more complex hydrolyzable lipids to be studied in Section 9.3. The triacylglycerols include lard (pork fat), tallow (beef fat), butterfat—all animal fats—as well as such plant oils (vegetable oils) as olive oil, cottonseed oil, corn oil, peanut oil, soybean oil, coconut oil, and linseed oil.

The name *triglycerides* is common in the older scientific literature on triacylglycerols.

In a particular fat or oil, fatty acids predominate, others either are absent or are present in trace amounts, and virtually all of the molecules are triacylglycerols. Data on the fatty acid compositions of several fats and oils are listed in Table 9.2. Oleic acid (C_{18}; one alkene group) is very common among both the fats and oils. Notice particularly, however, that

TABLE 9.2
Fatty Acids Obtained from Neutral Fats and Oils

Type of Lipid	Fat or Oil	Average Composition of Fatty Acids (%)					
		Myristic Acid	Palmitic Acid	Stearic Acid	Oleic Acid	Linoleic Acid	Others
Animal fats	Butter	8–15	25–29	9–12	8–33	2–4	*a*
	Lard	1–2	25–30	12–18	48–60	6–12	*b*
	Beef tallow	2–5	24–34	15–30	35–45	1–3	*b*
Vegetable oils	Olive	0–1	5–15	1–4	67–84	8–12	
	Peanut	—	7–12	2–6	30–60	20–38	
	Corn	1–2	7–11	3–4	25–35	50–60	
	Cottonseed	1–2	6–10	2–4	20–30	50–58	
	Soybean	1–2	6–10	2–4	20–30	50–58	*c*
	Linseed	—	4–7	2–4	14–30	14–25	*d*
Marine oils	Whale	5–10	10–20	2–5	33–40	—	*e*
	Fish	6–8	10–25	1–3	—	—	*e*

[a] Also, 3–4% butyric acid, 1–2% caprylic acid, 2–3% capric acid, 2–5% lauric acid.

[b] Also, linolenic acid, 1%. [c] Also, linolenic acid, 5–10%. [d] Also, linolenic acid, 45–60%.

[e] Large percentage of other highly unsaturated fatty acids (see Special Topic 9.2).

■ An acyl group has the general structure:

$$\mathrm{R\overset{\overset{\displaystyle O}{\|}}{C}—}$$

the vegetable oils tend to incorporate more of the acyl groups of the unsaturated fatty acids, like those of oleic and linoleic acid, than do the animal fats. Thus vegetable oils have more double bonds per molecule and so are often described as *polyunsaturated*. The saturated fatty acyl units of palmitic and stearic acids are far more common in animal fats, which are thus sometimes called the *saturated fats*.

The three acyl units in the more common triacylglycerol molecules present in a given fat or oil are contributed by two or three *different* fatty acids. Fats and oils are thus mixtures of different molecules that share common structural features. Although we cannot give, for example, *one* structure for cottonseed oil, we can describe what is probably a fairly typical molecule, like that of structure **1**.

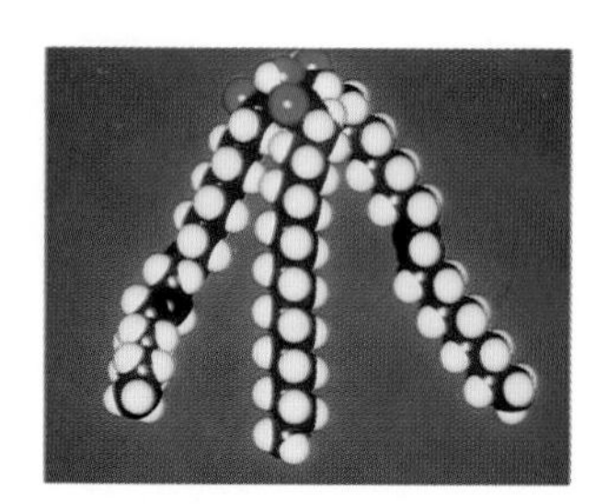

Model of structure **1**.

$$\begin{array}{l} CH_2—O—\overset{\overset{O}{\|}}{C}(CH_2)_7CH{=}CH(CH_2)_7CH_3 \\ | \\ CH—O—\overset{\overset{O}{\|}}{C}(CH_2)_{16}CH_3 \\ | \\ CH_2—O—\overset{\overset{O}{\|}}{C}(CH_2)_7CH{=}CHCH_2CH{=}CH(CH_2)_4CH_3 \end{array}$$

1

Plant Waxes Are Simple Esters with Long Hydrocarbon Chains The waxes occur as protective coatings on fruit and leaves as well as on fur, feathers, and skin. Nearly all **waxes** are esters of long-chain monohydric alcohols and long-chain monocarboxylic acids in both of which there is an *even* number of carbons. As many as 26 to 34 carbon atoms can be incorporated into *each* of the alcohol and the acid units, which makes the waxes almost totally hydrocarbon-like.

$$R—O—\overset{\overset{O}{\|}}{C}—R'$$

Alcohol unit — Fatty acyl unit

Components of waxes

Any particular wax, like beeswax, consists of a mixture of similar compounds that share the kind of structure shown above. In molecules of lanolin (wool fat), however, the alcohol portion is contributed by steroid alcohols, which have large ring systems that we'll study in Section 9.4. Waxes exist in sebum, a secretion of human skin that helps keep the skin supple.

■ Lanolin is used to make cosmetic skin lotions.

PRACTICE EXERCISE 2

One particular ester in beeswax can be hydrolyzed to give a straight-chain primary alcohol with 26 carbons and a straight-chain carboxylic acid with 28 carbons. Write the structure of this ester.

9.2

CHEMICAL PROPERTIES OF TRIACYLGLYCERIDES

Triacylglycerols can be hydrolyzed, saponified, and hydrogenated.

Triacylglycerols Can Be Hydrolyzed When we need the chemical energy of the triacylglycerols stored in our fat tissue, a special enzyme (a lipase) catalyzes their complete hydrolysis. The fatty acid molecules, bound to proteins (albumins) in the blood, are then sent to the liver. In general,

■ The solubility of free fatty acids in water is extremely low, only about 10^{-6} mol/L, so to be transported in blood fatty acids must be carried on protein molecules.

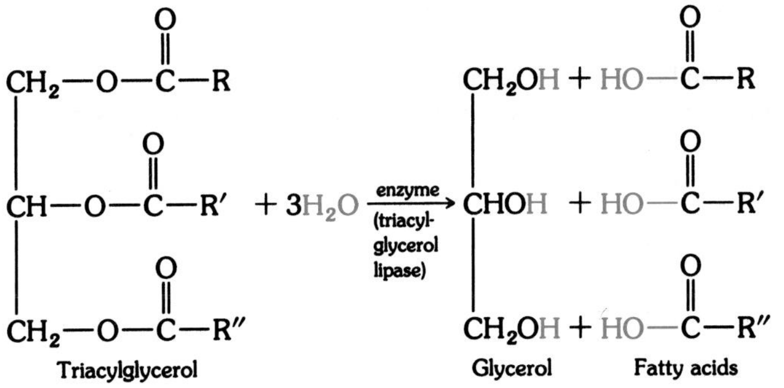

$$\begin{array}{l} CH_2-O-\overset{O}{\overset{\|}{C}}-R \\ | \\ CH-O-\overset{O}{\overset{\|}{C}}-R' \\ | \\ CH_2-O-\overset{O}{\overset{\|}{C}}-R'' \end{array} + 3H_2O \xrightarrow[\text{(triacylglycerol lipase)}]{\text{enzyme}} \begin{array}{l} CH_2OH + HO-\overset{O}{\overset{\|}{C}}-R \\ | \\ CHOH + HO-\overset{O}{\overset{\|}{C}}-R' \\ | \\ CH_2OH + HO-\overset{O}{\overset{\|}{C}}-R'' \end{array}$$

Triacylglycerol — Glycerol — Fatty acids

A specific example is

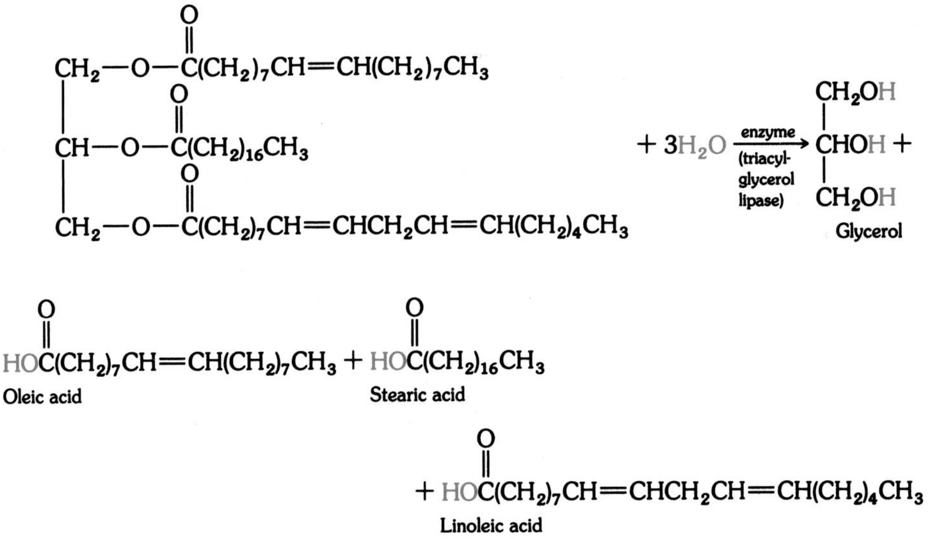

$$\begin{array}{l} CH_2-O-\overset{O}{\overset{\|}{C}}(CH_2)_7CH=CH(CH_2)_7CH_3 \\ | \\ CH-O-\overset{O}{\overset{\|}{C}}(CH_2)_{16}CH_3 \\ | \\ CH_2-O-\overset{O}{\overset{\|}{C}}(CH_2)_7CH=CHCH_2CH=CH(CH_2)_4CH_3 \end{array} + 3H_2O \xrightarrow[\text{(triacylglycerol lipase)}]{\text{enzyme}} \begin{array}{l} CH_2OH \\ | \\ CHOH \\ | \\ CH_2OH \end{array} +$$

Glycerol

$$HO\overset{O}{\overset{\|}{C}}(CH_2)_7CH=CH(CH_2)_7CH_3 + HO\overset{O}{\overset{\|}{C}}(CH_2)_{16}CH_3$$

Oleic acid — Stearic acid

$$+ HO\overset{O}{\overset{\|}{C}}(CH_2)_7CH=CHCH_2CH=CH(CH_2)_4CH_3$$

Linoleic acid

SPECIAL TOPIC 9.3 HOW DETERGENTS WORK

Soap Water is a very poor cleansing agent because it can't penetrate greasy substances, the "glues" that bind soil to skin and fabrics. When just a little soap is present, however, water cleans very well, especially warm water. Soap is a simple chemical, a mixture of the sodium or potassium salts of the long-chain fatty acids obtained by the saponification of fats or oils.

Detergents Soap is just one kind of detergent. All detergents are surface-active agents that lower the surface tension of water. All consist of ions or molecules that have long hydrocarbon portions plus ionic or very polar sections at one end. The accompanying structures illustrate these features and show the varieties of detergents that are available.

Although soap is manufactured, it is not called a synthetic detergent. This term is limited to detergents that are not soap, that is, not the salts of naturally occurring fatty acids obtained by the saponification of lipids. Most synthetic detergents are salts of sulfonic acids, but others have different kinds of ionic or polar sites. The great advantage of synthetic detergents is that they work in hard water and are not precipitated by the hardness ions—Mg^{2+}, Ca^{2+}, and the two ions of iron. These ions form messy precipitates ("bathtub ring") with the anions of the fatty acids present in soap. The anions of synthetic detergents do not form such precipitates.

Figure 1 shows how detergents work. In Figure 1*a* we see the hydrocarbon tails of the detergent work their way into the hydrocarbon environment of the grease layer. ("Like dissolves like" is the principle at work here.) The ionic heads stay in the water phase, and the grease layer becomes pincushioned with electrically charged sites. In Figure 1*b* we see the grease layer breaking up, aided with some agitation or scrubbing. Figure 1*c* shows a magnified view of grease globules studded with ionic groups and, being like-charged, these globules repel each other. They also tend to dissolve in water, so they are ready to be washed down the drain.

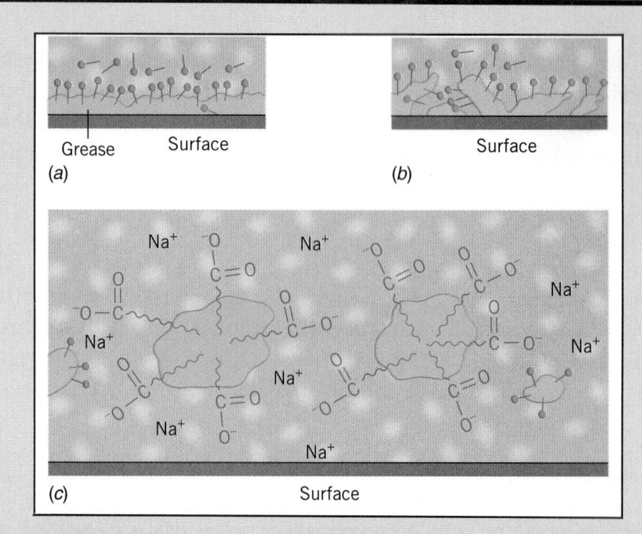

FIGURE 1

$CH_3(CH_2)_{14}CO_2^-Na^+$

Soap—an anionic detergent

$CH_3(CH_2)_{13}OSO_3^-Na^+$

A sodium alkyl sulfate—an anionic detergent

$CH_3(CH_2)_8$—C_6H_4—$SO_3^-Na^+$

A sodium alkylbenzenesulfonate—an anionic detergent

$CH_3(CH_2)_{11}\overset{+}{N}(CH_3)_3Cl^-$

A triethylalkylammonium ion—a cationic detergent

$CH_3(CH_2)_8O(CH_2CH_2O)_nH$

A nonionic detergent

When we *digest* triacylglycerols, hydrolysis is not complete. The digestive enzyme (pancreatic lipase) takes the hydrolysis only to monoacylglycerols, fatty acids, and some diacylglycerols.

Soaps Are Made by the Saponification of Triacylglycerols The saponification of the ester links in triacylglycerols by the action of a strong base (e.g., NaOH or KOH) gives glycerol and a mixture of the salts of fatty acids. These salts are soaps, and how they exert their detergent action is described in Special Topic 9.3. In general,

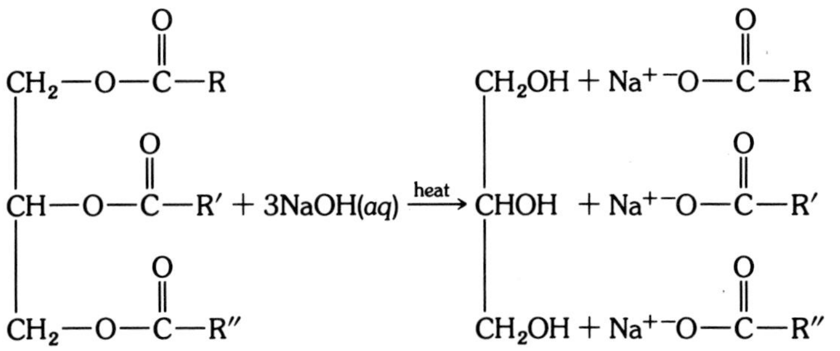

$$
\begin{array}{l}
\mathrm{CH_2-O-\overset{\overset{\large O}{\|}}{C}-R} \\
| \\
\mathrm{CH-O-\overset{\overset{\large O}{\|}}{C}-R'} + 3\mathrm{NaOH}(aq) \xrightarrow{\text{heat}} \\
| \\
\mathrm{CH_2-O-\overset{\overset{\large O}{\|}}{C}-R''}
\end{array}
\quad
\begin{array}{l}
\mathrm{CH_2OH} + \mathrm{Na^+\ ^-O-\overset{\overset{\large O}{\|}}{C}-R} \\
| \\
\mathrm{CHOH} + \mathrm{Na^+\ ^-O-\overset{\overset{\large O}{\|}}{C}-R'} \\
| \\
\mathrm{CH_2OH} + \mathrm{Na^+\ ^-O-\overset{\overset{\large O}{\|}}{C}-R''}
\end{array}
$$

Triacylglycerol — Glycerol — Mixture of sodium salts of fatty acids (soap)

PRACTICE EXERCISE 3

Write a balanced equation for the saponification of **1** with sodium hydroxide.

The Hydrogenation of Vegetable Oils Gives Solid Shortenings and Margarine

When hydrogen is made to add to some of the double bonds in vegetable oils, the oils become like animal fats, both physically and structurally. One very practical consequence of such partial hydrogenation is that the oils change from being liquids to solids at room temperature. Many people prefer solid, lard-like shortening for cooking, instead of a liquid oil. Therefore the manufacturers of such "hydrogenated vegetable oils" as Crisco and Spry take inexpensive, readily available vegetable oils, like corn oil and cottonseed oil, and catalytically add hydrogen to some (not all) of the alkene groups in their molecules. We say that the double bonds become *saturated*. Unlike natural lard, the vegetable shortenings have no cholesterol.

■ Except for the absence of cholesterol, hydrogenated vegetable oils are chemically and nutritionally identical to animal fats.

PRACTICE EXERCISE 4

Write the balanced equation for the complete hydrogenation of the alkene links in structure **1**.

The chief lipid material in margarine is produced from vegetable oils in the same way. The hydrogenation is done with special care so that the final product can melt on the tongue, one property that makes butterfat so pleasant. (If all the alkene groups in a vegetable oil were hydrogenated, instead of just some of them, the product would be just like beef or mutton fat, relatively hard materials that would not melt on the tongue.)

The popular brands of peanut butter, those with peanut oils that do not separate, are made by the partial hydrogenation of the oil in real peanut butter. The peanut oil changes to a solid, hydrogenated form at room temperature and therefore it cannot separate.

9.3

PHOSPHOLIPIDS

Phospholipid molecules have very polar or ionic sites in addition to long hydrocarbon chains.

Phospholipids are esters of either glycerol or sphingosine, which is a long-chain, dihydric amino alcohol with one double bond.

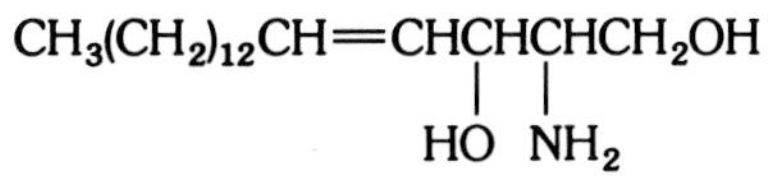

$$\mathrm{CH_3(CH_2)_{12}CH{=}CH\underset{\underset{\large HO}{|}}{C}H\underset{\underset{\large NH_2}{|}}{C}HCH_2OH}$$

Sphingosine

The phospholipids all have very polar but small molecular parts that are extremely important in the formation of cell membranes. We will survey their structures largely to demonstrate how they are both polar and hydrocarbon-like.

The Glycerophospholipids (Phosphoglycerides) Have Phosphate Units plus Two Acyl Units The **glycerophospholipids** occur in two broad types, the *phosphatides* and the *plasmalogens*. Both are based on glycerol esters. Molecules of the **phosphatides** have two ester bonds from glycerol to fatty acids plus one ester bond to phosphoric acid. The phosphoric acid unit, in turn, is joined by a phosphate ester link to a small alcohol molecule. Without this link, the compound is called *phosphatidic acid*.

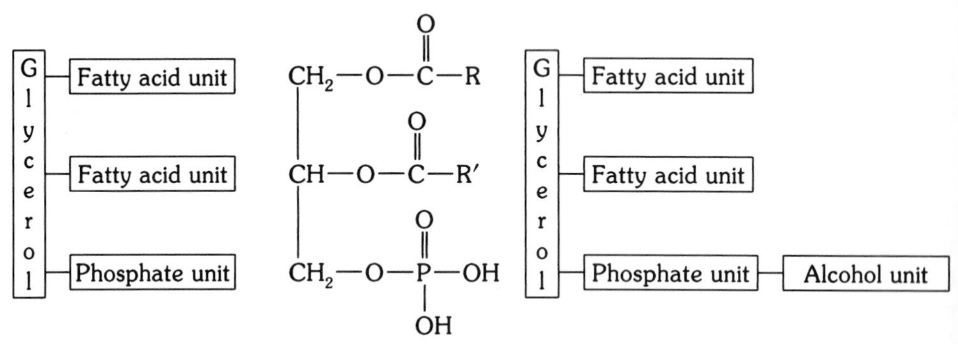

Phosphatidic acid components   Phosphatidic acid   Phosphatide components

Three particularly important phosphatides are esters between phosphatidic acid and either choline, ethanolamine, or serine, forming, respectively, phosphatidylcholine (lecithin), **2**, phosphatidylethanolamine (cephalin), **3**, and phosphatidylserine, **4**.

$$HOCH_2CH_2\overset{+}{N}(CH_3)_3 \qquad HOCH_2CH_2NH_2 \qquad HOCH_2CH(NH_3^+)CO_2^-$$

Choline   Ethanolamine   Serine

As the structures of **2, 3,** and **4** given below show, one part of each phosphatide molecule is very polar because it carries full electrical charges. These charges are partly responsible for the phosphatides being somewhat more soluble in water than triacylglycerols. The remainder of a phosphatide molecule is nonpolar and hydrocarbon-like, so phosphatides can be extracted from animal matter by relatively nonpolar solvents.

A typical phosphatide.

■ *Cephalin* is from the Greek *kephale*, head. Cephalin is found in brain tissue.

$$\begin{array}{l} CH_2OC(=O)R \\ | \\ CH{-}OC(=O)R' \\ | \\ CH_2OP(=O)(O^-)OCH_2CH_2\overset{+}{N}(CH_3)_3 \end{array}$$

2
Phosphatidylcholine (lecithin)

$$\begin{array}{l} CH_2OC(=O)R \\ | \\ CH{-}OC(=O)R' \\ | \\ CH_2OP(=O)(O^-)OCH_2CH_2\overset{+}{N}H_3 \end{array}$$

3
Phosphatidylethanolamine (cephalin)

$$\begin{array}{l} CH_2OC(=O)R \\ | \\ CH{-}OC(=O)R' \\ | \\ CH_2OP(=O)(O^-)OCH_2CH(NH_3^+)CO_2^- \end{array}$$

4
Phosphatidylserine

These three are the most common hydrolyzable lipids used to make animal cell membranes.

When pure, lecithin is a clear, waxy solid that is very hygroscopic. In air, it is quickly attacked by oxygen, which makes it turn brown in a few minutes. Lecithin is a powerful emulsifying agent for triacylglycerols, and this is why egg yolks, which contain lecithin, are used to make the emulsions found in mayonnaise, ice cream, candies, and cake dough.

■ *Lecithin* is from the Greek *lekitos*, egg yolk—a rich source of this phospholipid.

The Plasmalogens Have Both Ether and Ester Groups The **plasmalogens,** as we said, make up another family of glycerophospholipids, and they occur widely in the membranes of both nerve and muscle cells. They differ from the phosphatides by the presence of an unsaturated *ether* group instead of an acyl group at one end of the glycerol unit.

Glycerol — Fatty alcohol unit / Fatty acid unit / Phosphate unit — Amino alcohol unit

Plasmalogen components

Ether group: $CH_2OCH{=}CHR$

$CH{-}OC(=O)R'$

$CH_2OP(=O)(O^-)OCH_2CH_2{-}\overset{+}{N}(CH_3)_3$ or $-\overset{+}{N}H_3$

Plasmalogens

Plasmalogen molecules, like phosphatides, also carry electrically charged positions as well as long hydrocarbon chains.

The Sphingolipids Are Based on Sphingosine, Not Glycerol The two types of sphingosine-based lipids or **sphingolipids** are the *sphingomyelins* and the *cerebrosides,* and they are also important constituents of cell membranes, particularly those of nerve cells. The sphingomyelins are phosphate diesters of sphingosine. Their acyl units occur as acylamido parts, and they come from unusual fatty acids that are not found in neutral fats. Like the molecules of the phosphatides, those of the sphingolipids have two nonpolar tails.

The cerebrosides are not phospholipids. Instead they are **glycolipids,** lipids with a sugar unit (i.e., galactose or glucose) and not a phosphate ester system. The sugar unit, with its many OH groups, provides a strongly polar site, and it is usually a D-galactose or a D-glucose unit, or an amino derivative of these.

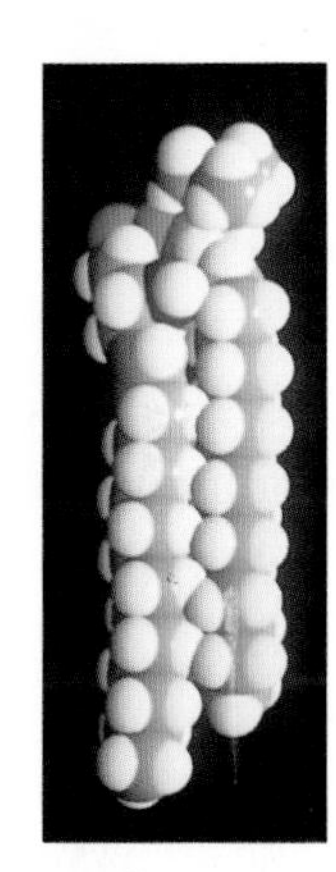

A sphingomyelin

Sphingosine — OH / Fatty acylamido unit / Phosphate—alcohol or Glycoside unit

Sphingolipid components

$CH_3(CH_2)_{12}CH{=}CHCHOH{-}CHNHC(=O)R{-}CH_2OP(=O)(O^-)OCH_2CH_2\overset{+}{N}(CH_3)_3$

Sphingomyelins

β-D-galactose unit

$CH_3(CH_2)_{12}CH{=}CHCHOH{-}CHNHC(=O)R{-}CH_2{-}O{-}$(β-D-galactose)

Cerebrosides

The cerebrosides are particularly prevalent in the membranes of brain cells.

9.4

STEROIDS

■ Steroid alcohols are called *sterols.*

Cholesterol and other steroids are nonhydrolyzable lipids.

Steroids are high-formula-mass aliphatic compounds whose molecules include a characteristic four-ring feature called the steroid nucleus. It consists of three six-membered rings and one five-membered ring, as seen in structure **5**. Several steroids are very active, physiologically.

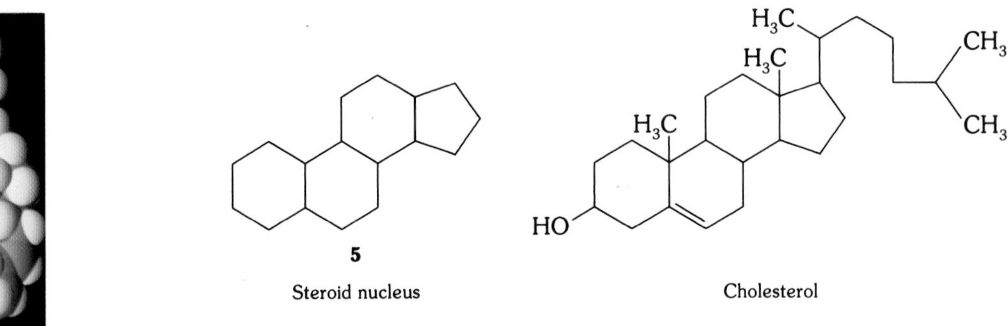

Steroid nucleus

Cholesterol

Cholesterol

Table 9.3 lists several steroids and their functions, and you can see the diverse roles they have in the body.

■ Cholesterol is the chief constituent in gallstones.

Cholesterol Molecules Are Components of Cell Membranes Cholesterol is an unsaturated steroid alcohol that makes up a significant part of the membranes of the cells of animals. The membrane of a human red blood cell (erythrocyte), for example, has about 25% cholesterol by mass, and so is particularly rich in cholesterol. Cholesterol is the body's raw material for making bile salts and steroid hormones, including the sex hormones listed in Table 9.3. Little cholesterol is found in plants, but they have compounds with similar structures.

Cholesterol enters the body via the diet, but up to 800 mg per day can normally be synthesized in the liver from two-carbon acetate units. Cholesterol made in the liver is used to make the bile salts like sodium cholate (see Table 9.3) or is converted to esters of cholesterol.

Sodium cholate
(a bile salt)

Cholesteryl ester
(R is long-chain)

Bile salts are secreted into the intestinal tract where they function as powerful surface active agents (detergents). They aid both in the digestion of dietary lipids and also in the absorption of the fat-soluble vitamins and the fatty acids from the digestive tract (eventually) into circulation. Cholesterol is put into circulation in the bloodstream as components of *very-low-density lipoproteins* (or VLDL). Esters of cholesterol occur in higher density complexes. The relationships between cholesterol, esters of cholesterol, the various kinds of lipoproteins, and the risk of heart disease will be discussed when we study the metabolism of lipids. We need to know more about proteins, genes, and enzymes before we can discuss this topic.

■ *Lipoprotein* molecules are combinations of lipids and proteins.

TABLE 9.3
Important Steroids

Vitamin D_3 Precursor

Irradiation of this derivative of cholesterol by ultraviolet light opens one of the rings to produce vitamin D_3. Meat products are sources of this compound.

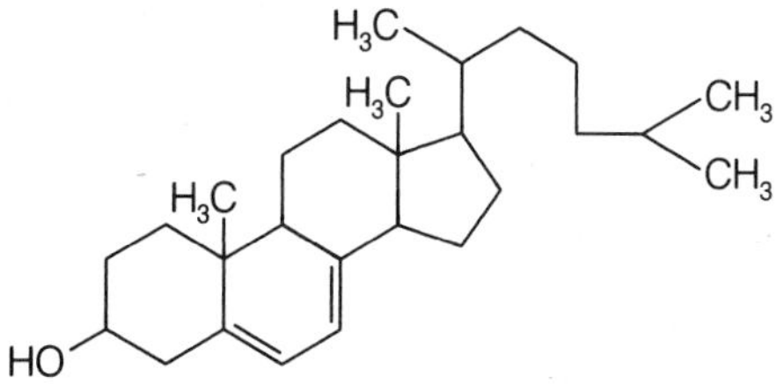

7-Dehydrocholesterol

Ultraviolet radiation

Vitamin D_3 is an antirachitic factor. Its absence leads to rickets, an infant and childhood disease characterized by faulty deposition of calcium phosphate and poor bone growth.

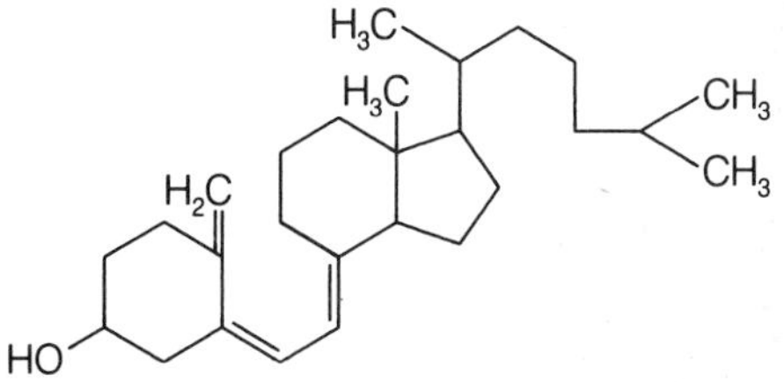

Vitamin D_3

Adrenocortical Hormone

Cortisol is one of the 28 hormones secreted by the cortex of the adrenal gland. Cortisone, very similar to cortisol, is another such hormone. When cortisone is used to treat arthritis, the body changes much of it to cortisol by reducing a keto group to the 2° alcohol group that you see in the structure of cortisol.

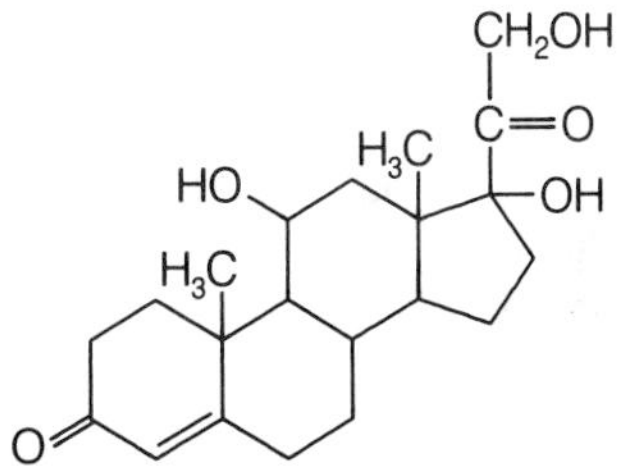

Cortisol

Sex Hormones

Estradiol is a human estrogenic hormone.

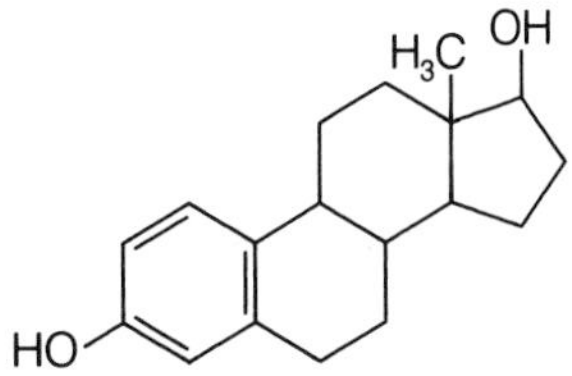

Estradiol

Progesterone, a human pregnancy hormone, is secreted by the corpus luteum.

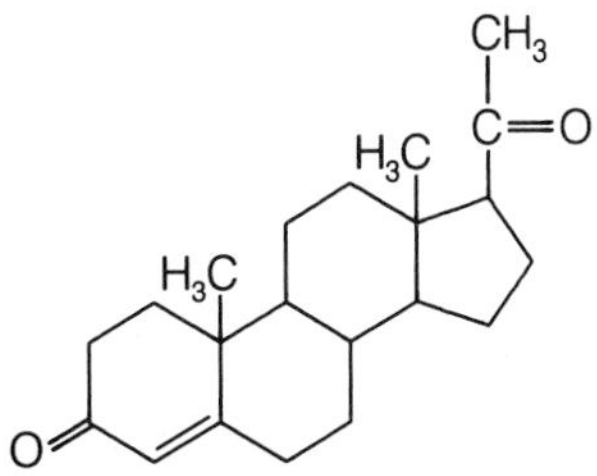

Progesterone

(Continued)

(Continued)

Testosterone, a male sex hormone, regulates the development of reproductive organs and secondary sex characteristics.

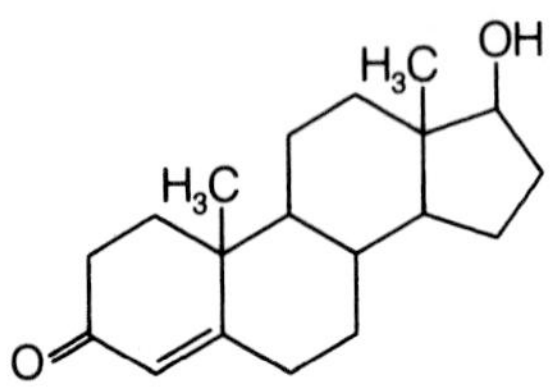

Testosterone

Androsterone is another male sex hormone.

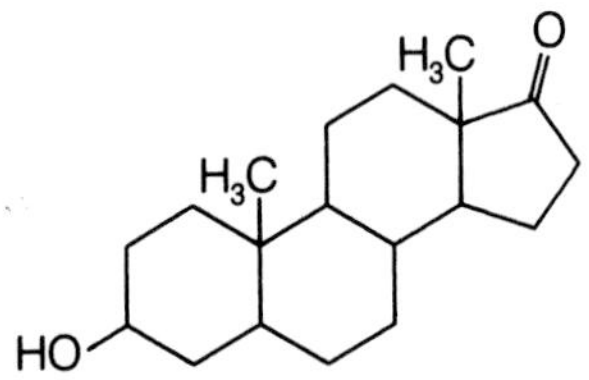

Androsterone

Synthetic Hormones in Fertility Control

Most oral contraceptive pills contain one or two synthetic, hormonelike compounds. (Synthetics must be used because the real hormones are broken down in the body.)

Synthetic Estrogens

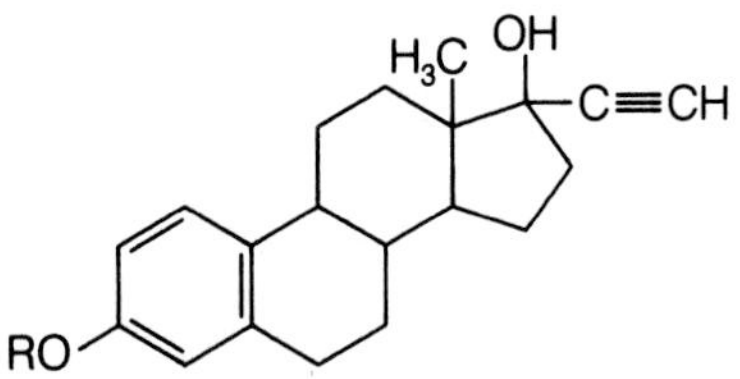

R=H, ethynylestradiol
R=CH_3, mestranol

Synthetic Progestin

The most widely used pills have a combination of an estrogen (20 to 100 μg if mestranol and 20 to 50 μg if ethinylestradiol) plus a progestin (0.35 to 2.5 mg depending on the compound).

A relatively new birth control technology, one that prevents implantation of a fertilized ovum in the uterus, has been developed. The compound is an antiprogesterone called mifepristone, or RU 486, and how it works is described in Special Topic 10.2 (page 263).

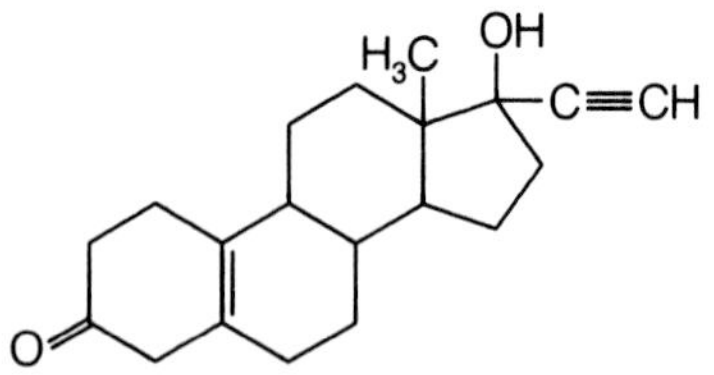

Norethynodrel

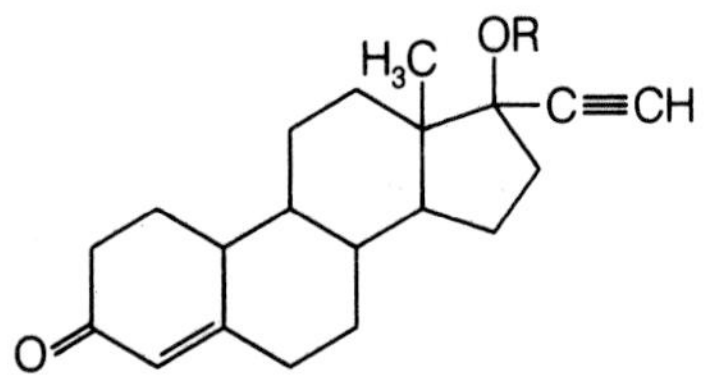

R=H, norethindrone
R=$COCH_3$, norethindrone acetate

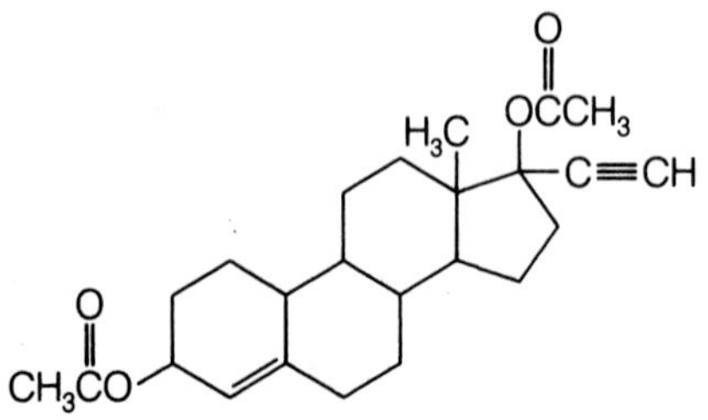

Ethynodiol diacetate

9.5

CELL MEMBRANES—THEIR LIPID COMPONENTS

Cell membranes consist mainly of a lipid bilayer plus molecules of proteins and cholesterol.

Cell membranes are made of both lipids and proteins. The lipid components are what keep a cell's insides within and its outsides without. The protein components perform services such as accepting hormone molecules and relaying them or their "messages" inside; for providing passages—closable molecular channels—for small ions and molecules; and for acting as "pumps" to move solutes across a cell membrane. We can only introduce the general features of animal cell membranes in this section. The services performed by the membrane proteins will be described in more detail after we have studied proteins and enzymes.

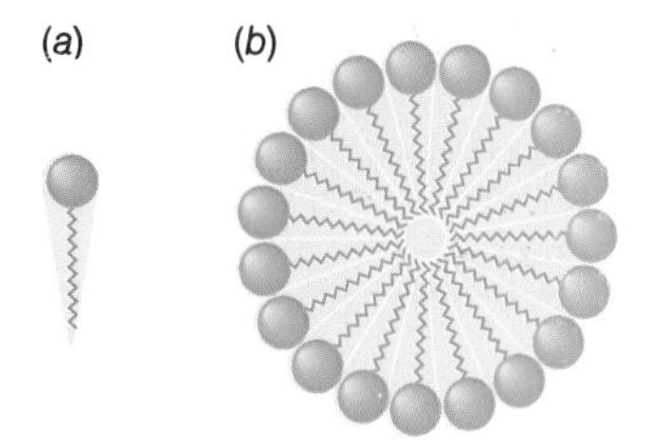

FIGURE 9.2
Detergent micelle. (*a*) Space-filling requirements of an amphipathic detergent or soap molecule with one hydrophobic tail (wavy line) and a hydrophilic head (blue sphere). (*b*) Micelle in water. The hydrophobic tails gather together as the hydrophilic heads have maximum exposure to the aqueous medium.

Both Hydrophilic and Hydrophobic Groups Are Necessary for Cell Membranes

The principal lipids of animal cell membranes are not triacylglycerols but more complex lipids, like the phospholipids and glycolipids, as well as cholesterol. The molecules of the hydrolyzable lipids in a membrane possess a part that is either very polar or fully ionic plus two nonpolar "tails." The polar or ionic sites are called **hydrophilic groups,** because they are able to attract water molecules. Hydrophilic groups force molecules to take up positions in membranes in such a way that the groups can be in contact with the water in body fluids both inside and outside the cell. In phospholipids, the hydrophilic groups are the phosphate diester units, which have ionic sites. In a glycolipid, the sugar unit with its many OH groups is the hydrophilic group.

The nonpolar, hydrocarbon sections of membrane lipids are called **hydrophobic groups** because they are water-avoiding. Hydrophobic groups tend to force molecules to become positioned *within* the membrane so that the groups are out of contact with water as much as possible. Substances like the phospholipids or glycolipids, with both hydrophilic and hydrophobic groups, are called **amphipathic compounds.** Soaps and detergents are also examples of amphipathic compounds.

■ *Hydrophilic*—from the Greek *hydor*, water, and *philos*, loving.
Hydrophobic—from the Greek *phobikos*, hating.

The molecules of amphipathic compounds, when mixed in the right proportion with water, spontaneously become grouped into *micelles.* A **micelle** is a globular aggregation in which hydrophobic contacts are minimized and hydrophilic interactions are maximized. Figure 9.2 shows how a micelle forms when the amphipathic molecules have a *single* hydrocarbon "tail," like a detergent or soap molecule.

In the Lipid Bilayer of Cell Membranes, Hydrophobic Groups Intermingle between the Membrane Surfaces As we said, molecules of phospholipids and glycolipids have two hydrocarbon "tails." These force a micelle made of such lipids to take up an extended disklike shape (see Fig. 9.3). Further extension of the shape shown in Figure 9.3 would produce two rows of molecules or a **lipid bilayer** arrangement, a sheetlike array that consists of two layers of lipid molecules aligned side by side. This is the basic architecture of an animal cell membrane (see Fig. 9.4). The hydrophobic "tails" of the lipid molecules intermingle in the center of the bilayer away from water molecules. In a sense, these "tails" dissolve in each other, following the "like-dissolves-like" rule. The hydrophilic "heads" stick out into the aqueous phase in contact with water. These water-avoiding and water-attracting properties, not covalent bonds, are the major "forces" that stabilize the membrane.

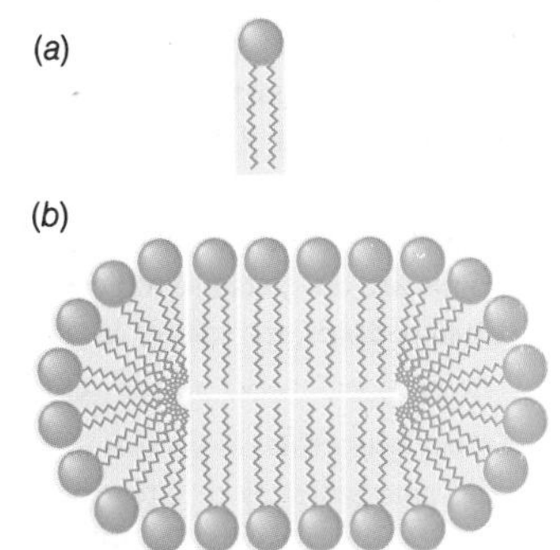

FIGURE 9.3
Phosphoglyceride micelle in water. (*a*) Space-filling requirements for an amphipathic phosphoglyceride, which has two hydrophobic tails (wavy lines) and a hydrophilic head (blue sphere). (*b*) A disklike micelle whose "wall" for the most part (top and bottom segments) is a lipid bilayer.

Cholesterol Molecules Also Help to Stabilize Membranes Cholesterol molecules are somewhat long and flat. In the lipid bilayer, they occur with their long axes lined up side by side with the hydrocarbon chains of the other lipids. The cholesterol OH groups are hydrogen-bonded to O atoms of ester groups in the membrane lipid molecules. Because the cholesterol units are relatively rigid, much more so than the fatty acid chains, cholesterol molecules help to keep a membrane from being too fluidlike.

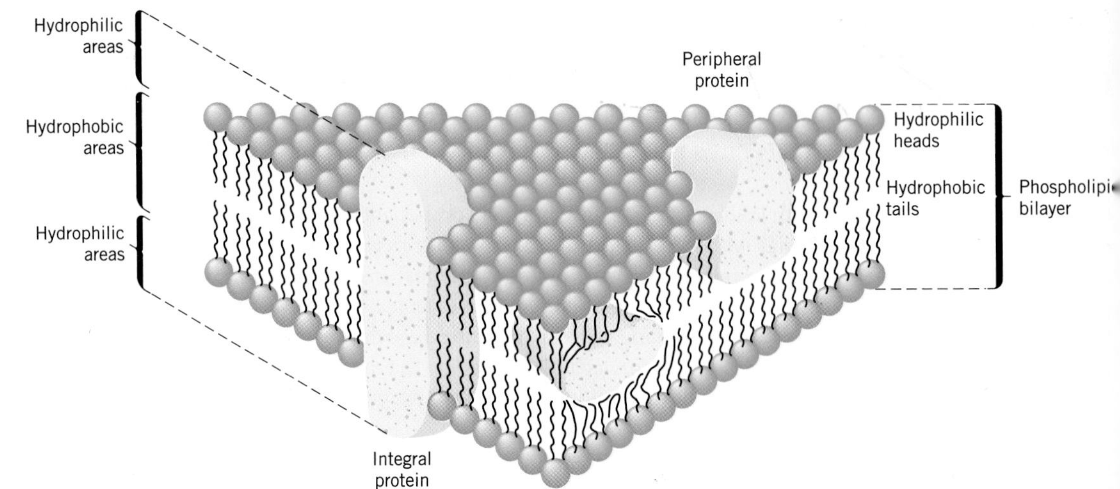

FIGURE 9.4
Cell membrane.

The Lipid Bilayer is Self-Sealing If a pin were stuck through a cell membrane and the pulled out, the lipid layer would close back spontaneously. This flexibility is allowed because, a we said, no covalent bonds hold neighboring lipid molecules to each other. Only the net force of attraction that we imply when we use the terms *hydrophobic* and *hydrophilic* are at work Yet the bilayer is strong enough to hold a cell together, and it is flexible enough to let things in and out. Water molecules move back and forth easily through the membrane, but other mole cules and ions are vastly less free to move. Their migrations depend on the protein component of the membrane, also indicated in Figure 9.4. We must, therefore, postpone further discussion of membranes until we have learned more about proteins, particularly such families of protein as enzymes and receptors.

SUMMARY

Lipids Lipids are ether-extractable substances in animals and plants, and they include hydrolyzable esters and nonhydrolyzable compounds. The esters are generally of glycerol or sphingosine with their acyl portions contributed by long-chain carboxylic acids called fatty acids. The fatty acids obtained from lipids by hydrolysis generally have long chains of even numbers of carbons, seldom are branched, and often have one or more alkene groups. The alkene groups are cis. Because molecules of all lipids are mostly hydrocarbon-like, lipids are soluble in nonpolar solvents but not in water.

Triacylglycerols Molecules of neutral fats, those without electrically charged sites or sites that are similarly polar, are esters of glycerol and a variety of fatty acids, both saturated and unsaturated. Vegetable oils have more double bonds per molecule than animal fats. The triacylglycerols can be hydrogenated, hydrolyzed, and saponified.

Waxes Molecules of the waxy coatings on leaves and fruit, or in beeswax or sebum, are simple esters between long-chain alcohols and fatty acids.

Glycerophospholipids Molecules of the glycerophospholipids are esters both of glycerol and of phosphoric acid. A second ester bond from the phosphate unit goes to a small alcohol molecule that car also have a positively charged group. Thus this part of a glycero phospholipid is strongly hydrophilic. The two types of glycerophos pholipids are the phosphatides and the plasmalogens. Both are vital to animal cell membranes. In phosphatide molecules there are two fatty acyl ester units besides the phosphate system. In plasmalo gens, there is one fatty acyl unit and a long-chain, unsaturated ether unit in addition to the phosphate system.

Sphingomyelins Sphingomyelins are esters of sphingosine, a di hydric amino alcohol. They also have a strongly hydrophilic phos phate system.

Glycolipids Also sphingosine-based, the glycolipids use a mono saccharide instead of the phosphate-to-small-alcohol unit to provide the hydrophilic section. Otherwise, they resemble the sphingo myelins.

Steroids Steroids are nonhydrolyzable lipids with the steroid nu cleus of four fused rings (three being C-6 rings and one a C-5 ring). Several steroids are sex hormones, and oral fertility control drugs mimic their structure and functions. Cholesterol, the raw material used by the body to make bile salts and other steroids, is manufac tured in the liver. Cholesterol is carried in circulation as cholesteryl esters in lipoprotein complexes. Cholesterol molecules are essential components of animal cell membranes.

Animal cell membranes A double layer of phospholipids or glycolipids plus cholesterol and assemblies of protein molecules make up the lipid bilayer part of a cell membrane. The hydrophobic tails of the amphipathic lipids intermingle within the bilayer, away from the aqueous phase. The hydrophilic heads are in contact with the aqueous medium. Cholesterol molecules help to stiffen the membrane.

REVIEW EXERCISES

The answers to Review Exercises whose numbers are in color are found in Appendix A. The answers to the other Review Exercises are found in the Study Guide that accompanies this book. The more challenging questions are marked with asterisks.

Lipids in General

9.1 Crude oil is soluble in ether, yet it isn't classified as a lipid. Explain.

9.2 Cholesterol has no ester group, yet we classify it as a lipid. Why?

9.3 Ethyl acetate has an ester group, but it isn't classified as a lipid. Explain.

9.4 What are the criteria for deciding if a substance is a lipid?

Fatty Acids

9.5 What are the structures and the names of the two most abundant *saturated* fatty acids?

9.6 Write the structures and names of the *unsaturated* fatty acids that have 18 carbons each and that have no more than three double bonds. Show the correct geometry at each double bond.

9.7 Write the equations for the reactions of palmitic acid with
(a) NaOH(*aq*)
(b) CH_3OH (when heated in the presence of acid)

*9.8 What are the equations for the reactions of oleic acid with each substance?
(a) Br_2
(b) KOH(*aq*)
(c) H_2 (in the presence of a catalyst)
(d) CH_3CH_2OH (heated with an acid catalyst)

9.9 Which of the following acids, **A** or **B**, is more likely to be obtained by the hydrolysis of a lipid? Explain.

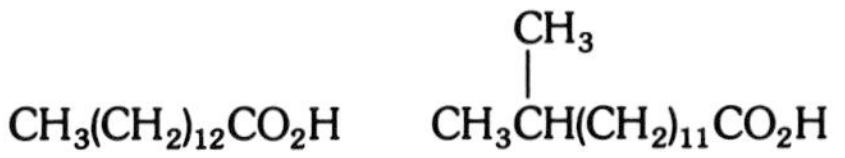

$CH_3(CH_2)_{12}CO_2H$ (**A**)

$CH_3CH(CH_3)(CH_2)_{11}CO_2H$ (**B**)

9.10 Without writing structures, state what kinds of chemicals the prostaglandins are.

Triacylglycerols

9.11 Write the structure of a triacylglycerol that involves linolenic acid, linoleic acid, and palmitic acid, besides glycerol.

9.12 What is the structure of a triacylglycerol made from glycerol, stearic acid, oleic acid, and palmitic acid?

9.13 Write the structures of all the products that would form from the complete hydrolysis of the following lipid. (Show the free carboxylic acids, not their anions.)

Write the structures of all the products that would form from the complete hydrolysis of the following lipid. (Show the free carboxylic acids, not their anions.)

$$CH_2OC(=O)(CH_2)_7CH{=}CHCH_2CH{=}CH(CH_2)_4CH_3$$
$$CHOC(=O)(CH_2)_{12}CH_3$$
$$CH_2OC(=O)(CH_2)_7CH{=}CH(CH_2)_7CH_3$$

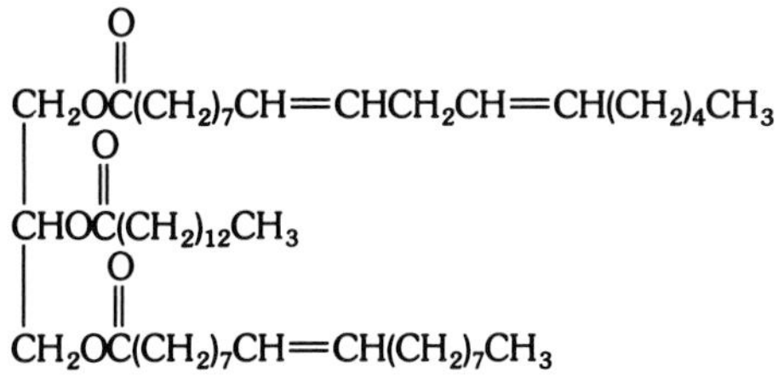

9.14 Write the structures of the products that are produced by the saponification (by NaOH) of the triacylglycerol whose structure was given in Review Exercise 9.13.

*9.15 The hydrolysis of a lipid produced glycerol, lauric acid, linoleic acid, and oleic acid in equimolar amounts. Write a structure that is consistent with these results. Is there more than one structure that can be written? Explain.

***9.16** The hydrolysis of 1 mol of a lipid gives 1 mol each of glycerol and oleic acid and 2 mol of lauric acid. The lipid molecule is chiral. Write its structure. Is more than one structure (constitution) possible? Explain.

9.17 What is the structural difference between the triacylglycerols of the animal fats and the vegetable oils?

9.18 Products such as corn oil are advertised as being "polyunsaturated." What does this mean in terms of the structures of the molecules that are present? Corn oil is "more polyunsaturated" than what?

9.19 What chemical reaction is used to make margarine?

9.20 Lard and butter are chemically almost the same substances, so what is it about butter that makes it so much more desirable a spread for bread than, say, lard or tallow?

Waxes

9.21 One component of beeswax has the formula $C_{36}H_{72}O_2$. When it is hydrolyzed, it gives $C_{18}H_{36}O_2$ and $C_{18}H_{38}O$. Write the most likely structure of this compound.

*9.22 When all the waxes from the leaves of a certain shrub are separated, one has the formula of $C_{60}H_{120}O_2$. Its structure is **A, B,** or **C**. Which is it most likely to be? Explain why the others can be ruled out.

$CH_3(CH_2)_{56}CO_2CH_2CH_3$ (**A**)

$CH_3(CH_2)_{29}CO_2(CH_2)_{28}CH_3$ (**B**)

$CH_3(CH_2)_{28}CO_2(CH_2)_{29}CH_3$ (**C**)

Phospholipids

9.23 Why are the phosphatides and plasmalogens both called glycerophospholipids?

9.24 What site in a glycerophospholipid carries a negative charge? What atom carries a positive charge?

9.25 In general terms, how are the sphingomyelins and cerebrosides structurally alike? How are they structurally different?

9.26 What structural unit provides the most polar groups in a molecule of a glycolipid? (Name it.)

9.27 Phospholipids are not classified as neutral fats. Explain.

9.28 Phospholipids are common in what part of a cell?

9.29 What are the names of the two types of sphingosine-based lipids?

9.30 Are the sugar units that are incorporated into the cerebrosides bound by glycosidic links or by ordinary ether links? How can one tell? Which kind of link is more easily hydrolyzed (assuming an acid catalyst)?

*9.31 The complete hydrolysis of 1 mol of a phospholipid gave 1 mol each of the following compounds: glycerol, linolenic acid, oleic acid, phosphoric acid, and the cation, $HOCH_2CH_2N(CH_3)_3{}^+$.
(a) Write a structure of this phospholipid that is consistent with the information given.
(b) Is the substance a glycerophospholipid or a sphingolipid? Explain.
(c) Are its molecules chiral or not? How can you tell?
(d) Is it an example of a lecithin or a cephalin? How can you tell?

***9.32** When 1 mol of a certain phospholipid was hydrolyzed, there was obtained 1 mol each of lauric acid, oleic acid, phosphoric acid, glycerol, and $HOCH_2CH_2NH_2$.
(a) What is a possible structure for this phospholipid?
(b) Is it a sphingolipid or a phosphoglyceride? Explain.
(c) Can its molecules exist as enantiomers or not? Explain.
(d) Is it a cephalin or a lecithin? Explain.

Steroids

9.33 What is the name of the steroid that occurs as a detergent in our bodies?

9.34 What is the name of a vitamin that is made in our bodies from a dietary steroid by the action of sunlight on the skin?

9.35 Give the names of three steroidal sex hormones.

9.36 What is the name of a steroid that is part of the cell membranes in animal tissues?

9.37 What is the raw material used by the body to make bile salts? How does the body use the bile salts?

9.38 How does the body carry cholesterol in circulation in the bloodstream?

Cell Membranes

9.39 Describe in your own words what is meant by the *lipid bilayer* structure of cell membranes.

9.40 How do the hydrophobic parts of phospholipid molecules avoid water in a lipid bilayer?

9.41 Besides lipids, what kinds of substances are present in a cell membrane?

9.42 What kinds of forces hold a cell membrane together?

9.43 What functions do the proteins of a cell membrane serve?

The Prostaglandins (Special Topic 9.1)

9.44 Name the fatty acid that is used to make the prostaglandins.

9.45 What effect is aspirin believed to have on prostaglandins, and how is this related to aspirin's medicinal value?

The Omega-3 Fatty Acids and Heart Disease (Special Topic 9.2)

9.46 What is it about the structure of linolenic acid that lets us call it an omega-3 acid?

9.47 What source of the omega-3 acids is relatively rich in the C-20 and C-22 acids?

9.48 Why have the omega-3 acids aroused the interests of people in nutrition and in medicine?

Detergent Action (Special Topic 9.3)

9.49 Which is the more general term, soap or detergent? Explain.

9.50 What kind of chemical is soap?

9.51 For household laundry work, which product is generally preferred, a synthetic detergent or soap? Why?

9.52 Why are soap and sodium alkyl sulfates called *anionic* detergents?

9.53 Explain in your own words how a detergent can loosen oils and greases from fabrics.

Additional Exercises

9.54 Examine the following structure.

$$CH_2OC(=O)(CH_2)_7CH{=}CHCH_2CH{=}CH(CH_2)_5CH_3$$
$$CHOC(=O)(CH_2)_{11}CH_3$$
$$CH_2OC(=O)(CH_2)_7CH{=}CH(CH_2)_8CH_3$$

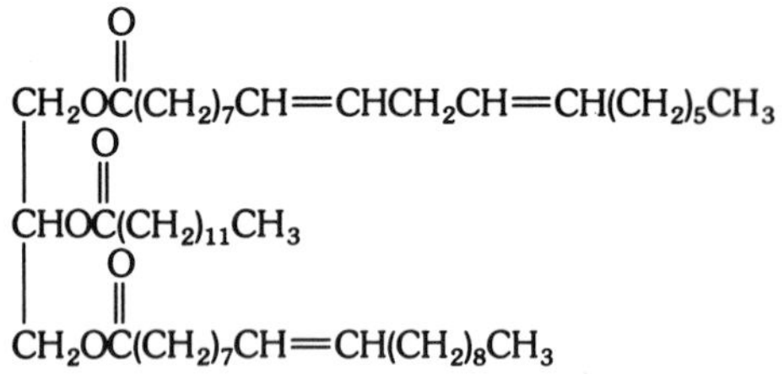

(a) Is it a triester of glycerol?
(b) Does it have hydrophobic groups?
(c) What are its hydrophilic functional groups?
(d) What would form if all of its alkene groups were hydrogenated?
(e) Is this molecule likely to be found among naturally occurring triacylglycerols? Explain.

*9.55 Examine the following structure and answer the questions that follow.

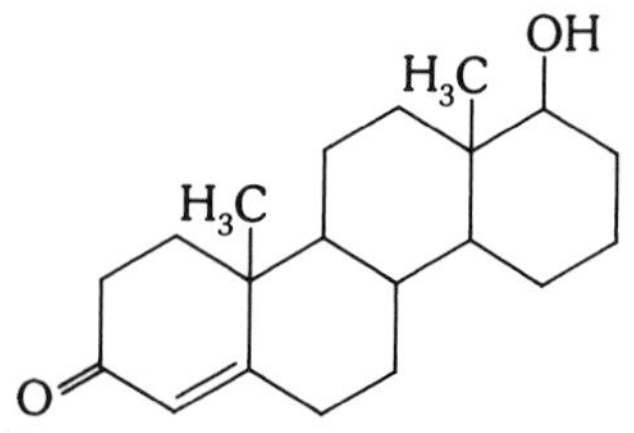

OH
H_3C
H_3C
O

(a) Is this compound amphipathic? Explain.
(b) Is it a member of the steroid family? How can one tell?

PROTEINS

10

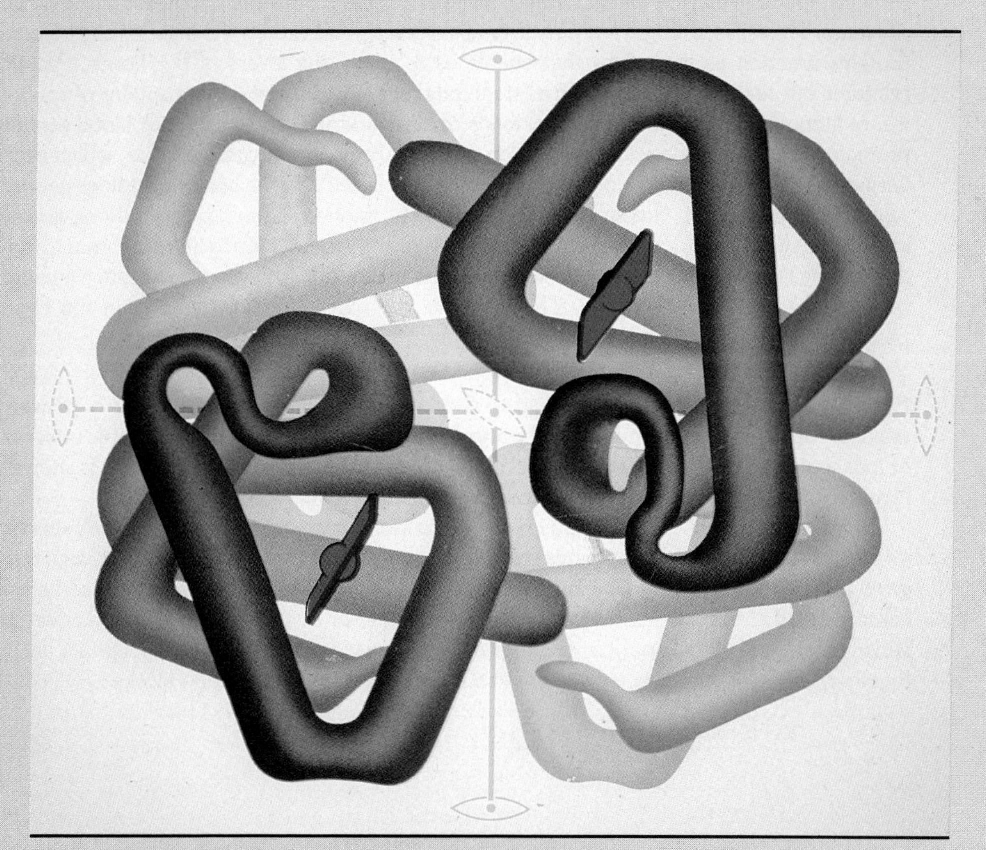

A representation of the hemoglobin molecule by artist Irving Geis. The flat red units are oxygen-holding molecules of heme, which are integral parts of hemoglobin (and which give the color to blood). Hemoglobin illustrates all four levels of complexity in protein structures, the principal topic of this chapter. (Illustration copyrighted © by Irving Geis.)

10.1

AMINO ACIDS. THE BUILDING BLOCKS OF PROTEINS

Living things select from among the molecules of about twenty α-amino acids to make the polypeptides in proteins.

Proteins, found in all cells and in virtually all parts of cells, constitute about half of the body's dry mass. They give strength and elasticity to skin and blood vessels. As muscles and tendons, proteins function as the cables that enable us to move the levers of our bones. Proteins reinforce our teeth and bones much as steel rods reinforce concrete. The proteins of antibodies, of hemoglobin, and of the various kinds of albumins and globulins in our blood serve as protectors and as the long-distance haulers of substances, like oxygen or lipids, which otherwise do not dissolve well in blood. Other proteins form parts of the communications network of our nervous system. Nearly all enzymes, some hormones and neurotransmitters, and cell membrane receptors are proteins that direct and control repair, construction, communication, and energy conversion in the body. No other class of compounds is involved in such a variety of functions, all essential to life. They deserve the name *protein,* taken from the Greek *proteios,* "of the first rank."

■ The chief elements in proteins are C, H, O, N, and S.

Polypeptides Are Made from α-Amino Acids The dominant structural units of **proteins** are high-formula-mass polymers called **polypeptides.** Metal ions and small organic molecules or ions are often present as well. The relationship of these parts to whole proteins is shown in Figure 10.1. Many proteins, however, are made entirely of polypeptides.

The monomer units for polypeptides are **α-amino acids,** which have the general structure given by **1**. Twenty such compounds make up the "standard" set (see Table 10.1), and in any given polypeptide some are used many times. Hundreds of **amino acid residues,** sometimes called **peptide units,** each derived from one or another of the various α-amino acids, are joined together in a single polypeptide molecule. Before we can study how polypeptides are put together, however, we must learn more about their α-amino acid building blocks.

■ With few exceptions, the amino acids not in the "standard" set consist of modifications of "standard" amino acid molecules made *after* a polypeptide has been put together.

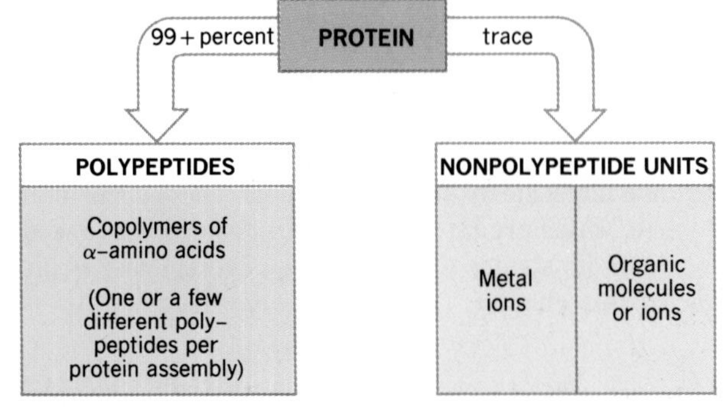

FIGURE 10.1
Components of proteins. Some proteins consist exclusively of polypeptide molecules, but most also have nonpolypeptide units such as small organic molecules or metal ions, or both.

joined together in a single polypeptide molecule. Before we can study how polypeptides are put together, however, we must learn more about their α-amino acid building blocks.

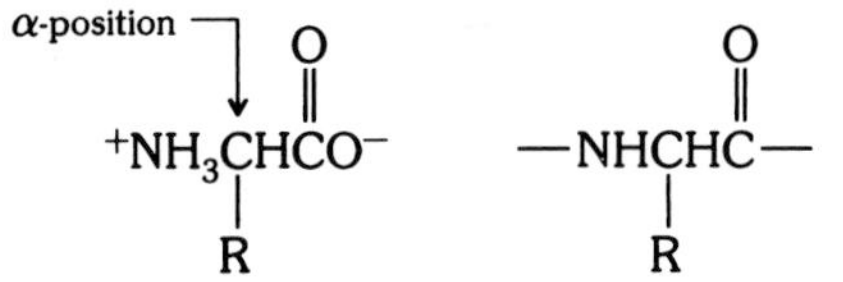

$$\overset{+}{N}H_3\underset{\underset{R}{|}}{C}H\overset{\overset{O}{\|}}{C}O^- \qquad -NH\underset{\underset{R}{|}}{C}H\overset{\overset{O}{\|}}{C}-$$

α-position (arrow to the central carbon)

α-Amino acids, general structural features — **1**

Amino acid residue (peptide unit)

The same set of 20 standard amino acids (Table 10.1) is used by all species of plants and animals. A few others are present in certain tissues as well as in some strains of bacteria.

The α-Amino Acids Are Based on a Dipolar Form of α-Amino Acetic Acid but Have Different Sidechains at the α-Position As you can see in Table 10.1, the amino acids differ in their **R** groups, called **sidechains,** located at the α-position in **1**.

■ Hereafter, when we say "amino acid," we'll mean α-amino acid.

In the solid state, amino acids exist entirely in the form shown by **1**, which is called a **dipolar ion** or a **zwitterion.** It is an electrically neutral particle but has a positive and a negative charge on different sites. Because dipolar ions are exceedingly polar, amino acids, like salts, have melting points that are considerably higher than those of most molecular compounds, and they tend to be much more soluble in water than in nonpolar solvents.

■ α-Amino acids melt around 300 °C but their simple esters, which cannot be dipolar ions, generally melt around 100 °C.

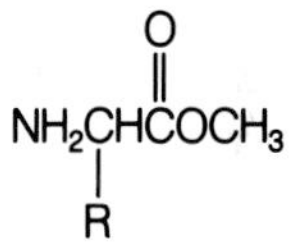

$$NH_2\underset{\underset{R}{|}}{C}H\overset{\overset{O}{\|}}{C}OCH_3$$

Methyl ester of α-amino acid

Structure **1** is actually an *internally neutralized molecule*. We can imagine that **1** started out with a regular amino group, NH_2, and an ordinary carboxyl group, CO_2H. But then the amino group, a proton acceptor, took a proton from the carboxyl group, a proton donor, to give the dipolar ionic form, **1**. Of course, **1** has its own (weaker) proton donating group, NH_3^+, and its own (also weaker) proton-accepting group, CO_2^-, so these dipolar ions can neutralize acids or bases of sufficient strength, like H_3O^+ and OH^-. Because polypeptides generally have at least one base-neutralizing NH_3^+ group and one acid-neutralizing CO_2^- group, *proteins are able to serve as buffers.*

For amino acids to exist in water as dipolar ions, **1**, the pH has to be about 6 to 7. If we make the pH much lower (more acidic) or much higher (more basic), the form of the amino acid changes. If we add enough strong acid like HCl*(aq)*, for example, to a solution of an amino acid to lower the pH to about 1, the CO_2^- groups take on protons. They change to CO_2H groups shown in structure **2**. Now the amino acid molecules are positively charged (cations) and can migrate to a negatively charged electrode, the cathode, in an electrolysis experiment.

■ In an electrolysis experiment the negatively charged electrode is called the *cathode,* so positively charged ions *(cations)* are attracted to it. Similarly, the positively charged electrode, the *anode,* attracts *anions,* ions bearing negative charge.

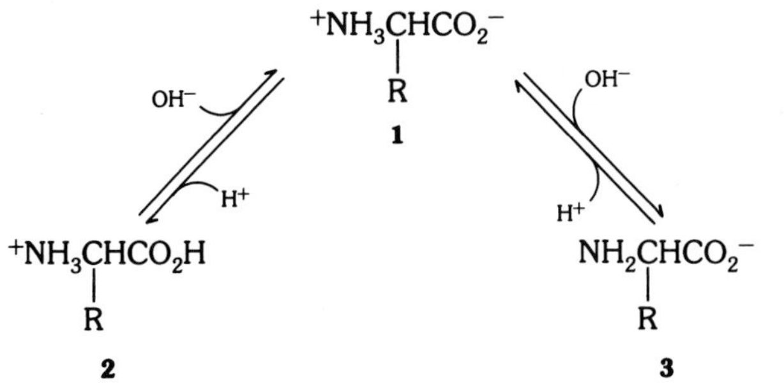

$$\underset{\mathbf{2}}{\overset{+}{N}H_3\underset{\underset{R}{|}}{C}HCO_2H} \underset{OH^-}{\overset{H^+}{\rightleftharpoons}} \underset{\mathbf{1}}{\overset{+}{N}H_3\underset{\underset{R}{|}}{C}HCO_2^-} \underset{H^+}{\overset{OH^-}{\rightleftharpoons}} \underset{\mathbf{3}}{NH_2\underset{\underset{R}{|}}{C}HCO_2^-}$$

If we add enough strong base to an amino acid solution to raise the pH to about 11, then the amino acid molecules transfer protons from NH_3^+ groups to OH^- ions, which changes NH_3^+ groups to NH_2 groups as seen in structure **3**. Because **3** is a negatively charged ion (an anion), it can migrate to the positively charged electrode in an electrolysis apparatus (the anode).

TABLE 10.1
Amino Acids: $^{+}NH_3CHCO_2^{-}$ (with R on the CH carbon)

Type	Sidechain, R	Name	Symbol 3-Letter	Symbol 1-Letter	pI
Side chain is nonpolar	$-H$	Glycine	Gly	G	5.97
	$-CH_3$	Alanine	Ala	A	6.00
	$-CH(CH_3)_2$	Valine	Val	V	5.96
	$-CH_2CH(CH_3)_2$	Leucine	Leu	L	5.98
	$-CH(CH_3)CH_2CH_3$	Isoleucine	Ile	I	6.02
	$-CH_2C_6H_5$	Phenylalanine	Phe	F	5.48
	$-CH_2-$ (3-indolyl, N–H)	Tryptophan	Trp	W	5.89
	$^{-}OC(=O)-CH-CH_2$ / H_2N^{+} / CH_2-CH_2 (ring) (complete structure)	Proline	Pro	P	6.30
Side chain has a hydroxyl group	$-CH_2OH$	Serine	Ser	S	5.68
	$-CH(OH)CH_3$	Threonine	Thr	T	5.64
	$-CH_2-C_6H_4-OH$	Tyrosine	Tyr	Y	5.66
Side chain has a carboxyl group (or an amide group)	$-CH_2CO_2H$	Aspartic acid	Asp	D	2.77
	$-CH_2CH_2CO_2H$	Glutamic acid	Glu	E	3.22
	$-CH_2CONH_2$	Asparagine	Asn	N	5.41
	$-CH_2CH_2CONH_2$	Glutamine	Gln	Q	5.65
Side chain has a basic amino group	$-CH_2CH_2CH_2CH_2NH_2$	Lysine	Lys	K	9.74
	$-CH_2CH_2CH_2NHC(=NH)NH_2$	Arginine	Arg	R	10.76
	$-CH_2-$ (4-imidazolyl, N–H)	Histidine	His	H	7.59
Side chain contains sulfur	$-CH_2SH$	Cysteine	Cys	C	5.07
	$-CH_2CH_2SCH_3$	Methionine	Met	M	5.74

A pH Exists for Each Amino Acid, Its Isoelectric Point, at Which No Net Migration in an Electric Field Occurs In an aqueous solution of an amino acid, a dynamic equilibrium exists between **1**, **2**, and **3**. If now a current is passed between electrodes dipping into such a solution, cations of form **2** migrate to the cathode. Anions of form **3** migrate to the anode. Neutral molecules, **1**, migrate nowhere. A molecule with *equal* numbers of positive and negative charges, like **1**, is said to be an **isoelectric molecule**, and it cannot migrate in an electric field.

Because the equilibrium is *dynamic*, a migrating cation like **2** could flip a proton to some acceptor, become **1** and isoelectric (neutral), and stop moving to the cathode. In another instant, it (now **1**) could shed another proton, become **3**, and so made to turn around and head for the anode. Similarly, an anion on its way to the anode might pick up a proton, become neutral and also stop dead. Then it might take another proton, become **2**, and get turned around. In the meantime, an isoelectric molecule, **1**, might either donate or accept a proton, become electrically charged, and start its own migration. The question is, what overall *net* migration occurs and how is this net effect influenced by the pH of the solution?

Although much coming and going occurs in an amino acid solution, the net molar concentrations of the species stay the same at equilibrium. If either **2** or **3** is in any molar excess because of the pH, then some *net* migration occurs toward one electrode or the other. When the net molar concentration of **2**, for example, is greater than that of **3**, some statistical net movement to the cathode occurs.

Remember, however, that these equilibria can be shifted by adding acid or base. By carefully adjusting the pH, in fact, we can so finely tune the concentrations at equilibrium that *no net migration occurs*. At the right pH, the rates of proton exchange are such that each unit that is not **1** spends an equal amount of time as **2** and as **3**. (And the concentrations of **2** and **3** are very low.) Thus, any net migration to one electrode is blocked.

■ These shifts of H^+ ions illustrate Le Châtelier's principle at work.

The pH at which no net migration of an amino acid can occur in an electric field is called the **isoelectric point** of the amino acid, and the symbol of this pH value is **pI** (see Table 10.1, last column). Now let's see what the concepts of isoelectric molecules and pI values have to do with proteins.

Proteins, Like Amino Acids, Have Isoelectric Points As we will soon see, all proteins have NH_3^+ or CO_2^- groups or can acquire them by a change in the pH of the surrounding medium. Whole protein molecules, therefore, can also be isoelectric at the right pH. *Each protein thus has its own isoelectric point.* Now think of what can happen if the pH is changed from the pI value. The entire electrical condition of a huge protein molecule can be made either cationic or anionic almost instantly, at room temperature, by adding strong acid or base—by changing the pH of the medium.

Such changes in the electrical charge of a protein have serious consequences at the molecular level of life. Being electrically charged can dramatically affect chemical reactions, for example, or greatly alter protein solubility. If proteins are to serve their biological purposes, some must not be allowed to go into solution and others must not be permitted to precipitate. We'll return to this concept in this and later chapters, but the discussion focuses our attention again on how important it is that an organism control the pH values of its fluids.

■ Casein, the protein in milk, precipitates when milk turns sour because a change in pH causes the casein molecules to become isoelectric.

We will next survey the types of side chains in amino acids and how they affect the properties of polypeptides and proteins. These properties include how a polypeptide molecule will spontaneously fold and twist into its distinctive and vital final shape. You should memorize the structures of a minimum of five amino acids that illustrate the types we are about to study; glycine, alanine, cysteine, lysine, and glutamic acid are suggested. How to use Table 10.1 to write their structures is described in the following example.

EXAMPLE 10.1
Writing the Structure of an Amino Acid

What is the structure of cysteine?

■ The amino acid proline is the only one of the standard 20 in which the α-amino group is itself joined to one end of the sidechain.

$H_2\overset{+}{N}$—(ring)—CO_2^-

Proline

ANALYSIS Regard any α-amino acid as having two structural features, a unit common to *all* α-amino acids plus a unique side chain at the α-position. So write the common unit first and then add the side chain.

SOLUTION The common unit is

$$^{+}NH_3\underset{|}{C}HCO_2^-$$

Next, either look up or recall from memory the side chain for the particular amino acid. For cysteine, this is CH_2SH, so simply attach this group to the α-carbon. Cysteine is

$$^{+}NH_3\underset{\underset{CH_2SH}{|}}{C}HCO_2^-$$

■

PRACTICE EXERCISE 1

Write the structures of the dipolar ionic forms of glycine, alanine, lysine, and glutamic acid. ■

Several Amino Acids Have Hydrophobic Sidechains The first amino acids in Table 10.1, including alanine, have essentially nonpolar, hydrophobic sidechains. When a long polypeptide molecule folds into its distinctive shape, these hydrophobic groups tend to be folded next to each other as much as possible rather than next to highly polar groups or to water molecules in the solution. This water avoidance by nonpolar sidechains is called the **hydrophobic interaction** of the sidechains and two factors are at work. One is the strong tendency of water molecules to form hydrogen bonds between each other and thus "reject" the presence of molecules or groups that cannot themselves "offer" hydrogen-bonding capabilities to water. The water molecules, in a sense, club together forcing nonpolar groups to stay away and be by themselves. The other factor in hydrophobic interactions is that of *London forces* of attraction. These are attractive forces between nonpolar groups made possible by temporary dipoles.

Some Amino Acids Have Hydrophilic OH Groups on Their Sidechains The second set of amino acids in Table 10.1 consists of those whose sidechains carry alcohol or phenol OH groups, which are polar and hydrophilic. They can donate and accept hydrogen bonds. As a long polypeptide chain folds into its final shape, these side chains tend to stick out into the surrounding aqueous phase to which they are attracted by hydrogen bonds.

■ Aspartame, a popular artificial sweetener, has an aspartic acid residue.

$$^{+}NH_3\underset{\underset{CH_2CO_2^-}{|}}{C}H\overset{\overset{O}{\|}}{C}—NH\underset{\underset{CH_2C_6H_5}{|}}{C}H\overset{\overset{O}{\|}}{C}OCH_3$$

Methyl ester

Aspartic acid residue — Phenylalanine residue

Aspartame (NutraSweet ®)

Two Amino Acids Have Carboxyl Groups on Their Sidechains The sidechains of aspartic and glutamic acid carry proton donating CO_2H groups. Because body fluids are generally slightly basic, the protons available from side chain CO_2H groups have been neutralized, so these groups actually occur mostly as CO_2^- groups on protein molecules in body fluids. The aqueous medium would have to be made quite acidic to prevent this and force protons back onto the sidechain CO_2^- groups. This is why the pI values of aspartic and glutamic acid are lower (more acidic) than the pI values of amino acids with nonpolar side chains.

Aspartic acid and glutamic acid often occur as asparagine and glutamine in which their side chain CO_2H groups have become amide groups, $CONH_2$, instead. These are also polar, hydrophilic groups, *but they are not electrically charged*. They are neither proton donors nor proton acceptors, so the pI values of asparagine and glutamine are higher than those of aspartic or glutamic acids.

PRACTICE EXERCISE 2

Write the structure of aspartic acid (in the manner of **1**) with the sidechain carboxyl (a) in its carboxylate form and (b) in its amide form.

Lysine, Arginine, and Histidine Have Basic Groups on Their Sidechains The extra NH_2 group on lysine makes its sidechain basic and hydrophilic. Remember that an amine in water makes the pH of the solution greater (more basic) than 7 because of the OH^- ion generated in the following equilibrium.

$$RNH_2(aq) + H_2O \rightleftharpoons RNH_3^+(aq) + OH^-(aq)$$

The addition of OH^- to this equilibrium shifts the equilibrium to the left in accordance with Le Châtelier's principle. (The extra base pulls H^+ from the NH_3^+ group). A solution of lysine has to be made basic, therefore, the prevent its sidechain NH_2 group from existing in the protonated form, NH_3^+, and so affect the net charge on the lysine molecule. This is why the pI value of lysine, 9.47, is relatively high. Arginine and histidine have similarly basic sidechains.

PRACTICE EXERCISE 3

Write the structure of arginine in the manner of **1**, but with its sidechain amino group in its protonated form. (Put the extra proton on the =NH unit, not the NH_2 unit of the side chain.)

PRACTICE EXERCISE 4

Is the sidechain of the following amino acid hydrophilic or hydrophobic? Does the sidechain have an acidic, basic, or neutral group?

$$\begin{array}{l} {}^+NH_3CHCO_2^- \\ \quad\quad\;\; | \\ \quad\quad CH_2CH_2CONH_2 \end{array}$$

Cysteine and Methionine Have Sulfur-Containing Sidechains The sidechain in cysteine has an SH group. As we studied in Section 3.6, molecules with this group are easily oxidized to disulfide systems, and disulfides are easily reduced to SH groups:

$$2RSH \underset{\text{(reduction)}}{\overset{\text{(oxidation)}}{\rightleftharpoons}} RS{-}SR + H_2O$$

Cysteine and its oxidized form, cystine, are interconvertible by oxidation and reduction, a property of far-reaching importance in some proteins.

$$\begin{array}{ccc} H{-}SCH_2CHCO_2^- & \xrightarrow[\;H_2O\;]{\text{(O) (oxidation)}} & SCH_2CHCO_2^- \\ + \quad | & & | \quad\quad | \\ NH_3^+ & & | \quad NH_3^+ \\ H{-}SCH_2CHCO_2^- & \xleftarrow[\;2(H)\;]{\text{(reduction)}} & SCH_2CHCO_2^- \\ | & & | \\ NH_3^+ & & NH_3^+ \\ \text{Cysteine (two molecules)} & & \text{Cystine} \end{array}$$

The **disulfide link** contributed by cystine is especially prevalent in the proteins that have a protective function, such as those in hair, fingernails, and the shells of certain crustaceans.

■ Some D-amino acid residues occur in bacterial cell membranes, which helps these disease-causing agents to survive in higher animals, where the enzymes for attacking polypeptides work only with L-forms.

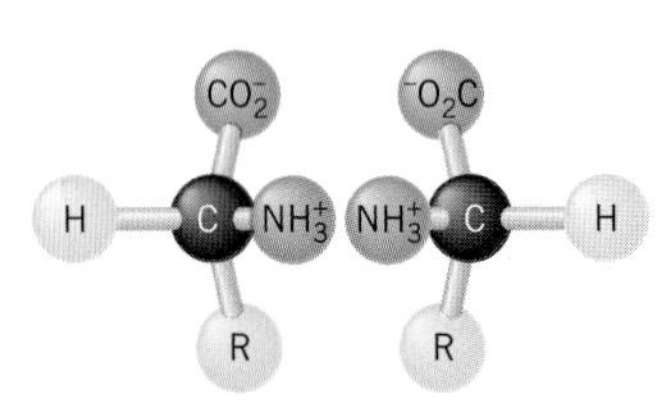

FIGURE 10.2
The two possible enantiomers of α-amino acids whose molecules have just one stereocenter. The absolute configuration on the right, which is in the L-family, represents virtually all the naturally occurring α-amino acids.

The α-Position in All Amino Acids except Glycine Is a Stereocenter All the amino acids except glycine consist of chiral molecules and can exist as pairs of enantiomers. For each possible pair of enantiomers, however, nature supplies just one of the two (with a few rare exceptions). All the naturally occurring amino acids, moreover, belong to the same optical family, the L-family. What this means is illustrated in Figure 10.2. It also means that all the proteins in our bodies, including all enzymes, are made from L-amino acids and are all chiral.

10.2 OVERVIEW OF PROTEIN STRUCTURE

Protein molecules can have four levels of complexity.
Protein structures are more complicated by far than those of carbohydrates or lipids, and every aspect of their structures is vital at the molecular level of life. We'll begin, therefore, with a broad overview of protein architecture.

Protein Structure Involves Four Features There are four possible levels of complexity in the structures of proteins. Disarray at any level almost always renders the protein biologically useless. A protein having the structure and overall shape that it normally possesses in a living system and that permit it to function biologically is called a **native protein.** The same protein when made to lose its molecular shape even while retaining its original molecular constitution (covalent structure), is a **denatured protein.** (We'll return to denaturing agents in Section 10.7.)

The first and most fundamental level of protein structure, the **primary structure,** concerns only the *sequence of amino acid residues* in the polypeptide(s) of the protein. However, it is this sequence that ultimately determines the three-dimensional structure of a protein and therefore determines how a protein can function.

The next level of protein complexity, the **secondary structure,** also concerns just individual polypeptides. It entails noncovalent forces, particularly the hydrogen bond, and it consists of the particular way in which a long polypeptide strand has coiled or in which strands have intertwined or lined up side to side.

The **tertiary structure** of a polypeptide concerns the further coiling, bending, kinking, or twisting of secondary structures. If you've ever played with a coiled door spring, you know that the coil (secondary structure) can be bent and twisted (tertiary structure). By and large, noncovalent forces such as hydrogen bonds and hydrophobic interactions stabilize these shapes. When polypeptides have disulfide bonds, they form from SH groups *after* the polypeptide has been synthesized in the cell, so the S—S covalent bond is usually classified as a feature of tertiary structure, not primary structure. In many polypeptides, attractions and repulsions between electrically charged sites on sidechains are also involved in determining the overall shape of a polypeptide.

Finally, we deal with those proteins having still another complexity, **quaternary structure.** A protein's quaternary structure forms by a coming together of two or more polypeptides, often with other relatively small molecules or ions that aggregate in a precise manner to form one grand whole. Each polypeptide unit—now actually a *subunit* of the protein—has all previous levels of structure.

10.3 PRIMARY STRUCTURES OF PROTEINS

The backbones of all polypeptides of all plants and animals have a repeating series of N—C—C(=O) units.

The Peptide Bond (Amide Bond) Joins Amino Acid Residues Together in a Polypeptide The **peptide bond** is the covalent bond that forms when amino acids are put together in a cell to make a polypeptide. It's nothing more than an amide system, carbonyl-to-nitrogen. To illustrate it and to show how polypeptides acquire their primary structure, we will begin by simply putting just two amino acids together to form a *dipeptide*.

Suppose that glycine acts at its carboxyl end and alanine acts at its amino end such that, by a series of steps (not given in detail), a molecule of water splits out and a carbonyl-to-nitrogen bond, a peptide bond, is created.

■ *Simple* amides have the following structure (see also Section 6.3).

$$R{-}\overset{O}{\overset{\|}{C}}{-}NH_2 \quad \text{(Amide bond)}$$

$$^{+}NH_3CH_2\overset{O}{\overset{\|}{C}}{-}O^- + H{-}\overset{+}{N}(H)_2CH(CH_3)\overset{O}{\overset{\|}{C}}O^- \xrightarrow[\text{by several steps}]{\text{in the body,}} {}^{+}NH_3CH_2\overset{O}{\overset{\|}{C}}{-}NHCH(CH_3)\overset{O}{\overset{\|}{C}}O^- + H_2O$$

Glycine (Gly)   Alanine (Ala)   Glycylalanine (Gly-Ala) **4** (Peptide bond)

■ How a cell causes a peptide bond to form involves a cell's nucleic acids and is described in Chapter 13.

Of course, there is no reason why we could not picture the roles reversed so that alanine acts at its carboxyl end and glycine at its amino end. This results in a different dipeptide (but an isomer of the first).

$$^{+}NH_3CH(CH_3)\overset{O}{\overset{\|}{C}}{-}O^- + H{-}\overset{+}{N}(H)_2CH_2\overset{O}{\overset{\|}{C}}O^- \xrightarrow[\text{by several steps}]{\text{in the body,}} {}^{+}NH_3CH(CH_3)\overset{O}{\overset{\|}{C}}{-}NHCH_2\overset{O}{\overset{\|}{C}}O^- + H_2O$$

Alanine (Ala)   Glycine (Gly)   Alanylglycine (Ala-Gly) **5** (Peptide bond)

The product of the union of any two amino acid residues by a peptide bond is called a **dipeptide**, and all dipeptides have the following features:

$$^{+}NH_3CH(R^1)\overset{O}{\overset{\|}{C}}{-}NHCH(R^2)\overset{O}{\overset{\|}{C}}O^-$$

Dipeptide (Peptide bond)

Structures **4** and **5** differ only in the sequence in which the sidechains, H and CH_3, occur on α-carbons. This is fundamentally how polypeptides also differ—in their sequences of sidechains.

■ The *di* in *di*peptide signifies that *two* amino acid residues are present and not the number of peptide bonds.

EXAMPLE 10.2
Writing the Structure of a Dipeptide

What are the two possible dipeptides that can be put together from alanine and cysteine?

ANALYSIS Both dipeptides must have the same backbone, so we write two of these first. We follow the convention that such backbones are always written in the N to

C—left to right—direction, but this is only a convention:

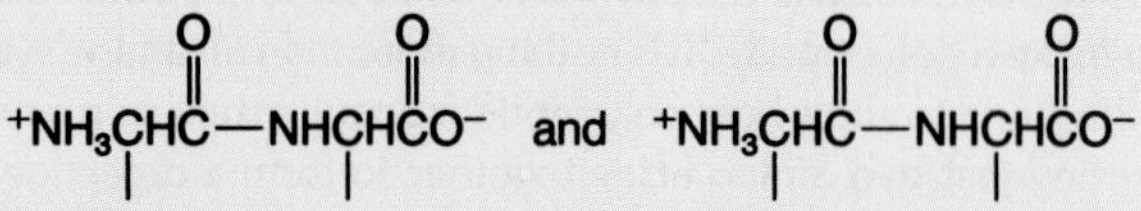

$$^{+}NH_3CH(—)C(=O)—NHCH(—)CO^{-} \quad \text{and} \quad ^{+}NH_3CH(—)C(=O)—NHCH(—)CO^{-}$$

Then either from memory or by using a table, we know that the two sidechains are CH_3 for alanine and CH_2SH for cysteine. We simply attach these in their two possible orders to make the finished structures.

SOLUTION

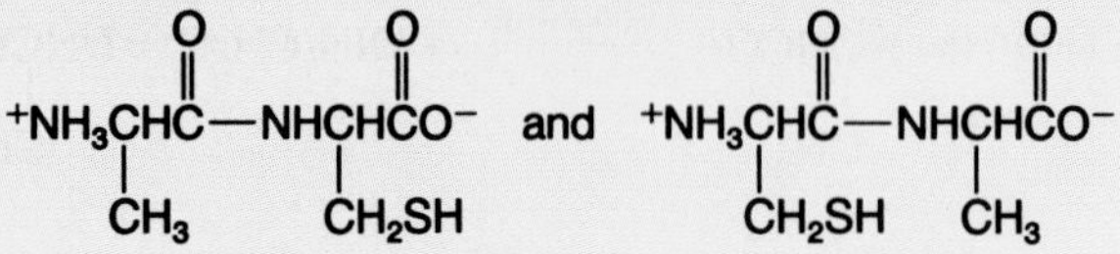

$$^{+}NH_3CH(CH_3)C(=O)—NHCH(CH_2SH)CO^{-} \quad \text{and} \quad ^{+}NH_3CH(CH_2SH)C(=O)—NHCH(CH_3)CO^{-}$$

Ala-Cys Cys-Ala

It would be worthwhile at this time simply to memorize the easy repeating sequence in a dipeptide, because it carries forward to higher peptides:

nitrogen – carbon – carbonyl – nitrogen – carbon – carbonyl

(Remember that the "carbon" in "nitrogen-carbon-carbonyl" is the *alpha* carbon. Remember also that the direction — left-to-right means nitrogen-carbon-carbonyl — is conventional, not a law of nature.)

PRACTICE EXERCISE 5

Write the structures of the two dipeptides that can be made from alanine and glutamic acid.

■ Biochemists find it easier to use single letter symbols for the amino acid residues when they want to compare the amino acid sequences in several similar polypeptides.

Three-Letter Symbol for Amino Acid Residues Simplify the Writing of Polypeptide Structures Each amino acid has been assigned a three-letter symbol, given in the third from the last column in Table 10.1. To use three-letter symbols to write a polypeptide structure, we have to follow certain rules. The convention is that a series of three-letter symbols, each separated by a hyphen, represents a polypeptide structure, provided that the first symbol (reading left to right) is the free amino end, NH_3^+, and the last symbol has the free carboxylate end, CO_2^-. The structure of the dipeptide **4**, for example, can be rewritten as Gly-Ala, and its isomer **5** as Ala-Gly. In both, the backbones are identical. In some applications, it is more convenient to use single-letter symbols, which are given in the next to the last column of Table 10.1.

Dipeptides still have NH_3^+ and CO_2^- groups, so a third amino acid can react at either end. In general,

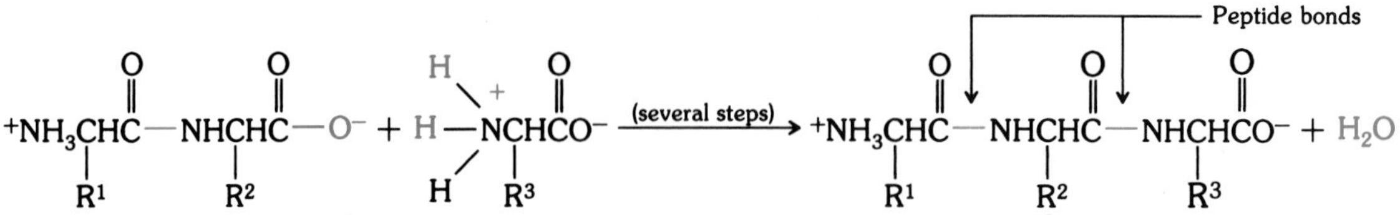

$$^{+}NH_3CH(R^1)C(=O)—NHCH(R^2)C(=O)—O^{-} + H—\overset{+}{N}H_2CH(R^3)CO^{-} \xrightarrow{\text{(several steps)}} {}^{+}NH_3CH(R^1)C(=O)—NHCH(R^2)C(=O)—NHCH(R^3)CO^{-} + H_2O$$

Peptide bonds

A tripeptide

If we start with glycine, alanine, and phenylalanine, the tripeptide, Gly-Ala-Phe, would be only one of six possible tripeptides involving these three different amino acids. The set of all possible sequences for a tripeptide made from Gly, Ala, and Phe is as follows:

■ A polypeptide is actually a *copolymer*, because the monomer units are not identical.

Gly-Ala-Phe Ala-Gly-Phe Phe-Gly-Ala

Gly-Phe-Ala Ala-Phe-Gly Phe-Ala-Gly

Each of these tripeptides still has groups at each end, NH_3^+ and CO_2^-, that can interact with still another amino acid to make a tetrapeptide. And this product would still have the end groups from which the chain could be extended even further. You can see how a repetition of this pattern many hundreds of times can produce a long polymer, a polypeptide.

The Sequence of Sidechains on the Repeating N—C—C(=O) Backbone Is the *Primary* Structure of All Polypeptides All polypeptides have the following skeleton in common. They differ in length (n) and in the kinds and sequences of sidechains.

$$\alpha\text{-positions}$$

$$^{+}NH_3—\underset{|}{CH}—\overset{O}{\overset{\|}{C}}\left[NH—\underset{|}{CH}—\overset{O}{\overset{\|}{C}}\right]_n NH—\underset{|}{CH}—CO_2^-$$

N-terminal amino acid residue

C-terminal amino acid residue

Polypeptide "backbone" (n can equal several thousand)

■ Notice, for later reference, the designations *N-terminal unit* and *C-terminal unit* for the residues with the free α-NH_3^+ and the free α-CO_2^- groups, respectively.

The peptide bond, as we said, is the chief covalent bond that holds amino acid residues together in polypeptides. The disulfide bond is the only other covalent bond that affects the amino acid residues. As we said, the disulfide bond generally becomes established *after* polypeptide molecules have been put together, so it usually is called a tertiary structural feature.

As the number of amino acid residues increases, some used several times, the number of possible polypeptides increases rapidly. For example, if 20 different amino acids are incorporated, each used only once, there are 2.4×10^{18} possible isomeric polypeptides! (And a polypeptide with only 20 amino acid residues in its molecule is a very small polypeptide.) The maximum number of ways by which 20 *different* amino acid residues can be joined is 20^{100} or about 1.3×10^{130}. The estimated total number of atoms of all kinds and combinations everywhere in the entire universe is only(!) 9×10^{78}. You can see that the statistical possibilities for having many *different* polypeptides exceeds by far the available supply of atoms and makes possible the huge number of unduplicated small and large variations not only between species but among individuals within a species who have ever lived or ever will live.

The *Peptide Group* Is Trans-planar *Rotation about the peptide bond is not free* despite its appearing to be a single bond. The unshared electron pair on the N atom of the NH group interacts with the C and O atoms of the carbonyl group to give the C—N peptide bond enough of the character of a double bond to prevent free rotation about it. Thus, in what we can call the *peptide group* of atoms—from one α-carbon through the carbonyl–nitrogen unit to the next α-carbon—the four atoms lie in the same plane (see Fig. 10.3). The H on N of the NH group and the O on C of the carbonyl group are held trans to each other, on opposite sides of the backbone. The rigidity of the peptide group places some constraints on the flexibility of the polypeptide chain and so affects its overall shape.

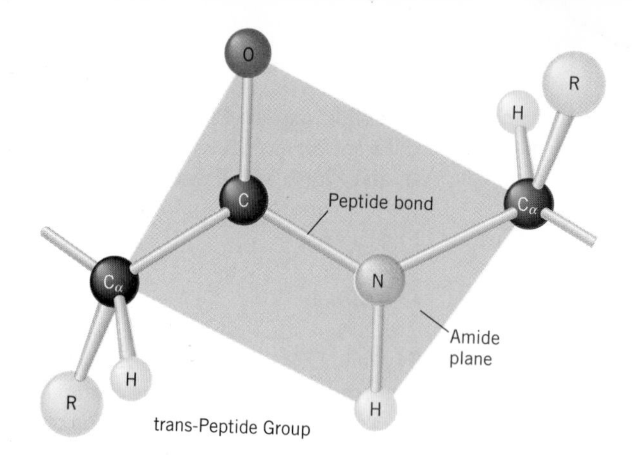

FIGURE 10.3
The trans nature of the *peptide group* showing how one C_α atom, the carbonyl carbon atom, the nitrogen atom, and the next C_α atom in the polypeptide chain are all in the same plane. The H in NH is trans to the O in C=O.

10.4

SECONDARY STRUCTURES OF PROTEINS

The α-helix and the β-pleated sheet are two important kinds of secondary protein structures.

Once a cell puts together a polypeptide, largely noncovalent forces, like the hydrogen bond and hydrophobic interactions, help to determine how a polypeptide twists into a particular native shape. The hydrophobic interactions are largely what "drive" the formation of an overall shape, and hydrogen bonds help to stabilize it.

■ The turns of a right-handed helix or screw are in the same direction taken as the fingers of your right hand curl when your thumb points along the axis of the helix in the direction in which the helix advances. Wood screws have right-handed helices.

The α-Helix Is a Major Secondary Structure of Polypeptides One of the most common configurations is the *α-helix,* a coiled configuration of a polypeptide strand discovered by Linus Pauling and R. B. Corey (see Fig. 10.4). In the **α-helix,** the polypeptide backbone coils as a right-handed screw, which permits all of its sidechains to stick to the outside of the coil.

■ Hydrogen bond

—N—H•••O=C<

Hydrogen Bonds Stabilize α-Helices In α-helices, hydrogen bonds extend from the H atoms of polar NH units in peptide groups to oxygen atoms of polar carbonyl units four residues farther along the backbone. Individually, single hydrogen bonds are weak forces of attraction, but they add up much like the individual forces that hold a zipper strongly shut. Generally, in very long polypeptides, only *segments* of the molecules, not entire lengths, are in an α-helix configuration. Coils that are about 11 residues long are common, but as many as 53 residues have been observed in the α-helix portion of one polypeptide molecule. Uncoiled portions of a polypeptide strand or a segment having a pleated sheet array (discussed below) occur between α-helix segments.

■ The protein in the cornea of the eye is also a member of the collagen family.

A Left-Handed Helix Characterizes the Individual Polypeptide Strands in Collagen The collagens, the most abundant proteins in vertebrates, are a family of extracellular proteins that give strength to bone, teeth, cartilage, tendons, skin, blood vessels, and certain ligaments. Glycine contributes a third of the amino acid residues in collagen's polypeptide subunits. Another 15–30% of the residues are contributed by proline and 4-hydroxyproline, an amino acid not on the list of 20 standard amino acids. In addition, there are residues from 3-hydroxyproline and 5-hydroxylysine, also not on the list of 20.

$H_2\overset{+}{N}$ ... CO_2^- ... OH — 4-Hydroxyproline

$H_2\overset{+}{N}$ ... CO_2^- ... OH — 3-Hydroxyproline

$^+NH_3CHCO_2^-$
$CH_2CH_2CHCH_2NH_2$
OH
— 5-Hydroxylysine

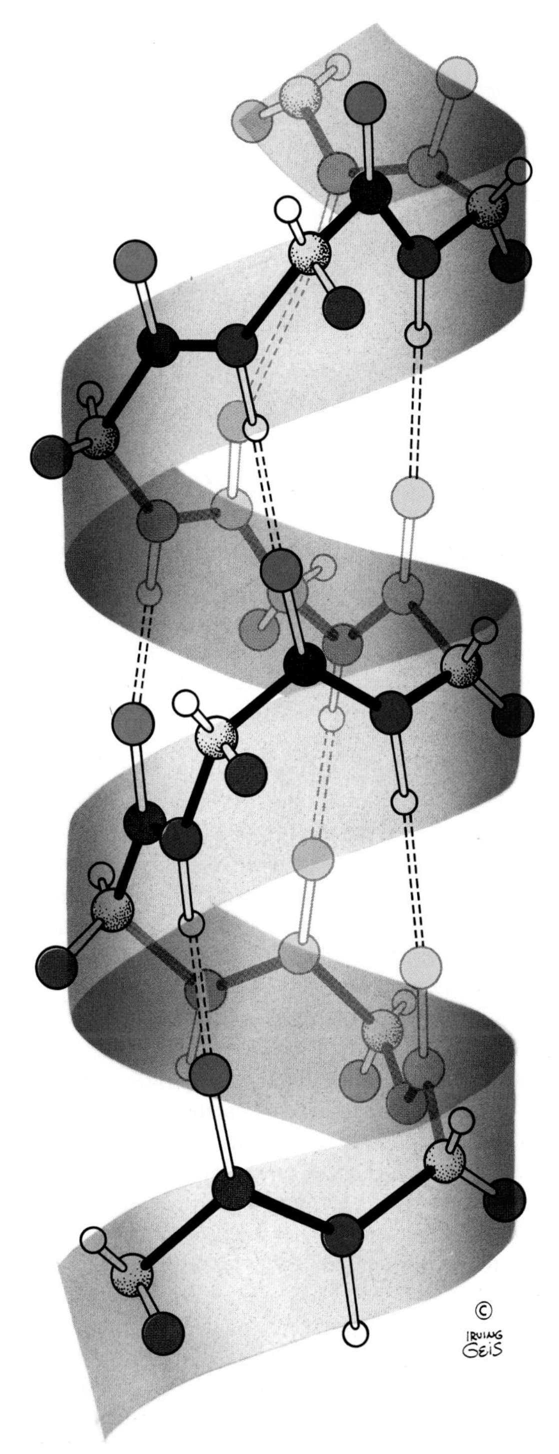

FIGURE 10.4
The α-helix. The polypeptide backbone follows the spiraling ribbon as a right-handed helix. The oxygen atoms of carbonyl groups are in red; nitrogen atoms of the peptide NH group are in dark blue with the H atoms of NH in white. Sidechain groups, R, are represented here only by simple spheres in purple. Note how they project to the *outside* of the ribbon. Dashed lines show how hydrogen bonds extend from each NH group's H atom to a carbonyl group's O atom four residues along the backbone.

In some types of collagen, monosaccharide molecules are incorporated (as glycosides of side chain OH groups).

The ring systems in the molecules of proline and its hydroxylated relatives limit strand flexibility. They restrict the coiling of collagen polypeptides to *left-handed* helices. There is a further level of structure to collagen, tertiary structure, which we'll consider shortly.

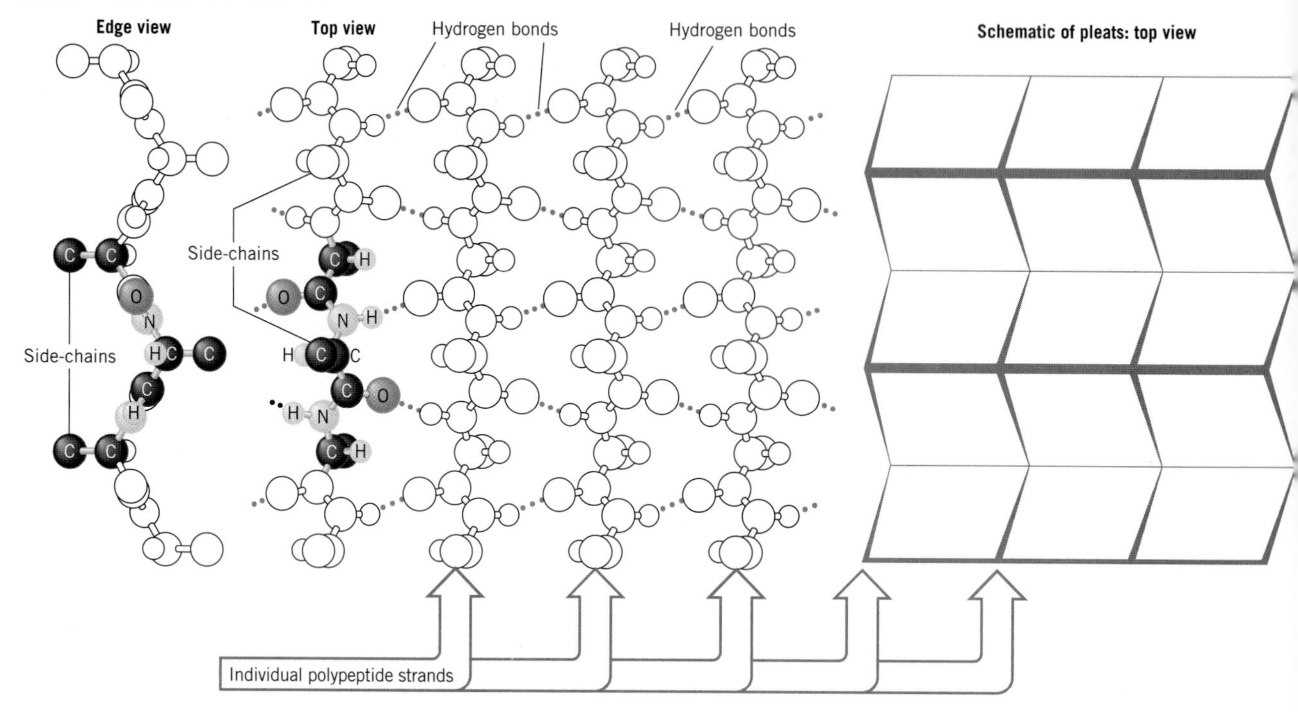

FIGURE 10.5
The β-pleated sheet.

■ A collagen fibril only 1 mm in diameter can hold a suspended mass as large as 10 kg (22 lb).

Vitamin C Is Essential to the Synthesis of Collagen The hydroxylated derivatives of lysine and proline are made with the help of an enzyme that requires ascorbic acid (vitamin C). The reactions that put OH groups on proline rings (or lysine side chains) occur *after* the initial polypeptide is made. Thus vitamin C is essential to growing children for the formation of strong bones and teeth as well as all other tissues that rely on collagen. When an adult's diet is deficient in ascorbic acid, wounds do not heal well, blood vessels become fragile, and an overall vitamin-deficiency condition called *scurvy* results.

The β-Pleated Sheet Is a Side-by-Side Array of Polypeptide Units Pauling and Corey also discovered that adjacent segments in some polypeptides line up side by side, to form a sheet-like array that is somewhat pleated and called the **β-pleated sheet** (see Fig. 10.5). The sidechains project above and below the surface of the sheet. This is another kind of secondary structure in which hydrogen bonds hold things together. As few as two segments of a polypeptide strand and as many as 15 have been found in the same pleated sheet, with each strand ranging from 6 to 15 residues long.

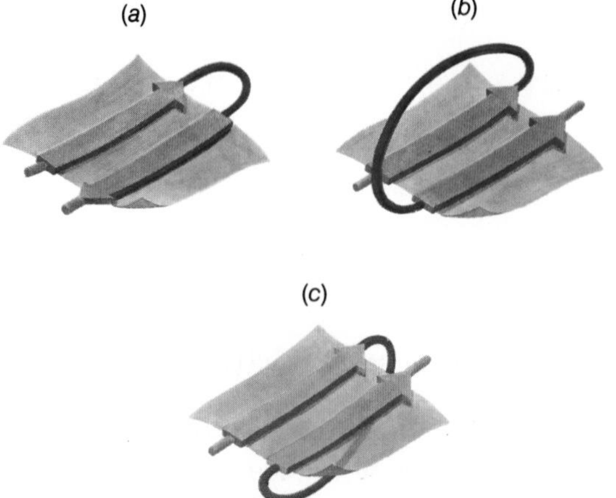

FIGURE 10.6
Segments of polypeptide strands can become adjacent to and parallel with each other in more than one way. (*a*) A hairpin loop brings the next segment into an antiparallel arrangement. (*b*) A back-and-over loop allows the next segment to have a parallel alignment. (*c*) A left-handed crossover loop also permits a parallel alignment. The loops are made of segments of the polypeptide chain.

When a hairpin turn carries the polypeptide chain from one segment of a pleated sheet to the next, the strands run in opposite directions. Other kinds of turns are possible, however, but are less common (see Fig. 10.6). The pleated sheet is a feature in portions of the polypeptide strands of many proteins and is the dominant feature in fibroin, silk protein (seen in Fig. 10.5).

■ Linus Pauling won the 1954 Nobel prize in chemistry for discovering the α-helix and β-sheet configurations.

10.5

TERTIARY STRUCTURES OF PROTEINS

Tertiary structures are the results of folding, bending, and twisting of secondary structures.

Once primary and secondary structures are in place, the final shaping of a protein occurs. All these activities happen spontaneously in cells, sometimes in a matter of seconds after the polypeptide molecule has been made and sometimes it takes several minutes. The "rules" followed by the polypeptide to give these shapes are still not fully known.

Tertiary Protein Structure Involves the Folding and Bending of Secondary Structure When α-helices take shape, their sidechains tend to project outward where, in an aqueous medium, they can be in contact with water molecules. Even in water-soluble proteins, however, as many as 40% of the sidechains are hydrophobic. Because such groups cannot break up the hydrogen-bonding networks among the water molecules, an entire α-helix or β-sheet undergoes further twisting and folding until the hydrophobic groups, as much as possible, are tucked to the inside, away from the water, and the hydrophilic groups stay exposed to the water. Thus the final shape of the polypeptide, its **tertiary structure,** emerges in response to simple molecular forces set up by the water-avoiding and the water-attracting properties of the sidechains. In fact, these hydrophobic interactions on which *tertiary* structure depends sometimes determine the best secondary structure for the polypeptide.

Disulfide Bonds Can Give Loops in Polypeptides or Join Two Strands Together Polypeptides that are to have disulfide bonds receive them by the oxidation of SH groups during the development of tertiary structure. If the SH group on the side chain of cysteine appears on two neighboring polypeptide molecules, then mild oxidation is all it takes to link the two molecules by a disulfide bond. This cross-linking, facilitated by proper enzymes, can also occur between parts of the same polypeptide molecule, in which case a closed loop results.

■ The letter symbols for cystine are

Cys | Cys and C | C

Ionic Bonds Also Stabilize Tertiary Structures Another force that can stabilize a tertiary structure is the attraction between a full positive and a full negative charge, each occurring on a particular sidechain. At the pH of body fluids, the sidechains of both aspartic acid and glutamic acid carry CO_2^- groups. The sidechains of lysine and arginine carry NH_3^+ groups. These oppositely charged groups can attract each other, like the attraction of oppositely charged ions in an ionic crystal. The attraction is an *electrostatic attraction,* and is sometimes called a **salt bridge.**

Hydrophobic Interactions Significantly Affect Polypeptide Shape In the tertiary structure of myoglobin, the oxygen-holding protein in muscle tissue, about 75% of the single polypeptide molecule consists of α-helix segments (see Fig. 10.7*a*). Virtually all of the hydrophobic groups of myoglobin are folded inside where they avoid water molecules as much as

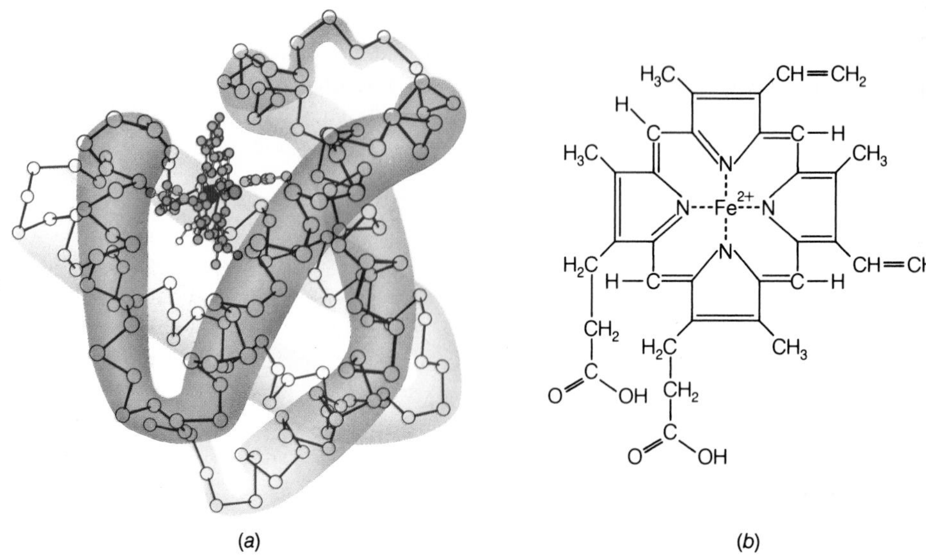

FIGURE 10.7
(*a*) Myoglobin (sperm whale), a polypeptide with 153 amino acid residues. The tube-like forms outline the eight segments that are in an α-helix. The flat, purple structure is the heme unit and the red circle is an oxygen molecule. Only the atoms that make up the backbone of the chain are indicated (by circles). The sidechains have been omitted (*b*) Heme molecule with its Fe^{2+} ion. (Illustration copyrighted © by Irving Geis.)

possible, and its hydrophilic groups are on the outside. A nonprotein molecule, heme, completes the native structure of myoglobin (Fig. 10.7*b*).

■ "Prosthetic" is from the Greek *prosthesis,* an addition.

Polypeptides Often Incorporate Prosthetic Groups into Their Tertiary Structures A nonprotein, organic compound that associates with a polypeptide, like heme in myoglobin, is called a **prosthetic group.** It is often the focus of the protein's biological purpose. Heme is the actual oxygen holder in myoglobin. Heme also serves the same function in **hemoglobin,** the oxygen carrier in blood. The heme molecule is held in the folded globin molecule by electrostatic attractions between two electrically charged sidechains and the Fe^{2+} ion in heme.

***β*-Sheets Are Often Twisted as Well as Pleated** Many polypeptides incorporate both helices and sheets within the same structure. In Figure 10.8, we see an artist's representation of two views of the molecule of a 247-residue enzyme, triose phosphate isomerase. Each broad, flat arrow is a segment of the chain that is in a β-sheet arrangement with a neighboring segment, also shown as a broad arrow. There are eight segments of the strand in the β-sheet regions of this enzyme, and the entire sheet is itself twisted to form a barrel-like configuration (called a β-barrel). α-Helix units occur between the strands involved in the β-sheet.

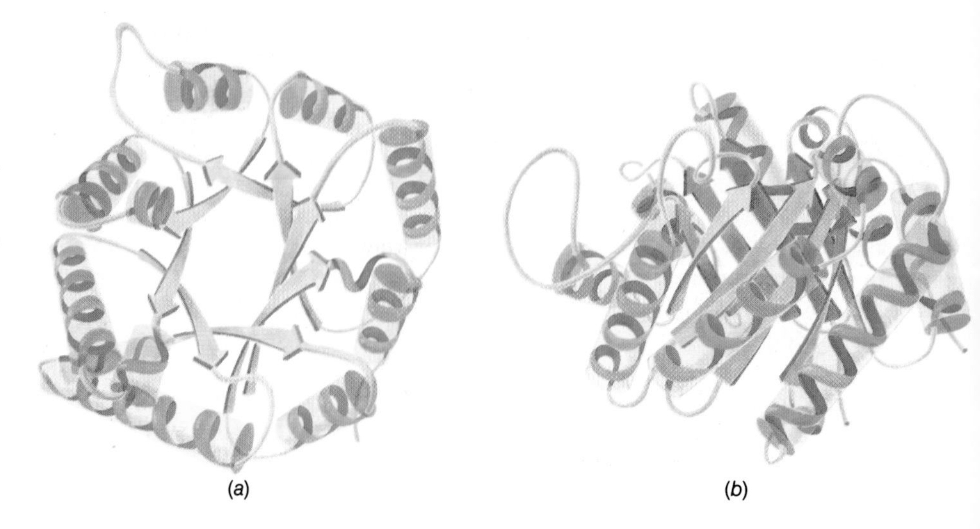

FIGURE 10.8
The molecule of the enzyme triose phosphate isomerase illustrates how both α-helix segments and β-sheet arrays can be incorporated together. The β-sheet consists of 8 parallel segments, but as indicated by the twists of the arrows, the sheet is itself twisted. The top view (*a*) and the side view (*b*) both indicate how this twist creates a barrel-like cylindrical structure. (From Voet/Voet, reprinted with permission from *Biochemistry* © 1990 by John Wiley & Sons, Inc.)

10.6

QUATERNARY STRUCTURES OF PROTEINS

For many proteins, the native form emerges only as two or more polypeptides assemble into a quaternary structure.

Proteins, like myoglobin, have finished shapes at the tertiary level. They are made up of single polypeptide molecules, sometimes with prosthetic groups. Many proteins, however, are aggregations of two or more polypeptides, and these aggregations constitute **quaternary structures.** One molecule of the enzyme phosphorylase, for example, consists of two tightly aggregated molecules of the same polypeptide. If the two become separated, the enzyme can no longer function. Individual molecules of polypeptides that make up an intact protein molecule are called the protein's *subunits.*

Hemoglobin has four subunits, two of one kind (designated α-subunits) and two of another (called the β-subunits), each subunit supporting a heme molecule (see Fig. 10.9). A combination of hydrophobic and electrostatic interactions as well as hydrogen bonds hold the subunits together. These forces do not work unless each subunit has the appropriate primary, secondary, and tertiary structural features. If even one amino acid residue is wrong, the results can be very serious, as in the example of sickle-cell anemia, described in Special Topic 10.1.

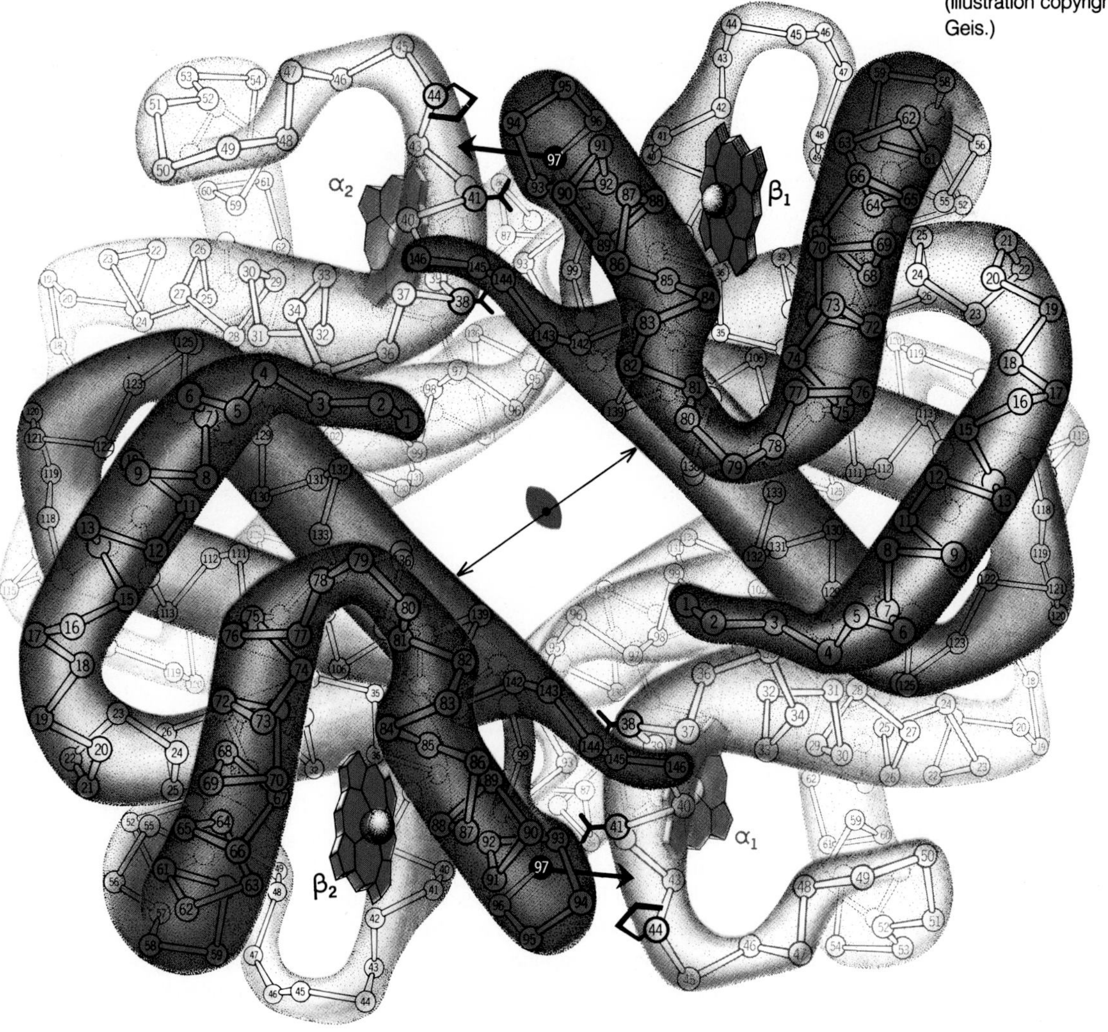

FIGURE 10.9
Hemoglobin. Four polypeptide chains, each with one heme molecule represented here by the colored, flat plates that contain spheres (Fe^{2+} ions), are nestled together. Only the atoms of the backbones are shown (by numbered circles). The central cavity, indicated by the double-headed arrow, has enough room to hold an organic anion not shown here, 2,3-bisphosphoglycerate (BPG), until hemoglobin starts to load up with oxygen molecules. (Illustration copyrighted © by Irving Geis.)

SPECIAL TOPIC 10.1

SICKLE-CELL ANEMIA AND ALTERED HEMOGLOBIN

The decisive importance of the primary structure to all other structural features of a polypeptide or its associated protein is illustrated by the grim story of sickle-cell anemia. This inherited disease is widespread among those whose roots are in equatorial regions of central and western Africa.

In its mild form, where only one parent carried the genetic trait, the symptoms of sickle-cell anemia are seldom noticed except when the environment has a low partial pressure of oxygen, as at high altitudes. In the severe form, when both parents carried the trait, the infant usually dies by the age of 2 unless treatment is begun early. The problem is *an impairment in blood circulation traceable to the altered shape of hemoglobin in sickle-cell anemia.* The altered shape is particularly a problem after the hemoglobin has delivered oxygen and is on its way back to the heart and lungs for more.

The fault at the molecular level lies in a β-subunit of hemoglobin. One of the amino acid residues should be glutamic acid, but is valine instead. Thus instead of a sidechain CO_2^- group, which is electrically charged and hydrophilic, there is an isopropyl sidechain, which is neutral and hydrophobic. Normal hemoglobin, symbolized as HbA, and sickle-cell hemoglobin, HbS, therefore have different patterns of electrical charges. Both have about the same solubility in well-oxygenated blood, but oxygen-free molecules of HbS clump together inside red cells and precipitate. This deforms the cells into a telltale sickle shape (see Figure 1). The distorted cells are harder to pump through capillaries, where the cells often create plugs. Sometimes the red cells split open. Any of these events place a strain on the heart. The error in one sidechain seems insignificant, but it is far from small in human terms.

The sickle-cell trait offers some resistance to malaria, which almost certainly explains why the trait survives largely where this tropical disease is most common. Normally, the mosquito-borne parasite that causes malaria resides within a red blood cell. However, the parasite cannot survive very long inside a sickled cell. The parasite has a high need for potassium ion, but the membrane of a sickled cell allows too much potassium ion to get through and escape. Thus people with the sickle-cell trait are statistically more likely than individuals unprotected from malaria to live long enough to bear children and so pass the trait to their offspring.

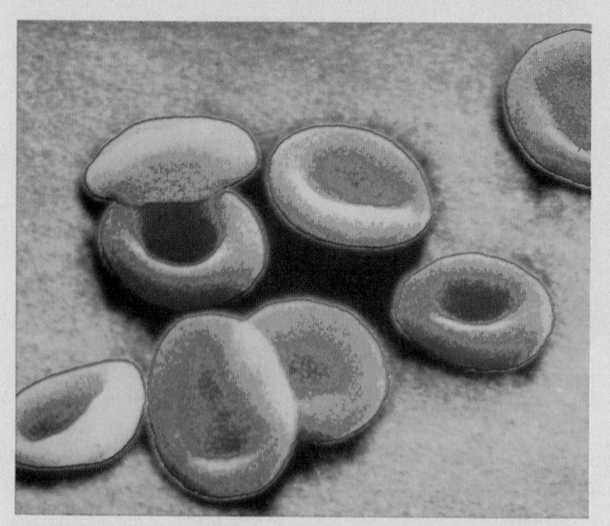

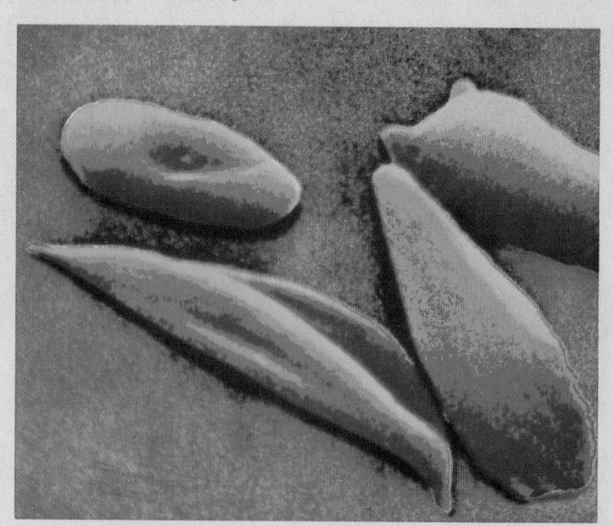

FIGURE 1 Electron micrographs of a normal red blood cell *(left)*, and a sickle cell *(right)*.

Covalent Cross-Linkages Occur in Collagen The polypeptide units in collagen, each with about 1000 amino acid residues and each in a left-handed helix, assemble in units of three molecules each. The three left-handed helices wrap around each other in a relatively open *right*-handed helix of helices to form the **triple helix,** cable-like system called *tropocollagen* (see Fig. 10.10). Between the polypeptide strands of the triple helix *covalently bonded* molecular bridges are erected by a series of reactions that cause lysine sidechains to link together. The details are beyond the scope of our study, but *covalent* cross-links are better able than hydrogen

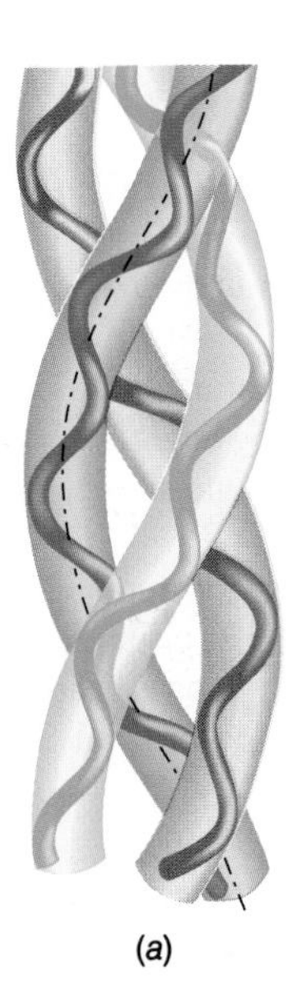

(a) (b)

FIGURE 10.10
The triple helix of collagen. (*a*) Schematic drawing. (Illustration copyrighted © by Irving Geis.) (*b*) Electron micrograph of collagen fibrils from the skin. A fibril is an orderly aggregation of collagen molecules aligned side by side by overlapping each other in a regularly repeating manner that produces the banded appearance. (Micrograph courtesy of Jerome Gross, Massachusetts General Hospital.)

bonds or hydrophobic interactions to resist forces that would work to undo the tertiary structure of collagen. With aging, additional covalent cross-links develop between collagen strands, leading to less muscular flexibility and agility.

A microfiber or *fibril* of collagen forms when individual tropocollagen cables overlap lengthwise. The mineral deposits in bones and teeth become tied into the protein at the gaps between the heads of tropocollagen molecules and the tails of others.

■ Meat from old animals is tougher because of their more highly cross-linked collagen.

10.7

COMMON PROPERTIES OF PROTEINS

Even small changes in the pH of a solution can affect a protein's solubility and its physiological properties.

Although proteins come in many diverse biological types, they generally have similar chemical properties toward ordinary substances because they have similar functional groups.

Protein Digestion Is Hydrolysis The digestion of a protein is nothing more than the hydrolysis of its peptide bonds to give a mixture of amino acids. Different digestive enzymes —all in the family of *proteases*—handle the cleavage of peptide bonds according to the nature of the sidechains nearby. The hydrolysis of a tripeptide illustrates digestion.

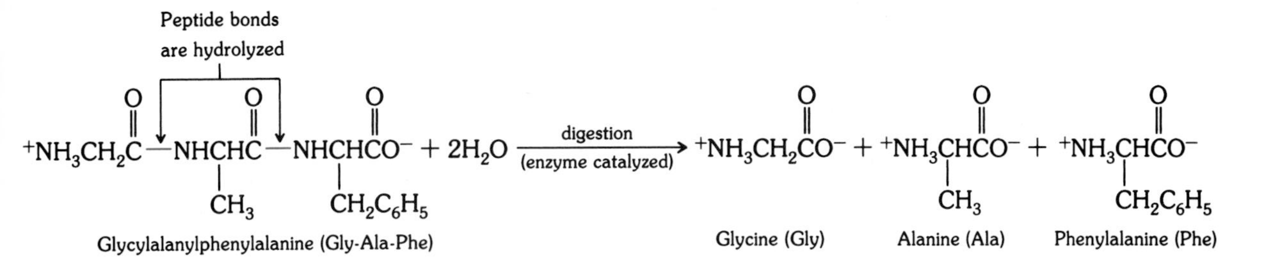

Protein *Denaturation* Is the Loss of Protein Shape Peptide bonds are not hydrolyzed when a protein is denatured. All that has to happen is some disruption of secondary or higher structural features. **Denaturation** is the disorganization of the overall molecular shape of a native protein. It can occur as an unfolding or uncoiling of helices, or as the separation of subunits. Because native proteins have their overall shapes in an aqueous environment where water molecules are intimately involved with hydrophobic interactions, even the removal of water can cause the denaturation of many proteins.

Usually, denaturation is accompanied by a major loss of solubility in water. When egg white is whipped or is heated, for example, as when you cook an egg, the albumin molecules unfold and become entangled among themselves. The system no longer blends with water—it's insoluble—and it no longer allows light to pass through.

Table 10.2 has a list of several reagents or physical forces that cause denaturation, together with brief explanations of how they work. How effectively a given denaturing agent is depends on the kind of protein. The proteins of hair and skin and of fur or feathers quite strongly resist denaturation because they are rich in disulfide links.

In recent years, several proteins have been discovered that, after denaturation, can be *renatured.* Often the denaturation of such proteins results by the cleavage of disulfide links. When an enzyme that handles the oxidation of SH groups back to S—S groups is presented to such denatured proteins, they are restored to their native forms. The original S—S linkages are faithfully remade.

Protein Solubility Depends Greatly on pH Because some sidechains as well as the end groups of polypeptides bear electrical charges, an entire polypeptide molecule can bear a net charge. Because these charged groups are either proton donors or proton acceptors, the net

TABLE 10.2
Denaturing Agents for Proteins

Denaturing Agent	How the Agent May Operate
Heat	Disrupts hydrophobic interactions and hydrogen bonds by making molecules vibrate too violently. Produces coagulation, as in the frying of an egg.
Microwave radiation	Causes violent vibrations of molecules that disrupt hydrogen bonds and hydrophobic interactions.
Ultraviolet radiation	Probably operates much like the action of heat (e.g., sunburning).
Violent whipping or shaking	Causes molecules in globular shapes to extend to longer lengths and then entangle (e.g., beating egg white into meringue).
Soaps	Probably affect hydrogen bonds and salt bridges.
Organic solvents (e.g., ethanol, acetone, 2-propanol)	May interfere with hydrogen bonds because these solvents can also form hydrogen bonds or can disrupt hydrophobic interactions. Quickly denature proteins in bacteria, killing them (e.g., the disinfectant action of 70% ethanol).
Strong acids and bases	Disrupt hydrogen bonds and salt bridges. Prolonged action leads to actual hydrolysis of peptide bonds.
Salts of heavy metals (e.g., salts of Hg^{2+}, Ag^{+}, Pb^{2+})	Cations combine with SH groups and form precipitates. (These salts are all poisons.)

charge is easily changed by changing the pH. For example, CO_2^- groups become electrically neutral CO_2H groups when they pick up protons as a strong acid is added.

Suppose that the net charge on a polypeptide is 1 −, and that one extra CO_2^- is responsible for it. When acid is added, we might imagine the following change, where the elongated shape is the polypeptide system.

$$\text{(polypeptide: } NH_3^+,\ CO_2^-,\ CO_2^-\text{)} + H_3O^+ \longrightarrow \text{(polypeptide: } NH_3^+,\ CO_2^-,\ CO_2H\text{)} + H_2O$$

net charge: 1− net charge: 0

Now the product polypeptide is isoelectric and neutral. On the other hand, a polypeptide might have a net charge of 1 +, caused by an excess of one NH_3^+ group. The addition of OH^- can cause the polypeptide to become isoelectric and neutral.

Like each amino acid, each protein has a characteristic pH, its isoelectric point, at which its net charge is zero and at which it cannot migrate in an electric field. One major significance of this is that polypeptide molecules that are neutral can aggregate and clump together to become particles of enormous size that simply drop entirely out of solution (see Fig. 10.11). *A protein is least soluble in water when the pH equals the protein's isoelectric point.* Therefore, whenever a protein must be *in solution* to work, as is true for many enzymes, the pH of the medium must be kept away from the protein's isoelectric point. Buffers in body fluids have the task of ensuring this.

An example of the effect of pH on solubility, mentioned earlier, is given by casein, milk protein, whose pI value is 4.7. As milk turns sour, the pH drops from its normal value of 6.3–6.6 to 4.7, and more and more casein molecules become isoelectric, denature, clump together, and separate as curds. As long as the pH of milk is something *other than* the pI for casein, the protein remains colloidally dispersed.

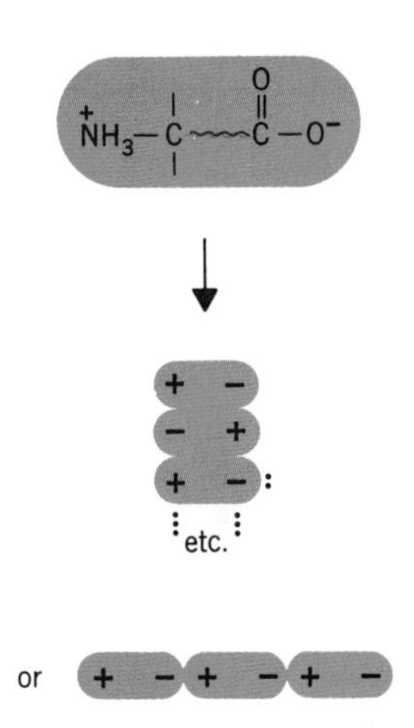

FIGURE 10.11
Several isoelectric protein molecules (top) can aggregate into very large clusters that no longer dissolve in water.

10.8

CELL MEMBRANES REVISITED—GLYCOPROTEIN COMPONENTS

Glycoproteins provide "recognition sites" on the surfaces of cell membranes.

In Section 9.5 we introduced the general features of cell membranes, giving particular attention to their lipid components. With the knowledge about protein structure gained in this chapter and about carbohydrates learned in Chapter 8, we can take a second, deeper look at membranes. We'll also learn how glycoproteins are involved in giving shock absorbancy to the cartilage that cushions bone joints.

Membrane Proteins Help to Maintain Concentration Gradients If the cell membrane were an ordinary dialyzing membrane, any kind of small molecule or ion could move freely back and forth between the interior of the cell and any surrounding fluid. The working of cells, however, demands that only some things be let in from this fluid and that others be let out. This means that many concentration *gradients* occur between the cell interior and whatever fluid is outside. A **gradient** is an unevenness in the value of some physical property throughout a system. A *concentration gradient* exists in a solution, for example, when one region of the solution has a higher concentration of solute than another, such as the variations in sugar concentration in unstirred coffee just after you add sugar. Gradients are generally unstable systems compared to systems with the same components but more thoroughly mixed up. Because of the random motions of ions and molecules in liquids and gases, the natural tendency is for gradients eventually to disappear in such media and for solute concentrations to become uniform. Yet living cells maintain a number of gradients.

Ion	Concentration (mmol/L) Plasma	Cells
Na^+	135–145	10
K^+	3.5–5.0	125

As the data in the table in the margin show, both sodium ions and potassium ions have quite different concentrations in the fluids on the inside of a cell as compared to fluids on the outside. Thus between the inside and the outside of a cell there is a considerable concentration gradient for both of these ions. *This gradient must be maintained against nature's spontaneous tendency to remove concentration gradients.*

Here is where some of the proteins in cell membranes carry out a vital function. One kind of assembly of membrane protein molecules can move sodium ions against their gradient. When too many sodium ions move to the inside of a cell, they are "pumped" back out to the external fluid by a special molecular machinery called the *sodium–potassium pump.* The same pump can move potassium ions back inside a cell. This movement of any solute *against* its concentration gradient requires chemical energy and is an example of **active transport.** Other reactions in cells supply the chemical energy that lets it work.

Gap Junctions Enable Substances to Move Directly from One Cell to Another In the cells of most tissues of multicelled organisms, membrane proteins provide a route for the *direct* movements of ions and molecules from one cell to another. These routes are through **gap junctions,** tubules fashioned from membrane proteins that "rivet" cells together (see Fig. 10.12). So many of such junctions occur in some tissues that the entire tissue is interconnected from within. Gap junctions in bone tissue, for example, enable bone cells at some distance from capillaries to receive nourishment and to remove wastes. Heart muscle is able to contract *synchronously* because gap junctions allow ions to move easily between cells. The gaps are large enough to allow certain ions, like Ca^{2+}, and certain relatively small molecules (up to formula masses of about 1200) to pass, but are not large enough for macromolecules like proteins and nucleic acids to get through.

Calcium ion appears to control the diameters of gap junctions. When the concentration of this ion is very low ($<10^{-7}$ *M*), the channels are fully open. As $[Ca^{2+}]$ increases, the gaps close, and they become completely shut when $[Ca^{2+}]$ reaches about 5×10^{-5} *M*. One consequence of this control is that if part of an interconnected mass of cells is injured, the closure of gaps limits the damage that might otherwise happen.

Some Proteins in Cell Membranes Are Receptors for Hormones and Neurotransmitters A **receptor molecule** is one whose unique shape enables it to fit only to the molecule of a compound that it is supposed to receive, its *substrate.* It is thus able to "recognize" the molecules of just one compound from among the hundreds whose molecules bump against it. This is roughly how specific hormones are able to find and stop only at the

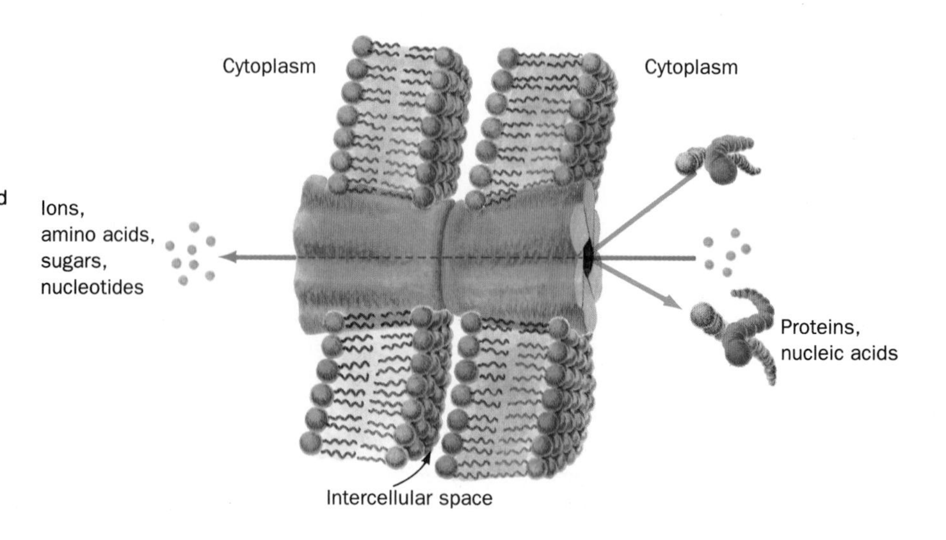

FIGURE 10.12
Gap junctions. A protein-fashioned channel (in brown) between two cells enables some small particles to pass directly from one cell to another. The blue spheres each with two wavy tails are phospholipid molecules of the cell membranes.

SPECIAL TOPIC 10.2

MIFEPRISTONE (RU 486)—RECEPTOR BINDING OF A SYNTHETIC ANTIPREGNANCY COMPOUND

When released from an ovarian follicle, the natural female hormone, progesterone, acts to prepare the system for pregnancy both by inhibiting further production of ova (egg cells) and by preparing the uterus for the implantation of the fertilized ovum. The action of progesterone involves the binding of its molecules to protein receptors within cells of the lining of the uterus (the endometrium).

Antiprogestins (Table 9.3) mimic the work of progesterone in that they cause a pseudopregnant state ("false pregnancy state") and so suppress the production of ova. Fertilization cannot occur without an ovum, of course, and so the synthetic progestin-containing medications are birth control pills.

Progesterone

Mifepristone (RU 486)

RU 486 (RU for Roussel-Uclaf, a French pharmaceutical company) was first prepared in 1980. It was soon discovered to be a strong binder to the progesterone binding sites on the receptor protein for progesterone. This action blocks the normal action of progesterone, and RU 486 is thus an *antiprogesterone.* If RU 486 is taken during the five to six day postcoital period (the period immediately following intercourse), its blocking action prevents pregnancy by suppressing the implantation of a fertilized ovum in the uterus. If used within 72 hours after unprotected intercourse, its failure rate is very low. It can thus be used as a "morning after" pill.

RU 486, followed by the use of two prostaglandin-like compounds, also induces abortion. Thus RU 486 is also an abortifacient (an abortion inducing agent). Under medical supervision, the use of RU 486 for this purpose has a 96% success record. The failures include continued pregnancy, only partial expulsion of the fetus, and the need for procedures to stem uterine bleeding.

That RU 486 "prevents pregnancy" is a controversial statement, because some view the onset of pregnancy as occurring at the moment of fertilization. Others regard pregnancy as not starting until the fertilized ovum has become implanted in the uterus. The controversy thus involves the question, "When does *pregnancy* begin?" and the not identical question, "When does *human life* begin?" Around these questions have surged some of the stormiest waters of the prolife–prochoice controversy.

cells where they are meant to stop, by being able to fit to unique receptors. An antiprogesterone birth control agent, RU 486, acts as a hormone-mimic, binding to a receptor protein (see Special Topic 10.2).

This fitting to a membrane protein also helps neurotransmitter molecules to become quickly attached to the right spot on the membrane of the next nerve cell after moving from one nerve cell to the next across the very narrow gap between them. Once a receptor molecule in a cell membrane accepts its unique substrate, further biochemical changes occur. An enzyme or a gene in the cell might be activated, for example.

■ Neurotransmitters are organic molecules that help carry nerve signals from the end of one nerve cell to the beginning of the next.

Membrane Proteins Are Glycoproteins Essentially all cells are "sugar-coated." The sugars are not the ordinary sugars of nutrition but are related to monosaccharides and *oligosaccharides*. The latter are carbohydrates that can be hydrolyzed to three or more — up to a few dozen — monosaccharide molecules. Some of the membrane carbohydrate units are

■ We learned in Section 9.3 that cerebrosides are glycolipids.

covalently joined to lipids—the **glycolipids**—of the membrane. Other membrane carbohydrates are bound to proteins, forming the **glycoproteins** of membranes (see Fig. 10.13). Most proteins are glycoproteins; several thousand have been identified and the list is growing.

The oligosaccharides of glycoproteins generally contain nitrogen or sulfur as additional elements. Nitrogen is present as an amine or an amide group. Sulfur occurs in SO_3^- groups joined to O atoms of carbohydrate systems or to N atoms of amino sugars. Amino sugars are those in whose molecules an OH group is replaced by an NH_2 group. Perhaps the most common amino sugar built into glycoproteins is D-glucosamine in the form of its *N*-acetyl derivative—*N*-acetyl-D-glucosamine. It and systems like it are usually joined to a polypeptide at an asparagine residue by what is called an *N-link*.

■ The *N* signifies that the acetyl group is attached to *nitrogen.*

N-Link

Polypeptide "backbone"

Asparagine residue of the polypeptide

β-D-Glucosamine

N-Acetyl-β-D-glucosamine

N-Acetyl-β-D-glucosamine unit N-linked to a polypeptide unit

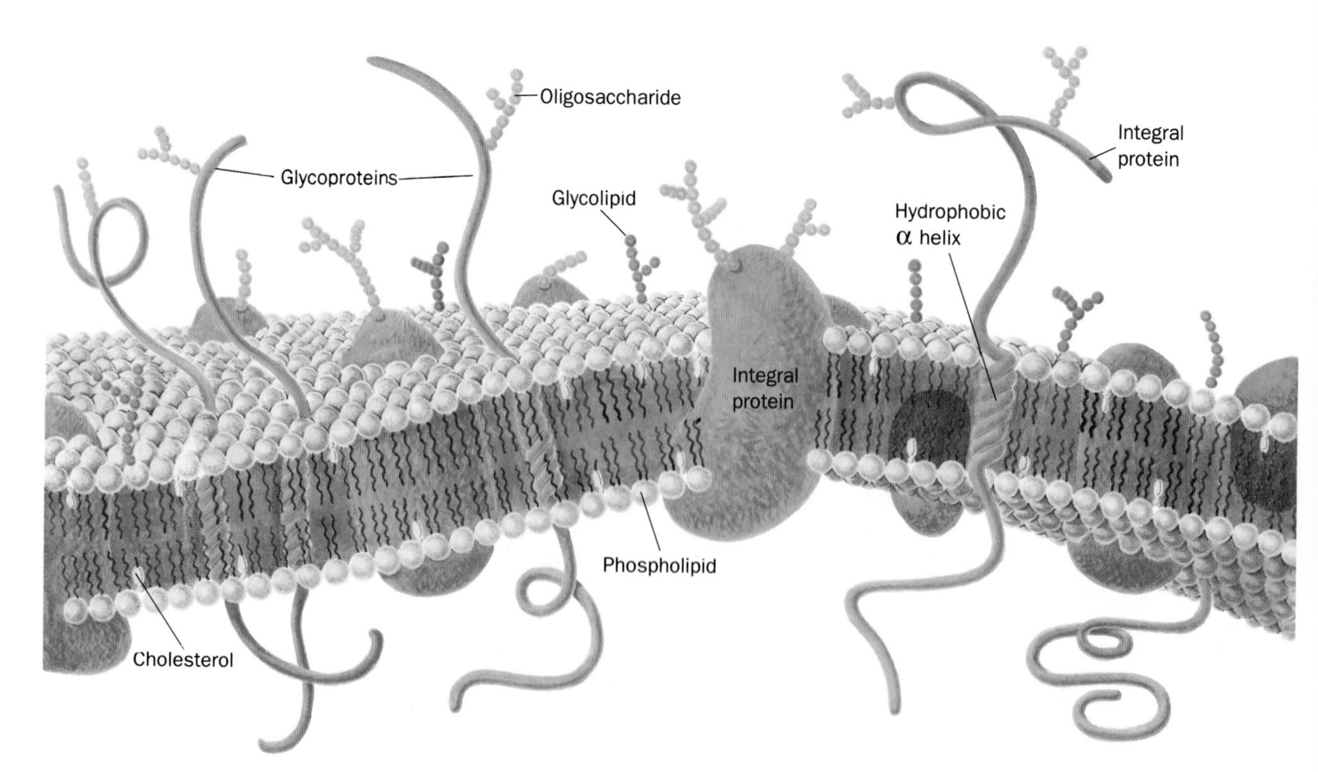

FIGURE 10.13
Glycoproteins as structural units in a cell membrane. The blue spheres each with two wavy tails are phospholipid molecules. Shown in yellow are cholesterol molecules. Chains of green beads represent glycolipids. Chains of yellow beads attached to polypeptides and proteins (in orange) are oligosaccharide units.

TABLE 10.3
Glycosaminoglycans—Their Repeating Disaccharide Units[a]

Hyaluronate monomer system	A component of ground substance particularly in connective tissue, in fluid that lubricates joints (synovial fluid), and the vitreous humor of the eye. Depending on the location, from 250 to 25,000 of these disaccharide units are joined into the polymer. (Note the N-acetyl-D-glucosamine unit on the right.)
Chondroitin-4-sulfate monomer system	A major component of cartilage and other connective tissues (after the Greek *chondros,* cartilage). There is also a 6-sulfate relative of this system. (Note that the unit on the right is derived from D-galactose, not D-glucose.)
Heparin monomer system	Occurs not in connective tissue, like the above, but in mast cells, cells that line the walls of arteries, particularly in the lungs, liver, and skin. Heparin inhibits the formation of blood clots, and its release from mast cells when an injury occurs prevents clotting from going too far. Heparin is widely used postsurgically to control clotting.

[a] In order to show the ring structures of the individual sugar units in their conventional array, distortions of connecting bonds must sometimes be tolerated in such structures.

One General Kind of Glycoprotein Gives Resiliency to Cartilage Table 10.3 gives the structures and chief uses of three of the several members of carbohydrate polymers known as the *glycosaminoglycans*[1] or *mucopolysaccharides.* Two of the glycosaminoglycans form the gel-like material called **ground substance** found in cartilage and other extracellular spaces and in which fibers of collagen and another fibrous protein called *elastin* are embedded. The fibers give the *tensile strength* to cartilage, the strength to withstand stretching tension without breaking. The ground substance holding the fibers gives *flexibility* and shock absorbancy to cartilage for reasons described next.

With molecules having so many polar groups with hydrogen bonding ability, it isn't surprising that glycosaminoglycan molecules are "sticky," and that the substances are thick, slimy, viscous materials resembling mucus. It also is not surprising that their molecules can attract and hold large numbers of water molecules of hydration. This is what gives ground

[1] In the term glycosaminoglycan, "glycos-" is from glycose, the generic name for monosaccharides; "glycan" is the generic name of all polysaccharides. The "gly-" part of both glycose and glycan is replaced by the prefix of a specific monosaccharide when naming specific polysaccharides. Thus starch is a *glucan* because it is a polymer of glucose.

substance the spongy, resilient nature needed for the cartilage in bone joints. As cartilage is squeezed during bumps and jolts, water is forced out. When the pressure is released, water rushes back in. In fact, this "tidal flow" of water is what carries metabolic wastes away and brings in nutrients to cartilage tissue, a tissue that lacks blood vessels. The movement of water caused by flexing the joints makes it easy to understand why the cartilage in joints becomes somewhat fragile, even brittle, during long periods of no exercise.

■ In carbohydrate chemistry, amylose and cellulose are examples of "straight chain" polysaccharides; amylopectin and glycogen are "branched chain."

Multifunctional Monosaccharide Units Have Numerous Ways of Combining into Oligosaccharides Glycoproteins include a large subfamily of N-linked oligosaccharides whose structures include several monosaccharides in both "straight chain" and "branched chain" connections. Many of the linkages between monosaccharide units involve oxygen atoms at ring positions not used for linking purposes by the nutritionally important disaccharide or polysaccharides that we studied in Chapter 8. Table 10.3, in fact, illustrates ($1 \rightarrow 3$) linkages in the first two structures. In Chapter 8 we saw mostly ($1 \rightarrow 4$) linkages, but ($1 \rightarrow 6$) linkages are also present in amylopectin and glycogen. The possibilities for a number of ($1 \rightarrow n$) linkages having varying geometries give living systems far more options for joining monosaccharide units than for joining amino acid units. Two identical amino acids, for example, can be joined in only one way by a peptide bond, but two identical monosaccharides can be linked to form 11 different disaccharides. Someone has calculated that only four *different* monosaccharide units can be linked in over 35,000 unique tetrasaccharides. *Thus an almost unlimited variety of structurally different oligosaccharides is available for making glycoproteins.* This momentous fact at the molecular level of life is behind the almost incredible spectrum of biological properties, observed throughout the living world. Different oligosaccharides, N-linked to polypeptides in a cell membrane, make possible the unique abilities of the membranes not only of species but also of individuals within species to interact discriminately among substances cruising near cells. The existence of the ABO blood groups; the inability of the sperm of one species to fertilize the egg of any other; the action of bacteria normally at just one tissue and not at any others; the existence of bacterial infections in animals that do not affect humans (and vice versa); the ability of hormones to be snared only by their own "target" tissues; these and many other properties can be traced to the occurrence of so many different oligosaccharides residing on the surfaces of cell membranes.

■ The varying glycoproteins that provide this uniqueness are called **glycoforms.**

The Linkage of Oligosaccharides to Proteins Largely Occurs at Protein Surfaces, Not in Their Interiors The N-links of oligosaccharides to polypeptides occur most often where the polypeptide strand is following a bend between segments of secondary structure, like β-sheets or α-helices. The oligosaccharide units, therefore, have little if any direct effect on tertiary structure but project, instead, from protein surfaces. This is why we could say near the start of this section that cell membranes are "sugar coated." The "sugar" consists of oligosaccharide units extending into the surrounding spaces from the glycoproteins that make up parts of cell membranes. Oligosaccharide units contribute to the adhesion between cells, and they have critical functions in all of the activities that depend on the "recognition" of hormones and neurotransmitters by a cell. In the next chapter we'll learn about the mechanism of recognition at the molecular level of life.

■ The oligosaccharide units also are involved in the actions of toxins, viruses, and bacteria.

10.9

CLASSES OF PROTEINS

Three criteria for classifying proteins are their solubility in aqueous systems, their compositions, and their biological functions.

We began this chapter with hints about the wide diversities of the kinds and uses of proteins. Now that we know about their structures, we can better understand how so many types of proteins with so many functions are possible. The following three major classifications of proteins and several examples give substance to the chapter's introduction.

Proteins Can Be Classified According to Solubility When proteins are classified by their solubilities, two families are the **fibrous proteins** and the **globular proteins.**

Fibrous Proteins

1. **Collagens** occur in bone, teeth, tendons, skin, blood capillaries, cartilage, and some ligaments. When such tissue is boiled with water, the portion of its collagen that dissolves is called *gelatin*.
2. **Elastins,** which have elastic, rubber-like qualities, are also in cartilage and are found in stretchable ligaments, the walls of large blood vessels like the aorta, the lungs, and the necks of grazing animals. Elastin, like collagen, is rich in glycine residues and proline, but not in hydroxyproline. Elastin chains are cross-linked by covalently bonded units that are largely responsible for elastin's elasticity.
3. **Keratins** occur in hair, wool, animal hooves, horns, nails, porcupine quills, and feathers. The keratins are rich in disulfide links, which contribute to the unusual stabilities of these proteins to environmental stresses.
4. **Myosins** are the proteins in contractile muscle.
5. **Fibrin** is the protein of a blood clot. During clotting, fibrin forms from its precursor, fibrinogen, by an exceedingly complex series of reactions.

■ When meat is cooked, some of its collagen changes to gelatin, which makes the meat easier to digest.

■ Elastin is not changed to gelatin by hot water.

Globular Proteins Globular proteins are soluble in water or in water that contains salts.

1. **Albumins** are present in egg white and in blood. In the blood, the albumins are buffers, transporters of water-insoluble molecules of lipids or fatty acids, and carriers of metal ions, like Cu^{2+} ions, that are insoluble in aqueous media at pH values higher than 7.
2. **Globulins** include antibodies, factors of the body's defenses against diseases. In addition, enzymes, many transport proteins, and receptor proteins are globulins.

Proteins Can Be Classified According to Biological Function Perhaps no other system more clearly dramatizes the importance of proteins than classifying them by their biological function.

1. Enzymes. The biological catalysts.
2. Contractile muscle. With stationary filaments, myosin, and moving filaments, actin.
3. Hormones. Such as growth hormone, insulin, and others.
4. Neurotransmitters. Such as the enkephalins and endorphins.
5. Storage proteins. Those that store nutrients that the organism will need such as seed proteins in grains, casein in milk, ovalbumin in egg white, and ferritin, the iron-storing protein in human spleen.
6. Transport proteins. Those that carry things from one place to another. Hemoglobin and the serum albumins are examples already mentioned. Ceruloplasmin is a copper-carrying protein.
7. Structural proteins. Proteins that hold a body structure together, such as collagen, elastin, keratin, and glycoproteins in cell membranes.
8. Protective proteins. Those that help the body to defend itself. Examples are the antibodies and fibrinogen.
9. Toxins. Poisonous proteins. Examples are snake venom, diphtheria toxin, and clostridium botulinum toxin (a toxic substance that causes some types of food poisoning).

SUMMARY

Amino acids About 20 α-amino acids supply the amino acid residues that make up a polypeptide. The molecules of all but one (glycine) are chiral and in the L-family. In the solid state or in water at a pH of roughly 6 to 7, amino acids exist as dipolar ions or zwitterions. Isoelectric points are the pH values of solutions in which amino acid (or protein) molecules are isoelectric. For amino acids without acidic or basic sidechains, the pI values are in the range of 6 to 7. Amino acids with CO_2H groups on sidechains have lower pI values. Those with basic sidechains have higher pI values. Several amino acids have hydrophobic sidechains, but the sidechains in others are strongly hydrophilic. The SH group of cysteine opens the possibility of disulfide cross-links between or within polypeptide units.

Polypeptides Amino acid residues are held together by peptide (amide) bonds, so the repeating unit in polypeptides is —NH—CH—CO—. Each amino acid residue has its own sidechain. This repeating system with a unique sequence of sidechains constitutes the primary structure of a polypeptide.

Once the primary structure is fashioned, the polypeptide coils and folds into higher features—secondary and tertiary—that are stabilized largely by hydrophobic interactions and hydrogen bonds. The most prominent secondary structures are the α-helix—a right-handed helix—and the β-pleated sheet. Individual polypeptides in collagen, which has an abundance of glycine, proline, and hydroxylated proline residues, are in a left-handed helix. Disulfide bonds form from SH groups on cysteine residues as many proteins assume their tertiary structure.

Proteins Many proteins consist just of one kind of polypeptide. Many others have nonprotein, organic groups—prosthetic groups—or metal ions. And still other proteins—those with quaternary structure—involve two or more polypeptides whose molecules aggregate in definite ways, stabilized by hydrophobic interactions, hydrogen bonds, and salt bridges. Thus the terms *protein* and *polypeptide* are not synonyms, although for some specific proteins they turn out to be.

Because of their higher levels of structure, proteins can be denatured by agents that do nothing to peptide bonds. A few denatured proteins can be renatured, but this is uncommon. The acidic and basic sidechains of polypeptides affect protein solubility, and when a protein is in a medium whose pH equals the protein's isoelectric point, the substance is least soluble. The amide bonds (peptide bonds) of proteins are hydrolyzed during digestion.

Membrane proteins—glycoproteins Incorporated into the lipid bilayer membranes of cells are proteins (and lipids) with attached oligosaccharide units of widely varying structure. Some of the proteins of a cell membrane provide conduits by which active transport processes can maintain concentration gradients. Other proteins provide gap junctions for direct movements, cell-to-cell, of certain dissolved species. The oligosaccharides of the membrane proteins stick out away from the membrane surface. They do not appear to affect the overall shapes of their attached polypeptides, and they serve as cell-recognition features for molecules moving near the cell. Certain oligosaccharides, the glycosaminoglycans, make up ground substance, which gives elasticity and shock absorbancy to cartilage.

REVIEW EXERCISES

The answers to Review Exercises whose numbers are in color are found in Appendix A. The answers to the other Review Exercises are found in the Study Guide that accompanies this book. The more challenging questions are marked with asterisks.

Amino Acids

10.1 One of the following structures is *not* of an amino acid on the list of standard 20. Which one is not on the list? How can you tell without looking at Table 10.1?

A: $NH_2CH_2CH_2CH_2CH_2CHCO_2^-$ with NH_3^+ on the CH

B: $^+NH_3CH_2CHCO_2^-$ with CH_3 on the CH

C: $^+NH_3CHCO_2^-$ with CH_2CO_2H on the CH

10.2 The following amino acid is on the standard list of 20.

$^+NH_3CHCO_2^-$ with $CH_2CH(CH_3)_2$ on the CH

(a) What part of its structure would be its amino acid *residue* in the structure of a polypeptide? (Write the structure of this residue.)

(b) With the aid of Table 10.1, write the name and the three-letter symbol of this amino acid.

(c) Is its sidechain hydrophobic or hydrophilic?

10.3 What structure will nearly all the molecules of glycine have at a pH of about 1?

10.4 What structure will most of the molecules of alanine have at a pH of about 12?

10.5 Pure alanine does not melt, but at 290 °C it begins to char and decompose. However, the ethyl ester of alanine, which has a free NH_2 group, has a low melting point, 87 °C. Write the structure of this ethyl ester, and explain this large difference in melting point.

***10.6** The ethyl ester of alanine (Review Exercise 10.5) is a much stronger base—more like ammonia—than alanine. Explain this.

10.7 Which of the following amino acids has the more hydrophilic sidechain? Explain.

A: $^+NH_3CHCO_2^-$ with $CH_2CH_2CH_2NHCNH_2$ on the CH (C=NH)

B: $^+NH_3CHCO_2^-$ with $CH_3CHCH_2CH_3$ on the CH

10.8 Which of the following amino acids has the more hydrophobic sidechain? Explain.

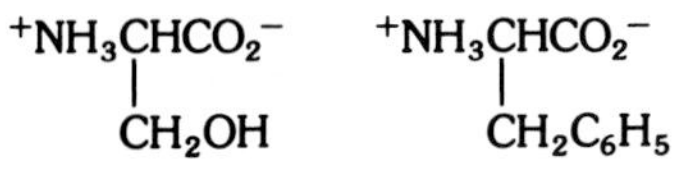

$$\underset{\textbf{A}}{\overset{+}{N}H_3\underset{|}{C}HCO_2^- \quad CH_2OH} \qquad \underset{\textbf{B}}{\overset{+}{N}H_3\underset{|}{C}HCO_2^- \quad CH_2C_6H_5}$$

A: $^+NH_3CHCO_2^-$ with sidechain CH_2OH

B: $^+NH_3CHCO_2^-$ with sidechain $CH_2C_6H_5$

10.9 Glutamic acid can exist in the following form.

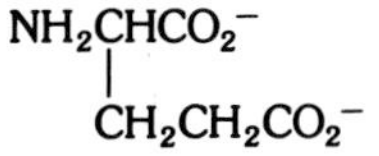

$NH_2CHCO_2^-$ with sidechain $CH_2CH_2CO_2^-$

(a) Would this form predominate at a pH of 2 or a pH of 10? Explain.

(b) To which electrode, the anode or the cathode—or to neither—would aspartic acid in this form migrate in an electric field?

10.10 When it is said that a substance is poorly soluble in water because of a *hydrophobic interaction*, what does "hydrophobic interaction" mean?

10.11 What kind of a reactant is required to convert cysteine into cystine: an acid, base, oxidizing agent, or reducing agent?

***10.12** Write two equilibrium equations that show how glycine, in its isoelectric form, can serve as a buffer.

10.13 Complete the following Fischer projection formula to show correctly the absolute configuration of L-serine.

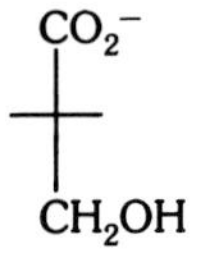

CO_2^- (top), CH_2OH (bottom) of Fischer cross

Polypeptides

***10.14** Each of the following structures has an amide linkage. Each can be hydrolyzed to glycine and lysine. The amide linkage in each of the two structures, however, cannot properly be called a *peptide bond*. This is true of which structure? Why?

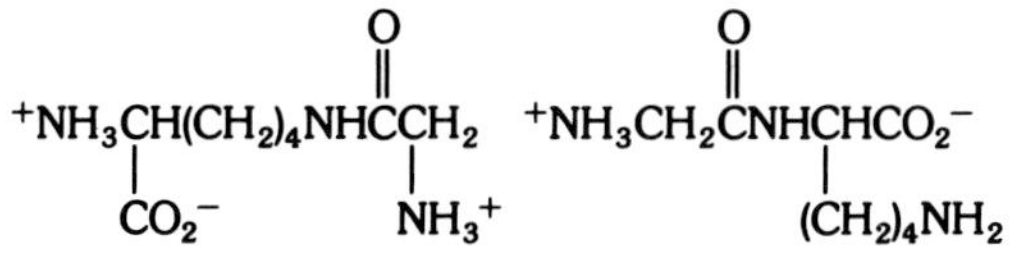

A: $^+NH_3CH(CH_2)_4NHCCH_2$ (C=O on the amide carbon; CO_2^- on the first CH; NH_3^+ on the terminal CH_2)

B: $^+NH_3CH_2CNHCHCO_2^-$ (C=O on the amide carbon; $(CH_2)_4NH_2$ on the CH)

10.15 Write both the conventional and the condensed structure (three-letter symbols) of the dipeptides that can be made from lysine and cysteine.

10.16 What are the condensed structures of the dipeptides that can be made from glycine and glutamic acid? (Do not use the three-letter symbols.)

10.17 Using three-letter symbols, write the structures of all of the tripeptides that can be made from lysine, glutamic acid, and cysteine.

10.18 Write the structures in three-letter symbols of all of the tripeptides that can be made from glycine, cysteine, and alanine.

10.19 What is the conventional structure of Val-Ile-Phe?

***10.20** Write the conventional structure for Val-Phe-Ala-Gly-Leu.

***10.21** Write the conventional structure for Asp-Lys-Glu-Thr-Tyr.

***10.22** Compare the sidechains in the pentapeptide of Review Exercise 10.20 (call it **A**) with those in the following, which we can call **B**

Lys-Glu-Asp-Thr-Ser

(a) Which of the two, **A** or **B**, is the more hydrocarbon-like?

(b) Which is probably more soluble in water? Explain.

***10.23** Compare the sidechains in the pentapeptide of Review Exercise 10.21, which we'll label **C**, with those in Phe-Leu-Gly-Ala-Val, which we can label **D**. Which of the two would tend to be less soluble in water? Explain.

***10.24** If the tripeptide Gly-Cys-Ala were subjected to mild oxidizing conditions, what would form? Write the structure of the product using three-letter symbols.

10.25 What is meant by a *peptide group?* Describe its geometry.

10.26 What atom or groups are *trans* to each other in a trans-planar peptide group?'

Higher Levels of Protein Structure

10.27 Which *level* of polypeptide complexity concerns the molecular "backbone" and the sequence of sidechains?

10.28 What is meant by *native* protein?

10.29 To what level of protein complexity is the disulfide bond normally assigned?

10.30 The disulfide bond is a *covalent* bond. Why isn't it assigned to the primary level of polypeptide structure?

10.31 An enzyme consists of two polypeptide chains associated together in a unique manner. To what level of protein structure is this detail assigned?

10.32 Does the trans-planar nature of the peptide group enlarge or reduce the *range of geometrical options* available to a polypeptide?

10.33 Describe the specific geometrical features of an α-helix structure. What force of attraction stabilizes it? Between what two kinds of sites in the α-helix does this force operate? How do the sidechains become positioned in the α-helix?

10.34 Give a brief description of the secondary structure of an individual polypeptide strand in collagen.

10.35 What function does ascorbic acid (vitamin C) perform in the formation of strong bones?

10.36 Describe the structure and geometry of tropocollagen. How is tropocollagen made into a collagen fibril?

10.37 Bridges between the polypeptide strands in collagen have what principal feature: a hydrophobic interaction, an electrostatic attraction (salt bridge), a disulfide system, or some other kind of covalent linkage?

10.38 What specific force of attraction stabilizes a β-pleated sheet? Where do the sidechains take up positions?

10.39 Does an α-helix or a β-sheet describe the *entire* secondary structure of a polypeptide? If not, how do these features occur?

10.40 What factors affect the bending and folding of α-helices in the presence of an aqueous medium?

10.41 What is meant by a salt bridge?

10.42 When is the disulfide bond normally put into place during the formation of a protein?

10.43 In what way does the hemoglobin represent a protein with quaternary structure (in general terms only)?

10.44 How do myoglobin and hemoglobin compare (in general terms only)?
(a) Structurally—at the quaternary level
(b) Where they are found in the body
(c) In terms of their prosthetic group(s)
(d) In terms of their functions in the body

Properties of Proteins

***10.45** What products form when the following polypeptide is completely digested?

$$^{+}NH_3CHCONHCHCONHCHCONHCHCONHCH_2CO_2^{-}$$

(side chains, in order: CH_2OH, CH_3, $CH(CH_3)_2$ [drawn as CH with H_3C and CH_3], $(CH_2)_4NH_2$)

10.46 Explain why a protein is least soluble in an aqueous medium that has a pH equal to the protein's pI value.

10.47 What is the difference between the *digestion* and the *denaturation* of a protein?

10.48 Some proteins can be denatured by a reducing agent but then completely renatured by a mild oxidizing agent. What functional groups are involved?

Cell Membranes

10.49 What is meant by "gradient" in the term *concentration gradient?*

10.50 Which has the higher level of Na^{+}, plasma or cell fluid?

10.51 Does cell fluid or plasma have the higher level of K^{+}?

10.52 In which fluid, plasma or cell fluid, would the level of sodium ion increase if the sodium ion gradient could not be maintained?

10.53 What does the sodium–potassium pump do?

10.54 What does "active" refer to in the term *active transport?*

10.55 What is a *gap junction* and what services does it perform?

10.56 The concentration of what species appears to control the size of the opening of a gap junction?

Glycoproteins and Cell Membranes

10.57 What does the prefix "glyco-" refer to in *glycoprotein?*

10.58 In general terms only, in what structural way does an oligosaccharide differ from a mono- or a disaccharide? From a polysaccharide?

10.59 The term "glycan" refers to what?

10.60 A "D-glucosaminoglucan" would be made of what monomer?

10.61 What is *ground substance?*

10.62 What kinds of substances provide tensile strength to cartilage and what substance gives cartilage its resiliency and shock-absorbing properties?

10.63 What role does the hydrogen bond play in the ability of cartilage tissue to carry out its functions?

10.64 What structural fact about monosaccharides (including the amino sugars) makes possible the huge variety of possible oligosaccharides?

Type of Proteins

10.65 What experimental criterion distinguishes between fibrous and globular proteins?

10.66 What is the relationship between collagen and gelatin?

10.67 How are collagen and elastin alike? How are they different?

10.68 What experimental criterion distinguishes between the albumins and the globulins?

10.69 What is fibrin and how is it related to fibrinogen?

10.70 What general name can be given to a protein that carries a carbohydrate molecule?

Sickle-Cell Anemia (Special Topic 10.1)

10.71 What is the primary *structural* fault in the hemoglobin of sickle-cell anemia?

10.72 What happens in blood cells in sickle-cell anemia that causes their shapes to become distorted?

10.73 What problems are caused by the distorted shapes of the red cells?

Mifepristone (RU 486) (Special Topic 10.2)

10.74 What is meant by a "receptor protein?"

10.75 In general terms only, how does RU 486 work in the post-coital period?

Additional Exercises

10.76 When an oligosaccharide unit is cleaved from its glycoprotein, the overall *shape* of the protein section is largely unchanged. Explain.

10.77 Write the structure of a pentapeptide that would hydrolyze to give only alanine.

***10.78** Consider the following structure.

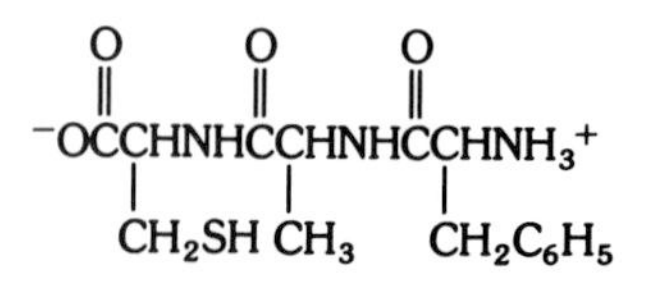

(a) If a polypeptide were *partially* hydrolyzed, could a molecule of this structure possibly form in theory? Explain.
(b) What is the three-letter symbol of the N-terminal residue?
(c) How would the structure of this compound be represented using three-letter symbols and following the rules for writing such a structure?
(d) Would a mild reducing agent have any affect on this compound? If so, write the structure of the product.

ENZYMES, HORMONES, AND NEUROTRANSMITTERS

11

In the most tense situations requiring the keenest reflexes, many of the body's hormones participate in assuring the needed energy supply. You can be sure that the adrenaline is flowing in this surfer, catching the curl off the northern coast of Oahu, Hawaii.

11.1 ENZYMES

Enzymes are biological catalysts whose activities often depend on cofactors made from B vitamins or metal ions.

Virtually all enzymes are proteins. A few have been discovered that are made of RNA, but they are exceptions. In this chapter we will only study enzymes that are proteins.

■ All enzyme molecules and most substrate molecules are chiral; they have handedness.

Enzymes Are Very Specific for the Substrates They Accept Enzyme specificity is one property that a theory of enzyme action has to explain. *In vivo,* a given enzyme acts on just one substrate. The digestive enzymes are even specific about which part of a substrate they hydrolyze. Some digestive enzymes specialize, for example, in hydrolyzing peptide bonds adjacent to particular sidechains on amino acid residues as the enzyme works to split long polypeptides into shorter molecules. Other digestive enzymes work exclusively to cleave C-terminal amino acid residues and still others to split off N-terminal residues. The starch hydrolyzing enzyme that catalyzes the hydrolysis of $\alpha(1 \rightarrow 4)$ oxygen bridges in starch does not work on the $\alpha(1 \rightarrow 6)$ bridges. Enzymes, as we said, are unusually specific *in vivo*.

Enzymes Work Best over Relatively Narrow Ranges of pH and Temperature Because nearly all enzymes are proteins, they are constrained by the general properties of proteins. The electrical charges on proteins, for example, vary and are sometimes differently distributed depending on the availability of acids or bases in the surrounding media, as we learned in Section 10.7. Changes in electric charge distribution can even affect the overall shape of a protein. We'll soon see that this explains why enzymes are active only in the pH range normal to their environment in the body. Pepsin, the gastric protein-digesting enzyme in the acidic medium of the stomach, is most active at a pH of about 2. Fumarase, an enzyme involved within cells in metabolizing acetyl units (from the breakdown of sugar or fatty acids), works best at a pH of just below 7 (see Fig. 11.1).

■ Many enzymes have been isolated for use in catalyzing chemical reactions *in vitro.*

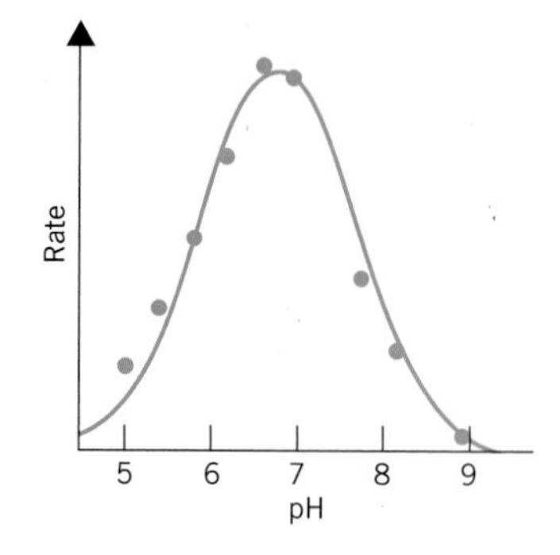

FIGURE 11.1
The effect of pH on the rate of an enzyme-catalyzed reaction, the enzyme being fumarase.

It is generally true that the rates of most reactions slow down as the temperature of the reacting mixture decreases and that reaction rates increase with increasing temperature. The rate-slowing effect of a colder temperature is also observed with enzyme catalysis. However, because enzymes are proteins, the opposite effect is not observed; the rates of enzyme-catalyzed reactions do *not* increase steadily with temperature. This is because enzymes (mostly proteins) are denatured by heat. Thus the efficiency of an enzyme will reach a peak over a small range of temperatures.

Enzymes Display Remarkable Rate Enhancements Enzymes, like all catalysts, affect the *rates* of reactions by providing a way for the reaction to occur that involves a lower energy barrier (a lower energy of activation) that the uncatalyzed reaction can take. Even small reductions of energy barriers can cause spectacular increases in rates. The enzyme carbonic anhydrase (CA) is an example. It catalyzes the interconversion of bicarbonate ion and protons with carbon dioxide and water:

$$CO_2 + H_2O \xrightleftharpoons{\text{carbonic anhydrase}} HCO_3^- + H^+$$

In actively metabolizing cells, where the supply of CO_2 is relatively high, this equilibrium shifts to the right. In blood that circulates into the lungs, where exhaling keeps the supply of CO_2 low, this equilibrium must shift to the left, *and the same enzyme participates in this change.* Each molecule of carbonic anhydrase aids in the conversion of 600,000 molecules of CO_2 *each second!* This is ten million times faster than the uncatalyzed reaction, which makes the speed of action of carbonic anhydrase among the highest of all known enzymes.

■ The equilibrium *must* shift to the right to make HCO_3^- and H^+ when the supply of CO_2 is high—a consequence of Le Châtelier's principle.

Enzymes Get Equilibria Established Extremely Rapidly In a chemical equilibrium, like that involving carbonic anhydrase, it is important to remind ourselves that the catalyst speeds up *equilibration*. It accelerates *both* the forward and the reverse reactions. Whether the equilibrium shifts to the right or to the left doesn't depend on the catalyst at all. It depends strictly on the inherent equilibrium constants; on the relative concentrations of reactants and products; on whether other reactions feed substances into the equilibrium or continuously remove them; and on the temperature. All the catalyst does is to make whatever shift is mandated by these conditions occur very rapidly.

Most Enzymes Consist of Polypeptides plus Cofactors The molecules of most enzymes include a nonpolypeptide component called a **cofactor.** The polypeptide is called the **apoenzyme,** but without the cofactor there is no enzymatic activity.

The cofactor of some enzymes is simply a metal ion. Zn^{2+} is the metal ion in carbonic anhydrase, for example. Fe^{2+} occurs in the cytochromes, a family of enzymes involved in biological oxidations. Most of the trace metal ions of nutrition are enzyme cofactors.

In other enzymes the cofactor is an organic molecule or ion called a **coenzyme.** Some enzymes have both a coenzyme and a metal ion cofactor.

B Vitamins Are Used to Make Coenzymes Thiamine diphosphate, a coenzyme with structure **1**, is a diphosphate ester of thiamine, a B vitamin, shown here in its fully ionized form.

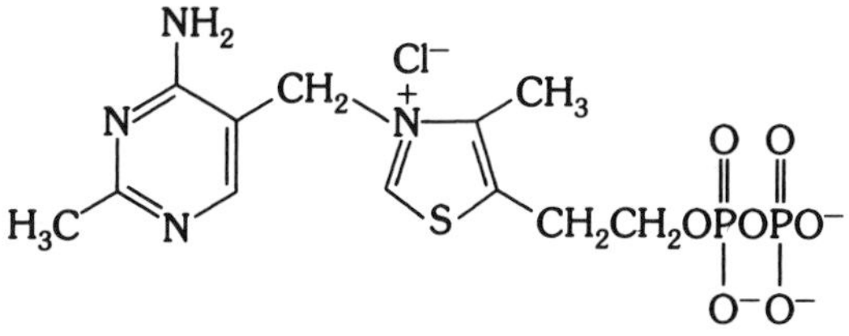

1

Thiamine diphosphate

■ When the diet is deficient in thiamine, often called vitamin B_1, a disease called *beriberi* results.

When pure, this coenzyme is a triprotic acid, but at the pH of body fluids it is ionized approximately as shown.

Nicotinamide's other name is *niacin.* A deficiency of this vitamin leads to *pellegra.*

Nicotinamide, another B vitamin, is part of the structure of nicotinamide adenine dinucleotide, **2a**, another important coenzyme. Mercifully, its long name is usually shortened to NAD^+ (or, sometimes, just NAD). The bottom half of the NAD^+ molecule is from adenosine monophosphate, AMP, a phosphate ester described on page 292. The upper half of NAD^+ is almost like its lower portion except that a molecule of nicotinamide has replaced the two-ring heterocyclic unit below it.

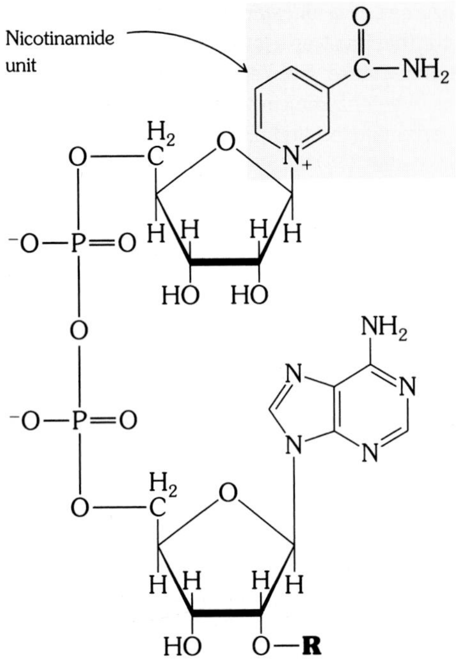

Nicotinamide unit

Riboflavin unit

Adenine unit

2

a NAD^+ **R**=H
b $NADP^+$ **R**=OPO_3^{2-}

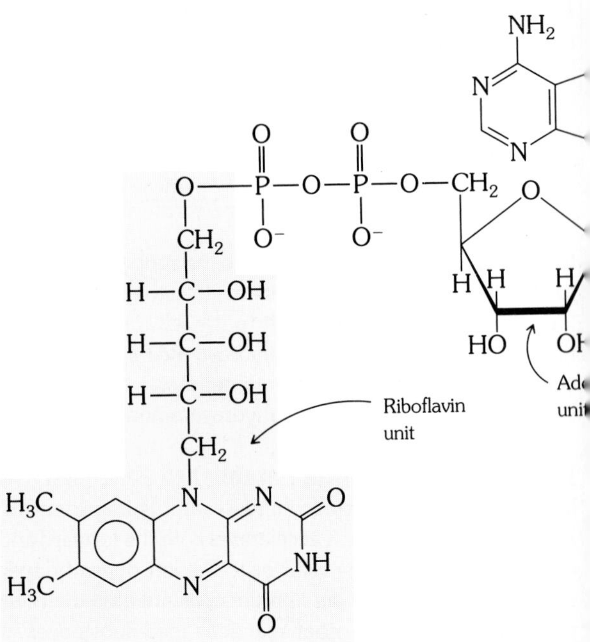

3

FAD

The P in $NADP^+$ refers to the extra phosphate ester unit.

Nicotinamide occurs in yet another important coenzyme, nicotinamide adenine dinucleotide phosphate, **2b**, a phosphate ester of NAD^+. Its name is usually shortened to $NADP^+$ (or, sometimes, just NADP). Both NAD^+ and $NADP^+$ are coenzymes in major biological oxidation–reduction reactions.

Quite often equations that involve enzymes with recognized coenzymes are written with the symbol of the coenzyme as a reactant or as a product. NAD^+, for example, is the cofactor for the enzyme that catalyzes the body's oxidation of ethyl alcohol to acetaldehyde. It serves, in fact, as the actual acceptor of the hydride ion, $H:^-$, given up by ethyl alcohol. The details are reserved to Special Topic 22.1, but the overall equation is written simply as follows.

$$\underset{\text{Ethanol}}{CH_3CH_2OH} + NAD^+ \longrightarrow \underset{\text{Ethanal}}{CH_3CH{=}O} + \underset{\text{Reduced form of } NAD^+}{NAD{:}H} + \underset{\text{Hydrogen ion (buffered)}}{H^+}$$

When the symbol of a coenzyme is used in an equation, remember that it stands for the entire enzyme that bears the coenzyme. In this reaction, the NAD^+ unit in the enzyme accepts $H:^-$ from the alcohol, and we can write this part of the reaction by the following equation.

$$NAD^+ + H:^- \longrightarrow NAD{:}H$$

By accepting the *pair of electrons* in $H:^-$, NAD^+ is reduced, and $NAD:H$ (usually written as NADH) is called the *reduced form* of NAD^+. $NADP^+$ can also accept hydride ion, and its reduced form is written as NADPH.

SPECIAL TOPIC 11.1

HOW NAD^+ AND FAD (OR FMN) PARTICIPATE IN ELECTRON TRANSFER

In the nicotinamide unit of both NAD^+ and $NADP^+$ there is a positive charge on the nitrogen atom, an atom that is also part of an aromatic ring. Despite how rich this ring is in electrons, the positive charge on N makes the ring an electron acceptor. Without the charge, the ring would otherwise be very similar to the ring in benzene, and it would strongly resist any reaction that disrupted its closed-circuit electron system. The positive charge on N, however, places the ring of the nicotinamide unit on an energy teeter-totter that the cell can tip either way without too much energy cost.

As seen in the following equations, when the donor of an electron pair, such as an oxidizable alcohol, arrives at the enzyme, the donor can transfer an electron pair (as $H{:}^-$) to the ring. Although the ring will temporarily lose its aromatic character, the positive charge of the ring nitrogen atom encourages this transfer of negative charge. In the next step of this metabolic pathway, the pair of electrons, still as $H{:}^-$, transfers to a riboflavin unit, also shown in the equations below. As this occurs, the ring of the nicotinamide unit recovers its stable, benzene-like nature.

Eventually, the pair of electrons winds up on an oxygen atom when, in the last step of this long metabolic pathway, oxygen that has been supplied by the air we breathe is reduced to water. We'll return to this in Chapter 14.

In the first reaction sequence, the alcohol is oxidized to an aldehyde as NAD^+ is reduced to NADH.

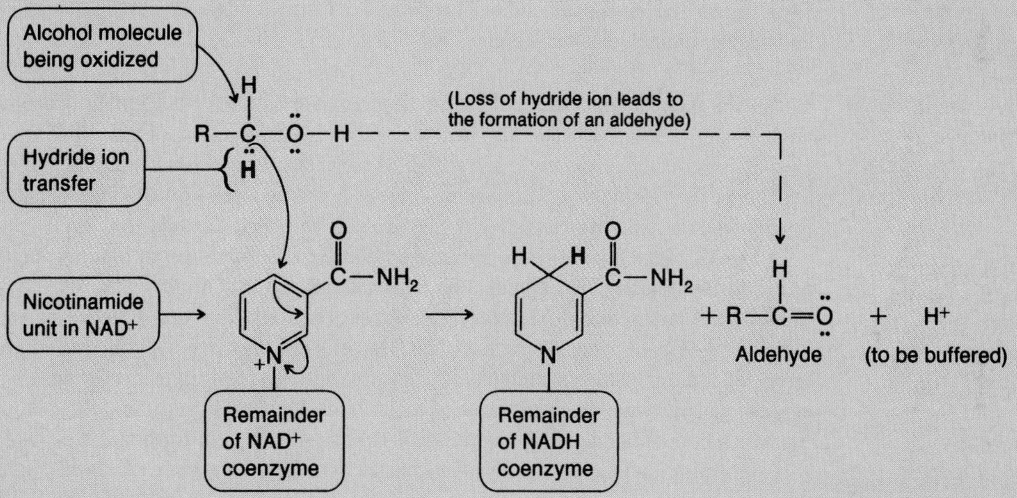

In the second reaction, NADH transfers $H{:}^-$ to FMN (or FAD). The reaction thus reoxidizes NADH to NAD^+ as FMN is reduced to $FMNH_2$ (or as FAD is reduced to $FADH_2$).

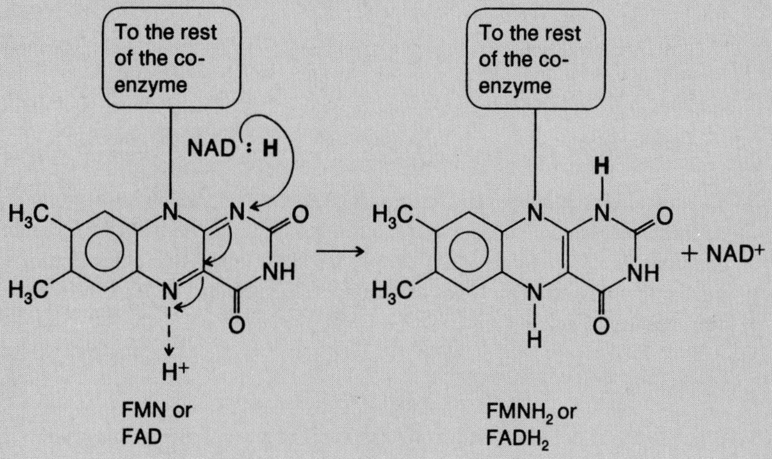

We learned earlier that we may not call something a catalyst unless it undergoes no *permanent* change, but the foregoing examples seem to contradict this. In the body, however, a reaction that alters an enzyme is followed at once by one that regenerates the enzyme. The NADH produced by the oxidation of ethyl alcohol, for example, is recovered in the next step. One of the possible enzymes for the next step has the cofactor FAD (for flavin adenine dinucleotide), **3**. FAD incorporates still another B vitamin, riboflavin. FAD can accept $H:^-$ from NAD:H, for example, change to $FADH_2$ (the second H is H^+ from the buffer), and so regenerate NAD^+. This is also shown in Special Topic 11.1, but the overall reaction is

■ Riboflavin is vitamin B_2.

$$NADH + FAD + H^+ \longrightarrow NAD^+ + FADH_2$$

The FAD-containing enzyme is, of course, now in its reduced form, $FADH_2$. $FADH_2$ passes on its load of hydrogen and electrons in yet another step and so is reoxidized and restored to FAD. The steps continue, but we'll stop here. The main points are that B vitamins are key parts of coenzymes, and that the catalytic activities of the associated enzymes directly involve the parts of the molecules contributed by these vitamins.

Flavin mononucleotide or FMN is a near relative of FAD that contains riboflavin. FMN is FAD minus the adenosine unit. The reduced form of FMN is $FMNH_2$, and FMN is also involved in biological oxidations.

■ Whenever we see *-ase* as a suffix in the name of any substance or type of reaction, the word is the name of an enzyme.

■ The digestive enzymes *trypsin, chymotrypsin,* and *pepsin,* all peptidases, have old (nonsystematic) names that do not end in *-ase.*

■ ATP is adenosine triphosphate, an important, high-energy triphosphate ester (see page 151).

Enzymes Are Named after Their Substrates or Reaction Types Nearly all enzymes have names that end in *-ase*. The prefix is either from the name of the substrate or from the kind of reaction. For example a **hydrolase** catalyzes hydrolysis reactions. An *esterase* is a hydrolase that aids the hydrolysis of esters. A *lipase* works on the hydrolysis of lipids. A *peptidase* or a *protease* catalyzes the hydrolysis of peptide bonds.

An **oxidoreductase** handles a redox equilibrium. Sometimes an oxidoreductase is called an *oxidase* when the favored reaction is an oxidation and a *reductase* when the reaction is a reduction. A **transferase** catalyzes the transfer of a group from one molecule to another, and a *kinase* is a special transferase that handles phosphate groups. Other broad categories of enzymes are the **lyases,** which catalyze elimination reactions that form double bonds; **isomerases,** which cause the conversion of a compound into an isomer; and **ligases,** which cause the formation of bonds at the expense of chemical energy in triphosphates, like ATP.

An International Enzyme Commission has developed a system of classifying and naming enzymes that places considerable chemical information into the enzyme's name. The names of the principal reactants, separated by a colon, are written first and then the name of the kind of reaction is written as a prefix to *-ase*. For example, in moving from left to right in the following equilibrium, an amino group transfers from the glutamate ion to the pyruvate ion:

■ This reaction, incidentally, is an example of how the body can make an amino acid—here, alanine—from other substances.

$$\underset{\text{Glutamate ion}}{^-O_2CCH_2CH_2\underset{\displaystyle NH_3^+}{\underset{|}{C}}HCO_2^-} + \underset{\text{Pyruvate ion}}{CH_3\overset{\displaystyle O}{\overset{\|}{C}}CO_2^-} \rightleftharpoons \underset{\alpha\text{-Ketoglutarate ion}}{^-O_2CCH_2CH_2\overset{\displaystyle O}{\overset{\|}{C}}CO_2^-} + \underset{\text{Alanine}}{CH_3\underset{\displaystyle NH_3^+}{\underset{|}{C}}HCO_2^-}$$

The systematic name for the enzyme is *glutamate : pyruvate aminotransferase*. In all but formal publications, such a cumbersome (but unambiguous) name is seldom used. This enzyme, for example, is often referred to simply as GPT (after the older name, glutamate : pyruvate transaminase) and sometimes as alanine transaminase or ALT. We'll largely stick with common names of enzymes.

PRACTICE EXERCISE 1

What is the most likely substrate for each of the following enzymes?

(a) sucrase **(b)** glucosidase **(c)** protease **(d)** esterase

Enzymes Often Occur as a Family of Similar Compounds Called *Isoenzymes* with Identical Functions Identical reactions are often catalyzed by enzymes with identical cofactors but slightly different apoenzymes. These variations are called **isoenzymes** or **isozymes.**

Here "iso-" signifies the same catalytic function, not identical molecular formulas.

Creatine kinase or CK, for example, consists of two polypeptide chains labeled *M* (for skeletal muscle) and *B* (for brain). It occurs as three isoenzymes. All catalyze the transfer of a phosphate group in the following equilibrium:

$$NH_2C(=NH_2^+)N(CH_3)CH_2CO_2^- + ATP \underset{}{\overset{\text{creatine kinase}}{\rightleftharpoons}} {}^-OP(=O)(O^-)NHC(=NH_2^+)N(CH_3)CH_2CO_2^- + ADP$$

Creatine — Creatine phosphate

When supplies of ATP are low and those of ADP are therefore high, the *reverse* of this equilibrium becomes a major path for making more ATP in muscle cells.

One CK isoenzyme, called CK(*MM*), has two *M* units and occurs in skeletal muscle. Another, CK(*BB*), has two *B* units and occurs in brain tissue. The third, CK(*MB*), has one *M* and one *B* polypeptide, and it is present almost exclusively in heart muscle, where it accounts for 15% to 20% of the total CK activity. The rest is contributed by CK(*MM*).

We have given this much detail about creatine kinase because this and similar enzymes have an extraordinarily important function in clinical analysis and diagnosis, as we'll see later in the chapter.

Bond to be broken
Substrate
δ+ δ−
Active site
δ− δ+
Apoenzyme
Enzyme
(*a*)

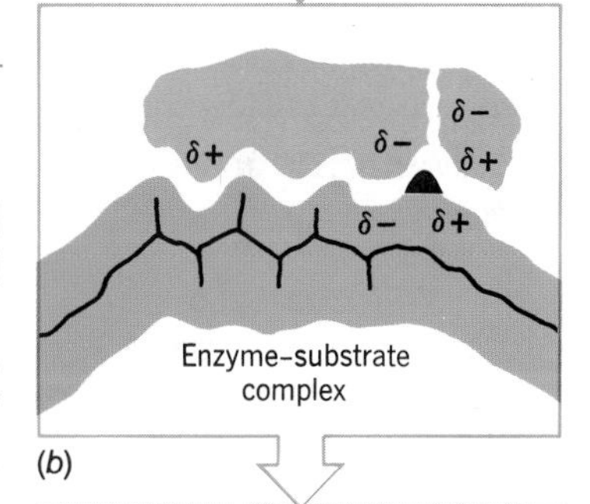

(*b*)

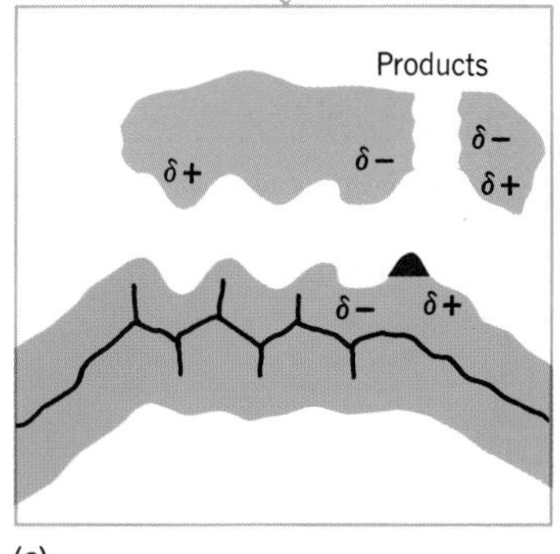

(*c*)

FIGURE 11.2
The lock-and-key model for enzyme action. (*a*) The enzyme and its substrate fit together to form an enzyme–substrate complex. (*b*) A reaction, such as the breaking of a chemical bond, occurs. (*c*) The product molecules separate from the enzyme.

11.2

THE ENZYME–SUBSTRATE COMPLEX

The chirality and flexibility of an enzyme and the sidechains of its amino acid residues allow only the enzyme's substrate to fit to it and to become activated for a reaction.

When an enzyme catalyzes a reaction of a substrate, molecules of each must momentarily fit to each other. This temporary combination is called an **enzyme–substrate complex.** It is part of a series of chemical equilibria that carry the substrate through a number of changes until the products of the overall reaction form.

$$E + S \rightleftharpoons E{-}S \rightleftharpoons E{-}S^{*} \rightleftharpoons E{-}P \rightleftharpoons E + P$$

Enzyme; Substrate; Enzyme–substrate complex; Substrate–activated $E{-}S$ complex; Enzyme–product complex; Enzyme (recovered); Product

The first equilibrium is the binding of the enzyme to the substrate like the fitting of a key (the substrate molecule) to a tumbler lock (the enzyme), so this theory is often called the **lock-and-key theory** of enzyme action. Shaped pieces that fit together are said to have *complementary shapes;* or we say that there is *complementarity* between the two shapes (see Fig. 11.2).

For an enzyme–substrate complex to form, there must be two kinds of complementarity. The first is what we have already implied—*geometrical complementarity:* a square peg fits a square hole better than to a round hole, for example. The other is *physical complementarity,* which concerns factors other than shape—hydrophobic interactions, hydrogen bonds, and electrical charges of *opposite* nature nestling *nearest* each other as the complex forms.

The chiral natures of enzymes and substrates place severe constraints on lock-and-key fitting and contribute to the high degree of specificity of enzymes. One illustration of the significance of chirality to fitting was observed in the early 1990s with a certain protease, HIV protease. The natural form of the protease, like all enzymes, is made from amino acids that are

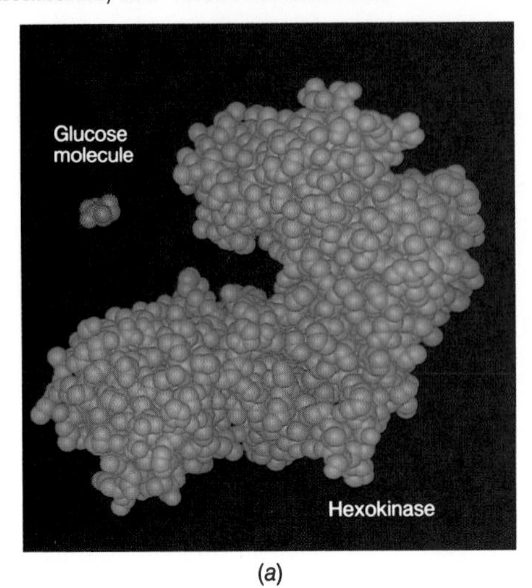

(a)

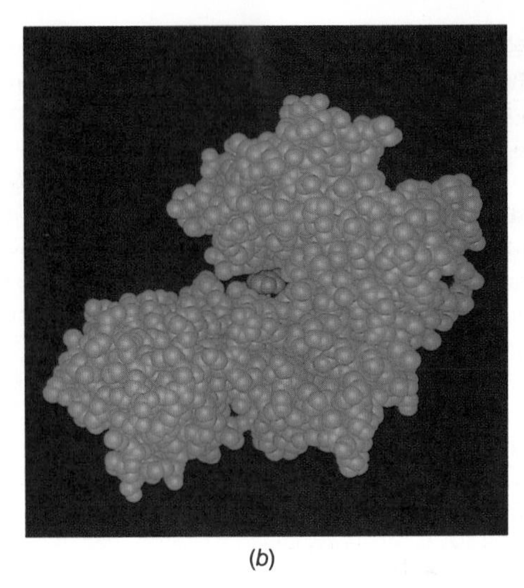

(b)

FIGURE 11.3
Induced fit theory. (a) A molecule of an enzyme, hexokinase, has a gap into which a molecule of its substrate, glucose, can fit. (b) The entry of the glucose molecule induces a change in the shape of the enzyme molecule, which now surrounds the substrate entirely. (Courtesy of T. A. Steitz, Department of Molecular Biophysics and Biochemistry, Yale University, New Haven, CT.)

all in the L-family. Likewise, the natural substrate protein for HIV protease is made of all L-family amino acids. Both the protease and the substrate protein made from all D-amino acids have been synthesized in the lab. The all-D enzyme worked only with the all-D substrate; the all-L (natural) enzyme cleaved only the all-L substrate. Another example of the significance of chirality to complementarity is the ability of trypsin, a digestive protease, to affect only substrates made of L-amino acids. The enzymes involved in the metabolism of glucose are similarly effective only with D-glucose, not L-glucose units.

■ Hormone uptake only by specific cells also depends on a flexible lock-and-key kind of recognition.

To get the substrate to fit to the enzyme depends on some flexibility in the enzyme molecule much as a lock flexibly adapts as the key is inserted. As the substrate molecule nestles onto the enzyme, the molecular groups of the substrate induce the enzyme molecule to adjust its shape to achieve the best fit (see Fig. 11.3). The initial contact with substrate and enzyme may cause changes in tertiary structure in the polypeptide of the protein. Such changes, which induce stress in the polypeptide, force the enzyme to modify its shape further. This **induced fit** model of how enzymes work describes the true nature of what is traditionally called the "lock-and-key" interaction.

Proteins, as we have indicated, consist of huge molecules, and not *all* parts of their molecules are ever *directly* involved in catalysis. Some of the enzyme's amino acid residues have sidechains (or groups of sidechains) with shapes and polar sites complementary to the substrate and so are called the enzyme's *binding sites.* Other groups on the enzyme, called *catalytic sites,* handle actual catalytic work in the complex. As stated in Section 11.1, catalytic sites are often supplied by molecules of coenzymes, and these must be bound to the apoenzyme to become integral parts of whole enzyme.

The Activation of the Substrate Changes It into Its Transition State The fit achieved by the enzyme and the substrate molecules in the $E-S$ complex is not perfect. However, the intermolecular forces that cause the complex to *begin* its formation in the first place now continue to work. These forces distort and stretch chemical bonds in the substrate to improve the fit. The chemical energy for this distortion is generally provided by the *gain in overall stability achieved in the complex.* The result of such changes is the conversion of the initial enzyme–substrate complex, $E-S$, into a substrate-activated complex, $E-S^*$. In $E-S^*$, the fit between enzyme and substrate is as good as it can be, and the substrate molecule has reached a unique condition of both shape and internal energy called its *transition state.* The perfecting of the enzyme–substrate fit as the transition state *forms,* rather than the initial, somewhat imperfect fitting of enzyme to substrate, largely accounts for the high catalytic power of an enzyme.

Whether $E-S^*$ collapses to return to the reactants or to proceed to the products depends on how much reactant and product concentrations are building up or declining, all in accordance with Le Châtelier's principle. If product forms (Fig. 11.2*b*), we might suppose that its molecule has a different distribution of electrical charges than that of the reactant molecule (see Fig. 11.2*c*). The enzyme–product complex, $E\text{–}P$, does not hold together, and the product molecule P slips off. The enzyme is then ready to receive another substrate molecule.

This has been a very broad and vastly simplified view of enzyme catalyses intended almost entirely to fortify one point: theories of how enzymes work start with the idea of induced fitting based on both geometrical and physical complementarity. An increasing number of detailed mechanisms that explain how specific enzymes or teams of enzyme work is accumulating. For an application of the principle of complementarity to antibodies, antigens, and the ABO blood groups, see Special Topic 11.2.

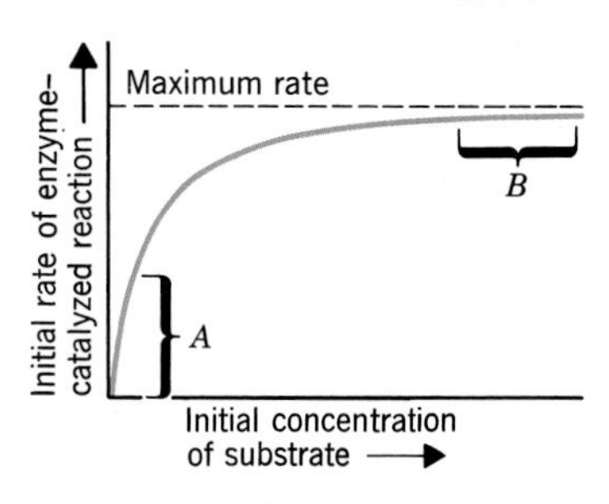

FIGURE 11.4
Initial rates of an enzyme-catalyzed reaction plotted versus the initial concentrations of substrates when the concentration of the enzyme is fixed in each experiment. The sections labeled *A* and *B* are discussed in the text.

11.3

KINETICS OF SIMPLE ENZYME–SUBSTRATE INTERACTIONS

At high substrate concentrations, an enzyme's rate enhancement levels off.
We know that reaction rates are sensitive to the *concentrations* of reactants. In many reactions between two species, if you double the initial concentration of one species, holding the other's constant, the rate doubles. Many enzyme-catalyzed reactions are like this if we treat their enzymes as actual (although temporary) reactants. At some fixed initial enzyme concentration, $[E_o]$, doubling the concentration of the substrate, $[S]$, doubles the reaction rate, V. What is significant about enzyme-catalyzed reactions is that this rate enhancement cannot be indefinitely extended to higher and higher initial substrate concentrations. Eventually, at some higher initial substrate concentration, no further rate acceleration occurs. The rate levels off.

We can see what this means with the aid of Figure 11.4, a plot of initial rates versus initial values of $[S]$. Imagine a series of experiments in all of which the molar concentration of the **enzyme, $[E_o]$, is the same. We assume a simple reaction, meaning one for which the enzyme is able to handle only *one* substrate molecule at a time. We will vary only the initial concentration of the substrate, $[S]$, from experiment to experiment. As we said, in most ordinary reactions, the initial rate would double each time we doubled the initial concentration of one reactant.**

■ Many enzymes consist of two or more polypeptide subunits *each of which has binding and catalytic sites and all of which become involved in the overall reaction.*

In our series of experiments, we do observe something like this, but only in those trials that have *low* initial values of $[S]$, as in part *A* of the plot in Figure 11.4. Here, a small increase in $[S]$ does cause a proportionate increase in initial rate. The curve rises steadily.

In succeeding experiments, however, at higher and higher initial values of $[S]$, the initial rates respond less and less until, in part *B* of the plot, the initial rates are constant *regardless of the value of* $[S]$. The reason for the leveling off is that we now have enough substrate molecules to saturate all of the active sites of all the enzyme molecules. Any additional substrate molecules have to wait their turns, so to speak.

The relationship between V, $[E_o]$, and $[S]$ for the kind of reaction just described is given by the following equation.

$$V = \frac{k[E_o][S]}{K_M + [S]} \tag{11.1}$$

■ In the early part of this century, biochemists Leonor Michaelis and Maude Menten were pioneer scientists in the field of enzyme kinetics.

The symbol k stands simply for a proportionality constant. The symbol K_M is another constant, called the *Michaelis constant*, and it has a particular value for a specific enzyme-catalyzed reaction. Notice what happens when $[S]$ has very small values, approaching zero. At very low values of $[S]$ the denominator becomes essentially identical with K_M, and Equation 11.1 becomes

$$V = \frac{k}{K_M}[E_o][S] \tag{11.2}$$

The ratio, k/K_M is a constant, being a ratio of four other constants. Equation 11.2, in other words, says that the velocity of the reaction is directly proportional to $[S]$ at *low* values of $[S]$. "Directly proportional" translates into a *straight line* or *linear* plot of initial rate versus $[S]$,

SPECIAL TOPIC 11.2
MOLECULAR COMPLEMENTARITY AND IMMUNITY, AIDS, EVEN THE ABO BLOOD GROUPS

The *immune system,* as large and complex as the nervous system, is the body's array of defenses against *pathogens*—disease-causing microorganisms and viruses. We cannot in a brief Special Topic do justice to the immune system, of course, but we can take note of some of the ways in which it shares basic operating principles with the enzyme–substrate reaction. The concept of the fitting of a substrate to an enzyme by means of geometrical and physical complementarity is also at the molecular base of the body's immune system as well as the existence of blood type groups.

When a pathogen has penetrated the first line of defense, the physical barriers of skin and mucous membranes, white blood cells known as *lymphocytes* go into action. They all begin life in bone marrow, but not all mature there. Two kinds of immunity involving two kinds of lymphocytes are recognized. One is *cellular immunity,* and it is handled by *T lymphocytes* or *T-cells* (after *t*hymus tissue, where T-cells mature). Cellular immunity handles viruses that have gotten inside cells, as well as parasites, fungi, and foreign tissue.

AIDS The human immunodeficiency virus (HIV) is able to destroy certain kinds of T-cells, the *helper T-cells.* This renders the immune system deficient in its ability to handle infections and results in AIDS, acquired immune deficiency syndrome. Relatively non-lethal diseases normally handled routinely by the body thus become lethal in AIDS victims.

Antigen–Antibody Reaction The second kind of immunity is *humoral immunity* (after an old word for fluid, *humor*). Humoral immunity is the responsibility of the *B lymphocytes* or *B-cells* (because they mature in *b*one marrow). B-cells act mostly against bacterial infections but also against those workings of viral infections that occur outside of cells. We'll limit the continuing discussion to the work of B-cells.

B-cells carry and manufacturer *antibodies,* glycoproteins that are able, by an interaction like that between substrate and enzyme, to attract and take antigens out of circulation and defeat the spread of the pathogen. An *antigen* is any molecular species or any pathogen that induces the immune system to make antibodies as well as gives the immune system a molecular-cellular memory for the antigen. Thus, at a later invasion of the same antigen, the immune system is poised for a far more rapid defensive response than it initially had. A *vaccine* is able to start the initial defensive response leading to the molecular memory for the antigen without causing the disease itself.

Figure 1 represents an antibody that is *dipolar;* it has two cross-linking molecular groups. The antigen in Figure 1 is represented by a unit that can become bound to at least three antibody-binding sites. *The antibody protein is specific for just one antigen.* As you can see, the interaction of antibody with its antigen essentially "polymerizes" the entire system into one vast "copolymer." The product is now in a far less soluble form and it bears molecular markings that are recognized by other white cells (phagocytes) that engulf the "polymer" and destroy it. There are some antigen–antibody complexes that are destroyed by a series of interacting proteins called the *complement system.* By tying up the antigens, the antibodies prevent the spread of the infection and thus allow the system the time needed to destroy the antigens.

The ABO Blood Groups The surfaces of red blood cells carry projecting oligosaccharide units of glycolipids. There are differences, however, among individuals in the structures of these sugar residues. One of the consequences of these differences is the existence of blood group systems, one being the *ABO system.* You might

which is approximately what we see in Figure 11.4. At low initial values of [*S*], the initial rates lie nearly on a straight line. The curve bends, as you can see, at higher values of [*S*]. As [*S*] becomes high enough, the K_M term in the denominator in Equation 11.1 is overwhelmed by the [*S*] term. Eventually, at sufficiently high values of [*S*], the [*S*] term in the *numerator* can be canceled by the entire denominator (which is now almost entirely contributed by [*S*] anyway), leaving the following simple expression as the equation for the velocity, where V_m means the *maximum* velocity.

$$V_m = k[E_o] \tag{11.3}$$

But $[E_o]$ is a *constant,* the concentration of the enzyme (in any form, free or combined in

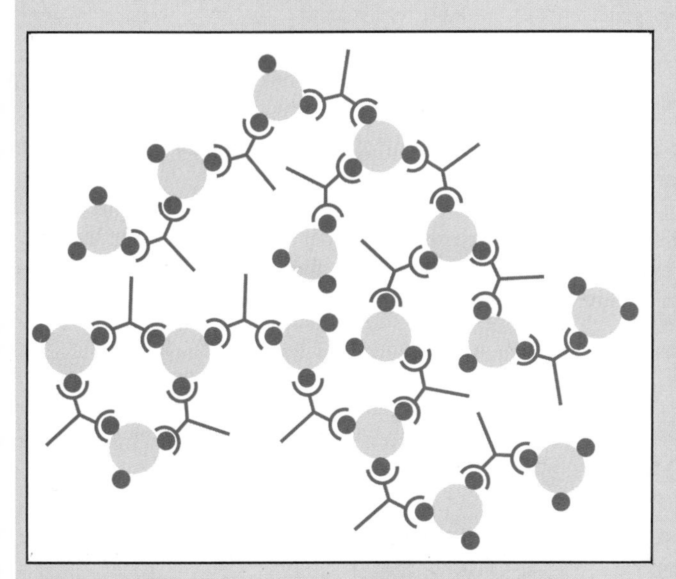

FIGURE 1
Molecules of the antibody, in green, cross-link with antigen particles, in red, to form a mass resembling a huge copolymer. (From Voet/Voet, reprinted with permission from *Biochemistry* © 1990 by John Wiley & Sons, Inc.)

have type A, type B, or type O blood. Some have a combination type, AB blood.

If you are of type A and by a transfusion are given blood from a type B person, the red cells from the type B blood will clump together (agglutinate) likely causing a blockage of blood capillaries that could be fatal. Thus your type A blood is able to "see" something in type B blood as a foreign material. Type A blood contains in the serum portion (the liquid minus the cells) an antibody against type B blood. What specifically is the antigen in the type B blood is the molecular unit at the tip of an oligosaccharide joined to a glycolipid of the (type B) red cell. This is why each kind of red cell is described as carrying an *antigen,* one of three types, A, B, and H.

If you are of type A, your serum includes anti-B antibodies. People with type B blood have anti-A antibodies. Those with AB blood have neither anti-A nor anti-B antibodies. AB blood type people are able to accept transfusions from people of any blood type. However, in all but emergency situations, transfusions are normally done using type AB blood. Type O people carry the H "antigen" on their red cells, and their blood has *both* anti-A and anti-B antibodies. Type O people, therefore, can *receive* transfusions only from individuals with type O blood. At the same time, type O people can *give* transfusions to all types—they are universal donors—because the antigen on the red cells in type O people actually has no "enemies" in other types of blood, no antibodies that can attack and agglutinate type O red cells when they are transfused into people of other blood types.

The H antigen in type O blood is given the name antigen because it is the precursor to the A and B antigens of other blood types. Type A individuals make type A antigen by adding an *N*-acetylgalactosamine residue to the tip of a glycolipid on the red cell. Type B people make type B antigen by adding a galactose residue to the same glycolipid. Type O individuals simply lack the enzymes needed for these transformations. The differences among these enzymes are thought to involve single amino acid residue substitutions in the enzymes' polypeptides. Table 1 summarizes donor–acceptor relationships for the blood types.

Table 1 Blood Type Acceptor-Donor Options

If Your Blood Type Is	You Can Accept Blood from One of This Type	You Can Donate Blood to One of This Type
O	O	O, A, B, AB
A	A or O	A or AB
B	B or O	B or AB
AB	AB, A, B, O	AB

complex) and k is, of course, also a constant. So the maximum velocity, V_m, *must* be a constant. Thus at high values of [S], the rate of the enzyme catalyzed reaction must level off at a maximum, as you can see it does in Figure 11.4. If we substitute the expression from Equation 11.3 into Equation 11.1, we obtain what is called the *Michaelis-Menten equation* for the rate of a simple enzyme catalyzed reaction.

■ The $E-S$ complex is sometimes called the Michaelis complex to honor the work of Leonor Michaelis.

$$V = \frac{V_m[S]}{K_M + [S]} \tag{11.4}$$

The reason for a *constant* rate at high values of [S] is that all enzyme molecules are now saturated with substrate and are in the form of the complex $E-S$. Further catalysis depends on the freeing of enzyme from enzyme–product complexes. At one specific value of [S], $[S] = K_M$, Equation 11.4 reduces to give

$$V = \frac{1}{2} V_m \qquad (11.5)$$

This equation gives meaning to the Michaelis constant, K_M; K_M is the substrate concentration at which the reaction rate is one-half of the maximum rate.

Our discussion involving Equations 11.1–11.5 concerned the response of rate to concentrations of enzyme and substrate when the enzyme carries only one active catalytic site. Let's now turn our attention to enzymes with more than one site; they offer the system numerous pathways for the regulation of enzyme action.

11.4

THE REGULATION OF ENZYMES

Enzymes are switched on and off by initiators, effectors, inhibitors, genes, poisons, hormones, and neurotransmitters.

A cell cannot be allowed to do everything at once. Some of its possible reactions have to be shut down while others occur. One way to keep a reaction switched off is to prevent its enzyme from being made in the first place. Another but harmful way is to deny the cell the amino acids, vitamins, and minerals that it needs to make enzymes. Hormones and neurotransmitters are natural regulators of enzymes, and we'll learn about them in the next section. In this section we study several other means to control enzymes.

■ The prevention of enzyme synthesis often involves the regulation of genes and their work of directing the synthesis of polypeptides, topics for Chapter 13.

Some Enzymes Display an Initial Resistance to the Formation of an Enzyme–Substrate Complex One of the very significant features of many enzymes is that the initial sharp rise of the curve in Figure 11.4 does not occur when the concentration of the substrate, [S], is low. It's as though the enzyme is inactive at low concentrations. For these enzymes, the plots of initial values of [S] versus initial rates look more like the curve in Figure 11.5. The plot has a lazy "s" shape, so it's called a *sigmoid plot* (after *sigma,* Greek for S). The rate increases very slowly with initial substrate concentration, then the rate takes off in the normal response of rate to concentration, and finally the rate levels off in the usual way.

Sigmoid rate plots are found among enzymes that remain inactive *until a sufficient concentration of substrate forces them into active forms.* What the curve suggests is something about *enzyme activation,* as we'll learn next.

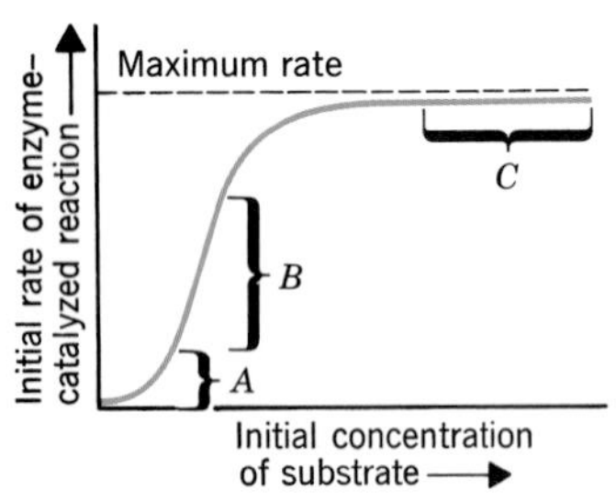

FIGURE 11.5
Initial rates of enzyme-catalyzed reactions plotted versus initial values of substrate concentrations, at fixed enzyme concentration, when an allosteric effect is observed. Sections labeled *A* and *B* of the curve—a sigmoid curve—are discussed in the text.

The Catalytic Sites in Some Enzymes Are Given Active Shapes by Reactions at Other Sites Enzymes with sigmoid rate curves (Fig. 11.5) consist of two or more polypeptide units *each* with a catalytic site that normally is in an *inactive* configuration, even in the presence of some substrate. Inactive catalytic sites, however, are eventually activated by the substrate, but only when the initial concentration of the substrate is high enough.

Let's suppose, for simplicity, that our enzyme is made of just two polypeptide chains. It has two catalytic sites, one on each polypeptide. We'll represent each site by a geometric shape, as shown in Figure 11.6. We have to suppose that the shape of each site is not quite complementary to the substrate, that the substrate must itself induce the correct fit, because the enzyme's response is sluggish at low substrate concentration. We're in region *A* of the sigmoid curve (Fig. 11.5), the region of the slower than normal rate.

When a substrate molecule induces a conformational change in the polypeptide unit to enable it to fit to one catalytic site, it simultaneously induces the other polypeptide to adopt a new shape *causing the second catalytic site to become active.* Now the same enzyme molecule can accept the second substrate molecule *much more easily than the first.* The enzyme is now being used at maximum efficiency, and the rate of reaction takes off, putting the system into part *B* of Figure 11.5.

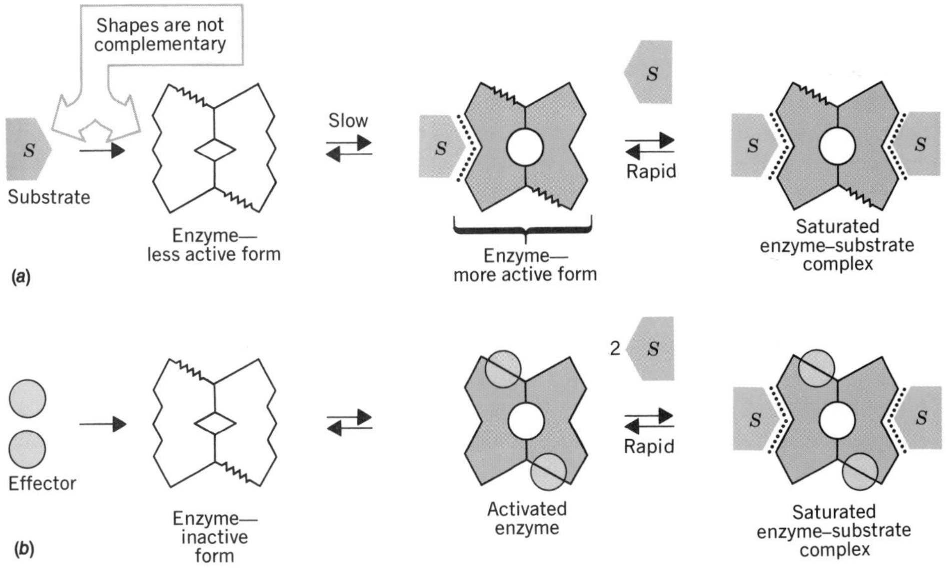

FIGURE 11.6 Allosteric activations. (*a*) Allosteric activation by substrate. (*b*) Allosteric activation by effector.

This phenomenon in which one active site is activated by an event that occurs elsewhere on the enzyme is called **allosteric activation.** The enzyme's subunits cooperate with each other to cause full activation, but this doesn't happen unless the level of substrate concentration has climbed high enough to start the process. *The activity of an enzyme subject to allosteric activation is thus regulated by how much its services are needed.* The "need," of course, increases as the concentration of substrate increases, because something has to be done about the increasing level of substrate.

■ *Allo-*, other; *-steric,* space —the other space or the other site.

In the next chapter, we'll see how a similar allosteric effect is caused by oxygen when it interacts with hemoglobin, which is not an enzyme, and how this enables hemoglobin to operate at 100% efficiency, or very nearly so.

Allosteric Activation Can Be Caused by Effectors Instead of Substrates The catalytic sites of some enzymes are activated by substances called **effectors** that are not substrates. When their molecules bind allosterically to the enzyme, meaning at a location distinct from the catalytic site, they force a configurational change that activates the enzyme (see Figure 11.6*b* where circles represent the effector molecules). The effector might, for example, be a molecule whose own metabolism *needs the products* made by the enzyme it activates.

Nerve Signals Can Indirectly Tell an Effector to Work Two of the very important effectors are *calmodulin,* a protein found in most cells, and *troponin,* present in muscle cells. Neither works as an effector, however, until it has itself been activated. The activator is calcium ion, and cells control their calcium ion levels by active transport mechanisms mediated by nerve signals.

Normally, the level of calcium ion that moves freely in solution in the cytosol is only about 10^{-7} mol/L. It must be kept extremely low, because the cytosol contains phosphate ion which forms an insoluble salt with Ca^{2+}. The level of Ca^{2+} just outside the cell is about 10^{-3} mol/L, very considerably higher. Despite the concentration gradient, which nature would normally erase by diffusion, the calcium ions stay outside until something changes the cell's permeability to them. Nerve signals do this.

■ The cytosol is the *solution* in the cytoplasm and does not include the organelles in the cytoplasm.

The overall sequence is roughly as follows. A nerve signal opens protein channels in the cell membrane for Ca^{2+} ions and they enter the cell, where they bind to calmodulin or troponin. The effector is thus activated and it then activates an enzyme to cause some chemical work or, in muscles, to cause muscle contraction. When the signal is over, the channels close, and Ca^{2+} ions are pumped back out through other portals. The effector is thus inactivated. This mechanism thus connects nerve signals to specific chemical activities in cells.

■ Other kinds of proproteins are known, like *proinsulin,* the precursor to the hormone *insulin,* a blood sugar regulator. The conversion of proinsulin to insulin entails the removal of a 33-residue polypeptide unit.

Some Enzymes Are Activated by the Removal of a Polypeptide Unit Several digestive enzymes are first made in inactive forms called **zymogens** or **proenzymes.** Their polypeptide strands have several more amino acid residues than the enzyme, and these extra units cover over the active site. Then, when the active enzyme is needed, a complex process is launched that clips off the extra units and, by exposing the active site, activates the enzyme.

One of the functions of enteropeptidase, a compound released in the upper intestine when food moves in from the stomach, is to convert the zymogen trypsinogen into the enzyme trypsin, which helps to digest proteins. Trypsin is activated by the deletion of a small polypeptide unit in trypsinogen. When no food is present, there is no need for trypsin, but when food enters, enteropeptidase comes in as well, and then trypsin is activated.

■ Kinases are the enzymes for this phosphorylation, and they must themselves be activated (usually by Ca^{2+}) only when needed.

Phosphorylation Activates Some Enzymes Glycogen phosphorylase, the enzyme that catalyzes the hydrolysis of glucose units from glycogen, is made in an inactive state. However, when a serine sidechain (CH_2OH) is changed to its phosphate ester, $CH_2OPO_3^{2-}$, the enzyme is activated.

■ Inhibitors are *negative effectors.*

Inhibitors Can Keep Enzymes Switched Off until They Are Needed Some substances, called **inhibitors,** bind reversibly to the enzyme and prevent it from working. In **allosteric inhibition,** molecules of the inhibitor bind to the enzyme somewhere other than the active site. This affects the shape of the active site or a binding site, and the enzyme–substrate complex cannot form (see Fig. 11.7*a*). Then if some reaction changes the inhibitor so that it no longer sticks to the enzyme, the catalyst becomes active.

In **competitive inhibition,** the inhibitor is a nonsubstrate molecule with a shape similar enough to that of the true substrate *that it can compete with the substrate for attachment to the active site.* When the inhibitor molecules lock to the enzyme's active sites, but do not undergo a reaction and leave, then the enzyme has become useless to the true substrate (see Fig. 11.7*b*).

A competitive inhibitor doesn't have to be a product of the enzyme's own work. It can be something else the cell makes, or it could be the molecules of a medication. What its molecules must do is *resemble* those of the normal substrate enough to bind to the active site of the enzyme.

Sometimes an inhibitor is the *product* of the reaction being catalyzed or one of the products produced later in a series of connected reactions. Now we have **feedback inhibition.** As the level of such a product increases, its molecules "feed back" with increasing success as inhibitors to one of the enzymes in the series of reactions that helped to make the product. A *feedback inhibitor works as an allosteric, noncompetitive inhibitor.* The amino acid isoleucine, for example, is a feedback inhibitor. It is made from another amino acid, threonine, by a series

FIGURE 11.7
Enzyme inhibition. (*a*) Allosteric inhibition. (*b*) Competitive inhibition.

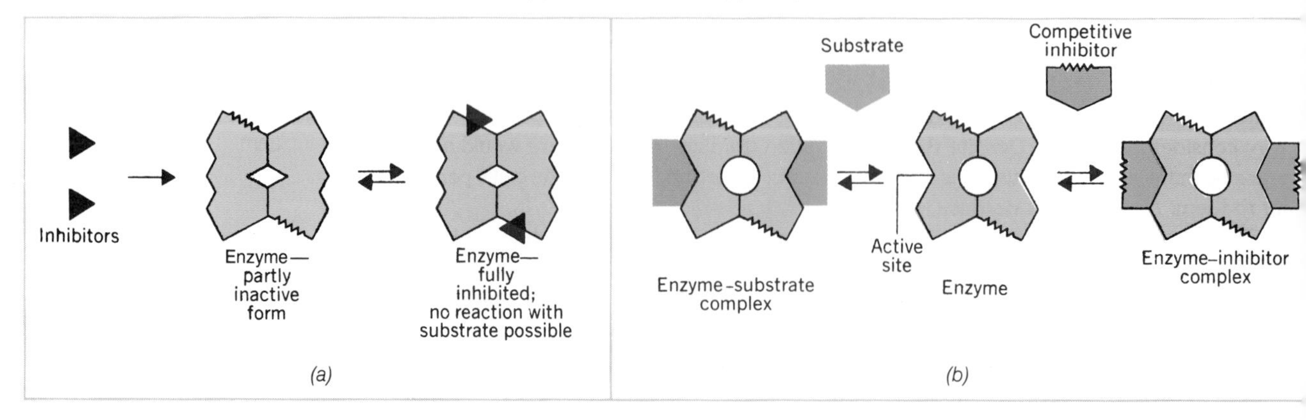

of steps, each with its own enzyme. As more and more isoleucine is made, its molecules more and more inhibit the enzyme involved with threonine in the first step, E_1, of the series:

$$
\begin{array}{ccc}
{}^{+}NH_3CHCO_2^- & \xrightarrow{E_1}\xrightarrow{E_2}\xrightarrow{E_3}\xrightarrow{E_4}\xrightarrow{E_5} & {}^{+}NH_3CHCO_2^- \\
| & & | \\
CHOH & \text{Inhibition of } E_1 \text{ by molecules of isoleucine} & CHCH_3 \\
| & & | \\
CH_3 & & CH_2CH_3 \\
\text{Threonine} & & \text{Isoleucine}
\end{array}
$$

■ Both threonine and isoleucine have two tetrahedral stereocenters. Can you spot them, and can you tell how many stereoisomers there are of each?

The beautiful feature of feedback inhibition is that the system for making a product shuts down automatically when enough product is made. Then, as the cell consumes this product, it eventually uses even product molecules that have been serving as inhibitors. The result is that when the product concentration has dropped very low, the enzyme needed to make more is released from its bondage.

Feedback inhibition is very common in nature. It helps to maintain a condition of **homeostasis** in which disturbances to systems by stimuli are minimized, because the stimulus is able to start a series of events that restore the system to the original state. Body temperature is a condition maintained by homeostatic mechanisms. The body does this so well that even small changes in temperature tell us that something is wrong.

A familiar homeostatic mechanism in the home is the work of a furnace controlled by a thermostat. When the room becomes hot enough (the desired "product"), the thermostat trips and the furnace shuts off. In time, the temperature drops, the thermostat trips back, and the furnace restarts.

Inhibition by Antibiotics A broad family of compounds called **antimetabolites** includes some made by bacteria and fungi and called **antibiotics.** Antibiotics are substances that inhibit or prevent the normal metabolism of a disease-causing bacterial system. Some antibiotics work by inhibiting an enzyme that the bacterium needs for its own growth. Both the sulfa drugs and penicillin work in this way.

■ An antimetabolite is called an *antibiotic* when it is the product of the growth of a fungus or a natural strain of bacteria.

Poisons Often Cause Irreversible Enzyme Inhibition The most dangerous **poisons** are effective even at very low concentrations because they are powerful inhibitors of enzymes, themselves not requiring high concentrations. By inhibiting an enzyme, "a little goes a long way." The cyanide ion, for example, forms a strong complex with one of the metal ion cofactors in an enzyme needed for our use of oxygen.

Enzymes that have SH groups are denatured and deactivated by such poisonous heavy metal ions as Hg^{2+}, Pb^{2+}, Cu^{2+}, and Ag^{+}.

Nerve gases and their weaker cousins, the organophosphate insecticides, inactivate enzymes of the nervous system.

PRACTICE EXERCISE 2

The following overall change is accomplished by a series of steps, each with its own enzyme.

$$
\begin{array}{ccc}
{}^{2-}O_3POCH_2\underset{\displaystyle OH}{\underset{|}{C}}H\overset{\displaystyle O}{\overset{\|}{C}}OPO_3^{2-} & \longrightarrow\longrightarrow\longrightarrow & {}^{2-}O_3POCH_2\underset{\displaystyle OPO_3^{2-}}{\underset{|}{C}}HCO_2^- \\
\text{1,3-Bisphosphoglycerate (1,3-BPG)} & & \text{2,3-Bisphosphoglycerate (2,3-BPG)}
\end{array}
$$

One of the enzymes in this series is inhibited by 2,3-BPG. What kind of control is exerted by 2,3-BPG on this series? (Name it.)

11.5

ENZYMES IN MEDICINE

The specificity of the enzyme for its substrate and the slight differences in properties of isoenzymes provide several unusually sensitive methods of medical diagnosis.

Enzymes that normally work only inside cells are not found in the blood except at extremely low concentrations. When cells are diseased or injured, however, their enzymes spill into the bloodstream. Much can be learned about the disease or injury by detecting such enzymes and measuring their levels.

Enzyme Assays of Blood Use Substrates as Chemical "Tweezers" Despite the enormous complexity of blood and the very low levels of enzymes in it, enzyme assays are relatively easy to carry out. The substrate for the enzyme is used to find its own enzyme, and the specificity of the enzyme–substrate system ensures that it will find nothing else. If no enzyme is present to match the substrate, nothing happens. Otherwise, the extent of the reaction of the substrate measures the concentration of the enzyme. In this section, we learn about some examples of this medical technology.

■ GPT is the transaminase introduced on page 276.

Viral Hepatitis Is Detected by the Appearance of GPT and GOT in Blood Heart, muscle, kidney, and liver tissue all contain the enzyme glutamate : pyruvate aminotransferase or GPT, which we introduced earlier. The liver, however, has about three times as much GPT as any other tissue, so the appearance of GPT in the blood generally indicates liver damage or a virus infection of the liver, such as viral hepatitis.

The level of another enzyme, glutamate : oxaloacetate aminotransferase or GOT, also increases in viral hepatitis, but the GPT level goes much higher than the GOT level. The ratio of GPT to GOT in the serum of someone with viral hepatitis is typically 1.6, compared with a level of 0.7 to 0.8 in healthy individuals. (Notice that we speak here of *ratios,* not absolute amounts, which are normally very low.)

■ The popular term for this set of events is *heart attack.*

Heart Attacks Cause Increased Levels of Three Enzymes in Blood Serum A *myocardial infarction* (MI) is the withering of a portion of the heart muscle following some blockage of the blood vessels that supply it with oxygen and nutrients. Such blockage can be caused by deposits, by hardening, or by a clot. If the patient survives, the withered muscle becomes scar tissue, and the outlook for a reasonably active life is generally good, particularly

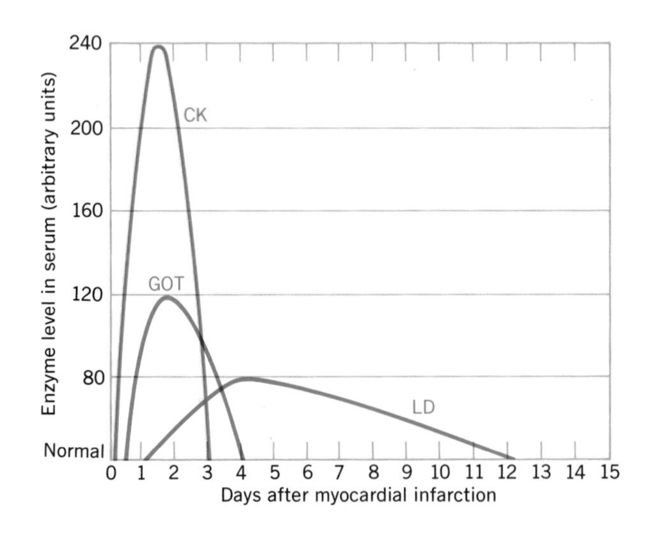

FIGURE 11.8
The concentrations of three enzymes in blood serum increase after a myocardial infarction. Here CK is creatine kinase; GOT is glutamate : oxaloacetate aminotransferase; and LD is lactate dehydrogenase.

TABLE 11.1
Clinical Report: Myocardial Infarction

DAY I		DAY II		DAY III	
DATE	5-25	DATE	5-26	DATE	5-27
CK	51	CK	552	CK	399
CK (MB)	Negative	CK (MB)	Moderate Positive	CK (MB)	Weak Positive
GOT	19	GOT	91	GOT	117
LD	88	LD	151	LD	247
LD ISOENZYME		LD ISOENZYME		LD ISOENZYME	
% of Total LD activity		% of Total LD activity		% of Total LD activity	
LD_1	28.3 %	LD_1	33.8 %	LD_1	37.9 %
LD_2	32.9 %	LD_2	32.3 %	LD_2	32.8 %
LD_3	19.7 %	LD_3	16.3 %	LD_3	15.7 %
LD_4	11.6 %	LD_4	9.1 %	LD_4	7.6 %
LD_5	7.5 %	LD_5	8.5 %	LD_5	6.0 %
	CB		C.B		B.L.

Normal Range

CK	Male	5–75 mU/ml
	Female	5–55 mU/ml
GOT		5–20 mU/ml
LD		30–100 mU/ml

LD Isoenzymes Normals

LD_1	14–29%
LD_2	29–40%
LD_3	18–28%
LD_4	7–17%
LD_5	3–16%

One International Unit (U) of activity is the reaction under standard conditions of one micromole per minute of a particular substrate used in the test. (Data courtesy of Dr. Gary Hemphill, Clinical Laboratories, Metropolitan Medical Center, Minneapolis, Minn.)

if treatment is started promptly. A diagnosis of an infarction of exceptionally high reliability can be made by the analysis of the serum for several enzymes and isoenzymes.

When a myocardial infarction occurs, the serum levels of three enzymes normally confined inside heart muscle cells begin to rise. See Figure 11.8. These enzymes are CK (page 277), GOT (just described) and LD. LD stands for lactate dehydrogenase, which catalyzes the formation of the oxidation–reduction equilibrium between lactate and pyruvate.

$$\underset{\text{Pyruvate}}{CH_3\overset{\overset{\displaystyle O}{\|}}{C}CO_2^-} + NAD{:}H + H^+ \xrightleftharpoons[\text{(an NAD}^+\text{ enzyme)}]{\text{lactate dehydrogenase (LD)}} \underset{\text{Lactate}}{CH_3\overset{\overset{\displaystyle OH}{|}}{C}HCO_2^-} + NAD^+$$

A clinical report from a patient with a myocardial infarction is shown in Table 11.1. You can see how sharply the levels of these three enzymes rose between day I and day II. The line

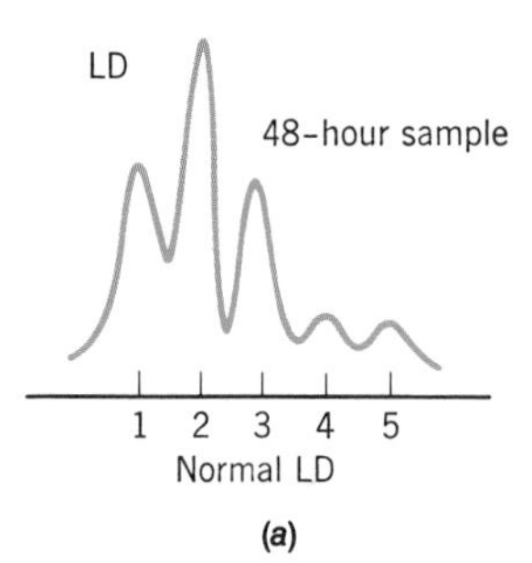

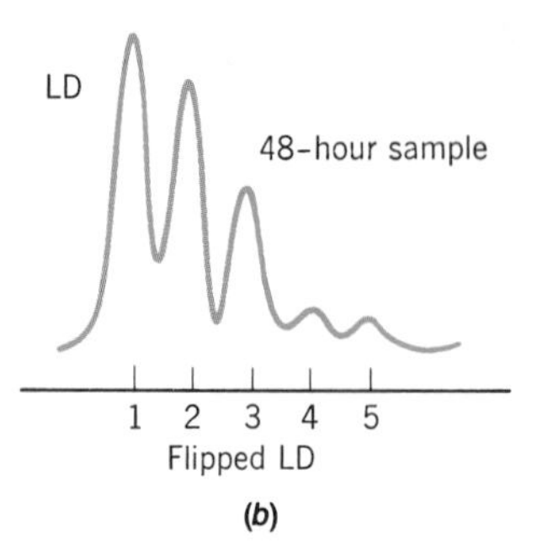

FIGURE 11.9 The lactate dehydrogenase isoenzymes. (*a*) The normal pattern of the relative concentrations of the five isoenzymes. (*b*) The pattern after a myocardial infarction. Notice the reversal in relative concentration between LD_1 and LD_2. This is the $LD_1 - LD_2$ flip.

of data labeled "CK(MB)" and the data in the columns headed by "LD ISOENZYME" provided the clinching evidence for the infarction. As we learned on page 277, the CK enzyme occurs as three isoenzymes, CK(*MM*), CK(*BB*), and CK(*MB*). The technique that uses a chemical substrate to determine the serum CK level can't tell these isoenzymes apart, because all three catalyze the reaction with the substrate.

To be sure that the increase in serum CK level is caused by injury to the *heart* tissue, the clinical chemist has to use a special technique to analyze specifically for the CK(*MB*) isoenzyme common to heart muscle tissue. As you can see in the chart, the patient's serum CK(*MB*) level did increase.

As additional confirmation of an infarction, the serum LD fraction is further separated by a special method into the five LD isoenzymes, and each is individually analyzed. The *relative* serum concentrations of these five differ distinctively between a healthy person (Fig. 11.9*a*) and one who has suffered an infarction (Fig. 11.9*b*). Of particular importance are the relative levels of the LD_1 and the LD_2 isoenzymes.

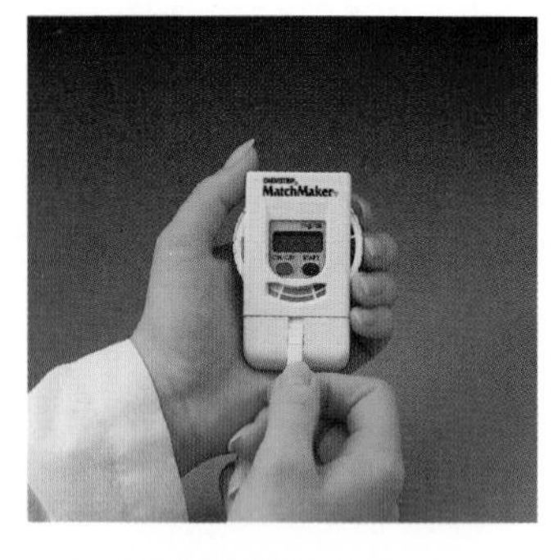

In the CHEMSTRIP MatchMaker device, the blood glucose level is measured by the intensity of a dye produced enzymatically and converted into mg of glucose per deciliter (100 mL) of blood.

Normally the LD_1 level is less than that of the LD_2, but following a myocardial infarction what is called an "$LD_1 - LD_2$ flip" occurs. The relative concentrations of LD_1 and LD_2 become reversed and the level of LD_1 rises higher than that of LD_2. When both the CK(*MB*) band and the $LD_1 - LD_2$ flip occur, the diagnosis of a myocardial infarction is essentially 100% certain.

The Blood Glucose Level Can Be Determined Enzymatically The regular determination of the level of glucose in blood is important to people with diabetes, because when this level is poorly managed, several complications can occur. One commercially available test uses a combination of chemicals, including enzymes, that react with blood glucose to generate a dye. The intensity of the resulting dye is proportional to the blood glucose level.

Similar enzyme-based tests are available to measure the serum levels of urea, triacylglycerols, bilirubin (a breakdown product of hemoglobin), and other compounds.

Enzymes Can Be Immobilized on Solid Supports In some applications, enzymes are physically immobilized onto the surfaces of extremely tiny, inert plastic beads, which makes it easier to separate the products from the enzymes. Immobilized enzymes last longer, are less sensitive to temperature, and are less vulnerable to oxygen. Immobilized enzymes, for example, are used in filtering systems to remove bacteria and viruses from air and water.

The enzyme *heparinase* is immobilized onto plastic beds that are then used to catalyze the breakdown of the heparin (page 265) added to blood sent through a hemodialysis machine. The added heparin inhibits the clotting of the blood outside the body, but it has to be removed before the blood goes back into the body. The heparinase catalyzes this removal.

■ The level of urea in the blood is called the BUN level (for blood urea nitrogen). A high BUN level indicates a kidney disorder.

Enzymes in Electrode Tips Make Possible Several Serum Assays Clinical chemists have a variety of electrodes made with immobilized enzymes for analyses that resemble the use of pH electrodes to measure pH. The specificity of the enzyme, immobilized on the electrode's tip, is a key factor in this technology. The level of urea in blood, for example, can be measured as a function of the concentration of the ammonium ions produced when urea is hydrolyzed. The tip of the urea electrode is lightly coated with a polymer that immobilizes urease, the enzyme that catalyzes the hydrolysis of urea.

$$NH_2{-}\overset{\overset{\displaystyle O}{\|}}{C}{-}NH_2 + H_2O + 2H^+ \xrightarrow{\text{urease}} 2NH_4^+ + CO_2$$

When this electrode is dipped into blood serum containing urea, urea and water migrate into the polymer, the urea is promptly hydrolyzed (thanks to the urease), and the electrode helps to register the appearance of the newly formed ammonium ion. It is relatively easy to correlate the response of the electrode to the concentration of the urea.

Enzyme electrodes for the determination of glucose, uric acid, tyrosine, lactic acid, acetylcholine, cholesterol, and other substances have been developed.

The removal of pollutants from water supplies is receiving the attention of specialists in immobilized enzymes. Both the nitrate and nitrite ions are pollutants that get into the ground water by the excessive use of nitrogen fertilizers. One problem with the nitrite ion is that it alters the hemoglobin into forms that cannot carry oxygen, a problem particularly dangerous to infants. Enzymes that reduce nitrate and nitrite ions to nitrogen have been successfully immobilized onto an electrode system that works effectively to remove these ions from polluted water.

■ The nitrate ion is the more common of the two, but NO_3^- is converted to NO_2^- in the body. A level of NO_3^- in water exceeding 50 mg/L is unacceptable for drinking water.

A Natural Blood-Clot-Dissolving Enzyme Can Be Activated by Other Enzymes

Three of the enzymes available to aid in dissolving blood clots that cause myocardial infarctions are streptokinase, tissue plasminogen activator (TPA), and a modified streptokinase called APSAC (for acylated plasminogen–streptokinase-activator complex).

■ Tissue plasminogen activator made by genetic engineering (Section 13.6) is referred to as *recombinant* tissue plasminogen activator and symbolized as TPA.

If you cut yourself, there is set in motion a huge cascade of enzyme-catalyzed reactions that bring about the formation of fibrin from a circulating polypeptide, fibrinogen. The long, stringy fibrin molecules form a brush-heap mat that entraps water and puts a seal, a blood clot, on the cut. After the wound heals, the clot must be dissolved. No part of a blood clot must break loose and circulate to the heart, because it will be stopped by tiny capillaries in heart muscle tissue. Such a blockage is one cause of a myocardial infarction. Clots in the lungs are also very serious.

■ About 150,000 people a year die from clots that start in their lungs.

To dissolve the fibrin of a clot, the body normally converts a zymogen, *plasminogen,* into the enzyme *plasmin.* We described the chemistry of zymogen activation in the previous section. Plasminogen actually is absorbed out of circulation by the fibrin as the clot forms. Its eventual activation, therefore, occurs exactly where its active form, plasmin, is needed. Plasmin, a protease, then catalyzes the hydrolysis of fibrin, and the clot "dissolves."

One of the interesting facts about plasminogen is that it becomes more susceptible to activation when bound to fibrin than when it is simply in circulation. In time, a circulating *tissue plasminogen activator* does what its name implies. It catalyzes the conversion of plasminogen to plasmin at the site of the blood clot.

Therapy for a myocardial infarction is intended to open blocked capillaries as rapidly as possible before the oxygen starvation of surrounding heart muscle tissue spreads the damage too widely. *Plasminogen activation therapy* has became the most commonly used method for treating clot-related infarctions. Obviously, the sooner this therapy is applied following a myocardial infarction, the better the chances that long-term heart muscle damage will be slight.

■ This therapy is called *thrombolytic therapy* because it *lyses* (breaks down) *thrombi* (blood clots).

Plasminogen activation therapy involves introducing into circulation one of the plasminogen-activating enzymes. Among the most commonly used are streptokinase (with aspirin) and tissue plasminogen activator (TPA). When therapy is started within the first four hours of a myocardial infarction, TPA therapy appears to have a small edge in effectiveness. There is a huge difference in cost, however; in the early 1990s, TPA cost about $2200 per dose, but the cost of streptokinase was about $200 per dose.

11.6

CHEMICAL COMMUNICATION—AN INTRODUCTION

Hormones and neurotransmitters are the chief methods by which cells communicate with one another.

Like any complex organism, the body is made up of highly specialized parts. Information, therefore, must flow among the parts to maintain a well-coordinated system. This flow is handled by chemical messengers, hormones or neurotransmitters, sent in response to a variety of signals.

■ Greek *hormon,* arousing.

Hormones are compounds made in specialized organs, the endocrine glands, secreted into the bloodstream, and usually sent some distance away where they launch responses in their particular **target tissue** or **target cells.** The distance might be as close as another neighboring tissue or as far away as 15 to 20 cm. The signal for releasing a hormone might be something conveyed by one of our senses, such as light or an odor, or it might be a stress, or a variation in the level of a particular substance in the blood or in another fluid. Insulin, for example, is released when the level of glucose in blood increases.

Neurotransmitters are chemicals made in nerve cells, called *neurons,* and sent to the next nerve cells. Thus, the distinctions between hormones and neurotransmitters concern differences in how far they go to exert their action. How these two kinds of messengers cause what they do bears many close similarities, however, as we will see.

■ We can imagine the unique oligosaccharide unit of one of the membrane's glycoproteins being like a fishing line dangling a unique lure at its tip that is attractive only to one kind of hormone or neurotransmitter molecule (refer back to Figure 10.13, page 264).

Target Cell Receptors Identify Chemical Messengers At a target cell, a hormone or neurotransmitter delivers its messages by binding to a cell **receptor.** Each receptor, a unique glycoprotein, has molecules so structured that it can accept just the messenger intended for it. It's another example of a lock-and-key mechanism at work to make interactions very specific. Sometimes molecules of toxic substances, virus particles, and even dangerous bacteria "recognize" a glycoprotein on the cells of one particular tissue and cause wholly unwanted changes to occur. Other receptor-like proteins recognize complementary molecular units on neighboring cells, lock to them, and so bind cells together. Such binding helps to control cell division. In cancerous tissue, cell-to-cell binding is weakened and cancer cells more easily proliferate and even break off, enter into circulation in blood or lymph, and so spread the cancer. Receptor-like glycoproteins on sperm cells are able to lock to molecular units on only the ova of the same species, thanks to a lock-and-key mechanism. Antibodies "recognize" antigens in the body's immune response by a lock-and-key mechanism. As evidence that such responses involve cellular *surfaces,* sometimes even killed bacteria or deactivated virus particles cause the body's immune system to develop antibodies. Once initiated, the body retains immunity for a long time.

■ When immunity is not life-lasting, as with the influenza virus, it is because the disease-causing material undergoes mutations.

Chemical Messengers Enter Cells by Four Major Mechanisms The formation of a receptor–messenger complex changes the receptor structure, so that now it is activated to do something. It might be to activate a gene, or an enzyme, or to alter the permeability of a cell membrane so that certain ions or small molecules can move across it. In neurons, the activation of a receptor sends the nerve signal on. We'll consider specific examples later. We offer only a broad overview here.

Figure 11.10 outlines the principal ways by which signals enter cells. Some hormones, once "recognized" by the target cell, move directly through the cell membrane, enter the cytosol, and find a receptor inside the cell. Steroid hormones work in this way. See [1] in Figure 11.10. They bind to receptors close to or inside the cell nucleus, where they induce changes in the way the cell uses DNA.

Polypeptide hormones cannot migrate directly through cell membranes, so they bind to receptors that are integral parts of the membrane. See [2] in Figure 11.10. Insulin and growth hormone work in this way.

Neurotransmitters also bind to membrane-bound receptors, and this opens channels through the membrane for metal ions, [3]. We have already seen how movements of the

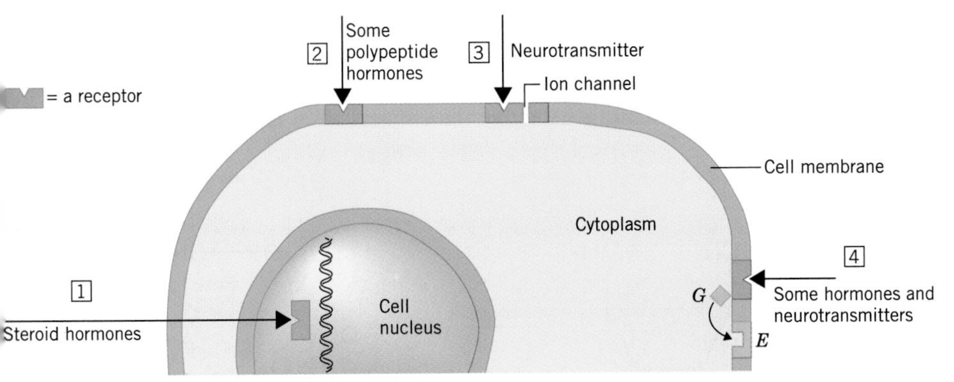

FIGURE 11.10
The ways by which hormones get chemical messages into cells

calcium ion can affect calmodulin or troponin and so activate a series of enzyme-catalyzed reactions.

Some receptors, in accepting neurotransmitters, hormones, or even light photons activate a small polypeptide of a G-protein complex, [4] in Figure 11.10. This leads to the activation of an enzyme (*E* in Figure 11.10). Remarkably, a large variety of cells share just a few mechanisms for taking advantage of the action of the G-protein. We'll study two, the cyclic AMP and the inositol phosphate cascades.

■ The G in G-protein stands for *guanyl nucleotide binding protein.* Of its three small polypeptide subunits, only one actually migrates to activate adenylate cyclase.

The Formation and Hydrolysis of Cyclic AMP Is a Major Mechanism by Which Many Cells Pass On Messages Cyclic nucleotides, particularly 3′,5′-cyclic AMP, are important secondary chemical messengers. How cyclic AMP works is sketched in Figure 11.11.

■ *Cyclic* refers to the *extra* ring of the phosphate diester system.

At the top of the figure we see a hormone—it could just as well be a neurotransmitter—that can combine with a receptor molecule at the surface of a cell. Because of the requirements of fitting substrate and receptor molecules together, the hormone bypasses all cells that do not have a matching receptor. The complex that forms between hormone and receptor activates the G-protein molecule bound on the cytosol side of the lipid bilayer. The G-protein finds a

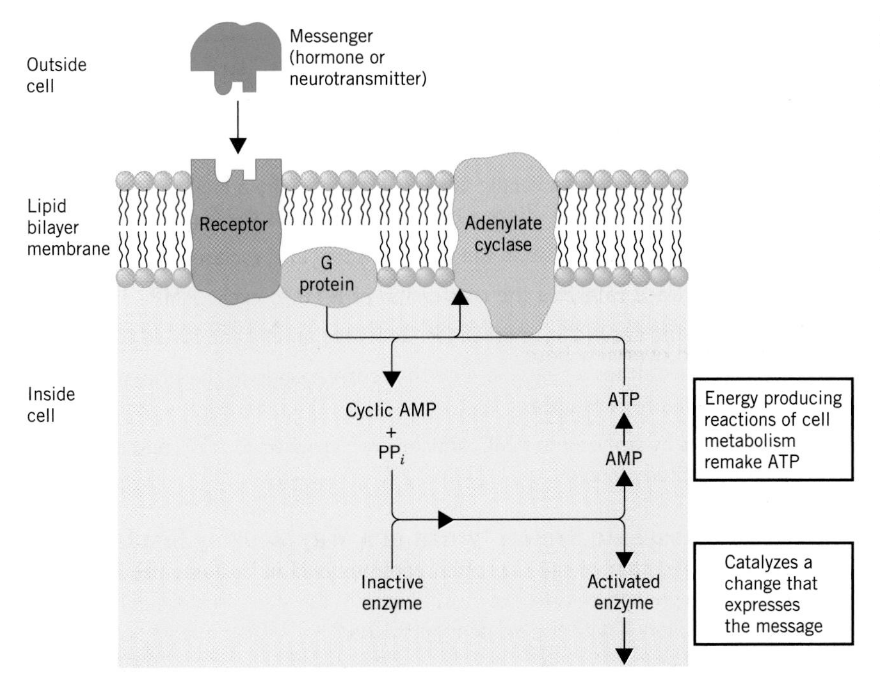

FIGURE 11.11
The activation of the enzyme adenylate cyclase by a hormone (or a neurotransmitter). The hormone–receptor complex releases a unit of the G-protein, which activates this enzyme. It then catalyzes the formation of cyclic AMP, which, in turn, activates an enzyme inside the cell.

molecule of an inactive form of the enzyme adenylate cyclase and activates it. This enzyme is an integral part of the cell membrane.

Once adenylate cyclase is activated, the "message" is on the inside of the cell membrane, because now the enzyme promptly catalyzes the conversion of ATP into cyclic AMP and diphosphate ion, PP_i.

■ The complete structure of ATP is shown on page 151.

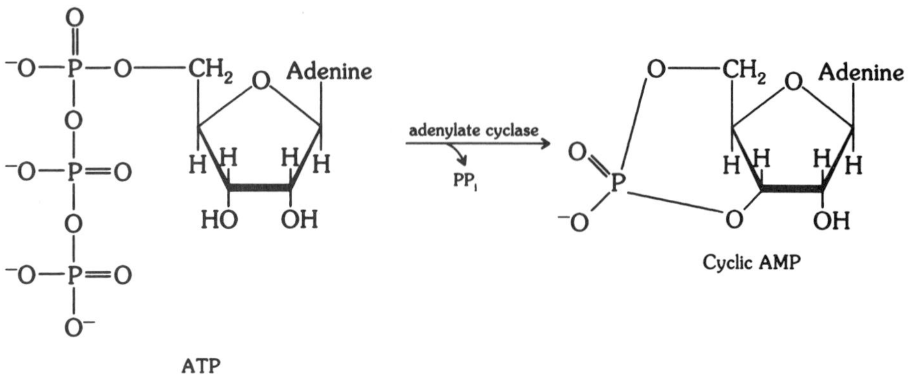

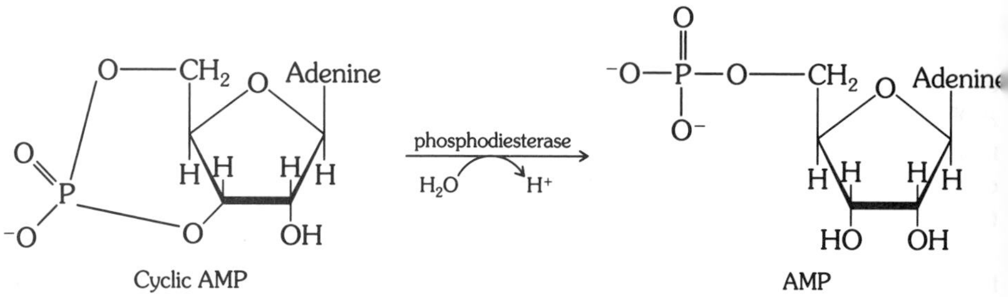

■ E. W. Sutherland, Jr., an American scientist, won the 1971 Nobel prize in physiology and medicine for his work on cyclic AMP.

The newly formed cyclic AMP now activates an enzyme, which, in turn, catalyzes a reaction. This last event is what the original message was all about.

Finally, an enzyme called *phosphodiesterase* catalyzes the hydrolysis of cyclic AMP to AMP, and this shuts off the cycle. Energy-producing reactions in the cell will now remake ATP from the AMP.

Let's summarize the steps in this remarkable chemical cascade.

1. A signal releases the hormone or neurotransmitter.
2. It travels to its target cell, next door for a neurotransmitter but some farther distance away for a hormone.
3. The primary messenger molecule finds its target cell by a lock-and-key mechanism and binds to a receptor, which alters a polypeptide unit of the G-protein.
4. The altered G-protein activates the enzyme adenylate cyclase.
5. Adenylate cyclase catalyzes the conversion of ATP to cyclic AMP.
6. Cyclic AMP, the secondary messenger, activates an enzyme inside the cell.
7. The enzyme catalyzes a reaction, one that corresponds to the primary message of the hormone or neurotransmitter.
8. Cyclic AMP is hydrolyzed to AMP, which is reconverted to ATP, and the system returns to the preexcited state.

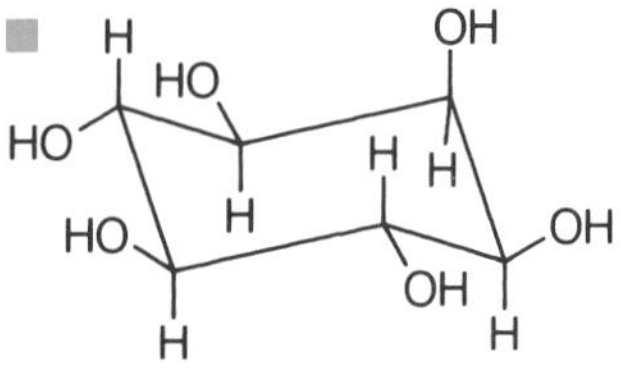

■ Hormones that work through the cyclic AMP cascade that we will encounter later include epinephrine, glucagon, norepinephrine, and vasopressin.

The Inositol Phosphate System Works in a Way Roughly Similar to the Cyclic AMP System Another of the G-protein communications systems used by cells involves phosphate and trisphosphate esters of inositol, one of the stereoisomers of hexahydroxycyclohexane. We can follow it with the aid of Figure 11.12.

■ Inositol

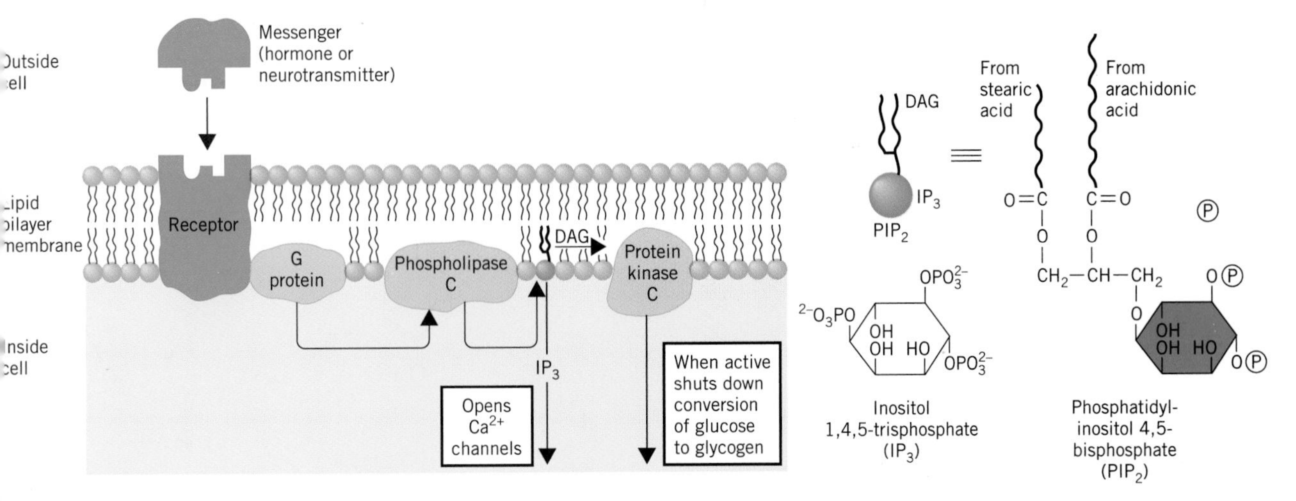

FIGURE 11.12
The inositol phosphate system that takes a message from a hormone or neurotransmitter and activates two enzymes to work cooperatively.

The primary messenger binds to the receptor, the G-protein is altered, and it activates an enzyme called phospholipase C. The pattern so far, as you can see, is quite similar to the cyclic AMP system. Phospholipase C now catalyzes the hydrolysis of a phosphate ester bond in a membrane-bound phospholipid, which we'll call PIP_2 (because its full name makes eyes glaze over). The products are two fragments *that are both messengers*. One we'll call IP_3 (an inositol trisphosphate) and the other DAG (a diacylglycerol).

■ There are several variations of the G-protein.

DAG, being mostly hydrocarbon-like, stays in the lipid bilayer of the cell membrane where it activates a membrane-bound enzyme, protein kinase C. As we'll study under the metabolism of glucose, this enzyme aids the cell in developing a higher glucose level, which might be needed for energy.

■ The phosphoinositol cascade mediates the following activities:

glycogenolysis in the liver
insulin secretion
smooth muscle contraction
platelet aggregation

The other fragment, IP_3, causes the rapid release of Ca^{2+} ion from intracellular storage systems. This ion might, for example, activate the contraction of a muscle. Of course, this is work requiring energy, so you can see that the two fragments, DAG and IP_3, work together. IP_3 tells the cell to do some work, DAG sees to the energy supply, and both are responses to one signal at the cell receptor.

Interestingly, malfunctions in the inositol phosphate system may be responsible for manic-depressive illness, one of the major psychiatric disorders. Lithium ion (as lithium carbonate) is used to treat this condition, and its action is evidently to shut down the inositol phosphate network in affected cells.

The inositol phosphate system may also be a target for the action of chemicals, including errant genes, that cause tumors and cancer. Protein kinase C has a function in cell division and in the control of the proliferation of cells, so chemicals that interact with this enzyme in the wrong way can affect cell division.

■ The cholera toxin is able to lock the G-protein into its active form, so adenylate cyclase cannot be shut off. This stimulates so much active transport of Na^+ ion into the gut, along with water, that massive diarrhea kills the victim.

11.7 HORMONES AND NEUROTRANSMITTERS

Interventions in the work of hormones and neurotransmitters are the bases of the action of a number of drugs, both licit and illicit.

Structurally, Hormones Come in Four Broad Types The principal endocrine glands of the human body and the major hormones they secrete are too numerous to catalog in detail. It is impossible to do justice to a subject as vast as hormones in one section of one chapter, so

TABLE 11.2
Neurotransmitters

Monoamines	
Acetylchloline	$(CH_3)_3\overset{+}{N}CH_2CH_2OCCH_3$ (C=O)
Dopamine	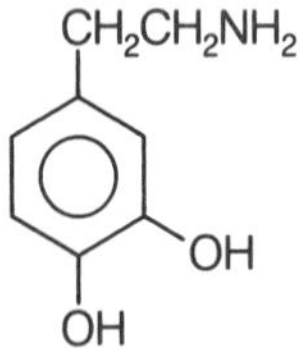 $CH_2CH_2NH_2$, OH, OH
Norepinephrine	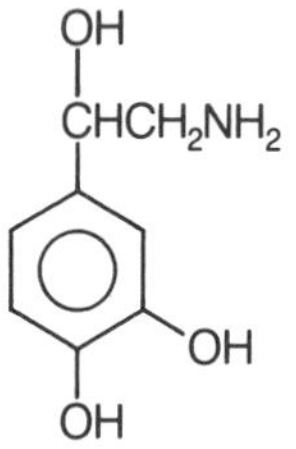 OH, $CHCH_2NH_2$, OH, OH
Serotonin	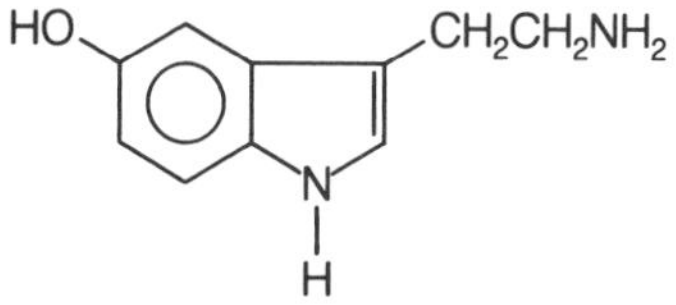 HO, $CH_2CH_2NH_2$, N–H
Amino Acids	
Glycine	$^{+}NH_3CH_2CO_2^{-}$
γ-Aminobutyric acid (GABA)	$^{+}NH_3CH_2CH_2CH_2CO_2^{-}$
Glutamic acid	$^{+}NH_3CHCO_2^{-}$ $CH_2CH_2CO_2H$
Neuropeptides	
Met-Enkephalin	Tyr-Gly-Gly-Phe-Met
Leu-Enkephalin	Tyr-Gly-Gly-Phe-Leu
β-Endorphin	Tyr-Gly-Gly-Phe-Met-Thr-Ser-Glu-Lys-Ser Gln-Thr-Pro-Leu-Val-Thr-Leu-Phe-Lys-Asn Ala-Ile-Val-Lys-Asn-Ala-His-Lys-Gly-Gln
Substance P	Arg-Pro-Lys-Pro-Gln-Gln-Phe-Phe-Gly-Leu-Met-NH_2
Angiotensin II	Asp-Arg-Val-Tyr-Ile-His-Pro-Phe-NH_2
Somatostatin	Ala-Gly-Cys-Lys-Asn-Phe Phe Cys Trp Ser-Thr-Phe-Thr

what follows is a very broad sketch of a few chemical aspects of hormone action. In later chapters we will mention specific hormones where they are particularly relevant to the metabolic activity being studied. Let's first consider some features of hormone molecules. They come in four general types.

Some hormones, the steroid hormones, are made from cholesterol and so have largely hydrocarbon-like molecules. This feature enables them to slip easily through the lipid bilayers of their target cells. Inside they find their final receptors, and the hormone–receptor complexes move to DNA molecules where they bind and affect the transcriptions of genetic messages. The sex hormones like estradiol, progesterone, and testosterone work in this way.

Many growth factors as well as insulin, oxytocin, and thyroid-stimulating hormone consist of polypeptides or proteins. These are able to alter the permeabilities of their target cells to the migrations of small molecules. The growth factors, for example, help get amino acids inside cells where they are needed for growth. Insulin helps to get glucose inside its target cells. We'll have much more to say about insulin in a later chapter. Either the absence of insulin or the absence (or inactivity) of insulin receptors results in the disease diabetes mellitus ("diabetes"). Several neurotransmitters are also polypeptides, and they alter the permeability of a neuron membrane to Ca^{2+} and Na^{+}. The cross-membrane movements of these ions are involved in the electrical signal that flows down a neuron.

The prostaglandins (Special Topic 9.1, page 226) are now classified as hormones, as *local hormones* because they work where they are made. The *eicosanoids* is the technical name for this family of compounds.

■ From *eicosane*, a C-20 hydrocarbon.

Finally, a number of hormones are relatively simple amino compounds made from amino acids. These include epinephrine (page 165) and thyroxine. Some of these are also neurotransmitters.

Neurotransmitters Move across the Narrow Synaptic Gap from One Neuron to the Next A partial list of neurotransmitters is given in Table 11.2. Some are nothing more than simple amino acids. Others are β-phenylethylamines or catecholamines (Special Topic 6.1, page 165), and many are polypeptides.

Each nerve cell has a fiber-like part called an *axon* that reaches to the face of the next neuron or to one of its filament-like extensions called *dendrites*. A nerve impulse consists of a traveling wave of electrical charge that sweeps down the axon as small ions migrate at different rates between the inside and the outside of the neuron. The problem is how to get this impulse launched into the next neuron so that it can continue along the length of the nerve fiber. This is solved by a *chemical* communication from one neuron to the next. Neurotransmitters are the chemicals involved, and they are made from amino acids within the neuron and stored in sacs, called *vesicles,* located near the ends of the axons.

■ The traveling wave of electrical charge moves rapidly, but still not as rapidly as electricity moves in electrical wires.

Between the terminal of an axon and the end of the next neuron, there is a very narrow, fluid-filled gap called the *synapse*. Neurotransmitters move across the synapse when the electrical wave causes them to be released from their vesicles.

When neurotransmitter molecules lock into their receptors on the other side of the synapse, adenylate cyclase (or phospholipase C) is activated. (We'll use the adenylate cyclase system to illustrate the process.) See Figure 11.13. Now the formation of cyclic AMP is catalyzed, and newly formed cyclic AMP initiates whatever change is programmed by the chemicals in the target neuron. An enzyme then deactivates adenylate cyclase by catalyzing the release of the neurotransmitter molecule.

If the newly released neurotransmitter were a hormone, it would be swept away in the bloodstream, but it's not. It's still a neurotransmitter in the synapse, so unless the system wants it to act again, it must be removed or deactivated. A number of options are open, depending on the neurotransmitter.

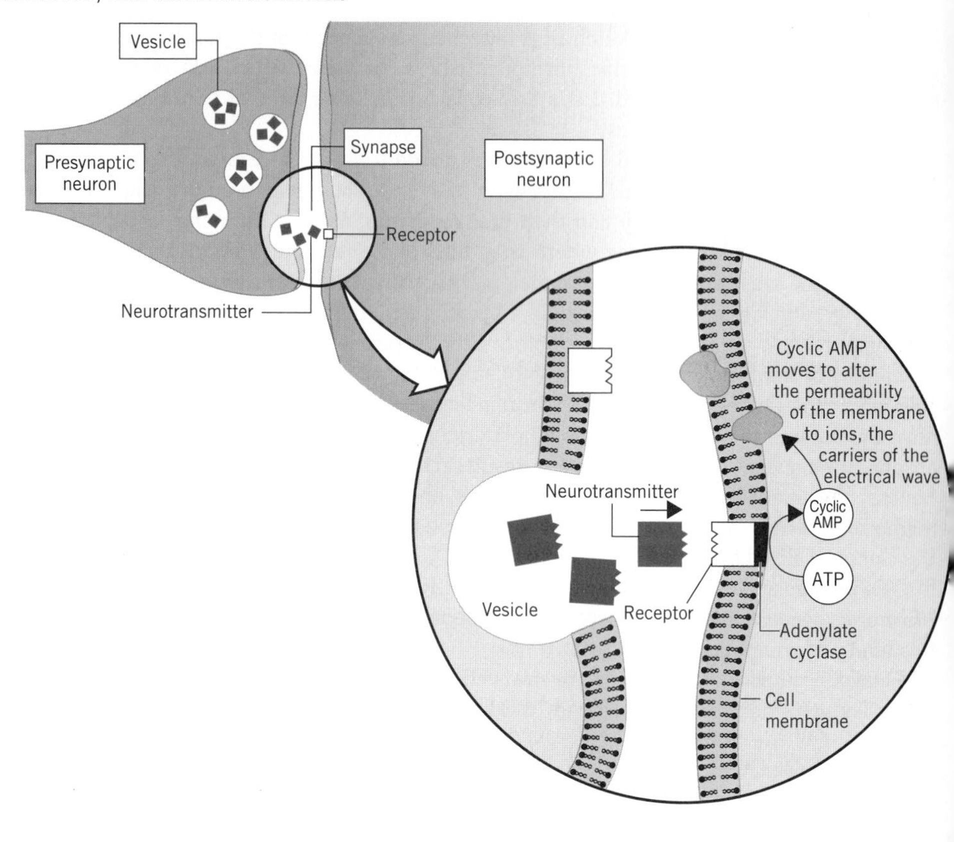

FIGURE 11.13 In neurotransmission, the neurotransmitter molecules, released from vesicles of the presynaptic neuron, travel across the synapse. At the postsynaptic neuron, they find their receptors, and the cyclic AMP system similar to that shown in Figure 11.11 becomes activated.

■ The ANS nerves handle the signals that run the organs that have to work autonomously (without conscious effort), such as the heart and the lungs.

The Neurotransmitter Acetylcholine Is Swiftly Hydrolyzed One method used to remove the neurotransmitter is to break it up by a chemical reaction. Acetylcholine, for example, a neurotransmitter in the autonomic nervous system or ANS, is catalytically hydrolyzed to choline and acetic acid. The enzyme is choline acetyltransferase.

$$\underset{\text{Acetylcholine}}{(CH_3)_3\overset{+}{N}CH_2CH_2O\overset{O}{\overset{\|}{C}}CH_3} + H_2O \xrightleftharpoons[\text{acetyltransferase}]{\text{choline}} \underset{\text{Choline}}{(CH_3)_3\overset{+}{N}CH_2CH_2OH} + \underset{\text{Acetic acid}}{HO\overset{O}{\overset{\|}{C}}CH_3}$$

Within two milliseconds (2×10^{-3} s) of the release of acetylcholine in the synapse, all its molecules are broken down. The synapse is now cleared for a fresh release of acetylcholine from the presynaptic neuron if the signal for its release continues. If the signal does not come, then the action is shut down.

■ The ANS nerves that use acetylcholine are called the *cholinergic nerves.*

The Botulinum Toxin Prevents a Neuron from Making Acetylcholine The botulinum toxin is an extremely powerful toxic agent made by the food-poisoning botulinum bacterium and works by preventing the *synthesis* of acetylcholine. Without this neurotransmitter, the cholinergic nerves of the ANS can't work.

Nerve Poisons Deactivate Choline Acetyltransferase The nerve gases act by deactivating the enzyme needed to break acetylcholine down after it has done its work. The absence of this enzyme means, therefore, that the signal transmitted by acetylcholine can't be turned off. It continues unabated until the heart fails, usually in a minute or two.

An antidote for nerve gas poisoning, atropine, works by blocking the receptor protein for acetylcholine, so despite the continuous presence of this neurotransmitter, it isn't able to complete the signal-sending work. This tones the system down, and other processes slowly restore the system to normal. (Given the extreme speed with which nerve gases work, atropine must obviously be used very promptly.)

Some organophosphate insecticides are mild nerve poisons and work in the same way. Other blockers of the receptor protein for acetylcholine are some local anesthetics like nupercaine, procaine, and tetracaine.

Drugs that block the action of a neurotransmitter are called **antagonists** to the neurotransmitter. The neurotransmitter itself is sometimes referred to as an **agonist.**

Some Neurotransmitters Are Reabsorbed by the Presynaptic Neuron Norepinephrine, another neurotransmitter, is deactivated by being reabsorbed by the neuron that released it, where it is then degraded. (Some is also deactivated right within the synapse.) The degradation of norepinephrine is catalyzed by enzymes called the **monoamine oxidases** or MAO.

■ The nerves that use norepinephrine are called the *adrenergic nerves* (after an earlier name for norepinephrine, noradrenaline).

Drugs that Inactivate the Monoamine Oxidases Are Used to Treat Depression One place where norepinephrine works is in the brain stem, where mood regulation is centered. If for any reason the monoamine oxidases are inactivated, then an excess of norepinephrine builds up in brain stem cells, and some spills back into the synapse and sends signals on. In some mental states, like depression, an abnormally low level of norepinephrine develops, so now one would want to inactivate the monoamine oxidases. This would leave what norepinephrine there is to carry on its work. Thus, some of the antidepressant drugs, such as iproniazid, work by inhibiting the monoamine oxidases.

■ O, CNHNHCH$(CH_3)_2$, N

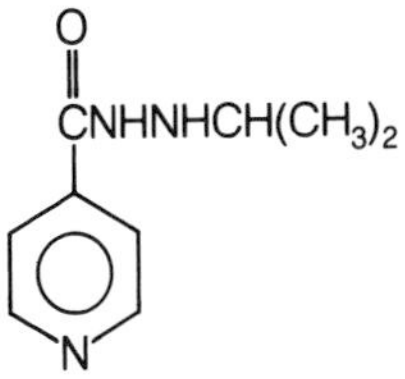

Iproniazid

Other antidepressants, like amitriptyline (e.g., Elavil) and imipramine (e.g., Tofranil), inhibit the reabsorption of norepinephrine by the presynaptic neuron. Without this reabsorption into the degradative hands of the monoamine oxidases, the level of norepinephrine and its signal-sending work stay high.

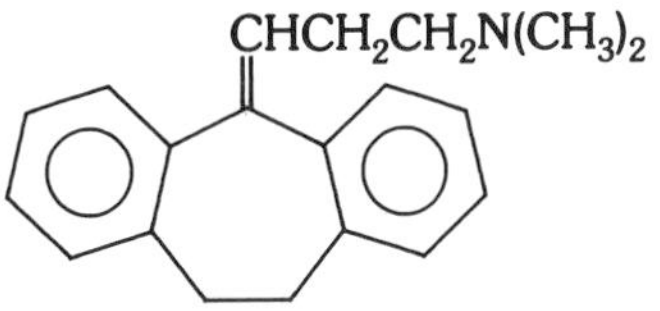

$CHCH_2CH_2N(CH_3)_2$

Amitriptyline (Elavil)

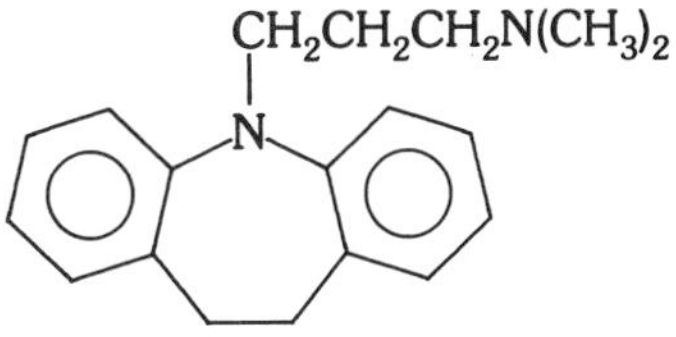

$CH_2CH_2CH_2N(CH_3)_2$, N

Imipramine (Tofranil)

Norepinephrine is both a hormone and a neurotransmitter. The adrenal medulla secretes it into the bloodstream in emergencies when it must be made available to all the nerve tissues that use it.

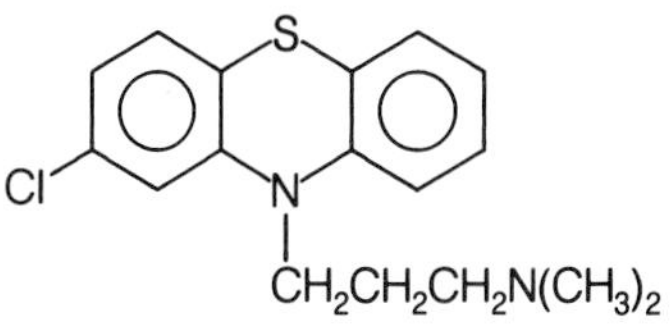

■ S, Cl, N, $CH_2CH_2CH_2N(CH_3)_2$

Chloropromazine

Dopamine Excesses Occur in Schizophrenia Dopamine, like norepinephrine, is also a monoamine neurotransmitter. It occurs in neurons of the midbrain that are involved with feelings of pleasure and arousal as well as with the control of certain movements.

In schizophrenia the neurons that use dopamine are overstimulated, because either the releasing mechanism or the receptor mechanism is overactive. Drugs commonly used to treat schizophrenia such as chlorpromazine (e.g., Thorazine) and haloperidol (Haldol) bind to dopamine receptors and thus inhibit its signal-sending work.

F, HO, O, N, Cl

Haloperidol

Amphetamine Abuse Causes Schizophrenia-like Symptoms Stimulants like the amphetamines (cf. Special Topic 6.1, page 165) work by triggering the release of dopamine into the arousal and pleasure centers of the brain. The effect is therefore a "high." But it's easy to abuse the amphetamines. When this occurs, the same kind of overstimulation associated with schizophrenia results, with resulting symptoms as delusions of persecution, hallucinations, and other disturbances of the thought processes.

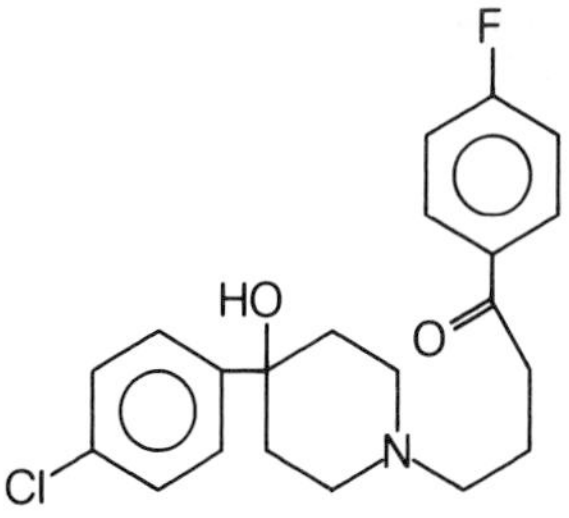

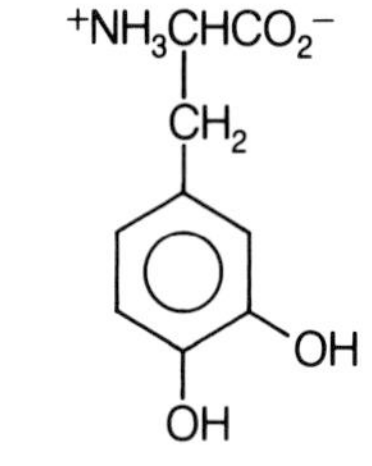

$^{+}NH_3CHCO_2^{-}$ | CH_2 | OH | OH

L-DOPA

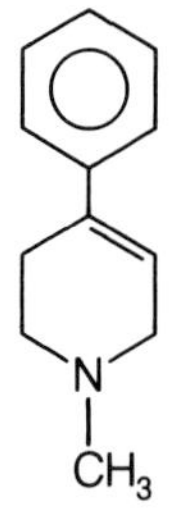

CH_3 N | CH_3 N^{+}

MPTP 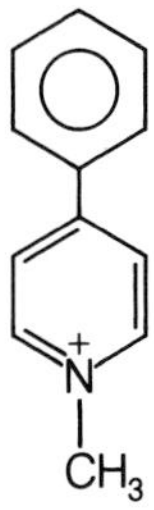 MPP^{+}

Dopamine-Releasing Neurons Have Degenerated in Parkinson's Disease When the dopamine-using neurons in the brain degenerate, as in Parkinson's disease, an extra supply of dopamine itself is then needed to compensate. This is why a compound called L-DOPA (*l*evorotatory *d*ihydr*o*xy*p*henyl*a*lanine) is used. The neurons that still work can use it to make extra dopamine.

In the mid-1980s, by an accidental discovery, it was found that a contaminant in street heroin, called MPTP for short, rapidly destroys the same cells that degenerate in Parkinson's disease. The active agent is actually a metabolic breakdown product called MPP^{+}. The cells affected by Parkinson's disease apparently do not die at once but first go into a dormant state. The rejuvenation of such cells has long been sought. In the early 1990s, scientists, working with experimental rats, found that a substance called *brain-derived neurotrophic factor* or BDNF protects the dormant cells, stimulates them into recovery, and protects susceptible cells.

In 1993, another neurotrophic factor, GDNF (for glial cell line-derived neurotrophic factor) was also found to be able to promote the survival of neurons that release dopamine. Another line of research into treating Parkinson's disease involves transplants of healthy fetal brain tissue. Preliminary results of the use of this technique on human patients, made public in 1992, were promising.

$^{+}NH_3CH_2CH_2CH_2CO_2^{-}$
GABA

GABA Inhibits Nerve Signals The normal function of some neurotransmitters is to *inhibit* signals instead of to initiate them. Gamma-aminobutyric acid (GABA) is an example, and as many as a third of the synapses in the brain have GABA available.

The inhibiting work of GABA can be made even greater by mild tranquilizers such as diazepam (e.g., Valium) and chlordiazepoxide hydrochloride (Librium), as well as by ethanol. The augmented inhibition of signals reduces anxiety, affects judgment, and induces sleep. Of course, you've probably heard of the widespread abuse of Valium and Librium, to say nothing of alcohol.

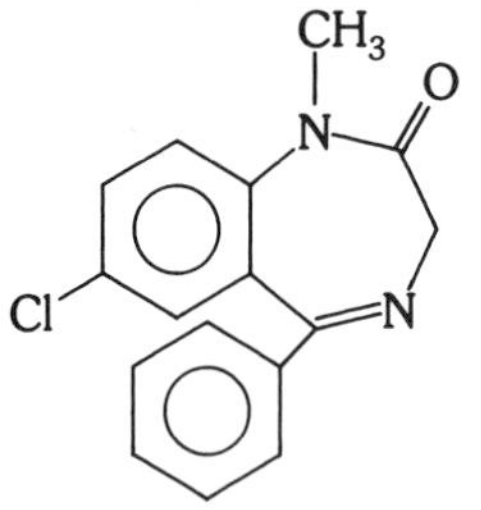

CH_3 O N Cl N | CH_3 NH_2Cl N Cl N O

Diazepam (Valium)

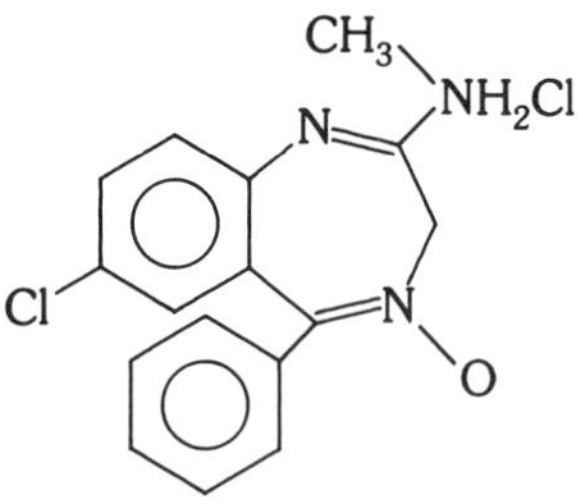

Chlordiazepoxide hydrochloride (Librium)

Greek, *chorea*, dance.

GABA Is Deficient in Huntington's Chorea The victims of Huntington's chorea, a hereditary neurological disorder, suffer from speech disturbances, irregular movements, and a steady mental deterioration, all related to a deficiency in GABA. Unhappily, GABA can't be administered in this disease, because it can't move out of circulation and into the regions of the brain where it works.

En- or *end-*, within; *kephale*, brain; *-orph-*, from morphine.

Several Polypeptides Act as Painkilling Neurotransmitters As we said, some neurotransmitters are relatively small polypeptides. See Table 11.2. One type includes the *enkephalins*; another consists of the *endorphins*. Both types are powerful pain inhibitors. One compound, dynorphin, is the most potent painkiller yet discovered, being 200 times stronger than morphine, an opium alkaloid that is widely used to relieve severe pain. Sites in the brain that strongly bind molecules of morphine also bind those of the enkephalins, so these natural painkillers are now often referred to as the body's natural opiates.

The Enkephalins Inhibit the Release of Substance P Substance P is a pain-signaling, polypeptide neurotransmitter. According to one theory, when a pain-transmitting neuron is activated, it releases substance P into the synapse. However, butting against such neurons are other neurons that can release enkephalins. And these, when released, inhibit the work of substance P. In this way, the intensity of the pain signal is toned down. The action of enkephalin might explain the delay of pain that sometimes occurs during an emergency when the brain and the body must continue to function to escape the emergency.

Substance P might be involved in the link between the nervous system and the body's immune system. It is known that some forms of arthritis flare up under stress, and stress deeply involves the nervous system. But arthritis is generally regarded as chiefly a disease of the immune system. In tests on rats with arthritis, flare-ups of the arthritis could be induced by injections of substance P. What interests scientists about this is the possibility of controlling the severity of arthritis by somehow diminishing levels of substance P in the affected joints.

One of the interesting developments in connection with the endorphins is that acupuncture, a pain-alleviating procedure developed in China many centuries ago, might work by stimulating the production and release of endorphins.

Many Neurotransmitters Exert More than One Effect Several neurotransmitters can be received by more than one kind of receptor. For example, at least three types of receptors for the opiates have been identified thus far. Such receptor multiplicity may explain how some neurotransmitters have multiple effects. Thus not only do opiates reduce pain, but they affect emotions, they induce sleep, and they affect the appetite, with each of the opiate receptors handling a different one of these functions.

Calcium Channel Blockers Are Drugs that Reduce the Vigor of Heart Muscle Contractions As we have often seen, calcium ions are major secondary chemical messengers, and neurotransmitters are able to open channels for calcium ions through cell membranes. Heart muscle tissue receives such signals at a rate that paces the heart as its muscles contract and relax during the heartbeat. Calcium ions are what finally deliver the message to contract. Then the cell pumps them back out and the muscle relaxes until another cycle starts.

Drugs like nifedipine (Procardia, Adalat), diltiazem (Cardizem), and verapamil (Isoptin) find calcium ion channels in heart muscle and block them. Not all are blocked, of course, so the effect is to reduce the migrations of Ca^{2+} through cell membranes. These calcium channel blockers (also called calcium antagonists or slow channel blocking agents) thus make each heart muscle contraction less vigorous. This reduces the risk of heart attacks in people known to be at risk, like those who experience angina pectoris and cardiac arrhythmias (heartbeat irregularities).

Nitric Oxide Is a Retrograde Messenger that Might Be Involved in the Storage of Memory Learning involves putting experiences, information, and data into memory in such a way that they can be retrieved and applied in new situations. If life has a molecular basis, then there must be a molecular basis to memory, too. What this basis is has long intrigued neuroscientists. According to one broad model of how some types of learning take place, a nerve cell that receives a signal associated with something to be remembered manufactures a *retrograde messenger*. This moves back to the signal-sending cell and strengthens the connection between the two cells. Each additional time that the same signal is sent, the signal is further strengthened.

■ "Retrograde" means moving or directed backward.

One of the candidates for retrograde messenger is nitric oxide, NO. Its tiny molecules are able to slip easily through cell membranes, which is an essential property for any retrograde messenger. It is known that cultured brain cells can make nitric oxide when certain receptors are stimulated, but when a binder of NO is present, experimental rats are unable to learn certain spatial tasks. The discovery of a neurotransmitter role for an otherwise noxious gas, NO, means that not all chemical messenger work is done by a common mechanism. NO is a gas and can diffuse, as we said, from cell to cell relatively easily in several directions. The

■ Abnormal NO systems have been found in some people with hypertension and in others with high blood cholesterol levels.

"classical" neurotransmitters must be stored and then released on signal. Next, they must find a receptor in a membrane in order to influence events within the cell. NO is able to slip directly through a membrane and so needs no membrane-bound receptor. (One enzyme that NO is known to stimulate directly is guanylate cyclase, an enzyme similar to adenylate cyclase.)

Carbon Monoxide Is Probably a Retrograde Messenger In the early 1990s, evidence was found that carbon monoxide, another noxious gas, might also be a brain messenger and possibly involved in a long-term learning-linked process. (One is certainly tempted to say, "Of all things!") In relatively large concentrations when carried into the body in inhaled air, CO binds so strongly to heme units in hemoglobin that the latter cannot carry oxygen, which causes death. Yet in small concentrations manufactured within the brain, CO activates an enzyme involved in learning. This is an area of rapidly moving research driven in part by intense curiosity about *how* we remember.

What we have done in this section is look at some *molecular* connections between conditions of the nervous system and particular chemical substances. This whole field is one of the most rapidly moving areas of scientific investigation today, and during the next several years we may expect to see a number of dramatic advances both in our understanding of what is happening and in the strategies of treating both mental and heart diseases.

■ Solomon Snyder (The Johns Hopkins University) has been a pioneer in studying how the brain works, biochemically.

SUMMARY

Enzymes Enzymes are the catalysts in cells. Some consist wholly of one or more polypeptides, and other enzymes include a cofactor besides the polypeptides. The cofactor can be an organic coenzyme, a metal ion, or both. Some coenzymes are phosphate esters of B vitamins and, in these examples, the vitamin unit usually furnishes the enzyme's active site. Because they are mostly polypeptide in nature, enzymes are vulnerable to all of the conditions that denature proteins. The name of an enzyme, which almost always ends in *-ase,* usually discloses either the identity of its substrate or the kind of reaction it catalyzes.

Some enzymes occur as small families called isoenzymes in which the polypeptide components vary slightly from tissue to tissue in the body. An enzyme is very specific both in the kind of reaction it catalyzes and in its substrate. Enzymes make possible reaction rates that are substantially higher than the rates of uncatalyzed reactions.

Induced fitting When an enzyme–substrate complex forms, the active site is brought together with the part of the substrate that is to react. Binding sites on the enzyme guide the substrate molecule in and induce a change in the conformation of the enzyme molecule to produce the best fit of the substrate. The recognition of the enzyme by the substrate occurs as a flexible, lock-and-key model that involves complementary shapes and electrical charges.

Enzyme kinetics An enzyme is a reactant in the initial phase of the reaction as it functions catalytically. At a sufficiently high initial substrate concentration, the active sites on all the enzyme molecules become saturated by substrate molecules, so at still higher initial substrate concentrations, there is little further increase in the initial rate. The rate behavior of enzyme-catalyzed reactions for which the enzyme has one catalytic site is described by the Michaelis-Menten equation. At low substrate concentration, the equation produces a straight-line plot of rate versus substrate concentration. At high substrate concentration, the equation shows that the rate must reach a maximum value.

Some enzymes seem to respond sluggishly to small increases in substrate concentration when the latter is very low. These display a sigmoid rate curve, and they require activation by substrate molecules before they produce their dramatic rate enhancements.

Regulation of enzymes Enzymes that display a sigmoid rate curve can be activated allosterically either by their own substrates or by effectors. Some enzymes are activated by genes. Some enzymes that are parts of the membranes of cells (or small bodies within cells) are activated by the interaction between a hormone or a neurotransmitter and its receptor protein. The work of many of these is to cause changes in the calcium ion level in a cell.

Other enzymes, such as certain digestive enzymes, exist as zymogens (proenzymes) and are activated when some agent acts to remove a small part that blocks the active site. The kinase enzymes are activated by being phosphorylated.

Enzymes can be inhibited by a non-product inhibitor or by competitive feedback that involves a product of the enzyme's action. Some inhibitors act allosterically; they bind to the enzyme at some location other than the catalytic site. Some of the most dangerous poisons bind to active sites and irreversibly block the work of an enzyme, or they carry enzymes out of solution by a denaturant action. Many antibiotics and other antimetabolites work by inhibiting enzymes in pathogenic bacteria.

Medical uses of enzymes The serum levels of many enzymes rise when the tissues or organs that hold these enzymes are injured or diseased. By monitoring these serum levels, and by looking for certain isoenzymes, we can diagnose diseases — for example, viral hepatitis and myocardial infarctions.

When a blood clot threatens or causes a heart attack, any one of three enzymes—streptokinase, ASPAC, or recombinant tissue plasminogen activator (TPA)—can be used to initiate the hydrolysis of the fibrin of the clot.

Enzymes are also used in analytical systems that measure concentrations of substrates, such as in tests for glucose. In some analytical systems, enzymes are immobilized on electrodes where they catalyze a reaction that produces a product; the electrode then senses and measures this product.

Chemical communication with hormones and neurotransmitters The carbohydrate tails of membrane glycoproteins project away from the membrane surface and provide recognition sites for chemical messengers (and other particles). The messenger molecule and the receptor protein form a complex. One common response to the formation of the complex is the release of a G-protein. In the adenylate cyclase cascade, this activates adenylate cyclase, which then triggers the formation of cyclic AMP. In turn, cyclic AMP sets off other events, such as the activation of an enzyme that catalyzes a reaction, one that is ultimately what the "signal" of the neurotransmitter was all about.

When the release of a G-protein is followed by the inositol phosphate cascade, the G-protein molecule activates phospholipase C, which breaks up a phospholipid in the membrane into two enzyme activators, PIP_2 and DAG. PIP_2 initiates the rapid release of Ca^{2+} from cytoplasm stores to cause muscle contraction, and DAG helps to keep the cell's glucose level high so that its metabolism can supply the energy.

Hormones Endocrine glands secrete hormones, and these primary chemical messengers travel to their target cells in the blood, where they activate a gene, or an enzyme, or affect the permeability of a cell membrane. They recognize their own target cells by binding to specific receptor proteins.

The steroid hormones can move into a cell to its nucleus and there find a receptor. Polypeptide hormones bind to membrane-bound receptors to initiate their action.

Neurotransmitters In response to an electrical signal, vesicles in an axon release a neurotransmitter that moves across the synapse. Its molecules bind to a receptor protein on the next neuron, and then the pattern is much like that of hormones. The result, however, is to open channels through the cell membrane for the migration of ions.

Neurotransmitters include amino acids, monoamines, and polypeptides. Some neurotransmitters *activate* some response in the next neuron, whereas others *deactivate* some activity. A number of medications work by interfering with neurotransmitters or with the opening of calcium ion channels.

REVIEW EXERCISES

The answers to Review Exercises whose numbers are in color are found in Appendix A. The answers to the other Review Exercises are found in the Study Guide that accompanies this book. The more challenging questions are marked with asterisks.

Nature of Enzymes

11.1 What are (a) the function and (b) the composition, in general terms only, of an enzyme?

11.2 To what does *specificity* refer in enzyme chemistry?

11.3 Define and distinguish among the following terms.
(a) apoenzyme (b) cofactor (c) coenzyme

11.4 Write the equation for the equilibrium catalyzed by carbonic anhydrase. What is particularly remarkable about the enzyme?

11.5 What in general does an enzyme do to an equilibrium?

Coenzymes

11.6 What B vitamin is involved in the $NAD^+/NADH$ system?

11.7 The active part of either FAD or FMN is furnished by which vitamin?

11.8 Complete and balance the following equation.

$$CH_3\overset{\displaystyle OH}{\overset{|}{C}}HCH_3 + NAD^+ \longrightarrow CH_3\overset{\displaystyle O}{\overset{\|}{C}}CH_3 + \_\_\_ + \_\_\_$$

11.9 Complete and balance the following equation.

$$\_\_\_ + NADH + FAD \longrightarrow NAD^+ + \_\_\_$$

11.10 In what structural way do NAD^+ and $NADP^+$ differ? What formula can be used for the reduced form of $NADP^+$?

Kinds of Enzymes

11.11 What *kind* of reaction does each of the following enzymes catalyze?
(a) an oxidase (b) transmethylase
(c) hydrolase (d) oxidoreductase

11.12 What is the difference between lactose and lactase?

11.13 What is the difference between a hydrolase and hydrolysis?

11.14 What are isoenzymes (in general terms)?

11.15 What are the three isoenzymes of creatine kinase? Give their symbols and state where they are principally found.

Theory of How Enzymes Work

11.16 What name is given to the part of an enzyme where the catalytic work is carried out?

11.17 How is enzyme specificity explained?

11.18 What is the induced-fit theory?

11.19 The Michaelis-Menten equation applies to an enzyme having how many catalytic sites?

*11.20 Which relationship, $V \propto [S]$ or $V \propto [E_o]$, applies under each of the following conditions?
(a) A relatively high concentration of substrate.
(b) A relatively low concentration of substrate.

11.21 When the velocity of an enzyme-catalyzed reaction obeying the Michaelis-Menten equation equals half of the maximum velocity, what does the Michealis constant compare to?

Enzyme Activation and Inhibition

11.22 What does *allosteric* mean?

11.23 If the plot of initial reaction rate versus initial substrate concentration at constant [E] has a sigmoid shape, what does this signify about the active site(s) on the enzyme?

11.24 How does a substrate molecule activate an enzyme whose rate curve is sigmoid?

11.25 How does an effector differ from a substrate in causing allosteric activation?

11.26 What are the names of two important effectors? Which one is used in muscle cells?

11.27 What are the approximate concentrations of calcium ion in the cytosol and the fluid just outside a cell? Why doesn't simple diffusion wipe out this concentration gradient?

11.28 Why must the concentration of Ca^{2+} be so low in the cytosol?

11.29 Does the concentration of Ca^{2+} in the cytosol measure the amount of Ca^{2+} in the whole cytoplasm? Explain.

11.30 What does Ca^{2+} do to calmodulin or troponin?

11.31 When Ca^{2+} combines with troponin, what happens with respect to other proteins in the cell? What then happens to Ca^{2+}?

11.32 What is the relationship of a zymogen to its corresponding enzyme? Give an example of an enzyme that has zymogen.

11.33 What is enteropeptidase and what does it do?

11.34 What is plasmin and in what form does in normally circulate in the blood?

11.35 What role does a phosphorylation reaction have in connection with some enzymes?

11.36 How does competitive inhibition of an enzyme work?

11.37 Feedback inhibition of an enzyme works in what way?

11.38 Why is feedback inhibition an example of a homeostatic mechanism?

***11.39** How do competitive inhibition and allosteric inhibition differ?

11.40 How do the following poisons work?
(a) CN^- (b) Hg^{2+}
(c) nerve gases or organophosphate insecticides

11.41 What are antimetabolites, and how are they related to antibiotics?

11.42 In broad terms, how does penicillin work?

Enzymes in Medicine

11.43 If an enzyme such as CK or LD is normally absent from blood, how can a *serum* analysis for either tell anything? (Answer in general terms.)

11.44 What is the significance of CK(*MB*) in serum in trying to find out whether a person has had a heart attack and not just some painful injury in the chest region?

11.45 What CK isoenzyme would increase if the injury in Review Exercise 11.44 were to skeletal muscle?

11.46 What is the LD_1-LD_2 flip, and how is it used in diagnosis?

11.47 How is immobilized urease used?

11.48 How is immobilized heparinase used?

11.49 Describe an example of an immobilized enzyme on an electrode tip and the function it serves.

11.50 What three enzymes are available to help dissolve a blood clot? Which one occurs in human blood, and how is it obtained for therapeutic uses?

11.51 What substance makes up most of a blood clot, and what happens to it when TPA works?

Chemical Communication

11.52 What are the names of the sites of the synthesis of (a) hormones and (b) neurotransmitters?

11.53 What do the lock-and-key and induced fit concepts have to do with the work of hormones and neurotransmitters?

11.54 In what general ways do hormones and neurotransmitters resemble each other?

11.55 What general name is given to the substance on a target cell that recognizes a hormone or neurotransmitter?

11.56 Name the two kinds of "enzyme cascades" studied in this chapter.

11.57 Name the small polypeptide that both systems (Review Exercise 11.56) use to activate something inside the cell.

11.58 What function does adenylate cyclase have in the work of at least some hormones?

11.59 How is cyclic AMP involved in the work of some hormones and neurotransmitters?

11.60 After cyclic AMP has caused the activation of an enzyme inside a cell, what happens to the cyclic AMP that stops its action until more is made?

***11.61** In what ways does the inositol "cascade" resemble the cyclic AMP cascade?

11.62 When the G-protein of the inositol cascade completes its work, it has produced *two* new messengers. In general terms, what does each one do next?

Hormones

11.63 What are the four broad types of hormones?

11.64 What structural fact about the steroid hormones makes it easy for them to get through a cell membrane?

11.65 In each case, what substance (or kind of substance) can enter a target cell more readily following the action of the hormone?
(a) insulin (b) growth hormone (c) a neurotransmitter

11.66 Why are the prostaglandins called *local hormones?*

Neurotransmitters

11.67 What happens to acetylcholine after it has worked as a neurotransmitter? What is the name of the enzyme that catalyzes this change? In chemical terms, what does a nerve gas poison do?

11.68 How does atropine counter nerve gas poisoning?

11.69 How does a local anesthetic such as procaine affect the functioning of acetylcholine as a neurotransmitter?

11.70 How does the botulinum toxin work?

11.71 What, in general terms, are the monoamine oxidases, and in what way are they important?

11.72 What does iproniazid do chemically in the neuron-signaling that is carried out by norepinephrine?

11.73 In general terms, how do antidepressants such as amitriptyline or imipramine work?

11.74 Which neurotransmitter is also a hormone, and what is the significance of this dual character to the body?

11.75 The overactivity of which neurotransmitter is thought to be one biochemical problem in schizophrenia?

11.76 How do the schizophrenia-control drugs chlorpromazine and haloperidol work?

11.77 How can the amphetamines, when abused, give schizophrenia-like symptoms?

11.78 How does L-dopa work in treating Parkinson's disease?

11.79 How does a neurotrophic factor like BDNF work in treating Parkinson's disease?

11.80 Which common neurotransmitter in the brain is a signal inhibitor? How do such tranquilizers as Valium and Librium affect it?

11.81 Why is enkephalin called one of the body's own opiates? How does it appear to work?

11.82 What does substance P do?

***11.83** How do the calcium channel blockers reduce the risk of a heart attack?

11.84 What is meant by the term *retrograde messenger?*

11.85 What physical property of nitric oxide or carbon monoxide enables them to act in a retrograde manner?

Work of NAD^+, FMN, or FAD (Special Topic 11.1)

11.86 When NAD^+ is involved in the oxidation of an alcohol, what small pieces leave the alcohol molecule? Where does each go?

11.87 When FMN or FAD is involved in the oxidation of an alcohol, what small pieces leave the alcohol molecule? Where does each go?

11.88 What specifically makes the NAD^+ unit an attractor of $H{:}^-$?

11.89 What molecule finally gets the electrons of $H{:}^-$ at the end of the entire metabolic pathway?

Complementarity and Immune Responses (Special Topic 11.2)

11.90 What is a pathogen? (In general terms.)

11.91 What two kinds of immunity are recognized and how do they differ?

11.92 The HIV particle attacks what kind of immunity and in what way?

11.93 What is an antibody? An antigen?

11.94 At the molecular level, what aspects of molecular structure explain the high specificity of the immune response?

11.95 In what kind of immunity are the B-cells operative, and what kind of substance do B-cells eventually make to counter an alien material?

11.96 Why is a glycolipid on a red blood cell of an A-type individual referred to as an antigen and not an antibody?

11.97 At the molecular level involving their red blood cells, in what specific ways do the A, B, and O types differ?

11.98 What is present in the blood of an A-type person that makes receiving blood from a B-type dangerous?

11.99 Why is it that O-types can donate blood to people of any type but can receive blood only from other O-types?

Additional Exercises

11.100 How does the plot of initial rate versus initial [S] at constant [E] look when (a) an allosteric effect is occurring and (b) no allosteric effect is observed? (Draw pictures.)

***11.101** If you drink enough methanol, you will become blind or die. One strategy to counteract methanol poisoning is to give the victim a nearly intoxicating drink of ethanol. As the ethanol floods the same enzyme that attacks the methanol, the methanol gets a lessened opportunity to react and it is slowly and relatively harmlessly excreted. Otherwise, it is oxidized for formaldehyde, the actual poison from an overdose of methanol:

$$\underset{\text{Methanol}}{CH_3OH} \xrightarrow{\text{dehydrogenase}} \underset{\text{Formaldehyde}}{CH_2O}$$

What kind of enzyme inhibition might ethanol be achieving here? (Name it.)

***11.102** Truffles are edible, potato-shaped fungi that grow underground in certain parts of France, and are highly prized by gourmet cooks and gourmands. Pigs are used to locate truffles buried as much as one meter below the surface because truffles carry traces of a steroid, androsten-16-en-3-ol, which is a powerful sex attractant for pigs. A sex attractant is a species-specific chemical compound made and released in trace amounts by a female member of a species toward which a male member experiences a powerful sexual response. A male pig, of course, does not initially know that it is a truffle emitting the attractant. (It

appears that all sex attractants for humans are *nonchemical* being better understood, perhaps, as public relations activities.)

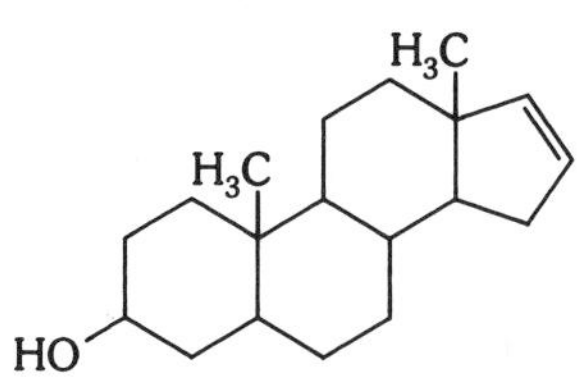

Androsten-16-en-3-ol

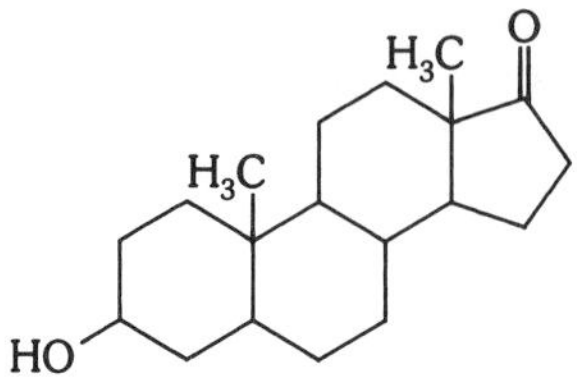

Androsterone

Notice the structural similarity between androsten-16-en-3-ol and a human sex hormone, androsterone.

(a) In terms of what general theory would we explain how androsten-16-en-3-ol has a particular specificity in pigs and not in humans but androsterone has a specificity in humans and not in pigs?

(b) In order to convert androsten-16-en-3-ol into androsterone *in vitro,* what specific *series* of changes in functional groups would have to be carried out. Your answer would begin something like "First, we have to change such and such a group into. . . . . Then this new group would have to be changed . . ." (All needed changes involve one-step reactions we have studied.)

***11.103** The structure of isoleucine is given on page 285. Draw the Fischer projection structures of its stereoisomers. (See Special Topic 7.1.)

***11.104** Referring to the equation given on page 285 for the conversion of threonine to isoleucine, what is the maximum number of milligrams of isoleucine that could be made from 150.0 mg of threonine?

EXTRACELLULAR FLUIDS OF THE BODY

12

High-altitude work, such as climbing along the southeast buttress of Fitzroy in Patagonia, places intense demands on the buffer system of the blood. The tendency is to overbreathe and so remove carbon dioxide too rapidly from the body. This leads to alkalosis and possibly fatal high-altitude sickness. The acid-base status of the blood is the major topic of this chapter.

12.1

DIGESTIVE JUICES

The end products of the digestion of the nutritionally important carbohydrates, neutral fats and oils, and proteins are monosaccharides, fatty acids, monoacylglycerols, and amino acids.

Life engages two environments, the outside environment we commonly think of, and the internal environment, which we usually take for granted. When healthy, our bodies have nearly perfect control over the internal environment, and so we are able to handle large changes outside more or less well: large temperature fluctuations, chilling winds, stifling humidity, and a fluctuating atmospheric tide of dust and pollutants.

■ The fluids of the internal environment make up about 20% of the mass of the body.

Cells Exist in Contact with Interstitial Fluids and Blood The **internal environment** consists of all the **extracellular fluids,** those that aren't actually inside cells. About three-quarters consists of the **interstitial fluid,** the fluid in the spaces or interstices *between* cells. The blood makes up nearly all the rest. The lymph, cerebrospinal fluid, digestive juices, and synovial fluids (lubricants of joints) are also extracellular fluids.

The chemistry occurring inside cells, in the *intracellular fluid,* has been and will continue to be a major topic of our study. Here we will focus on two of the extracellular fluids, the digestive juices and the blood.

The Digestive Tract Is a Convoluted Tube Running through the Body with Access to Several Solutions of Hormones and Hydrolases The principal parts of the digestive tract are given in Figure 12.1. The **digestive juices** are dilute solutions of electrolytes and hydrolytic enzymes (or zymogens) either in the cells lining the intestinal tract or in solutions that enter the tract from various organs.

The intestinal tract also includes specialized cells that manufacture and release hormones in response to various signals — the presence of arriving nutrients, or the distension of the wall of the tract, or a change in pH. Some hormones are secreted into the tract itself, and others enter the bloodstream. Taken as a whole, the digestive tract is itself a vast endocrine gland, the largest we have. The lower part of the stomach releases *gastrin,* which helps to stimulate the release of gastric juice. Cells in the upper intestinal tract release the hormone *cholecystokinin.* It acts to modulate the release of material from the stomach into the intestinal tract, which means that digestion proceeds at a sufficiently slow rate to be complete. Cholecystokinin also tells the gallbladder to release bile, and it makes the pancreas release pancreatic enzymes. *Secretin* is another hormone of the upper intestinal tract, and it makes the pancreas release bicarbonate ion for neutralizing incoming gastric acid.

■ Dextrins are partial breakdown products of amylopectin, a component of starch.

Saliva Provides α-Amylase, a Starch-Splitting Enzyme The flow of **saliva** is stimulated by the sight, smell, taste, and even the thought of food. Besides water (99.5%), saliva includes a food lubricant called *mucin* (a glycoprotein) and an enzyme, *α-amylase.* This enzyme catalyzes the partial hydrolysis of starch to dextrins and maltose, and it works best at the pH of saliva, 5.8–7.1. Proteins and lipids pass through the mouth essentially unchanged.

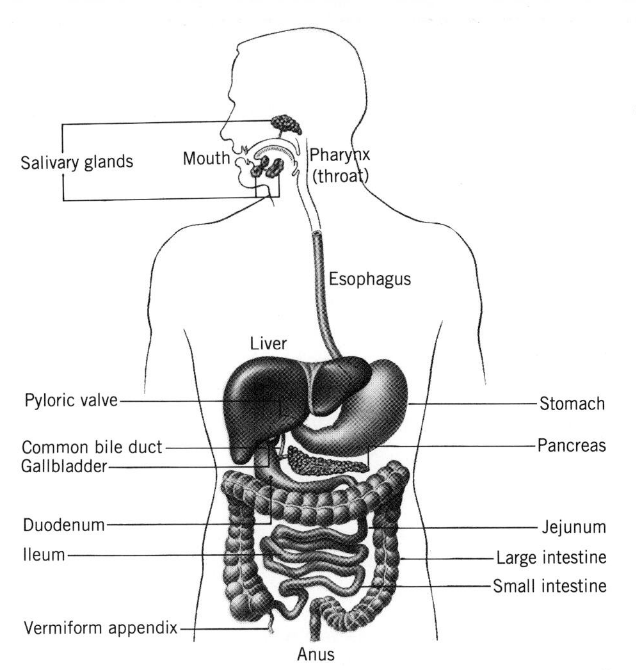

FIGURE 12.1
Organs of the digestive tract

Gastric Juice Starts the Digestion of Proteins with Pepsin When food is on the way to the stomach, the neurotransmitter acetylcholine is produced in gastric cells and histamine is manufactured. When food arrives in the stomach and distends it, the cells in the gastric lining release gastrin. These three compounds, gastrin, acetylcholine, and histamine, stimulate the release of the fluids that combine to give **gastric juice.** One kind of gastric cell secretes mucin, which coats the stomach to protect it against its own digestive enzymes and its acid. Mucin is continuously produced and only slowly digested. If for any reason its protection of the stomach is hindered, part of the stomach itself could be digested, and this would lead to an ulcer.

■ Histamine is made in the body from the amino acid histidine.

NH_2CH_2–CH_2– (imidazole ring, NH, N)

Histamine

Another kind of gastric cell (the parietal cells) secretes hydrogen ion at a concentration of roughly 0.15 mol/L (pH 0.8 or over a million times more acidic than blood). A K^+-H^+ ion "pump" (using ATP chemical energy) takes K^+ ion out of stomach fluids and puts H^+ into the fluids. The K^+ is sent back out along with Cl^- ion. Because chloride ion is the chief anion in gastric juice, the overall effect is the secretion of hydrochloric acid. Histamine is the specific stimulator of the K^+-H^+ ion pump. One of the most common and successful medications for the treatment of stomach ulcers is cimetidine (Tagamet®), and it acts to competitively inhibit the histamine receptor of the K^+-H^+ ion pump. By preventing histamine binding, cimetidine shuts down the K^+-H^+ ion pump, thus preventing the secretion of acid and so allowing the stomach to heal the ulcer.

■ $SCH_2CH_2NHC(=NCN)NHCH_3$ / CH_2 / H_3C (imidazole ring, NH, N)

Cimetidine

The stomach acid coagulates proteins and activates a protease. Protein coagulation retains the protein in the stomach longer for exposure to the protease, *pepsin*.

Gastric cells also secrete the zymogen, *pepsinogen*. Pepsinogen is changed into pepsin by the action of hydrochloric acid and traces of pepsin. The optimum pH of pepsin is about 2, reflecting how relatively acidic stomach fluid becomes as it mixes with incoming food. Pepsin catalyzes the only important digestive work in the stomach, the hydrolysis of some of the peptide bonds of proteins to make shorter polypeptides.

■ A *protease* is an enzyme that catalyzes the digestion of proteins.

Adult gastric juice also has a lipase, but it does not start its work until it arrives in the higher pH medium of the upper intestinal tract.

The gastric juice of infants is less acidic than the adult's. To compensate for the protein-coagulating work normally done by the acid, infant gastric juice contains rennin, a powerful protein coagulator. Because the pH of an infant's gastric juice is higher than that in the adult, its lipase gets an early start on lipid digestion.

The churning and digesting activities in the stomach produce a liquid mixture called *chyme*. This is released in portions through the pyloric valve into the duodenum, the first 12 inches of the upper intestinal tract.

Pancreatic Juice Furnishes Several Zymogens and Enzymes As soon as chyme appears in the duodenum, hormones (cholecystokinin and secretin) are released that circulate to the pancreas and induce this organ to release two juices. One is almost entirely dilute sodium bicarbonate, which neutralizes the acid in chyme. The other juice is the one usually called **pancreatic juice.** It carries enzymes or zymogens that become involved in the digestion of practically everything in food. It contributes an *α-amylase* similar to that present in saliva, a *lipase, nucleases,* and zymogens for protein-digesting enzymes.

■ The nucleases include ribonuclease (RNase) and deoxyribonuclease (DNase).

■ Enteropeptidase used to be called enterokinase.

The conversion of the proteolytic zymogens to active enzymes begins with a "master switch" enzyme called *enteropeptidase,* which we mentioned in Section 11.4. It is released from cells that line the duodenum when chyme arrives, and it then catalyzes the formation of trypsin from its zymogen, trypsinogen.

$$\text{Trypsinogen} \xrightarrow{\text{enteropeptidase}} \text{trypsin}$$

Trypsin then catalyzes the change of the other zymogens into their active enzymes.

■ These proteases must exist as zymogens first or they will catalyze the self-digestion of the pancreas, which does happen in acute pancreatitis.

$$\text{Procarboxypeptidase} \xrightarrow{\text{trypsin}} \text{carboxypeptidase}$$

$$\text{Chymotrypsinogen} \xrightarrow{\text{trypsin}} \text{chymotrypsin}$$

$$\text{Proelastase} \xrightarrow{\text{trypsin}} \text{elastase}$$

Trypsin, chymotrypsin, and *elastase* catalyze the hydrolysis of large polypeptides to smaller ones. *Carboxypeptidase,* working in from C-terminal ends of small polypeptides, carries the action further to amino acids and di- or tripeptides.

Bile Salts Are Powerful Surfactants Necessary to Manage Dietary Triacylglycerols and Fat-Soluble Vitamins In order to digest triacylglycerols, the lipase in pancreatic juice needs the help of the powerful detergents in bile, called the *bile salts.* These help to emulsify water-insoluble fatty materials and so greatly increase the exposure of lipids to water and lipase. The digestion of triacylglycerols gives the anions of fatty acids and monoacylglycerols (plus some diacylglycerols).

■ The structure of a typical bile salt was given on page 234.

Bile is a juice that enters the duodenum from the gallbladder. Its secretion is stimulated by a hormone released when chyme contains fatty material. Bile is also an avenue of excretion, because it can carry cholesterol and breakdown products of hemoglobin. These and further breakdown products constitute the bile pigments, which give color to feces.

The bile salts also assist in the absorption of the fat-soluble vitamins (A, D, E, and K) from the digestive tract into the blood. This work reabsorbs some bile pigments, some of which eventually leave the body via the urine. Thus the bile pigments are responsible for the color of both feces and urine.

Cells of the Intestines Carry Several Digestive Enzymes The term **intestinal juice** embraces not only a secretion but also the enzyme-rich fluids found inside certain kinds of cells that line the duodenum and jejunum. The secretion of some of these cells delivers an amylase and enteropeptidase, which we just described.

■ These intestinal cells last only about two days before they self-digest. They are constantly being replaced.

The other enzymes in this region work within their cells as digestible compounds are already being absorbed. An *aminopeptidase,* working inward from N-terminal ends of small polypeptides, hydrolyzes them to amino acids. The hydrolase enzymes *sucrase, lactase,* and *maltase* handle the digestion of disaccharides: sucrose to fructose and glucose; lactose to galactose and glucose; and maltose to glucose. An intestinal lipase and enzymes for the hydrolysis of nucleic acids are also present.

As the anions of fatty acids and mono- and diacylglycerols migrate into the cells of the duodenal lining, they are reconstituted into triacylglycerols, which are taken up by the lymph system and then delivered to the blood in special complexes of lipids and proteins called chylomicrons. (We'll study what happens to them and to dietary cholesterol in Chapter 16.)

Some Vitamins and Essential Amino Acids Are Made in the Large Intestine No digestive functions are performed in the large intestine. Microorganisms in residence there, however, make vitamins K and B, plus some essential amino acids. These are absorbed by the body, but their contribution to overall nutrition in humans is not large.

Water and sodium chloride are reabsorbed from the ileum, and undigested matter (including fiber), microorganisms, and some water make up the feces.

12.2

BLOOD AND THE ABSORPTION OF NUTRIENTS BY CELLS

The balance between the blood's pumping pressure and its colloidal osmotic pressure tips at capillary loops.

The circulatory system, Figure 12.2, is one of our two main lines of chemical communication between the external and internal environments. All of the veins and arteries together are called the **vascular compartment.** The **cardiovascular compartment** includes this plus the heart.

■ The nervous system with its neurotransmitters is the other line of communication.

The Blood Moves Nutrients, Oxygen, Messengers, Wastes, and Disease Fighters throughout the Body Blood in the pulmonary branches moves through the lungs where waste carbon dioxide is exchanged for oxygen. The oxygenated blood then moves to the rest of the system via the systemic branches.

■ About 8% of the body's mass is blood. In the adult, the blood volume is 5 to 6 L.

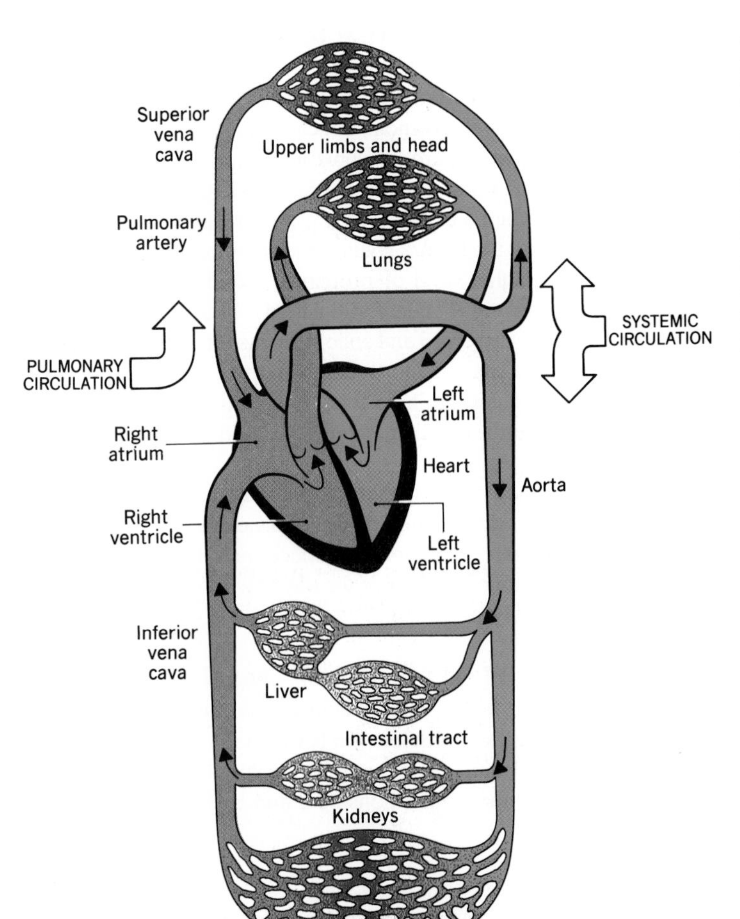

FIGURE 12.2
Human circulatory system. When oxygen-depleted venous blood (blue areas) returns to the heart, it is pumped into the capillary beds of the alveoli in the lungs to reload oxygen and get rid of carbon dioxide. Then the oxygenated blood (red areas) is distributed by the arteries throughout the body, including the heart muscle.

FIGURE 12.3
Major components of blood

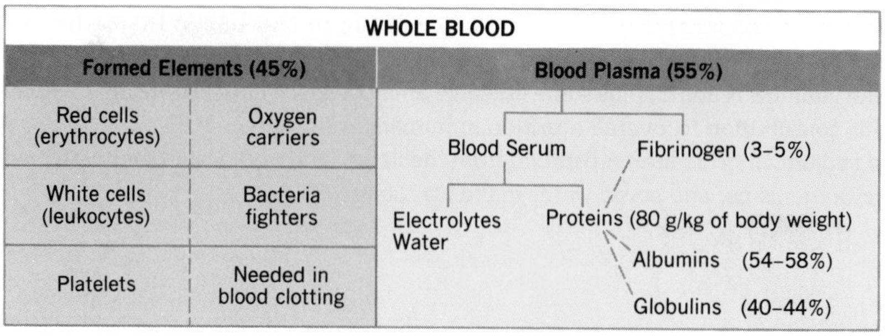

WHOLE BLOOD		
Formed Elements (45%)		**Blood Plasma (55%)**
Red cells (erythrocytes)	Oxygen carriers	Blood Serum; Fibrinogen (3–5%)
White cells (leukocytes)	Bacteria fighters	Blood Serum: Electrolytes, Water; Proteins (80 g/kg of body weight)
Platelets	Needed in blood clotting	Proteins: Albumins (54–58%); Globulins (40–44%)

At the intestinal tract, the blood picks up the products of digestion. Most of these are immediately monitored at the liver and many alien chemicals are modified for elimination. In the kidneys, the blood replenishes its buffer supplies and eliminates waste nitrogen, mostly as urea. The pH of blood and its electrolyte levels depend largely on the kidneys. At endocrine glands, the blood picks up hormones whose secretions are often in response to something present in the blood.

White cells in blood provide protection against infection; red cells or **erythrocytes** carry oxygen and waste bicarbonate ion; and platelets are needed for blood clotting and other purposes. The blood also carries several zymogens that participate in the blood-clotting mechanism.

■ About a quarter of the plasma proteins are replaced each day.

The Proteins in Blood Are Vital to Its Colloidal Osmotic Pressure The principal types of substances in whole blood are summarized in Figure 12.3. Among the proteins, the **albumins** help carry hydrophobic molecules, like fatty acids and steroid hormones, and they contribute 75–80% of the osmotic effect of the blood. Some **globulins** carry ions (e.g., Fe^{2+} and Cu^{2+}) that otherwise are insoluble when the pH is greater than 7. Some globulins are antibodies that help to protect against infectious disease. **Fibrinogen** is converted to an insoluble form, **fibrin,** when a blood clot forms, as we discussed in Section 11.5.

■ The osmolarity of plasma is about 290 mOsm/L.

Figure 12.5 gives the level of solutes, including proteins, in various components of the major body fluids. The higher level of protein in blood is the principle reason why blood has a higher osmotic pressure than interstitial fluid.[1] The *total* osmotic pressure of blood is caused, of course, by all the dissolved and colloidally dispersed solutes. The small ions and molecules, however, can move back and forth between the blood and the interstitial compartment. The large protein molecules can't do this, so it is their presence that gives to blood a higher effective osmotic pressure than interstitial fluid. The margin of difference is the blood's colloidal osmotic pressure.

As a consequence of the higher osmotic pressure of blood, water tends to flow into the blood from the interstitial compartment. This cannot be allowed to happen everywhere and continually, however, or the interstitial spaces and then the cells would eventually become too dehydrated to maintain life. Before we see how this problem is managed, we need to survey some of the **electrolytes** in blood.

The Chief Ions in Blood Are Na^+, Cl^-, and HCO_3^- Figure 12.4 gives the levels of the electrolytes in body fluids. The sodium ion is the chief cation in both the blood and the interstitial fluid and the potassium ion is the major cation inside cells. A sodium–potassium pump, a special ATP-run protein complex, maintains these gradients. Both ions are needed to maintain osmotic pressure relationships, and both are a part of the regulatory system for

[1] As a reminder and a useful memory aid, high solute concentration means high osmotic pressure; and solvent flows in osmosis or dialysis from a region where the solute is dilute to the region where it is concentrated. The "goal" of this flow is to even out the concentrations everywhere.

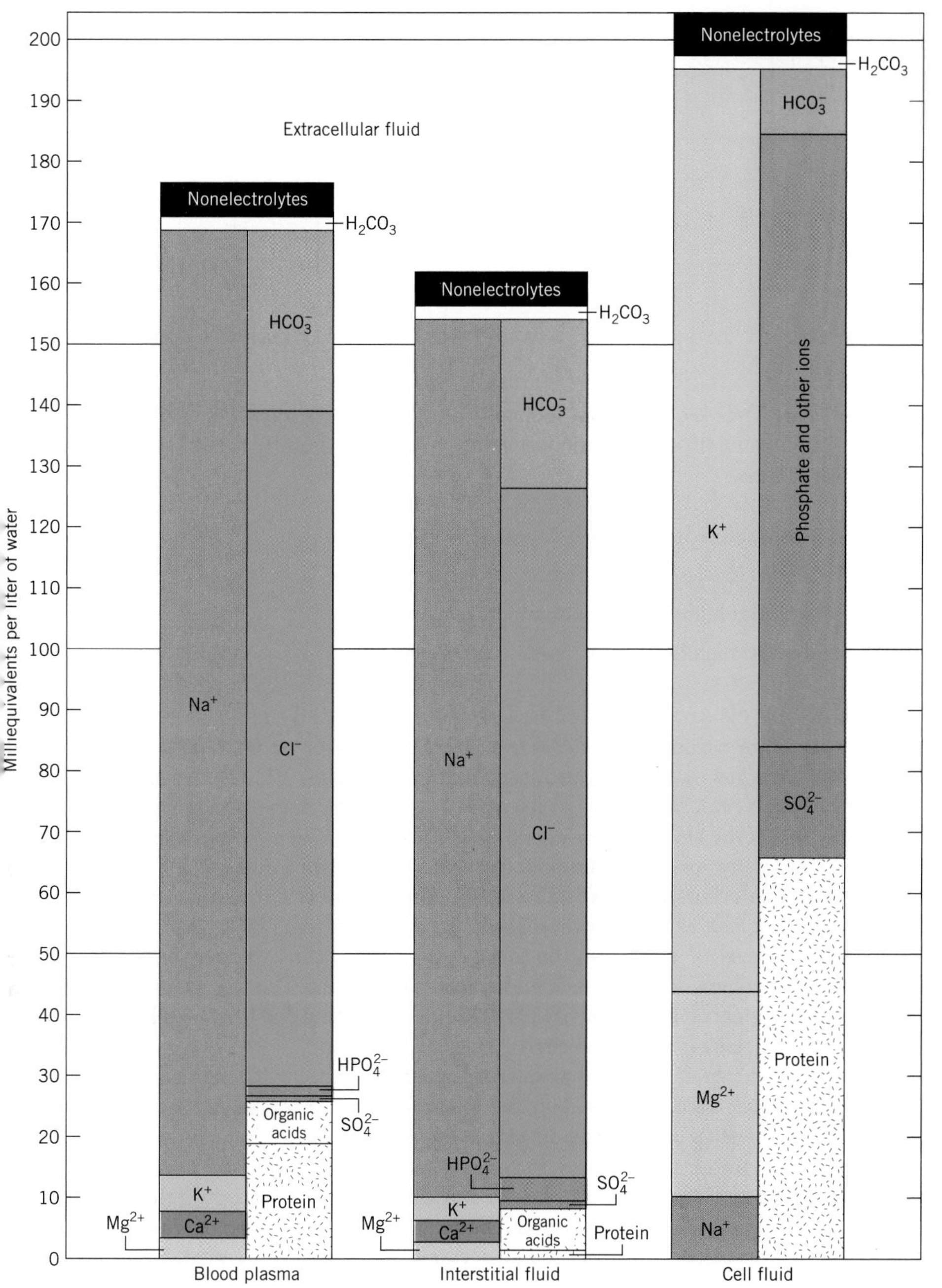

FIGURE 12.4 Electrolyte composition of body fluids. (Adapted by permission from J. L. Gamble, *Chemical Anatomy, Physiology and Pathology of Extracellular Fluids*, 6th ed. Harvard University Press, Cambridge, MA, 1954.)

■ The **milliequivalent**, abbreviated **meq**, is a unit often used by clinical chemists for a particular amount of substance and is analogous to the *millimole* (10^{-3} mol). Think of 1 meq of an ion as equal to the number of milligrams of the species that contains 1 millimole of electrical charges. Thus 1 mmol of Ca^{2+} contains 2 meq of positive electrical charge; 19 mmol HCO_3^- contains 19 meq of negative charge.

acid–base balance. Both ions are also needed for the smooth working of the muscles and the nervous system.

Changes in the concentrations of sodium and potassium ion in blood can lead to serious medical emergencies, so a special vocabulary has been developed to describe various situations. We will see here how a technical vocabulary can be built on a few word parts, and we will use some of these word parts in later chapters, too. For example, we use *-emia* to signify "in the blood." *Hypo-* indicates a condition of a low concentration of something, and *hyper-* is the opposite, a condition of a high concentration of something. We can specify this something by a

■ Another word part is *-uria*, of the urine. Thus *glucosuria* means glucose in the urine.

■ The symbol "meq" is defined in the margin of the preceding page.

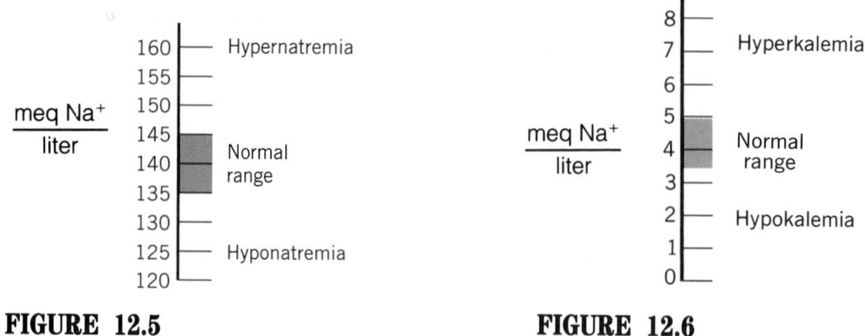

FIGURE 12.5
The sodium ion level in the blood

FIGURE 12.6
The potassium ion level in the blood

word part, too. Thus *-nat-* signifies sodium (from the Latin *natrium* for sodium), and *-kal-* designates potassium (from the Latin *kalium* for potassium). Putting these together gives us the following terms.

Hyponatremia: low level of sodium ion in blood
Hypokalemia: low level of potassium ion in blood
Hypernatremia: high level of sodium ion in blood
Hyperkalemia: high level of potassium ion in blood

■ You'll see it called the "sodium level," not sodium *ion* level, in nearly all references, but sodium *ion* level is always intended.

Na^+ and K^+ Levels are Regulated in Tandem Figure 12.5 shows the normal range for the sodium ion level in blood. Figure 12.6 does the same thing for the potassium ion level. Table 12.1 gives the normal ranges for these cations in various units.

If our kidneys cannot make urine or if we drink water faster than the kidneys can handle it, the sodium level of the blood decreases and we will display the signs of hyponatremia: flushed skin, fever, and a dry tongue (and a noticeable decrease in urine output).

Hypernatremia occurs from excessive losses of water under circumstances in which sodium ions are not lost, as in diarrhea, diabetes, and even in some high-protein diets.

Because the level of potassium ion in blood is so low, small changes are particularly dangerous. Severe hyperkalemia leads to death by heart failure. This danger exists in victims with crushing injuries, severe burns, or heart attacks—anything that breaks cells open so that they spill their K^+ ions into general circulation.

At the other extreme, severe hypokalemia, caused by any unusual losses of body fluids including prolonged, excessive sweating, can lead to death by heart failure. Both body fluids *and electrolytes* have to be replaced during severe exercise.

■ Experienced backpackers use salt tablets that contain not only NaCl but also some KCl to supply K^+. Beverages like Gatorade® similarly resupply the body with electrolytes.

The serum levels of sodium ions and potassium ions are regulated in tandem by the kidneys. If the intake of Na^+ is high, there will be a loss of K^+ from the body. If the intake of K^+ is high, there will be a loss of Na^+ from the body. It's the total positive charge that must be maintained so as to balance the total negative charge. One reason we can't tolerate seawater (3% NaCl) and will die if we drink it is that it upsets the sodium–potassium balance in the body.

TABLE 12.1
Sodium and Potassium Ions in the Human Body

Area	Na^+	K^+
Total body	2700–3000 meq	3200 meq
Plasma level	135–145 meq/L	3.5–5.0 meq/L
Intracellular level	10 meq/L	125 meq/L
Mass of 1 meq	23.0 mg	39.1 mg

TABLE 12.2
Calcium and Magnesium Ions in the Human Body

Area	Ca^{2+}	Mg^{2+}
Total body	6×10^4 meq (1.2 kg)	1000 meq (24 g)
Plasma level	4.2–5.2 meq/L	1.5–2.0 meq/L
Intracellular level	*a*	35 meq/L
Mass of 1 meq	20.0 mg	12.2 mg

[a] Free in solution in the cytosol, about 10^{-7} mol/L.

Mg^{2+} Is Second to K^+ as a Cation Inside Cells The normal ranges for the levels of calcium ion and magnesium ion in the body are given in Table 12.2. The level of magnesium ion in the blood is even lower than that of K^+ (Fig. 12.4), so small variations can mean large trouble. Hypomagnesemia, for example, is observed when the kidneys aren't working properly, or in alcoholism, or in untreated diabetes. Some of its signs are muscle weakness, insomnia, and cramps in the legs or the feet. Injections of isotonic magnesium sulfate solution are sometimes used to restore Mg^{2+} to the serum.

On the opposite side, hypermagnesemia can lead to cardiac arrest, and it can be brought on by the overuse of magnesium-based antacids such as milk of magnesia, $Mg(OH)_2$.

Nearly All the Body's Ca^{2+} Is in Bones and Teeth The Ca^{2+} not in bones and teeth is absolutely vital, as we explained in the previous chapter. It is an important secondary chemical messenger involved in activating enzymes and in muscle contraction.

■ It's common among the elderly to suffer the loss of Ca^{2+} from bones, a condition called *osteoporosis*.

Hypocalcemia can be brought on by vitamin D deficiency, the overuse of laxatives, an impaired activity of the thyroid gland, and even by hypomagnesemia. Hypercalcemia, on the other hand, is caused by the opposite conditions: an overdose of vitamin D, the overuse of calcium ion-based antacids, or an overactive thyroid. In severe hypercalcemia, the heart functions poorly.

Cl^- and HCO_3^- Provide Almost All the Negative Charge to Balance the Cationic Charges in Blood The chloride ion in blood helps to maintain osmotic pressure relationships, the acid–base balance, and the distribution of water in the body. It has a function in oxygen transport. The bicarbonate ion is the chief acid-neutralizing buffer in blood and the principal form in which waste CO_2 is carried.

■ In blood, the normal range for $[Cl^-]$ is 100 to 106 meq/L (1 meq $Cl^- = 35.5$ mg).

In hypochloremia, there is an excessive loss of Cl^- from the blood. For every Cl^- lost, the blood either must lose one (+) charge, like Na^+, or retain one extra (−) charge on some other anion. Electrical neutrality dictates these simple facts. The chief ion retained is HCO_3^-. Because this ion tends to raise the pH of a fluid, *a condition of hypochloremia can cause the pH of blood to increase—alkalosis*.

By the same token, hyperchloremia, a rise in the level of Cl^-, could mean that HCO_3^- has to be dumped, which would mean a loss of base. Thus *hyperchloremia can cause a decrease in blood pH—acidosis*.

Fluids That Leave the Blood Must Return in Identical Volume The blood vessels undergo extensive branching until the narrowest tubes called the capillaries are reached. Blood enters a capillary loop (Fig. 12.7) as arterial blood, but it leaves on the other side of the loop as venous blood.

During the switch, fluids and nutrients leave the blood and move into the interstitial fluids and into the tissue cells themselves. *In the same volume* the fluids must return to the blood, but now they must carry the wastes of metabolism. The rate of this diffusion of fluids throughout the body is sizable, about 25 to 30 L/s. Some fluids return to circulation by way of the lymph ducts, which are thin-walled, closed-end capillaries that bed in soft tissue.

■ The lymph system makes antibodies and it has white cells that help defend the body against infectious diseases.

FIGURE 12.7
The exchange of nutrients and wastes at capillaries. As indicated at the top, on the arterial side of a capillary loop the blood pressure counteracts the pressure from dialysis and osmosis, and fluids are forced to leave the bloodstream. On the venous side of the loop, the blood pressure has decreased below that of dialysis and osmosis, so fluids flow back into the bloodstream. On the top right and the bottom is shown how a normal red cell distorts as it squeezes through a capillary loop. Red cells in sickle-cell anemia do not pass through as smoothly. The bottom drawing also shows how some fluids enter the lymph system.

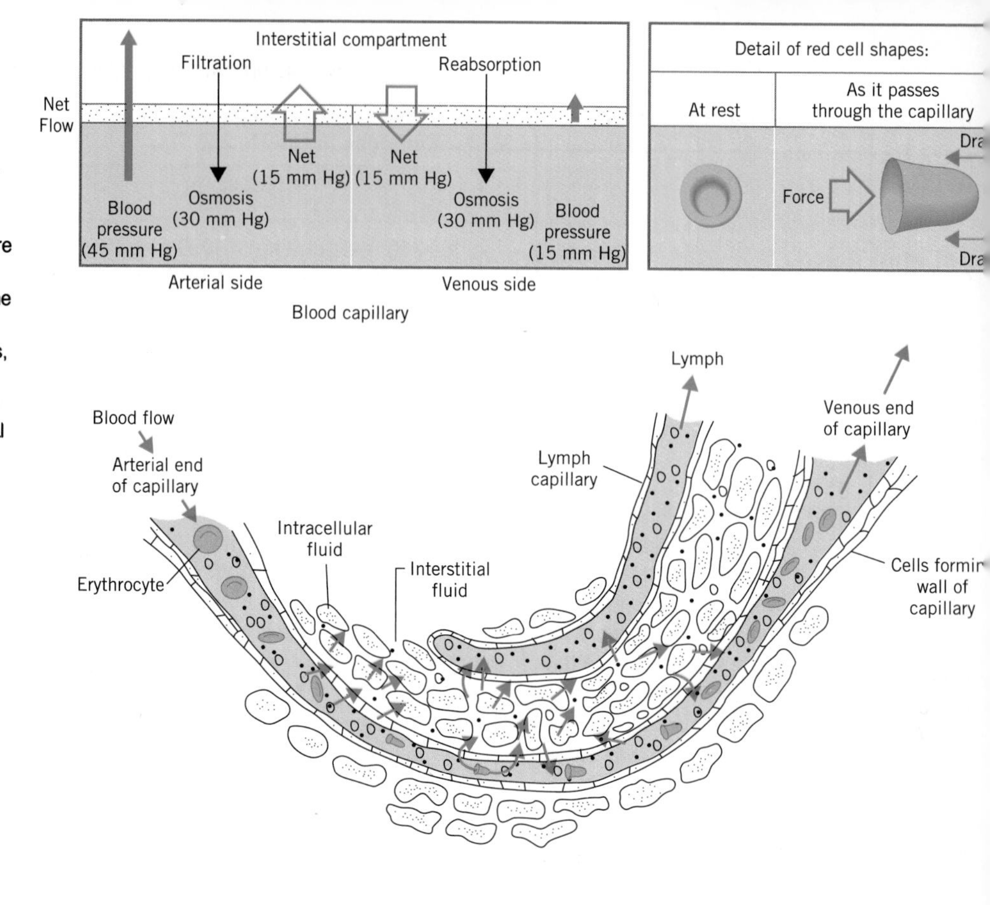

Blood Pressure Overcomes Osmotic Pressure on the Arterial Side of a Capillary Loop Interstitial fluids have a natural tendency to dialyze into the bloodstream because the blood has the higher colloidal osmotic pressure. On the arterial side of a capillary loop, however, the blood pressure is sufficiently high to overcome this. Water and dissolved solutes are forced, instead, from the blood *into* the interstitial space. While they are there, exchanges of chemicals occur. Nutrients move into cells, and cells get rid of wastes.

Osmotic Pressure Overcomes Blood Pressure on the Venous Side of a Capillary Loop As blood emerges from the thin constriction of a capillary into the venous side, its pressure drops. Now it is too low to prevent the natural diffusion of fluids back into the bloodstream. By this time, of course, the fluids are carrying waste products. These relationships are illustrated in Figure 12.7. It shows how the colloidal osmotic pressure contributed by the dispersed macromolecules in blood, particularly the albumins, makes the difference in determining the direction of diffusion.

Blood Loses Albumins in the Shock Syndrome When the capillaries become more permeable to blood proteins, as they do in such trauma as sudden severe injuries, major surgery, and extensive burns, the proteins migrate out of the blood. Unfortunately, this protein loss also means the loss of the colloidal osmotic pressure that helps fluids to return from the tissue areas to the bloodstream. As a result, the total volume of circulating blood drops quickly, and this drastically reduces the blood's ability to carry oxygen and to remove carbon dioxide. The drop in blood volume and the resulting loss of oxygen supply to the brain send the victim into traumatic **shock.**

■ The prompt restoration of blood volume is mandatory in the treatment of shock.

Blood Also Loses Proteins in Kidney Disease and Starvation Sometimes the proteins in the blood are lost at malfunctioning kidneys. The effect, although gradual, is a slow but unremitting drop in the blood's colloidal osmotic pressure. Fluids accumulate in the interstitial regions. Because this takes place more slowly and water continues to be ingested, there is no sudden drop in blood volume as in shock. The victim appears puffy and water-logged, a condition called **edema.**

■ Greek, *oidema*, swelling.

Edema can also appear at one stage of starvation, when the body has metabolized its circulating proteins to make up for the absence of dietary proteins.

Any obstruction in the veins can also cause edema, as in varicose veins and certain forms of cancer. Now it is the venous blood pressure that rises, creating a back pressure that reduces the rate at which fluids can return to circulation from the tissue areas. The localized swelling that results from a blow is a temporary form of edema caused by injuries to the capillaries.

12.3

BLOOD AND THE EXCHANGE OF RESPIRATORY GASES

The binding of oxygen to hemoglobin is allosteric, and it is affected by the pH, pCO_2, and the pO_2 of the blood.

The carrier of oxygen in blood is **hemoglobin,** a complex protein found inside erythrocytes. It consists of four subunits, each with one molecule of heme, an organic group that holds an iron(II) ion, the actual oxygen-binding unit. Two of the subunits are identical and have the symbol α. The other two are also identical and have the symbol β.

■ Each red cell carries about 2.8×10^8 molecules of hemoglobin.

When hemoglobin is oxygen free, it is called *deoxyhemoglobin;* in this state the molecule has a cavity in which a small organic anion nestles. This is the **2,3-bisphosphoglycerate** ion, called simply **BPG,** and it has an important function in oxygen transport.[2]

■

$^{2-}O_3POCH_2CHCO_2^-$ (with OPO_3^{2-} on the CH carbon)

BPG (2,3-bisphosphoglycerate)

The Subunits of Hemoglobin Cooperate in Its Oxygenation We define the **oxygen affinity** of the blood as the percent to which the blood has all of its hemoglobin molecules saturated with oxygen. A fully laden hemoglobin molecule carries four oxygen molecules and is called **oxyhemoglobin.** For the maximum efficiency in moving oxygen from the lungs to tissues that need it, all hemoglobin molecules should leave the lungs in the form of fully loaded oxyhemoglobin. Let's see what factors ensure this.

■ The structure in Figure 10.9, page 257, is actually deoxyhemoglobin, and the central cavity for BPG was pointed out.

First, the partial pressure of oxygen is higher in the lungs than anywhere else in the body; pO_2 is 100 mm Hg in freshly inhaled air in alveoli and only about 40 mm Hg in oxygen-depleted tissues. Because of this partial pressure gradient, oxygen naturally migrates from the lungs into the bloodstream. It's as though the higher partial pressure *pushes* oxygen into the blood. Second, newly arrived oxygen actually reacts with hemoglobin, binding to its Fe^{2+} ions to form oxyhemoglobin, so this tends to *pull* oxygen into the blood.

An allosteric effect also helps to load hemoglobin with oxygen. Figure 12.8 shows a plot of oxygen affinity versus the oxygen partial pressure. It has the sigmoid shape that in Chapter 11

■ We first learned about the allosteric effect on page 283.

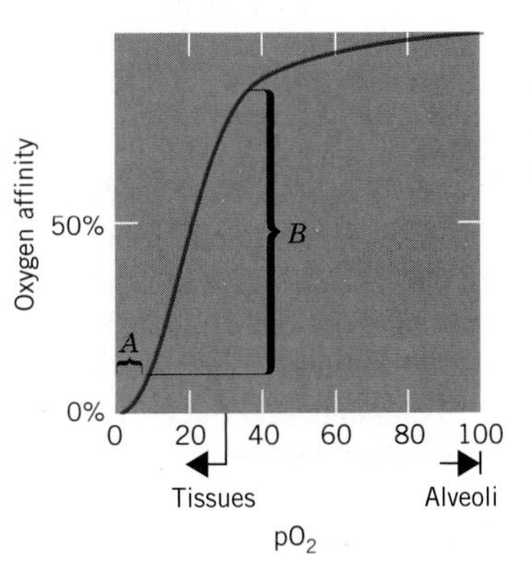

FIGURE 12.8
Hemoglobin-oxygen dissociation curve. Regions *A* and *B* are discussed in the text.

[2] The name 2,3-diphosphoglycerate (DPG) has been supplanted by 2,3-bisphosphoglycerate (BPG). "Diphospho-" signifies a diphosphate ester, an ester of diphosphoric acid (page 149). "Bisphospho" signifies two ("bis") monophosphate ester groups.

we learned to associate with the allosteric effect among enzymes. At low values of pO_2, in region *A* of the plot, the ability of the blood to take up oxygen rises only slowly with increases in pO_2. But eventually the oxygen affinity takes off, and rises very steeply with still more increases in pO_2, in region *B* of the plot. Eventually, the oxygen affinity starts to level off. It almost seems that a small "molecular dam" thwarts the efforts of oxygen molecules to be joined to hemoglobin at low partial pressures of oxygen.

■ This natural resistance is needed in working cells where we want no restrictions on the deoxygenation of blood.

What is thought to be happening is as follows. We'll represent deoxyhemoglobin by structure **1**, below. Each circle in **1** is a polypeptide subunit with its heme unit but without oxygen. The oval figure centered within the structure represents one BPG anion. When the first O_2 molecule manages to bind (**1 → 2**), it induces a change in the shape of the affected subunit, which we have represented as a change from a circle to a square:

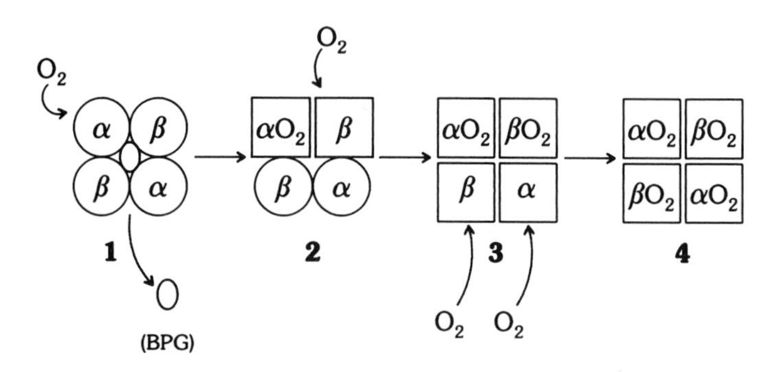

This change tends to squeeze out the BPG unit, and this "breaks the dam." The next oxygen molecule enters more readily (**2 → 3**). As its affected subunit changes its shape, the remaining two subunits allosterically change their shapes and become very receptive to the third and fourth molecules of oxygen. These two flood into this hemoglobin molecule (**3 → 4**) far more readily than either would bind to a completely deoxygenated hemoglobin molecule. Thus if a hemoglobin molecule accepts just one oxygen molecule, it's almost certain that it will soon accept three more to become fully oxygenated. Few if any partially oxygenated hemoglobin molecules leave the lungs.

■ Carbon monoxide binds 150 to 200 times more strongly to hemoglobin than does oxygen and thus prevents the oxygenation of hemoglobin and causes internal suffocation.

For the sake of the remaining discussion, we'll simplify what we have just described by letting the symbol HHb represent an entire hemoglobin molecule. The first H in HHb stands for a potential hydrogen ion, and we overlook the fact that more than one is actually present in hemoglobin. (We're now also overlooking the fact that *four* molecules of O_2 bind to one of hemoglobin.) With this in mind, we can represent the oxygenation of hemoglobin as the *forward* reaction in the following equilibrium where oxyhemoglobin is represented as the anion, HbO_2^-:

■ About 20% of a smoker's hemoglobin is more or less permanently tied up by carbon monoxide.

$$\underset{\text{Hemoglobin}}{HHb} + O_2 \rightleftharpoons \underset{\text{Oxyhemoglobin}}{HbO_2^-} + H^+ \qquad (12.1)$$

Two facts indicated by this equilibrium are that HHb is a weak acid and that it becomes a stronger acid as it become oxygenated. Being stronger, it is produced in its ionized state as HbO_2^- and H^+. *The presence of* H^+ *in equilibrium 12.1 means that the equilibrium can be shifted one way or another simply by changing the pH,* a fact of enormous importance at the molecular level of life, as we'll see.

To understand the oxygenation of hemoglobin, we have to see how various stresses shift equilibrium 12.1 to the *right* in the lungs. One stress, as we've already noted, is the relatively high value of pO_2 (100 mm Hg) in the alveoli. This stress acts on the left side of 12.1, so it helps to shift the equilibrium to the right.

$$\underset{\substack{\text{In the}\\\text{red cell}}}{HHb} + \underset{\substack{\text{From}\\\text{air}}}{O_2} \longrightarrow \underset{\substack{\text{In the}\\\text{red cell}}}{HbO_2^-} + \underset{\substack{\text{In the}\\\text{red cell}}}{H^+}$$

Another stress that isn't evident from equation 12.1 is the removal of H^+ *as it forms*. The red cell in the lungs is carrying waste HCO_3^-. The newly forming H^+ ions are therefore promptly equilibrated with HCO_3^- and CO_2 by the help of carbonic anhydrase. (We write only the forward reaction of this equilibrium here because this is how the equilibrium shifts when both H^+ and HCO_3^- are high.)

■ The stimulation of HHb to bind O_2 caused by the removal of H^+ is called the **Bohr effect** after Christian Bohr, a Danish scientist (and the father of nuclear physicist Niels Bohr).

$$\underset{\text{In the red cell}}{H^+} + \underset{\text{In the red cell}}{HCO_3^-} \xrightarrow{\text{carbonic anhydrase}} \underset{\text{In the red cell}}{CO_2} + H_2O \qquad (12.2)$$

■ As we learned in the last chapter, carbonic anhydrase is one of the body's fastest working enzymes, and it has to be fast because the red cells are always on the move.

This switch from the appearance of H^+ as a product to its disappearance as a reactant is called the **isohydric shift.**

We see here one of the beautiful examples of coordinated activity in the body, the coupling of the uptake of oxygen to the release of the carbon dioxide to be exhaled. The neutralization of the H^+ ions produced from the uptake of O_2 produces CO_2. The loss of CO_2 from the red cell by exhaling pulls this and all previous equilibria to the right in the lungs.

$$\underset{\text{In the red cell}}{CO_2} \xrightarrow{\text{exhaling}} \underset{\text{In exhaled air}}{CO_2}$$

Thus the uptake of O_2 as hemoglobin oxygenates, which pushes the equilibria to the right, simultaneously produces a chemical (CO_2) whose loss pulls the same equilibria to the right.

Let's now see how this waste is picked up at cells that have produced it and is carried to the lungs; and let's also see how this works cooperatively with the *release* of oxygen at cells that need it.

The Hemoglobin Subunits and BPG Cooperate in the Deoxygenation of Oxyhemoglobin in Metabolizing Tissues Consider, now, a tissue that has done some chemical work, used up some oxygen, and made some waste carbon dioxide. When a fully oxygenated red cell arrives in such a tissue, some of the events we just described reverse themselves.

We can think of this reversal as beginning with the diffusion of waste CO_2 from the tissue into the blood. An impetus for this diffusion is the higher pCO_2 (50 mm Hg) in active tissue versus its value in blood (40 mm Hg). Once the CO_2 arrives in the blood, it moves inside a red cell where it encounters carbonic anhydrase. Equation 12.2 is therefore run in reverse. It was part of the equilibrium managed by carbonic anhydrase, and it shifts to the left as more and more CO_2 arrives. In other words, the equilibrium now shifts to *make* HCO_3^-.

■ CO_2 molecules diffuse in body fluids 30 times more easily than O_2 molecules, so the partial pressure gradient for CO_2 need not be as steep as that for O_2.

$$H_2O + \underset{\text{From the working tissue}}{CO_2} \xrightarrow{\text{carbonic anhydrase}} \underset{\text{In the red cell}}{HCO_3^-} + H^+ \qquad (12.2\text{—reversed})$$

Of course, this generates hydrogen ions, and if you'll look back to Equation 12.1, which is also part of an equilibrium, you will see that an increase in the level of H^+ (caused by the influx of waste CO_2) can only make Equation 12.1 run in reverse.

$$\underset{\text{Just made in the red cells at working tissue}}{H^+} + \underset{\text{In the red cell; just arrived at tissue}}{HbO_2^-} \longrightarrow \underset{\text{In the red cell}}{HHb} + \underset{\text{Will diffuse into the tissue needing oxygen}}{O_2} \qquad (12.1\text{—reversed})$$

This reaction, another isohydric shift, not only neutralizes the acid generated by the arrival of waste CO_2, but also helps to force oxyhemoglobin to give up its oxygen. Notice the cooperation. The tissue that needs oxygen has made CO_2 and, hence, it has indirectly made the H^+ that is required to release this needed oxygen from newly arrived HbO_2^-.

The deoxygenation of HbO_2^- is also aided by the BPG anions that were pushed out when HbO_2^- formed. These anions are still inside the red cell, and as soon as an O_2 molecule leaves oxyhemoglobin, a BPG anion starts to move back in. The changes in shapes of the hemoglobin subunits now operate in reverse, and all oxygen molecules smoothly leave. It's all or nothing again, and the efficiency of the unloading of oxygen is so high that if one O_2 molecule leaves, the other three follow essentially at once. BPG helps this to happen. Partially deoxygenated hemoglobin units do not slip through and go back to the lungs.

■ The high negative charge on BPG keeps it from diffusing through the red cell's membrane.

BPG and Hemoglobin Levels Are Higher in People Living at High Altitudes It's interesting that those who live and work at high altitudes, such as the populations in Nepal in the Himalayan Mountains or the people in the Andes Mountains in Bolivia, have 20% higher levels of BPG in their blood and more red blood cells than those who live at sea level.

The extra red cells give them more hemoglobin to help them carry more oxygen per milliliter of blood, and the extra BPG increases the efficiencies of both loading and unloading oxygen. When lowlanders take trips to high altitudes, their bodies start to build more red cells and to make more BPG so that they can function better where the partial pressure of the atmospheric oxygen is lower. Those who patiently wait during the few days that it takes for these events to occur before they set off on strenuous backpacking expeditions are less likely to suffer high altitude sickness, a condition that can cause death.

■ No conditioning at a low altitude can get the cardiovascular system ready for a low pO_2 at a high altitude.

To summarize the chemical reactions we have just studied, we can write the following equations. The cancel lines show how we can arrive at the overall net results.

Oxygenation: (These reactions occur in the lungs.)

$$HHb + O_2 \longrightarrow HbO_2^- + \cancel{H^+}$$

HHb	O_2	HbO_2^-	$\cancel{H^+}$
In the red cell	From air	Will go in red cell to tissue	In the red cell

■ CA is carbonic anhydrase.

$$\cancel{H^+} + HCO_3^- \xrightarrow{CA} CO_2 + H_2O$$

$\cancel{H^+}$	HCO_3^-	CO_2	H_2O
Just made	In red cell (but from tissues)	In red cell (but will be exhaled)	

■ **Net effect of oxygenating hemoglobin:**

$$\mathbf{HHb + O_2 + HCO_3^- \longrightarrow HbO_2^- + CO_2 + H_2O} \quad (12.3)$$

HHb	O_2	HCO_3^-	HbO_2^-	CO_2	H_2O
In the red cell	From air	In red cell (but from tissues)	Will go in red cell to tissue	Will leave the lungs in exhaled air	

Deoxygenation: (These reactions occur wherever tissues are low in oxygen.)

$$CO_2 + H_2O \xrightarrow{CA} HCO_3^- + \cancel{H^+}$$

CO_2	H_2O	HCO_3^-	$\cancel{H^+}$
Waste from tissues		In red cell (in blood still within tissue but will go to the lungs)	In the red cell

$$\cancel{H^+} + HbO_2^- \longrightarrow HHb + O_2$$

$\cancel{H^+}$	HbO_2^-	HHb	O_2
Just made	In red cell (in blood within tissues)	In red cell (will return to the lungs)	Goes into tissue needing it

■ **Net effect of deoxygenating oxyhemoglobin:**

$$\mathbf{CO_2 + H_2O + HbO_2^- \longrightarrow HHb + HCO_3^- + O_2} \quad (12.4)$$

CO_2	H_2O	HbO_2^-	HHb	HCO_3^-	O_2
Waste from tissues		In red cell (in blood with tissues)	In red cell (will go to the lungs)	Goes in blood to the lungs	Goes into tissue

These summarizing equations omit two features. They do not show the importance of BPG and of the concentration of H^+ in both the loading and the unloading of oxygen. Concerning the concentration of H^+, Figure 12.9 shows the plots of oxygen affinity versus pO_2 at two different values of pH, one relatively low (pH 7.2) compared with the normal value of 7.35, and the other relatively high (pH 7.6). You may recall that the pO_2 in the vicinity of oxygen-starved cells is around 30 to 40 mm Hg. Notice in Figure 12.9 that in this range of the partial pressure of oxygen, the blood's ability to hold oxygen is much less at the lower pH of 7.2 than it is at a pH of 7.6. In actively metabolizing cells, there is a localized drop in pH caused chiefly by the presence of the CO_2 that these cells have made. This drop in pH caused by CO_2 cannot help but to assist in the deoxygenation of HbO_2^-.

Thus precisely where O_2 should be unloaded there is a chemical signal (a lower pH) that makes it happen. It's an altogether beautiful example of how a set of interrelated chemical equilibria shift in just the directions that are required for health and life. Figure 12.9 will also help us understand in what ways both acidosis and alkalosis are serious threats.

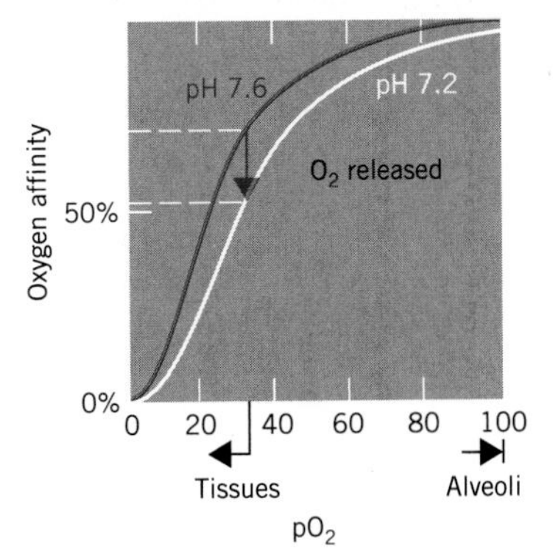

FIGURE 12.9
Hemoglobin–oxygen dissociation curves at two different values of the pH of blood

Some CO_2 Is Carried to the Lungs on Hemoglobin There is another chemical reaction involving waste carbon dioxide that we must now mention. Not all the CO_2 made by metabolizing cells winds up as HCO_3^-. Some CO_2 reacts with the hemoglobin that has just been freed by deoxygenation:

$$CO_2 + HHb \longrightarrow Hb{-}CO_2^- + H^+ \qquad (12.5)$$

Waste from tissue	Just released by deoxygenation of HbO_2^-	Carbaminohemoglobin (in red cells)

This is actually the forward reaction of an equilibrium. The product, $Hb{-}CO_2^-$, is called **carbaminohemoglobin,** and it is one form in which some of the waste CO_2 travels in the blood back to the lungs.

Notice in Equation 12.5 that this reaction also produces H^+, just as did the reaction of water and waste CO_2. Thus, whether waste CO_2 is changed to HCO_3^- and H^+ or to $Hb{-}CO_2^-$ and H^+, either fate helps to generate hydrogen ions that are needed to react with HbO_2^- and make it unload O_2.

When the red cell reaches the lungs, where H^+ will now be *generated* by

$$HHb + O_2 \longrightarrow HbO_2^- + H^+ \qquad (12.1, again)$$

the reaction of Equation 12.5 will be forced to shift into reverse, because the added H^+ provides this kind of stress. Of course, this releases the CO_2 where it can be exhaled, and it gets hemoglobin ready to take on more oxygen.

Chloride Ion Is Also Needed to Deoxygenate Hemoglobin Still another factor we have not mentioned thus far and that helps to unload oxygen from oxyhemoglobin is the chloride ion. Hemoglobin, HHb, binds chloride ion, and it binds it better than does oxyhemoglobin. The following equilibrium exists along with all the others.

$$HHb(Cl^-) + O_2 \longrightarrow HbO_2^- + Cl^- + H^+ \qquad (12.6)$$

Therefore to *unload* oxygen (do the reverse of 12.6), chloride ion must be available *inside* the red cell. The red cell obtains it by a mechanism called the *chloride shift.* Let's see how it occurs.

As chloride ions are drawn into the red cell to react with some of the HHb that is released in active tissue, negative ions have to move out so that there is electrical charge balance. The negative ions that leave are the newly forming bicarbonate ions. From 60% to 90% of all waste CO_2 returns to the lungs as the bicarbonate ion, but most of this makes the trip *outside* red cells.

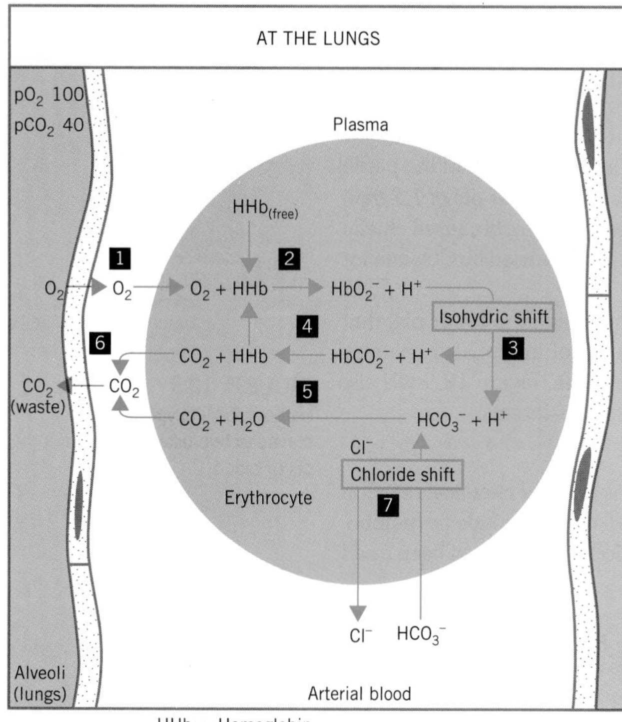

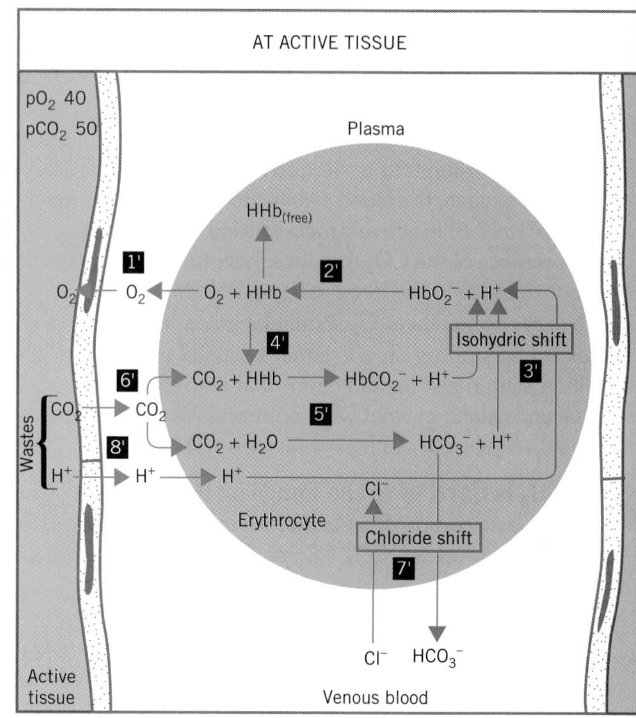

HHb = Hemoglobin
HbO_2^- = oxyhemoglobin [actually, $Hb(O_2)_4$]

$HbCO_2^-$ = Carbaminohemoglobin [actually, $Hb(CO_2)_4^{n-}$ where $n-$ = 1– to 4–]

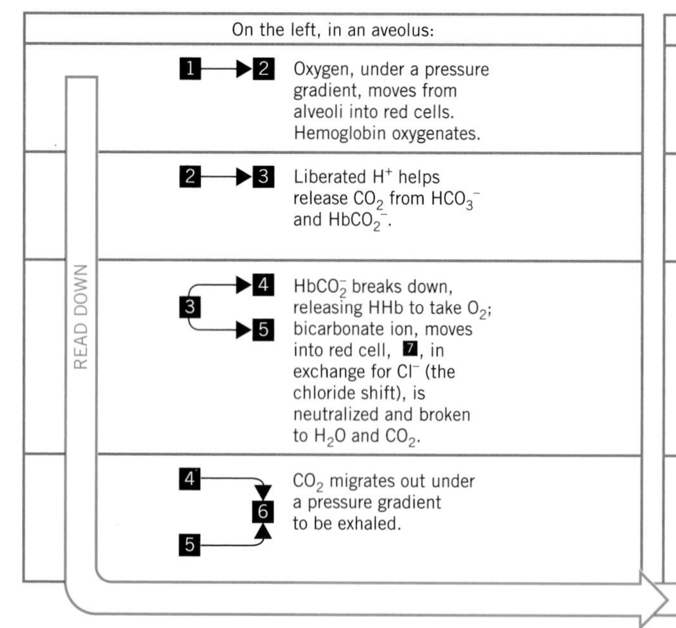

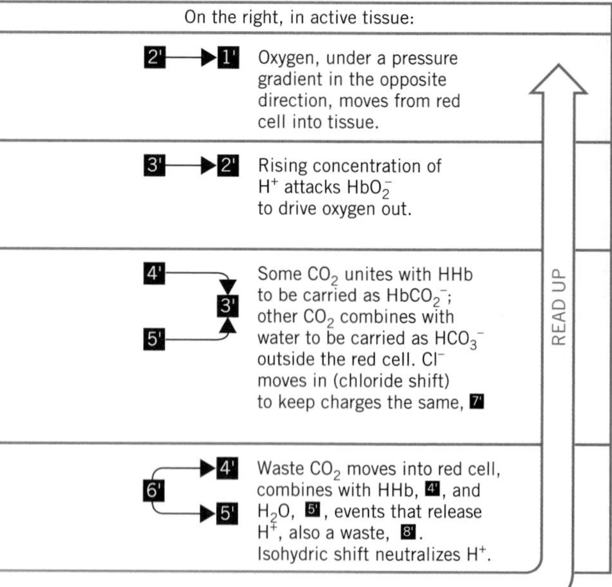

FIGURE 12.10
Oxygen and carbon dioxide exchange in the blood

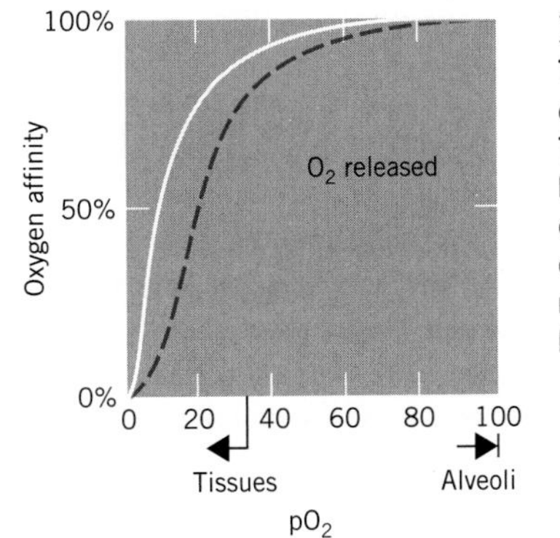

FIGURE 12.11 The myoglobin–oxygen dissociation curve is a solid line. The dashed line is the curve for hemoglobin. Over the entire range of pH in tissues that might need oxygen, the oxygen affinity of myoglobin is greater than that of hemoglobin.

For every chloride ion that enters the red cell, one HCO_3^- ion leaves, and this switch is called the **chloride shift.** When the red cell gets back to the lungs, the various new chemical stresses make all the equilibria, including the chloride shift, run in reverse.

If you're bewildered by all these equilibria and how they are made to shift in the correct directions, you're almost certainly not alone among your classmates. This isn't easy material, but it is so much at the heart of so many aspects of health and one's ability to have an active life that the effort to master it is very worthwhile. As you make this effort, get the key equilibria down and memorized. For each one ask: What are the stresses that can make it shift, and where in the body is each stress important, in the lungs or at actively metabolizing cells?

■ For those entering careers in nursing and respiratory therapy, there is no other single topic of such career-lasting importance than the chemistry of respiration and its associated electrolyte balance.

The key stresses are the following: the relative partial pressures of O_2, the relative partial pressures of CO_2, and the changes in the levels of H^+ caused by the influx of CO_2 or by its loss by means of exhaling.

After you have studied the various equilibria from the stress point of view, and you can write all the equilibria and discuss the influence of various stresses, then you might find Figure 12.10 a useful way to review. Follow the direction of the large U-shaped arrow that curves around the legend, and use the boxed numbers in the legend to follow the events in the figure. Notice particularly that the reactions that occur in the red cell when it is in metabolizing tissue are just the reverse of those that happen when the cell is in the lungs.

■ For those who simply want to understand the respiratory demands (and limitations) of active sports, including skiing at altitude, the chemistry of respiration is the key.

Myoglobin Binds O_2 More Strongly than Hemoglobin Myoglobin is a heme-containing protein in red muscle tissue such as heart muscle. Its function is to bind and store oxygen for the needs of such tissue. Unlike hemoglobin, myoglobin has only one polypeptide unit and only one heme unit per molecule. Moreover, there is no allosteric effect when it binds oxygen, as the shape of the myoglobin–oxygen dissociation curve indicates (see Fig. 12.11).

Myoglobin's oxygen affinity is greater than that of hemoglobin, especially in the range of pO_2 associated with actively metabolizing tissues. Consequently, *myoglobin is able to take oxygen from oxyhemoglobin:*

$$HbO_2^- + HMy \longrightarrow HHb + MyO_2^-$$

■ HMy = myoglobin

This ability is vital to heart muscle which, as much as the brain, must have an assuredly continuous supply of oxygen. When oxymyoglobin, MyO_2^-, gives up its oxygen for the cell's needs, it can at once get a fresh supply from the circulating blood. Not only does this cell now have CO_2 and H^+ available to deoxygenate HbO_2^-, it also has the superior oxygen affinity of its own myoglobin to draw more O_2 into the cell.

Fetal Hemoglobin Binds Oxygen More Strongly than Adult Hemoglobin The hemoglobin in a fetus is slightly different from that of an adult, and it has a higher oxygen affinity than adult hemoglobin. This helps to ensure that the fetus successfully pulls oxygen from the mother's oxyhemoglobin to satisfy its own needs.

12.4

ACID–BASE BALANCE OF THE BLOOD

The proper treatment of acidosis or of alkalosis depends on knowing if the underlying cause is a metabolic or a respiratory disorder.

Acid–base balance in the blood exists when the pH of blood is in the range of 7.35 to 7.45. A decrease in pH, acidosis, or an increase, alkalosis, is serious and requires prompt attention because all the equilibria that involve H^+ in the oxygenation or the deoxygenation of blood are sensitive to pH. If the pH falls below 6.8 or rises above 7.8, life is not possible.

■ Acidosis is sometimes called *acidemia*, and alkalosis is called *alkalemia*.

Disturbances in Either Metabolism or Respiration Can Upset the Blood's Acid–Base Balance In general, acidosis results from either the retention of acid or the loss of base by the body, and these can be induced by disturbances in either metabolism or respiration. Similarly, alkalosis results either from the loss of acid or from the retention of base, and some disorder in either metabolism or respiration can be the underlying cause.

A malfunction in respiration can be caused by any kind of injury to the *respiratory centers*. These are units in the brain that sense changes in the pH and pCO_2 of the blood and instruct the lungs to breathe either more rapidly or more slowly. Another cause of a malfunction in respiration is any kind of injury or disease of the lungs.

What we'll do in this section is study four situations, metabolic and respiratory acidosis as well as metabolic and respiratory alkalosis. We will learn how the values of pH, pCO_2, and serum $[HCO_3^-]$ change in each situation. (Normal values are given in the margin.)

■ Normal values (arterial blood):

$pH = 7.35 - 7.45$
$pCO_2 = 35–45$ mm Hg
$[HCO_3^-] = 19–24$ meq/L

(1.0 meq HCO_3^- = 61 mg HCO_3^-. The unit "meq" was defined in the margin on page 311.)

Metabolic Acidosis Receives a Respiratory Compensation, Hyperventilation In **metabolic acidosis,** the lungs and the respiratory centers are working, and the problem is metabolic. Acids are being produced faster than they are neutralized, or they are being exported too slowly.

Excessive loss of base, such as from severe diarrhea, can also result in metabolic acidosis. (In diarrhea, the alkaline fluids of the duodenum leave the body, and as base migrates to replace them, there will be a depletion of base somewhere else, such as in the blood, at least for a period of time.)

As the pH of the blood falls and the molar concentration of H^+ rises, there are parallel *but momentary* increases in the values of pCO_2 and $[HCO_3^-]$. The value of pCO_2 starts to increase because the carbonate buffer, working hard to neutralize the extra H^+, manufactures CO_2:

$$\underset{\text{Produced by acidosis}}{H^+} + HCO_3^- \longrightarrow H_2O + CO_2$$

The kidneys work harder during this situation to try to keep up the supply of HCO_3^-.

The chief compensation for metabolic acidosis, however, involves the respiratory system. The respiratory centers, which are sensitive to changes in pCO_2, instruct the lungs to blow the CO_2 out of the body. The lungs, in other words, hyperventilate. As the equation given just above shows, the loss of each molecule of CO_2 means a net neutralization of one H^+ ion.

Hyperventilation, however, is overdone. So much CO_2 is blown out that pCO_2 actually decreases; a low arterial pCO_2 is called **hypocapnia.** Thus as the blood pH decreases, so too do the values of pCO_2 (from hyperventilation) and $[HCO_3^-]$ (from the reaction with H^+). We can summarize the range of a number of clinical situations that involve metabolic acidosis as follows.

Clinical Situations of Metabolic Acidosis

Lab Results:	pH↓ (7.20); pCO_2↓ (30 mm Hg); [HCO_3^-]↓ (12 meq/L)
Typical Patient:	An adult male comes to the clinic with a severe infection. He does not know that he has diabetes.
Range of Causes:	Diabetes mellitus; severe diarrhea (with loss of HCO_3^-); kidney failure (to export H^+ or to make HCO_3^-); prolonged starvation; severe infection; aspirin overdose, alcohol poisoning.
Symptoms:	Hyperventilation (because the respiratory centers have told the lungs to remove excess CO_2 from the blood); increased urine output (to remove H^+ from the blood); thirst (to replace water lost as urine); drowsiness; headache; restlessness; disorientation.
Treatment:	If the kidneys function, use isotonic HCO_3^- intravenously to restore HCO_3^- level, thereby neutralizing H^+ and raising pCO_2. In addition, restore water. In diabetes, use insulin therapy. If the kidneys do not function, hemodialysis must be tried.

■ (↓) means a decrease from normal and (↑) means an increase. Some typical values are in parentheses. Note that some changes do not necessarily bring values outside the normal ranges.

Respiratory Acidosis Is Compensated by a Metabolic Response In **respiratory acidosis,** either the respiratory centers or the lungs have failed, and the lungs are hypoventilating *because they cannot help it.* The blood now cannot help but retain CO_2. An increase in arterial pCO_2 is called **hypercapnia.** The retention of CO_2 functionally means the retention of acid, because CO_2 can neutralize base by the following equation.

$$HO^-(aq) + CO_2(aq) \longrightarrow HCO_3^-(aq) \quad (12.7)$$

The decrease in the level of base lowers the pH of the blood and gives rise to respiratory acidosis.

Clinical Situations of Respiratory Acidosis

Lab Results:	pH↓ (7.21); pCO_2↑ (70 mm Hg); [HCO_3^-]↑ (27 meq/L)
Typical Patient:	Chain smoker with emphysema or anyone with chronic obstructive pulmonary disease.
Range of Causes:	Emphysema, severe pneumonia, asthma, anterior poliomyelitis, or any cause of shallow breathing such as an overdose of narcotics, barbiturates, or general anesthesia; severe head injury.
Symptoms:	Shallow breathing (which is involuntary)
Treatment:	Underlying problem must be treated; possibly intravenous sodium bicarbonate; possibly hemodialysis.

The body responds metabolically as best it can to respiratory acidosis by using HCO_3^- to neutralize the acid, by making more HCO_3^- in the kidneys, and by exporting H^+ via the urine.

Metabolic Alkalosis Also Receives a Respiratory Compensation, Hypoventilation In **metabolic alkalosis,** the system has lost acid, or it has retained base (HCO_3^-), or it has been given an overdose of base (e.g., antacids). Metabolic alkalosis can also be caused by a kidney-associated decrease in the serum levels of K^+ or Cl^-. The loss of these ions means the retention of Na^+ and HCO_3^- ions, because these work in tandem and oppositely. The loss of acid could be from prolonged vomiting, which removes the gastric acid. This is followed by an effort to borrow serum H^+ to replace it, and the pH of the blood increases.

Whatever the cause, the respiratory centers sense an increase in the level of base in the blood (as the level of acid drops), and they instruct the lungs to retain the most readily available neutralizer of base it has, namely CO_2, which removes OH^- according to equation 12.7. To help retain CO_2 to neutralize base, the lungs hypoventilate. Thus metabolic alkalosis leads to hypoventilation.

■ Compensation by hypoventilation is obviously limited by the fundamental need of the body for some oxygen.

Notice carefully that hypoventilation alone cannot be used to tell whether the patient has metabolic *alkalosis* or respiratory *acidosis*. Either condition means hypoventilation. But one condition, respiratory acidosis, could be treated by intravenous sodium bicarbonate, a base. This would aggravate metabolic alkalosis.

You can see that the lab data on pH, pCO_2, and $[HCO_3^-]$ must be obtained to determine which condition is actually present. Otherwise, the treatment used could be just the opposite of what should be done.

■ An overdose of "bicarb" ($NaHCO_3$), from a too aggressive use of this home remedy for "heartburn," can cause metabolic alkalosis.

Clinical Situations of Metabolic Alkalosis

Lab Results:	pH↑ (7.53); pCO_2↑ (56 mm Hg); $[HCO_3^-]$↑ (45 meq/L)
Typical Patient:	Postsurgery patient with persistent vomiting.
Range of Causes:	Prolonged loss of stomach contents (vomiting or nasogastric suction); overdose of bicarbonate or of medications for stomach ulcers; severe exercise, or stress, or kidney disease (with loss of K^+ and Cl^-); overuse of a diuretic.
Symptoms:	Hypoventilation (to retain CO_2); numbness, headache, tingling; possibly convulsions.
Treatment:	Isotonic ammonium chloride (a mild acid), intraveneously with great care; replace K^+ loss.

■ Ammonium ion acts as a neutralizer as follows:

$$NH_4^+ + OH^- \longrightarrow NH_3 + H_2O$$

Respiratory Alkalosis Is Compensated Metabolically by a Reduced Bicarbonate Level In **respiratory alkalosis,** the body has lost acid usually by some involuntary hyperventilation such as hysterics, prolonged crying, or overbreathing at high altitudes, or by the mismanagement of a respirator. The respiratory centers have lost control, and the body expels CO_2 too rapidly. The loss of CO_2 means the loss of a base neutralizer from the blood. Hence, the level of base rises; the pH increases. To compensate, the kidneys excrete base, HCO_3^-, so the serum level of HCO_3^- decreases.

■ Tissue that gets too little O_2 is in a state of *hypoxia.* If it gets none at all, it is in a state of *anoxia.*

Extreme respiratory alkalosis can occur to high mountain climbers, like climbers of Mount Everest (8848 m, 29,030 ft). At its summit, the barometric pressure is 253 mm Hg and the pO_2 of the air is only 43 mm Hg (as compared to 149 mm Hg at sea level). Hyperventilation brings a climber's arterial pCO_2 down to only 7.5 mm Hg (compared to a normal of 40 mm Hg) and the blood pH is above 7.7!

Clinical Situations of Respiratory Alkalosis

Lab Results:	pH↑ (7.56); pCO_2↓ (23 mm Hg); $[HCO_3^-]$↓ (20 meq/L)
Typical Patient:	Someone nearing surgery and experiencing anxiety.
Range of Causes:	Prolonged crying; rapid breathing at high altitudes; hysterics; fever; disease of the central nervous system; improper management of a respirator.
Symptoms:	Hyperventilation (that can't be helped); numbness, headache, tingling. Convulsions may occur.
Treatment:	Rebreathe one's own exhaled air (by breathing into a sack); administer carbon dioxide; treat underlying causes.

Take careful notice that hyperventilation alone cannot be used to tell what the condition is. Either metabolic *acidosis* or respiratory *alkalosis* is accompanied by hyperventilation, but the treatments are opposite in nature.

■ People working in emergency care situations get the requisite lab data rapidly, and they must be able to interpret the data on the spot.

Combinations of Primary Acid–Base Disorders Are Possible We have just surveyed the *four primary acid–base disorders.* Combinations of these are often seen, and healthcare professionals have to be alert to the ways in which the lab data vary in such combinations. Someone with diabetes, for example, might also suffer from an obstructive pulmonary disease. Diabetes causes metabolic acidosis and a *decrease* in $[HCO_3^-]$. The pulmonary disease causes respiratory acidosis with an *increase* in [HCO_3^-]. In combination, then, the lab data on bicarbonate level will not be in the expected pattern for either. We will not carry the study of such complications further. We mention them only to let you know that they exist. There are standard ways to recognize them.[3]

12.5

BLOOD AND THE FUNCTIONS OF THE KIDNEYS

Both filtration and chemical reactions in the kidneys help to regulate the electrolyte balance of the blood.

Diuresis is the formation of urine in the kidneys, and it is an integral part of the body's control of the electrolyte and buffer levels in blood.

Urea Is the Chief Nitrogen Waste Exported in the Urine Huge quantities of fluids leave the blood by diffusion each day at the hundreds of thousands of filtering units in each kidney. Substances in solution but not those in colloidal dispersions (e.g., proteins) leave in these fluids. Then active transport processes in kidney cells pull all of any escaped glucose, any amino acids, and most of the fluids and electrolytes back into the blood. Most of the wastes are left in the urine that is being made.

■ The net urine production is 0.6 to 2.5 L/day.

Urea is the chief nitrogen waste (30 g/day), but creatinine (1 to 2 g/day), uric acid (0.7 g/day), and ammonia (0.5 g/day) are also excreted with the urine. If the kidneys are injured or diseased and cannot function, wastes build up in the blood, which leads to a condition known as *uremic poisoning.*

■ *Ur-*, of the urine; *-emia,* of the blood. *Uremia* means substances of the urine present in the blood.

The Hormone Vasopressin Helps Control Water Loss A nonapeptide hormone, **vasopressin,** instructs the kidneys to retain or excrete water and thus helps to regulate the overall levels of solutes in blood. The hypophysis, where vasopressin is made, releases it when the osmotic pressure of blood increases by as little as 2%. At the kidneys, vasopressin promotes the reabsorption of water, and therefore it is often called the *antidiuretic hormone* or ADH.

■ The monitors in the hypophysis of the blood's osmotic pressure are called *osmoreceptors.*

An osmotic pressure that is higher than normal (hypertonicity) means a higher concentration of solutes and colloids in blood. The released vasopressin therefore helps the blood to retain water and thus keeps the solute levels from going still higher. In the meantime, the thirst mechanism is stimulated to bring in water to dilute the blood.

Conversely, if the osmotic pressure of blood decreases (becomes hypotonic) by as little as 2%, the hypophysis retains vasopressin. None reaches the kidneys, so the water that has left the bloodstream at the "filtering units" does not return as much. Remember that a low osmotic pressure means a low concentration of solutes, so the absence of vasopressin at the kidneys when the blood is hypotonic lets urine form. This reduces the amount of water in the blood and thereby raises the concentrations of its dissolved matter. You can see that with the help of vasopressin a normal individual can vary the intake of water widely and yet preserve a stable, overall concentration of substances in blood.

■ In *diabetes insipidus,* vasopressin secretion is blocked, and unchecked diuresis can make from 5 to 12 L of urine a day.

■ The inosital phosphate system (Figure 11.12, page 293) mediates the chemical signals of vasopressin.

The Hormone Aldosterone Helps the Blood Retain Sodium Ion The adrenal cortex makes **aldosterone,** a hormone that works to stabilize the sodium ion level of the blood. This steroid hormone is secreted if the blood's sodium ion level drops. When aldosterone arrives at the kidneys, it initiates reactions that make sodium ions that have left the blood return again. Of course, to keep things isotonic in the blood, the return of sodium ions also requires the return of water.

[3] See, for example, H. Valtin and F. J. Gennari, *Acid–Base Disorders, Basic Concepts and Clinical Management,* 1987. Little, Brown and Company, Boston.

Conversely, if the sodium ion level of the blood increases, then aldosterone is not secreted, and sodium ions that have migrated out of the blood at a filtering unit are permitted to stay out. They remain in the urine being made, together with some extra water.

The Kidneys Make HCO_3^- for the Blood's Buffer System We have seen that breathing is the body's most direct means of controlling acid as it removes or retains CO_2. The kidneys are the body's means of controlling base, as they make or remove HCO_3^-.

The kidneys also adjust the blood's levels of HPO_4^{2-} and $H_2PO_4^-$ the anions of the phosphate buffer. Moreover, when acidosis develops, the kidneys can put H^+ ions into the urine. Some neutralization of these ions by HPO_4^{2-} and by NH_3 takes place, but the urine becomes definitely more acidic as acidosis continues, as we've mentioned before.

■ Urine taken after several hours of fasting normally has a pH of 5.5 to 6.5.

■ In severe acidosis, the pH of urine can go as low as 4.

Figure 12.12 shows the various reactions that take place in the kidneys, particularly during acidosis. (The numbers in the following boxes refer to this figure.) The breakdown of metabolites, [1], makes carbon dioxide, which enters the equilibrium whose formation is catalyzed by carbonic anhydrase, [2]. The ionization of carbonic acid, [3], makes both bicarbonate ion and hydrogen ion. The bicarbonate ion goes into the bloodstream, [4], but the hydrogen ion is put into the tubule, [5], where urine is accumulating. This urine already contains sodium ions and monohydrogen phosphate ions, but to make step [5] possible, *some* positive ion has to go with the HCO_3^-. Otherwise, there would be no net electrical balance. The kidneys have the ability to select Na^+ to go with HCO_3^- at [4]. The kidneys can make Na^+ travel one way and H^+ the other. Newly arrived H^+ can be buffered by HPO_4^{2-} in the developing urine, [6]. Moreover, the kidneys have an ability not generally found in other tissues to synthesize ammonia and use it to neutralize H^+, [7]. Thus the ammonium ion also appears in the urine.

The Kidneys Excrete Organic Anions When acidosis has a metabolic origin, the serum level of the anions of organic acids increases. Organic acids are made at accelerated rates in metabolic disorders, like diabetes or starvation, and are the chief cause of the pH change in metabolic acidosis. The base in the blood buffer has to neutralize them.

The kidneys let organic anions stay in the urine, but only by letting increasing quantities of water stay, too. There is a limit to how concentrated the urine can become, so as solutes stay

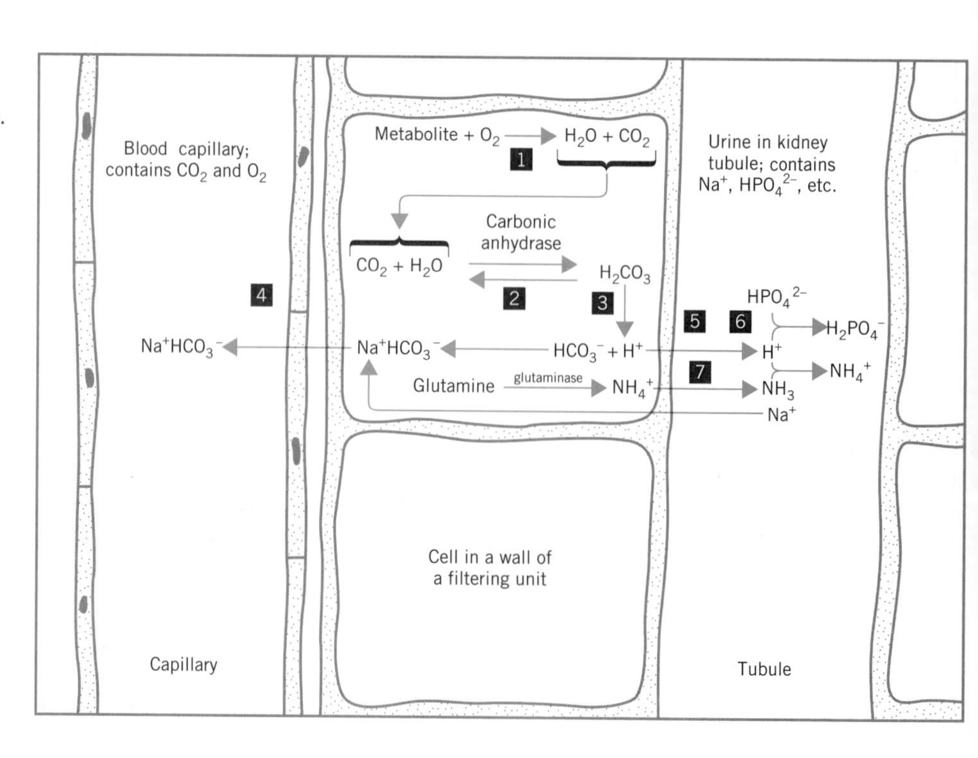

FIGURE 12.12
Acidification of the urine. The numbers refer to the text discussion.

in the urine, water must also stay. Someone with metabolic acidosis, therefore, can experience a general dehydration as the system borrows water from other fluids to make urine. The thirst mechanism normally brings in replacement water, so the individual drinks copious amounts of fluids.

The Kidneys Can Export HCO_3^- In alkalosis, the kidneys can put bicarbonate ion into the urine, and it no longer uses HPO_4^{2-} to neutralize H^+. Both actions raise the pH of the urine, and in severe alkalosis it can go over a pH of 8.

The Kidneys Also Help to Regulate Blood Pressure If the blood pressure drops, as in hemorrhaging, the kidneys secrete a trace of renin into the blood. Renin is an enzyme that acts on one of the zymogens in blood, angiotensinogen, to convert it to the enzyme, angiotensin I. This, in turn, helps to convert still another protein in blood to angiotensin II, a neurotransmitter.

Angiotensin II is the most potent vasoconstrictor known. When it makes blood capillaries constrict, the heart has to work harder, and this makes the blood pressure increase. This helps to ensure that some semblance of proper filtration continues at the kidneys.

Angiotensin II also triggers the release of aldosterone, which we've already learned helps the blood to retain water. This is important because the maintenance of the overall blood volume is needed to sustain a proper blood pressure.

SUMMARY

Digestion The release of digestive juices is under the control of nerve signals and such digestive hormones as gastrin, cholecystokinin, and secretin. α-Amylase in saliva begins the digestion of starch. Pepsin in gastric juice starts the digestion of proteins. In the duodenum, trypsinogen (from the pancreas) is activated by enteropeptidase (from the intestinal juice) and becomes trypsin, which helps to digest proteins. Enteropeptidase also activates chymotrypsin (from chymotrypsinogen) and carboxypeptidase (from procarboxypeptidase). These also help to digest proteins. The pancreas supplies an important lipase, which, with the help of the bile salts, catalyzes the digestion of hydrolyzable lipids. The bile salts also aid in the absorption of the fat-soluble vitamins, A, D, E, and K.

Intestinal juice supplies enzymes for the digestion of disaccharides, nucleic acids, small polypeptides, and lipids.

The products of the digestion of proteins are amino acids; of carbohydrates: glucose, fructose, and galactose; and of the triacylglycerols: anions of fatty acids and monoacylglycerols (and some diacylglycerols). Complex lipids and nucleic acids are also hydrolyzed.

Blood Proteins in blood give it a colloidal osmotic pressure that assists in the exchange of nutrients at capillary loops. Albumins are carriers for hydrophobic molecules and serum-soluble metallic ions. Gamma globulin helps to defend the body against bacterial infections. Fibrinogen is the precursor of fibrin, the protein of a blood clot.

Among the electrolytes, anions of carbonic and phosphoric acid are involved in buffers, and all ions are involved in regulating the osmotic pressure of the blood. The chief cation in blood is Na^+, and the chief cation inside cells is K^+. The balances between Na^+ and K^+ as well as between Ca^{2+} and Mg^{2+} are tightly regulated. Ca^{2+} is vital to the operation of muscles, including heart muscle. Mg^{2+} is involved in a number of enzyme systems.

The blood transports oxygen and products of digestion to all tissues. It carries nitrogen wastes to the kidneys. It unloads cholesterol and heme breakdown products at the gallbladder. And it transports hormones to their target cells. Lymph, another fluid, helps to return some substances to the blood from tissues.

Sudden failure to retain the protein in blood leads to an equally sudden loss in blood volume and a condition of shock. Slower losses of protein, as in kidney disease or starvation, lead to edema.

Respiration The relatively high pO_2 in the lungs helps to force O_2 into HHb. This creates HbO_2^- and H^+. In an isohydric shift, the H^+ is neutralized by HCO_3^-, which is returning from working tissues that make CO_2. The resulting CO_2 leaves during exhaling. Some of the H^+ also converts $HbCO_2^-$ to HHb and CO_2.

In deoxygenating HbO_2^- at cells that need oxygen, the influx of CO_2 makes HCO_3^- and H^+. The H^+ then moves (isohydric shift) to HbO_2^- and breaks it down to HHb and O_2. Both Cl^- and BPG help to ease the last of the four O_2 molecules out of oxyhemoglobin. The low oxygen affinity of blood in tissues where pCO_2 is high helps in the release of oxygen also.

In red muscle tissue, myoglobin's superior oxygen affinity ensures that such tissue can obtain oxygen from the deoxygenation of oxyhemoglobin. Fetal hemoglobin also has an oxygen affinity superior to that of adult hemoglobin.

Acid–base balance The body uses the bicarbonate ion of the carbonate buffer to inhibit acidosis by irreversibly removing H^+ when the lungs release CO_2. HCO_3^- is replaced by the kidneys, which can also put excess H^+ into the urine. Dissolved CO_2 in the blood's carbonate buffer works to control alkalosis by neutralizing OH^-. Metabolic acidosis, with hyperventilation, and metabolic alkalosis, with hypoventilation, arise from dysfunctions in metabolism. Respiratory acidosis, with hypoventilation, and metabolic alkalosis, with hyperventilation, occur when the respiratory centers or the lungs are not working.

Diuresis The kidneys, with the help of hormones and changes in blood pressure, blood osmotic pressure, and concentrations of ions, monitor and control the concentrations of solutes in blood. Vasopressin tells the kidneys to keep water in the bloodstream. Aldosterone tells the kidneys to keep sodium ion (and therefore water also) in the bloodstream. A drop in blood pressure tells the kidneys to release renin, which activates a vasoconstrictor, and aldosterone, which helps to raise blood pressure and retain water. In acidosis, the kidneys transfer H^+ to the urine and replace some of the HCO_3^- lost from the blood. In alkalosis the kidneys put some HCO_3^- into urine.

REVIEW EXERCISES

The answers to Review Exercises whose numbers are in color are found in Appendix A. The answers to the other Review Exercises are found in the Study Guide that accompanies this book. The more challenging questions are marked with asterisks.

Digestion

12.1 What are the names of the two chief extracellular fluids?

12.2 Name the fluids that have digestive enzymes or digestive zymogens.

12.3 Describe how the flow of the acid component of gastric juice is controlled.

12.4 Cimetidine is described as a competitive enzyme inhibitor. What does this mean? How does cimetidine work to aid in the healing of an ulcer?

12.5 What is the result of the work of cholecystokinin? Of secretin?

12.6 What enzymes or zymogens are there, if any, in each of the following?
(a) saliva (b) gastric juice
(c) pancreatic juice (d) bile
(e) intestinal juice

12.7 Name the enzymes and the digestive juices that supply them (or their zymogens) that catalyze the digestion of each of the following.
(a) large polypeptides (b) triacylglycerols
(c) amylose (d) sucrose
(e) di- and tripeptides (f) nucleic acids

12.8 What are the end products of the digestion of each of the following?
(a) proteins
(b) carbohydrates
(c) triacylglycerols

12.9 What functional groups are hydrolyzed when each of the substances in Review Exercise 12.8 is digested? (Refer back to earlier chapters if necessary.)

12.10 In what way does enteropeptidase function as a "master switch" in digestion?

12.11 What would happen if the pancreatic zymogens were activated within the pancreas?

12.12 What services do the bile salts render in digestion?

12.13 What does mucin do (a) for food in the mouth and (b) for the stomach?

12.14 What is the catalyst for each of the following reactions?
(a) pepsinogen → pepsin
(b) trypsinogen → trypsin
(c) chymotrypsinogen → chymotrypsin
(d) procarboxypeptidase → carboxypeptidase
(e) proelastase → elastase

12.15 Rennin does what for an infant?

12.16 Why is gastric lipase unimportant to digestive processes in the adult stomach but useful in the infant stomach?

12.17 In terms of where they work, what is different about intestinal juice compared to pancreatic juice?

12.18 What secretion neutralizes chyme, and why is this work important?

12.19 What happens to the molecules of the monoacylglycerols and fatty acids that form from digestion?

12.20 In a patient with a severe obstruction of the bile duct the feces appear clay-colored. Explain why the color is light.

12.21 The cholesterol in the diet undergoes no reactions of digestion. Explain.

Substances in Blood

12.22 In terms of their general composition, what is the greatest difference between blood plasma and interstitial fluid?

12.23 What is the largest contributor to the net osmotic pressure of the blood as compared to the interstitial fluid?

12.24 What is fibrinogen? Fibrin?

12.25 What services are performed by albumins in blood?

12.26 In what two different regions are Na^+ and K^+ ions mostly found? What are the chief functions of these ions?

12.27 In hypernatremia, the sodium ion level of blood is above what value?

12.28 The sodium ion level of blood is below what value in hyponatremia?

12.29 What causes the hyperkalemia in crushing injuries?

12.30 Above what level is the blood described as hyperkalemic?

12.31 Excessive drinking of water tends to cause what condition, hyponatremia or hypernatremia?

12.32 The overuse of milk of magnesia can lead to what condition that involves Mg^{2+}?

12.33 Inside cells, what is a function that Mg^{2+} serves?

12.34 Where is most of the calcium ion in the body?

12.35 What does Ca^{2+} do in cells?

12.36 What condition is brought on by an overdose of vitamin D, hypercalcemia or hypocalcemia?

12.37 Injections of magnesium sulfate would be used to correct which condition, hypomagnesemia or hypermagnesemia?

12.38 What is the principal anion in both the blood and the interstitial fluid?

12.39 What is the normal range of concentration of Cl^- in blood? Explain how hypochloremia leads to alkalosis.

Exchange of Nutrients at Capillary Loops

12.40 What two opposing forces are at work on the arterial side of a capillary loop? What is the net result of these forces, and what does the net force do?

12.41 On the venous side of a capillary loop there are two opposing forces. What are they, what is the net result, and what does this cause?

12.42 Explain how a sudden change in the permeability of the capillaries can lead to shock.

12.43 Explain how each of the following conditions leads to edema.

(a) kidney disease (b) starvation
(c) a mechanical blow

Exchange of Respiratory Gases

12.44 What are the respiratory gases?

12.45 What compound is the chief carrier of oxygen to actively metabolizing tissues?

12.46 The binding of oxygen to hemoglobin is said to be *allosteric*. What does this mean, and why is it important?

*12.47 Write the equilibrium expression for the oxygenation of hemoglobin. In what direction does this equilibrium shift when:

(a) The pH decreases?
(b) The pO_2 decreases?
(c) The red cell is in the lungs?
(d) The red cell is in a capillary loop of an actively metabolizing tissue?
(e) CO_2 comes into the red cell?
(f) HCO_3^- ions flood into the red cell?

***12.48** Using chemical equations, describe the isohydric shift when a red cell is (a) in actively metabolizing tissues and (b) in the lungs.

*12.49 In what two ways does the oxygenation of hemoglobin in red cells in alveoli help to release CO_2?

***12.50** In what way does waste CO_2 at active tissues help to release oxygen from the red cell?

*12.51 In what way does extra H^+ at active tissue help release oxygen from the red cell?

12.52 Where is carbonic anhydrase found in the blood, and what function does it have in the management of the respiratory gases in (a) an alveolus and (b) actively metabolizing tissues?

12.53 How does BPG help in the process of oxygenating hemoglobin?

12.54 In what way is BPG involved in helping to deoxygenate HbO_2^-?

12.55 What are some changes involving the blood that occur when the body remains at a high altitude for a period of time, and how do these changes help the individual?

12.56 How would the net equations for the oxygenation and the deoxygenation of blood be changed to include the function of BPG?

12.57 What are the two main forms in which waste CO_2 moves to the lungs?

12.58 How is oxygen affinity affected by pCO_2, and how is this beneficial?

12.59 What is the chloride shift and how does it aid in the exchange of respiratory gases?

12.60 In what way is the superior oxygen affinity of myoglobin over hemoglobin important?

12.61 Fetal hemoglobin has a higher oxygen affinity than adult hemoglobin. Why is this important to the fetus?

Acid–Base Balance of the Blood

12.62 Construct a table using arrows (↑) or (↓) and typical lab data that summarize the changes observed in respiratory and metabolic acidosis and alkalosis. The column headings should be as follows:

Condition	pH	pCO_2	$[HCO_3^-]$

*12.63 With respect to the *directions* of the changes in the values of pH, pCO_2, and $[HCO_3^-]$ in both respiratory acidosis and metabolic acidosis, in what way are the two types of acidosis the same? In what way are they different?

12.64 Hyperventilation is observed in what two conditions that relate to the acid–base balance of the blood? In one, giving carbon dioxide is sometimes used, and in the other, giving isotonic HCO_3^- can be a form of treatment. Which treatment goes with which condition and why?

12.65 In what two conditions that relate to the acid–base balance of the blood is hypoventilation observed? Isotonic ammonium chloride or isotonic sodium bicarbonate are possible treatments. Which treatment goes with which condition, and how do they work?

12.66 In which condition relating to acid–base balance does hyperventilation have a beneficial effect? Explain.

*12.67 Hyperventilation is part of the *cause* of the problem in which condition relating to the acid–base balance of the blood?

***12.68** Hypoventilation is the body's way of helping itself in which condition that relates to the acid–base balance of the blood?

***12.69** In which condition that concerns the acid–base balance of the blood is hypoventilation part of the *problem* rather than the cure?

12.70 How can a general dehydration develop in metabolic acidosis?

***12.71** Which condition, metabolic or respiratory acidosis or alkalosis, results from each of the following situations?
(a) hysterics
(b) overdose of bicarbonate
(c) emphysema
(d) narcotic overdose
(e) diabetes
(f) overbreathing at a high altitude
(g) severe diarrhea
(h) prolonged vomiting
(i) cardiopulmonary disease
(j) barbiturate overdose

***12.72** Referring to Review Exercise 12.71, which is happening in each situation, hyperventilation or hypoventilation?

12.73 Why does hyperventilation in hysterics cause alkalosis?

12.74 Explain how emphysema leads to acidosis.

12.75 Prolonged vomiting leads to alkalosis. Explain.

12.76 Uncontrolled diarrhea can cause acidosis. Explain.

12.77 Respiratory alkalosis causes hypocapnia or hypercapnia?

12.78 For each 1 °C above normal human body temperature, the rate of CO_2 production increases by 13%. If the rate of breathing does not increase, what results—hypocapnia or hypercapnia?

Blood Chemistry and the Kidneys

12.79 If the osmotic pressure of the blood has increased, what, in general terms, has changed to cause this?

12.80 How does the body respond to an increase in the osmotic pressure of the blood?

12.81 If the sodium ion level of the blood falls, how does the body respond?

12.82 What is the response of the kidneys to a decrease in blood pressure?

12.83 Alcohol in the blood suppresses the secretion of vasopressin. How does this affect diuresis?

12.84 In what ways do the kidneys help to reduce acidosis?

Additional Exercises

12.85 Monoacylglycerols are able to migrate through membranes of the cells of the intestinal tract that absorb them. Glycerol, however, is unable to accomplish this movement. How might we explain these relative abilities to migrate through a cell membrane?

***12.86** When the gallbladder is surgically removed, lipids of low formula mass are the only kinds that can be easily digested. Explain.

***12.87** It has been reported that some long-distance Olympic runners have trained at high altitudes and then had some of their blood withdrawn and frozen. Days or weeks after returning to lower altitudes and just prior to a long race, they have used some of this blood to replace an equal volume of what they are carrying. This is supposed to help them in the race. How would it work?

***12.88** Aquatic diving animals are known to have much larger concentrations of myoglobin in their red muscle tissue than humans. How is this important to their lives?

NUCLEIC ACIDS

13

Only a hippopotamus can have a hippopotamus baby (which is a good thing when you think about it). The molecular basis for such good things is presented in this chapter.

13.1
HEREDITY AND THE CELL

The chemicals associated with living things are organized in structural units called cells, and the instructions for making these chemicals are carried on molecules of nucleic acids.

We have learned that nearly every reaction in a living organism requires its special catalyst, and that these catalysts are called enzymes. The set of enzymes in one organism is not exactly the same as in another, although many enzymes in different species are quite similar. A major chemical requirement of reproduction is that each organism transmits to its offspring the capacity to possess the set of enzymes that are unique to the organism.

Nucleic Acids Carry Instructions for Making Enzymes In reproduction, an organism does not duplicate the enzymes themselves and then pass them on directly to its offspring. Instead, it sends on the *instructions* for making its unique enzymes from amino acids. It does this by duplicating the compounds of a different family, the *nucleic acids*. These then direct the synthesis of enzymes (and all other proteins) in the offspring. Certain nucleic acids called **ribozymes** are themselves also able to function as enzymes. Our purpose in this chapter is to study how nucleic acids carry genetic instructions and act on them, but we have to learn something about the biological context first.

■ T. R. Cech and S. Altman shared the 1989 Nobel prize in chemistry for discovering ribozymes. It came as a huge surprise in the 1980s to learn that the body uses anything other than proteins as enzyme catalysts.

Every Cell Carries Nucleic Acids The cell is the smallest unit of an organism that has life and can duplicate itself in the organism. Cells of different systems vary widely in shape and size, but they generally have similar parts. Figure 13.1 outlines the major features of an animal cell. The cell boundary is the cell membrane, which we studied in Sections 9.5 and 10.6. Everything enclosed by it is called *protoplasm,* which contains several discrete particles or cellular bodies. Prominent among them are the **mitochondria,** which are tiny *organelles* (subcellular bodies) where adenosine triphosphate (ATP) is made for the cell's needs for chemical energy.

The part of the cell outside of its nucleus is called the *cytoplasm.* (The liquid portion itself is called the *cytosol,* as we have learned.) One kind of particle in the cytoplasm is called a **ribosome,** which consists mostly of nucleic acids and polypeptides. Ribosomes have essential functions in the synthesis of polypeptides, including the polypeptides of enzymes. Polypeptide synthesis, however, is under the primary control of nucleic acids *inside* the cell's nucleus. Nucleic acids thus occur in both the cytoplasm and nucleus.

Genes Are the Fundamental Units of Heredity The nucleus has its own membrane, and inside is a web-like network of protein. The nucleus also contains twisted and intertwined filaments of nucleoprotein called *chromatin.* Chromatin is like a strand of pearls, each pearl made of proteins called *histones* around which are tightly coiled one of the kinds of nucleic acid called DNA for short. DNA also links the "pearls." Each "pearl" is called a *nucleosome.* We'll study the structure of DNA in the next section, but DNA molecules have portions that

■ The histones are not just spools for wrapping DNA strands; they contribute to the regulation of DNA activity.

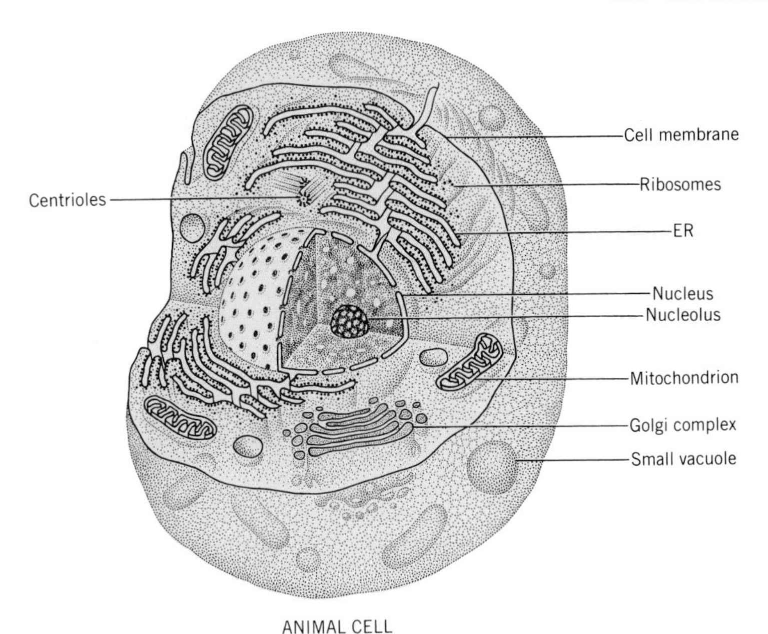

FIGURE 13.1
Model of a generalized animal cell. Although cells differ greatly from tissue to tissue, most have the features shown here. ER stands for endoplasmic reticulum. (From G. C. Stephens and B. B. North, *Biology, A Contemporary Perspective,* 1974. John Wiley and Sons. Used by permission.)

constitute individual **genes,** the fundamental units of heredity. A single gene, for example, contains the information for making a single enzyme, so unique genes translate into unique enzymes. A major goal of this chapter is to learn how the chemical structure of a gene enables it to function in this way.

Prior to Cell Division, Genes Replicate When cell division begins, the chromatin strands thicken and become rod-like bodies that accept staining agents and so can be seen under a microscope. These discrete bodies are called **chromosomes.** The thickening of chromatin into chromosomes is caused by the synthesis of new DNA and histones. The new chromosomes, including their DNA, are exact copies of the old, if all goes well as it does to a remarkable extent. This reproductive duplication of DNA is called **replication,** so by the replication of DNA, the genetic message of the first cell is passed to each of the two new cells.

■ There are 23 matched pairs of chromosomes in the human cell for a total of 46 chromosomes.

When two germ cells, a sperm and an ovum, unite at conception to form the single cell from which the entire organism will grow, DNA from both germ cells combine. Genetic characteristics of both parents thus are passed on to the offspring. Every cell made from the first cell has the entire set of genes, but obviously most genes in the older organism are turned off most of the time. Genes that are behind the formation of fingernails, for example, must not operate in heart muscle cells! Another goal of our study is to learn how (some) genes might be regulated.

■ Every human cell has between 50,000 and 100,000 genes.

13.2

THE STRUCTURE OF NUCLEIC ACIDS

Genetic information is carried by the sidechains of a twisted, double-stranded polymer called the DNA double helix.

The nucleic acids that store and direct the transmission of genetic information are polymers nicknamed DNA and RNA. **DNA** is **deoxyribonucleic acid** and **RNA** is **ribonucleic acid.** The monomer molecules for the nucleic acids are called **nucleotides.**

Nucleotides Are Monophosphate Esters of Pentoses to Which Heterocyclic Bases Are Joined Unlike the monomers of proteins, the nucleotides can be further hydrolyzed. As outlined in Figure 13.2, the hydrolysis of a representative mixture of nucleotides produces

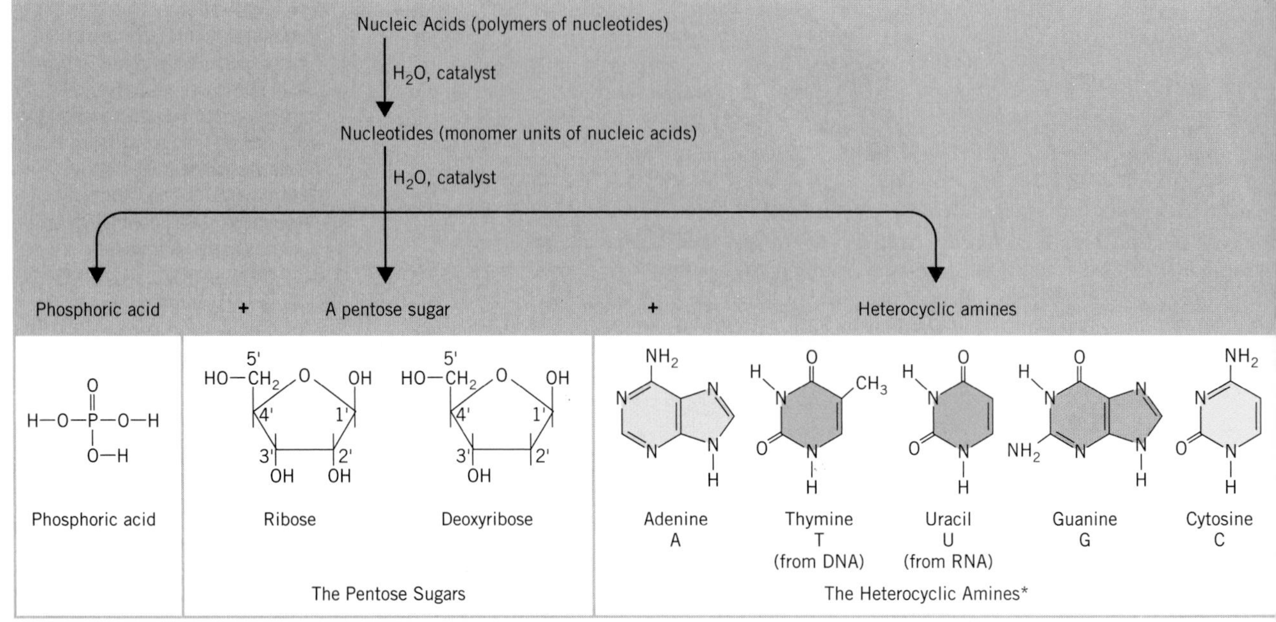

*These are the five principal heterocyclic amines obtainable from nucleic acids. Others, not shown, are known to be present. Although they differ slightly in structure, they are informationally equivalent to one or another of the five shown here.

FIGURE 13.2
The hydrolysis products of nucleic acids

three kinds of products: inorganic phosphate, a pentose sugar, and a group of heterocyclic amines called the **bases,** which have single-letter symbols:

Bases from DNA		Bases from RNA	
Adenine	A	Adenine	A
Thymine	T	Uracil	U
Guanine	G	Guanine	G
Cytosine	C	Cytosine	C

Three bases are thus common to both DNA and RNA, and one base is different. The sugar unit is also different. It is a pentose in both, but the hydrolysis of RNA gives ribose—hence the R in RNA—and that of DNA gives deoxyribose—which lends the D to DNA.

In the structure of a typical nucleotide of RNA, adenosine monophosphate or AMP, the base adenine (A) is joined to the hemiacetal carbon (C-1′) of the pentose; the C-5′ OH group of the pentose has been changed into a ester of phosphoric acid (see Fig. 13.3). (The prime, as in 5′, refers to the numbering of the pentose ring; unprimed numbers are used to number positions on the rings of the bases.) All nucleotides are monophosphate esters with structures

■ Ring carbon number 1′ is the carbon of the aldehyde group when the pentose ring is open.

FIGURE 13.3
A typical nucleotide, AMP, and the smaller units from which it is assembled. The phosphate ester group forms by the splitting out of water between an OH group of phosphoric acid and an H atom at the 5′ OH group of the ribose unit. The splitting out of water between the 1′ OH group of the ribose unit and the H atom on a ring nitrogen of adenine (A) joins adenine to the sugar unit. Similar structures could be drawn with the other bases of Figure 13.2. In the formation of each such nucleotide, the H atom used in this splitting out of water is the one attached to the ring nitrogen drawn as the lowest in the structure of the base.

Adenine

Phosphoric acid

Ribose

$(-2H_2O)$
(several steps)

Adenine unit (A)

Ribose unit

Adenosine monophosphate "AMP"
(Adenosine 5′-phosphate)

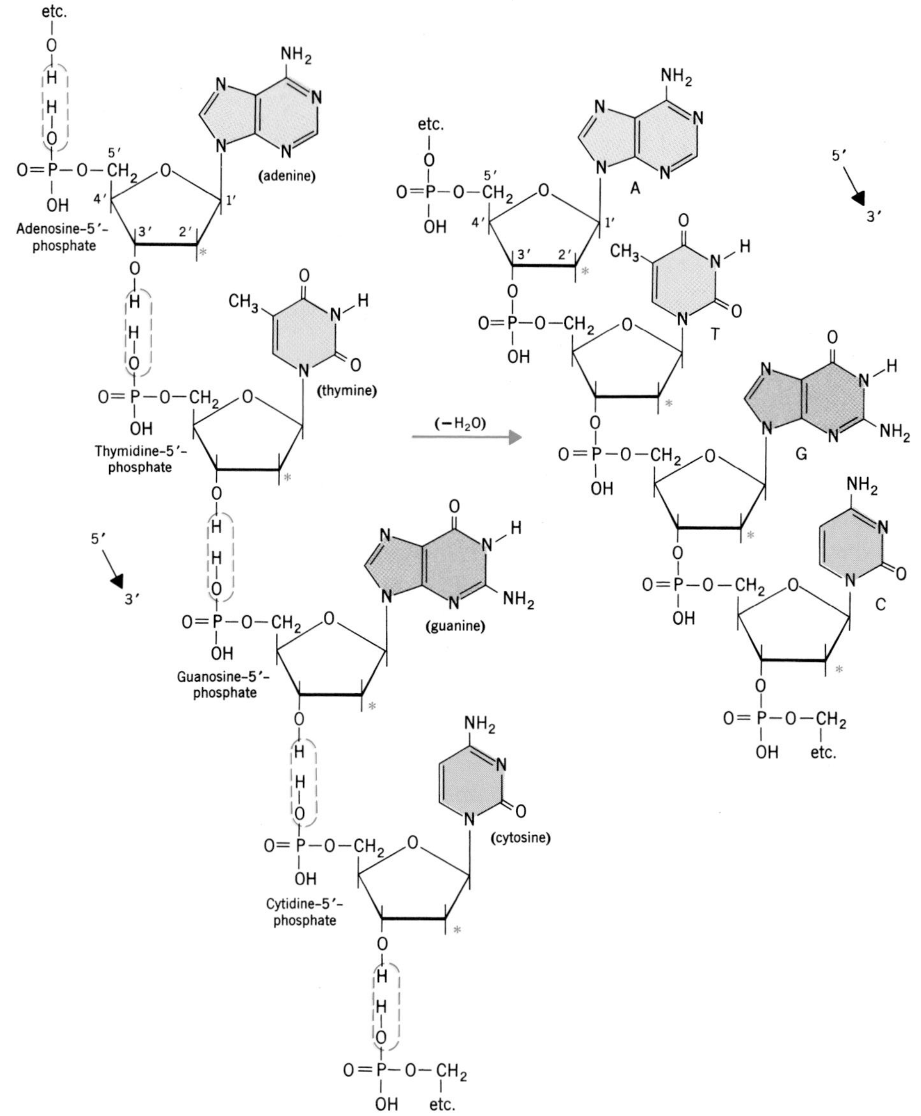

FIGURE 13.4
The relationship of a nucleic acid chain to its nucleotide monomers. On the right is a short section of a DNA strand. On the left are the nucleotide monomers from which it is made (after many steps). The colored asterisks by the pentose units identify the 2′ positions of these rings where there would be another OH group if the nucleic acid were RNA (assuming that uracil also replaced thymine).

like that of adenosine monophosphate. In making its nucleotides, the cell splits out two molecules of water as indicated in Figure 13.3 (but by *several* steps that we won't discuss).

When the OH at C-2′ is replaced by H, and the pentose is deoxyribose, the resulting nucleotide is one for DNA. The pattern for all nucleotides, therefore, is as follows.

■ Some 20 enzymes are required, the chief being DNA polymerase, discovered in 1958 by Arthur Kornberg (Nobel prize, 1959).

$$\begin{array}{c} \text{base} \\ | \\ \text{HO—phosphate—pentose—OH} \end{array}$$

The Bases Project from the Backbones of Nucleic Acids Figure 13.4 shows the general scheme of how nucleotides are linked in nucleic acids. Many steps and many enzymes are required but, in the overall result, water splits out between an OH group at a C-3′ ring position of one nucleotide and the phosphate unit of the next nucleotide. The result of the splitting out of water is a *phosphodiester* system. When this linking is repeated thousands of times, a nucleic acid is the product. A nucleic acid is thus a copolymer for which there are four monomers. If the sequence at the top end of the chain in Figure 13.4 did not continue, the top would be regarded as the *starting* point of the polymer chain. It's called the 5′ end after the number of the first position encountered on a pentose ring. At the opposite end, the chain ends at a C-3′ OH group. Thus the chain in Figure 13.4 is said to be running, top to bottom

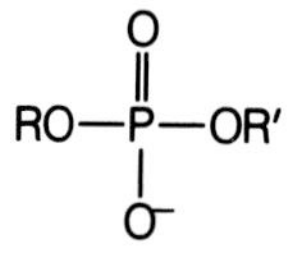

■ $RO—P(=O)(O^-)—OR'$

Phosphodiester (anion form)

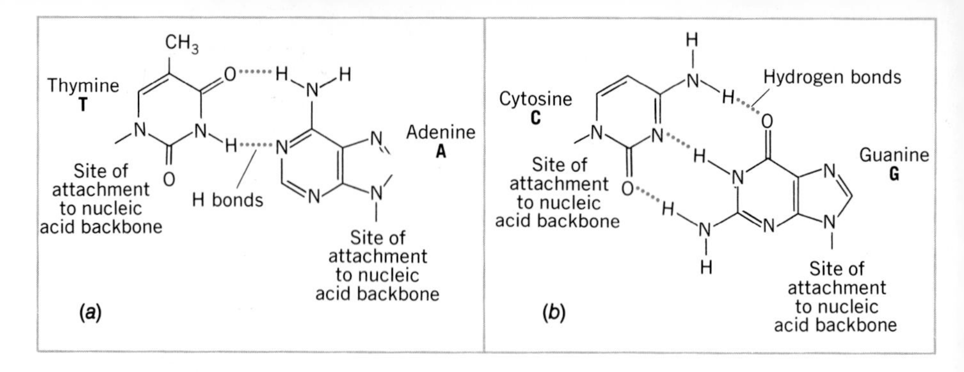

FIGURE 13.5
Hydrogen bonding between base pairs. (*a*) Thymine (T) and adenine (A) form one base pair between which are two hydrogen bonds. (*b*) Cytosine (C) and guanine (G) form another base pair between which are three hydrogen bonds. Adenine can also base-pair to uracil (U).

(beginning to end), in the 5′→3′ direction. The bases are in the sequence of A to T to G to C, 5′→3′, so the *structural formula* of the portion of the DNA polymer in Figure 13.4 can be highly condensed simply as ATGC.

What are shown in Figure 13.4 as free OH groups attached to the P atoms of the chain are moderately strong proton-donating groups, similar to the OH groups in phosphoric acid. The pH of cellular fluid is slightly higher (more basic) than 7, however, so these acidic groups have given up protons (which are neutralized by the buffer), and nucleic acids exist *in vivo* as multiply charged anions.

The repeating pattern for the backbone of any nucleic acid, DNA or RNA, thus shows alternating phosphate and pentose units, and projecting from each pentose is one of the bases. The backbone holds the system together, and the bases—their selection and sequence—carry the genetic information. *The distinctiveness of any one nucleic acid is in the sequence of its side chain bases.* There are 24 different sequences possible for just the four different bases of DNA.

ATGC	TAGC	GCAT	CATG
ATCG	TACG	GCTA	CAGT
AGTC	TGAC	GATC	CTAG
AGCT	TGCA	GACT	CTGA
ACTG	TCAG	GTAC	CGAT
ACGT	TCGA	GTCA	CGTA

Many genes involve thousands of bases (actually thousands of *pairs* of bases, as we will see), each base obviously used many times, and you can begin to see how the uniqueness of a given gene rests altogether on the sequences of the DNA bases.

Pairs of Bases Are Attracted to Each Other by Hydrogen Bonds The sidechain bases have functional groups so arranged geometrically that they fit to each other in pairs by means of hydrogen bonds, a phenomenon called **base pairing** (see Fig. 13.5). *The locations and geometries of the functional groups of the bases permit only certain base pairs to exist.* In DNA, G and C always form a pair, and A and T form another pair. In RNA, G and C always pair, and U and A always pair. Neither G nor C ever pairs with A, T, or U.

■ Crick and Watson shared the 1962 Nobel prize in medicine and physiology with Maurice Wilkins.

DNA Occurs as Paired Strands Twisted into a Double Helix In 1953, Francis Crick of England and James Watson of the United States proposed a structure for DNA, the **DNA double helix,** that made possible an understanding of how heredity works at the molecular level. Using X-ray data obtained by Rosalind Franklin, they deduced that two complementary DNA molecules line up side by side to form a double strand that then twists into a right-handed helix. The strands will not separate except by unwinding the helix. DNA in its double helix form is called **duplex DNA.**

■ Two irregular objects are *complementary* when one fits to the other, as your right hand would fit to its impression in clay.

The idea of complementary strands is a key feature of the double helix just as it was in our study of enzymes and substrates or receptors and hormones (or neurotransmitters). In DNA chemistry, *strand complementarity* also refers to the kind of fitting of one whole strand to the

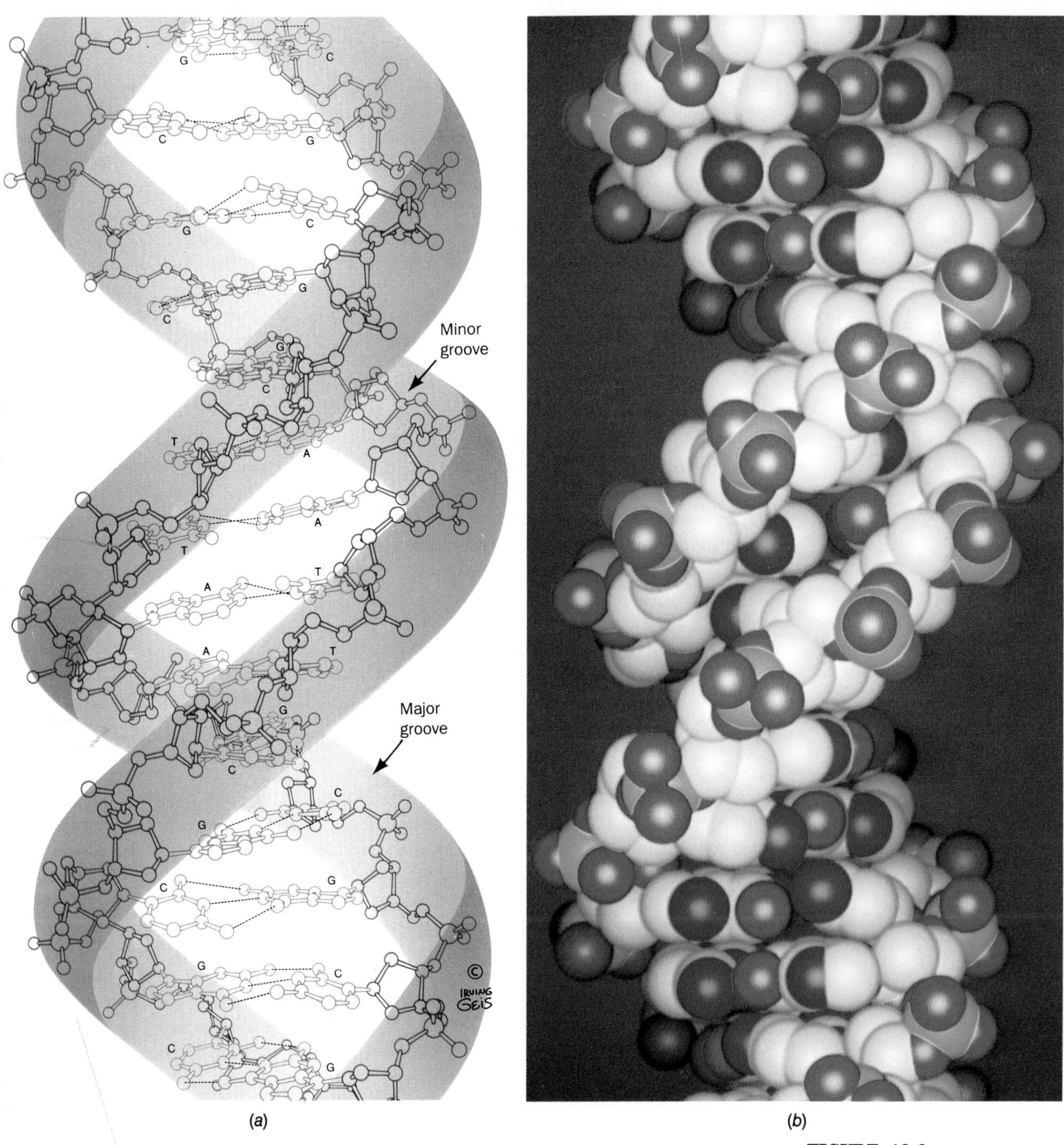

FIGURE 13.6
The structure of duplex DNA—the native form (B-DNA)—seen here in (*a*) a ball-and-stick drawing and (*b*) a computer-generated space-filling model. (Drawing copyrighted © by Irving Geis. Computer graphics courtesy of Robert Stodola, Fox Chase Cancer Center.)

other. This occurs by base pairing *between the strands,* an idea suggested by two significant 1 : 1 mole ratios discovered earlier by Erwin Chargaff. He had found that A and T are always present in a 1 : 1 ratio in DNA *in all species,* and that G and C also occur in a 1 : 1 mole ratio in all species studied. As we have seen, A pairs to T and G pairs to C, so each pair *must* be in a 1 : 1 ratio. Crick and Watson made sense of the simple 1 : 1 ratios by proposing that A pairs to T *between two complementary strands,* and that G pairs to C also between the strands.

Whenever adenine (A) is attached to one strand, then thymine (T) is attached opposite it on the complementary strand. And whenever guanine (G) projects from one strand, then cytosine (C) is opposite it on the complementary strand. A molecular model of the DNA double helix discovered by Crick and Watson is shown in Figure 13.6. The system resembles a spiral staircase in which the steps, which are nearly perpendicular to the long axis of the spiral, consist of the base pairs.

■ The mole ratios of (A + T) to (C + G) vary between species.

Largely Hydrophobic Forces Stabilize the DNA Double Helix Hydrogen bonds occur between the base pairs, but these bonds are no longer regarded as the primary source of the stability of the double helix structure. Hydrogen bonds determine which bases can pair, but *hydrophobic interactions stabilize duplex DNA*. The rings themselves of the bases are planar and unsaturated, like benzene rings, and relatively nonpolar despite the ring N atoms. In an aqueous medium there is a natural tendency for these rings to stack closely together because they cannot offer much hydrogen-bonding alternatives to water molecules. Water excludes them, so they stack together, which is exactly what the DNA duplex structure portrays. Around the edge of each spiraling backbone project the negatively charged, anionic sites of the phosphate units, giving these strongly hydrophilic groups full access to the cellular fluid. What may not be immediately obvious from Figure 13.6 is that the two DNA strands run in opposite directions.

■ The *in vivo* form of DNA is sometimes called B-DNA. Other forms are A-DNA and Z-DNA. They differ in coiling geometries.

Duplex DNA Has a Major and a Minor Groove The structure of duplex DNA in Figure 13.6 is the *in vivo* form of DNA. Notice that it has two grooves, one major and one minor. These are binding sites for molecules of proteins involved in several gene functions, including gene repression and gene activation. They are also sites for the initial molecular interactions of certain drugs, antibiotics, carcinogens (cancer-causing agents), and poisons. Polypeptides with net positive charges would be attracted to these spiraling duplex DNA grooves, and complementary fitting to the DNA becomes a factor in protein–DNA interactions.

Figure 13.6 shows only a short segment of duplex DNA, a sequence too short to show that duplex DNA is further twisted and coiled into superhelices. The looping and coiling are necessary to fit the cell's DNA into its nucleus. A typical human cell nucleus, for example, is only about 10^{-7} m across, but if all of its DNA double helices were stretched out, they would measure a little over 1 m, end to end.

■ There are an estimated 3 billion base pairs in one nucleus of a human cell but only about 5% are a part of genes. The functions of the remaining 95% are not yet fully known.

In DNA Replication, the Bases on Each Strand Guide the Formation of New Complementary Strands When DNA replicates, the cell makes an exact complementary strand for each of the two original strands, so two identical double helices are made. The built-in guarantee that each new strand will be the complement of one of the old is the requirement that A pairs to T and C to G. A very general picture that shows how this works is in Figure 13.7. Realize that this figure explains only one of the many aspects of replication, how base pairing ensures two new complementary strands. Many of the details of how replication occurs without everything becoming hopelessly tangled in the nucleus are understood, but we must leave them to more advanced references. One feature of DNA replication correctly shown by Figure 13.7 is that the synthesis of replica DNA strands occurs as the parent strand unwinds. The formation of the replicas does not await the complete unwinding of the parent before the replication commences.

In Higher Organisms, Sections of DNA Molecules Called *Exons* Collectively Carry the Message of One Gene Although a single gene can itself be thought of as a continuous sequence of nucleotide units, each with its sidechain, *the gene does not occur as a continuous sequence within a DNA system of chromatin*. In their locations in chromatin, virtually all individual genes are divided or split. This means that a gene is made up of *sections*, called **exons** of a DNA chain. Interrupting the exons and separating them are other, usually longer sections of the DNA molecule called **introns.** The gene for the β-subunit of human hemoglobin, for example, consists of 990 bases, but two intron units, 120 and 550 bases long, interrupt the gene, as illustrated in Figure 13.8. Some introns are short as 50 base pairs in length and others as long as 200,000 base pairs. Introns make up an average of roughly 80% of a human structural gene (that part of a gene that does not include regulator and control sites).

■ *Exon* refers to the part that is expressed and *intron* to the segments that *int*errupt the exons.

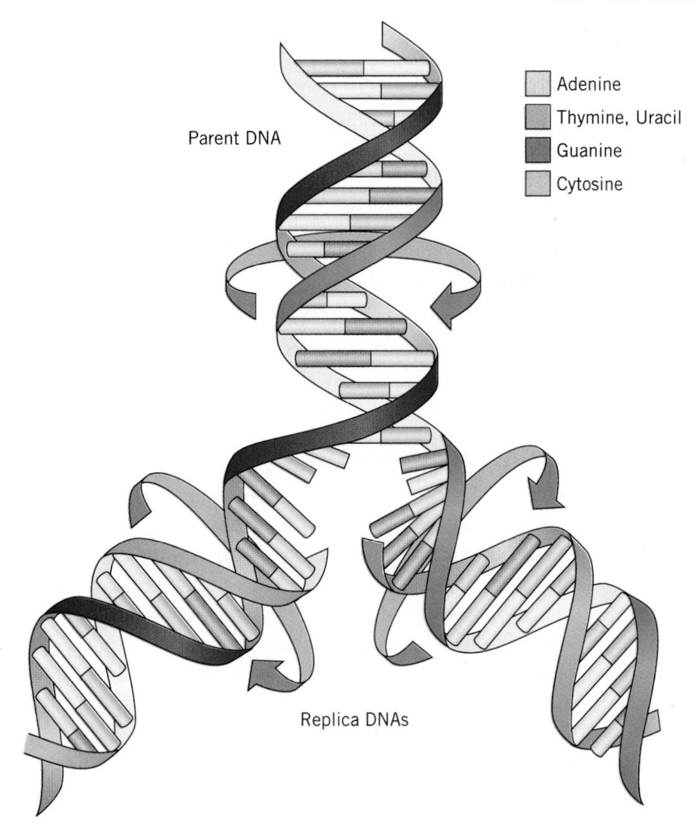

FIGURE 13.7
The replication of DNA.

A single gene can have as many as 50 intron segments, but not all DNA molecules in higher organisms have introns. A few human genes do exist with completely continuous sequences of bases. Just what purposes are served by the introns is currently under considerable speculation. They may serve a protective role or they may have a regulatory function.

■ There are no introns in human genes that code for histones or for α-interferon.

We'll see later in this chapter how the exons of one gene manage to get the gene's message together. For the present, we can consider that a **gene** is a particular section of a DNA strand minus all the introns in this section. A gene, in other words, is a specific series of bases strung in a definite sequence along a DNA backbone.

Gene Sequencing Has Been Automated Chemists have developed automatic gene sequencing instruments that determine the sequences of bases in individual genes. The genes of one human being have to be similar, of course, to those of another for the two to be members of the same species. However, every individual human being that has ever lived or ever will live, except for identical twins, has genes unique in some ways. This uniqueness of individuals is so analogous to the uniqueness of fingerprints that genetic "fingerprints" often are introduced as evidence in certain criminal trials (see Special Topic 13.1).

■ American chemists Richard Roberts and Phillip Sharp shared the 1993 Nobel prize in physiology and medicine for their independent discovery in 1977 of "split genes."

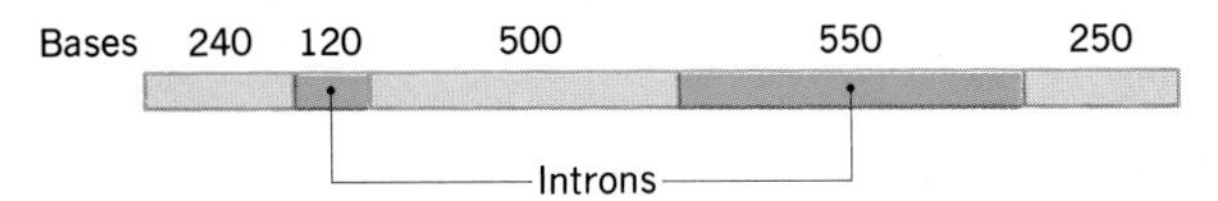

FIGURE 13.8
The gene for the β-subunit of hemoglobin is a split gene with two long intron sections.

SPECIAL TOPIC 13.1
DNA TYPING—"GENETIC FINGERPRINTING"—AND CRIME PROSECUTION

In the entire set of human genes, the human *genome,* there are many regions consisting of nucleotide sequences repeated in tandem. These regions, called *minisatellites,* all have a common core sequence, but the *number* of repeated sequences in the minisatellites varies from individual to individual. These variations, which can be measured, are so considerable between individuals that they are the basis of a major new technique in discovering the truth concerning a suspect in certain crimes, particularly rape.

Suppose, for example, a sample of semen can be obtained from the sperm left by a rapist. The DNA is "amplified" or cloned by means of the *polymerase chain reaction* to convert it into a sample large enough for the analysis. To apply the polymerase chain reaction, the DNA is first denatured by warming the sample, which causes the strands of duplex DNA to unwind and separate. (Denatured DNA renatures itself spontaneously if the temperature of the system is carefully controlled over a sufficient length of time.) A sample of the enzyme DNA polymerase is then added together with specially prepared, short DNA "primer" strands, and the mixture is incubated. During this period, cloning of the DNA occurs and new but identical copies of duplex DNA molecules form.

The sample of cloned DNA is next hydrolyzed with the use of a specific enzyme, called a *restriction enzyme.* The enzyme is able to catalyze the breakage (hydrolysis) of a DNA strand *only at specific sites,* which releases the core segments described earlier. The fragments of DNA can then be separated and made to bind (by base pairing) to short, radioactively labeled, specially made DNA that has been designed to be able to bind to core segments. Now the labeled DNA fragments are separated by a special technique so that the fragments occur as thin, parallel deposits spread out on a film. Their locations on the film can be detected by the ability of their atomic radiations to affect photographic film. The radiations darken the film only in characteristic locations and produce a series of 30 to 40 dark bands, roughly analogous to the bar codes on groceries that are used for pricing at checkout counters. Each individual has a unique genetic "bar code." Because every cell in the body has the entire genome, a single hair of a rape suspect can provide the same fundamental genome as comes from a sample of semen or blood. Thanks to the polymerase chain reaction, enough DNA material can be made from the tiniest of samples to measure the person's genetic "bar code" and compare it to that obtained from the semen sample. Figure 1 shows how one suspect was trapped.

When the two "bar codes" match and there has been sufficient care in making sure that the samples are from *different* people and that the cloning work has been carefully done, the jury has evidence considered by nearly all scientists to be as powerful as fingerprints for a conviction. If "bar codes" do not match, the district attorney looks for another suspect. DNA typing evidence is thus as powerful an ally of the innocent as it is an enemy of the guilty. If a crime site specimen is old or has been subjected to harsh environmental conditions and has deteriorated, it will either give a true test or none at all. It does not give a false test that will lead to the conviction of an innocent person.

The use of DNA typing is going through rigorous court challenges. Does every individual actually have a completely unique genome? (The DNA "fingerprints" of identical twins are the same.) Is the evidence from DNA typing truly like fingerprint evidence? (Members of a South American tribe founded by a few individuals and now excessively inbred have very similar DNA fingerprints.) What are the statistical probabilities that two DNA samples from *different* people will be different? Enough doubts have been raised over these questions in the minds of some judges that (as of early 1993) the court systems of Massachusetts and Guam would not admit DNA "fingerprints" as evidence unless or until these uncertainties are resolved. The use of real fingerprints went through similar challenges. During the period of the challenges to DNA typing, recognizing that the technique is an extraordinarily powerful tool to establish the truth, scientists have been at work to remove the doubts, improve the technology, and establish the reliability of every procedural step from the finding of raw evidence to the court appearance.

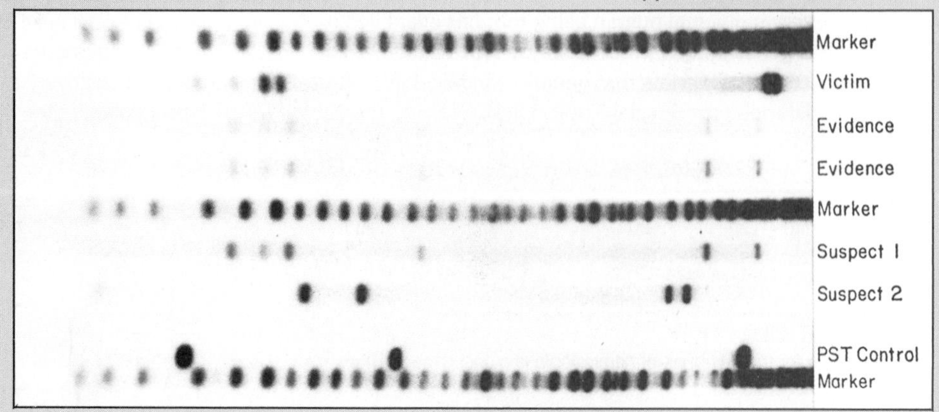

FIGURE 1
DNA fingerprinting. The banding pattern of the DNA of suspect 1 matches that of the evidence. Neither the rape victim's DNA nor that of suspect 2 does.

13.3 RIBONUCLEIC ACIDS

■ *Molecular biology* is the hybrid science that encompasses all of the chemistry of nucleic acids, genes, chromosomes, and genetic expressions in protein syntheses.

Triplets of bases on messenger RNA correlate with individual amino acids through the *genetic code*.

The general scheme that relates DNA to polypeptides is illustrated in Figure 13.9. We'll refer to it often, but first we need to learn more about RNA and its various types, particularly the four that participate in expressing genes in higher organisms.

Ribosomal RNA (rRNA) Is the Most Abundant RNA We have already mentioned the small particles in the cytoplasm, called ribosomes (Fig. 13.1). Each forms two subunits, as shown in Figure 13.9, which come together to form a complex with messenger RNA, another type that we'll study soon.

■ Over 50 kinds of polypeptides exist in one ribosome.

Ribosomes contain proteins and a ribonucleic acid called **ribosomal RNA,** abbreviated **rRNA.** Except in a few viruses, rRNA is single-stranded, but its molecules often have hairpin loops in which base pairing occurs. The rRNA of a ribosome is itself made from longer versions by losing some of its chain pieces. This loss is catalyzed by a region of the very RNA being groomed, so such RNA regions are actually working as enzymes, the ribozymes we mentioned on page 332.

■ Other ribozymes are now known that catalyze reactions of substrates other than themselves.

Ribosomes are the sites of polypeptide synthesis. However, the ribosome does not know how to arrange the amino acid residues of the polypeptide in the right order. The rRNA of the

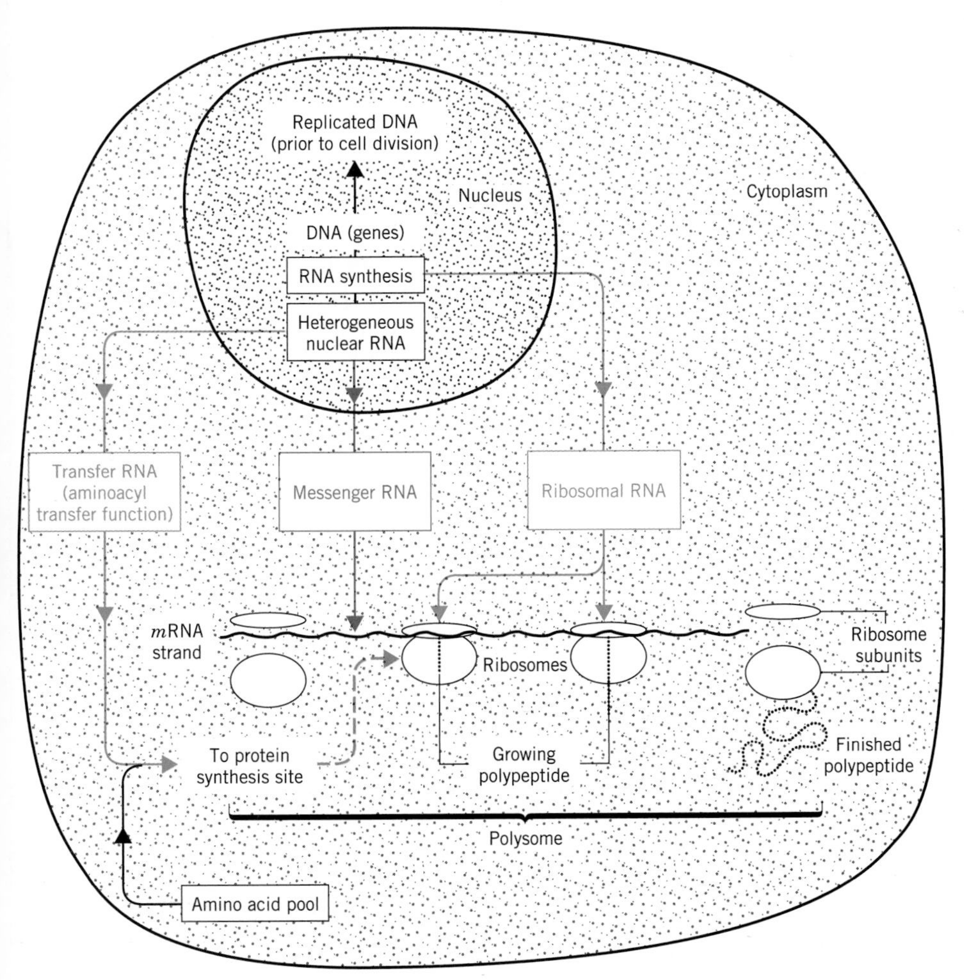

FIGURE 13.9
The relationships of nuclear DNA to the various RNAs and to the synthesis of polypeptides

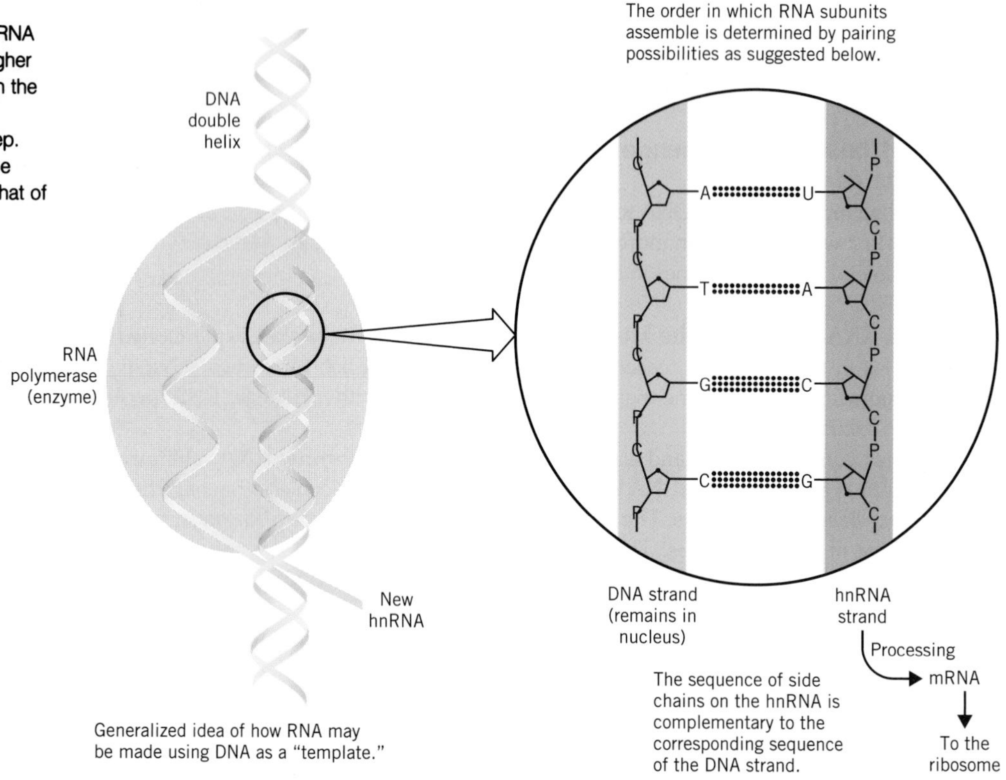

The order in which RNA subunits assemble is determined by pairing possibilities as suggested below.
DNA double helix
RNA polymerase (enzyme)
New hnRNA
Generalized idea of how RNA may be made using DNA as a "template."
A U
T A
G C
C G
DNA strand (remains in nucleus)
hnRNA strand
Processing
mRNA
To the ribosome
The sequence of side chains on the hnRNA is complementary to the corresponding sequence of the DNA strand.

FIGURE 13.10
DNA-directed synthesis of hnRNA in the nucleus of a cell in a higher organism. The shaded oval on the left represents a complex of enzymes that catalyze this step. (Notice that the direction of the hnRNA strand is opposite to that of the DNA strand.)

ribosome cannot itself direct this aspect of putting a polypeptide together. Another kind of RNA called *messenger RNA* is the director of polypeptide synthesis, and messenger RNA is made from another kind of RNA that we will study.

Heterogeneous Nuclear RNA (hnRNA) Is Complementary to DNA—Exons and Introns When a cell uses a gene to direct the synthesis of a polypeptide, its first step is to use the single DNA strand bearing the polypeptide's gene to guide the assemblage of a complementary molecule of RNA, called **heterogeneous nuclear RNA,** abbreviated **hnRNA.**[1] Figure 13.10 gives a very broad picture of how sidechain bases on DNA guide the nucleotides with sidechain bases for hnRNA into the correct sequence during the assembly of hnRNA. Uracil (U) is now used instead of thymine (T), so when a DNA strand has an adenine (A) side chain, then uracil not thymine takes the position opposite it on the complementary RNA. The enzyme involved is *RNA polymerase*.

■ Part of the grooming of hnRNA molecules installs at their 3′ ends a long poly-A tail (about 200 adenosine units long), and at the other end a nucleotide triphosphate "cap."

The next general step in gene-directed polypeptide synthesis is the processing of hnRNA to the next kind of RNA that we must study, messenger RNA or mRNA.

Each Messenger RNA (mRNA) Is Complementary to One Gene Molecules of hnRNA have large sections complementary to the introns of the DNA, and these sections must be deleted. Special enzymes catalyze reactions that snip these pieces out of hnRNA and splice together just the units corresponding to the exons of the divided gene (see Fig. 13.11). The result is a much shorter RNA molecule called **messenger RNA,** or **mRNA.**

■ A small family of nuclear ribonucleoproteins, snRNPs (or "snurps"), helps this resplicing process.

In mRNA we have a sequence of bases complementary just to the gene's exons, so this mRNA now carries the unsplit genetic message. We have now moved the genetic message

[1] Heterogeneous nuclear RNA has been called *primary transcript RNA (ptRNA)*, or simply *primary transcript*, and sometimes *pre-messenger RNA (pre-mRNA)*.

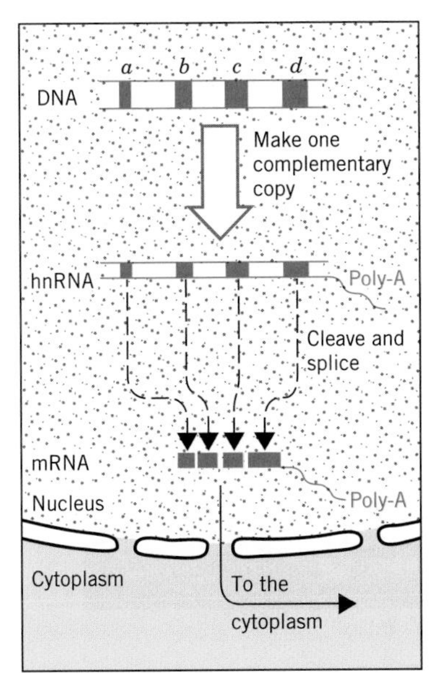

FIGURE 13.11
The RNA made directly at a DNA strand is hnRNA. Only the segments made at sites *a*, *b*, *c*, and *d*—the exons—are needed to carry the genetic message to the cytoplasm. The hnRNA is processed, therefore, and its segments that matched the introns of the gene are snipped out. Then the segments that matched the exons are rejoined to make the mRNA strand.

from a split gene on DNA to a molecule of mRNA, and the name for this overall process is **transcription.**

Triplets of Bases on mRNA Are Genetic Codons Each group of three adjacent bases on a molecule of mRNA constitutes a unit of genetic information called a **codon** (taken from the word *code*). Thus it is a sequence of *codons* on the mRNA backbone more than a sequence of individual bases that now carries the genetic message. (We'll explain shortly why *three* bases per codon are necessary.)

Once they are made, the mRNAs move from the nucleus to the cytoplasm where they attach ribosomes. Many ribosomes can be strung like beads along one mRNA chain, and such a collection is called a *polysome* (short for *polyribosome*).

■ F. Jacob and J. Monod (Nobel prizes, 1965) conceived the idea that a messenger RNA must exist.

Ribosomes are traveling packages of enzymes intimately associated with rRNA. Each ribosome moves along its mRNA chain while the codons on the mRNA guide the synthesis of a polypeptide. To complete this system, we need a way to bring individual amino acids to the polysome's polypeptide assembly sites. For this, the cell uses still another type of RNA.

Transfer RNA (tRNA) Molecules Can Recognize Both Codons and Amino Acids The substances that carry aminoacyl units to mRNA *in the right order for a particular polypeptide* are a collection of similar compounds called **transfer RNA** or **tRNA.** Their molecules are small, each typically having only 75 nucleotides. As seen in Figure 13.12, they are single-stranded but with hairpin loops.

■ $NH_2CH(R)C(=O)—$

Aminoacyl unit

We're now dealing with the molecular basis of *information,* so we can use language analogies. On a human level, we use language to convey information, and language involves words built from a common alphabet. We are aware that many languages exist among human societies and that the world knows several alphabets. To communicate between languages, we have to translate. The same need for translation occurs at the level of genes and polypeptides. tRNA is the master translator in cells.

tRNA is able to work with two "languages," the nucleic acid or genetic and the polypeptide. The nucleic acid language is expressed in an alphabet of 4 letters, the four bases, A, T (or U), G, and C. The polypeptide language has an alphabet of 20 letters, the side chains on the 20 amino acids. To translate from a 4-letter language to a 20-letter language requires that the 4 nucleic acid letters be used in groups of a minimum of 3 letters. Then there will be enough combinations of letters for the larger alphabet of the amino acids. Thus there can be at least one nucleic acid "word," built of 3 letters, for each of the 20 amino acids. This is exactly how

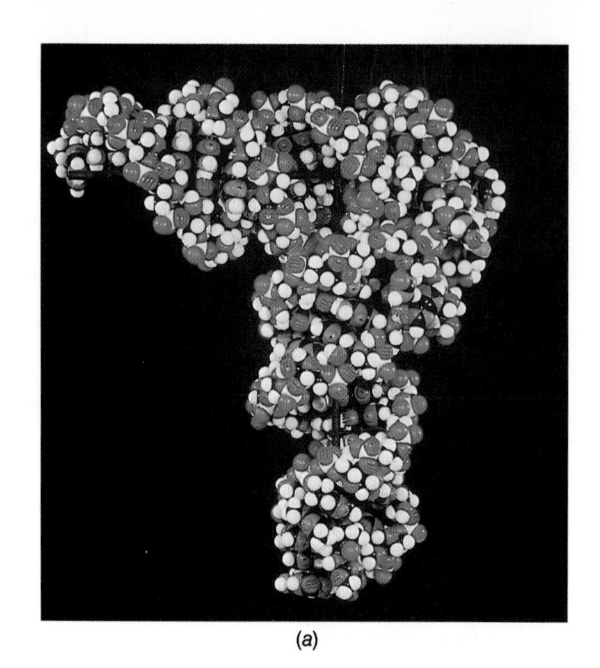

(a)

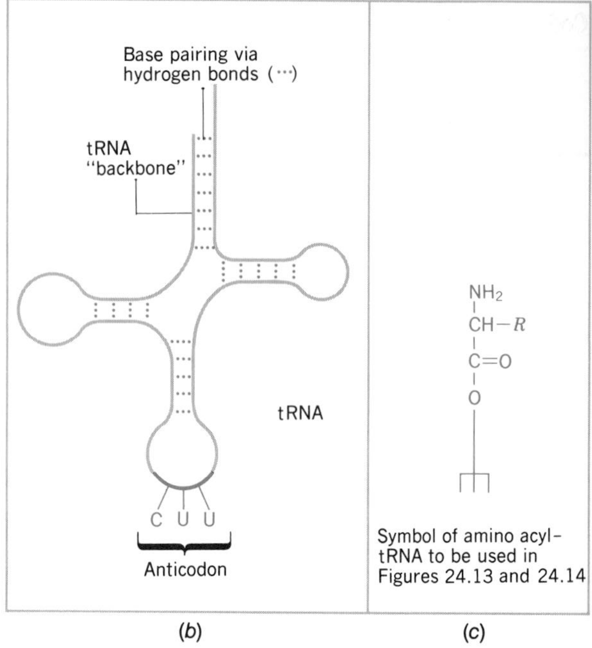

(b) (c)

FIGURE 13.12
Transfer RNA (tRNA). (*a*) The tRNA for phenylalanine. Its anticodon occurs at the tip of the base, and the place where the phenylalanyl residue can be attached is at the upper left point. (*b*) Highly schematic representation of the model to highlight the occurrence of double-stranded regions. (*c*) Symbol of the aminoacyl-tRNA unit that will be used in succeeding figures. (Molecular model courtesy of Academic Press/Molecular Design Incorporated.)

tRNA is structured for its work of translating between the nucleic acid and the polypeptide languages. It is able to connect the three-letter codon "words" aligned along an mRNA backbone to a matching alignment of sidechains of individual amino acids in a polypeptide being made.

One part of a tRNA molecule can recognize a codon because it carries a triplet of bases complementary to the codon. This triplet on tRNA is called an **anticodon.** There are many individual kinds of tRNA, each carrying a particular anticodon. In the tRNA molecule in Figure 13.12, the triplet CUU is its anticodon.

Another functional group of each tRNA molecule, an OH group at a terminal ribose unit, can attach a particular aminoacyl unit (by an ester bond). We can use the symbol tRNA-aa for this new compound, where we use "aa" for the aminoacyl group. The anticodon on each tRNA-aa molecule is unique. A given tRNA-aa molecule, therefore, can be brought into alignment only with one codon of mRNA at a polysome. This ensures that the attached aa unit is brought into place in the correct sequence.

To be able to work with the polypeptide language, each tRNA is able to recognize the amino acid that "belongs" to it. Complementary fitting is at work along with special enzymes.

As each tRNA-aa molecule docks at its unique location at the mRNA chain, the growing polypeptide chain tethered "next door" is transferred to it. You can see that a unique series of codons *can allow the polypeptide chain to grow only with an equally unique sequence of amino acid residues.* The pairing of the triplets of bases between the codons and anticodons can permit only one sequence.

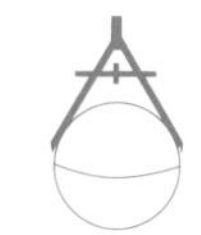

■ UUU and UUC are called *synonyms* for Phe.

The Genetic Code Is the Correlation between Codons and Amino Acids Table 13.1 displays the known correlations of codons with amino acids, the **genetic code.** Most amino acids are associated with more than one codon, which apparently minimizes the harmful effects of genetic mutations. (These, in molecular terms, are small changes in the structures of genes.) Phenylalanine, for example, is coded either by UUU or by UUC. Alanine is coded by any one of the four triplets: GCU, GCC, GCA, or GCG. Only two amino acids go with single codons, tryptophan (Trp) and methionine (Met).

TABLE 13.1
Codon Assignments

First	Second				Third
	U	C	A	G	
U	Phenylalanine	Serine	Tyrosine	Cysteine	U
	Phenylalanine	Serine	Tyrosine	Cysteine	C
	Leucine	Serine	CT[a]	CT	A
	Leucine	Serine	CT	Tryptophan	G
C	Leucine	Proline	Histidine	Arginine	U
	Leucine	Proline	Histidine	Arginine	C
	Leucine	Proline	Glutamine	Arginine	A
	Leucine	Proline	Glutamine	Arginine	G
A	Isoleucine	Threonine	Asparagine	Serine	U
	Isoleucine	Threonine	Asparagine	Serine	C
	Isoleucine	Threonine	Lysine	Arginine	A
	Methionine[b]	Threonine	Lysine	Arginine	G
G	Valine	Alanine	Aspartic acid	Glycine	U
	Valine	Alanine	Aspartic acid	Glycine	C
	Valine	Alanine	Glutamic acid	Glycine	A
	Valine	Alanine	Glutamic acid	Glycine	G

[a] The codon CT is a signal codon for chain termination.
[b] The codon for methionine, AUG, serves also as the codon for *N*-formylmethionine, the chain-initiating unit in polypeptide synthesis in bacteria and mitochondria.

The Genetic Code Is Almost Universal for All Plants and Animals A few single-celled species have been found with codon assignments not given in Table 13.1. Moreover, some genes occur in human mitochondria (Fig. 13.1), and mitochondrial genes have unique codons. Apart from these exceptions, the genetic code of Table 13.1 is shared from the lowest to the highest forms of life in both the plant and animal kingdoms. Once again we see a remarkable kinship with nature.

■ Some (but not all) of the enzymes present in a mitochondrion are made *within* this organelle under the direction of mitochondrial nucleic acids. According to one theory, mitochondria evolved from single-celled organisms.

mRNA Codons Also Relate to DNA Triplets It's important to remember that a codon cannot appear on a strand of mRNA unless a complementary triplet of bases was on an exon unit of the original DNA strand. For example, there could not be the UUC codon on mRNA unless the DNA strand had the triplet AAG, because G pairs with C and A of DNA pairs with U of mRNA.

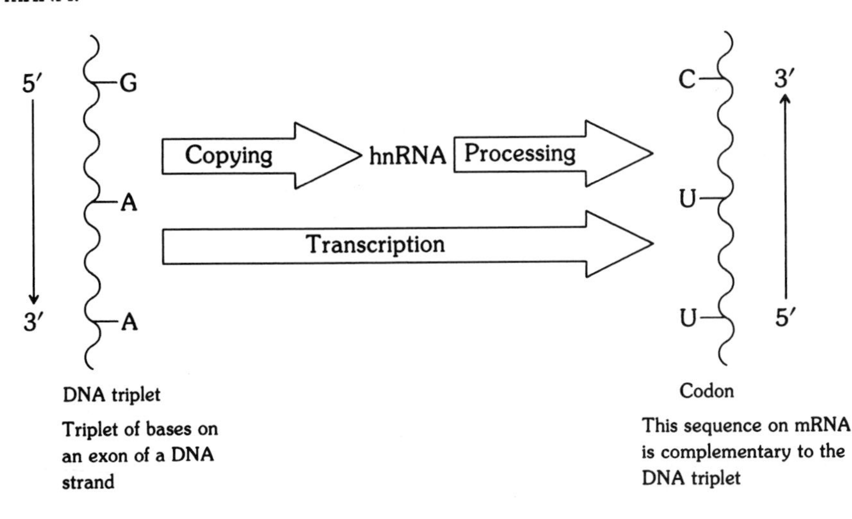

The Direction in Which a Codon Triplet Is Written Has Structural Meaning As shown here, a DNA strand and the RNA strand made directly from it run in opposite directions. To avoid confusion in writing codons on a horizontal line, scientists use the following conventions. The 5′ end of a codon is written on the left end of the three-letter symbol, and the direction, left to right, is 5′ to 3′. (See also Fig. 13.4.) This is why the codon given above is written as UUC, not as CUU. Opposite this codon is the triplet GAA on the DNA strand, which is also written from the 5′ to 3′ end. To give another example, the complement to the mRNA codon, AAG, is the DNA triplet, CTT.

PRACTICE EXERCISE 1

Using Table 13.1, what amino acids are specified by each of the following codons on an mRNA molecule?

(a) CCU **(b)** AGA **(c)** GAA **(d)** AAG

PRACTICE EXERCISE 2

What amino acids are specified by the following base triads on DNA?

(a) GGA **(b)** TCA **(c)** TTC **(d)** GAT

13.4

mRNA-DIRECTED POLYPEPTIDE SYNTHESIS

tRNA molecules carry aminoacyl groups, one by one, to a strand of mRNA attached to a ribosome where the amino acid residues are joined by peptide bonds to make a polypeptide.

In the previous section we saw how a particular genetic message can be transcribed from the exons of a split gene to a series of codons on mRNA. In this section we will learn in broad terms how the next general step is accomplished, the mRNA-directed assemblage of a polypeptide. This step is called **translation.**

Genetic Translation Follows Transcription We will now use the fork-like symbol in part *c* of Figure 13.12 for tRNA-aa. The three tines of the fork stand for the anticodon triplet. We will assume that all the needed tRNA-aa combinations have been assembled and are waiting like so many spare parts to be used at the polypeptide assembly line. The cell must invest considerable amounts of chemical energy, like that of ATP, to carry out all of these steps, and we also assume that this energy is available as needed.

A polypeptide is started with an N-terminal methionine residue. After the polypeptide has been made, this residue is left in place only if methionine is supposed to be the N-terminal unit. Otherwise, methionine will be removed, and the second aminoacyl group will be the N-terminal unit. By anchoring the start of polypeptide synthesis in one triplet, AUG for methionine, the remaining triplets on the mRNA strand cannot overlap each other. Thus a sequence such as AUGCCGAGU . . . must have CCG as the second triplet and AGU as the third, and so forth. Although there seems to be a GCC "triplet" in this sequence, it cannot be read as a triplet because the starting triplet preempts the first G in the series.

$^{+}NH_3CHCO_2^{-}$ with side chain $CH_2CH_2SCH_3$

Methionine, Met

The principal steps in genetic translation are the following.

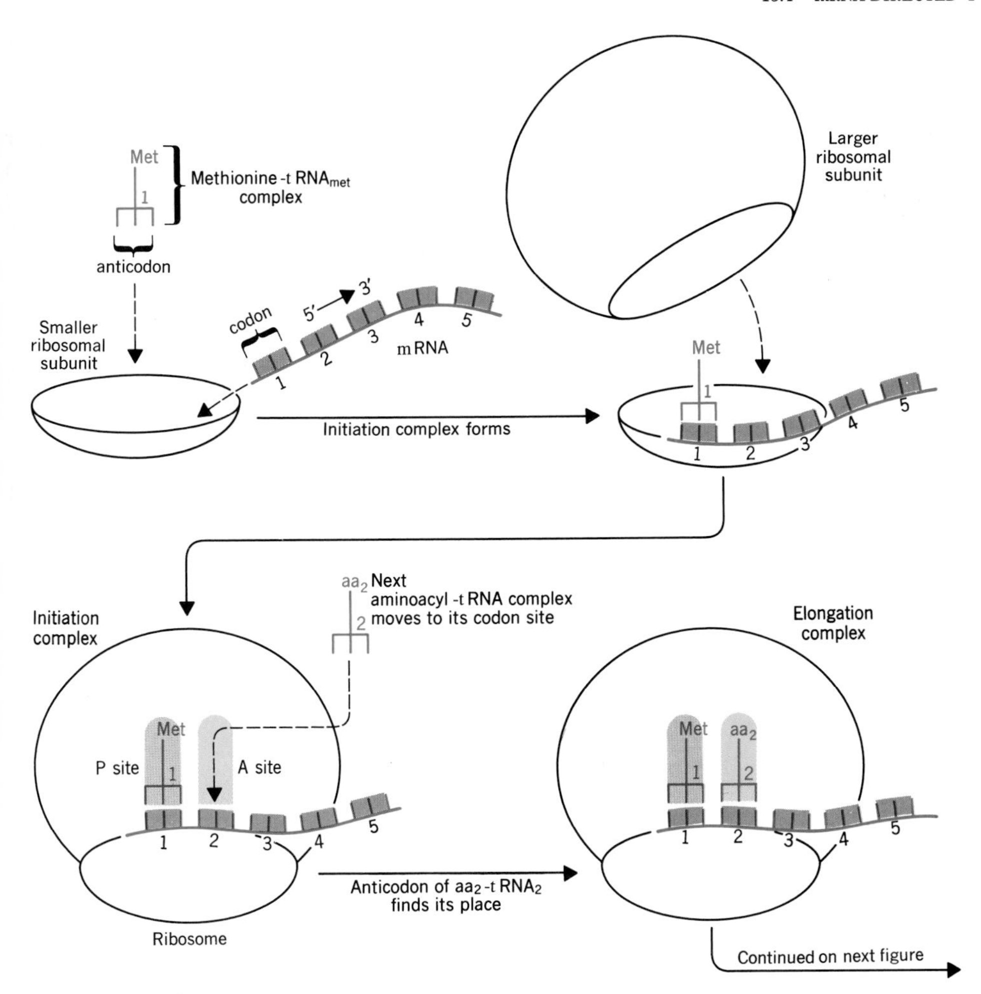

FIGURE 13.13
Formation of the elongation complex at the beginning of the synthesis of a polypeptide

1. **Formation of the elongation complex** As shown in Figure 13.13, the elongation complex is made of several pieces: two subunits of the ribosome, plus the first aminoacyl-tRNA unit (which is Met-$tRNA_1$), plus the mRNA molecule beginning at its first codon end. The Met-$tRNA_1$ comes to rest with its anticodon matched to the first codon on mRNA and with the bulk of the tRNA molecule in contact with a portion of the ribosome's surface called the P site. This is a site where there are enzymes that can catalyze the transfer of a growing polypeptide chain to a newly arrived aminoacyl unit.

   Now the second tRNA unit, tRNA-aa_2, which holds the second aminoacyl group, aa_2, has to find the mRNA codon that matches its own anticodon, and it does this at another site on the ribosome called the A site.

   We now have two aminoacyl units, Met and aa_2, aligned side by side, one over the P site and the second over the A site, held in place with the aid of the hydrogen bonds of the matched codon–anticodon pairs. The work of completing the elongation complex is now complete, and actual chain lengthening can start.

■ The "P" in P site refers to the peptidyl transfer site. The A site is the aminoacyl binding site. (A third site, not shown, is the E site, which temporarily holds the departing tRNA.)

2. **Elongation of the polypeptide chain** There is now a series of repeating steps, illustrated in Figure 13.14. The methionine residue moves from its tRNA to the newly

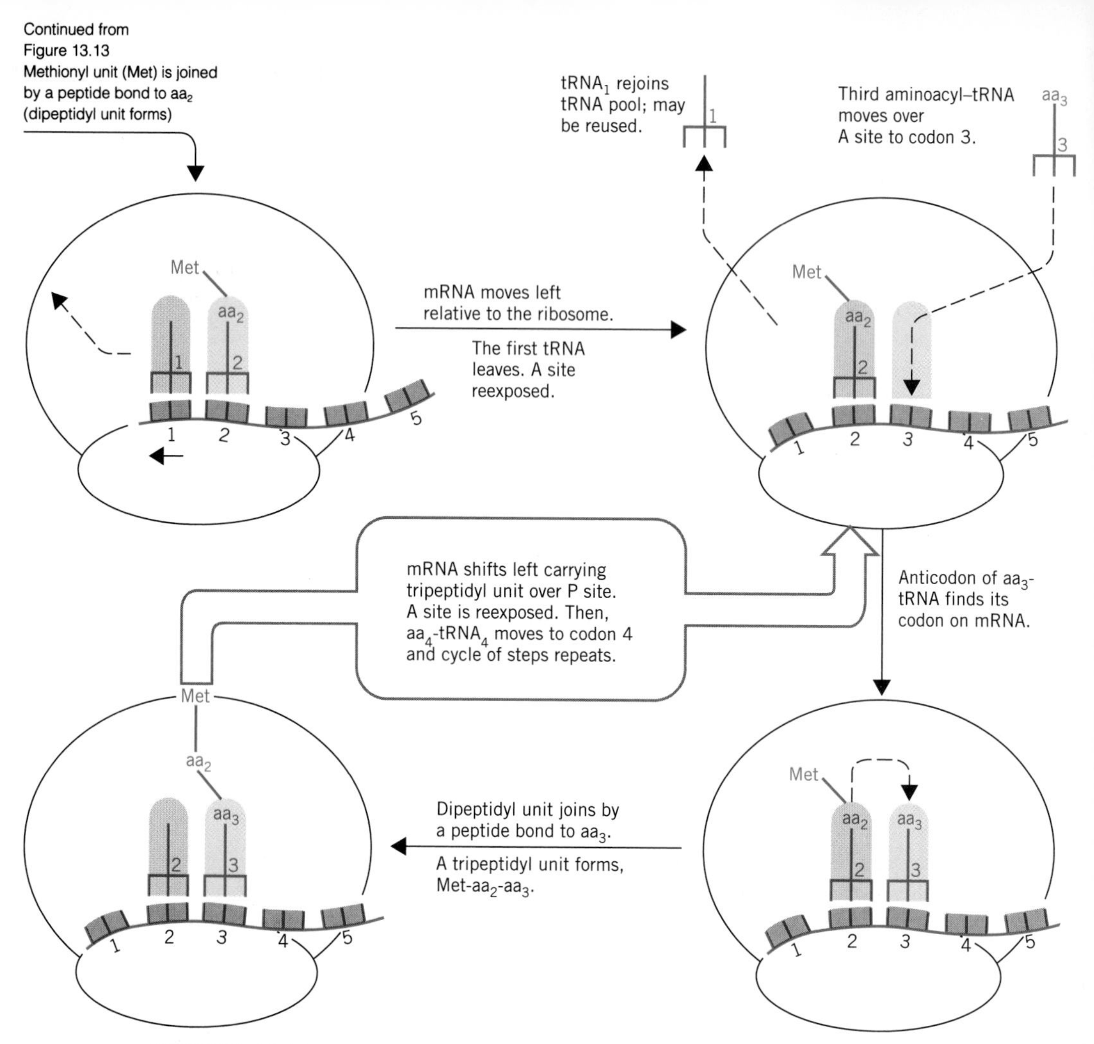

FIGURE 13.14
The elongation steps in the synthesis of a polypeptide. The dipeptidyl unit, Met-aa_2, formed by the process of Figure 13.13, now has a third amino acid residue, aa_3, added to it.

arrived, second aminoacyl unit. This makes the first peptide bond, and it takes place by acyl transfer, much as we discussed when we studied the synthesis of amides in an earlier chapter.

■ This kind of translocation of the growing polypeptide to the next amino acid is what is blocked by the toxin of the diphtheria bacillus.

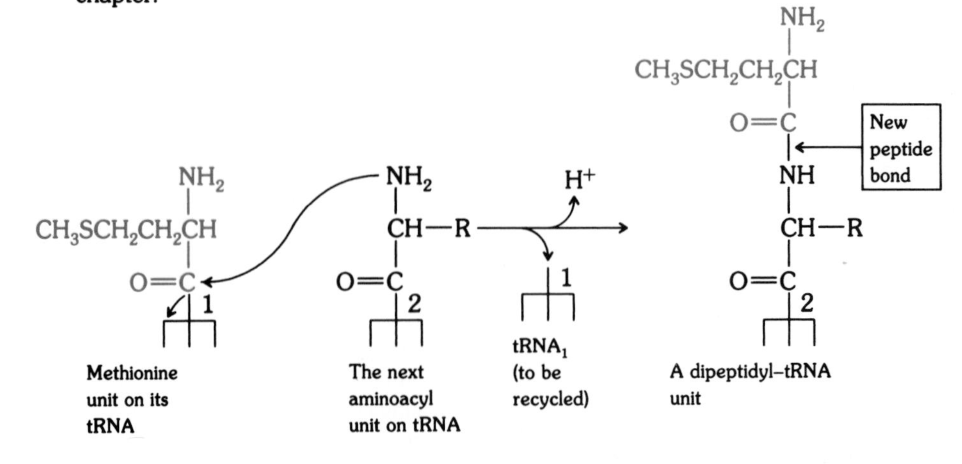

The mRNA unit now shifts one codon over—leftward as we have drawn it. (It's actually a relative motion of the mRNA and the ribosome.) This movement positions what is now a *dipeptidyl*–tRNA unit over the P site. The third aminoacyl–tRNA now finds its anticodon-to-codon matching at the third codon of the mRNA strand, which is over the recently vacated A site.

Elongation now occurs; the dipeptide unit transfers to the amino group of the third amino acid. Another peptide bond forms. And a tripeptidyl system has been made.

The cycle of steps can now occur again, starting with a movement of the mRNA chain relative to the ribosome that shifts this tripeptidyl–tRNA and positions it over the P site. The fourth amino acid residue is carried to the mRNA; the tripeptidyl unit transfers to it to make a tetrapeptidyl unit, and so forth. This cycle of steps continues until a special chain-terminating codon is reached.

■ Polypeptide synthesis can occur at several ribosomes moving along the mRNA strand at the same time.

3. **Termination of polypeptide synthesis** Once a ribosome has moved down to one of the chain-terminating codons of the mRNA strand (UAA, UAG, or UGA), the polypeptide synthesis is complete, and the polypeptide is released. The ribosome can be reused, and the polypeptide acquires its higher levels of structure. The folding and aligning of polypeptides into these higher structural levels are often directed by folding catalysts, enzymes that guide the polypeptides toward the correct overall aggregations and geometries. Because they have a chaperone function, these enzymes are actually called *chaperonines*.

■ In mammals, it takes only about one second to move each amino acid residue into place in a growing polypeptide.

A *Repressor* Can Keep a Gene Switched Off until an *Inducer* Acts As we have said, every cell in the body carries the entire set of genes. In any given tissue, therefore, most must be permanently switched off. It wouldn't do, for example, for a cell to be making a protein-digesting enzyme that would then catalyze the destruction of its own proteins.

Because so many steps occur between the divided gene and the finished polypeptide, there are a large number of points at which the cell can control the overall process. We briefly discuss only one way this is done, one classic example, and we have to rely heavily on Figure 13.15. It comes from discoveries involving a bacterium called *Escherichia coli,* or *E. coli,* a one-celled organism found in our intestinal tracts.

E. coli are able to obtain all the carbon atoms they need from the metabolism of lactose, milk sugar. They must first hydrolyze this disaccharide to galactose and glucose, and the enzyme β-galactosidase is essential for this reaction. The enzyme, of course, is not needed until lactose is available, so the gene for the enzyme is switched off until lactose molecules arrive. We must leave some details to more advanced treatments. Our aim is only to illustrate in broad outline one of the most important kinds of activities at the molecular level of life, the control of gene expression. What we describe here is an *inducible gene* and just one strategy for the control of genes.

The DNA segment responsible for the structure of β-galactosidase is in a region of the DNA strand called the *structural gene* (see Fig. 13.15). Next to it, on the same DNA strand, is a segment of DNA to which a **repressor** molecule can bind. Because this DNA unit is part of the switch for the operation of the structural gene, it is named the *operator site.* Immediately next to it, down the same DNA chain, is a small segment called the *promoter site.* It holds in readiness the enzyme, *DNA polymerase,* needed to transcribe the structural gene. But this enzyme cannot work until the operator site is switched on. The structural gene plus the repressor and operator sites, taken together, make up a unit called the *lac operon.*

The repressor is a polypeptide made at the direction of its own gene, a *regulator gene,* located just above the promoter site in Figure 13.15. When receptor molecules are present, they bind to the DNA of the operator site. *This binding of the repressor is what prevents gene transcription and translation.*

Now suppose that some molecules of lactose appear. The enzyme made with the aid of the structural gene is now needed. The first of the lactose molecules to arrive are altered slightly (by steps we will not discuss) into **inducer** molecules. These have shapes that enable them to fit in some way to the repressor molecule at the operator site. As they attach to the repressor, the shape of the repressor becomes changed so that it no longer can stick to the operator gene, *and it drops off.* DNA polymerase is now released to work with the structural gene to help to make the *mRNA* needed for the manufacture of β-galactosidase.

FIGURE 13.15
Repression and induction of the enzyme β-galactosidase in *E. coli*. The inducer is lactose, whose hydrolysis requires this enzyme. The first lactose molecules to arrive bind to and remove the repressor from the operator gene. Now the structural gene is free to direct the synthesis of mRNA coded to make the enzyme. The repressor is made at the direction of another gene *(top)*, the regulator gene.

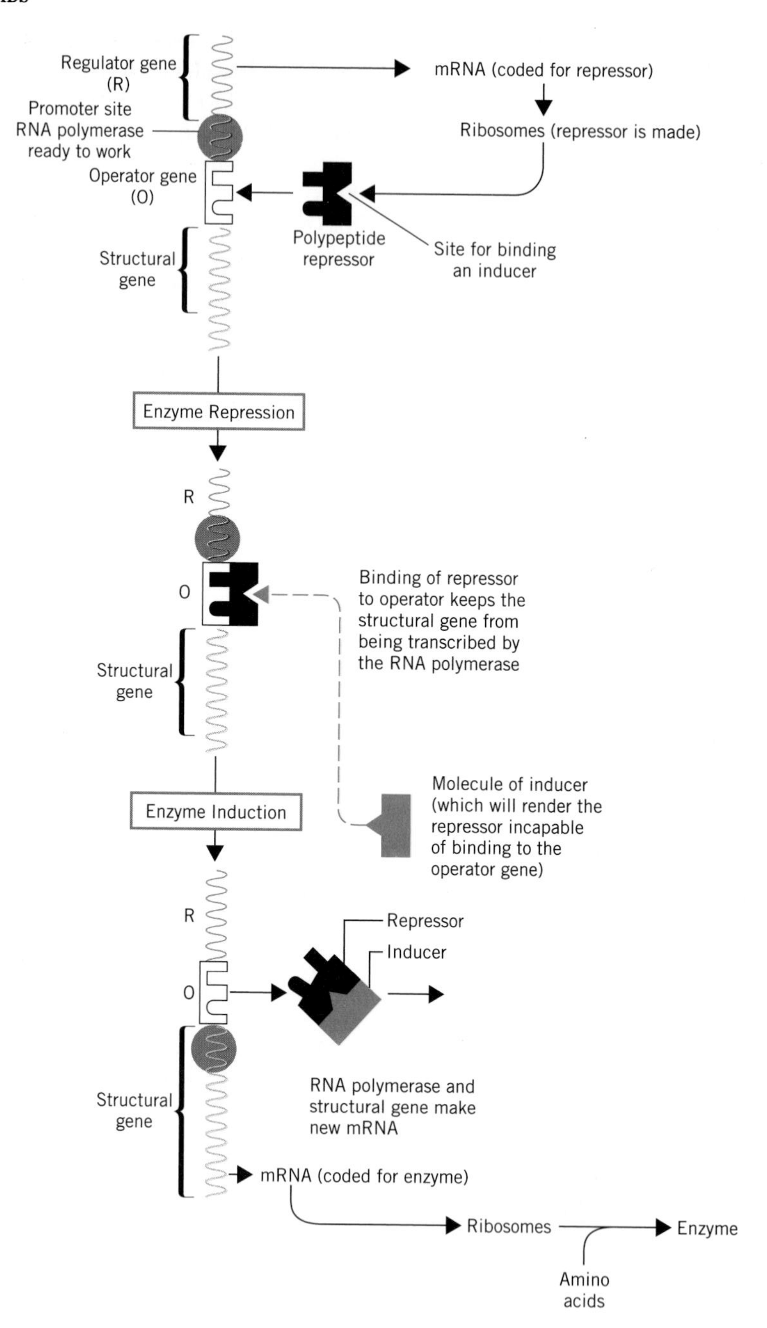

The overall process we have described is called **enzyme induction,** and β-galactosidase is one of many *inducible enzymes*. Don't let the beauty of it be smothered by the details. If no substrate (lactose in this example) is in the cell, an enzyme isn't needed, and the genetic machinery for making it stays switched off. Only when the enzyme is needed, signaled by the arrival of its substrate, is it manufactured. Many human enzymes no doubt work in the same way. Without thousands of events at the molecular level of life happening *automatically,* that is, without human thought, we couldn't imagine minds free enough for any higher thoughts.

Many Antibiotics Kill Bacteria by Interfering with Genetic Translation or Transmission Bacteria, like us, have to manufacture polypeptides to stay alive and multiply. Inhibiting the synthesis of bacterial polypeptides at any one of several steps in the overall process will kill the bacteria. Streptomycin, for example, inhibits the initiation of polypeptide synthesis. Chloramphenicol inhibits the ability to transfer newly arrived aminoacyl units to the elongating strand. The tetracyclines inhibit the binding of tRNA-aa units when they arrive at the ribosome. Actinomycin binds tightly to DNA. Erythromycin, puromycin, and cycloheximide interfere with elongation.

■ These inhibiting activities render the *enzymes* for the various steps inactive.

X Rays and Atomic Radiations Can Damage Genes Atomic radiations, particularly X rays and gamma rays, go right through soft tissue where they create unstable ions and radicals (particles with unpaired electrons). New covalent bonds can form from radicals. Even side-chain bases might be annealed together. If such events were to happen to DNA molecules, the polypeptide eventually made at their direction might be faulty. The DNA made by replication might then be seriously altered. If the initial damage to the DNA is severe enough, replication won't be possible and the cell involved is reproductively dead. This, in fact, is the *intent* when massive radiation doses are used in cancer therapy — to kill cancer cells.

Chemicals that are used in cancer therapy often mimic radiations by interfering with the genetic apparatus of cancer cells. Such chemicals are called **radiomimetic substances.**

13.5 VIRUSES

Viruses take over the genetic machinery of host cells to make more viruses. They are at the borderline of living systems and are generally regarded only as unique packages of dead chemicals, except when they get inside their host cells. They then seem to be living things, because they reproduce. They are a family of materials called **viruses.**

Viruses Consist of Nucleic Acids and Proteins Viruses are agents of infection made of nucleic acid molecules surrounded by overcoats of protein molecules. Unlike a cell, a virus has either RNA or DNA, but not both. Viruses must use host cells to reproduce because they can neither synthesize polypeptides nor generate their own energy for metabolism. The simplest virus has only four genes and the most complex has about 250.

■ Viruses are intracellular parasites.

Each kind of virus has something on its surface that is complementary to something on the surface of the host cell. The glycoproteins of membranes are involved. Thus each virus has one particular kind of host cell and does not attack all kinds. At least, this theory helps us understand how viruses are so unusually selective. A virus that attacks, for example, the nerve cells in the spinal cord has no effect on heart muscle cells. A large number of viruses exist that do not affect any kind of human cell. Many viruses attack only plants.

■ A complete virus particle *outside* its host cell is called a *virion.*

The protein overcoat of some viruses includes an enzyme that catalyzes the breakdown of the cell membrane of the host cell. When such a virus particle sticks to the surface of a host cell, its overcoat catalyzes the opening of a hole into the cell. Then the viral nucleic acid squirts into the cell, or the whole virus might move in. Each virus that works this way evidently has its own unique membrane-dissolving enzyme.

Once a virus particle or the parts of one gets inside its host cell one of two possible fates awaits it. It might become turned off and change into a *silent gene;* or it might take over the genetic machinery of the cell and reproduce so much of itself that it bursts the host cell walls. The new virus particles that spill out then infect neighboring host cells, and in this way the infection spreads. A virus that has become a silent gene might later be activated. Some cancer-causing agents, including ultraviolet light, may initiate cancer by this mechanism.

Most viruses contain RNA, not DNA, so the manufacture of more of their RNA must somehow be managed without the direction of DNA.

RNA Viruses Either Carry or Make Enzymes for Synthesizing More RNA RNA viruses have to solve a major problem if they are to infect a host cell. Host cells normally (in health) have no enzyme that can direct the synthesis of a copy of an RNA molecule *from the* *instructions of another RNA.* In healthy host cells, copies of RNA molecules are made by the direction of DNA, not RNA, like the synthesis of hnRNA directed by DNA.

Two basic solutions to this problem occur involving two different enzymes. One is called *RNA replicase,* and it can catalyze the manufacture of RNA *from the directions encoded on RNA.* Some viruses carry this enzyme. Others direct the host cell to synthesize RNA replicase. Either way, once RNA replicase is inside the host cell, it handles the manufacture of the *mRNA* needed to make more viral RNA and protein, so new virus particles can form.

■ David Baltimore and Howard Temin shared the 1975 Nobel prize in physiology and medicine (together with Renato Dulbecco) for the discovery of reverse transcriptase.

The second way by which RNA directs the synthesis of RNA occurs with viruses that carry a DNA polymerase enzyme. This enzyme, called *reverse transcriptase,* directs the synthesis of viral DNA, which subsequently is used to code for more viral RNA. Reverse transcriptase can use RNA information to make DNA. This is unusual because it's normally the other way around; DNA information is used to make RNA.

Four Basic Strategies Are Used by RNA Viruses to Make More RNA Figure 13.16 outlines the strategies used by four kinds of RNA viruses to manufacture more of their own RNA. The (+) and (−) signs denote single-stranded RNA molecules of opposite complementarity. A (±) sign denotes a double-stranded nucleic acid. By convention, the messenger RNA that the virus must make is designated (+)*mRNA.* Refer to this figure as we briefly discuss four kinds of RNA viruses.

The polio virus contains a single-stranded RNA molecule. Inside its host cell, it functions as a messenger RNA at the host's ribosomes for the synthesis of both overcoat proteins and molecules of RNA replicase. This enzyme then synthesizes (−)RNA molecules which, in turn, direct the synthesis of the (+)mRNA that now takes over the host cell's genetic machinery.

The rabies virus has (−)RNA molecules, but they are not messengers in their host cells. This virus carries its own RNA replicase into the host cells where it directs the synthesis of (+)RNA. This is then used to make the new (−)RNA molecules required for additional virus particles.

■ In the 1919 worldwide influenza epidemic, 20 million people died.

The influenza virus also has (−)RNA, each molecule bearing 10 genes. During infection, segments of the RNA exist that can become resorted as intact, new (−)RNA forms. In this way the influenza virus is able to change into new strains that make the job of immunizing large populations against influenza for long periods particularly difficult.

■ *Reo* in reovirus is from **r**espiratory, **e**nteric **o**rphan—a virus in search of a disease.

The reovirus of Figure 13.16 is present in the intestinal and respiratory tracts of mammals without causing disease. Its RNA is double-stranded RNA, or (±)RNA, and it carries its own RNA polymerase. It can use this on both (+) and (−) strands to make more viral (±)RNA.

The retroviruses (Fig. 13.16) form a family of viruses with (+)RNA that cannot make more RNA without first making double-stranded DNA. To do this, retroviruses carry reverse transcriptase. Thus in retroviruses, the flow of information goes from RNA to DNA. (This explains the *retro-* prefix; it suggests a reversal or retrograde action.) Reverse transcriptase uses retroviral (+)RNA to direct the formation of (−)DNA, which directs the formation of (±)DNA. This DNA is incorporated into the host cell's collection of genes and then directs the synthesis of the (+)mRNA needed to make more retrovirus particles.

■ *Oncogenic* means cancer-inducing.

Cancer-Causing Viruses Transform Normal Genes in Host Cells The retroviruses include the only known cancer-causing RNA viruses, technically termed the *oncogenic RNA viruses.* (Several DNA-based viruses also cause cancer.) They transform host cells so that they

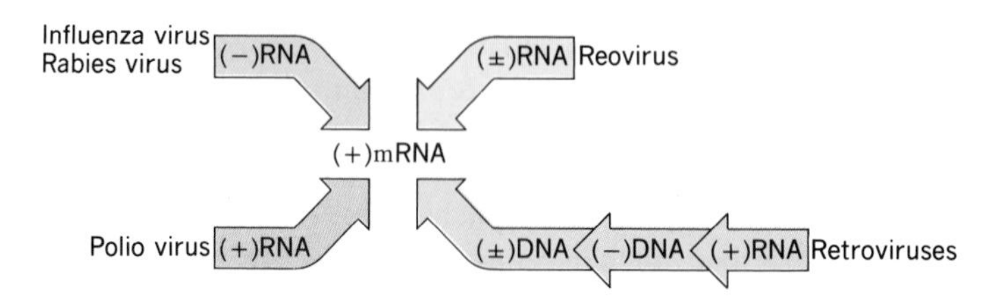

FIGURE 13.16
Overall strategies used by RNA viruses to make their messenger RNA

SPECIAL TOPIC 13.2 
INTERFERON

The term *interferon* refers to a family of similar polypeptides that are chiefly characterized by an ability to inhibit viruses. There are at least three types in humans, designated as α-, β-, and γ-interferon. The interferons are glycoproteins that have about 150 amino acid residues per molecule. α-Interferon is from white blood cells; β-interferon is from certain connective tissue (fibroblasts); and γ-interferon is from cells of the immune system (lymphocytes).

Viruses are potent stimuli for the induction of interferon synthesis in humans. When an invading virus particle first encounters a white blood cell, particularly the type made in lymph tissue, the mechanism in the cell for making the mRNA coded for interferon is switched on. In a short time, the cell is manufacturing and releasing interferon, which then acts as a signal to other cells to make interferon, too. By binding to its host cell, α-interferon directly activates an enzyme in the cell called tyrosine kinase. This enzyme then brings together three polypeptide subunits and so activates a transcription factor that enters the cell nucleus where it completes its work. Circulating cells called killer cells are activated to attack and destroy virus particles. In this way, interferon works to inhibit a viral infection.

The discovery of interferon in 1957 came about when two scientists, Alick Isaacs and Jean Lindenmann, wondered why victims of one viral disease never seemed to come down with a second viral disease at the same time. It seemed to them that something in the first viral attack triggered a mechanism that provided protection against an attack by a different virus. The search for this "something" led to the discovery of interferon.

In the earliest clinical trials, interferon provided relief and sometimes a cure to a few patients who had rapidly acting cancer, such as osteogenic sarcoma (bone cancer), multiple myeloma, melanoma (a form of skin cancer), breast cancer, and some forms of leukemia and lymphoma. However, it takes nearly 25,000 pints of blood to make 100 mg of interferon, at a cost of a few billion dollars, so large-scale clinical tests really depend on interferon made by recombinant DNA technology.

The initial excitement that interferon would be the cure-all, free of side effects, for any kind of viral disease or any kind of cancer has faded. Yet tests with experimental animals continue to be promising, and several medical groups and biotechnology companies are continuing their testing efforts.

In 1988, the U.S. Food and Drug Administration approved the use of α-interferon for the treatment of genital warts. In a clinical trial, it eliminated these warts in over 40% of the patients, and in another 24% the warts were reduced in size. About 8 million Americans suffer from this sexually transmitted virus, which has been linked to cervical cancer. Other treatments were usually ineffective. Pregnant women were advised against this treatment because it could induce abortions.

grow chaotically and continuously. They do this by changing normal genes in the host cell to *oncogenes,* genes that henceforth are able to continue the cancerous growth.

The Host Cell of the AIDS Virus Is Part of the Human Immune System The acquired immunodeficiency syndrome, AIDS, is also caused by a retrovirus, the human immunodeficiency virus or HIV. One reason why this virus is so dangerous is that its host cell, the T4 lymphocyte, is a vital part of the human immune system, as we mentioned in Special Topic 11.2, page 280. By destroying T4 lymphocytes, HIV exposes the body to other infectious diseases, like pneumonia, or to certain rare types of cancer. One of the strategies being used to retard the development of AIDS, if not cure it in those with this syndrome, is to offer the HIV virus a nucleotide that can bind to its reverse transcriptase but, once bound, inhibit the further work of this enzyme. Thus AZT, a nucleotide with a modified sugar unit, has been a commonly used nucleotide medication against AIDS. Used by itself, however, and not as part of a combination of drugs, AZT does not delay the onset of AIDS in patients known to be infected with the HIV virus.

O, HN, CH_3, O, N, $HOCH_2$, O, N_3

AZT

Some Viral Infections Produce Interferons, Which Fight Further Infection One of the many features of the body's defense against some viral infections is a small family of polypeptides called the *interferons,* described further in Special Topic 13.2. Supplies of interferons are now available by genetic engineering, which we will study next.

13.6

RECOMBINANT DNA TECHNOLOGY AND GENETIC ENGINEERING

Single-celled organisms can be made to manufacture the proteins of higher organisms.

Human insulin, human growth hormone, and human interferons are now being manufactured by a technology that involves the production of *recombinant DNA.* With the aid of Figure 13.17, we'll learn how this technology works. It represents one of the important advances in scientific technology of this century. It has permitted the *cloning,* the synthesis of identical copies, of a number of genes. The use of recombinant DNA to make genes and the products of such genes is called **genetic engineering.**

■ The term *cloning* is used for the operation that places new genetic material into a cell where it becomes a part of the cell's gene pool. The new cells that follow this operation are called *clones.*

Genes Alien to Bacteria Can Be Inserted into Bacterial Plasmids Bacteria generally make polypeptides using the same genetic code as humans. There are some differences in the machinery, however. An *E. coli* bacterium, for example, has DNA not only in its single chromosome but also in large, circular, supercoiled DNA molecules called **plasmids.** Each plasmid carries just a few genes, but several copies of a plasmid can exist in one bacterial cell. Each plasmid can replicate independently of the chromosome.

The plasmids of *E. coli* can be removed and given new DNA material, such as a new gene, with base triplets for directing the synthesis of a particular polypeptide. It can be a gene completely alien to the bacteria, like the subunits of human insulin, or human growth hormone, or human interferon. The DNA of the plasmids is snipped open by special enzymes called *restriction enzymes* absorbed from the surrounding medium. This medium can also contain naked DNA molecules, such as those of the gene to be cloned. Then, with the aid of a DNA-knitting enzyme called *DNA ligase,* the new DNA combines with the open ends of the plasmid. This recloses the plasmid loops. The DNA of these altered plasmids is called **recombinant DNA.** The altered plasmids are then allowed to be reabsorbed by bacterial cells.

The remarkable feature of bacteria with recombinant DNA is that when they multiply, the plasmids in the offspring also have this new DNA. When these multiply, still more altered plasmids are made.

Between their cell divisions, the bacteria manufacture the proteins for which they are genetically programmed, including the proteins specified by the recombinant DNA. In this way, bacteria can be tricked into making the *human* proteins we have mentioned. The technology isn't limited to bacteria; yeast cells work, too.

People who rely on the insulin of animals, like cows, pigs, or sheep, sometimes experience allergic responses. They are also vulnerable to the availability of the pancreases of these animals, because they once were the *only* sources of insulin. The ability to make *human* insulin, therefore, has been a welcome development for diabetics.

If an interferon should prove effective in treating various types of viruses, hepatitis, and possibly even certain kinds of cancer, the ability to make this unusually rare and costly substance in large quantities at low cost by recombinant DNA technology will truly be a major technological advance. In fact, just the clinical tests alone would be too costly without the availability of manufactured interferon.

■ Kary B. Mullis of the United States was a cowinner of the 1993 Nobel Prize in chemistry for discovering the polymerase chain reaction. He shared the prize with Michael Smith of Canada, who discovered how to make specific changes in the molecular coding of DNA so that custom-made proteins can be synthesized by living organisms.

Recombinant DNA technology has interacted with archeology in an interesting way. In the early 1980s scientists were able to remove segments of DNA molecules from an Egyptian mummy and from an extinct horse-like animal (a quagga) and reproduce these segments using bacteria. No segment was long enough to include a complete gene, but the technique no doubt will be developed as another tool for the study of evolutionary history. To amplify faithfully (clone) the amount of "archaeologic" DNA from a tiny sample, the *polymerase chain reaction* is used (see Special Topic 13.1).

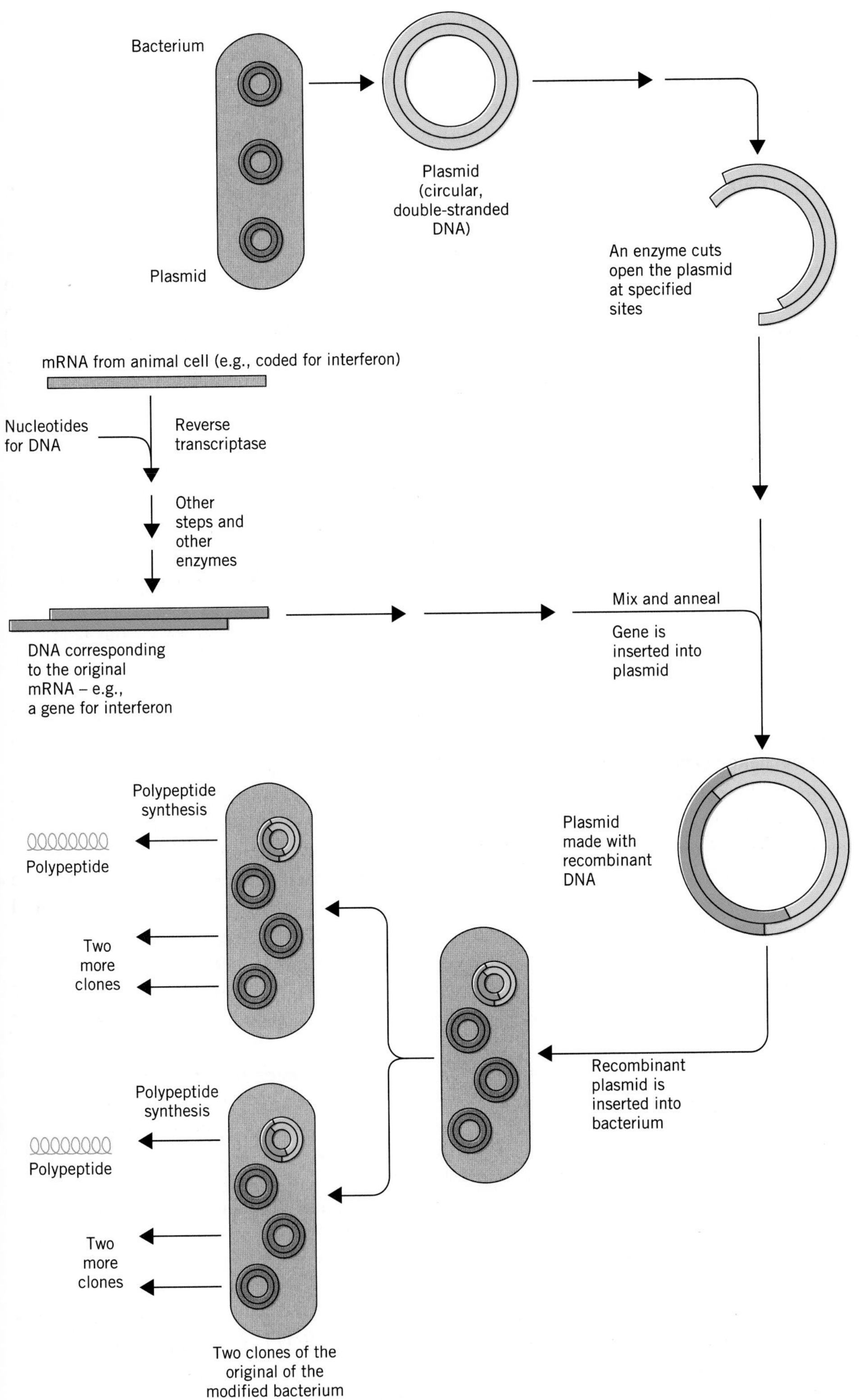

FIGURE 13.17
Recombinant DNA is made by inserting a DNA strand, coded for some protein not made by the bacteria, into the circular DNA of the bacterial plasmid.

Recombinant DNA Can Be Inserted into Cells of Higher Organisms Sometimes, before it will function, a desired polypeptide has to be "groomed" by a cell *after* it has been made by genetic translation. It might have to be attached to a carbohydrate molecule, for example. Bacteria lack the enzymes for such grooming work, so cells of higher organisms are used. When these cells are large enough, the new DNA can be inserted directly into them using glass pipets of extremely small diameters (0.1 μm). Although only a small fraction of such inserted DNA becomes taken up into the cell's chromosomes, it can be enough when amplified by successive cell divisions.

Because viruses are able to get inside cells, some have been used to carry new DNA along. Retroviruses can be customized for this purpose, for example (Section 13.7).

Genetic Engineering Offers Major Advances in Medicine One of the hopes of genetic engineering research is to have ways to correct genetic faults. As we will study in the next section, a number of undesirable conditions are caused by flawed or absent genes. Dwarfism, for example, is caused by a lack of growth hormone, a relatively small polypeptide. In experiments with mice, genetic engineering has successfully introduced growth hormone into mice, with dramatic effects on the mouse size.

In another medical application of genetic engineering, the smallpox vaccine is being remodeled to provide altered forms that might give immunity to many other diseases, ranging from malaria to influenza.

■ This hormone, called the atrial natriuretic factor (ANF), was discovered in the mid 1980s, but had been long-suspected.

A polypeptide hormone made by the heart, which reduces blood pressure, can be manufractured by genetic engineering and used by victims of high blood pressure. As we mentioned in Section 11.5, the clot-dissolving enzyme called tissue plasminogen activator has been genetically engineered for use in reducing the damage to heart tissue following a sudden heart attack.

The alteration of a sticky protein that mussels use to cling to underwater surfaces is being studied to find an adhesive that can be used after surgery.

The list of potential applications of genetic engineering to health problems grows yearly. It seems likely that kidney dialysis patients will need fewer blood transfusions if a blood-cell-producing substance, erythropoietin, can be made by this technology. Hemophiliacs who lack a blood-clotting factor may have it available. The synthesis of a number of drugs by genetic engineering is being studied, some that might be used against cancer.

Agriculture Is Affected by Genetic Engineering In agriculture, genetic engineering is being tried for developing pest-resistant strains of plants, even plants that manufacture their own fertilizer. Gene manipulations have developed cows that yield much larger quantities of milk. Pigs with leaner meat have been produced. The list of similar applications is certain to grow.

The list of problems for society will also grow. In an age of large milk surpluses, some ask, who needs cows that can make more? What of the dairy farmers with huge investments already? These and questions like them are serious enough that if you follow newspapers and news magazines, you no doubt will read of the debates. There has never been a technology yet that had no associated problems. The extent to which these problems are accepted depends on human values expressed through laws and the acceptance of laws.

13.7

HEREDITARY DISEASES AND GENETIC ENGINEERING

About 4000 inherited disorders in humans are caused directly or indirectly by flawed genes.

In Cystic Fibrosis, a Defective Gene Makes a Defective Membrane Protein The victims of cystic fibrosis overproduce a thick mucus in the lungs and the digestive tract, which clogs these systems and which often leads to death in children. About one person in 20 carries the defective gene associated with this disease, and it hits about one in every 1000 newborns. The gene that is defective in cystic fibrosis normally directs the synthesis of a transmembrane protein that lets chloride ion pass out of a cell. For those with the defective gene, this passage of chloride ion is impaired. When Cl^- passes unimpaired out of a cell *it also takes water along,* probably for osmotic pressure reasons. When water is kept in diminished supply outside of the cells of the lungs and airways, however, the mucus thickens and does not flow properly. The thickened mucus makes breathing more difficult and provides a breeding ground for the bacteria that cause a certain form of pneumonia.

■ The membrane protein is called the CFTR protein after cystic fibrosis, transmembrane conductance regulator.

Gene Therapy for Cystic Fibrosis Is Being Tried In late 1992, teams of scientists received the permission of a Federal panel to attempt *gene therapy* for cystic fibrosis. The general plan is to take advantage of the ability of retroviruses (page 352) to insert DNA into a chromosome of the host cell. If the piggybacked DNA is a replacement for the DNA of a defective gene, the host cell will thereby acquire the DNA needed to counter the inherited defect. The *surface* of the altered retrovirus particle is left the same; it still has the original proteins by means of which the particle "recognizes" and sticks to the host cell. Only the retroviral particle interior is altered so that the virus particles cannot manufacture more of themselves in the host and so cause serious infection. Instead, the altered virus particles must launch the desired genetic correction based on the correct DNA. One of the common cold viruses is the designated delivery vehicle for gene therapy against cystic fibrosis. The material is simply squirted into the airways. If corrected genes become installed in only 10 percent of the airway cells, it is thought that the defects in Cl^- and water flows will be rendered nonlethal.

Gene Therapy Has Worked against Enzyme Deficiencies The very first effort to cure a human disease using gene therapy began in 1990. It involved two little girls, ages 4 and 9, born with a defective ADA gene, which left them with almost no natural immunity. The children were injected with white blood cells that had been given the correct gene. Two years later, the girls had functioning immune systems and instead of having to lead very isolated lives were in public school. Sickle-cell anemia is another disease caused by a defective gene, as we described in Special Topic 10.1.

■ The ADA gene makes the enzyme *adenosine deaminase,* which is vital to the immune system.

Albinism Is a Genetic Defect Albinism, the absence of pigments in the skin and the irises of the eyes, is caused by a defect in a gene that directs the synthesis of an enzyme that is needed to make these pigments. The pigments absorb the ultraviolet rays in sunlight, radiation that can induce cancer, so victims of albinism are more susceptible to skin cancer.

PKU Is a Genetic Disease Treated Nutritionally Phenylketonuria, or PKU disease, is a brain-damaging genetic disease in which abnormally high levels of the phenylketo acid called phenylpyruvic acid occur in the blood. This condition causes permanent brain damage in the newborn. Because of a defective gene, an enzyme needed to handle phenylalanine is not made, and this amino acid is increasingly converted to phenylpyruvic acid.

$$C_6H_5CH_2\overset{\overset{\displaystyle O}{\|}}{C}CO_2H \qquad C_6H_5CH_2\underset{\underset{\displaystyle NH_3^+}{|}}{C}HCO_2^-$$

Phenylpyruvic acid Phenylalanine

PKU can be detected by a simple blood test within four or five days after birth. If the infant's diet is kept very low in phenylalanine, it can survive the critical danger period and experience no brain damage. The infant's diet should include no aspartame, a low-calorie sweetener

■ There is enough phenylalanine in just one slice of bread to be potentially dangerous to a PKU infant.

(contained in NutraSweet®), because it is hydrolyzed by the digestive processes to give phenylalanine.

$$\underset{\text{Aspartame}}{{}^{+}NH_3\underset{\underset{CO_2^-}{|}}{\underset{|}{CH_2}}{\overset{|}{C}}HC(=O)-NH\underset{\underset{C_6H_5}{|}}{\underset{|}{CH_2}}{C}HC(=O)OCH_3} + 2H_2O \longrightarrow \underset{\text{Aspartic acid}}{{}^{+}NH_3\underset{\underset{CO_2^-}{|}}{\underset{|}{CH_2}}CHCO_2^-} + \underset{\text{Phenylalanine}}{{}^{+}NH_3\underset{\underset{C_6H_5}{|}}{\underset{|}{CH_2}}CHCO_2^-} + \underset{\text{Methanol}}{CH_3OH}$$

Maintaining a low-phenylalanine diet, of course, is not the easiest and best solution, so genetic engineers are working to correct the fundamental gene defect.

The *Human Genome Project* Aims to Map All Human Genes and Determine Their DNA Sequences The entire complement of genetic information of a species is called its **genome.** The *Human Genome Project,* formally launched in 1990, intends to discover the "map" of every human chromosome and to determine the sequence of bases in every human gene. It's perhaps the largest single project ever attempted in molecular biology, and its total cost may come within range of (but will probably be less than) what it cost to put an astronaut on the moon. If the genome is a library, if the chromosomes are book sections, and if the genes are individual books, then the Human Genome Project's goal in *mapping* chromosomes is to create the card catalog. "In which section of the library is the gene for cystic fibrosis located?" Answer: on chromosome 7. "In which particular part (shelf) of the chromosome 7 is the gene responsible for cystic fibrosis located?" Answer: in region q31. Now comes the *sequencing* question; it's not the same as the mapping question. "What is the sequence of bases in the cystic fibrosis gene?" The answer is known, but we can't give it here; the gene involves over 6000 bases. "What's wrong with this gene to make it cause cystic fibrosis?" For about 70 percent of cystic fibrosis mutations, there is a deletion (an absence) of three bases in exon number 10. So much human suffering from so small a molecular defect!

■ In early 1993, the location of the gene for Huntington's disease, a deadly neurological disorder (page 298) was found near the tip of chromosome 4.

SUMMARY

Hereditary information The genetic apparatus of a cell is mostly in its nucleus and consists of chromatin, a complex of DNA and proteins (histones). Strands of DNA, a polymer, carry segments that are individual genes. Chromatin replicates prior to cell division, and the duplicates segregate as the cell divides. Each new cell thereby inherits exact copies of the chromatin of the parent cell. If copying errors are made, the daughter cells (if they form at all) are mutants. They may be reproductively dead—incapable of themselves dividing—or they may transmit the mutant character to succeeding cells. The expression of this might be as a cancer, a tumor, or a birth defect. Atomic radiations, particularly X rays and gamma rays, are potent mutagens, but many chemicals, those that are radiomimetic, mimic these rays.

DNA Complete hydrolysis of DNA gives phosphoric acid, deoxyribose, and a set of four heterocyclic amines, the bases adenine (A), thymine (T), guanine (G), and cytosine (C). The molecular backbone of the DNA polymer is a series of deoxyribose units joined by phosphodiester groups. Attached to each deoxyribose is one of the four bases. The order in which triplets of bases occur is the cell's way of storing genetic information.

In higher organisms, a gene consists of successive groups of triplets, the exons, separated by introns. Thus the gene is a split system, not a continuous series of nucleotide units. DNA exists in cell nuclei as duplex DNA, double helices held to this geometry largely by hydrophobic forces. The base A always pairs with the base T and C always pairs with G. Using this structure and the faithfulness of base pairing, Crick and Watson explained the accuracy of replication. After replication, each new double helix has one of the parent DNA strands and one new, complementary strand.

RNA RNA is similar to DNA except that in RNA ribose replaces deoxyribose and uracil (U) replaces thymine (T). Four main types of RNA are involved in polypeptide synthesis. One is rRNA, which is in ribosomes. A ribosome contains both rRNA and proteins that have enzyme activity needed during polypeptide synthesis.

mRNA is the carrier of the genetic message from the nucleus to the site where a polypeptide is assembled. mRNA results from a chemical processing of the longer RNA strand, hnRNA, which is made directly under the supervision of DNA.

tRNA molecules are the smallest RNAs to participate directly in polypeptide synthesis. Their function is to convey aminoacyl units to the polypeptide assembly site. tRNAs recognize where they are to go by base pairing between an anticodon on the tRNA molecule and its complementary codon on mRNA. Both codon and anticodon consist of a triplet of bases.

Polypeptide synthesis Genetic information is first transcribed when DNA directs the synthesis of mRNA. Each base triplet on the exons of DNA specifies a codon on mRNA. The mRNA moves to the cytoplasm to form an elongation complex with subunits of a ribosome and the first and second tRNA-aa unit to become part of the developing polypeptide.

The ribosome then rolls down the mRNA as tRNA-aa units come to the mRNA codons during the moment when the latter are aligned over the proper enzyme site of a ribosome. Elongation of the polypeptide then proceeds to the end of the mRNA strand or to a chain-terminating codon. After chain termination, the polypeptide strand leaves, and it may be further modified to give it its final N-terminal amino acid residue. Chaperonine enzymes guide the polypeptides into their final geometries.

The whole operation can be controlled by a feedback mechanism in which an inducer molecule removes a repressor of the gene, thus letting the gene work.

Several antibiotics inhibit bacterial polypeptide synthesis, which causes the bacteria to die.

Viruses Viruses are packages of DNA or RNA encapsulated by protein. Once they get inside their host cell, virus particles take over the cell's genetic machinery, make enough new virus particles to burst the cell, and then repeat this in neighboring cells. Viruses are implicated in human cancer. Some viruses make new RNA under the direction of existing RNA, using RNA replicase. Others, the retroviruses, use RNA and reverse transcriptase to make DNA, which then directs the synthesis of new RNA.

Recombinant DNA Recombinant DNA is DNA made from bacterial plasmids and DNA obtained from another source and encoded to direct the synthesis of some desired polypeptide. The altered plasmids are reintroduced into the bacteria, where they become machinery for synthesizing the polypeptide (e.g., human insulin, or growth hormone, or interferon). Yeast cells can be used instead of bacteria for this technology.

Cells of higher organisms are also used as sites for inserting new DNA. In this kind of genetic engineering, microsyringes or tailored viruses have been used to insert the DNA. Many applications in medicine and agriculture exist.

Gene therapy Hereditary diseases stem from defects in DNA molecules that either prevent the synthesis of necessary enzymes or that make the enzymes in forms that won't work. The identification of the chromosomes bearing the defective genes has enabled the analyses of the genes themselves. In gene therapy, it is hoped that healthy genes can be substituted for those that are defective. Gene therapy for cystic fibrosis entails using a modified retrovirus to convey correct DNA material into the host cells without causing a viral infection.

REVIEW EXERCISES

The answers to Review Exercises whose numbers are in color are found in Appendix A. The answers to the other Review Exercises are found in the Study Guide that accompanies this book. The more challenging questions are marked with asterisks.

The Cell

13.1 What is the term used for each of the following?
(a) the *liquid* inside the cell but outside the nucleus
(b) the entire contents of the cell
(c) the region of the cell outside the nucleus
(d) the particle at which polypeptide synthesis occurs
(e) the particle in which ATP is synthesized
(f) the name of the chemical that makes up a gene
(g) the nucleoprotein material inside a cell nucleus
(h) the protein around which DNA strands are wound
(i) the fundamental unit of heredity

13.2 What is the relationship between a chromosome and chromatin?

13.3 The duplication of a gene occurs in what part of the cell?

13.4 In a broad, overall sense, what happens when DNA replicates?

Structural Features of Nucleic Acids

13.5 What is the general name for the chemicals that are most intimately involved in the storage and the transmission of genetic information?

13.6 The monomer units for the nucleic acids have what *general* name?

13.7 What are the names of the two sugars produced by the complete hydrolysis of all the nucleic acids in a cell?

13.8 What are the names and symbols of the four bases that are liberated by the complete hydrolysis of (a) DNA and (b) RNA?

13.9 How are all DNA molecules structurally alike?

13.10 How do different DNAs differ structurally?

13.11 How are all RNA molecules structurally alike?

13.12 What are the principal structural differences between DNA and RNA?

13.13 When DNA is hydrolyzed, the ratios of A to T and of G to C are each very close to 1 : 1, *regardless of the species investigated.* Explain.

13.14 What does base pairing mean, in general terms?

13.15 What is the chief stabilizing factor for the geometrical form taken by duplex DNA?

***13.16** If the AGGCTGA sequence appeared on a DNA strand, what would be the sequence on the DNA strand opposite it in a double helix?

13.17 The *accuracy* of replication is assured by the operation of what factors?

13.18 What is the relationship between a single molecule of single-stranded DNA and a single gene?

13.19 Suppose that a certain DNA strand has the following groups of nucleotides, where each lowercase letter represents a group several nucleotides long.

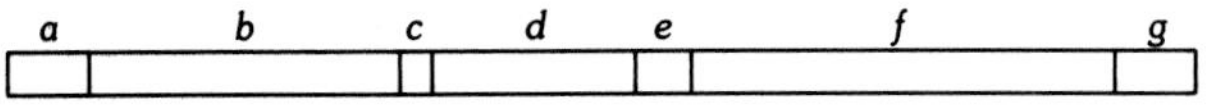

a	*b*	*c*	*d*	*e*	*f*	*g*

Which sections are likelier to be the introns? Why?

13.20 In general terms only, what particular contribution does a gene make to the structure of a polypeptide?

Ribonucleic Acids

13.21 What is the general composition of a ribosome, and what function does this particle have?

13.22 What is hnRNA, and what role does it have?

13.23 To which kind of RNA does the term "primary transcript" refer?

13.24 What is a codon, and what kind of nucleic acid is a continuous, uninterrupted series of codons?

13.25 What is an anticodon, and on what kind of RNA is it found?

13.26 Which triplet, ATA or CGC, cannot be a codon? Explain.

13.27 Which amino acids are specified by the following codons?
(a) UUU (b) UCC (c) ACA (d) GAU

***13.28** What are the anticodons for the codons of Review Exercise 13.27?

***13.29** Suppose that sections *x*, *y*, and *z* of the following hypothetical DNA strand are the exons of one gene.

3′ 5′

AAA	GAA TAT CTC	AGG	GGT	TGT CTA
x		*y*		*z*

What is the structure of each of the following substances made under its direction?
(a) the hnRNA
(b) the mRNA
(c) the tripeptide that is made using the given genetic information (Use the three-letter symbol format for the tripeptide structure, referring as needed to Table 10.1 for these symbols.)

Polypeptide Synthesis

13.30 Use the identifying letters to arrange the following symbols or terms in the correct order in which they are synthesized in going from a gene to an enzyme. (Place the letter of the first material to be involved on the left.)

hnRNA	duplex DNA	polypeptide	mRNA
A	**B**	**C**	**D**

___ < ___ < ___ < ___

earliest to be involved — last to appear

13.31 What is meant by translation, as used in this chapter? And what is meant by transcription?

***13.32** To make the pentapeptide, Met-Ala-Trp-Ser-Tyr,
(a) What do the sequences of bases on the mRNA strand have to be?
(b) What is the anticodon on the first tRNA to move into place?

13.33 In general terms, explain how the overlapping of triplets on mRNA is avoided.

13.34 The discussion about the *lac operon* was included to illustrate what aspect of the chemistry of polypeptide synthesis?

13.35 Not all of the DNA of the *lac operon* codes for the synthesis of β-galactosidase. What is the name given to the part that is so coded? What names are given to other segments of the DNA of the *lac operon* and what is their function?

13.36 In general terms, what is a repressor, and what does it do?

13.37 What does an inducer do (in general terms)?

13.38 How do some of the antibiotics work at the molecular level?

13.39 The genetic code is the key to translating between what two "languages"?

13.40 What is meant by the statement that the genetic code is universal? And is the code strictly universal?

13.41 In general terms, how do X rays cause cancer?

13.42 How do X rays and gamma rays work in cancer therapy?

13.43 What is a radiomimetic substance?

Viruses

13.44 What is a virus made of?

13.45 In general terms, how does a virus discriminate among all possible host cells and "find" just one kind of host cell?

13.46 In general terms, once a virus particle has joined to the membrane of its host cell, what must occur next if the viral infection is to advance?

13.47 In general terms, in the systems having it, what does RNA replicase do? What is true about a normal host cell that requires a virus to have RNA replicase?

13.48 In general terms and in connection with the work of (some) viruses, what does reverse transcriptase do?

13.49 Which kind of mRNA enables a virus to use the host cell to make more virus particles, (−)mRNA or (+)mRNA?

13.50 If the mRNA in a given virus is (−)mRNA, what must the virus–host cell system accomplish if the virus is to multiply?

13.51 If the mRNA in a given virus is (−)mRNA, what enzyme is carried into the host cell by the virus?

13.52 What does the prefix *retro-* signify in retrovirus?

13.53 What is meant by a *silent gene* and where does it come from?

13.54 Some viruses are called *oncogenic RNA viruses*. What does "oncogenic" mean here?

13.55 What is the full name of the HIV system?

13.56 What is the host cell of HIV and how does this fact make AIDS so dangerous?

13.57 What is the theory concerning the action of AZT?

Recombinant DNA

13.58 What is a plasmid, and what is it made of?

13.59 What is the name of the enzyme that can snip open the DNA of a plasmid?

13.60 Recombinant DNA is made from the DNA of two different kinds of sources. What are they?

13.61 Recombinant DNA technology is carried out to accomplish the synthesis of what kind of substance (in general terms)?

13.62 What does "genetic engineering" refer to?

Hereditary Diseases

13.63 At the molecular level of life, what kind of defect is the fundamental cause of a hereditary disease?

13.64 The defective gene in cystic fibrosis leads to an impairment of what specific activity of the cells of the affected tissues? Why does this activity result in a problem with mucus?

13.65 Only in the broadest terms, what is gene therapy mean to accomplish?

13.66 A retrovirus instead of some other kind of virus is being used in gene therapy for cystic fibrosis. Why?

13.67 In altering the retrovirus used in gene therapy for cystic fibrosis, only the interior of the virus is changed. Why is the surface of the virus particle left alone?

13.68 What is the molecular defect in PKU, and how does it cause the problems of the victims? How is it treated?

DNA Typing (Special Topic 13.1)

13.69 What fact about cells makes it possible to use cells from any part of the body of a suspect for DNA fingerprinting in a rape case for which a semen sample has been obtained?

13.70 What is meant by the *polymerase chain reaction*, and why is it used as part of the technology of DNA typing?

13.71 A restriction enzyme separates DNA molecules into pieces given what general name? How are they used to give the genetic "bar code"?

Interferon (Special Topic 13.2)

13.72 What kinds of compounds are the interferons and what are they able to do that protects us from disease?

13.73 What scientific observation led scientists to look for something that would do what the interferons do?

13.74 What kinds of infectious agents stimulate the production of interferons?

13.75 How are interferons made for clinical trials?

Additional Exercises

13.76 Compare the structures of uracil and thymine (Fig. 13.2).
(a) How do the structures differ?
(b) Can the structural difference affect the hydrogen-bonding capabilities of these two bases?
(c) How does the known behavior of uracil and thymine toward adenine bear on the answer to part (b)?

13.77 Write the structures of nucleotides that involve deoxyribose and each of the following bases.
(a) adenine (b) cytosine

***13.78** Consider the following compound.

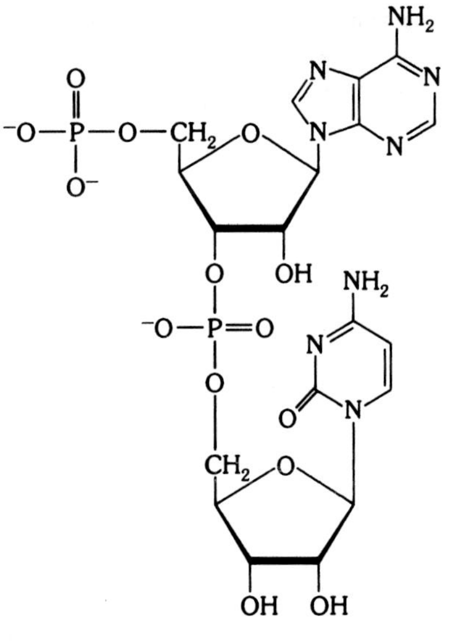

NH$_2$ N N O ⁻O—P—O—CH$_2$ O N N O⁻ O OH NH$_2$ ⁻O—P=O N O O N CH$_2$ O OH OH

(a) Is it a mononucleotide, a dinucleotide, or a higher nucleotide (an oligonucleotide)? How can you tell?
(b) Could it be obtained by the partial hydrolysis of DNA or RNA? How can you tell?
(c) Where is the 5′ end, at the bottom or top of the structure as written?
(d) In terms of the single-letter-symbols for bases, how is the structure of this compound written?

13.79 Consider the structure of AZT found on page 353.

(a) What is the name of its sidechain base?

(b) The sidechain base of AZT could form a hydrogen-bonded pair to which other sidechain base among those that occur among the nucleic acids?

(c) Could AZT become hydrogen-bonded to sidechains of *both* DNA and RNA? Or to only one of these two? (If so, which one?)

(d) If the N_3 group on the AZT molecule were replaced by an OH group, would the resulting molecule be a nucleotide? If so, would it be a nucleotide of RNA, DNA, both, or neither? If neither, what would have to be done to make the product of replacing N_3 by OH into a nucleotide?

BIOCHEMICAL ENERGETICS

14

When Gail Devers (USA) won a gold medal for the 100-m dash in the 1992 Olympic Games, her body mobilized an incredible burst of energy from chemicals, like glucose. In this chapter we'll study how the body handles such transformations at the molecular level.

14.1

ENERGY FOR LIVING—AN OVERVIEW

High-energy phosphates, such as ATP, are the body's means of trapping the energy from the oxidation of the products of digestion.

We cannot use solar energy directly, like plants. We cannot use steam energy, like a coal-fired electrical power plant. We need *chemical* energy for living. We obtain it from food, and we use it to make high-energy molecules. These then drive the chemical "engines" behind muscular work, signal sending, and chemical manufacture in tissue. Our principal source of chemical energy is the **catabolism** (i.e., the breaking down) of carbohydrates and fatty acids, although we can also use most of the amino acids obtained from proteins for energy.

■ *Catabolism* is from the Greek, *cata-*, down; *ballein,* to throw or cast.

Heats of Combustion Disclose the Energy Available from Catabolism The energy from the *combustion* of 1 mol of glucose to CO_2 and H_2O is exactly the same as the energy available from any other method to convert glucose to the same products, 673 kcal/mol (2.82×10^3 kJ/mol). The difference is that our cells don't *burn* glucose or they would receive their energy solely as heat. We oxidize glucose by a number of small steps, some of which make other high-energy compounds. Thus some of the energy available from changing glucose to CO_2 and H_2O is used to run chemical reactions that make high-energy compounds, and the rest of the energy of catabolism is released as heat.

If we were to burn 1 mol of palmitic acid, $CH_3(CH_2)_{14}CO_2H$, a typical fatty acid, we could obtain 2400 kcal (1.00×10^4 kJ) of energy. In the body, we can also change this compound to CO_2 and H_2O, and we obtain the same energy per mole. However, as with glucose, not all this energy is released as heat. The catabolism of palmitic acid occurs by a number of steps some of which make other high-energy compounds, like the high-energy phosphates.

Each Organophosphate Has a Potential for Transferring Its Phosphate Unit to Another Molecule In Section 5.6 we first learned about the existence of energy-rich triphosphate esters. These and many similar high-energy compounds are involved in the mobilization of chemical energy in the body. Table 14.1 gives the names and structures of the principal phosphates we'll encounter. They are arranged in the order of their **phosphate group transfer potentials,** their relative abilities to transfer a phosphate group to an acceptor, as in the following equation:[1]

■ The bond in color denotes the P—O bond that breaks in an energy-releasing reaction of a high-energy phosphate.

$$\underset{\text{Higher potential}}{R{-}O{-}PO_3^{2-}} + R'{-}O{-}H \longrightarrow R{-}O{-}H + \underset{\text{Lower potential}}{R'{-}O{-}PO_3^{2-}} \qquad (25.1)$$

The numbers in the last column of Table 14.1 are measures of the relative potentials that the compounds have to transfer their phosphate groups to water under standard conditions. (We don't need to know anything about these conditions, just that they supply a common

We can use the positions of compounds in Table 14.1 to predict whether a particular transfer is possible. A compound with a higher (more negative) phosphate group transfer
transfer is possible. A compound with a higher (more negative) phosphate group transfer

[1] We will usually write the phosphate unit as $O{-}PO_3^{2-}$, but its state of ionization varies with the pH of the solution. At physiological pH, the unit is mostly in the singly and doubly ionized forms, $O{-}PO_3H^-$ and $O{-}PO_3^{2-}$.

TABLE 14.1
Some Organophosphates in Metabolism

Organophosphate	Structure	Phosphate Group Transfer Potential
Phosphoenolpyruvate	$CH_2{=}C(OPO_3^{2-})CO_2^-$	−14.8
1,3-Bisphosphoglycerate	$^{2-}O_3POCH_2CH(OH)C(=O)OPO_3^{2-}$	−11.8
Creatine phosphate	$^{2-}O_2CH_2N(CH_3)C(=NH_2^+)NHPO_3^{2-}$	−10.3
Acetyl phosphate	$CH_3C(=O)OPO_3^{2-}$	−10.1
Adenosine triphosphate, ATP	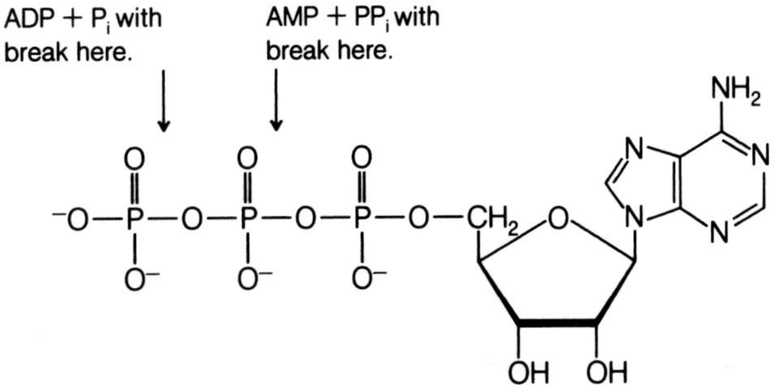	−7.3[a]
Glucose-1-phosphate	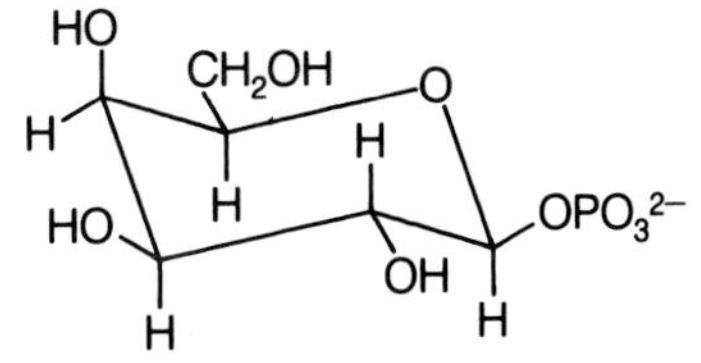	−5.0
Fructose-6-phosphate	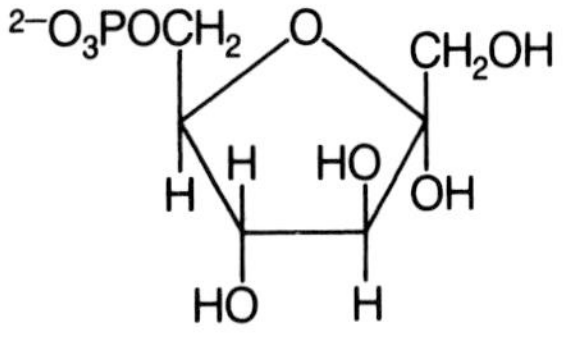	−3.8
Glucose-6-phosphate	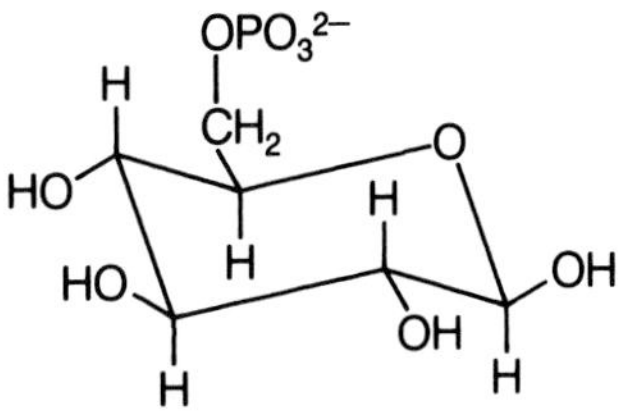	−3.3
Glycerol-1-phosphate	$HOCH_2CH(OH)CO_2OPO_3^{2-}$	−2.2

[a] This value applies whether ADP or AMP forms.

potential can, in principal, always be used to make one with a lower (less negative) potential, assuming that the right enzyme and any other needed reactants are available.

PRACTICE EXERCISE 1

Using Table 14.1, tell whether each reaction can occur.

(a) ATP + glycerol-3-phosphate → ADP + glycerol-1,3-diphosphate
(b) ATP + glucose → ADP + glucose-1-phosphate
(c) Glucose-1-phosphate + creatine → glucose + creatine phosphate

Adenosine Triphosphate (ATP) Is the Body's Chief Energy Broker One compound in Table 14.1, **adenosine triphosphate** or **ATP,** is so important that we must review what we learned about it in Section 5.6. From the lowest to the highest forms of life, ATP is universally used as the principal carrier of energy for living functions. it is the chief means used by the body to trap energy available by oxidations.

We saw in Section 5.6 our first example of how ATP can be used—to power muscle contraction. This activity can be simply written as a chemical reaction of ATP with the proteins in relaxed muscle (or with something within the proteins):

$$\text{"Relaxed" muscle} + \text{ATP} \longrightarrow \text{"contracted" muscle} + \text{ADP} + P_i$$

This reaction produces changes in tertiary structures of muscle proteins that cause the fibers made from these proteins to contract. Simultaneously, ATP changes to **ADP,** which is **adenosine diphosphate,** plus inorganic phosphate, P_i.

ATP has two phosphate bonds either of whose rupture can release much useful energy. Occasionally the second bond (see Table 14.1) is broken in some transfer of chemical energy, and the products then are **AMP, adenosine monophosphate,** and the inorganic diphosphate ion, PP_i.

Occasionally triphosphates other than ATP are the carriers of energy. Guanosine triphosphate, GTP, is an example that we'll encounter later in this chapter.

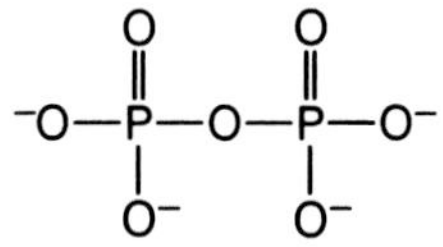

Diphosphate ion, PP_i

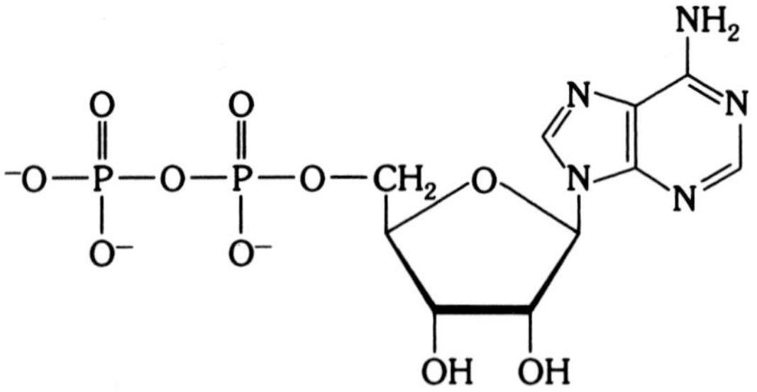

Adenosine diphosphate, ADP
(fully ionized form)

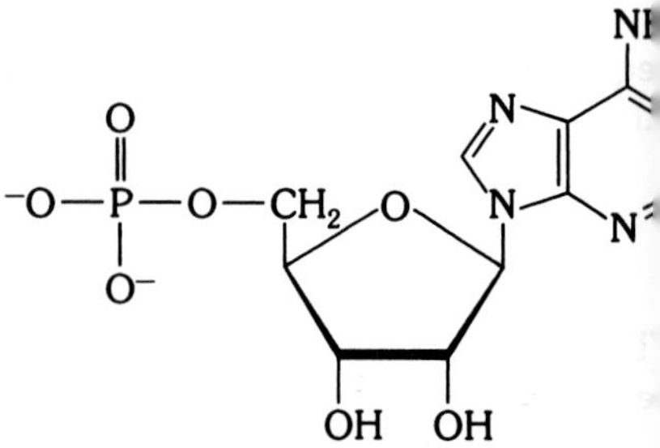

Adenosine monophosphate, AMP
(fully ionized form)

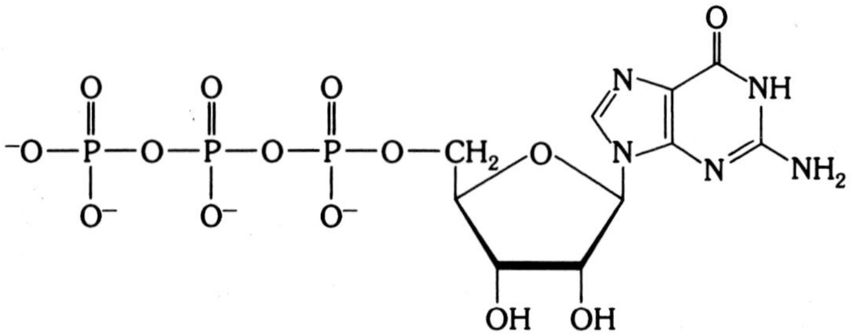

Guanosine triphosphate, GTP
(fully ionized form)

By convention, ATP and the phosphates higher than ATP in Table 14.1 are called **high-energy phosphates.** Notice that ATP is not at the head of the list. This means that ATP can be made from those above it in Table 14.1 by their transfer of phosphate to ADP. Then the ATP can, in turn, make any of the phosphates lower on the list. This intermediate position of the ATP and ADP is therefore one reason why this system is so useful as an energy broker. ADP can *accept* chemical energy from the phosphates with higher (more negative) potential and be changed itself to a high-energy phosphate (ATP). Then the newly made ATP can pass the chemical energy on by phosphorylations that make compounds lower on the list.

The Resynthesis of ATP Is a Major Goal of Catabolism Almost any energy-demanding activity of the body consumes ATP. The adult human *at rest* consumes about 40 kg (about 80 mol) of ATP per day yet, at any one instant, there is less than 50 g (0.1 mol) of ATP in existence in the body. When we engage in exercise, our rate of consumption of ATP can go as high as 0.5 kg per minute! Obviously, the rapid resynthesis of ATP is one of the highest priority activities of the body, and it occurs continuously. Virtually all biochemical energetics come down to the synthesis and uses of this compound.

■ Between the completely resting and the vigorously exercising states, the rate of ATP consumption can vary by a factor of 100.

When one of the compounds in Table 14.1 *above* ATP is first made by catabolism, and then its phosphate group is made to transfer to ADP to give ATP, the overall ATP synthesis is called **substrate phosphorylation** (where the substrate is ADP). The transfer is direct from an organic phosphate donor to ADP. In substrate phosphorylation, in other words, the phosphate group doesn't come directly from a phosphate ion in solution. Nonetheless, it is possible for inorganic phosphate to be joined *directly* to ADP. The direct use of phosphate ion happens in another kind of phosphorylation, **oxidative phosphorylation,** done by a series of reactions called the **respiratory chain,** which we'll study soon.

Creatine Phosphate Phosphorylates ADP in Muscles As you saw in Table 14.1, not all high-energy phosphates are triphosphates. Creatine phosphate, for example, is as important to muscle contraction as ATP. The ATP that is actually present in rested muscle can sustain muscle activity for only a fraction of a second. To provide for the *immediate* regeneration of ATP, muscle tissue makes and stores creatine phosphate (phosphocreatine) during periods of rest. Then, as soon as some ATP is used, and ADP plus P_i is made, creatine phosphate regenerates ATP.

■ The enzyme is *creatine kinase,* the CK enzyme we studied in Section 11.1.

What actually happens is that an increase in the supply of ADP shifts the following equilibrium to the right to raise the concentration of ATP. The high value of K_{eq} tells us that the forward reaction is favored.

$$\text{Creatine phosphate} + \text{ADP} \xrightleftharpoons{\text{creatine kinase, CK}} \text{creatine} + \text{ATP} \qquad K_{eq} = 162$$

Then, during periods of rest, when the ATP level is substantially raised by other reactions, this equilibrium shifts back to the left to resupply the creatine phosphate reserves.

Although muscle tissue has three to four times as much creatine phosphate as ATP, even this reserve cannot supply the high-energy phosphate needs of muscle for more than a 100-m to 200-m sprint. For a long sustained period of work, the body uses other methods to make ATP. We'll now take an overview of all of the routes to ATP, and then we will look closely at individual pathways.

All Bioenergetic Pathways Converge on the Citric Acid Cycle and the Respiratory Chain Our first interest is what *initiates* ATP synthesis. This process is under feedback control, and if the supply of ATP is high, no more needs to be made. Only as ATP is used up is first one mechanism and then another thrown into action. Generally, *it's an increase in the*

■ When the level of ATP drops and the levels of ADP + P_i rise, the rate of breathing is also accelerated.

Figure 14.1 is a broad outline of the metabolic pathways that can generate ATP. Think of the last one shown in the figure, the respiratory chain, as being at the bottom of a tub, nearest the drain through which ATP will leave for some use. As with water that leaves a tub, the first to leave is at the plug. This activates the motions of water at higher levels.

FIGURE 14.1
The major pathways for making ATP

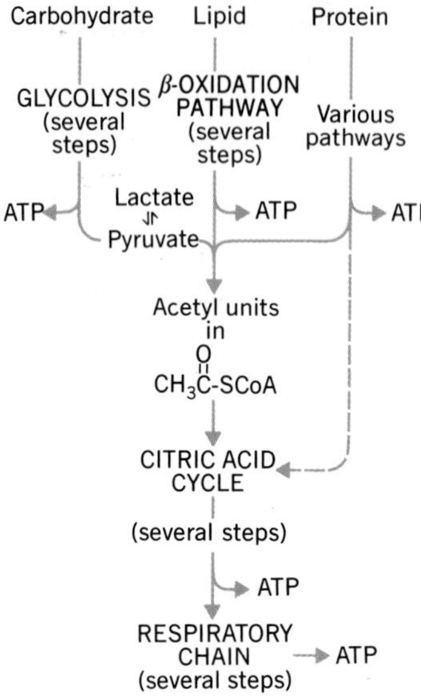

Carbohydrate Lipid Protein
GLYCOLYSIS (several steps)
β-OXIDATION PATHWAY (several steps)
Various pathways
ATP
Lactate ⇌ Pyruvate
ATP
ATP
Acetyl units in $CH_3\overset{O}{\overset{\|}{C}}\text{-SCoA}$
CITRIC ACID CYCLE
(several steps)
ATP
RESPIRATORY CHAIN (several steps)
ATP

■ Hans Krebs won a share of the 1953 Nobel prize in medicine and physiology for his work on the citric acid cycle.

By analogy, once the respiratory chain is launched, the next pathway to be thrown into action is the one second from the bottom in Figure 14.1, the **citric acid cycle.** The *chief purpose of the citric acid cycle is to supply the chemical needs of the respiratory chain.*

The citric acid cycle also requires a "fuel," and this need is filled by an acetyl derivative of an enzyme cofactor called coenzyme A. **Acetyl coenzyme A,** or acetyl CoA, is the "fuel" for the citric acid cycle. *The catabolism of molecules from all three major foods—carbohydrates, lipids, and proteins—can produce acetyl coenzyme A.*

Fatty acids are a major source of acetyl CoA, particularly during periods of sustained activities. A series of reactions called the **β-oxidation pathway** breaks fatty acids into acetyl units. We will study this pathway in Chapter 16.

Most amino acids can also be catabolized to acetyl units or to intermediates of the citric acid cycle itself, and we'll survey these reactions in Chapter 17.

Starting with glucose, an important pathway called **glycolysis** breaks glucose units down to the pyruvate ion and also makes some ATP by substrate phosphorylation. Then a short pathway converts pyruvate to acetyl CoA from which more ATP is made by oxidative phosphorylation.

Glycolysis Can Make ATP When a Cell Is Deficient in Oxygen One important aspect of glycolysis is that it makes some ATP by substrate phosphorylation *under low oxygen conditions* and independently of the respiratory chain. Even when a cell is temporarily starved for oxygen, glycolysis is able to make some ATP for a while. Glycolysis is thus a backup source of ATP for cells (temporarily) running low on oxygen.

When glycolysis has to run with insufficient oxygen, then its end product is the lactate ion, not the pyruvate ion. Once the cell obtains sufficient oxygen, however, lactate is converted to pyruvate.

■ The pyruvate ⇌ lactate equilibrium is discussed further in Chapter 16.

$$\underset{\text{Glucose}}{C_6H_{12}O_6} \xrightarrow[\text{not balanced)}]{\text{(Several steps;}} \underset{\text{Pyruvate}}{2CH_3\overset{O}{\overset{\|}{C}}CO_2^-} \underset{\text{With } O_2\text{, shifts to left}}{\overset{\text{If no } O_2\text{, shifts to right}}{\rightleftharpoons}} \underset{\text{Lactate}}{2CH_3\overset{OH}{\overset{|}{C}}HCO_2^-}$$

Glycolysis

$$\downarrow$$

$$\underset{\text{Acetyl coenzyme A}}{CH_3\overset{O}{\overset{\|}{C}}—S—CoA}$$

Thus glycolysis can end either with lactate or with pyruvate, depending on the oxygen supply. We will study details of glycolysis in the next chapter.

The full sequence of oxygen-consuming reactions from glucose to pyruvate ions to acetyl CoA and on through the respiratory chain is sometimes called the **aerobic sequence** of glucose catabolism. That part of the sequence that runs without oxygen and from glucose only as far as lactate ions is called the **anaerobic sequence** of glucose catabolism.

■ *Aerobic* signifies the use of air. *Anaerobic*, stemming from "not air," means in the absence of the use of oxygen.

Our interest in this chapter is in the citric acid cycle and the respiratory chain, pathways that can accept breakdown products from *any* food. Their reaction pathways converge on the citric acid cycle, which we will take up next.

14.2

THE CITRIC ACID CYCLE

Acetyl CoA is used to make the citrate ion, which is then broken down, bit by bit, to CO_2 as units of ($H:^-$ + H^+) are sent into the respiratory chain.

Figure 14.2 gives the reactions of the citric acid cycle, a series of reactions that break down acetyl groups.[2] The two carbon atoms of this group end up in molecules of CO_2, and the hydrogen of the acetyl group is fed into the respiratory chain by means of transfers of $H:^-$ and H^+. In the end, these pieces of H:H are irreversibly oxidized by oxygen (from respiration) to H_2O, a reaction of the respiratory chain that occurs in the innermost part of a mitochondrion.

■ The mitochondrion is an organelle (small body) in the cytoplasm (as we illustrated in Figure 13.1, page 333).

Coenzyme A Is the Common Carrier of Acetyl Units Before an acetyl group can enter the citric acid cycle, it must be made and attached to coenzyme A, which we represent as CoASH. In acetyl CoA, the acetyl residue replaces the H on HS of CoASH.

■ Pantothenic acid is another of the B vitamins.

$CH_3C(=O)SCoA$

Acetyl coenzyme A (acetyl CoA)

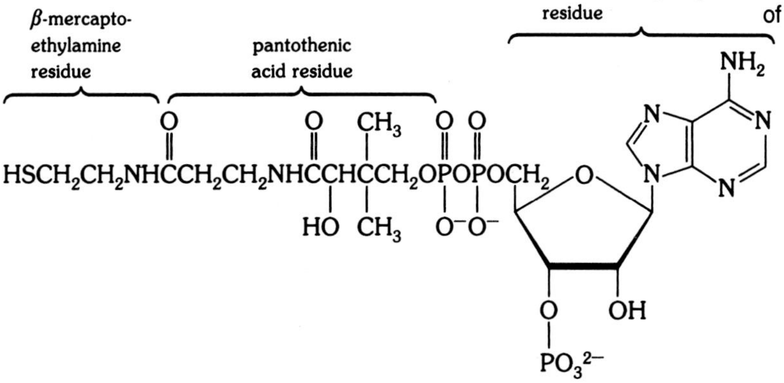

β-mercapto-ethylamine residue | pantothenic acid residue | adenosine 3′-phosphate residue

$HSCH_2CH_2NHC(=O)CH_2CH_2NHC(=O)CH(OH)C(CH_3)_2CH_2OP(=O)(O^-)OP(=O)(O^-)OCH_2$–(ribose with 3′-$OPO_3^{2-}$, 2′-OH, adenine with NH_2)

Coenzyme A (CoASH)

As we said, amino acid residues, fatty acids, and glucose are all sources of acetyl groups for acetyl coenzyme A. When glucose is the source, it is catabolized by many steps to pyruvate ion.

The conversion of pyruvate ion to acetyl coenzyme A involves both a decarboxylation and an oxidation. The overall equation for this complicated and irreversible change is as follows.

$$\underset{\text{Pyruvate}}{CH_3C(=O)CO_2^-} + \underset{\text{Coenzyme A}}{HSCoA} + NAD^+ \xrightarrow{\text{5 steps}} \underset{\text{Acetyl coenzyme A}}{CH_3C(=O)SCoA} + CO_2 + NADH$$

[2] You should be aware that the citric acid cycle goes by two other names as well: the **tricarboxylic acid cycle** and **Krebs' cycle.** You might encounter any of these names in other references.

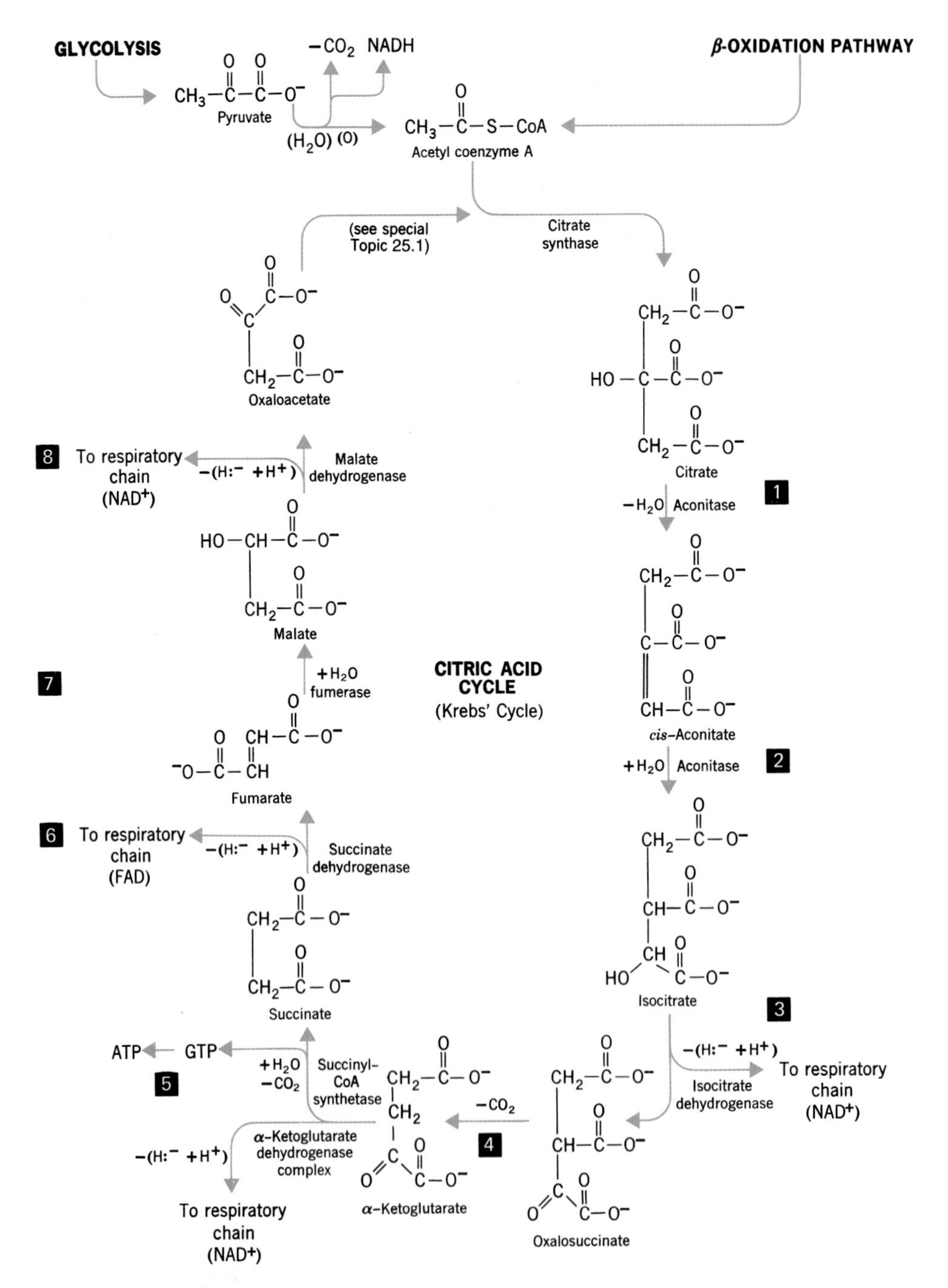

FIGURE 14.2
The citric acid cycle. The boxed numbers refer to the text discussion. The names of the enzymes for each step are given by the arrows.

We leave the details to more advanced treatments, but there are five actual steps catalyzed by a large, multienzyme complex called *pyruvate dehydrogenase*. Three B vitamins—thiamine, nicotinic acid (niacin), and riboflavin—are among the compounds needed to make the coenzymes for the complex.

Acetyl coenzyme A is a thioester, which means an ester in which an oxygen atom has been replaced by a sulfur atom. A thioester is far more reactive in transferring an acyl group than an ordinary ester. Thus acetyl CoA is a particularly active transfer agent for the acetyl group, and just such a transfer of an acetyl group launches one turn of the citric acid cycle.

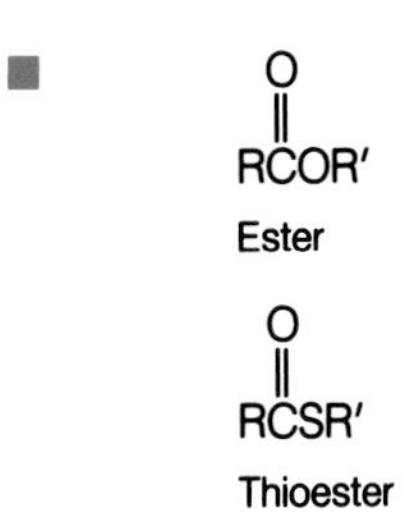

The Citric Acid Cycle Dismantles Acetyl Groups, Sending Hydrogen to the Respiratory Chain and CO_2 to Waste Disposal The enzymes for the citric acid cycle occur in mitochondria in close proximity to the enzymes of the respiratory chain. In the first step of the citric acid cycle, the acetyl group of acetyl CoA transfers to oxaloacetate ion, an ion that has two carboxylate groups and a keto group. The enzyme is *citrate synthase*. The acetyl unit of acetyl CoA adds across the keto group in a type of reaction that we have not studied before. For background to this reaction see Special Topic 14.1.

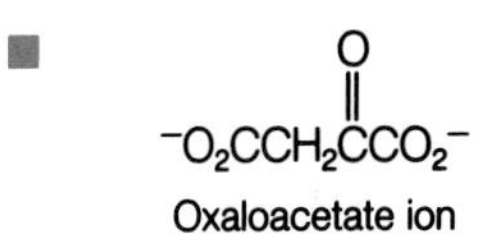

The product is the citrate ion. Now begins a series of reactions by which the citrate ion is degraded bit by bit until another oxaloacetate ion is regenerated. The numbers of the following steps match those in Figure 14.2.

■ At physiological pH, the acids in the cycle exist largely as their anions.

1. Citrate is dehydrated to give the double bond of *cis*-aconitate. This is the dehydration of an alcohol. The enzyme is *aconitase*.
2. Water adds to the double bond of *cis*-aconitate to give an isomer of citrate called isocitrate; aconitase also catalyzes this step. Thus the net effect of steps 1 and 2 is to switch the alcohol group in citrate to a different carbon atom. This changes the alcohol from being tertiary to one that is secondary, from an alcohol that cannot be oxidized to one that can.

■ The fluoroacetate ion, $FCH_2CO_2^-$, one of the most toxic of small-molecule poisons, inhibits aconitase.

3. The secondary alcohol group of isocitrate is dehydrogenated (oxidized) by *isocitrate dehydrogenase* to give the keto group of oxalosuccinate. The alcohol system gives up both H:⁻ and H^+.

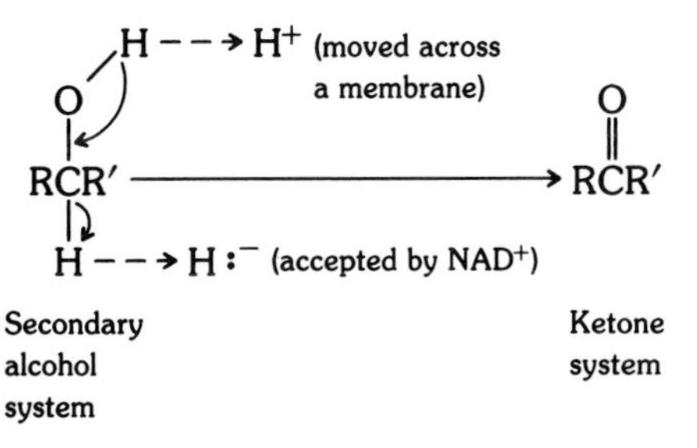

An NAD^+ coenzyme accepts the H:$^-$ unit; the released proton (H^+) is transferred across a membrane within the mitochondrion. The NAD^+ thereby changes to NAD:H (usually written simply as NADH).

What happens to the electron pair in NAD:H and to the H^+ is described in Section 14.3. Overall, the fates of NADH and H^+ are intimately tied to the synthesis of three molecules of ATP, so the step we have here described is the first delivery of chemical energy from the citric acid cycle to a triphosphate.

■ We studied how H:$^-$ transfers to NAD^+ in Section 11.1 and Special Topic 11.1.

4. Oxalosuccinate, still bound to isocitrate dehydrogenase, loses a carboxyl group—it decarboxylates—to give α-ketoglutarate.
5. α-Ketoglutarate now undergoes a very complicated series of reactions, all catalyzed by one team of enzymes called *α-ketoglutarate dehydrogenase*, a complex enzyme system resembling pyruvate dehydrogenase mentioned earlier. A carboxyl group is lost and another dehydrogenation occurs. The same kind of fate occurs to H:$^-$ and H^+ here as happened in step 3, so three more ATPs can now be made by the respiratory chain.

$^{-}O_2CCH_2CH_2CO_2^{-}$
Succinate ion

The product (not shown in Fig. 14.2) is the coenzyme A derivative of succinic acid. converted by the action of the enzyme *succinyl CoA synthetase* to the succinate ion, whi shown in Figure 14.2. The reaction also generates guanosine triphosphate, GTP, fron diphosphate. GTP is another high-energy triphosphate, which we mentioned on page 36 being similar to ATP. GTP is able to phosphorylate ADP to make one ATP. Thus the citric cycle includes one substrate phosphorylation.

$HCCO_2^{-}$ ‖ $^{-}O_2CCH$
Fumarate ion (trans geometry)

6. Succinate donates hydrogen to FAD (not to NAD^+) in a reaction catalyzed by *fumarate dehydrogenase,* and the fumarate ion forms. Both $H:^-$ and H^+ are accepted by FAD, which becomes $FADH_2$. $FADH_2$ also intersects with the respiratory chain, and its further involvement in this chain produces two ATPs (not three, which are possible from NADH).
7. Fumarate adds water to its double bond, and malate forms. The enzyme is *fumarase.* Note the 2° alcohol group in malate.

$^{-}O_2CCH_2CH(OH)CO_2^{-}$
Malate ion

8. Malate is oxidized by *malate dehydrogenase* to oxaloacetate. The 2° alcohol group of malate, in a reaction resembling step 3, gives up $H:^-$ to NAD^+ to form NADH. The chemical energy in NADH will be used to make three more ATP molecules by the respiratory chain.

One turn of the citric acid cycle is now complete and one molecule of the carrier, oxaloacetate, has been remade to enable another turn of the cycle. It can accept another acetyl group from acetyl coenzyme A.

SPECIAL TOPIC 14.1
CONDENSING ESTERS—A MAJOR C—C BOND-MAKING REACTION

The reaction by which an acetyl group enters the citric acid cycle is just one example of many reactions of two carbonyl compounds in which an acyl group of one becomes joined to an alpha position of another. The product is a much larger molecule. We'll first explain how this happens under laboratory conditions with a very simple system, two molecules of the same ester, ethyl acetate. Organic chemists call the reaction the *Claisen ester condensation.*

The Alpha Hydrogen of a Carbonyl Compound Is "Mobile" The acid ionization constant of an alkane is estimated to be about 10^{-40}; clearly, no one would call an alkane an acid! The ionization constant of the H atom attached to the α-position of the CH_3CO unit in ethyl acetate is about 10^{-18}, making it a billion trillion times as strong an acid (but still nothing we'd call an acid in water).

The reason for this greater acidity is the presence of the two nearby electronegative oxygen atoms. This greater acidity (mobility) of a hydrogen on a carbon alpha to oxygens is all we need to understand how the Claisen condensation and similar reactions in the body can occur.

With ethyl acetate, the reaction is as follows, where $B:^-$ is a powerful base, so powerful that the reaction cannot be run in water. (Usually it is run in ethyl alcohol.) In living systems, reactions like this use enzymes to handle necessary proton exchanges.

$$CH_3\overset{O}{\overset{\|}{C}}-OCH_2CH_3 + CH_3\overset{O}{\overset{\|}{C}}-OCH_2CH_3 \xrightarrow{B:^-} CH_3\overset{O}{\overset{\|}{C}}-CH_2\overset{O}{\overset{\|}{C}}-OCH_2CH_3 + HOCH_2CH_3$$

Ethyl acetoacetate

Notice that an acetyl group (in red) has been joined to the alpha carbon of the second ester molecule. A new C—C bond has been made. Let's see how a strong base handles this reaction.

SPECIAL TOPIC 14.1 CONTINUED

Step 1 The base takes an H^+ from the alpha H—C position of one ethyl acetate.

$$B\!:^- + H\text{—}CH_2\overset{\overset{\Large O}{\|}}{C}\text{—}OCH_2CH_3 \longrightarrow$$

Ethyl acetate

$$^-\!:CH_2\overset{\overset{\Large O}{\|}}{C}\text{—}OCH_2CH_3 + B\text{—}H$$

Anion of ethyl acetate

Step 2 The new anion attacks the carbonyl carbon of another ester molecule. (This carbon has a partial positive charge on it.)

$$CH_3\overset{\overset{\Large \ddot{O}:}{\|}}{C}\text{—}OCH_2CH_3 + {}^-\!:CH_2\overset{\overset{\Large O}{\|}}{C}\text{—}OCH_2CH_3 \longrightarrow$$

$$CH_3\underset{\underset{\large OCH_2CH_3}{|}}{\overset{\overset{\large :\ddot{O}:^-}{|}}{C}}\text{—}CH_2\overset{\overset{\Large O}{\|}}{C}\text{—}OCH_2CH_3$$

Step 3 The product of this reaction expels $CH_3CH_2O^-$, which takes H^+ from $B\!:\!H$ (and the base is thus regenerated).

$$CH_3\underset{\underset{\large OCH_2CH_3}{|}}{\overset{\overset{\large :\ddot{O}:^-}{|}}{C}}\text{—}CH_2\overset{\overset{\Large O}{\|}}{C}\text{—}OCH_2CH_3 \longrightarrow \quad B\!:\!H$$

$$CH_3\overset{\overset{\Large O}{\|}}{C}\text{—}CH_2\overset{\overset{\Large O}{\|}}{C}\text{—}OCH_2CH_3 + HOCH_2CH_3 + B\!:^-$$

Ethyl acetoacetate

This overall reaction is formally very similar to the initial reactions the body uses for several biosyntheses: to make long hydrocarbon chains from acetate units, and to make cholesterol and the sex hormones. So we'll meet this kind of reaction in later chapters.

Citrate Synthase Manages a Claisen-like Condensation of Acetyl CoA with Oxaloacetate If we take acetyl CoA and oxaloacetate through the same steps, we have the following in which the services of a base, still represented by $B\!:^-$, are provided by the enzyme citrate synthase. Realize that what follows is a considerable simplification of a series of steps in which the way charges are managed by the enzyme are not indicated.

Step 1 Acetyl CoA gives up a proton.

$$B\!:^- + H\text{—}CH_2\overset{\overset{\Large O}{\|}}{C}SCoA \longrightarrow BH + {}^-\!:CH_2\overset{\overset{\Large O}{\|}}{C}SCoA$$

Acetyl CoA

Step 2 The new anion attacks the keto group in oxaloacetate. The new C—C bond forms.

$$\underset{\underset{\large CH_2CO_2^-}{|}}{\overset{\overset{\large \ddot{O}\!:}{\|}}{C}}\text{—}CO_2^- + {}^-\!:CH_2\overset{\overset{\Large O}{\|}}{C}SCoA \longrightarrow {}^-\!:\ddot{O}\text{—}\underset{\underset{\large CH_2CO_2^-}{|}}{\overset{\overset{\large CH_2\overset{O}{\overset{\|}{C}}SCoA}{|}}{C}}\text{—}CO_2^-$$

Step 3 The CoA-S group is hydrolyzed, and a proton is donated so that the 3° alcohol group can form. (The CoA-S unit is similar to an ordinary ester. It is a thioester, and these hydrolyze more readily than ordinary esters. In fact this hydrolysis is part of the driving force of the overall reaction.)

$$B\text{—}H + {}^-\!:\ddot{O}\text{—}\underset{\underset{\large CH_2CO_2^-}{|}}{\overset{\overset{\large CH_2\overset{O}{\overset{\|}{C}}SCoA}{|}}{C}}\text{—}CO_2^- + H_2O \longrightarrow$$

$$B\!:^- + H\text{—}\ddot{O}\text{—}\underset{\underset{\large CH_2CO_2^-}{|}}{\overset{\overset{\large CH_2\overset{O}{\overset{\|}{C}}O^-}{|}}{C}}\text{—}CO_2^- + H\text{—}SCoA + H^+$$

Citrate ion

Thus a Claisen-like condensation launches the acetyl group of acetyl CoA into the citric acid cycle.

14.3

THE RESPIRATORY CHAIN

The flow of electrons to oxygen in the respiratory chain creates a proton gradient in mitochondria that drives the synthesis of ATP.

The term *respiration* refers to more than just breathing. It includes the chemical reactions that use oxygen in cells. We are now ready to learn specifically how oxygen is reduced to water and how the chemical energy of this event helps to drives the entire respiratory chain. The most basic statement we can write for what happens when water forms emphasizes how electrons and protons combine with an oxygen atom:

$$(:) \quad + \quad 2H^+ + \cdot\ddot{\underset{\cdot\cdot}{O}}\cdot \longrightarrow H{-}\ddot{\underset{\cdot\cdot}{O}}{-}H + \text{energy}$$

Pair of electrons — Pair of protons — Atom of oxygen — Molecule of water

■ The gain of e^- or of e^- carriers such as $H:^-$ is *reduction;* the loss of e^- or of $H:^-$ is *oxidation.*

Remember that when any particle, like the oxygen atom, gains electrons it is reduced, and the donor of electrons is oxidized. The electrons and protons for the reduction of oxygen come mostly from intermediates in the citric acid cycle. As you saw in the discussion of steps 3, 5, and 6 of this cycle, the electrons of C—H bonds are used, not O—H bonds, and the hydride ion $H:^-$ is often (but not always) the vehicle for carrying the electrons from one species to another.

The Mitochondrion Is the Cell's "Powerhouse" The respiratory chain is one long series of oxidation–reduction reactions. The flow of electrons from the initial donor is down an energy hill all the way to oxygen, and it is irreversible. As this flow occurs, other complex events take place to make ATP from ADP and P_i.

The cell's principal site of respiratory chain activity and ATP synthesis, a *mitochondrion,* is often dubbed the powerhouse of the cell (see Fig. 14.3). Some tissues have thousands of mitochondria in the cytoplasm of a single cell. A mitochondrion has two important mem-

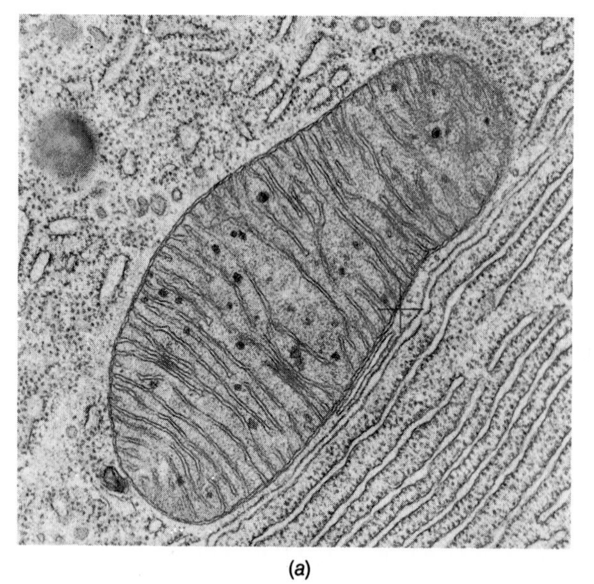

(a)

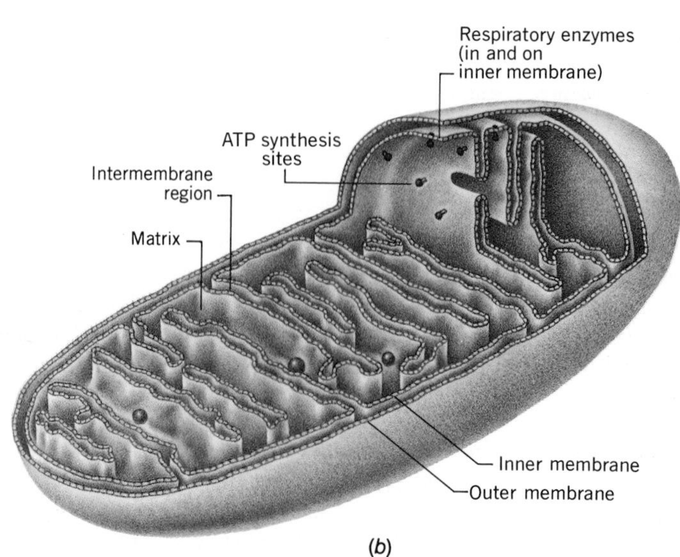

(b)

FIGURE 14.3

A mitochondrion. (*a*) Electron micrograph (×53,000) of a mitochondrion in a pancreas cell of a bat. (*b*) Perspective showing the interior. The respiratory enzymes are incorporated into the inner membrane. On the matrix side of this membrane are enzymes that catalyze the synthesis of ATP. (Micrograph courtesy of Dr. Keith R. Porter.)

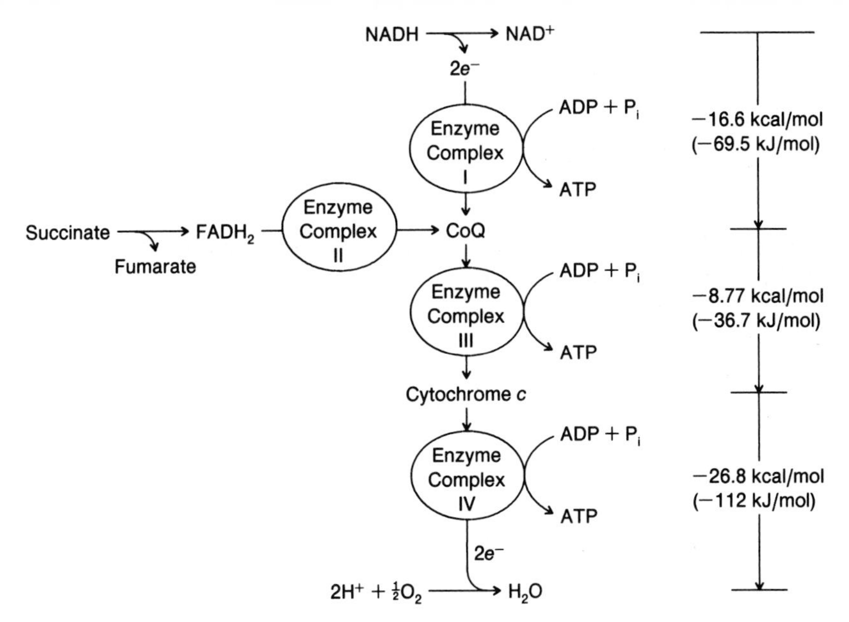

FIGURE 14.4
The enzyme complexes of the respiratory chain and the energy drop from NAD^+ to reduced oxygen (water). The figures for energy refer to the portion of the energy change that is not unavoidably dissipated as heat. The symbols for the enzymes are as follows.
NADH = enzyme with reduced nicotinamide cofactor
$FADH_2$ = enzyme with reduced riboflavin cofactor
CoQ = enzyme with coenzyme Q cofactor

branes, one outer and one inner. The inner membrane is very convoluted and has a surface area many times that of the outer membrane. The space between the two membranes is called the *intermembrane space.* The space deep inside the inner membrane is filled with a gel-like material called the *matrix*. It's less than 50% water and rich in soluble enzymes, cofactors, inorganic ions, and substrates.

■ A cell in the flight muscle of a wasp has about a million mitochondria.

The Chief Agents of Electron Transfer Are the Respiratory Enzymes Built into the inner membrane itself are numerous immobile enzyme clusters designated as enzyme complexes I, II, III, and IV. Also existing as parts of the inner membrane are enzymes capable of moving from one complex to another and named coenzyme Q (CoQ) and cytochrome *c*. Taken together, these enzymes are called the **respiratory enzymes,** and they operate the respiratory chain.

The Operation of the Respiratory Chain Moves H^+ Ions from the Matrix to the Intermembrane Space One very important property of the inner membrane is that *only at particular channels is it permeable to chemical species, particularly ions like* H^+. The operation of the respiratory chain transfers H^+ ions from the matrix to the intermembrane compartment. *This establishes an unstable proton gradient between the intermembrane region (more* H^+*) and the matrix (less* H^+*).* Eventually the fluid of the intermembrane space will have a pH 1.4 units lower (more acidic) than the matrix.

Remember, given the random motions of species in any fluid, that nature eventually destroys a gradient. However, the gradient-destroying flow of H^+ ion back into the matrix is permitted only at certain inner-membrane channels. The flowing protons, therefore, can return to the matrix only at certain locations, *places where the returning protons activate a complex enzyme system that makes and releases ATP*. This is basically how the cell connects activities of the respiratory chain with the synthesis of ATP. An outline of this understanding was first formulated by English scientist Peter Mitchell, and it is called the **chemiosmotic theory.**

■ Peter Mitchell received the 1978 Nobel prize in chemistry. His proposal was made in 1961 and inspired a flurry of controversy among scientists before intensive research led to its acceptance.

One Entry to the Respiratory Chain Is the Synthesis of NADH If we let MH_2 represent any metabolite that can donate $H\!:^-$ to NAD^+, we can write the following equation.

$$MH_2 + NAD^+ \longrightarrow M + NADH + H^+$$

■ Think of MH_2 as $M\!:\!H$ (with a second H attached to M), which becomes M when $H\!:^-$ and H^+ leave.

By this reaction, the electron pair has moved from MH_2 to NADH, which next interacts with enzyme complex 1.

Figure 14.4 gives the sequence by which the respiratory enzyme complexes interact and shows the reaction-favoring releases of chemical energy that impel the overall sequence.

Enzyme Complex I of the Respiratory Chain Oxidizes NADH and Reduces CoQ As indicated in Figure 14.4, enzyme complex I first accepts a pair of electrons (carried by H:$^-$) from NADH. The electron flow has now started toward oxygen.

The acceptor of H:$^-$ in complex I is FMN, which carries the B vitamin riboflavin as part of its coenzyme. The hydride unit, H:$^-$, in NADH transfers to FMN according to the following equation.

■ We studied the FMN enzyme in Section 11.1 and Special Topic 11.1.

$$\text{NADH} + \text{H}^+ + \text{FMN} \longrightarrow \text{FMNH}_2 + \text{NAD}^+$$

This restores the NAD^+ enzyme, and moves the electron pair one more step along the respiratory chain.

■ Think of NADH as NAD:H (with H attached to NAD) in which the electron pair dots are those that move along the respiratory chain.

What happens next is the transfer of just a pair of electrons, not H:$^-$, from $FMNH_2$ while the hydrogen *nucleus* of H:$^-$ leaves the carrier as H^+. *This transfer induces conformational changes in an inner membrane protein that cause protons to transfer from the matrix into the intermembrane region.* Thus the buildup of the H^+ gradient across the inner mitochondrial membrane begins. The two electrons are accepted by an iron-sulfur protein, which we'll symbolize by FeS–P. The iron in FeS–P occurs as Fe^{3+}. By accepting one electron, it is reduced to the Fe^{2+} state, so we need *two* units of FeS–P to handle the pair of electrons now about to leave $FMNH_2$. This step can be written as

■ There are actually three types of iron–sulfur proteins having different atom ratios of iron to sulfur. We do not indicate the actual net charge; regardless of the type, the oxidized and reduced forms all differ by a net charge of 1+.

$$\text{FMNH}_2 + 2\text{FeS–P} \longrightarrow \text{FMN} + 2\text{FeS–P}\cdot + 2\text{H}^+$$

where we use FeS–P· to represent the reduced form of the iron–sulfur protein in which Fe^{2+} occurs.

Before we go on, we will introduce a way to display these reactions that helps to emphasize the restoration of each enzyme to its original condition. For example, the last three reactions we have studied can be represented as follows:

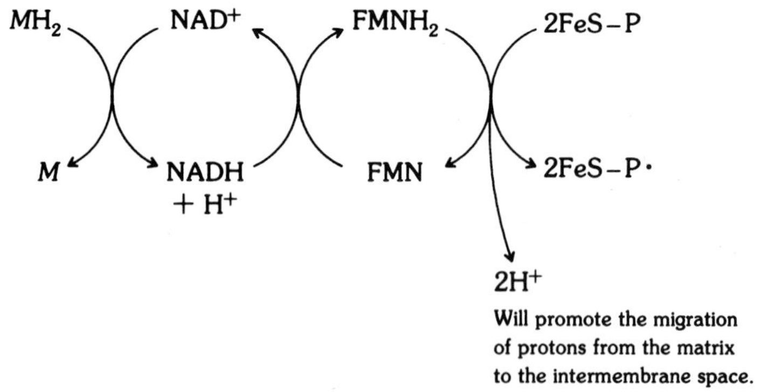

MH_2 → M; NAD^+ → NADH + H^+; NADH + H^+ → NAD^+; FMN → $FMNH_2$; $FMNH_2$ → FMN; 2FeS–P → 2FeS–P·; → $2H^+$

$2H^+$ Will promote the migration of protons from the matrix to the intermembrane space.

Reading from left to right, at the point where the first pair of curved arrows touch, we see that as MH_2 changes to M, it passes H:$^-$ to NAD^+ to change it to NADH. H^+ is also released. The touching of the second pair of curved arrows tells us that as NADH and H^+ are changed back to NAD^+ the system passes one unit of H:$^-$ plus H^+ across to FMN to change it to $FMNH_2$. At the third pair of curved arrows, the display shows $FMNH_2$ changing back to FMN as it passes two electrons to two FeS–P molecules and also as it expels two protons (H^+).

Between three and four protons, on the average, are moved from the matrix into the intermembrane space by each pair of H^+ ions released by respiratory chain oxidation. From which *specific* chemical species the protons actually migrate is still being studied.

The last work of enzyme complex I is to pass electrons from the reduced iron–sulfur complex to coenzyme Q. We will now drastically reduce the attention we give to specific detail.

Respiratory Enzyme Complexes III and IV Continue the Respiratory Chain As you have seen, the respiratory chain is quite complicated, and from this point on it becomes even more so. What follows next are further transfers of electrons through enzyme complexes III and IV, which involve a series of individual enzymes called the *cytochromes*. These are designated, in the order in which they participate, as cytochromes b, c_1, and c, which together with an iron–sulfur protein make up enzyme complex III. Cytochromes a and a_3 are parts of enzyme complex IV, often called *cytochrome c oxidase*.

■ *Cyto-*, cell; *-chrome*, pigment. The cytochromes are colored substances.

Complex IV is the enzyme complex that catalyzes the reduction of oxygen to water that we described earlier. It carries a copper ion that alternates between the Cu^{2+} and the Cu^{+} states.

Figure 14.5 is a diagram of the electron movement through complexes I, III, and IV. The net result of the operation of this, the main branch of the respiratory chain, starting with MH_2 and NAD^+, can be expressed by the following overall equation.

$$MH_2 + nH^+ + \tfrac{1}{2}O_2 \xrightarrow[\text{(enzyme complexes I, III, and IV)}]{\text{respiratory chain}} M + H_2O + nH^+$$

MH_2: From inside the inner membrane

nH^+: These protons help make the proton gradient across the inner membrane of the mitochondrion.

Complex II, indicated in Figure 14.4 (but not in Fig. 14.5), is just another way to enter the respiratory chain at coenzyme Q. The succinate ion of step 6 of the citric acid cycle is one substrate for complex II.

■ Complex II includes the riboflavin-based coenzyme, FAD.

Two Gradients Are Established, a Proton and a Plus Charge Gradient The operation of the respiratory chain pumps protons from the mitochondrial matrix to the intermembrane region *without also putting the equivalent of negatively charged ions there, too.* Two gradients are thus set up, a gradient of H^+ ions and a gradient of positive charge. The positive charge gradient could be erased either by the migration of negative ions to the intermembrane region *or by the migration of any kind of positive ion from this region into the matrix.* Calcium ions, for example, might move, *and precisely such movements of Ca^{2+} ions are involved with nerve signals and muscular work.* Thus the chemiosmotic theory helps us understand events other than ATP synthesis, like electrical signal sending in nerve tissue.

■ When a gradient of electrical charge is established by chemical species, it is called an *electrochemical gradient.*

■ Because chemical reactions create the gradients, we have the *chemi-* part of the term *chemiosmotic.*

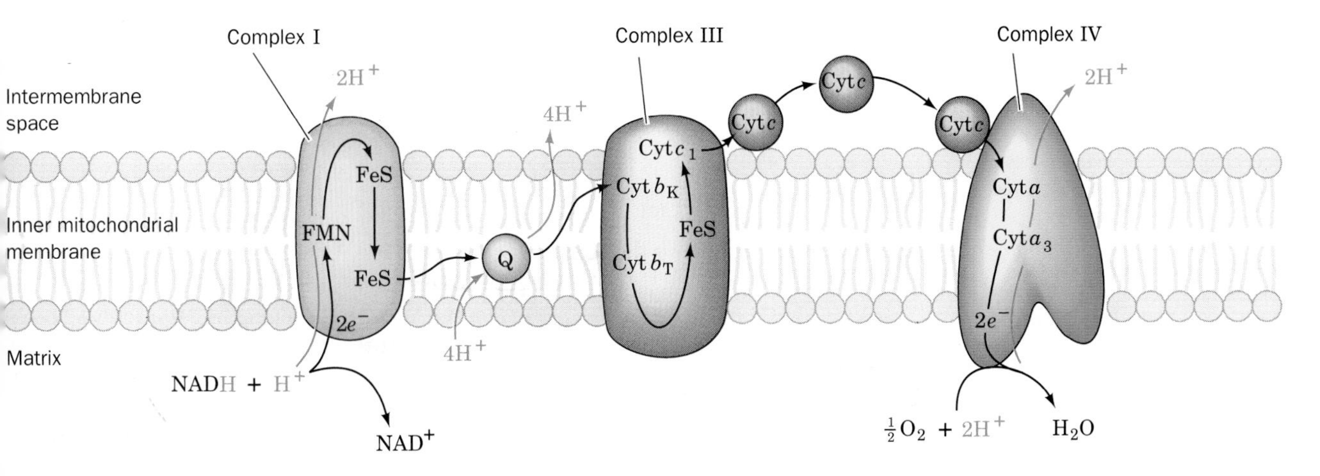

FIGURE 14.5
The electron flow (black) and the proton transfers (red) from NADH through enzyme complexes I, III, and IV of the respiratory chain. Q represents coenzyme Q, CoQ. Cyt *c* is cytochrome *c*. Complex II is not shown, but it transfers electrons from succinate ion to coenzyme Q. (From Voet and Voet, *Biochemistry*, John Wiley & Sons, Inc., 1990. Used by permission.)

The Proton Channels of the Inner Mitochondrial Membrane Are Part of an Enzyme for Making ATP from ADP and P_i Embedded in the inner mitochondrial membrane are complexes of proteins that form a tube through it. At least one tube exists for every unit of respiratory enzymes. The tube itself is called the F_0 component of a complex enzyme called the **proton-pumping ATPase.** On the matrix side of the inner membrane occurs the other component, called the F_1 component, at which the proton channel ends. In electron micrographs of the inner membrane, the F_1 components project like lollipops from the matrix-side surface (see Fig. 14.6). The F_1 components consist of eight polypeptides one of which is the proton gate and three of which are the actual sites where ADP and P_i are put together to make ATP.

■ The proton-pumping ATPase is also called the *proton-translocating ATPase* as well as the *$F_0F_1ATPase$.*

In a theory proposed by Paul Boyer, the linkage between the flow of protons through F_0 and the synthesis and release of ATP from F_1 is made as follows by a series of steps involving allosteric interactions (see Fig. 14.7). Each of the ATP-synthesizing F_1 subunits is capable of existing in any one of three configurations, open (O), loosely closed (L), and tightly closed (T). The O configuration has little ability to bind substrates, whether ADP plus P_i or ATP. The L configuration can bind ADP + P_i *but can do nothing catalytically to change these particles to ATP until protons arrive to cause a change in configuration of the protein shape from the L to the T state.* Now the formation of ATP is catalyzed, *but it is tightly held to the T-shaped subunit.* It cannot be released until the imbalance in proton concentration—the proton gradient—builds up sufficiently to drive protons through the F_0 channel and into F_1. The arriving protons cause conformational changes in all of the F_1 subunits. The *T* subunit with its bound ATP changes into an *O* subunit, which permits the ATP to drop off. The *L* subunit (with loosely bound ADP plus P_i) changes to the *T* form as a result of which another ATP is put together to await release by further surges of protons.

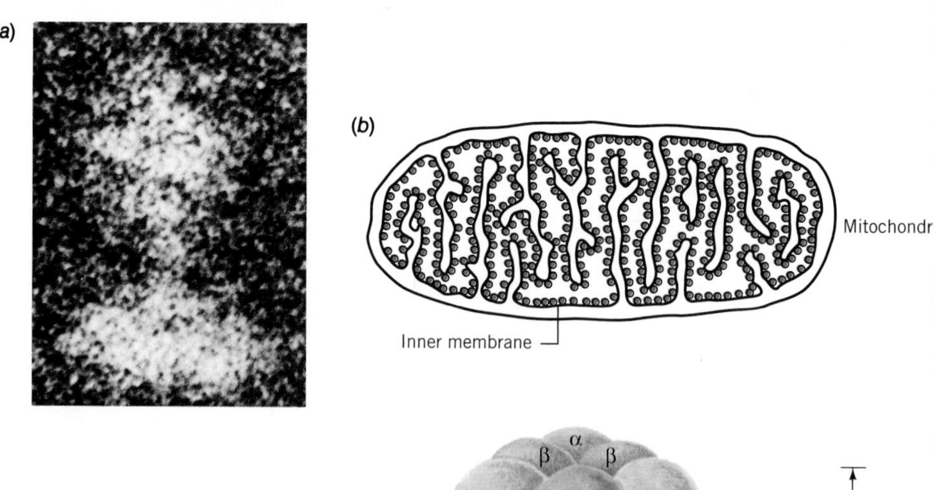

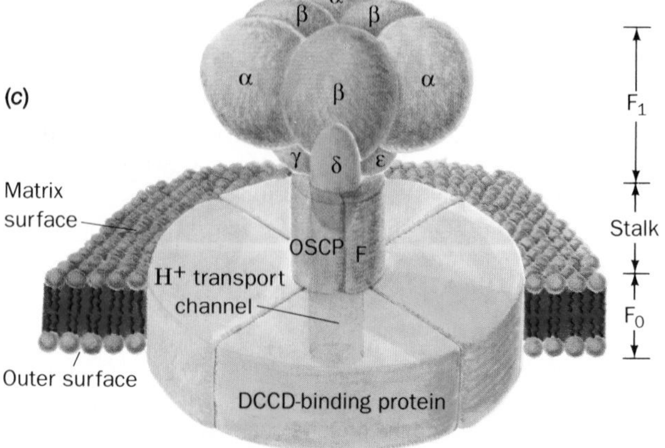

FIGURE 14.6
Proton-pumping ATPase. (*a*) Electron micrograph of an intact inner mitochondrial membrane showing the "lollipop" projections of the enzyme on the matrix side. [From Parsons, D. F., *Science* **140,** 985 (1963). Copyright © 1963 American Association for the Advancement of Science. Used by permission.] (*b*) Interpretive drawing of the mitochondrion with the same projections shown in part *a*. (*c*) Interpretive drawing of one proton-pumping ATPase complex. The unit labeled δ is the proton gate between the F_0 proton channel (beneath) and the other subunits of the F_1 complex (above). The three subunits labeled β are ATP synthesis sites. (Adapted with permission from Voet and Voet, *Biochemistry*, John Wiley & Sons, Inc., 1990. Used by permission.)

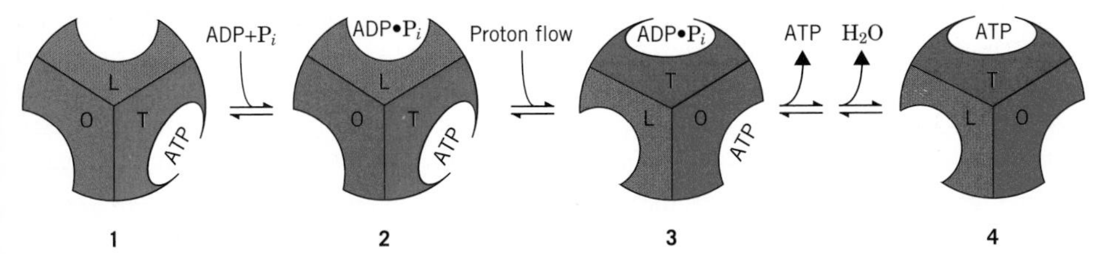

FIGURE 14.7
The synthesis and release of ATP from subunits of the F_0 portion of the proton-pumping ATPase complex. Each of the three β-subunits shown in Figure 14.6 is capable of being in the O, L, or T configuration discussed in the text. In set **1**, an ATP molecule, previously made, is tightly held. The system picks up ADP plus P_i, which become loosely held, as the system changes to **2.** Now, **2** to **3**, we imagine that protons in the gradient have entered the complex via F_0 to cause conformational changes that let ATP drop away, get the ADP plus P_i ready to become ATP (**3** to **4**) and let an old O site change to an L site in readiness to accept ADP and P_i. (Adapted with permission from Voet and Voet, *Biochemistry,* John Wiley & Sons, Inc., 1990. Used by permission.)

It is because a series of *oxidations* creates the proton gradient that makes this ATP synthesis possible that the overall synthesis is called **oxidative phosphorylation,** or sometimes *respiratory chain phosphorylation.* Each molecule of NADH that enters the chain at enzyme complex I can lead to a maximum of three ATPs. Each $FADH_2$ that enters at complex II can cause just two ATP molecules to form. Let's now go back over the major aspects of oxidative phosphorylation

Oxidative Phosphorylation

1. The synthesis of ATP occurs at an enzyme located on the matrix side of the inner mitochondrial membrane.
2. ATP synthesis is driven by a flow of protons that occurs from the intermembrane side to the matrix side of the inner membrane.
3. The flow of protons is through special channels in the membrane and down a concentration gradient of protons that exists across the inner membrane.
4. The energy that creates the proton gradient and the (+) charge gradient is provided by the flow of electrons between the respiratory chain enzymes that make up integral packages of the inner membrane.
5. The inner mitochondrial membrane is a closed envelope except for the special channels for the flow of protons and for special transport systems that let needed solutes move into or out of the innermost mitochondrial compartment.

■ This migration through a semipermeable membrane explains the *osmotic* part of the term *chemiosmotic.*

A Transport Protein in the Mitochondrial Membrane Moves ADP Inside as It Carries ATP Out Both ATP and ADP are highly charged species and so could not easily get through a lipid bilayer membrane. A transport protein of an inner mitochondrial membrane "pump" called *ATP–ADP translocase* solves this problem. The migration of newly made ATP to the intermembrane region and thence to the cytosol outside is coupled to the movement of ADP into the matrix. The expression "is coupled" means that the membrane pump that moves ADP in one direction (in) simultaneously moves ATP in the other (out).

The Rate of ATP Resynthesis Is Sensitive to the Molar Concentrations of ADP and ATP With considerable simplification of a complex phenomenon, it is as if ADP, P_i, and ATP are involved in the following equilibrium.

$$\text{ADP} + \text{P}_i \rightleftharpoons \text{ATP} \quad (14.2)$$

When the concentrations of ADP and P_i are high, a stress thereby exists that causes ATP to be made; equilibrium 14.2 shifts to the right—the direction predicted by Le Châtelier's principle. When the supply of ATP is relatively high, on the other hand, little more is (or need be) made. During our quiet times, when the least amounts of ADP and P_i are being made by ATP-consuming, energy-demanding activities, the machinery for making ATP operates very slowly—enough to maintain basal metabolism. Then, when our lives become more active, the rate of use of ATP accelerates, and ADP plus P_i are made. Their appearance shifts equilibrium 14.2 to the right, thus using them up. A question not yet fully resolved is the location in the cell of these shifts. Is the rate of ATP synthesis controlled by changes in concentrations in the cytosol or in the mitochondrial matrix? Or is the rate controlled at the ATP–ADP translocase protein of the mitrochondrial membranes? In any case, the rate depends on the supply of ADP, and ATP is made no more rapidly than actually needed.

■ Rotenone is a naturally occurring insecticide.

■ Remember that the respiratory chain also occurs in bacteria, so antibiotics that interfere with the chain can kill bacteria.

■ A natural and healthy mechanism for uncoupling the chain from ATP synthesis exists in the heat-generating, brown fat tissue that cold-adapted animals have (and which will be discussed further in Special Topic 16.2, page 411.)

Some Antibiotics and Poisons Inhibit Oxidative Phosphorylation One of the barbiturates, amytal sodium, and the powerful insecticide, rotenone, block the respiratory chain by interfering at enzyme complex I. The antibiotic antimycin A stops the chain at complex III. The cyanide ion blocks the chain at its very end, at cytochrome a_3 in complex IV. These agents, by inhibiting the chain, thus work to inhibit the creation of the proton gradient and in this manner inhibit ATP synthesis.

Other substances interfere with the *use* of the gradient. They do not prevent the respiratory chain from operating, but they cancel the effect of the gradient. They *uncouple* the chain from ATP synthesis. For example, 2,4-dinitrophenol increases the permeability of the inner membrane to protons so that they can use random routes to reenter the matrix. When the gradient is destroyed without the use of the F_0 proton channel, little if any ATP synthesis can occur. The chemical energy released by the operation of the chain is then converted to heat.

Oxidative Phosphorylation Makes a Maximum of Twelve ATP Molecules Using Intermediates of the Citric Acid Cycle The data in Table 14.2 show where ATP production is generated as the citric acid cycle runs. If, to the twelve ATPs made this way, we add the three ATPs made possible by the conversion of pyruvate to acetyl CoA, then fifteen ATPs can be made by the degradation of one pyruvate ion.

Keep in mind that pyruvate isn't the only raw material either for acetyl coenzyme A or for other metabolites (MH_2) that can fuel the respiratory chain. Also remember that the energy of the respiratory chain can be used to run other operations besides making ATP—for example, the operation of nerves. Thus Table 14.2 has to be viewed as giving the upper limits to ATP production from pyruvate by the respiratory chain.

TABLE 14.2
ATP Production by Oxidative Phosphorylation

Step	Receiver of ($H:^- + H^+$) in the Respiratory Chain	Molecules of ATP Formed
Isocitrate → α-ketoglutarate	NAD^+	3
α-Ketoglutarate → succinyl CoA	NAD^+	3
Succinyl CoA → succinate (via GTP)	—	1
Succinate → fumarate	FAD	2
Malate → oxaloacetate	NAD^+	3
Total ATP via citric acid cycle		12
Pyruvate → acetyl CoA	NAD^+	3
Total ATP from pyruvate via the citric acid cycle		15

SUMMARY

High-energy compounds Organophosphates whose phosphate group transfer potentials equal or are higher than that of ATP are classified as high-energy phosphates. A lower energy phosphate can be made by phosphate transfer from a higher energy phosphate, a process called substrate phosphorylation.

Citric acid cycle Acetyl groups from acetyl coenzyme A are joined to a four-carbon carrier, oxaloacetate, to make citrate. This six-carbon salt of a tricarboxylic acid then is degraded bit by bit as pieces of hydrogen, $(H\!:^- + H^+)$, are fed to the respiratory chain. Each acetyl unit leads to a maximum of twelve ATP molecules.

Fatty acids and glucose are important suppliers of acetyl groups for the citric acid cycle. In aerobic glycolysis, glucose units are broken to pyruvate units. The oxidative decarboxylation of pyruvate leads to the synthesis of one NADH (which eventually leads to the formation of three ATPs), and it supplies acetyl units to the citric acid cycle from which twelve more ATPs are made.

Respiratory chain A series of electron transfer enzymes called the respiratory enzymes occur together as groups called respiratory assemblies in the inner membranes of mitochondria. These enzymes process NADH or $FADH_2$, made by reduction reactions from metabolites (MH_2) and NAD^+ or FAD. NADH or $FADH_2$ then pass on electrons until cytochrome oxidase uses them (together with H^+) to reduce oxygen to water.

Oxidative phosphorylation As electrons flow from NADH to oxygen in the respiratory chain, protons are released. They cause conformational changes in proteins imbedded in the inner membrane of a mitochondrion. These changes cause protons to move from the mitochondrial matrix to the intermembrane side of the inner membrane, making a proton gradient and a positive charge gradient across the inner membrane. As protons flow back at the allowed F_0 conduits of the proton-pumping ATPase complex of this membrane, ATP is made and released by the F_1 complex of this enzyme. In some systems, other kinds of positive ions migrate, as in the operation of nerves. Various drugs and antibiotics can block the respiratory chain. Other chemicals are able to uncouple the work of the respiratory chain from the synthesis of ATP so that the operation of the chain converts chemical energy to heat. The rate at which ATP is made increases as the concentration of ADP increases.

REVIEW EXERCISES

The answers to Review Exercises whose numbers are in color are found in Appendix A. The answers to the other Review Exercises are found in the Study Guide that accompanies this book. The more challenging questions are marked with asterisks.

Energy Sources

14.1 What products of the digestion of carbohydrates, triacylglycerols, and the polypeptides can be used as sources of biochemical energy to make ATP?

14.2 The complete catabolism of glucose gives what products?

14.3 The identical products form when glucose is burned in open air as when it is fully catabolized in the body. How, then, do these two processes differ in their overall accomplishments?

14.4 What are the end products of the complete catabolism of fatty acids?

High-Energy Phosphates

14.5 Complete the following structure of ATP.

$$\text{Adenosine—O—}\overset{\overset{\displaystyle O}{\|}}{\underset{\underset{\displaystyle O^-}{|}}{P}}\text{—}$$

14.6 Write the structures of ADP and of AMP in the manner started by Review Exercise 14.5.

14.7 At physiological pH, what does the term *inorganic phosphate* stand for? (Give formulas and names.)

14.8 Phosphate X has a phosphate group transfer potential of −13 kcal/mol. What does this quantity refer to?

14.9 Why are creatine phosphate and ATP called high-energy phosphates but glycerol-3-phosphate is not?

*14.10 If the organophosphate of *M* has a *lower* (less negative) phosphate group transfer potential than that of *N*, in which direction does the following reaction tend to go spontaneously, to the left or to the right?

$$M\text{—}OPO_3^{2-} + N \overset{?}{\rightleftharpoons} N\text{—}OPO_3^{2-} + M$$

14.11 Using data in Table 14.1, tell (yes or no) whether ATP readily transfers a phosphate group to each of the following possible compounds. (Assume, of course, that the right enzyme is available.)
(a) glycerol (b) fructose (c) creatine (d) acetic acid

14.12 All of the possible phosphate transfers in Review Exercise 14.11 are classified as *substrate* phosphorylations. What does this mean?

14.13 What is the function of creatine phosphate in muscle tissue?

14.14 Whether or not creatine phosphate is used in muscle tissue is under *feedback control*. Explain.

Overview of Metabolic Pathways

14.15 In the general area of biochemical energetics, what is the purpose of each of the following pathways?
(a) respiratory chain (b) anaerobic glycolysis
(c) citric acid cycle (d) fatty acid cycle

14.16 What prompts the respiratory chain to go into operation?

14.17 Which tends to increase the rate of breathing, an increase in the body's supply of ATP or an increase in its supply of ADP? Explain.

14.18 In general terms, the intermediates that send electrons down the respiratory chain come from what metabolic pathway that consumes acetyl groups?

14.19 Arrange the following sets of terms in sequence in the order in which they occur or take place. Place the identifying letter of the first sequence of a set to occur on the left of the row of letters.

(a) Citric acid cycle (A) pyruvate (B) acetyl CoA (C)
respiratory chain (D) glycolysis (E)

(b) Citric acid cycle (A) β-oxidation (B) acetyl CoA (C)
respiratory chain (D)

14.20 The *aerobic sequence* begins with what metabolic pathway and ends with which pathway?

14.21 The β-oxidation pathway occurs to what kind of compound?

14.22 The anaerobic sequence begins and ends with what compounds?

14.23 What does it mean for the cell that anaerobic glycolysis is a "backup?"

Citric Acid Cycle

14.24 What makes the citric acid cycle start up?

14.25 What chemical unit is degraded by the citric acid cycle? Give its name and structure.

14.26 How many times is a secondary alcohol group oxidized in the citric acid cycle?

14.27 Water adds to a carbon–carbon double bond how many times in one turn of the citric acid cycle?

*14.28 The enzyme for the conversion of isocitrate to oxalosuccinate (Fig. 14.2) is stimulated by one of these two substances, ATP or ADP. Which one is the more likely activator? Explain.

14.29 What is the maximum number of ATP molecules that can be made from the use of respiratory chain phosphorylation to break down each of the following?
(a) pyruvate
(b) an acetyl group in acetyl CoA

*14.30 Glutamic acid, one of the amino acids, can be converted to α-ketoglutarate (Fig. 14.2). How many ATP molecules can be made from the entry of α-ketoglutarate into the citric acid cycle?

Respiratory Chain

14.31 Is the following (unbalanced) change a reduction or an oxidation? How can you tell?

$$CH_3CH(OH)CH_2CO_2^- \longrightarrow CH_3C(=O)CH_2CO_2^-$$

14.32 Which kind of enzyme would be more likely to cause the change given in Review Exercise 14.31, an enzyme like aconitase or an enzyme like isocitrate dehydrogenase? Explain.

14.33 What is missing in the following basic expression for what must happen in the respiratory chain?

$$\tfrac{1}{2}O_2 + 2H^+ \longrightarrow H_2O$$

14.34 What general name is given to the set of enzymes involved in electron transport?

*14.35 Write the following display in the normal form of a chemical equation.

$$CH_3CH(OH)CO_2^- \rightarrow CH_3C(=O)CO_2^- \qquad NAD^+ \rightarrow NAD{:}H + H^+$$

(a) Which specific species is oxidized? (Write its structure.)
(b) Which species is reduced?

*14.36 Write the following equation in the form of a display like that shown in Review Question 14.35.

$$^-O_2CCH_2CH_2CO_2^- + FAD \longrightarrow {}^-O_2CCH{=}CHCO_2^- + FADH_2$$

14.37 Arrange the following in the order in which they receive and pass on electrons.
FMN NAD$^+$ CoQ FeS–P

14.38 What does respiratory enzyme complex IV do?

14.39 What is FAD, and where is it involved in the respiratory chain?

14.40 Across which cellular membrane does the respiratory chain establish a gradient of H^+ ions? On which side of this membrane is the value of the pH lower?

14.41 According to the chemiosmotic theory, the flow of what particles most directly leads to the synthesis of ATP?

14.42 If the inner mitochondrial membrane is broken, the respiratory chain can still operate, but the phosphorylation of ADP that normally results stops. Explain this in general terms.

14.43 Complete and balance the following equation. (Use 1/2 as the coefficient of oxygen as shown.)

$$MH_2 + \underset{\text{From the matrix}}{H^+} + \tfrac{1}{2}O_2 \longrightarrow$$

14.44 What is the difference between substrate and oxidative phosphorylation?

14.45 Besides a gradient of H^+ ions, what other gradient exists in mitochondria as a result of the operation of the respiratory chain? In terms of helping to explain how cations other than H^+ move across a membrane, of what significance is this gradient?

14.46 Briefly describe the theory presented in this chapter that explains how a flow of protons across the inner mitochondrial membrane initiates the synthesis of ATP from ADP and P_i.

Ester Condensations (Special Topic 25.1)

14.47 Which would be more acidic, acetone or propane? Explain.

14.48 Suppose that ethyl propanoate went through the steps of the Claisen condensation. Write the structure of the final product.

Additional Exercises

*14.49 The conversion of pyruvate to an acetyl unit is both an oxidation and a decarboxylation.

(a) If *only* decarboxylation occurred, what would form from pyruvate? Write the structure of the other product in the following.

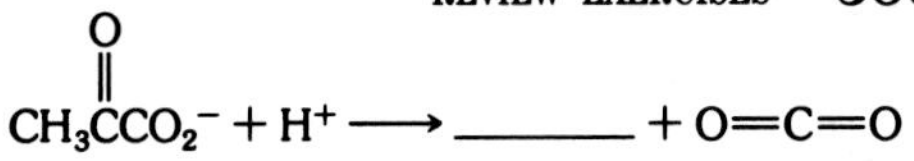

(b) If this product is oxidized, what is the name and the structure of the product of such oxidation?

(c) Referring to Figure 14.2, which specific compound undergoes an oxidative decarboxylation similar to that of pyruvate? (Give its name.)

*14.50 One cofactor in the enzyme assembly that catalyzes the oxidative decarboxylation of pyruvate requires thiamine, one of the B vitamins. Therefore in beriberi, the deficiency disease for this vitamin, the level of what substance can be expected to rise in blood serum (and for which an analysis can be made as part of the diagnosis of beriberi)?

*14.51 Consider the oxidation of one acetyl group via the citric acid cycle to two molecules of carbon dioxide.

(a) This overall change requires that how many *pairs* of electrons be transferred to the enzymes of the respiratory chain?

(b) In the transfer of these pairs of electrons, what else transfers?

(c) What is a word common to the names of the enzymes that catalyze these electron transfers?

METABOLISM OF CARBOHYDRATES

15

No doubt the act of mouth-watering has a molecular basis, but all we'll attempt to understand in this chapter is the metabolism of sugar and other carbohydrates.

15.1
GLYCOGEN METABOLISM
15.2
GLUCOSE TOLERANCE
15.3
THE CATABOLISM OF GLUCOSE
15.4
GLUCONEOGENESIS

15.1

GLYCOGEN METABOLISM

Much of the body's control over the blood sugar level is handled by its regulation of the synthesis and breakdown of glycogen.

The digestion of the starch and disaccharides in the diet gives glucose, fructose, and galactose, but their catabolic pathways all converge very quickly to that of glucose itself. Galactose, for example, is changed by a few steps in the liver to glucose-1-phosphate. Fructose is changed to a compound that occurs early in glycolysis. For these reasons, this chapter concentrates on the metabolism of glucose, but we'll begin with information about the glucose in circulation. The concentration of circulating glucose, as we soon will learn, is affected by several factors and demands involved in its distribution, storage, and use. We will go into details about the catabolism of glucose both by glycolysis and by the pentose phosphate pathway. Finally, we'll learn how glucose is synthesized in the body from smaller molecules.

■ mg/dL = milligrams per deciliter, where 1 dL = 100 mL.

■ *-glyc-*, sugar
-emia, in blood
hypo-, under, below
hyper-, above, over
renal, of the kidneys
-uria, in urine

A Special Vocabulary Exists to Describe Variations in the *Blood Sugar Level* The concentration of monosaccharides in whole blood, expressed in milligrams per deciliter (mg/dL), is called the **blood sugar level.** This is very nearly the same as the glucose level, because glucose is overwhelmingly the major monosaccharide. When determined after several hours of fasting, the plasma sugar level, called the **normal fasting level,** is 70 to 110 mg/dL (3.9 to 6.1 mmol/L).[1] In a condition of **hypoglycemia,** the plasma sugar level is *below* normal, and in **hyperglycemia** it is *above* the normal fasting level. (Sometimes you will see the term *normoglycemia* used for levels in the normal range.)

When the blood sugar level becomes too hyperglycemic (becomes too high), the kidneys are unable to put back into the blood all of the glucose that leaves them. Glucose then appears in the urine, a condition called **glucosuria.** The blood sugar level above which this happens is called the **renal threshold** for glucose, and it is in the range of roughly 160 to 180 mg/dL, and higher in some individuals.

Excess Blood Glucose Normally Is Withdrawn from Circulation When there is more than enough glucose in circulation to meet energy needs, the body does not eliminate the excess glucose but conserves its chemical energy. There are two ways to do this. One is to convert glucose to fat, and we'll study how this is done in the next chapter. The other is to synthesize glycogen, which we'll discuss here.

■ Glycogen (page 217), recall, is a polymer of glucose with both $\alpha(1\rightarrow4)$ and $\alpha(1\rightarrow6)$ bridges and resembling amylopectin, a component of starch.

Liver and muscle cells can convert glucose to glycogen by a series of steps called **glycogenesis** ("glycogen creation"). The liver holds 70 to 110 g of glycogen, and the muscles, taken as a whole, contain 170 to 250 g. When muscles need glucose, they take it back out of glycogen. When the blood needs glucose because the blood sugar level has dropped too much, the liver hydrolyzes as much of its glycogen reserves as needed and then puts the glucose into circulation. The overall series of reactions in either tissue that hydrolyzes glycogen is called **glycogenolysis** (lysis or hydrolysis of glycogen), a process controlled by several hormones.

Epinephrine Launches a Multiple-Enzyme Glycogenolysis Cascade When muscular work is begun, the adrenal medulla secretes the hormone **epinephrine.** Epinephrine (as

[1] The normal reference laboratory values or "normals" in this text are those used by the Massachusetts General Hospital and published in *The New England Journal of Medicine,* **327,** page 718 (Sept. 3, 1992).

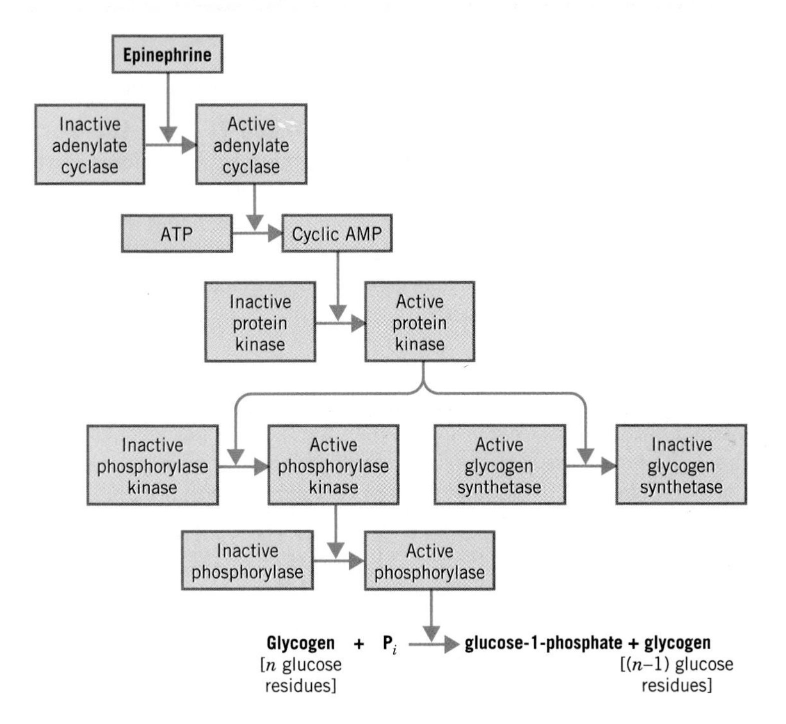

FIGURE 15.1
The epinephrine "cascade"

well as a close relative, norepinephrine) activates glycogenolysis by the steps outlined in Figure 15.1. At its target cells, epinephrine activates the enzyme *adenylate cyclase,* which catalyzes the conversion of some ATP to cyclic AMP as we studied in Section 11.6.

Cyclic AMP then activates another enzyme, and this still another enzyme, and so on in a cascade of events (see Fig. 15.1). Each molecule of epinephrine, by activating adenylate cyclase, triggers the formation of dozens of molecules of cyclic AMP. *Each* of these activates a **succeeding enzyme, and so on until the final enzyme appears not only suddenly but in a relatively large quantity. Thus one epinephrine molecule triggers the rapid mobilization of thousands of glucose units, and these are now ready to supply energy that the body needs.**

While the epinephrine "cascade" proceeds to release glucose from glycogen, the affected tissue simultaneously shuts down a team of enzymes called *glycogen synthetase* that otherwise would do the opposite, change glucose to glycogen. This is done at the step in Figure 15.1 where the enzyme called *active protein kinase* catalyzes the phosphorylation of *two* enzymes. **One is inactive phosphorylase kinase and the other is active glycogen synthetase. The first action continues the epinephrine cascade, until glucose units are obtained from glycogen. The second action, the phosphorylation of glycogen synthetase, *shuts down an enzyme that would help to remake glycogen*. The potential competition, therefore, cannot develop. It's a remarkable aspect of this system.**

The end product of glycogenolysis isn't actually glucose but glucose-1-phosphate. Cells that can do glycogenolysis also have an enzyme called *phosphoglucomutase,* which catalyzes the conversion of glucose-1-phosphate to its isomer, glucose-6-phosphate:

$$\text{Glucose-1-phosphate} \; (\text{C-1: } OPO_3^{2-}; \text{ C-6: } CH_2OH) \xrightleftharpoons{\text{phosphoglucomutase}} \text{Glucose-6-phosphate} \; (\text{C-1: } OH; \text{ C-6: } CH_2OPO_3^{2-})$$

Glucose-1-phosphate　　　　Glucose-6-phosphate

■ Epinephrine (adrenaline) and norepinephrine (noradrenaline)—see Special Topic 6.1, page 165—are called the "fight or flight" hormones because the brain causes their release in sudden emergencies.

■ An estimated 30,000 molecules of glucose are released from glycogen for each molecule of epinephrine that initiates glycogenolysis.

Glucose Is Trapped in the Muscle Cell When It Is in the Form of Glucose-6-Phosphate Glucose-6-phosphate, rather than glucose, is the form in which a glucose unit must be to enter a pathway that produces ATP. *It is also in a form that cannot migrate out of muscle cells.* It cannot be lost, therefore, from tissue needing it during exercise. Thus glycogenolysis in muscle tissue is an important supplier of energy. When the supply of muscle glycogen is low, muscle cells can take glucose from circulation, *trap it as glucose-6-phosphate,* and then convert this to glycogen.

■ The glucagon molecule has 29 amino acid residues.

Glucagon Activates Liver Glycogenolysis and Thus Affects the Blood Sugar Level The α-cells of the pancreas make a polypeptide hormone, **glucagon,** which helps to maintain a normal blood sugar level. When the blood sugar level drops, the α-cells release glucagon. Its target tissue is the liver, where it is a strong activator of glycogenolysis.

Glucagon works by a cascade process very similar to that initiated by epinephrine. Like epinephrine, glucagon activates adenylate cyclase. Unlike epinephrine, glucagon also inhibits glycolysis, so this action helps to keep the supply of glucose up. Glucagon, also unlike epinephrine, does not cause an increase in blood pressure or pulse rate, and it is longer acting than epinephrine.

Liver Cells Can Release Glucose to Circulation Glucose units released from liver glycogen as glucose-6-phosphate are converted to glucose by the enzyme *glucose-6-phosphatase* that the liver has but is not present in the muscles. This enzyme catalyzes the hydrolysis of glucose-6-phosphate to glucose and inorganic phosphate.

■ The letter P is often used to represent the whole phosphate group in the structures of phosphate ester intermediates in metabolism.

$$\text{Glucose-6-P} + H_2O \xrightarrow{\text{glucose-6-phosphatase}} \text{glucose} + P_i$$

Glucose can now leave the liver and so help raise the blood sugar level. During periods of fasting, therefore, the overall process in the liver from glucose-1-phosphate to glucose-6-phosphate to glucose is a major supplier of glucose for the blood. Glucagon, which triggers this, is thus an important regulator of the blood sugar level. The brain depends on the liver during fasting to maintain its favorite source of chemical energy, circulating glucose. (We are beginning to see in chemical terms how vital the liver is to the performance of other organs.) When circulating glucose is taken up by a brain cell, it is promptly trapped in the cell by being converted to glucose-6-phosphate.

Several hereditary diseases involve the glucose–glycogen interconversion, and some are discussed in Special Topic 15.1.

Human Growth Hormone Stimulates the Release of Glucagon Growth requires energy, so the action of glucagon that helps to supply a source of energy—glucose—aids in the work of the human growth hormone. In some situations, such as a disfiguring condition known as acromegaly, there is an excessive secretion of human growth hormone that promotes too high a level of glucose in the blood. This is undesirable because a prolonged state of hyperglycemia from any cause can lead to some of the same blood-capillary related complications observed when diabetes is poorly controlled.

■ Acromegaly is sometimes called *giantism* because the bone structures of victims are enlarged.

Insulin Strongly Lowers the Blood Sugar Level The β-cells of the pancreas make and release **insulin,** a polypeptide hormone. Its release is stimulated by an increase in the blood sugar level, such as normally occurs after a carbohydrate-rich meal. As insulin moves into action, it finds its receptors at the cell membranes of muscle and adipose tissue. The insulin–receptor complexes make it possible for glucose molecules to move easily into the affected cells, which lowers the blood sugar level.

■ Adipose tissue is the fatty tissue that surrounds internal organs.

Not all cells depend on insulin to take up glucose. Brain cells, red blood cells, and cells in the kidneys, the intestinal tract, and the lenses of the eyes take up glucose directly.

In one form of diabetes mellitus, type I diabetes, the β-cells have been destroyed, and the pancreas is unable to release insulin. Such individuals must receive insulin, usually by intravenous injection. If more insulin is put into circulation than needed for the management of the blood sugar level, this level falls too low and *insulin shock* results.

■ Life-saving first aid for someone in insulin shock is sugared fruit juice or candy to counter the hypoglycemia.

SPECIAL TOPIC 15.1
GLYCOGEN STORAGE DISEASES

A number of inherited diseases involve the storage of glycogen. For example, in **Von Gierke's disease,** the liver lacks the enzyme glucose-6-phosphatase, which catalyzes the hydrolysis of glucose-6-phosphate. Unless this hydrolysis occurs, glucose units cannot leave the liver. They remain as glycogen in such quantities that the liver becomes very large. At the same time, the blood sugar level falls, the catabolism of glucose accelerates, and the liver releases more and more pyruvate and lactate.

In **Cori's disease,** the liver lacks an enzyme needed to catalyze the hydrolysis of 1,6-glycosidic bonds, the bonds that give rise to the many branches of a glycogen molecule. Without this enzyme, only a partial utilization of the glucose in glycogen is possible. The clinical symptoms resemble those of Von Gierke's disease, but they are less severe.

In **McArdle's disease,** the muscles lack phosphorylase, the enzyme needed to obtain glucose-1-phosphate from glycogen (see Fig. 15.6). Although the individual is not capable of much physical activity, physical development is otherwise relatively normal.

In **Andersen's disease,** both the liver and the spleen lack the enzyme for putting together the branches in glycogen. Liver failure from cirrhosis usually causes death by age 2.

Hypoglycemia Can Make You Faint Your brain relies almost entirely on glucose for its chemical energy, so if hypoglycemia develops rapidly, you can become dizzy and may even faint. The brain consumes about 120 g/day of glucose, and a quick onset of hypoglycemia starves the brain cells. They do have the ability to switch over to other nutrients, but brain cells can't do this very rapidly.

Somatostatin Inhibits Glucagon and Slows the Release of Insulin The hypothalamus, a specific region in the brain, makes **somatostatin,** another hormone that participates in the regulation of the blood sugar level. When the β-cells of the pancreas secrete insulin, which helps to *lower* the blood sugar level, the α-cells should not at the same time release glucagon, which helps to *raise* this level. Somatostatin acts at the pancreas to inhibit the release of glucagon as well as to slow down the release of insulin. It thus helps to prevent a wild swing in the blood sugar level that insulin alone might cause.

Persistent Hyperglycemia Indicates Diabetes Whenever hyperglycemia develops and tends to persist, something is wrong with the mechanisms for withdrawing glucose from circulation. In an individual with diabetes under poor control and so with sustained hyperglycemia, some blood glucose combines with hemoglobin to give glycohemoglobin (or glycosylated hemoglobin). The measurement of the level of this substance has become the best way to determine the average blood sugar level of an individual, better even than direct measurements of blood glucose. The level of glycohemoglobin doesn't fluctuate as widely and quickly as the blood sugar level. Let's now examine in more detail the body's ability to manage a somewhat steady concentration of blood glucose.

■ Diabetes is a common cause of hyperglycemia, but there are other possible causes.

15.2
GLUCOSE TOLERANCE

The ability of the body to tolerate swings in the blood sugar level is essential to health.

Your **glucose tolerance** is the ability of your body to manage its blood sugar level within the normal range. We'll take an overview here of the many factors that contribute to glucose tolerance.

■ Carl Cori and Gerti Cori shared the 1947 Nobel prize in physiology and medicine.

The Cori Cycle Describes the Distributions and Uses of Glucose The strategies used by the body to maintain its blood sugar level within the normal range form a cycle of events called the **Cori cycle,** outlined in Figure 15.2.

At the bottom of the figure we see glucose as it enters the bloodstream from the intestinal tract. Its molecules either stay in circulation or are soon removed by various tissues. Two are shown in the figure, those of muscle and liver. Muscle cells can trap glucose molecules and use them either to make ATP by glycolysis or to replenish the muscle's glycogen reserves. Liver cells can similarly trap glucose, and the liver is able to release glucose back into the bloodstream when the blood sugar level must be raised.

When glucose is used in glycolysis, the end product is either pyruvate or lactate, depending on the oxygen supply, as we learned in the previous chapter. Either pyruvate or lactate can be used to make more ATP by means of the citric acid cycle and the respiratory chain. However, when extensive anaerobic glycolysis is carried out in a tissue, then the lactate level increases considerably.

■ *Neo,* new; *-neogenesis,* new creation:, *gluconeogenesis,* the synthesis of new glucose.

Because lactate still has C—H bonds, it continues to have useful chemical energy, so instead of simply excreting excess lactate at the kidneys, the body recycles it. It converts a fraction of it—about five-sixths—to glucose. This synthesis of glucose from smaller molecules is called **gluconeogenesis,** and it requires ATP energy. The remaining one-sixth of the lactate is catabolized to make the ATP needed for gluconeogenesis. (We go into more details about gluconeogenesis later in this chapter.) The recycling of lactate completes the Cori cycle.

We can see from all these processes that many factors affect the blood sugar level. Some tend to increase it and some do the opposite. Figure 15.3 summarizes them in a different kind of display.

Glucose Tolerance Can Be Measured by the Glucose Tolerance Test In the **glucose tolerance test,** the individual is given a drink that contains glucose, generally 75 g for an adult and 1.75 g per kilogram of body weight for children, and then the blood sugar level is checked at regular intervals.

Figure 15.4 gives typical plots of this level versus time. The lower curve is that of a person with normal glucose tolerance, and the upper curve is one whose glucose tolerance is typical of a person with diabetes. In both, the blood sugar level initially increases sharply. The healthy person, however, soon manages the high level and brings it back down with the help of a

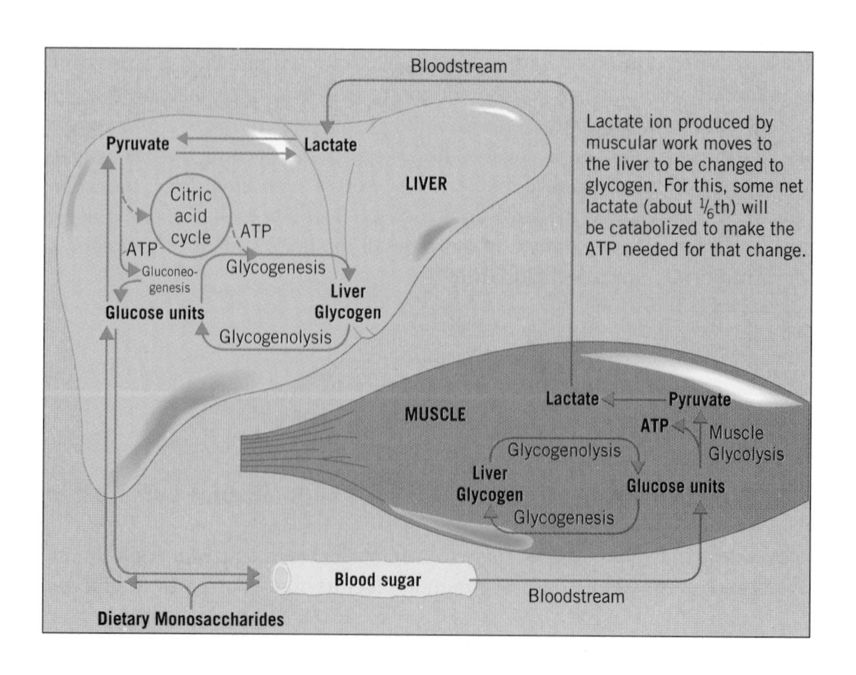

FIGURE 15.2
The Cori cycle

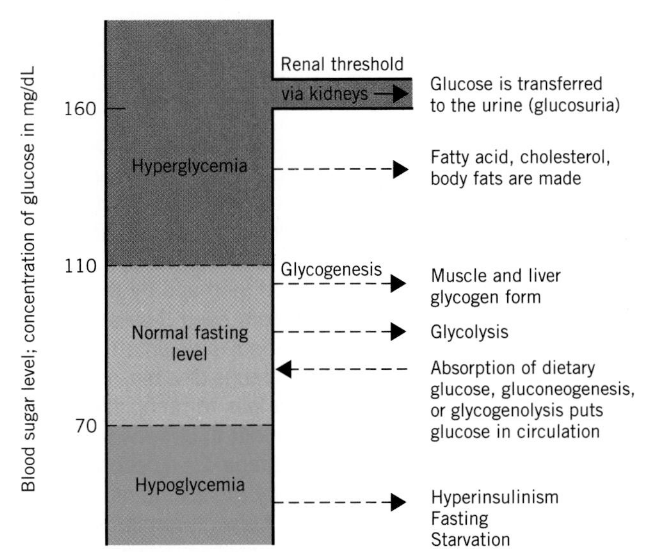

FIGURE 15.3
Factors that affect the blood sugar level

normal flow of insulin and somatostatin. In the diabetic, the level comes down only very slowly and remains essentially in the hyperglycemic range throughout.

Notice that in the normal individual, the blood sugar level can sometimes drop to a mildly hypoglycemic level. Such hypoglycemia is possible also in someone who has eaten a carbohydrate-rich breakfast. With glucose pouring into the bloodstream, the release of a bit more insulin than needed can occur. This leads to the overwithdrawal of glucose from circulation, and midmorning brings dizziness and sometimes even fainting (or falling asleep in class). The prevention isn't more sugared doughnuts but a balanced breakfast.

■ Unhappily, most people with midmorning sag go into another round of sugared coffee and sugared rolls. The glucose gives a short lift, but then an oversupply of insulin restores the mild hypoglycemia of the sag.

Glucose Tolerance Is Poor in Those with Diabetes The subject of glucose tolerance is nowhere of more concern than in connection with diabetes. **Diabetes** is defined clinically as a disease in which the blood sugar level persists in being much higher than warranted by the dietary and nutritional status of the individual. Invariably, a person with untreated diabetes has glucosuria, and the discovery of this condition often triggers the clinical investigations that are necessary to diagnose diabetes.

As discussed in Special Topic 15.2, there are two broad kinds of diabetes, type I and type II. Type I diabetics are unable to manufacture insulin (at least not enough), and they need daily insulin therapy to manage their blood sugar levels. Maintaining a relatively even and normal blood sugar level is the best single strategy that such diabetics have for the prevention of some of the vascular and neural problems that can complicate their health later. Most type II diabetics are able to manage their blood sugar levels by a good diet, weight control, exercise, and sometimes the use of oral medications.

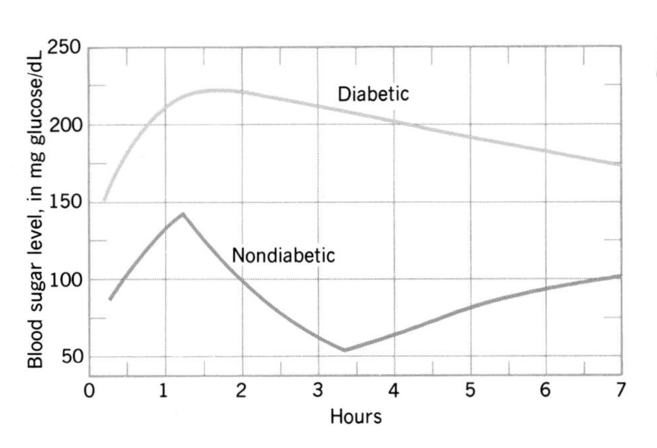

FIGURE 15.4
Glucose tolerance curves

SPECIAL TOPIC 15.2 DIABETES MELLITUS

The name for this disorder is from the Greek *diabetes,* to pass through a siphon, and *mellitus,* honey-sweet—meaning to pass urine that contains sugar. We'll call it diabetes for short. In severe, untreated diabetes, the victim's body wastes away despite efforts to satisfy a powerful thirst and hunger. To the ancients, it seemed as if the body were dissolving from within.

Between 1 and 6% of the United States population has diabetes, and almost as many others are believed to have this disease. It ranks third behind heart disease and cancer as a cause of death.

Between 10 and 25% of all cases of diabetes are of the severe, insulin-dependent variety in which the β-cells of the pancreas are unable to make and secrete insulin. This is **type I diabetes,** and insulin therapy is essential. It is also called **insulin-dependent diabetes mellitus** or **IDDM.** Most victims contract IDDM before the age of 40, often as adolescents, so IDDM has sometimes been called *juvenile-onset diabetes.*

The rest of all those with diabetes have a form called **type II diabetes,** or **non-insulin-dependent diabetes mellitus, NIDDM.** Most victims are able to manage their blood sugar levels by diet and exercise alone, without insulin injections. Their problem is not the lack of insulin but with a breakdown in the machinery for taking advantage of it at insulin's target cells. Most who contract NIDDM do so when they are over 40, so NIDDM has been called *adult-onset diabetes.*

Type I Diabetes Develops in Six Stages D. S. Eisenbarth, a diabetes specialist, divides the onset of type I diabetes into six stages. We'll review them as background for illustrations of equilibrium chemistry and factors that shift equilibria.

Stage 1 is thought to be an existing genetic defect most likely involving more than one gene. However, sets of identical twins are known in which only one twin becomes diabetic. In some way, the genetic problem must be related to a problem of the immune system.

Stage 2 is a triggering incident, like a viral infection. The mumps virus, for example, causes diabetes in some. Usually, the onset of virus-caused type I diabetes occurs slowly over a few years.

Stage 3 is the appearance in the blood of certain antibodies. (*Antibodies* are substances made by the immune system to counteract the effects of invading substances called *antigens* that are alien to the body.) Substances on the membranes of the pancreatic β-cells have been altered (perhaps by the virus) so that the body's immune system sees β-cells as foreign antigens and so makes antibodies against them. Type I diabetes is thus an autoimmune disease, *auto* because the body's immune system fails to recognize the proteins of its own body and sets out to destroy them and, therefore, itself.

Stage 4 is a period during which the pancreas loses its ability to secrete insulin. Stage 5 is diabetes and persistent hyperglycemia. Most of the pancreatic β-cells have disappeared. Stage 6 is the period following complete destruction of β-cells.

Immune-Suppressant Therapy Works if the Problem Is Caught in Its Early Stages Cyclosporine is an agent used to suppress the rejection of transplanted organs, like kidney transplants. When used in the early stages of the onset of type I diabetes, cyclosporine prevents insulin dependence in a significant fraction of individuals tested.

Insulin Receptors Are a Problem in Type II Diabetes The onset of type II diabetes is much slower than that for type I, and obesity, lack of exercise, and a sugar-rich diet are factors. Some scientists believe that sugar evokes such a continuous presence of insulin that the insulin-receptor proteins of target cells literally wear out faster than they can be replaced. When the weight is reduced, particularly in connection with physical exercise, the relative numbers of receptor proteins rebound.

Under the incessant demand to produce and secrete insulin, the β-cells can eventually give out. In a veritable epidemic of NIDDM, over 60% of the older adult population of the Pacific islet of Nauru has become diabetic following a change in lifestyle from active fishing and farming to a very sedentary life. (The discovery of huge phosphate reserves on Nauru Island made the Nauruans one of the world's wealthiest peoples.)

The conventional wisdom that places insulin and its receptors at the heart of the NIDDM problem has been recently challenged by the discovery that the β-cells secrete not only insulin but also another protein called *amylin.* Is it the relatively high level of amylin that suppresses the uptake of glucose by target cells? If so, would an amylin control drug be the answer to NIDDM? Needless to say, much research is in progress.

Glucosylation of Proteins May Cause the Long-Term Complications of Diabetes The immediate complications of IDDM are an elevated blood sugar level, metabolic acidosis, and eventual death from coma and uremic poisoning. Insulin therapy corrects these immediate problems, but it deals less well with the longer-term complications.

The continuous presence of a high level of blood glucose in both IDDM and NIDDM shifts certain chemical equilibria in favor of glucosylated compounds. The aldehyde group of the open form of glucose, for example, can react with amino groups, like those on sidechains of lysine residues, to form products called *Schiff bases.*

$$\underset{\text{Aldehyde group}}{-CH{=}O} + \underset{\text{Amino group}}{H_2N-} \rightleftharpoons \underset{\text{Schiff base system}}{-CH{=}N-} + H_2O$$

Hemoglobin, for example, gives this reaction, and a high level of glucose shifts this equilibrium to the right. The level of glucosylated hemoglobin thus increases. When the glucose level is brought down and kept within a normal range, the Schiff base level also declines.

The problem with the Schiff bases in the long term is that they undergo molecular rearrangements to more permanent products, called *Amadori compounds,* in which the C=N double bond has migrated to C=C positions. *After a time, the formation of the Amadori compounds is not reversible.*

When these reactions occur in the basement membrane of blood capillaries, they swell and thicken; the condition is called *microangiopathy.* (The basement membrane is the protein support structure that encases the single layer of cells of a capillary.) Microangiopathy is believed to lead to the other complications, most of which involve the vascular system or the neural networks: kidney problems, gangrene of the lower limbs, and blindness. Diabetes is the leading cause of new cases of blindness in the United States, and it is the second most common cause of blindness, overall.

Blindness from Diabetes May Also Reflect the Reduction of Glucose to Sorbitol Glucose is reduced by the enzyme *aldose reductase* to sorbitol. It's a minor reaction in cells of the lens of the eye, but *an abundance of glucose shifts equilibria in favor of too much sorbitol.* Sorbitol, unlike glucose, tends to be trapped in lens cells, and as the sorbitol concentration rises so does the osmotic pressure in the fluid. *This draws water into the lens cells, which generates pressure and leads to cataracts.*

Diet Control Is Mandatory The best single treatment of diabetes is any effort that keeps a strict control on the blood sugar level to avoid the episodes of upward surges followed by precipitous declines.

A Number of Technologies Are in Use or Are Being Tested People with IDDM gauge their insulin needs by blood tests for blood sugar levels. We mentioned an enzyme-based test in Section 11.5, but it requires a puncture to obtain a drop of blood. Another technology in a testing stage (Futrex Inc., Maryland) uses a hand-held, battery-driven source of infrared rays, which are focused onto the skin at the wrist or a fingertip. The meter converts the amount of light absorbed to a blood glucose level. With some diabetics, insulin pumps can be implanted much like heart pacemakers. These monitor the blood sugar level and release insulin according to the need. The use of insulin nasal sprays immediately before a meal is another approach being tested. Several groups of scientists are working on insulin pills, a technology made difficult by the fact that insulin, like any protein, is digested.

When stripped of neighboring cells, β-cells can be transplanted. They need not even be inserted into the receiver's pancreas, and they start to make insulin in a few weeks. Apparently the cells *adjacent* to the β-cells are responsible for inducing the body's immune-centered rejection process, so without such cells the β-cells are more safely transplanted.

Human β-cells work best, of course, but those from pigs and cows also appear to be usable provided that they are encapsulated in very small plastic spheres. These spheres have microscopic holes large enough to let insulin molecules escape but not large enough to let antibodies inside. This technique has cured type I diabetes in experimental animals. A test of this technique in a human was begun in 1993, and early indications were very promising.

15.3

THE CATABOLISM OF GLUCOSE

Glycolysis and the pentose phosphate sequence are the chief catabolic pathways open to glucose.

■ Greek, *glykos*, sugar or sweet; *-lysis*, dissolution.

Glycolysis, as we noted in the previous chapter, is a series of reactions that change glucose to pyruvate or to lactate while a small but important amount of ATP is made. Other monosaccharides eventually enter the same glycolysis pathway as glucose, as seen in Figure 15.5, so when we study glycolysis we cover most of monosaccharide catabolism.

Anaerobic Glycolysis Ends in Lactate When a cell receives oxygen at a rate slower than needed, glycolysis can still operate, but it ends in lactate, not pyruvate. The overall equation for this *anaerobic glycolysis,* or the **anaerobic sequence,** is

$$\underset{\text{Glucose}}{C_6H_{12}O_6} + 2ADP + 2P_i \xrightarrow[\text{glycolysis}]{\text{anaerobic}} \underset{\text{Lactate}}{2CH_3CH(OH)CO_2^-} + 2H^+ + 2ATP$$

Except during extensive exercise, glycolysis is operated with sufficient oxygen, and it is aerobic. Its overall equation is

$$\underset{\text{Glucose}}{C_6H_{12}O_6} + 2ADP + 2P_i + 2NAD^+ \longrightarrow \underset{\text{Pyruvate}}{2CH_3C(=O)CO_2^-} + 2ATP + 2NADH + 2H^+ + 2H_2O$$

The NADH produced by aerobic glycolysis is involved with the respiratory chain, so more ATP is made by using its $H:^-$ as NAD^+ is regenerated.

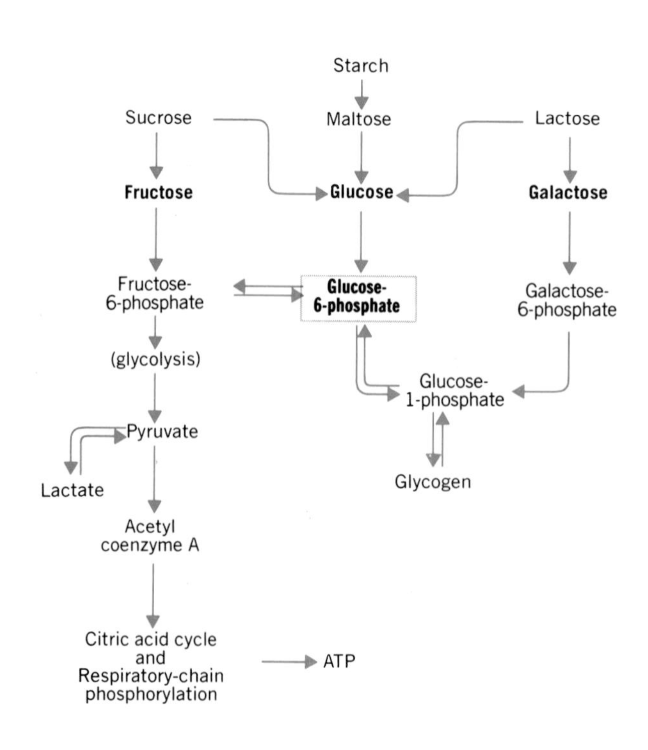

FIGURE 15.5
Convergence of the pathways in the metabolism of dietary carbohydrates

Glycolysis Begins with ATP Consumption but Then Generates More Figure 15.6 outlines the steps to pyruvate (or lactate) that can begin with either glucose or glycogen. The steps that lead to fructose-1,6-bisphosphate are actually up an energy hill, because they consume ATP. But this is like pushing a sled or bike up the short backside of a long hill, because the investment in energy is more than repaid by the long, downhill slide to lactate and more ATP. When glycogen is the source of glucose for glycolysis, the initial investment in ATP is slightly smaller.

■ The numbered steps in the discussion of glycolysis, below, refer to Figure 15.6.

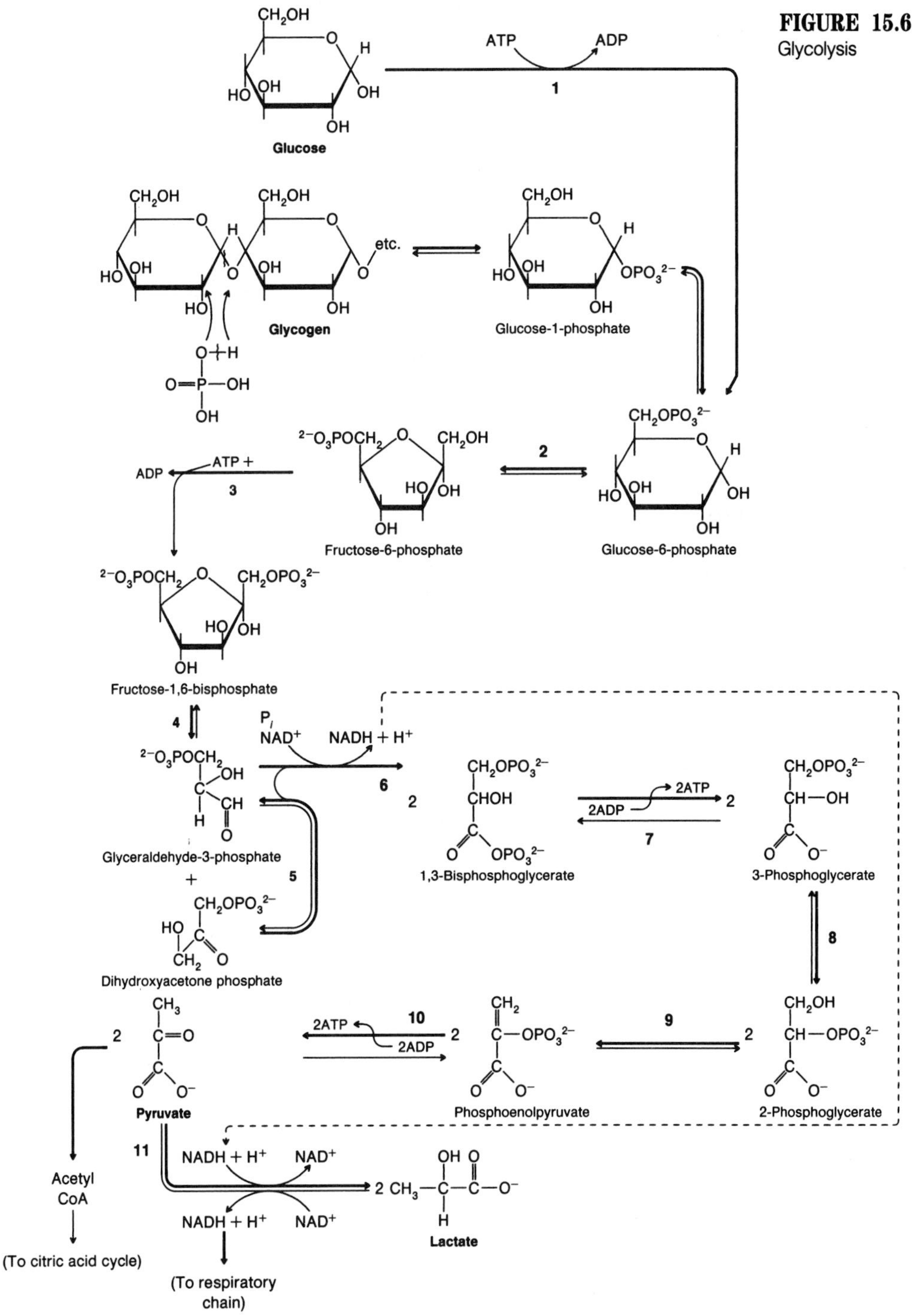

FIGURE 15.6
Glycolysis

A *kinase* is an enzyme that handles transfers of phosphate units between ATP and some other substrate.

1. Glucose is phosphorylated by ATP under catalysis by *hexokinase* to give glucose-6-phosphate.
2. Glucose-6-phosphate changes to its isomer, fructose-6-phosphate. The enzyme is *phosphoglucose isomerase*. This may seem to be a major structural change, but it involves little more than some shifts of bonds and hydrogens, as the arrows in the following sequence show:

These equilibria shift constantly to the right as long as later reactions continuously remove products as they form.

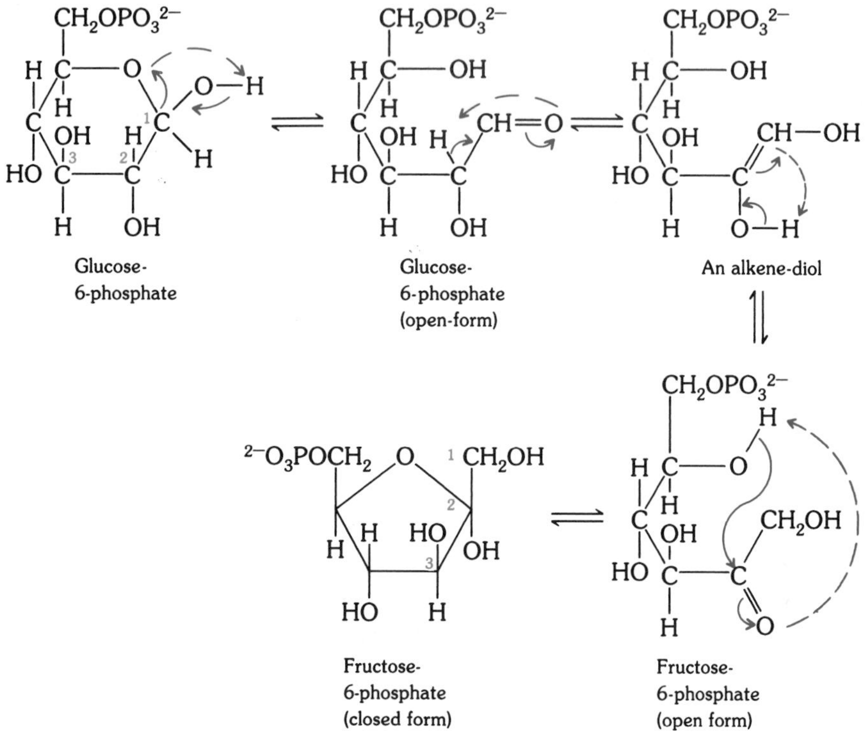

Glucose-6-phosphate ⇌ Glucose-6-phosphate (open-form) ⇌ An alkene-diol ⇌ Fructose-6-phosphate (open form) ⇌ Fructose-6-phosphate (closed form)

3. ATP phosphorylates fructose-6-phosphate to make fructose-1,6-bisphosphate. The enzyme is *phosphofructokinase*. This step is essentially irreversible, and it ends the energy-using phase of glycolysis.

All kinases require Mg^{2+} as a cofactor.

4. Fructose-1,6-bisphosphate breaks apart into two triose monophosphates. The reaction is catalyzed by *aldolase*. *This is actually a reverse aldol condensation*, a reaction whose we postponed from Chapter 4. The regular aldol condensation is discussed in Special 15.3. If you study this Special Topic now, it will be easier to see how we can visualize ste glycolysis, as follows, in terms of a few simple and reasonable shifts of electrons and prc

Sidechain groups on certain proton-donating or proton-accepting amino acid residues of the enzyme participate in shuttling protons around.

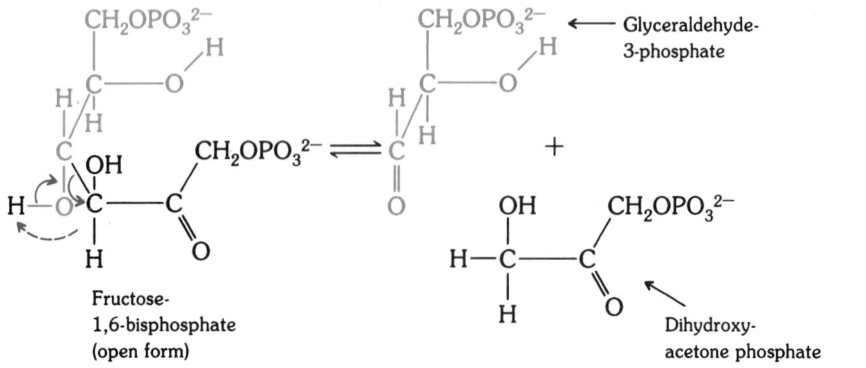

Fructose-1,6-bisphosphate (open form) ⇌ Glyceraldehyde-3-phosphate + Dihydroxyacetone phosphate

SPECIAL TOPIC 15.3
THE ALDOL CONDENSATION AND REVERSE ALDOL CONDENSATION

Reactions that make new carbon–carbon bonds or break them have special places in both organic chemistry and biological chemistry. In this Special Topic we will examine the *aldol condensation,* one that joins aldehydes or ketones together and that can be reversed. The term *aldol* signifies that the product has both an aldehyde group and an alcohol group. Step 4 of glycolysis, catalyzed by aldolase, is a reverse of this reaction. An example of the forward running of an aldol condensation occurs in gluconeogenesis. Let's see how the aldol condensation occurs in the molecule-building or forward direction. Each step is reversible, however, so we'll write each step as an equilibrium. We'll also show the reaction using a aldehyde, but the same principles apply to the condensation of ketones or to a mixed reaction involving an aldehyde and a ketone.

The reaction whose mechanism we will examine is

$$\underset{\text{(Two molecules of the same aldehyde)}}{RCH_2\overset{O}{\overset{\|}{C}}H + H{-}\overset{R}{\overset{|}{C}}H\overset{O}{\overset{\|}{C}}H} \underset{}{\overset{\text{aldol condensation}}{\rightleftharpoons}} \underset{\text{An aldol (a }\beta\text{-hydroxy aldehyde)}}{RCH_2\overset{OH}{\overset{|}{C}}H{-}\overset{R}{\overset{|}{C}}H\overset{O}{\overset{\|}{C}}H}$$

Notice that the reaction is simply the addition of one molecule of the aldehyde (the second) to another molecule (the first). *It always involves the H atom of an alpha position of the adding molecule, never a hydrogen attached at any other position on the chain.* By the same token, *the reverse aldol condensation requires that the OH group be beta to the carbonyl group.*

Step 1 Proton transfer from the aldehyde to a proton acceptor, B:⁻. (*In vivo,* sidechains of amino acid residues, like the NH_2 group of lysine, serve as carriers and transfer agents in proton shuttles.)

$$B{:}^- + \underset{\text{(Aldehyde with one }\alpha\text{-hydrogen)}}{H{-}\overset{R}{\overset{|}{C}}H\overset{O}{\overset{\|}{C}}H} \overset{\text{proton transfer}}{\rightleftharpoons} {}^-{:}\overset{R}{\overset{|}{C}}H\overset{O}{\overset{\|}{C}}H + B{:}H$$

We see here why only a hydrogen attached to the *alpha* position of the aldehyde can be involved. It's the only H atom on the chain with any acidity whatsoever, and it is only *very* weakly acidic. That it has any acidity at all is caused by the nearby carbonyl group, an electronegative group that can attract electron density in the anion of the aldehyde and so help to stabilize it.

Step 2 The anion of the aldehyde is attracted to the $\delta+$ charge on the permanently polarized carbonyl carbon of the intact aldehyde molecule. The new carbon–carbon bond forms in this step. The product is the anion of the aldol.

$$RCH_2\overset{O}{\overset{\|}{C}}H + {}^-{:}\overset{R}{\overset{|}{C}}H\overset{O}{\overset{\|}{C}}H \rightleftharpoons RCH_2\overset{O^-}{\overset{|}{C}}H{-}\overset{R}{\overset{|}{C}}H\overset{O}{\overset{\|}{C}}H$$

Step 3 The anion of the aldol is the conjugate base of an alcohol and so is a very strong base. It recovers a proton in the last step.

$$B{:}H \quad RCH_2\overset{O^-}{\overset{|}{C}}H{-}\overset{R}{\overset{|}{C}}H\overset{O}{\overset{\|}{C}}H \overset{\text{proton transfer}}{\rightleftharpoons} \underset{\text{Aldol product}}{RCH_2\overset{OH}{\overset{|}{C}}H{-}\overset{R}{\overset{|}{C}}H\overset{O}{\overset{\|}{C}}H} + B{:}^-$$

Ketones are able to engage in the same kind of reaction. All that would change in the mechanism given here is that instead of H on the carbonyl carbon there would be another R group.

In the reverse of an aldol condensation, each step is run in reverse, in turn. Thus the aldol product would first transfer H from its OH group to an acceptor to give the anion of the aldol. Next, the carbon–carbon bond would *break* as the two smaller fragments form. *It is precisely this kind of reaction that occurs in step 4 of glycolysis.* If you find the HO group that is *beta* to the keto group of fructose-1,6-bisphosphate (position 4 of the original ring), you will see how the reverse aldol condensation breaks the bisphosphate into two half-size fragments.

In gluconeogenesis (Fig. 15.7), you can see that glyceraldehyde-3-phosphate and dihydroxyacetone phosphate are joined by an aldol condensation to give fructose-1,6-bisphosphate.

5. Dihydroxyacetone phosphate, in the presence of *triose phosphate isomerase,* changes to its isomer, glyceraldehyde-3-phosphate.

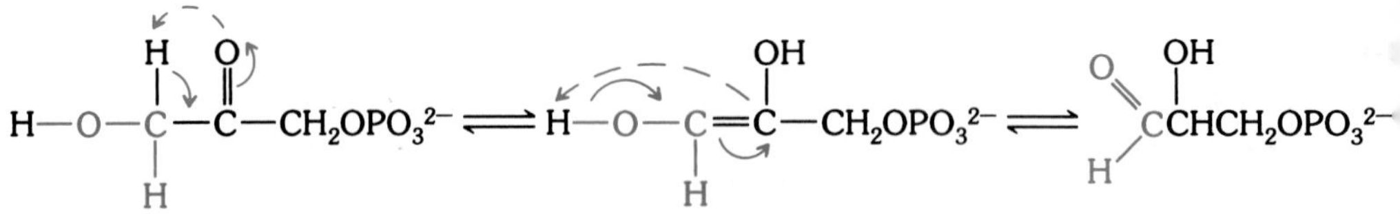

$$H-O-CH_2-C(=O)-CH_2OPO_3^{2-} \rightleftharpoons H-O-CH=C(OH)-CH_2OPO_3^{2-} \rightleftharpoons OHC-CH(OH)CH_2OPO_3^{2-}$$

Dihydroxy-acetone phosphate — An alkene-diol — Glyceraldehyde-3-phosphate

This change ensures that all the chemical energy in glucose will be obtained, because the main path of glycolysis continues with glyceraldehyde-3-phosphate. All dihydroxyacetone shuttles through glyceraldehyde-3-phosphate.

■ The continuous removal of glyceraldehyde-3-phosphate by its subsequent reaction shifts the dihydroxyacetone phosphate equilibrium to the right.

6. Glyceraldehyde-3-phosphate is simultaneously oxidized and phosphorylated. The enzyme, glyceraldehyde-3-phosphate dehydrogenase, has NAD^+ as a cofactor, and an SH group on the enzyme participates. Inorganic phosphate is the source of the new phosphate group. We can visualize how it happens as follows.

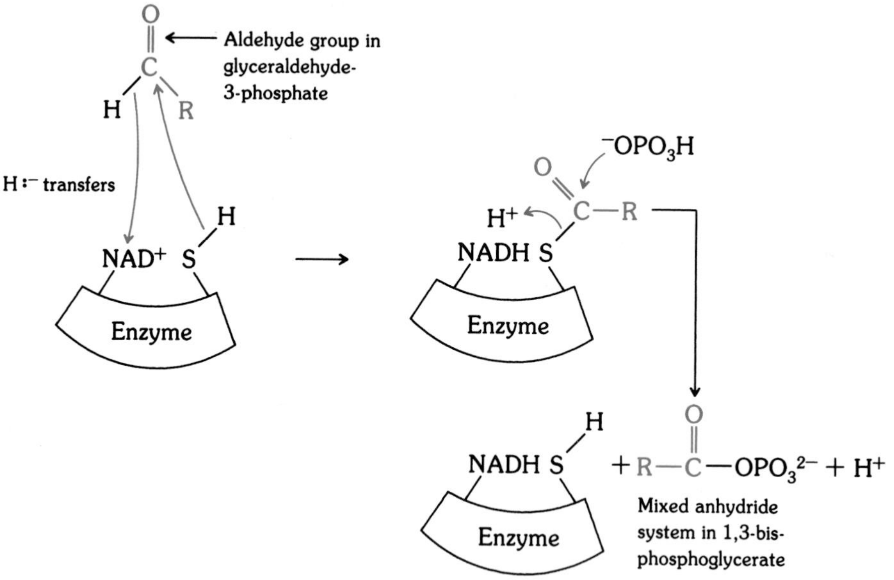

Aldehyde group in glyceraldehyde-3-phosphate

H:⁻ transfers

NAD^+ S—H / Enzyme $\longrightarrow$ H^+ NADH S—C(=O)—R ← $^-OPO_3H$ / Enzyme

NADH S—H / Enzyme $+ R-C(=O)-OPO_3^{2-} + H^+$

Mixed anhydride system in 1,3-bis-phosphoglycerate

7. 1,3-Bisphosphoglycerate has a higher phosphate group transfer potential than ATP. With the help of *phosphoglycerate kinase,* it transfers a phosphate to ADP, so this gives us back the original investment of ATP. (Remember that each glucose molecule with six carbons is processed through *two* three-carbon molecules, so *two* ATPs are made here for each original single glucose molecule.)

8. The phosphate group in 3-phosphoglycerate shifts to the 2-position, catalyzed by *phosphoglycerate mutase.*

9. The dehydration of 2-phosphoglycerate to phosphoenolpyruvate[2] is catalyzed by *enolase.* This step has been compared to cocking a huge bioenergetic gun. A simple,

[2] When a carbon atom of an alkene group holds an OH group, the compound is called an *enol* ("ene" + "-ol"). Enols are unstable alcohols that spontaneously rearrange into carbonyl compounds:

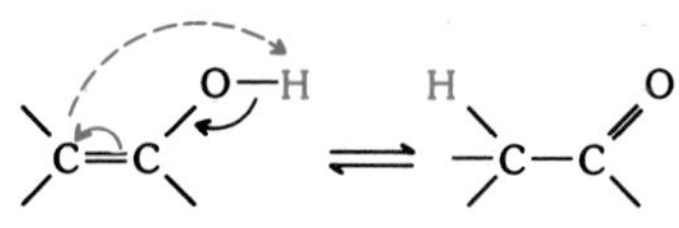

$$>C=C(-O-H)- \rightleftharpoons H-C(<)-C(=O)-$$

Enol form — Carbonyl form

low-energy reaction, dehydration of an alcohol group, converts a low-energy phosphate into the highest energy phosphate in all of metabolism, the phosphate ester of the enol form of pyruvate, phosphoenolpyruvate.

10. More ATP is made as the phosphate group in phosphoenolpyruvate transfers to ADP. The enzyme is *pyruvate kinase.* The enol form of pyruvate that is left behind promptly and irreversibly rearranges to the keto form of the pyruvate ion. The instability of enols (see footnote 2) is the driving force for this entire step.

11. If the mitochondrion is running aerobically when pyruvate is made, the pyruvate ion changes to acetyl coenzyme A, from which an acetyl group can enter the citric acid cycle (or be used in another way).

Anaerobic Glycolysis Provides a Way to Restore an Enzyme Vital to Continued Glycolysis in the Absence of Oxygen The enzyme at step 6 now has its coenzyme, NAD^+, in its reduced form, NADH. *As long as* $H:^-$ *is on an NADH unit, the enzyme lacks* NAD^+ *and so is "plugged." Further glycolysis, therefore, is blocked at step 6.* The block is continuously removed when the cell has sufficient oxygen because NADH simply gives its hydride unit to the respiratory chain, NAD^+ reappears, and glycolysis can run again. When the cell is deficient in oxygen, however, perhaps because of excessive work, glycolysis would shut down quickly without an alternative mechanism for changing NADH back to NAD^+. Exactly such an alternative is the conversion of pyruvate to lactate.

■ We studied the oxidative decarboxylation of pyruvate to acetyl coenzyme A, catalyzed by pyruvate dehydrogenase, in Section 14.2.

■ No reaction in the body is truly complete until its enzyme is fully restored.

When oxygen isn't available, the $H:^-$ in NADH is unloaded onto the keto group of pyruvate, which thereby changes to the 2° alcohol group in lactate. *This is why lactate is the end product of anaerobic glycolysis.* Lactate serves to store $H:^-$ made at step 6 until the cell once again becomes aerobic. The overall change from glucose to lactate by anaerobic glycolysis can be represented by the following equation. (Note the generation of acid.)

$$\text{Glucose} + 2\text{ADP} + 2\text{P}_i \longrightarrow 2\text{ lactate} + 2\text{ATP} + 2\text{H}^+$$

The importance of anaerobic glycolysis is that the cell can continue to make some ATP even when insufficient oxygen is available to run the respiratory chain. Of course, there are limits. The longer the cell operates anaerobically and the more that lactate accumulates, the more the cell runs an **oxygen debt.** Eventually the system has to slow down to let respiration bring back oxygen to metabolize lactate.

Excessive Exercise Causes Lactic Acid Acidosis By describing the production of lactate rather than lactic acid, we have obscured the generation of acid during glycolysis (but did note it in the equation, above). Extensive physical exercise that forces tissues to operate anaerobically can overtax the buffer. A form of metabolic acidosis called *lactic acid acidosis* is the result. When the pH of fluids in muscle tissue decreases, the muscles become tired and sore. The respiratory response to *metabolic acidosis* is hyperventilation, which blows out carbon dioxide and thus helps to remove acid from the body. During the cool down period following exercise, the body reestablishes the acid–base balance of the blood, and excess lactate ion is shuttled into the Cori cycle.

■ Hunters know that meat from animals run to exhaustion is sour.

The *Pentose Phosphate Pathway* Makes NADPH The biosyntheses of some substances in the body require a reducing agent. Fatty acids, for example, are almost entirely alkane-like, and alkanes are the most reduced types of organic compounds. The reducing agent used in the biosynthesis of fatty acids is NADPH, the reduced form of $NADP^+$.

■ $NADP^+$ is a phosphate derivative of NAD^+.

The body's principal route to NADPH is the **pentose phosphate pathway** of glucose catabolism. This complicated series of reactions (which we'll not study in detail) is very active in adipose tissue, where fatty acid synthesis occurs. Skeletal muscles have very little activity in this pathway.

The pentose phosphate pathway also goes by the names *hexose monophosphate shunt* and *phosphogluconate pathway.*

The oxidative reactions in the pentose phosphate pathway convert a hexose phosphate into a pentose phosphate, hence the name of the series. There are two broad sequences, one oxidative and the other nonoxidative. The overall equation for the oxidative sequence is

$$\text{Glucose-6-phosphate} + 2\text{NADP}^+ + H_2O \longrightarrow \text{ribose-5-phosphate} + 2\text{NADPH} + 2H^+ + CO_2$$

The ribose-5-phosphate can now be used to make the pentose systems in the nucleic acids, if they are needed. If not, ribose-5-phosphate undergoes a series of isomerizations and group transfers that make up the nonoxidative phase of the pentose phosphate pathway. These reactions have the net effect of converting three pentose units into two hexose units (glucose) and one triose unit (glyceraldehyde).

The hexose units can be catabolized by glycolysis, or they can be recycled into the oxidative series of the pentose phosphate pathway. Glyceraldehyde, as we'll soon see, can be converted to glucose and recycled as well. Thus glycolysis, the pentose phosphate reactions, and the resynthesis of glucose all interconnect, and specific bodily needs of the moment determine which pathway is operated.

Overall, with pentose recycled, the balanced equation for the complete oxidation of one glucose molecule via the pentose phosphate pathway is as follows.

$$6\ \text{Glucose-6-phosphate} + 12\text{NADP}^+ + 6H_2O \xrightarrow{\text{pentose phosphate pathway}} 5\ \text{glucose-6-phosphate} + 6CO_2 + 12\text{NADPH} + 12H^+ + P_i$$

15.4

GLUCONEOGENESIS

Some of the steps in gluconeogenesis are the reverse of steps in glycolysis. The overall scheme of gluconeogenesis, by which glucose is made from smaller molecules, is given in Figure 15.7. Excess lactate can be used as a starting material, as we have already mentioned. Just as important, several amino acids can be degraded to molecules that can be used to make glucose, too.

Most of our glucose needs are met by gluconeogenesis during periods of fasting.

You'll recall that the brain normally uses circulating glucose for energy. Therefore the ability of the body to manufacture glucose from noncarbohydrate sources such as amino acids—even those from the body itself—is an important backup during times when glucose either isn't in the diet (starvation) or cannot be effectively used (untreated diabetes).

Gluconeogenesis Is Not the Exact Reverse of Glycolysis There are three steps in glycolysis that cannot be directly reversed: steps 1, 3, and 10 of glycolysis (see again Figure 15.6). However, the liver and the kidneys have special enzymes that create bypasses.

The enzyme for this step requires the vitamin biotin.

In the bypass that gets back and around step 10 of glycolysis, the synthesis of phosphoenolpyruvate from pyruvate, carbon dioxide is used as a reactant; ATP energy is used to drive the steeply uphill reaction; and *pyruvate carboxylase* is the enzyme. Pyruvate is changed to oxaloacetate:

$$\underset{\text{Pyruvate}}{CH_3\overset{O}{\overset{\|}{C}}CO_2^-} + HCO_3^- + \text{ATP} \xrightarrow[\text{pyruvate carboxylase}]{\text{(in mitochondria)}} \underset{\text{Oxaloacetate}}{{}^-O_2CCH_2\overset{O}{\overset{\|}{C}}CO_2^-} + \text{ADP} + P_i$$

This reaction occurs in mitochondria, but the *use* of oxaloacetate in gluconeogenesis occurs in the cytosol. *By segregating anabolism reactions from their opposite, catabolism, the body is better able to control which pathways to operate.* After a complex shuttle mechanism gets

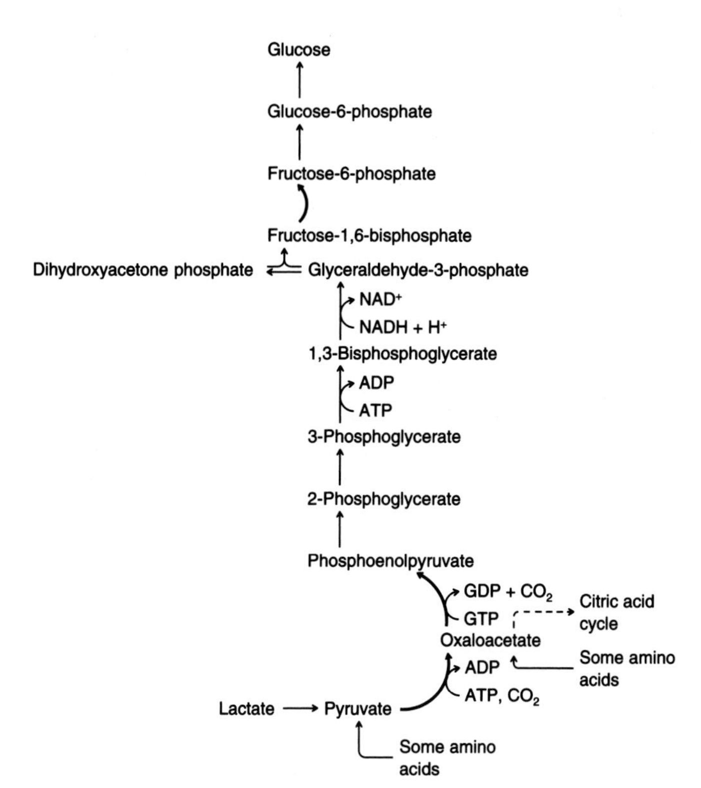

FIGURE 15.7
Gluconeogenesis. The straight arrows signify steps that are the reverse of corresponding steps in glycolysis. The heavy, curved arrows denote steps that are unique to gluconeogenesis.

oxaloacetate through the mitochondrial membranes, it reacts in the cytosol with another triphosphate, guanosine triphosphate (GTP), as carbon dioxide splits out and bonds rearrange:

$$^{-}\!:\!\ddot{O}-\overset{\overset{O}{\|}}{C}-CH_2-\overset{\overset{O}{\|}}{C}-CO_2^- + {}^{-}O-\underset{\underset{O^-}{|}}{\overset{\overset{O}{\|}}{P}}-O-\underset{\underset{O^-}{|}}{\overset{\overset{O}{\|}}{P}}-O-\underset{\underset{O^-}{|}}{\overset{\overset{O}{\|}}{P}}-O-\text{guanosine} \longrightarrow$$

Oxaloacetate

$$CO_2 + CH_2=\overset{\overset{O-PO_3^{2-}}{|}}{C}-CO_2^- + GDP$$

Phosphoenol-pyruvate

Because each glucose to be made requires two pyruvates, and because each pyruvate uses two high-energy phosphates in gluconeogenesis, this bypass costs the equivalent of four ATPs (2ATP + 2GTP) per glucose molecule to be made.

At the reversal of step 7 (see again Fig. 15.6), two more ATPs are used per molecule of glucose made. Thus a total of six ATPs are needed to make one glucose molecule by gluconeogenesis that starts from pyruvate. This may be compared with two ATPs that are produced by anaerobic glycolysis (that begins with glucose).

The bypasses to the reverses of steps 3 and 1 of glycolysis require only specific enzymes that catalyze the hydrolysis of phosphate ester groups, not high-energy boosters. Such enzymes are integral parts of the enzyme team for gluconeogenesis.

The other steps in gluconeogenesis are run as reverse shifts in equilibria that occur in the opposite direction in glycolysis.

SUMMARY

Glycogen metabolism The regulation of glycogenesis and glycogenolysis is a part of the machinery for glucose tolerance in the body. Hyperglycemia stimulates the secretion of insulin and somatostatin, and insulin helps cells of adipose tissue to take glucose from the blood. Somatostatin helps to suppress the release of glucagon (which otherwise stimulates glycogenolysis and leads to an increase in the blood sugar level).

When glucose is abundant, the body either replenishes its glycogen reserves or makes fat. In muscular work, epinephrine stimulates a cascade of enzyme activations that begins with the activation of adenylate cyclase and ends with the release of many glucose molecules from glycogen.

When glucose is in short supply, the body makes its own by gluconeogenesis from noncarbohydrate molecules, including several amino acids. In diabetes, some cells that are starved for glucose make their own, also. Such cells are unable to obtain glucose from circulation, so the blood sugar level is hyperglycemic to a glucosuric level. The glucose tolerance test is used to see how well the body handles an overload of glucose. In the management of diabetes, the maintenance of a reasonably steady blood sugar level in the normal range is vital. The measurement of the glycohemoglobin in circulation serves to monitor how well the normal range is kept over a long period of time.

Following strenuous exercise, when lactate is plentiful, the liver makes glucose from lactate. The many pathways that involve glycogen and glucose form a cycle of events called the Cori cycle.

Glycolysis Under anaerobic conditions, glucose can be catabolized to lactate ion. (Galactose and fructose enter this pathway, too.) When lactate is used to store $H:^-$, one of the enzymes in glycolysis can be regenerated in the absence of oxygen, and glycolysis can be run to make some ATP without the involvement of the respiratory chain. Then, when the cell is aerobic again, the $H:^-$ that this enzyme must shed to work again is given directly into the respiratory chain, and pyruvate instead of lactate becomes the end product of glycolysis.

Pentose phosphate pathway The body's need for NADPH to make fatty acids is met by catabolizing glucose through the pentose phosphate pathway.

Gluconeogenesis Most of the steps in gluconeogenesis are simply the reverse of steps in glycolysis, but there are a few that require rather elaborate bypasses. Special teams of enzymes and supplies of high-energy phosphates are used for these. Many amino acids can be used to make glucose by gluconeogenesis.

REVIEW EXERCISES

The answers to Review Exercises whose numbers are in color are found in Appendix A. The answers to the other Review Exercises are found in the Study Guide that accompanies this book. The more challenging questions are marked with asterisks.

Blood Sugar

15.1 What are the end products of the complete digestion of the carbohydrates in the diet?

15.2 Why can we treat the catabolism of carbohydrates as almost entirely that of glucose?

15.3 What is meant by *blood sugar level?* By *normal fasting level?*

15.4 What is the range of concentrations in mg/dL for the normal fasting level of whole blood?

***15.5** A level of 5.5 mmol/L for glucose in the blood corresponds to how many milligrams of glucose per deciliter?

15.6 What characterizes the following conditions?

(a) glucosuria (b) hypoglycemia
(c) hyperglycemia (d) glycogenolysis
(e) renal threshold (f) glycogenesis

15.7 What is gluconeogenesis? Which condition, hypoglycemia or hyperglycemia, would activate gluconeogenesis?

15.8 Explain how severe hypoglycemia can lead to disorders of the central nervous system.

Hormones and the Blood Sugar Level

15.9 When epinephrine is secreted, what soon happens to the blood sugar level?

15.10 At which one tissue is epinephrine the most effective?

15.11 What does epinephrine activate at its target cell?

15.12 Arrange the following in the correct order in which they work in epinephrine-initiated glycogenolysis. Place the identifying letters in their correct order, left to right, in the series.

phosphorylase kinase	phosphorylase	cyclic AMP
A	B	C
adenylate cyclase	protein kinase	
D	E	

15.13 One epinephrine molecule triggers the ultimate formation of roughly how many glucose units, 10^2, 10^3, or 10^4?

15.14 What might be the result if phosphorylase and glycogen synthetase were both activated at the same time?

15.15 What switches glycogen synthetase off when glycogenolysis is activated?

15.16 What is the end product of glycogenolysis, and what does phosphoglucomutase do to it?

15.17 Why can liver glycogen but not muscle glycogen be used to resupply blood sugar?

15.18 What is glucagon, what does it do, and what is its chief target tissue?

15.19 Which is probably better at increasing the blood sugar level, glucagon or epinephrine? Explain.

15.20 How does human growth hormone manage to promote the supply of the energy needed for growth?

15.21 What is insulin, where is it released, and what is its chief target tissue?

15.22 What triggers the release of insulin into circulation?

*15.23 If brain cells are not insulin-dependent cells, how can too much insulin cause insulin shock?

15.24 What is somatostatin, where is it released, and what kind of effect does it have on the pancreas?

Glucose Tolerance and the Cori Cycle

15.25 What is meant by *glucose tolerance?*

15.26 How is glucose trapped in muscle cells and what happens to it?

15.27 What are the main steps in the Cori cycle?

15.28 In general terms, what happens to excess lactate produced in muscles during exercise?

15.29 What is the purpose of the glucose tolerance test? How is it conducted?

15.30 Describe what happens when each of the following persons takes a glucose tolerance test.
(a) a nondiabetic individual
(b) a diabetic individual

15.31 Describe a circumstance in which hyperglycemia might arise in a nondiabetic individual.

Catabolism of Glucose

15.32 Fill in the missing substances and balance the following incomplete equation:

$$\underset{\text{Glucose}}{C_6H_{12}O_6} + 2\ ADP + \_\_\_\_ \longrightarrow \underset{\text{Lactate}}{C_3H_5O_3^-} + 2H^+ + \_\_\_\_$$

15.33 What particular significance does glycolysis have when a tissue is running an oxygen debt?

15.34 Why is the rearrangement of dihydroxyacetone phosphate into glyceraldehyde-3-phosphate important?

15.35 What happens to pyruvate (a) under aerobic conditions and (b) under anaerobic conditions?

15.36 What happens to lactate when an oxygen debt is repaid?

***15.37** What is the maximum number of ATPs that can be made by the complete catabolism of (a) one molecule of glucose and (b) one glucose residue in glycogen?

15.38 The pentose phosphate pathway uses $NADP^+$, not NAD^+. What forms *from* $NADP^+$, and how does the body use what forms (in general terms)?

Gluconeogenesis

15.39 In a period of prolonged fasting or starvation, what does the system do to try to maintain its blood sugar level?

*15.40 Amino acids are not excreted, and they are not stored in the same way that glucose residues are stored in a polysaccharide. What probably happens to the excess amino acids in a high-protein diet of an individual who does not exercise much?

15.41 The amino groups of amino acids can be replaced by keto groups. Which amino acids could give the following keto acids that participate in carbohydrate metabolism?
(a) pyruvic acid (b) oxaloacetic acid

*15.42 The amino acids glutamic acid, arginine, histidine, and proline can all be catabolized to α-ketoglutarate. What metabolic cycle in the body makes it possible for α-ketoglutarate to be used eventually in gluconeogenesis? Explain.

*15.43 The carbon atoms of succinyl CoA can wind up in glucose molecules. What metabolic pathway in the body enables succinyl CoA to connect to gluconeogenesis?

Glycogen Storage Diseases (Special Topic 15.1)

15.44 For each of the following diseases, name the defective enzyme, and state the biochemical and physiological consequences.
(a) Von Gierke's disease (b) Cori's disease
(c) McArdle's disease (d) Andersen's disease

Diabetes Mellitus (Special Topic 15.2)

15.45 What is the biochemical distinction between type I and type II diabetes?

15.46 Juvenile-onset diabetes is usually which type?

15.47 Adult-onset diabetes is usually which type?

15.48 Briefly state the six stages in the onset of type I diabetes.

15.49 Viruses that cause diabetes attack which target cells?

15.50 Type I diabetes is an autoimmune disease. What does this mean?

15.51 What are some explanations for the lack of glucose uptake in NIDDM?

15.52 Sustained hyperglycemia causes damage to which specific tissue, causing damage that might be responsible for other complications?

15.53 How is glucose involved in the formation of a Schiff base? What other kinds of compounds react with glucose in this way?

15.54 When the glucose level in blood drops, what happens to the level of glucosylated hemoglobin? Why?

15.55 What happens to the Schiff bases involving glucose if given enough time? Why is this serious?

15.56 Describe a theory that explains how the hydrogenation of glucose might contribute to blindness.

Aldol Condensation (Special Topic 15.3)

15.57 Write the products that would form in an aldol condensation starting with (a) ethanal, (b) propanal, and (c) acetone.

15.58 What products would form if the following compound underwent a reverse aldol condensation?

$$C_6H_5CH_2\overset{\displaystyle OH}{\overset{|}{C}}H-\underset{\underset{\displaystyle C_6H_5}{|}}{C}H\overset{\displaystyle O}{\overset{\|}{C}}H$$

Additional Exercises

*15.59 Suppose that a sample of glucose is made using some atoms of carbon-13 in place of the common isotope, carbon-12. Suppose further that this is fed to a healthy, adult volunteer and that all of it is taken up by the muscles.

(a) Will some of the original molecules be able to go back out into circulation? Explain.

(b) Can we expect any carbon-13 compounds to end up in the liver? Explain.

(c) Can we ever expect to see carbon-13 labeled glucose molecules in circulation again? Explain.

*15.60 Referring to data in the previous chapter, (a) how much ATP can be made from the chemical energy in one pyruvate? (b) The conversion of one lactate to one pyruvate transfers one $H:^-$ to NAD^+. How many ATP molecules can be made just from the operation of this step? (c) If all the possible ATP that can be obtained from the complete catabolism of lactate were made available for gluconeogenesis, how many molecules of glucose could be made? (Assume that ATP can substitute for GTP.)

METABOLISM OF LIPIDS 16

For an activity lasting much longer than a sprint, an athlete must draw on the chemical energy of lipids, the major topic of this chapter. This kayaker is probably thinking about other things, like survival, while negotiating Rainbow Falls on the Tuolumne River in California.

16.1

ABSORPTION AND DISTRIBUTION OF LIPIDS

Several lipoprotein complexes in the blood transport triacylglycerols, fatty acids, cholesterol, and other lipids from tissue to tissue.

The digestion of triacylglycerols, as we learned in Chapter 12, produces a mixture of the anions of long-chain fatty acids and monoacylglycerols (plus some diacylglycerols). As these move into the cells of the intestinal membrane, they become reconstituted into triacylglycerols. These plus the cholesterol in the diet make up most of the *exogenous lipid* material eventually delivered into circulation.

■ *Exogenous* means "from the outside" (e.g., from the diet).

Chylomicrons Carry Dietary Lipids and Cholesterol Lipids are virtually insoluble in water and, together with proteins, they are transported in blood in tiny particles called **lipoprotein complexes.** There are several kinds, each with its own function, and they are classified according to their densities (see Table 16.1), which range from 0.95 g/cm^3 to 1.21 g/cm^3.

■ The solubility in water of a long-chain fatty acid is typically less than 10^{-6} mol/L, but that of a complex of the fatty acid with a plasma albumin is about 2×10^{-3} mol/L.

The lipoprotein complexes with the lowest density are called *chylomicrons.* They are put together from exogenous lipid material but include 2% protein or less. After their assembly within the cells of the intestinal membrane, chylomicrons are delivered to the lymph, which carries them to the bloodstream. When they enter the capillaries embedded in muscle and adipose tissue (fat tissue), chylomicrons encounter binding sites and are held up. An enzyme, *lipoprotein lipase,* now catalyzes the hydrolysis of the chylomicrons' triacylglycerols to fatty

TABLE 16.1
Composition of Plasma Lipoproteins

Type of Complex[a]	Chief Constituents (in order of amount)	Density (in g/cm^3)
Chylomicrons and chylomicron remnants	Triacylglycerols and cholesterol from the diet	<0.95
VLDL	Triacylglycerols from the liver, cholesteryl esters, cholesterol	0.95–1.006
IDL	Cholesteryl esters, cholesterol, triacylglycerols (trace)	1.066–1.019
LDL	Cholesteryl esters, cholesterol, triacylglycerols (trace)	1.019–1.063
HDL	Cholesteryl esters, cholesterol	1.063–1.210

[a] VLDL = very-low-density lipoprotein complex, IDL = intermediate-density lipoprotein complex, LDL = low-density lipoprotein complex, HDL = high-density lipoprotein complex.

Data from M. S. Brown and J. L. Goldstein in *Harrison's Principles of Internal Medicine,* 11th ed (1987), page 1651. Edited by E. Braunwald, K. J. Isselbacher, R. G. Petersdorf, J. D. Wilson, J. B. Martin, and A. S. Fauci.

acids and monoacylglycerols, which are promptly absorbed by the nearby tissue. As the hydrolysis occurs, the chylomicrons shrink to *chylomicron remnants,* which still contain dietary cholesterol. The remnants break loose from the capillary surface, and circulate to the liver where they are absorbed (see Fig. 16.1). *The overall functions of the chylomicrons are thus to deliver exogenous fatty acids to muscle and adipose tissue and to carry dietary cholesterol to the liver.*

■ *Chyle* is the lymph from the intestines. (Lymph was discussed on page 313.)

Lipoprotein Complexes Transport Endogenous Lipids The liver is a site where breakdown products from carbohydrates are used to make fatty acids and, from them, triacylglycerols. Lipids made in the liver are called *endogenous* lipids ("generated from within"). Three similar lipoprotein complexes are used to transport endogenous lipids in the bloodstream: *very-low-density lipoproteins (VLDL), intermediate-density lipoproteins (IDL),* and *low-density lipoproteins (LDL).* These three complexes also carry cholesterol. Another complex, *high-density lipoprotein (HDL),* carries back to the liver the cholesterol released from tissues and no longer needed by them.

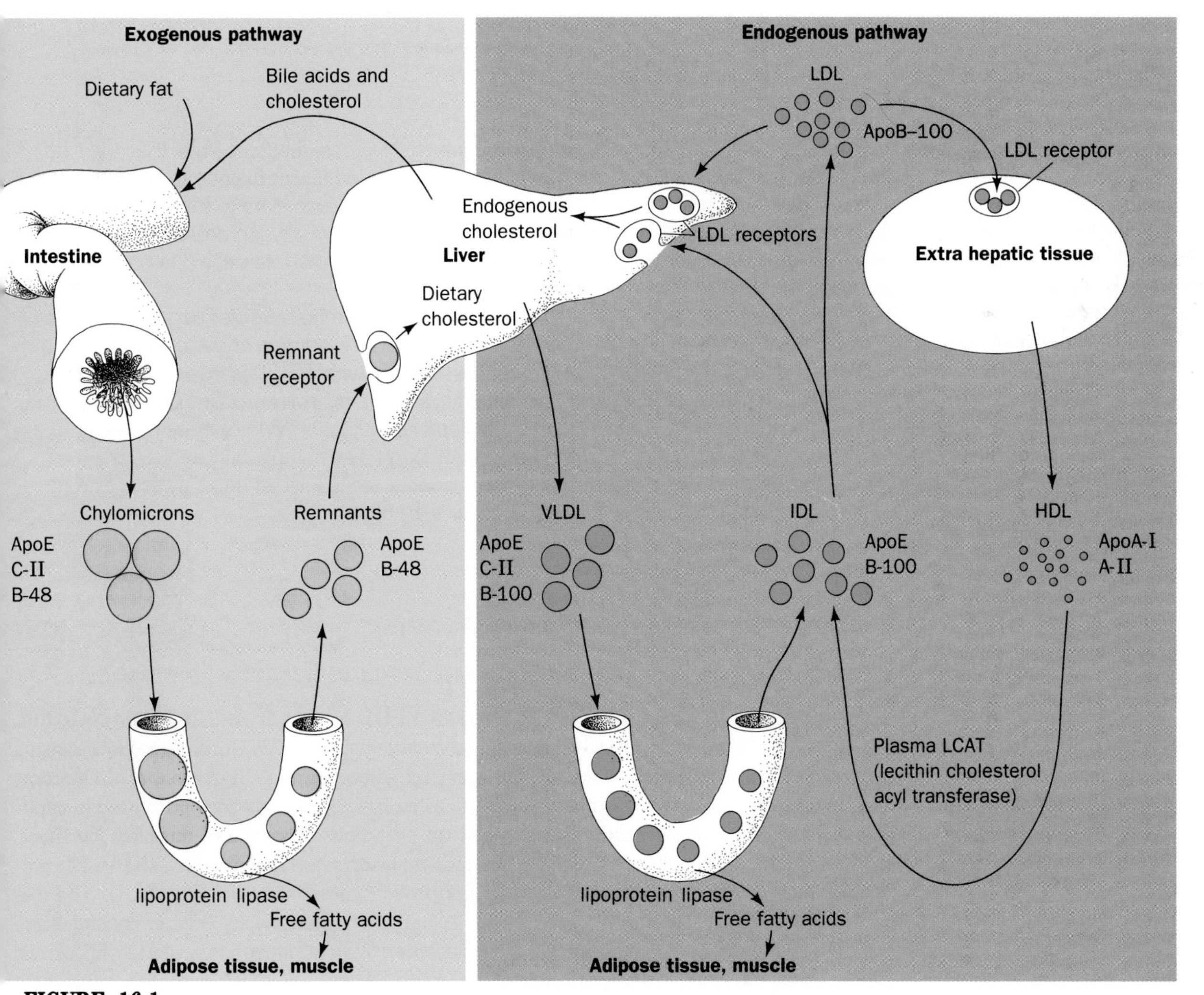

FIGURE 16.1
The transport of cholesterol and triacylglycerols by lipoprotein complexes. The abbreviations ApoE, C-II, B-48, and so forth refer to specific proteins involved with the lipoprotein complexes. [After M. S. Brown and J. L. Goldstein *in* E. Brunwald, K. J. Isselbacher, R. G. Petersdorf, J. B. Martin, and A. S. Fauci, Eds., *Harrison's Principles of Internal Medicine* (11th edition), p. 1652, McGraw-Hill (1987).]

SPECIAL TOPIC 16.1
"GOOD" AND "BAD" CHOLESTEROL AND LIPOPROTEIN RECEPTORS

The receptor proteins for LDL, whether on the liver or on extrahepatic tissue, have a crucial function. If they are reduced in number or are absent, there is little ability to remove cholesterol from circulation. The level of cholesterol in the blood, therefore, becomes too high. The result is *atherosclerosis,* a disease in which several substances, including collagen, elastic fibers, and triacylglycerols, but chiefly cholesterol and its esters, form plaques in the arterial wall (see Figure 1). Such plaques are the chief cause of heart attacks.

Inadequate LDL Receptors Cause Serum Levels of LDL, the "Bad Cholesterol," To Increase The ranges of serum cholesterol concentrations regarded as "desirable," "borderline high," and "risk" are given in Table 1. Some people have a genetic defect that bears specifically on the LDL receptors at the liver and causes elevated levels of LDL cholesterol. Two genes are involved. Those who carry two mutant genes have *familial hypercholesterolemia,* a genetically caused high level of cholesterol in the blood—3 to 5 times higher than average. Even on a zero-cholesterol diet, the victims have very high cholesterol levels. Their cholesterol along with other materials slowly comes out of the blood at valves and other sites, reduces the dimensions of the blood capillaries, and thus restricts blood flowage. Atherosclerosis has set in. Because elevated levels of LDL cholesterol are so often involved, LDL cholesterol is often called "bad cholesterol." Because the HDL complexes perform the desirable service of helping to carry cholesterol back to the liver, HDL cholesterol is sometimes called the "good cholesterol."

In atherosclerosis, the heart must work harder. Eventually arteries and capillaries in the heart itself become reduced so much in cross-sectional size that they no longer are able to bring sufficient oxygen to heart tissue. The plaques sometimes cause clots to form in capillaries of the heart and so block all oxygen delivery to affected heart tissue. Now a myocardial infarction ("heart attack") has occurred.

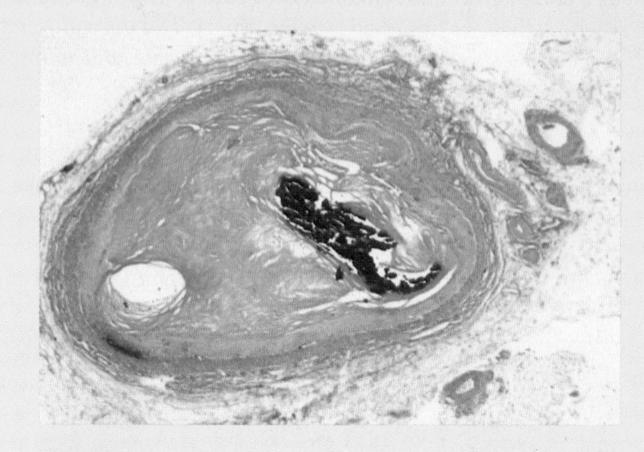

FIGURE 1
The buildup of plaque in this artery in the heart has almost closed it.

Victims of familial hypercholesterolemia generally have their first heart attacks as children and are dead by their early twenties. People with one defective gene and one normal gene for the LDL receptor proteins generally have blood cholesterol levels that are two or three times higher than normal. Although they number only about 0.5% of all adults, they account for 5% of all heart attacks among those younger than 60.

■ American scientists J. L. Goldstein and M. S. Brown shared the 1985 Nobel prize in medicine for their work on LDL receptor proteins and how they help to control blood cholesterol levels.

■ Cholesterol ($d = 1.05$ g/cm^3) is more dense than triacylglycerols (density of about 0.9 g/cm^3).

VLDL Changes to IDL and Then to LDL as VLDL Moves from the Liver to Other Tissues The liver packages triacylglycerols and cholesterol into VLDL, the very-low-density lipoprotein complexes, and releases them into circulation. During circulation, a VLDL complex undergoes somewhat continuous changes as its triacylglycerols are hydrolyzed in capillaries and the hydrolysis products are taken up by adipose tissue and muscles. By these processes, which strongly resemble the changes that occur to chylomicrons, the VLDL particles change to IDL—intermediate-density lipoprotein complexes (see Fig. 16.1). The loss of the lower-density triacylglycerols leaves a particle richer in the higher-density components, so the net density increases (see Table 16.1). With continued loss of triacylglycerols, the IDL change to low-density lipoprotein complexes, LDL (Fig. 16.1). While these activities take place, the VLDL lose some of their protein, and their cholesterol becomes largely esterified by fatty acids and changed to cholesteryl esters. Some LDL is reabsorbed by the liver, but *the main purpose of LDL is to deliver cholesterol to extrahepatic tissue* (tissue other

One Kind of Lipoprotein in HDL Is Undesirable It isn't only the *relative concentration* of HDL that confers protection against atherosclerosis; the *composition* of the HDL is also a factor. HDL is made of about equal parts of lipid and protein. The two most abundant proteins are apolipoprotein A-I and A-II (or apoA-I and apoA-II). When *both* are found in the same HDL particles, the HDL is less "good" than when only apoA-I is present. It is the apoA-I that confers on the HDL its ability to protect against atherosclerosis, but this ability deteriorates when apoA-II is also present (as shown in 1993 by experiments in mice). *Thus the ratio of HDL to LDL does not alone provide information concerning an individual's status regarding atherosclerosis. The composition of the HDL lipoproteins must also be considered.* Just how awaits further research.

Cholesterol Levels Can Be Reduced High blood cholesterol levels occur in many people besides those with genes for hypercholesterolemia, even in people with normal genes for the receptor proteins. The causes of their high cholesterol levels have not been fully unraveled. Smoking, obesity, and lack of exercise are known to contribute to the cholesterol problem. High-cholesterol foods also appear to be factors. There is some evidence that as the liver receives more and more cholesterol from the diet it loses more and more of its receptor proteins. This forces more and more cholesterol to linger in circulation. Some people, however, are able to sustain high-cholesterol and high-lipid diets with no ill effects. Eggs are 0.5% cholesterol (all in the yolk), so egg-less diets are commonly recommended for people with high cholesterol. Yet, one 88-year-old man had eaten 25 eggs a day for at least 15 years without having elevated blood cholesterol levels!

Changing from a typically American high-fat diet to a low-fat diet lowered the serum cholesterol levels of several dozen adults with borderline high levels by an average of only 5 percent. In the same study, reported in 1993, lovastatin brought about a 27 percent reduction in serum cholesterol. Thus the effect of diet on serum cholesterol levels among adults who are healthy, except for borderline high cholesterol levels, is relatively small. Serum cholesterol, of course, is not the only risk factor for heart disease, and the 1993 study in no way endorses a high-fat diet.

TABLE 1
Serum Levels of Cholesterol in All Forms

Form	Desirable	Borderline High	Risk
Total cholesterol	<200 mg/dL (<5.18 mmol/L)	200–239 mg/dL (5.18–6.19 mmol/L)	>239 mg/dL (>6.20 mmol/L)
LDL	<130 mg/dL	130–159 mg/dL	>160 mg/dL
HDL	<35 mg/dL (<0.91 mmol/L)		

Source: *New England Journal of Medicine*, Sept. 3, 1992, page 718.

than liver tissue) *to be used to make cell membranes and, in specialized tissues, steroid hormones.*

HDL Transports Cholesterol from Extrahepatic Tissue to the Liver High-density lipoprotein complexes—HDL—are cholesterol scavengers. When cells of extrahepatic tissue break down for any reason, their cholesterol molecules are picked up by HDL particles and changed to cholesteryl esters. En route to the liver, HDL undergoes some changes of its own. There is evidence that HDL can transfer cholesteryl esters to VLDL. By one means or another, some of the HDL becomes more like LDL before entering liver cells by means of LDL receptors (Fig. 16.1). Some evidence also exists for specific HDL receptors on liver cells. When receptors for cholesterol-bearing lipoprotein complexes are absent, defective, or in too few numbers, either at the liver or at extrahepatic tissue, there are serious consequences, which are discussed in Special Topic 16.1.

16.2

STORAGE AND MOBILIZATION OF LIPIDS

The high energy density of stored triacylglycerol makes it a choice form for storing energy.

The energy stored per gram of tissue or solution is called the *energy density* of the material. Isotonic glucose solution, for example, carries only about 0.2 kcal per gram. When the glucose is changed into glycogen, however, *and is no longer in solution,* there is little associated water. Now we can get more energy into storage in 1 g; the energy density of wet glycogen is about 1.7 kcal/g. Triacylglycerol, in sharp contrast, has an energy density of about 7.7 kcal/g.

■ Because lipids are water-insoluble, they attract the least amount of associated water in storage.

A 70-kg adult male has about 12 kg of triacylglycerol in storage. If he had to exist on no food, just water and a vitamin–mineral supplement, and if he needed 2500 kcal/day, this fat would supply his caloric needs for 43 days. Of course, during this time the body proteins would also be wasting away, and metabolic acidosis (page 322) would be a problem of growing urgency.

■ These data are for information; they're certainly not recommendations!

Adipose Tissue Is the Principal Lipid Storage Depot The chief depot for the storage of fatty acids is adipose tissue, a very metabolically active tissue. There are two kinds, brown and white. Both types are associated with internal organs, where they cushion the organs against mechanical bumps and shocks, and insulate them from swings in temperature. For a discussion of how brown adipose tissue uses the respiratory chain to generate heat, not ATP, see Special Topic 16.2. The discussion that continues concerns the metabolic activities of white adipose tissue. This tissue stores energy as triacylglycerols chiefly on behalf of the energy budgets of other tissues.

■ The relatively high concentration of mitochondria, which hold iron-containing enzymes (cytochromes), causes the color of brown fat tissue.

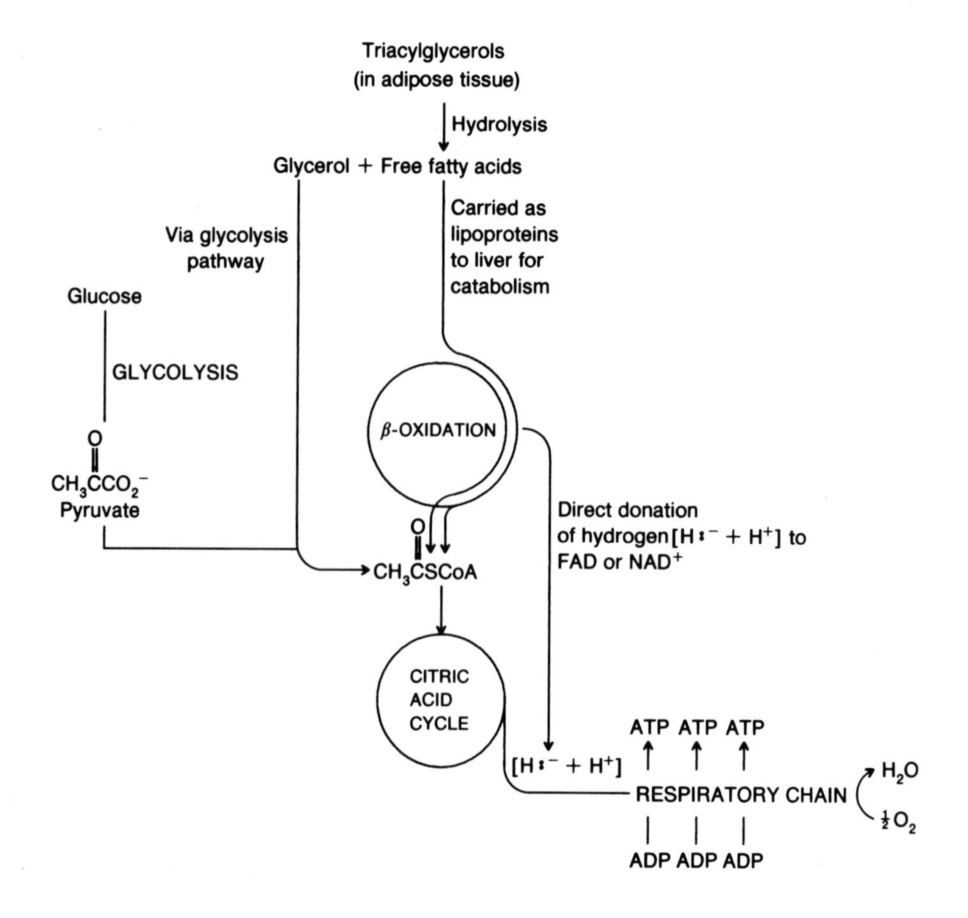

FIGURE 16.2
Pathways for the mobilization of energy reserves in the triacylglycerols of adipose tissue

SPECIAL TOPIC 16.2
BROWN FAT AND THERMOGENESIS

Both kinds of adipose tissue, white and brown, store triacylglycerols, but white adipose tissue does not metabolize them except to break them down to free fatty acids and glycerol. The fatty acids from white adipose tissue are then exported to other tissues for catabolism.

Respiratory Chain Oxidation and ATP Synthesis Are Uncoupled in Brown Adipose Tissue The triacylglycerols in brown adipose tissue are catabolized within this tissue for little other use than to generate heat. This is possible because cells of brown adipose tissue can switch off the capability of the inner mitochondrial membranes to accept and hold a proton gradient. Recall that the respiratory chain normally establishes a proton gradient across the inner mitochondrial membrane, and that as protons flow back through selected channels, they trigger the synthesis of ATP from ADP and P_i.

When the respiratory chain runs but cannot set up the proton gradient, the chemical energy released by the chain emerges only as heat. Such generation of heat is called *thermogenesis*. Two stimuli, both mediated by the neurotransmitter norepinephrine, trigger this heat-generating activity, exposure to cold (and the resulting shivering) and the ingestion of food.

The advantage to the body of thermogenesis induced by a cold outside temperature is that the body can oxidize its own fat to help keep itself warm.

The advantage of food-induced thermogenesis is that the individual is protected from getting fat by too much eating in relationship to exercise. Thermogenesis in brown adipose tissue removes fat by catabolism, rather than by exercise, as new calories (in the food) are imported.

A Body in Dietary Balance Uses Fatty Acids as Well as Glucose for Energy Some tissues, like heart muscle and the renal cortex, use breakdown products of fatty acids in preference to glucose for energy. Skeletal muscles can use both fatty acids and short molecules made from them for energy. In fact, most of the energy needs of resting muscle tissue are met by intermediates of fatty acid catabolism, not from glucose. Given enough time for adjustment, during starvation, for example, even brain cells can obtain some energy from fatty acid breakdown products.

Fatty acids, consequently, come and go from adipose tissue, and the balance between this tissue's receiving or releasing them is struck by the energy requirements elsewhere. In extreme distress, when either glucose is in very low supply (as in starvation) or what is available can't be used (as in diabetes), the body must turn almost entirely to its fatty acids for energy.

Figure 16.2 outlines the many steps involved in tapping the lipid reserves for their energy. (It also reminds us of the connection between glucose and ATP.) Triacylglycerol molecules in adipose tissue are first hydrolyzed to free fatty acids and glycerol. The cellular lipase needed for this is activated by a process involving cyclic AMP and the hormone epinephrine. The fatty acids are carried as albumin complexes to the liver, the chief site for their catabolism.

■ Epinephrine is also involved in glucose metabolism.

Insulin suppresses the cellular lipase that releases fatty acids from adipose tissue. Thus when insulin is in circulation, and its presence is linked to the presence of blood glucose, the fatty acids are less required for energy.

The glycerol that is produced when the fatty acids are released is changed to dihydroxyacetone phosphate, and it enters the pathway of glycolysis.

16.3 OXIDATION OF FATTY ACIDS

Acetyl groups are produced by the β-oxidation of fatty acids and are fed into the citric acid cycle and respiratory chain.

The degradation of fatty acids takes place inside mitochondria by a repeating series of steps known as the **β-oxidation pathway** (see Fig. 16.3).

■ The β-oxidation pathway has also been called the *fatty acid cycle*, but it isn't exactly a true *cycle*, like the citric acid cycle.

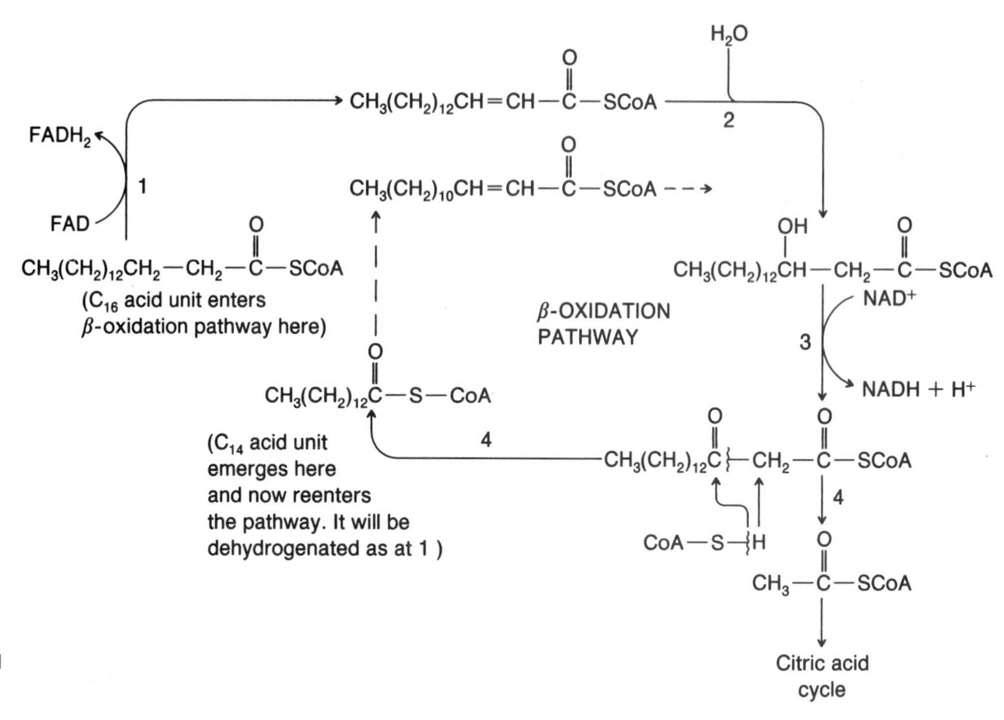

FIGURE 16.3
The β-oxidation pathway. The numbers refer to the numbered steps discussed in the text.

A Fatty Acid Is First Joined to Coenzyme A The β-oxidation pathway occurs in the mitochondrial matrix, and before a fatty acid can enter this pathway, it has to be joined to coenzyme A. It costs one ATP to do this, but now the fatty acyl unit is activated. The ATP itself breaks down to AMP and PP_i. *The subsequent hydrolysis of the diphosphate is the driving force for the overall change,* and this means that the actual cost in high-energy phosphate to commence the β-oxidation pathway is *two* high-energy phosphate bonds.

Getting a fatty acid inside the matrix of a mitochondrion from the cytosol begins with the formation of fatty acyl CoA in the cytosol. From this form, the fatty acyl unit is passed over to a protein of the mitochondrial membrane, migrates through, and is finally passed to coenzyme A inside. It's complicated, but the cell has now isolated the fatty acyl unit for *catabolism;* the steps by which a fatty acyl unit are made (its anabolism) occur in the cytosol. As we have said, nature segregates anabolism from catabolism and thereby is able to control both more easily.

■ Franz Knoop directed much of the research on the fatty acid cycle, so this pathway is sometimes called *Knoop oxidation.*

Fatty Acyl CoA Is Catabolized by Two Carbons at a Time The repeating sequence of the β-oxidation pathway consists of four steps as the result of which one molecule of $FADH_2$, one of NADH, and one of acetyl coenzyme A are made in addition to a fatty acyl unit with two less carbons.

The now shortened fatty acyl unit is carried again through the four steps, and the process is repeated until no more two-carbon acetyl units can be made. The $FADH_2$ and the NADH fuel the respiratory chain. The acetyl groups pass into the citric acid cycle, or they enter the general pool of acetyl coenzyme A that the body draws from to make other substances (e.g., cholesterol). Let's now look at the four steps in greater detail. The numbers that follow refer to Figure 16.3.

1. The first step is dehydrogenation. FAD accepts ($H:^- + H^+$) from the α- and the β-carbons of the fatty acyl unit of palmityl coenzyme A.

$$CH_3(CH_2)_{12}CH_2{-}CH_2\overset{O}{\overset{\|}{C}}SCoA + FAD \xrightarrow{\boxed{1}}$$

Palmityl coenzyme A

$$CH_3(CH_2)_{12}CH{=}CH\overset{O}{\overset{\|}{C}}SCoA + FADH_2 \longrightarrow [H{:}^- + H^+] \text{ (FAD) }$$

An α,β-unsaturated acyl derivative of coenzyme A

Transfers to respiratory chain

The enzyme is *acyl-CoA dehydrogenase.*

$FADH_2$ interacts with the respiratory chain at enzyme complex II (page 377).

2. The second step is hydration. Water adds to the alkene double bond formed by the previous step, and a 2° alcohol group forms.

$$CH_3(CH_2)_{12}CH{=}CH\overset{O}{\overset{\|}{C}}SCoA + H_2O \xrightarrow{\boxed{2}} CH_3(CH_2)_{12}\overset{OH}{\overset{|}{C}H}CH_2\overset{O}{\overset{\|}{C}}SCoA$$

A β-hydroxyacyl derivative of coenzyme A

The enzyme is *enoyl-CoA hydratase.*

3. The third step is another dehydrogenation—a loss of $(H{:}^- + H^+)$. This oxidizes the secondary alcohol to a keto group. Notice that these steps end in the oxidation of the β-position of the original fatty acyl group to a keto group, which is why the pathway is called *beta* oxidation.

$$CH_3(CH_2)_{12}\overset{OH}{\overset{|}{C}H}CH_2\overset{O}{\overset{\|}{C}}SCoA + NAD^+ \xrightarrow{\boxed{3}} CH_3(CH_2)_{12}\overset{O}{\overset{\|}{C}}CH_2\overset{O}{\overset{\|}{C}}SCoA + NADH + H^+$$

A β-keto acyl derivative of coenzyme A

NAD^+

Transfers to respiratory chain ⟵ $[H{:}^- + H^+]$

3-L-Hydroxyacyl-CoA dehydrogenase is the enzyme for this oxidation. One form of this enzyme has been found to be deficient in about 10% of infants who died of sudden infant death syndrome (SIDS), so without the enzyme there may be a fatal imbalance between glucose and fatty acid metabolism.

4. The fourth step breaks the bond between the α-carbon and the β-carbon. This bond has been weakened by the stepwise oxidation of the β-carbon, and now it breaks to release one unit of acetyl coenzyme A. (This bond-breaking reaction is little more than a *reverse* Claisen condensation; we studied the Claisen reaction in Special Topic 14.1, page 372)

$$CH_3(CH_2)_{12}\overset{O}{\overset{\|}{C}}{-}CH_2\overset{O}{\overset{\|}{C}}SCoA \xrightarrow{\boxed{4}} CH_3(CH_2)_{12}\overset{O}{\overset{\|}{C}}SCoA + CH_3\overset{O}{\overset{\|}{C}}SCoA$$

CoA—S—H

Myristyl coenzyme A

Acetyl coenzyme A

Transfers to citric acid cycle

12ATP ⟵ via respiratory chain

The enzyme is *thiolase* (or β-ketoacyl-CoA thiolase).

The remaining acyl unit, the original shortened by two carbons, now goes through the cycle of steps again: dehydrogenation, hydration, dehydrogenation, and cleavage. After seven such cycles, one molecule of palmityl coenzyme A is broken into eight molecules of acetyl coenzyme A.

One Palmityl Unit Yields 129 ATP Molecules Table 16.2 shows how the maximum yield of ATP from the oxidation of one unit of palmityl coenzyme A adds up to 131 ATPs. The

TABLE 16.2
Maximum Yield of ATP from Palmityl CoA

Intermediates Produced by Seven Turns of the Fatty Acid Cycle	ATP Yield per Intermediate	Total ATP Yield
7 $FADH_2$	2	14
7 NADH	3	21
8 Acetyl Coenzyme A	12	96
		131 ATP
Deduct two high-energy phosphate bonds for activating the acyl unit		−2
Net ATP yield per palmityl unit		129 ATP

net from palmitic acid is two ATPs fewer, or 129 ATPs, because the activation of the palmityl unit—joining it to CoA—requires this initial investment, as we mentioned earlier.

16.4

BIOSYNTHESIS OF FATTY ACIDS

Acetyl CoA molecules that are not needed to make ATP can be made into fatty acids.

Acetyl CoA stands at a major metabolic crossroad. It can be made from any monosaccharide in the diet, from virtually all amino acids, and from fatty acids. Once made, it can be shunted into the citric acid cycle where its chemical energy can be used to make ATP; or its acetyl group can be made into other compounds that the body needs. In this section we'll see how acetyl CoA can be made into long-chain fatty acids.

Fatty Acid Synthesis Begins with the Activation of Acetyl CoA Whenever acetyl CoA molecules are made within mitochondria but aren't needed for the citric acid cycle and respiratory chain, they are exported to the cytosol. The enzymes for the synthesis of fatty acids are found there, not within the mitochondria, illustrating again the general rule that the body segregates its sequences of catabolism from those of anabolism.

As might be expected, because fatty acid synthesis is in the direction of climbing an energy hill, the cell has to invest some energy of ATP to make fatty acids from smaller molecules. The first payment occurs in the first step in which the bicarbonate ion reacts with acetyl CoA to form malonyl CoA.

■ The enzyme for this step, *acetyl CoA carboxylase*, requires the vitamin biotin.

$$\underset{\text{Acetyl coenzyme A}}{CH_3\overset{O}{\overset{\|}{C}}SCoA} + HCO_3^- + ATP \longrightarrow \underset{\text{Malonyl coenzyme A}}{^-O\overset{O}{\overset{\|}{C}}CH_2\overset{O}{\overset{\|}{C}}SCoA} + 2H^+ + ADP + P_i$$

The extra carbonyl group in malonyl CoA activates the CH_2 unit, now between two carbonyl groups, for fatty acid synthesis.

The Growing Fatty Acyl Unit Is Moved from Enzyme to Enzyme by a Construction Boom Molecular Unit The enzyme that now builds a long-chain fatty acid is actually a huge complex of seven enzymes called *fatty acid synthase* (see Fig. 16.4). In the center of this

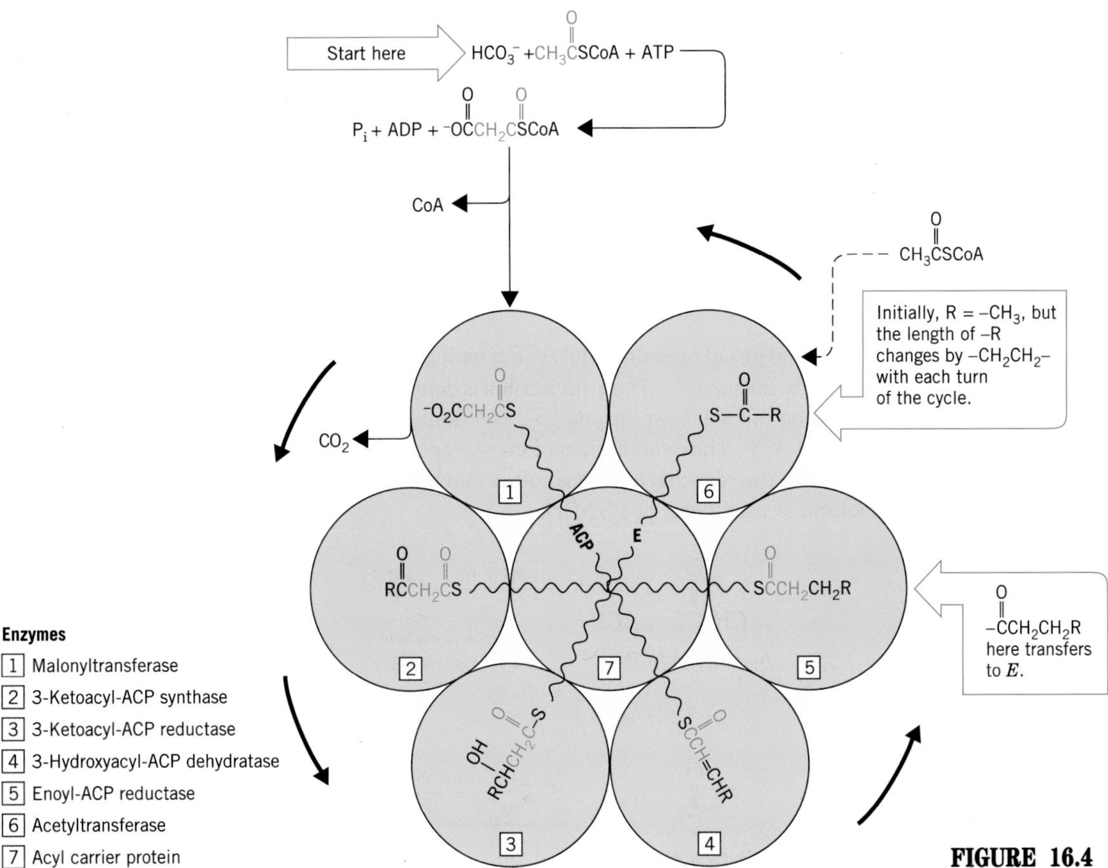

FIGURE 16.4
The synthesis of fatty acids. At the top, an acetyl group is activated and joined as a malonyl unit to an arm of the acyl carrier protein, ACP. Another acetyl group transfers from acetyl CoA to site **E**. In a second transfer, this acetyl group is then joined to the malonyl unit as CO_2 splits back out. This gives a β-ketoacyl system whose keto group is reduced to CH_2 by the next series of steps. One turn of the "cycle" adds a CH_2CH_2 unit to the growing acyl chain.

complex is a molecular unit long enough to serve as a swinging arm carrier. It's called the *acyl carrier protein,* or ACP, and like the boom of a construction crane, this arm swings from enzyme to enzyme in the synthase complex. Thus the arm brings what it carries over first one enzyme and then another, and at each stop a reaction is catalyzed that contributes to chain lengthening. Let's see how it works. The enzymes are represented as colored circles in Figure 16.4 and identified by numbered boxes used in the following discussion. The *names* of the enzymes involved are given in Figure 16.4.

The malonyl unit in malonyl CoA, just made, transfers to the swinging arm of ACP, placing the product over enzyme [1]:

$$^{-}O\overset{O}{\overset{\|}{C}}CH_2\overset{O}{\overset{\|}{C}}S{-}CoA + ACP \longrightarrow {}^{-}O\overset{O}{\overset{\|}{C}}CH_2\overset{O}{\overset{\|}{C}}S{-}ACP + CoA$$

Malonyl coenzyme A — Malonyl ACP

In the meantime, a similar reaction occurs over enzyme [6] to another molecule of acetyl CoA at a different unit of the synthase, a unit that we'll call simply *E*.

$$CH_3\overset{O}{\overset{\|}{C}}S{-}CoA + E \longrightarrow CH_3\overset{O}{\overset{\|}{C}}S{-}E + CoA$$

Acetyl *E*

Next, enzyme [1] catalyzes the transfer of the acetyl group of acetyl *E* to the malonyl group of malonyl ACP as carbon dioxide, the initial activator, is ejected. The loss of CO_2, in fact, is the driving force for this transfer reaction, a driving force initially put in place by energy from ATP. A four-carbon derivative of ACP, acetoacetyl ACP, forms. It's over enzyme [2]. The *E* unit is vacated.

■ Glucagon, epinephrine, and cyclic AMP—all stimulators of the use of glucose to make ATP—depress the synthesis of fatty acids in the liver. Insulin, however, promotes it.

$$\underset{\text{Acetyl } E}{CH_3\overset{O}{\overset{\|}{C}}S-E} + \underset{\text{Malonyl ACP}}{{}^-O\overset{O}{\overset{\|}{C}}CH_2\overset{O}{\overset{\|}{C}}S-ACP} \longrightarrow \underset{\text{Acetoacetyl ACP}}{CH_3\overset{O}{\overset{\|}{C}}CH_2\overset{O}{\overset{\|}{C}}S-ACP} + CO_2 + E$$

The ketone group in acetoacetyl ACP is next reduced by enzyme [3] to a secondary alcohol and passed to enzyme [4]. Then this alcohol is dehydrated by enzyme [4] to introduce a double bond, putting the unit over enzyme [5]. The double bond is next reduced over enzyme [5] to give butyryl ACP. The overall effect of these steps is to reduce the keto group to CH_2. Notice that NADPH, the reducing agent manufactured by the pentose phosphate pathway of glucose catabolism, is used here, not NADH.

$$\underset{\text{Acetoacetyl ACP}}{CH_3\overset{O}{\overset{\|}{C}}CH_2\overset{O}{\overset{\|}{C}}S-ACP} \xrightarrow[\text{(reduction of the keto group by enzyme 3)}]{NADPH + H^+ \;\; NADP^+} CH_3\overset{OH}{\overset{|}{C}}HCH_2\overset{O}{\overset{\|}{C}}S-ACP$$

(dehydration by enzyme 4) $\downarrow$ H_2O

$$\underset{\text{Butyryl ACP}}{CH_3CH_2CH_2\overset{O}{\overset{\|}{C}}S-ACP} \xleftarrow[\text{(reduction of the double bond at enzyme 5)}]{NADP^+ \;\; NADPH + H^+} CH_3CH{=}CH\overset{O}{\overset{\|}{C}}S-ACP$$

The butyryl group is now transferred to the *vacant E* unit of the synthase, the unit that initially held an acetyl group. Butyryl *E* instead of acetyl *E* is now over enzyme [6]. This ends one complete cycle of the β-oxidation pathway. To recapitulate, we have gone from two two-carbon acetyl units to one four-carbon butyryl unit.

The steps now repeat as indicated in Figure 16.4. A new malonyl unit is joined to the ACP. Then the newly made *butyryl* group is made to transfer to the malonyl unit as CO_2 is again ejected. This elongates the fatty acyl chain to six carbons in length, positions it on the swinging arm, and gets it ready for the several-step reduction of the keto group to CH_2. The swinging arm mechanism and the enzymes of the synthase complex go to work until the chain is that of the six-carbon acyl group, the hexanoyl group.

■ $$\underset{\text{Hexanoyl group}}{CH_3CH_2CH_2CH_2CH_2\overset{O}{\overset{\|}{C}}-}$$

In the next repetition of this series, the six-carbon acyl group will be elongated to an eight-carbon group, and the process will repeat until the chain is sixteen carbons long. Overall, the net equation for the synthesis of the palmitate ion from acetyl CoA is

$$8CH_3\overset{O}{\overset{\|}{C}}SCoA + 7ATP + 14NADPH \longrightarrow \underset{\text{Palmitate ion}}{CH_3(CH_2)_{14}CO_2^-} + 7ADP + 7P_i + 8CoA + 14NADP^+ + 6H_2O$$

■ Because the symbols ATP, ADP, and P_i are given without their electrical charges, we can't provide an electrical balance to equations such as this.

The process creates a heavy demand for NADPH—14 NADPH to make one palmitate ion. If acids with chains longer than the chain in the palmitate ion are needed, or acids with double bonds are to be made, additional steps or different pathways using different enzymes are taken.

■ The pentose phosphate pathway for the catabolism of glucose (page 399) is the body's chief supplier of NADPH.

16.5

BIOSYNTHESIS OF CHOLESTEROL

Excessive cholesterol can inhibit the formation of a key enzyme required in the multistep synthesis of cholesterol.

In addition to serving as a raw material for making fatty acids, acetyl CoA can be used to make the steroid nucleus. Cholesterol, an alcohol with this nucleus, is the end product of a long, multistep process, and is used by the body to make various bile salts and sex hormones and to be incorporated into cell membranes. In mammals, about 80% to 95% of all cholesterol synthesis takes place in cells of the liver and the intestines. We won't go into all the details, but we will go far enough to learn more about how the body normally controls the process. If sufficient cholesterol is provided by the diet, then the body's synthesis should be shut down. Let's see how this is done.

Steroid nucleus

Cholesterol

Cholesterol Is Made from Acetyl Units When the level of acetyl CoA builds up in the liver, the following equilibrium shifts to the right (in accordance with Le Châtelier's principle):

$$2CH_3\overset{O}{\overset{\|}{C}}SCoA \rightleftharpoons CH_3\overset{O}{\overset{\|}{C}}CH_2\overset{O}{\overset{\|}{C}}SCoA + CoASH$$

Acetyl CoA — Acetoacetyl CoA

■ The forward reaction is an example of the Claisen ester condensation (Special Topic 14.1, page 372), the joining of an acyl group of one ester molecule to the α-position of a second ester molecule.

When cholesterol synthesis is switched on, then acetoacetyl CoA combines with another acetyl CoA:

$$CH_3\overset{O}{\overset{\|}{C}}CH_2\overset{O}{\overset{\|}{C}}{-}SCoA + CH_3\overset{O}{\overset{\|}{C}}SCoA \xrightleftharpoons[\text{synthase}]{\text{HMG-CoA}} {}^{-}O\overset{O}{\overset{\|}{C}}CH_2\underset{CH_3}{\underset{|}{\overset{OH}{\overset{|}{C}}}}CH_2\overset{O}{\overset{\|}{C}}SCoA + CoASH \quad (16.1)$$

HMG–CoA (β-hydroxy-β-methylglutaryl CoA)

■ The reaction is the addition of one carbonyl compound (acetyl CoA) to the keto group of another, so it strongly resembles the aldol condensation (Special Topic 15.3, page 397).

The Reduction of HMG–CoA Commits the Cell to the Complete Cholesterol Synthesis Both a reduction and a hydrolysis occur in the next step, which is a complex change catalyzed by *HMG–CoA reductase*, a key enzyme.

$$HMG{-}CoA + 2NADPH + 2H^+ \xrightarrow[\text{reductase}]{\text{HMG-CoA}} HOCH_2CH_2\underset{CH_3}{\underset{|}{\overset{OH}{\overset{|}{C}}}}CH_2CO_2^- + 2NADP^+ + CoA{-}SH$$

Mevalonate

Mevalonate is next carried through a long series of reactions until cholesterol is made. As we said, we'll not take it that far, but consider, instead, how cholesterol synthesis is controlled.

Cholesterol Is a Natural Inhibitor of HMG-CoA Reductase The control of HMG–CoA reductase is the major factor in the overall control of the biosynthesis of cholesterol. Cholesterol itself is one inhibitor, and it works by inhibiting both the *synthesis* of the enzyme and the enzymatic activity of any of its existing molecules. In the presence of cholesterol, the

■ Fifteen Nobel prizes have gone to scientists who devoted the better parts of their careers to various aspects of cholesterol and its uses in the body.

enzyme isn't totally deactivated. There is just *less* of it free to do catalytic work. Thus if the diet is relatively rich in cholesterol, the body tends to make less of it. If the diet is very low in cholesterol, the body makes more.

■ On a low-cholesterol diet, an adult makes about 800 mg of cholesterol per day.

Lovastatin and Compactin Lower Blood Cholesterol Levels A drug that lowers the cholesterol level, lovastatin (Mevacor®), was approved in 1987 and has proved to be extraordinarily effective in suppressing the body's natural synthesis of cholesterol. It works partly as a competitive inhibitor of the enzyme HMG–CoA reductase, which is a key choke point in cholesterol synthesis. You can see from the resemblances of the highlighted parts of the structures below how lovastatin might compete with mevalonate for a position on the enzyme.

■ Individuals with severe hypercholesterolemia who take lovastatin show a decrease in serum cholesterol of 30%. When lovastatin and compactin are used in combination, the decrease is 50% to 60%.

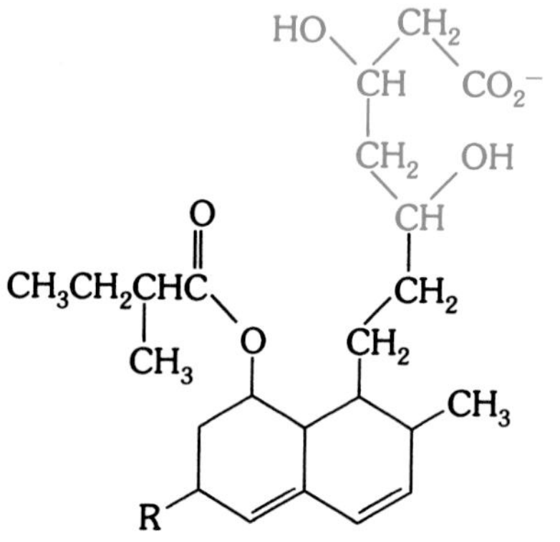

$R = H$ Compactin

$R = CH_3$ Lovastatin

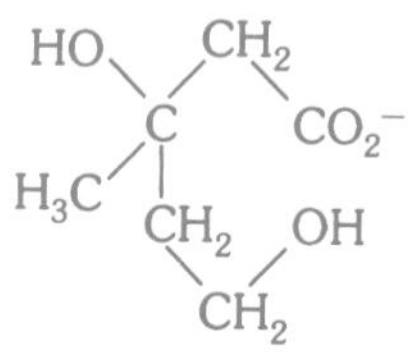

Mevalonate

Lovastatin also increases the synthesis of the mRNA responsible for making the liver LDL receptors. With more of these, the liver is better able to reabsorb cholesterol, recycle it, or export it in the bile. In some people, lovastatin also enhances the levels of the HDL units that carry cholesterol back to the liver for removal.

Carbohydrate and Lipid Metabolisms Are Intertwined Figure 16.5 provides a summary of much of what we have covered in this and the previous chapter about the chief uses of acetyl CoA and its relationship to carbohydrate and lipid catabolism. One point emphasized by this figure is that triacylglycerols can be made from any of the three dietary components: carbohydrates, lipids, and proteins.

16.6

KETOACIDOSIS

An acceleration of the fatty acid cycle tips some equilibria in a direction that leads to ketoacidosis.

Cells of certain tissues have to engage in gluconeogenesis in two serious conditions, starvation and uncontrolled diabetes mellitus. In starvation, the blood sugar level drops because of nutritional deficiencies, so the body (principally the liver) tries to compensate by making glucose. The consequences are fatal unless the underlying causes are treated.

The Level of Acetyl CoA Increases When Gluconeogenesis Is Accelerated If you look back to Figure 15.7 (page 401), you will see that gluconeogenesis consumes oxaloacetate, the carrier of acetyl units in the citric acid cycle (Fig. 14.2, page 370). When

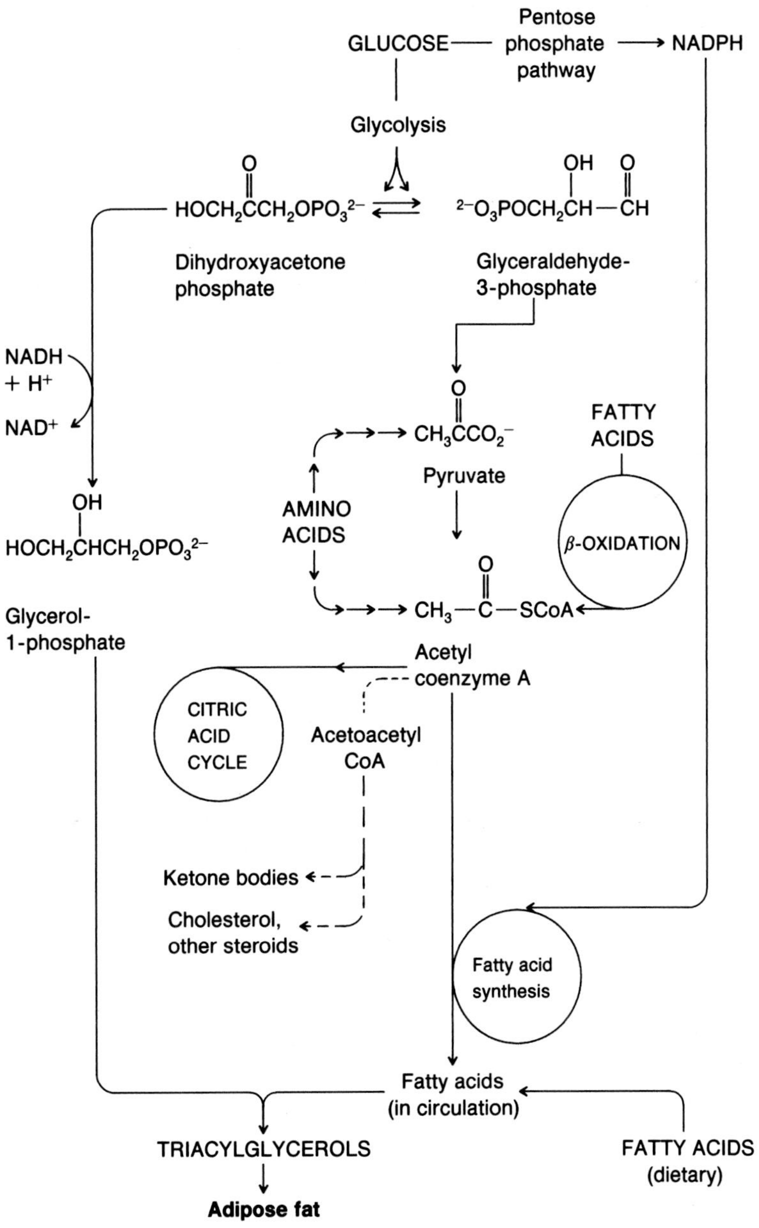

FIGURE 16.5
Principal sources of triacylglycerols for adipose tissue and the chief uses of acetyl CoA

oxaloacetate is diverted from the citric acid cycle, acetyl coenzyme A cannot put its acetyl group into the cycle. Yet acetyl coenzyme A continues to be made by the fatty acid cycle, *so acetyl CoA levels build up.*

As the supply of acetyl CoA increases in the liver, the following equilibrium shifts to the right to make acetoacetyl CoA.

■ Again we see applications of Le Châtelier's principle; the increase in the level of acetyl CoA is the stress and the equilibrium, by shifting to the right, relieves the stress.

$$\underset{\text{Acetyl CoA}}{2CH_3\overset{O}{\overset{\|}{C}}SCoA} \rightleftharpoons \underset{\text{Acetoacetyl CoA}}{CH_3\overset{O}{\overset{\|}{C}}CH_2\overset{O}{\overset{\|}{C}}SCoA} + CoASH$$

As the level of acetoacetyl CoA increases, equilibrium 16.1 shifts to the right to make more HMG–SCoA. As the level of HMG–SCoA increases (and little if any is being diverted to the synthesis of cholesterol), a liver enzyme splits it to acetoacetate ion and acetyl coenzyme A:

$$\text{MG–SCoA} \longrightarrow \underset{\text{Acetoacetate}}{CH_3\overset{\overset{O}{\|}}{C}CH_2\overset{\overset{O}{\|}}{C}O^-} + CH_3\overset{\overset{O}{\|}}{C}SCoA$$

The net effect of these steps, starting from acetyl CoA, is the following:

$$2CH_3\overset{\overset{O}{\|}}{C}SCoA + H_2O \longrightarrow \underset{\text{Acetoacetate}}{CH_3\overset{\overset{O}{\|}}{C}CH_2\overset{\overset{O}{\|}}{C}O^-} + 2CoASH + H^+$$

Notice the hydrogen ion. It makes the situation dangerous, and an increased synthesis of "new" glucose was the cause. As we said, conditions of starvation or untreated diabetes mellitus can be instigators of accelerated gluconeogenesis. Figure 16.6 outlines the chain of events that occurs in untreated diabetes, and you may wish to refer to this figure as you continue with the following discussion.

Accelerated Acetoacetate Production Leads to Acidosis The acid produced by the formation of acetoacetate must be neutralized by the buffer. Under an increasingly rapid production of acetoacetate and hydrogen ion, the blood buffer slowly loses ground. A condition of *acidosis* sets in. It is *metabolic* acidosis, because the cause lies in a disorder of metabolism. Because the chief species responsible for this acidosis has a keto group, the condition is often called **ketoacidosis.**

Blood Levels of the Ketone Bodies Increase in Starvation and Diabetes The *acetoacetate ion* is called one of the **ketone bodies.** The two others are *acetone* and the *β-hydroxybutyrate ion*. Both are produced from the acetoacetate ion. Acetone arises from acetoacetate by the loss of the carboxyl group:

$$\underset{\text{Acetoacetate}}{CH_3\overset{\overset{O}{\|}}{C}CH_2\overset{\overset{O}{\|}}{C}O^-} + H_2O \longrightarrow \underset{\text{Acetone}}{CH_3\overset{\overset{O}{\|}}{C}CH_3} + HCO_3^-$$

β-Hydroxybutyrate is a *ketone* body not because it has a keto group but because it is made from and is found together with a species that does.

β-Hydroxybutyrate is produced when the keto group of acetoacetate is reduced by NADH:

$$\underset{\text{Acetoacetate}}{CH_3\overset{\overset{O}{\|}}{C}CH_2\overset{\overset{O}{\|}}{C}O^-} + NADH + H^+ \longrightarrow \underset{\beta\text{-Hydroxybutyrate}}{CH_3\overset{\overset{OH}{|}}{C}HCH_2\overset{\overset{O}{\|}}{C}O^-} + NAD^+$$

The ketone bodies enter general circulation. Because acetone is volatile, most of it leaves the body via the lungs, and individuals with severe ketoacidosis have "acetone breath," the noticeable odor of acetone on the breath.

The vapor pressure of acetone at body temperature is nearly 400 mm Hg (about 0.53 atm), so it readily evaporates from the blood in the lungs.

Acetoacetate and β-hydroxybutyrate can be used in skeletal muscles to make ATP. Heart muscle uses these two for energy in preference to glucose. Even the brain, given time, can adapt to using these ions for energy when the blood sugar level drops in starvation or prolonged fasting. The ketone bodies are not in themselves abnormal constituents of blood. Only when they are produced at a rate faster than the blood buffer can handle them are they a problem.

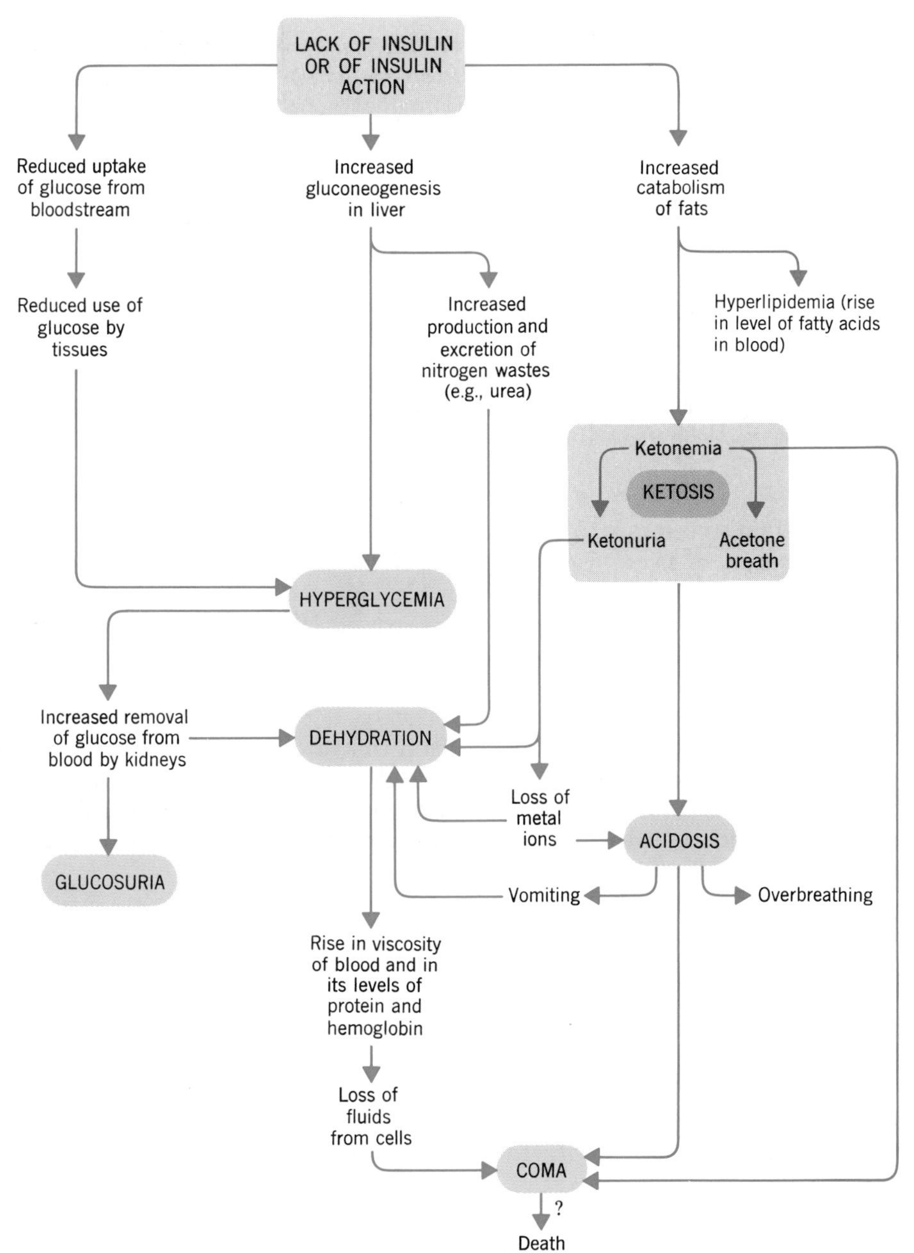

FIGURE 16.6
The principal sequence of events in untreated diabetes

The Conditions of Ketonemia, Ketonuria, and "Acetone Breath" Collectively Constitute Ketosis Normally, the levels of acetoacetate and β-hydroxybutyrate in the blood are, respectively, 2 μmol/dL and 4 μmol/dL. In prolonged, undetected, and untreated diabetes, these values can increase as much as 200-fold. The condition of excessive levels of ketone bodies in the blood is called **ketonemia.**

■ 1 μmol = 1 micromole = 10^{-6} mol.

As ketonemia becomes more and more advanced, the ketone bodies begin to appear in the urine, a condition called **ketonuria.** When there is a combination of ketonemia, ketonuria, and acetone breath, the overall state is called **ketosis.** The individual will be described as *ketotic.* As unchecked ketosis becomes more severe, the associated ketoacidosis worsens and the pH of the blood continues its fatal descent.

Condition	$[HCO_3^-]_{blood}$ in mmol/L
Normal	22–30
Mild acidosis	16–20
Moderate acidosis	10–16
Severe acidosis	<10

The Urinary Removal of Organic Anions Means the Loss of Base from the Blood To leave the ketone body anions in the urine, the kidneys have to leave positive ions with them to keep everything electrically neutral. Na^+ ions, the most abundant cations, are used. One Na^+ ion has to leave with each acetoacetate ion, for example. This loss of Na^+ is often referred to as the "loss of base" from the blood, although Na^+ is not a base. But the loss of one Na^+ stems from the appearance of one acetoacetate ion *plus one* H^+ *ion* that the blood had to neutralize. Thus each Na^+ that leaves the body corresponds to the loss of one HCO_3^- ion, the true base, consumed in neutralizing one H^+. Hence, the loss of Na^+ is taken as an indicator of the loss of this true base.

Another way to understand the urinary loss of Na^+ as the loss of base from the blood is that a Na^+ ion has to accompany a bicarbonate ion when it goes from the kidneys into the blood. The kidneys manufacture HCO_3^- ions normally in order to replenish the blood buffer system. The greater the number of Na^+ ions that have to be left in the *urine* in order to clear ketone bodies from the blood, the less the amount of true base, HCO_3^-, that can be put into the *blood*.

Diuresis Must Accelerate to Handle Ketosis The solutes that are leaving the body in the urine cannot, of course, be allowed to make the urine too concentrated. Otherwise, osmotic pressure balances are upset. Therefore increasing quantities of water must be excreted. To satisfy this need, the individual has a powerful thirst. Other wastes, such as urea, are also being produced at higher than normal rates, because amino acids are being sacrificed in gluconeogenesis. These wastes add to the demand for water to make urine.

Polyuria is the technical name for the overproduction of urine.

Internal Water Shortages in Ketosis Spell Dehydration of Critical Tissues If, during a state of ketosis, insufficient water is drunk, then water is simply taken from extracellular fluids. The blood volume therefore tends to drop, and the blood becomes more concentrated. It also thickens and becomes more viscous, which makes the delivery of blood more difficult.

Because the brain has the highest priority for blood flow, some of this flow is diverted from the kidneys to try to ensure that the brain gets what it needs. This only worsens the situation in the kidneys, and they have an increasingly difficult time clearing wastes. As the water shortage worsens, some water is borrowed from the intracellular supply. This, in addition to a combination of other developments, leads to coma and eventually death.

SUMMARY

Lipid absorption and distribution As fatty acids and monoacylglycerol move out of the digestive tract they become reconstituted as triacylglycerols. These, in addition to cholesterol and proteins, are packaged as chylomicrons. As chylomicrons move in the bloodstream, they unload some of their triacylglycerols and change to chylomicron remnants, which are taken up by the liver. The liver organizes cholesterol, triacylglycerols available to it, and proteins into VLDL and then sends these very-low-density lipoprotein complexes into circulation. Triacylglycerols are again unloaded where they are needed, and the VLDL become more dense, changing into intermediate-density complexes, IDL. Much of these are reabsorbed by the liver. Those that aren't become more dense and change over to low-density lipoprotein complexes, LDL. Most of the LDL finds its way to extrahepatic tissues to deliver cholesterol for cell membranes. Endocrine glands need cholesterol to make steroid hormones. LDL can be reabsorbed by the liver to recycle its lipids. Cholesterol not needed in extrahepatic tissue is carried back to the liver by high-density lipoprotein complexes, HDL. The liver can excrete excess cholesterol via the bile, or it makes bile salts.

Storage and mobilization of lipids The favorable energy density of triacylglycerol means that more energy is stored per gram of this material than can be stored by any other chemical system. The adipose tissue is the principal storage site, and fatty material comes and goes from this tissue according to the energy budget of the body. When the energy of fatty acids is needed, they are liberated from triacylglycerols and carried to the liver, the chief site of fatty acid catabolism.

Catabolism of fatty acids Fatty acyl groups, after being pinned to coenzyme A inside mitochondria, are catabolized by the β-oxidation pathway. By a succession of four steps—dehydrogenation, hydration of a double bond, oxidation of the resulting alcohol, and cleavage of the bond from the α- to the β-carbon—one cycle of the β-oxidation pathway removes one two-carbon acetyl group. The pathway then repeats another cycle as the shortened fatty acyl group continues to be degraded. Each cycle of reactions produces one $FADH_2$ and one NADH, which pass $H:^-$ to the respiratory chain for the synthesis of ATP. Each cycle of β-oxidation also sends one acetyl group into the citric acid cycle which, via the respiratory chain, leads to several more ATPs. The net ATP production is 129 ATPs per palmityl residue.

Biosynthesis of fatty acids Fatty acids can be made by a repetitive cycle of steps. It begins by building one butyryl group from two acetyl groups. The four-carbon butyryl group is attached to an acyl carrier protein that acts as a swinging arm on the enzyme complex. This arm moves the growing fatty acyl unit first over one enzyme and then another as additional two-carbon units are added. The process consumes ATP and NADPH.

Biosynthesis of cholesterol Cholesterol is made from acetyl groups by a long series of reactions. The synthesis of one of the enzymes is inhibited by excess cholesterol, which gives the system a mechanism for keeping its own cholesterol synthesis under control.

Ketoacidosis Acetoacetate, β-hydroxybutyrate, and acetone build up in the blood—ketonemia—in starvation or in diabetes. The first two are normal sources of energy in some tissues. When they are made faster than they can be metabolized, however, there is an accompanying increased loss of bicarbonate ion from the blood's carbonate buffer. This leads to a form of metabolic acidosis called ketoacidosis.

The kidneys try to leave the anionic ketone bodies in the urine, but this requires both Na^+ (for electrical neutrality) and water (for osmotic pressure balances). The excessive loss of Na^+ in the urine is interpreted as a loss of "base." With Na^+ leaving the body in the urine, less is available to accompany replacement HCO_3^-, made by the kidneys, when this base should be going into the blood. Under developing ketoacidosis, the kidneys have extra nitrogen wastes and, in diabetes, extra glucose that need to leave via the kidneys in the urine. This also demands an increased volume of water. Unless this is brought in by the thirst mechanism, it has to be sought from within. However, the brain has first call on blood flowage, so the kidneys suffer more. Eventually, if these events continue unchecked, the victim goes into a coma and dies.

REVIEW EXERCISES

The answers to Review Exercises whose numbers are in color are found in Appendix A. The answers to the other Review Exercises are found in the Study Guide that accompanies this book. The more challenging questions are marked with asterisks.

Absorption and Distribution of Lipids

16.1 What are the end products of the digestion of triacylglycerols?

16.2 What happens to the products of the digestion of triacylglycerols as they migrate out of the intestinal tract?

16.3 What are chylomicrons and what is their function?

16.4 What happens to chylomicrons as they move through capillaries of, say, adipose tissue?

16.5 What happens to chylomicron remnants when they reach the liver?

16.6 What are the two chief sources of cholesterol that the liver exports?

16.7 What do the following symbols stand for?
(a) VLDL (b) IDL
(c) LDL (d) HDL

16.8 The loss of what kind of substance from the VLDL converts them into IDL?

16.9 What happens to cause the increase in density between VLDL and LDL?

16.10 What tissue can reabsorb IDL complexes?

16.11 What is the chief constitutent of LDL?

16.12 In extrahepatic tissue, what two general uses await delivered cholesterol?

16.13 If the liver lacks the key receptor proteins, which specific lipoprotein complexes can't be reabsorbed?

16.14 Explain the relationship between the liver's receptor proteins for lipoprotein complexes and the control of the cholesterol level of the blood.

16.15 What is the chief job of the HDL?

***16.16** Why does HDL have a higher density than chylomicrons?

Storage and Mobilization of Lipids

16.17 With reference to the storage of chemical energy in the body, what is meant by *energy density?*

16.18 Arrange the following in their order of increasing quantity of energy that they store per gram.

Wet glycogen	Adipose lipids	Isotonic glucose
1	**2**	**3**

16.19 Briefly describe two conditions in which the body would have to turn to the fatty acids for energy.

16.20 How does insulin suppress the mobilization of fatty acids from adipose tissue?

16.21 Arrange the following processes in the order in which they occur when the energy in storage in triacylglycerols is mobilized.

β-Oxidation	Oxidative phosphorylation	Citric acid cycle
A	**B**	**C**

Lipoprotein formation	Lipolysis in adipose tissue
D	**E**

16.22 What specific function does β-oxidation have in obtaining energy from fatty acids?

16.23 What specific function does the citric acid cycle have in the use of fatty acids for energy?

***16.24** Name two hormones that activate the lipase in adipose tissue. Referring to the previous chapter, what does the presence of these hormones do for the blood sugar level?

*16.25 Explain how an increase in the blood sugar level indirectly inhibits the mobilization of energy from adipose tissue.

*16.26 When lipolysis occurs in adipose tissue, what happens to the glycerol?

Catabolism of Fatty Acids

16.27 How are long-chain fatty acids activated for β-oxidation?

*16.28 Write the equations for the four steps of β-oxidation as it operates on butyryl CoA. How many more cycles are possible after this one?

16.29 How is the FAD enzyme recovered from its reduced form, $FADH_2$, when β-oxidation operates?

16.30 How is the reduced form of the NAD^+ enzyme that is used in β-oxidation restored to its oxidized form?

16.31 Why is fatty acid catabolism called *beta* oxidation?

Biosynthesis of Fatty Acids

16.32 Where are the principal sites for each activity in a liver cell?
(a) fatty acid catabolism
(b) fatty acid synthesis

*16.33 Outline the steps that make butyryl ACP out of acetyl CoA.

16.34 What metabolic pathway in the body is the chief supplier of NADPH for fatty acid synthesis?

Biosynthesis of Cholesterol

16.35 The enzyme for the formation of which intermediate in cholesterol synthesis is the major control point in this pathway?

16.36 How does cholesterol itself work to inhibit the activity of the enzyme referred to in Review Exercise 16.35?

Ketoacidosis

16.37 What species is diverted from the citric acid cycle to gluconeogenesis?

*16.38 Why does the diversion of Review Exercise 16.37 lead to an increase in the level of acetyl CoA?

16.39 Two molecules of acetyl CoA can combine to give the coenzyme A derivative of what keto acid? Give its structure.

*16.40 In two steps, the compound of Review Exercise 27.39 gives one unit of a ketone body and one other significant species (besides recovered CoA). What is it? Why is it a problem?

16.41 Give the names and structures of the ketone bodies.

16.42 What is ketonemia?

16.43 What is ketonuria?

16.44 What is meant by acetone breath?

16.45 Ketosis consists of what collection of conditions?

16.46 What is ketoacidosis? What form of acidosis is it, metabolic or respiratory?

16.47 The formation of which particular compound most lowers the supply of HCO_3^- in ketoacidosis?

16.48 What are the reasons for the increase in the volume of urine that is excreted by someone with untreated type I diabetes?

*16.49 If the ketone bodies (other than acetone) can normally be used by heart and skeletal muscle, what makes them dangerous in starvation or in diabetes?

*16.50 Why does the rate of urea production increase in untreated, type I diabetes?

16.51 When a physician refers to the loss of Na^+ as the loss of *base*, what is actually meant?

Good and Bad Cholesterol (Special Topic 16.1)

16.52 What is atherosclerosis?

16.53 What is familial hypercholesterolemia and how do people get it?

16.54 Which of the lipoprotein complexes is sometimes called "bad cholesterol?" Explain why it is so designated.

16.55 Which lipoprotein complex carries "good cholesterol?"

16.56 Some scientists believe that a relatively high level of HDL is associated with protection against atherosclerosis, and that one's risk of having heart disease declines when the level of HDL is raised by exercise and losing weight. Why would a low level of HDL tend to be associated with heart disease?

16.57 Which substance is believed to be associated with the *most common* inherited risk factor for heart attack?

Brown Adipose Tissue (Special Topic 16.2)

16.58 With respect to the catabolism of fatty acids, how do white and brown adipose tissue differ?

16.59 In the ordinary operation of the respiratory chain, the chain is coupled to the synthesis of ATP because of what condition concerning mitochondria?

16.60 What happens to the energy released by the respiratory chain when it isn't used to make ATP?

16.61 What is meant by *thermogenesis?*

Additional Exercises

*16.62 Complete the following equations for one cycle of β-oxidation by which a six-carbon fatty acyl group is catabolized.

(a) $CH_3CH_2CH_2CH_2CH_2\overset{O}{\overset{\|}{C}}SCoA + FAD \rightarrow$ ______ + ______

(b) ______ $+ H_2O \rightarrow$ ______

(c) ______ $+ NAD^+ \rightarrow$ ______ $+ NADH + H^+$

(d) ______ $+ CoASH \rightarrow$ ______ + ______

*16.63 Myristic acid, $CH_3(CH_2)_{12}CO_2H$, can be catabolized by β-oxidation just like palmitic acid.
(a) How many units of acetyl CoA can be made from it?
(b) In producing this much acetyl CoA, how many times does $FADH_2$ form and then deliver its hydrogen to the respiratory chain?
(c) Referring again to part (a), how many times does NADH form as acetyl CoA is produced and then delivers its hydrogen to the respiratory chain?
(d) Calculate the maximum net number of molecules of ATP that can be made by means of the catabolism of one molecule of myristic acid. (Table 16.2 provides clues about this calculation.)

METABOLISM OF NITROGEN COMPOUNDS 17

This fellow needs *meat*, and no arguments, please. Carnivores rely on weaker, slower, or less alert animals to supply them their needed amino acids. How amino acids are broken down and used by the body is studied in this chapter.

17.1

SYNTHESIS OF AMINO ACIDS IN THE BODY

The body can manufacture a number of amino acids from intermediates that appear in the catabolism of nonprotein substances.

Amino acids, the end products of protein digestion, are rapidly transported across the walls of the small intestine. Some very small, simple peptides can also be absorbed. Once amino acids enter circulation, they become part of what is called the **nitrogen pool.**

The *Nitrogen Pool* Consists of All Nitrogenous Compounds Anywhere in the Body

■ The polypeptides in proteins that serve as enzymes have a particularly rapid turnover.

Figure 17.1 illustrates the various compartments of the nitrogen pool and how they are interrelated. Amino acids enter the nitrogen pool not only as digestive products but also as products of the breakdown of proteins in body fluids and tissues. These are undergoing constant turnover, fairly rapidly among the liver proteins and those in the blood and quite slowly among muscle proteins.

As indicated in Figure 17.1, individual amino acids can be used in any one of the following ways, depending on the body's needs of the moment.

1. the synthesis of new or replacement proteins
2. the synthesis of such nonprotein nitrogen compounds as heme, creatine, nucleic acids, and certain hormones and neurotransmitters
3. the production of ATP or of glycogen and fatty acids, substances with the potential for making ATP
4. the synthesis of any needed nonessential amino acids

■ In both starvation and diabetes the body draws from its amino acid pool to make glucose.

The Body Uses Both Essential and Nonessential Amino Acids to Make Proteins

We do not have to take in *all* of the 20 amino acids in the diet; we are able to make roughly half of them ourselves from other substances in food. Those that *must* be in the diet are called the

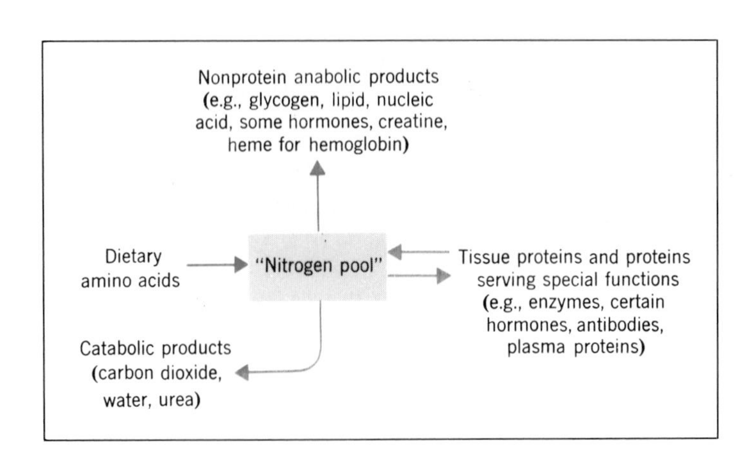

FIGURE 17.1
The nitrogen pool

essential amino acids. The others are called the *nonessential amino acids*. Be sure to remember that "nonessential" in this context refers only to a *dietary* need because, one way or another, the body must have on hand all of the amino acids whenever it needs to make polypeptides. We'll broadly study how the body makes nonessential amino acids next. Figure 17.2 gives an overview to which we'll refer as we go along.

The Reductive Amination of α-Ketoglutarate to Give Glutamate Makes the Ammonium Ion a Source of Nitrogen for Nonessential Amino Acids Many of the syntheses of nonessential amino acids outlined in Figure 17.2 depend on the availability of the glutamate ion. It is made from α-ketoglutarate by a reaction called **reductive amination,** in which the ammonium ion is the source of nitrogen and NADPH is the reducing agent. (In some cells, NADH enzymes can also work.) We need a reducing agent because the keto group (of

This reaction is the body's principal means for incorporating inorganic nitrogen (as NH_3) into organic compounds.

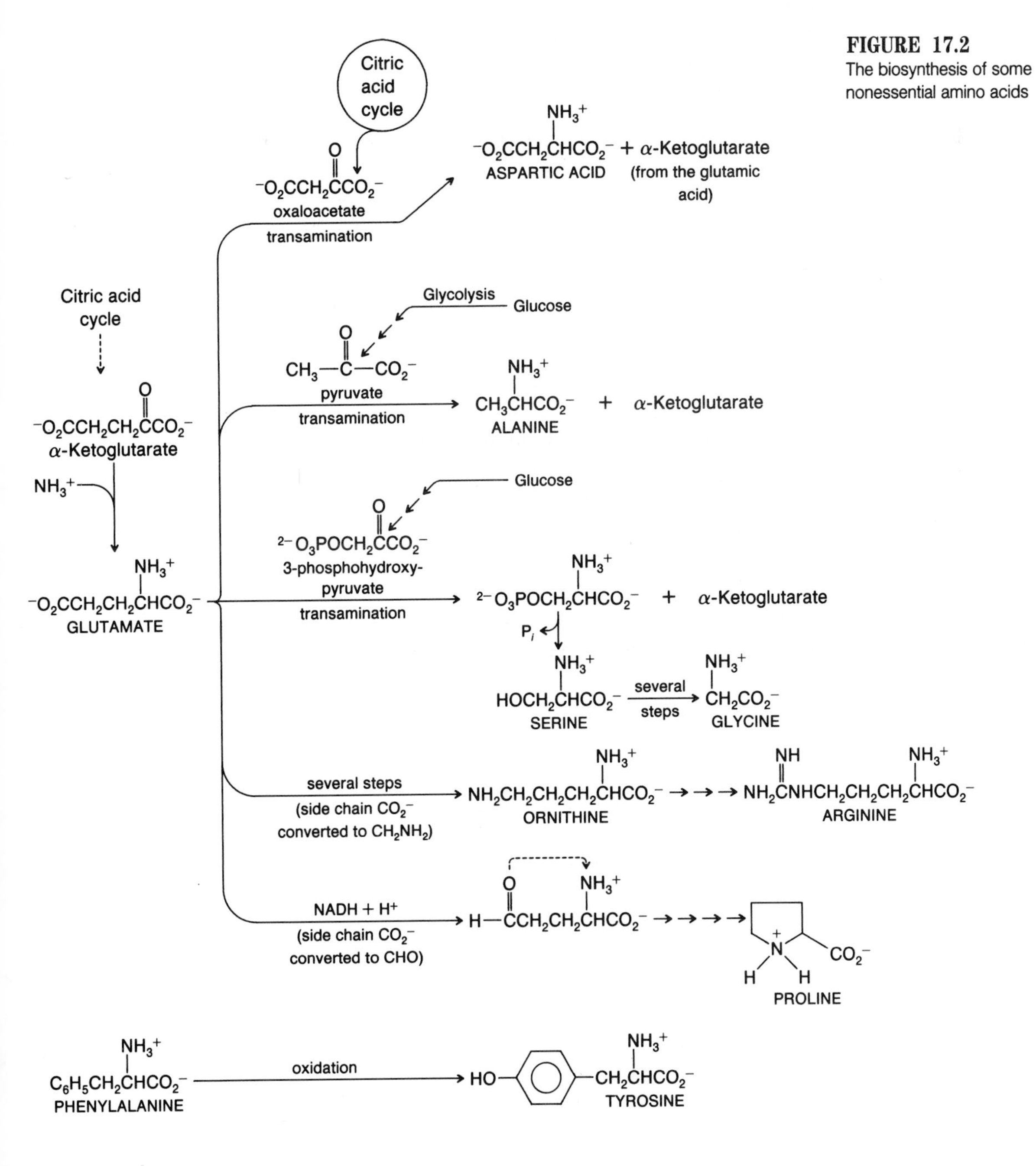

FIGURE 17.2
The biosynthesis of some nonessential amino acids

α-ketoglutarate) is in a higher oxidation state than the $CHNH_2$ unit in glutamate, the product. The overall result is

■ The enzyme is *glutamate dehydrogenase.*

$$\underset{\alpha\text{-Ketoglutarate}}{^{-}O_2CCH_2CH_2\overset{\overset{O}{\|}}{C}CO_2^-} + NH_4^+ + NADPH + H^+ \rightleftharpoons \underset{\text{Glutamate}}{^{-}O_2CCH_2CH_2\overset{\overset{NH_3^+}{|}}{C}HCO_2^-} + NADP^+ + H_2O$$

Glutamate Supplies the Amino Group to Make Many Nonessential Amino Acids

Special aminotransferase enzymes (sometimes called transaminases) are able to catalyze the transferral of an amino group from glutamate to the keto carbonyl group of a keto acid. The reaction is called **transamination.** The required ketone compounds can come from the catabolism of glucose, as you can see in Figure 17.2. We'll illustrate the general case of what happens in a transamination as follows.

■ The coenzymes for aminotransferases are compounds that collectively are called vitamin B_6.

$$\underset{\text{An }\alpha\text{-keto acid}}{R\overset{\overset{O}{\|}}{C}CO_2^-} + \underset{\text{Glutamate}}{^{-}O_2CCH_2CH_2\overset{\overset{NH_3^+}{|}}{C}HCO_2^-} \rightleftharpoons \underset{\text{An }\alpha\text{-amino acid}}{R\overset{\overset{NH_3^+}{|}}{C}HCO_2^-} + \underset{\alpha\text{-Ketoglutarate}}{^{-}O_2CCH_2CH_2\overset{\overset{O}{\|}}{C}CO_2^-}$$

17.2

CATABOLISM OF AMINO ACIDS

The breakdown products of amino acid catabolism eventually enter the pathways of catabolism of either carbohydrates or lipids.

There is no special mechanism for the storage of amino acids analogous to the storage system for glucose (glycogen) or for fatty acids (fat in adipose tissue). Amino acids in the nitrogen pool that aren't needed to make other amino acids or other nitrogen compounds are soon catabolized. Most of this work is done in the liver (indicating once again how important this organ is).

Amino Acids, When Stripped of Amino Groups, Can Be Used to Make Virtually Anything Else the Body Needs The ultimate end products of the complete catabolism of amino acids are urea, carbon dioxide, and water. On the way to these compounds, however, several intermediates form that can enter other pathways. In fact, all the pathways for the use of carbohydrates, lipids, and proteins are interconnected in one way or another (see Fig. 17.3).

■ 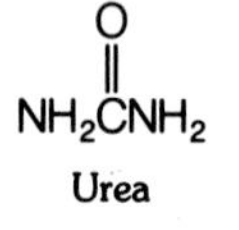

Urea

Notice in Figure 17.3 the central importance of two small molecules, acetyl CoA and pyruvate. Notice also that there is no route from acetyl CoA to pyruvate, which means that (in all animals, at least) *glucose cannot be made from fatty acids.* (We'll return to this shortly.)

We won't study in detail how each amino acid is catabolized, because each requires its own particular scheme, usually quite complicated. Early in each pathway, however, the amino acid gets rid of its nitrogen, which is shuttled into the synthesis of urea. The nonnitrogen fragment then eventually enters a pathway we have already studied. There are three kinds of reactions, besides transamination, that occur often: *oxidative deamination, direct deamination,* and *decarboxylation.* We'll study these and how they apply to certain selected amino acids.

Amino Groups Are Shuttled through the α-Ketoglutarate–Glutamate Switch toward Urea Synthesis One of the steps in the catabolic process that removes amino groups of amino acids is **oxidative deamination,** the name given to the reverse of reductive amination, studied on page 427. In the display that follows, the step at the first pair of curved

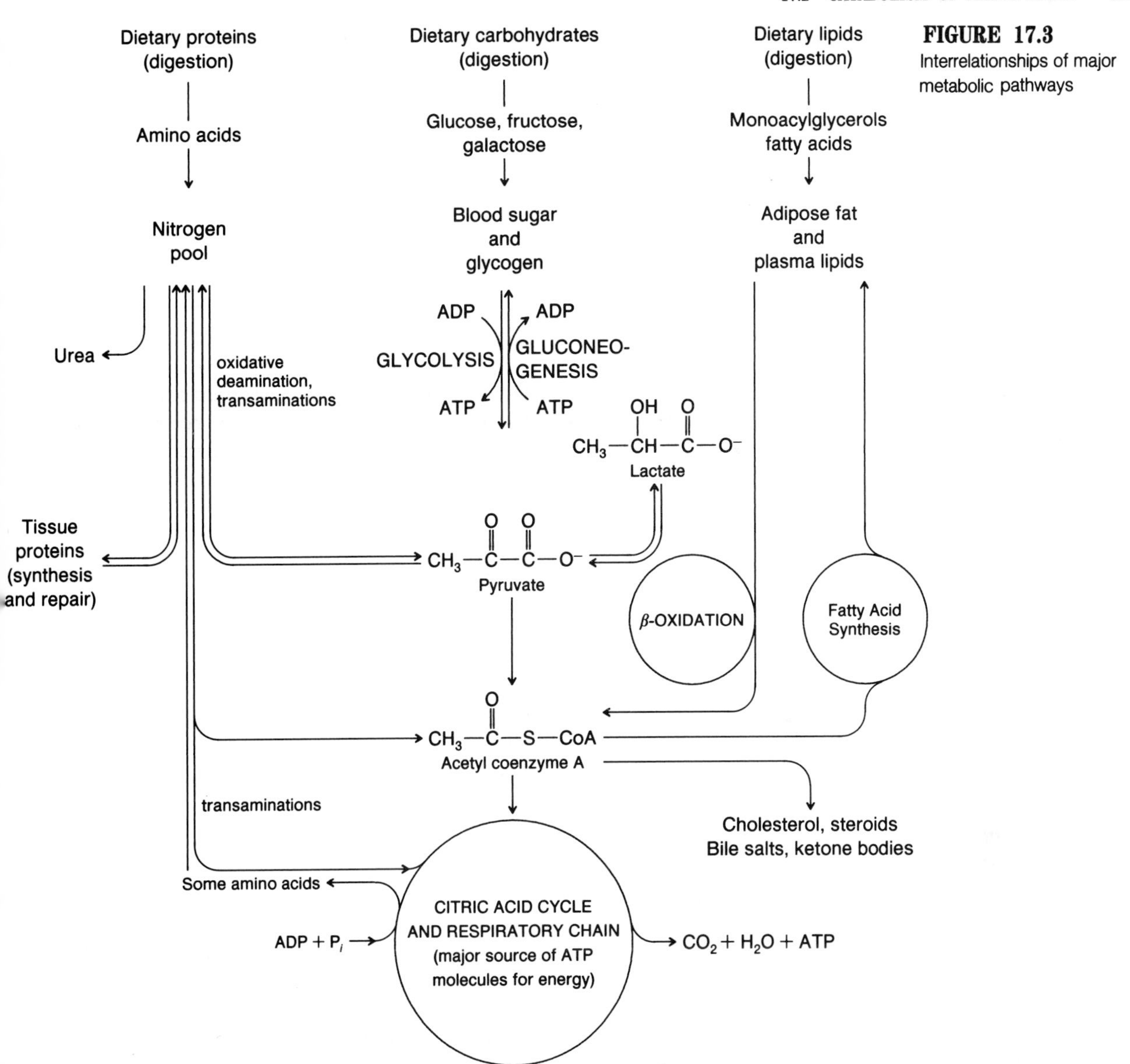

FIGURE 17.3 Interrelationships of major metabolic pathways

arrows is a transamination. Oxidative deamination occurs at the second pair, and the arrowheads point, left to right, in the direction of catabolism.

Most amino acids are deaminated by this route.

$$\underset{\alpha\text{-Amino acid}}{RCH(NH_3^+)CO_2^-} + \underset{\alpha\text{-Ketoglutarate}}{{}^-O_2CCH_2CH_2C(=O)CO_2^-} \rightleftharpoons \underset{\alpha\text{-Keto acid}}{RC(=O)CO_2^-} + \underset{\text{Glutamate}}{{}^-O_2CCH_2CH_2CH(NH_3^+)CO_2^-}$$

$$\text{Glutamate} + NAD^+ + H_2O \rightleftharpoons \alpha\text{-Ketoglutarate} + NADH + H^+ + NH_4^+ \xrightarrow[\text{steps}]{\text{several}} \text{Urea}$$

Notice that the nitrogen of the amino acid on the upper left ends up in urea, and that the α-ketoglutarate–glutamate pair provides a switching mechanism to convey this nitrogen in the right direction for catabolism.

Alanine Catabolizes to Pyruvate The transamination of alanine gives pyruvate, which can go into the citric acid cycle, be used in gluconeogenesis, or be used for the biosynthesis of fatty acids.

■ This display shows how excess alanine from the diet, alanine not needed that day to make proteins or other nitrogen compounds, can be used to make glucose, glycogen, or fatty acids according to other needs.

$$CH_3CH(NH_3^+)CO_2^- \text{ (Alanine)} \xrightarrow{\text{transamination}} CH_3C(=O)CO_2^- \text{ (Pyruvate)} \longrightarrow \text{Acetyl CoA} \longrightarrow \text{citric acid cycle}$$

Pyruvate → gluconeogenesis; Acetyl CoA → fatty acid synthesis

Aspartic Acid Catabolizes to Oxaloacetate The transamination of aspartic acid gives oxaloacetate, an intermediate in both gluconeogenesis and the citric acid cycle.

$$^-O_2CCH_2CH(NH_3^+)CO_2^- \text{ (Aspartic acid)} \xrightarrow{\text{transamination}} {}^-O_2CCH_2C(=O)CO_2^- \text{ (Oxaloacetate)} \longrightarrow \text{citric acid cycle}$$

Oxaloacetate → gluconeogenesis

Direct Deamination Removes Amino Groups without Oxidation Two amino acids, serine and threonine, have OH groups which make possible a nonoxidative loss of NH_3, called **direct deamination.** They are able to undergo the simultaneous loss of water and ammonia because their OH groups are strategically located on the carbon adjacent to the one that holds an amino group. Here's how direct deamination happens with serine.

■ During prolonged fasting or starvation, the body degrades its own proteins to make the brain's favorite source of energy, glucose, via gluconeogenesis.

$$HOCH_2CH(NH_3^+)CO_2^- \text{ (Serine)} \xrightarrow[-H_2O]{\text{(dehydration of alcohol)}} CH_2{=}C(NH_2^+H)CO_2^- \longrightarrow CH_3C(=NH)CO_2^- + H^+ \text{ (An imine)}$$

$$CH_3C(=NH)CO_2^- \xrightarrow[\text{(hydrolysis of the imine group)}]{H_2O} NH_3 + CH_3C(=O)CO_2^- \text{ (Pyruvate)}$$

The first step is the dehydration of the alcohol system of serine to give an unsaturated amine. This spontaneously rearranges into an imine, a compound with a carbon–nitrogen double bond. Water can add to this double bond, but the product spontaneously breaks up, so the net effect is the hydrolysis of the imine group to a keto group (in pyruvate) and ammonia. Thus serine breaks down to pyruvate, which, as we now well know, can send an acetyl group into the citric acid cycle, or can contribute an acetyl group to fatty acid synthesis, or can be used to make glucose.

■ Imine groups easily hydrolyze because they can add water and then split out ammonia.

Sustained Gluconeogenesis Necessarily Consumes Body Proteins Figure 17.4 broadly outlines how the reactions just surveyed are involved in the catabolism of several amino acids. Notice particularly that oxaloacetate occurs in two places, as an intermediate in the citric acid cycle and as the product of the oxidative deamination of aspartate. The oxaloacetate made from aspartate has two options: to be used to make ATP or to make

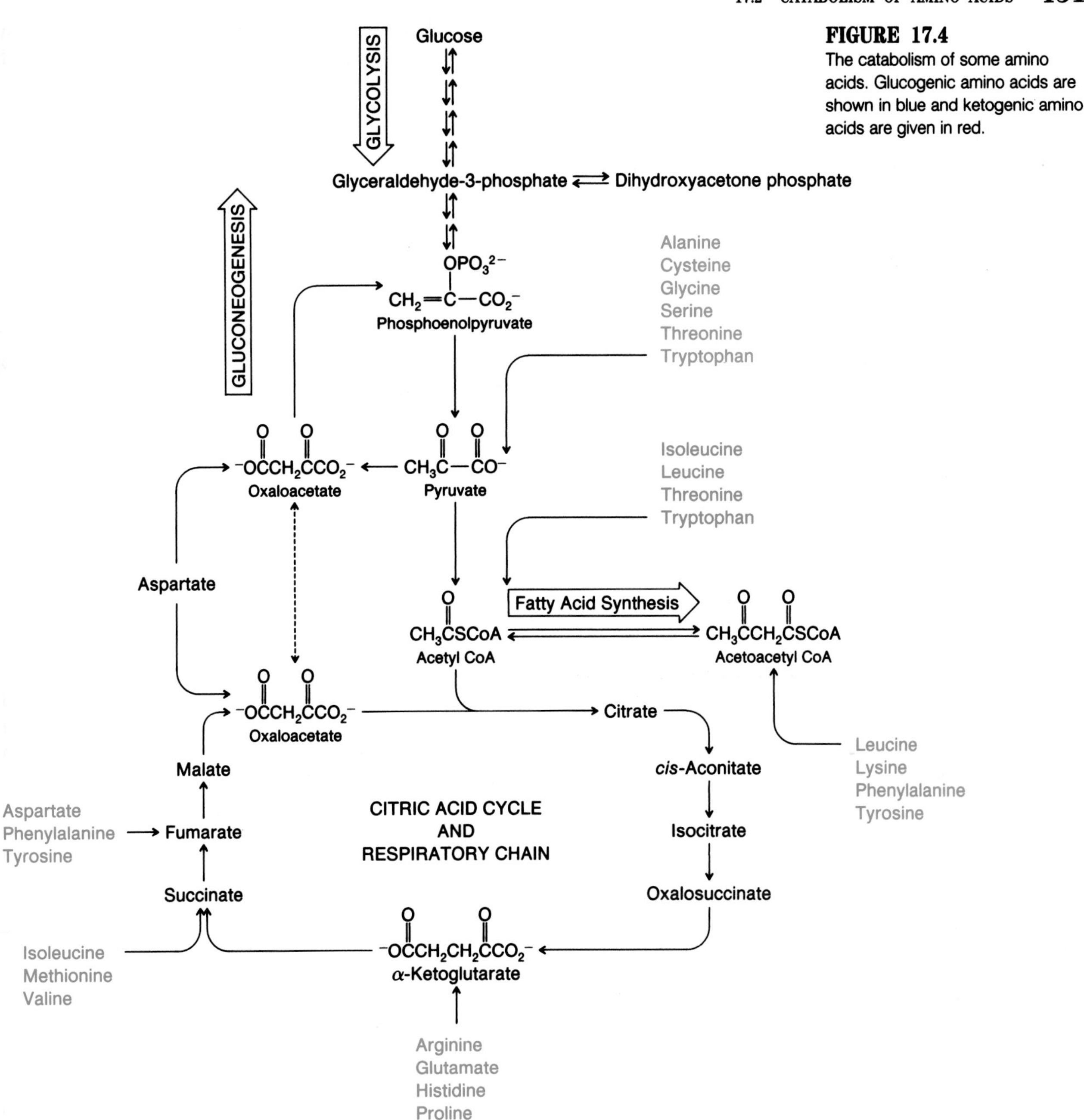

FIGURE 17.4
The catabolism of some amino acids. Glucogenic amino acids are shown in blue and ketogenic amino acids are given in red.

glucose by gluconeogenesis. Oxaloacetate thus connects the catabolism of amino acids to carbohydrate synthesis.

Because of the occurrence of oxaloacetate in the citric acid cycle, and because its carbons can originate in fatty acids (by way of acetyl coenzyme A), we might think that fatty acids could also be used to make glucose. Not so, at least *not for a net gain of glucose.* The removal of oxaloacetate from the citric acid cycle for gluconeogenesis means the removal of what carries acetyl units (of any origin) into the cycle. This leads to a backup of acetyl CoA (as we learned in Section 16.6) and a buildup of the ketone bodies.

For a *net gain* of glucose molecules via gluconeogenesis, the oxaloacetate of the citric acid cycle cannot be counted as available. *Only oxaloacetate made from amino acids can give a net gain of glucose this way.* This is why gluconeogenesis under conditions of starvation or diabetes necessarily breaks down body proteins. It needs some of their amino acids to make oxaloacetate to be able to make glucose.

The amino acids indicated in Figure 17.4 as sources of carbons for the synthesis of glucose are called the *glucogenic amino acids.* These are any that can be degraded to pyruvate, or to such citric acid cycle intermediates as α-ketoglutarate, succinate, fumarate, or oxaloacetate. Amino acids that can be degraded to acetyl CoA or acetoacetate can be used to make fatty acids and so are called the *ketogenic amino acids.* (Only leucine and lysine are exclusively ketogenic.) A few amino acids are both glucogenic and ketogenic.

The Decarboxylation of Some Amino Acids Leads to Neurotransmitters Some special enzymes can split out just the carboxyl groups from amino acids and so make amines. The reaction, called **decarboxylation,** is used to make some neurotransmitters and hormones. Dopamine, norepinephrine, and epinephrine are all made by steps that begin with the decarboxylation of dihydroxyphenylalanine, which the body makes from the amino acid tyrosine.

■ The symbol (O) signifies an oxidation step.

Tyrosine ($HOC_6H_4CH_2CH(NH_3^+)CO_2^-$) —(O)→ DOPA (dihydroxyphenylalanine) ($(HO)_2C_6H_3CH_2CH(NH_3^+)CO_2^-$) —($-CO_2$)→ Dopamine ($(HO)_2C_6H_3CH_2CH_2NH_2$) —(O)→ Norepinephrine ($(HO)_2C_6H_3CH(OH)CH_2NH_2$) —methylation→ Epinephrine ($(HO)_2C_6H_3CH(OH)CH_2NHCH_3$)

Tyrosine —($-CO_2$)→ Tyramine ($HOC_6H_4CH_2CH_2NH_2$) —(O)→ Dopamine

17.3 FORMATION OF UREA

Ammonia and amino groups are converted into urea by a complex cycle of reactions called the urea cycle.

Urea, as we have learned, is the chief nitrogen waste made by the body. Most of its nitrogen indirectly comes from amino acids, but some comes (also indirectly) from two of the sidechain bases of nucleic acids.

■ Recall that the amino group of glutamate can come from any other amino acid by way of another shuttle.

Oxaloacetate Shuttles Nitrogen from Glutamate to Aspartate and Then into the Synthesis of Urea There are two direct sources for the nitrogen atoms in urea. One is the ammonium ion produced by the oxidative deamination of glutamate.

The other nitrogen atom in urea comes from a specific amino acid, aspartate. However, because aspartate can be made from glutamate by a transamination, you can see that

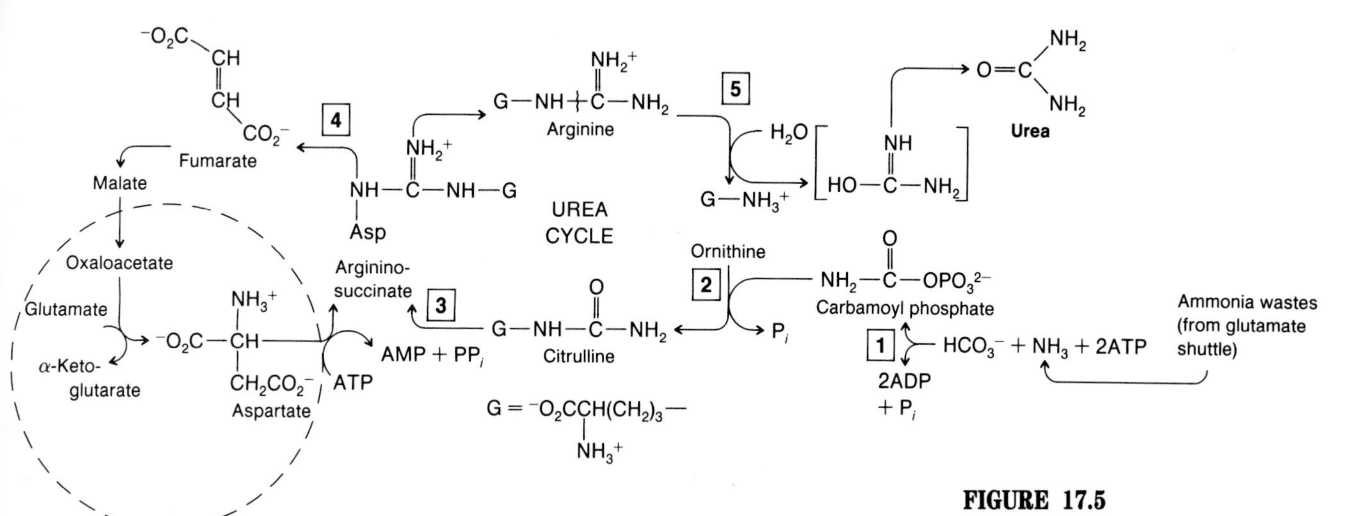

FIGURE 17.5
The urea cycle. The boxed numbers refer to the text discussion. The dashed-line circle is the aspartate–oxaloacetate shuttle also discussed in the text.

glutamate is close to being the direct source of both nitrogens in urea. Here is the shuttle from glutamate to aspartate just described.

$$^-O_2CCH_2CH_2CH(NH_3^+)CO_2^- \ (\text{Glutamate}) + {}^-O_2CC(=O)CH_2CO_2^- \ (\text{Oxaloacetate}) \longrightarrow {}^-O_2CCH_2CH_2C(=O)CO_2^- \ (\alpha\text{-Ketoglutarate}) + {}^-O_2CCH(NH_3^+)CH_2CO_2^- \ (\text{Aspartate})$$

$$\text{Aspartate} \longrightarrow NH_3 \xrightarrow{\text{To the urea cycle}}$$

The Urea Cycle Is the Only Way the Body Has to Make Urea

A series of reactions called the **urea cycle** manufactures urea from ammonium ion, carbon dioxide, and aspartate. Figure 17.5 displays its steps, and the boxed numbers in this figure refer to the following discussion.

■ The urea cycle is sometimes called the *Krebs ornithine cycle.*

1. Ammonia, with the help of ATP, reacts with CO_2 to form carbamoyl phosphate, a high-energy phosphate. In a sense, this is an activation of ammonia that launches it into the next step that takes it into the cycle.
2. The carbamoyl group transfers to the carrier unit, ornithine, as P_i is ejected. This consumes energy from high-energy phosphate. Citrulline forms.
3. Citrulline condenses with the alpha amino group of aspartate to give argininosuccinate.
4. Fumarate forms from the original aspartate as the amino group stays with the arginine that emerges. Fumarate is an intermediate in the citric acid cycle. By a transamination that involves glutamate, it is reconverted to aspartate.
5. Arginine is hydrolyzed. Urea forms and ornithine is regenerated to start another turn of the cycle.

■ This step occurs insides a mitochondrion.

■ Steps 2 through 5 occur in the cell's cytosol.

The overall result of the urea cycle is given by the following equation (which, as Figure 17.5 makes clear, is extremely simplified):

$$2NH_3 + H_2CO_3 \longrightarrow \underset{\text{Urea}}{NH_2\overset{O}{\overset{\|}{C}}NH_2} + 2H_2O$$

To do justice to the overall event, we must factor in the ATP consumption, as follows:

$$NH_4^+ + CO_2 + 3ATP + \text{aspartate} \longrightarrow \text{urea} + 2ADP + \text{fumarate} + 4P_i + AMP$$

At Elevated Levels, the Ammonium Ion Is Toxic If an infant is born without any one of the enzymes needed for the five steps of the urea cycle, it will die soon after birth. It will be unable, on its own, to clear ammonia from its blood. (Prior to birth, the mother's metabolism handles this.) Ammonia and the ammonium ion are toxic at sufficiently high levels.

■ There are many genetic defects involving enzymes for the body's utilization of amino acids.

Some inherited genetic defects produce enzymes for this cycle that have reduced activity. Such individuals have a condition called **hyperammonemia,** an elevated level of NH_3 in the blood. Infants that have this genetic defect improve on low-protein diets. If the level is not high enough to cause death, it can be expected to cause mental retardation. Periodic but unremembered episodes of bizarre behavior by one man—babbling, pacing, crying, glassy eyes—were not understood until it was found that he had a rare deficiency of the enzyme for step 2 of the urea cycle (ornithine transcarboxylase).

17.4

CATABOLISM OF OTHER NITROGEN COMPOUNDS

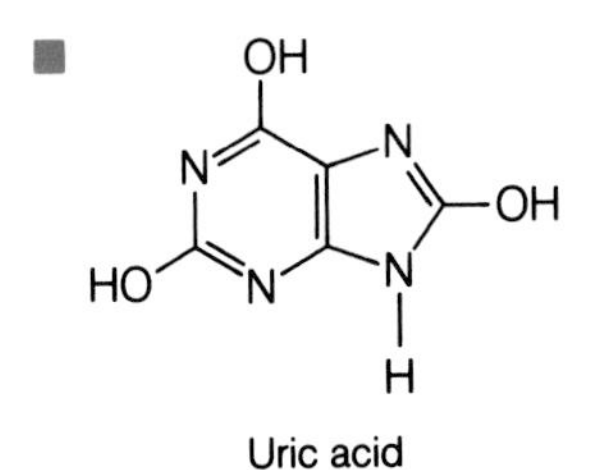

Uric acid

Uric acid and the bile pigments are other end products of the catabolism of nitrogen compounds.

The nitrogen of the purine bases of nucleic acids, adenine (A) and guanine (G), is excreted as uric acid, which also has the purine nucleus. After studying how uric acid forms, we'll see how defects in this pathway can lead to gout or to a particularly difficult disease of children, the Lesch-Nyhan syndrome.

Purine

The Catabolism of AMP Gives the Urate Ion The numbered steps in Figure 17.6 are discussed next to show how the adenine unit of AMP can be used to make uric acid.

1. A transamination removes the amino group of the adenine side chain in adenosine monophosphate, AMP.

FIGURE 17.6
The catabolism of adenosine monophosphate, AMP. The boxed numbers refer to the discussion in the text.

AMP (Adenosine monophosphate) → [1] → Inosine → [2] → Hypoxanthine + Ribose—P

Hypoxanthine → [3] → Xanthine; Hypoxanthine (recycle) → Nucleotides

Xanthine → [4] → Uric acid (Keto form ⇌ Phenolic form)

2. Ribose phosphate is removed and will enter the pentose phosphate pathway of carbohydrate catabolism. The product is hypoxanthine.
3. An oxidation produces xanthine. (The steps from guanine lead to xanthine, too.)
4. Another oxidation produces the keto form of uric acid, which exists partly in the form of a phenol. Actually, it's the salt of uric acid that forms, sodium urate, because the acid is neutralized by base in the buffer system.

Overproduction of the Urate Ion Causes Gout In the disease known as *gout,* the rate of formation of sodium urate is more rapid than its rate of elimination. Crystals of this salt precipitate in joints where they cause painful inflammations and lead to a form of arthritis. Kidney stones may form as this salt comes out of solution in this organ.

Just why the formation of sodium urate accelerates isn't well understood, but genetic factors are involved. Normally, some of the hypoxanthine made in step 2 is recycled back to nucleotide bases that are needed to make nucleic acids or high-energy phosphates. Some individuals with gout are known to have a partial deficiency of the enzyme system required for this recycling of hypoxanthine. Hence, most if not all of their hypoxanthine ends up as more sodium urate than normal.

■ Ethyl alcohol accelerates the synthesis of urate ion by increasing the rate of catabolism of adenosine monophosphate.

The Absence of One Enzyme Needed for Hypoxanthine Recycle Leads to Self-mutilating Behavior in Infants In Lesch-Nyhan syndrome, the enzyme for recycling hypoxanthine is totally lacking. The result is both bizarre and traumatic. Infants with this syndrome develop compulsive, self-destructive behavior at age 2 or 3. Unless their hands are wrapped in cloth, they will bite themselves to the point of mutilation. They act with dangerous aggression toward others. Some become spastic and mentally retarded. Kidney stones develop early, and gout comes later.

■ The lack of one enzyme usually means great personal and family suffering.

The Catabolism of Heme Produces the Bile Pigments Erythrocytes have life spans of only about 120 days. Eventually they split open. Their hemoglobin spills out and then is degraded. Its breakdown products are eliminated via the feces and, to some extent, in the urine. In fact, the characteristic colors of feces and urine are caused by partially degraded heme molecules called the **bile pigments.**

The degradation of heme begins before the globin portion breaks away. The heme molecule partly opens up to give a system that has a chain of four small rings called pyrrole rings. (This is why the bile pigments are sometimes called the *tetrapyrrole pigments.*)

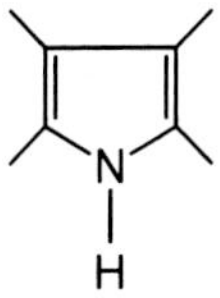

■ Pyrrole skeleton

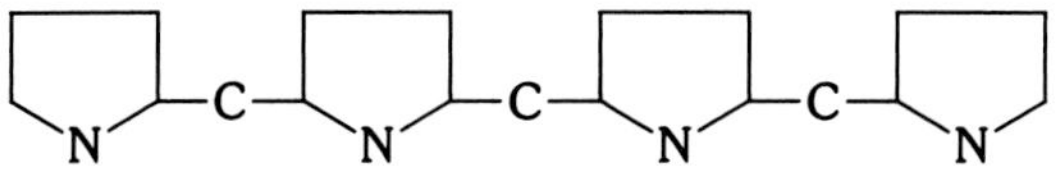

Carbon skeleton of the bile pigments

The rings have varying numbers of double bonds according to the state of oxidation of the pigment.

The slightly broken hemoglobin molecule, now called verdohemoglobin, then splits into globin, iron(II) ion, and a greenish pigment called **biliverdin** (Latin *bilis,* bile, + *virdus,* green). Globin enters the nitrogen pool. Iron is conserved in a storage protein called ferritin and is reused. Biliverdin is changed in the liver to a reddish-orange pigment called **bilirubin** (Latin *bilis,* bile + *rubin,* red). Bilirubin is not only made by the liver but is also removed from circulation by the liver, which transfers it to the bile. In this fluid it finally enters the intestinal tract.

The pathway from hemoglobin to bilirubin after the rupture of an erythrocyte, as well as the fate of bilirubin, is shown in Figure 17.7.

Bilirubin is the principal bile pigment in humans. In the intestinal tract, bacterial enzymes convert bilirubin to a colorless substance called *mesobilirubinogen.* This is further processed to form a substance known as **bilinogen,** which usually goes by other names that describe

SPECIAL TOPIC 17.1
JAUNDICE AND THE TETRAPYRROLE PIGMENTS

Jaundice (French, *jaune,* yellow) is a condition that is symptomatic of a malfunction somewhere along the pathway of heme metabolism. If bile pigments accumulate in the plasma in concentrations high enough to impart a yellowish coloration to the skin, the condition of *jaundice* is said to exist. Jaundice can result from one of three kinds of malfunctions.

Hemolytic jaundice results when hemolysis takes place at an abnormally fast rate. Bile pigments, particularly bilirubin, form faster than the liver can clear them. Hepatic diseases such as infectious hepatitis and cirrhosis sometimes prevent the liver from removing bilirubin from circulation. The stools are usually clay-colored, because the pyrrole pigments do not reach the intestinal tract.

Obstructions of bile ducts can prevent release of bile into the intestinal tract, and the tetrapyrrole pigments in bile cannot be eliminated. Under these circumstances, they tend to reenter general circulation. The kidneys remove large amounts of bilirubin, but the stools are usually clay-colored. As the liver works harder and harder to handle its task of removing excess bilirubin, it can weaken and become permanently damaged.

differences in destination rather than structure. Thus bilinogen that leaves the body in the feces is called *stercobilinogen* (Latin, *stercus,* dung). Some bilinogen is reabsorbed via the bloodstream, comes to the liver, and finally leaves the body in the urine. Now it is called *urobilinogen*. Some bilinogen is reoxidized to give **bilin,** a brownish pigment. Depending on its destination, bilin is called *stercobilin* or *urobilin*.

Special Topic 17.1 describes how the bile pigments are involved in jaundice.

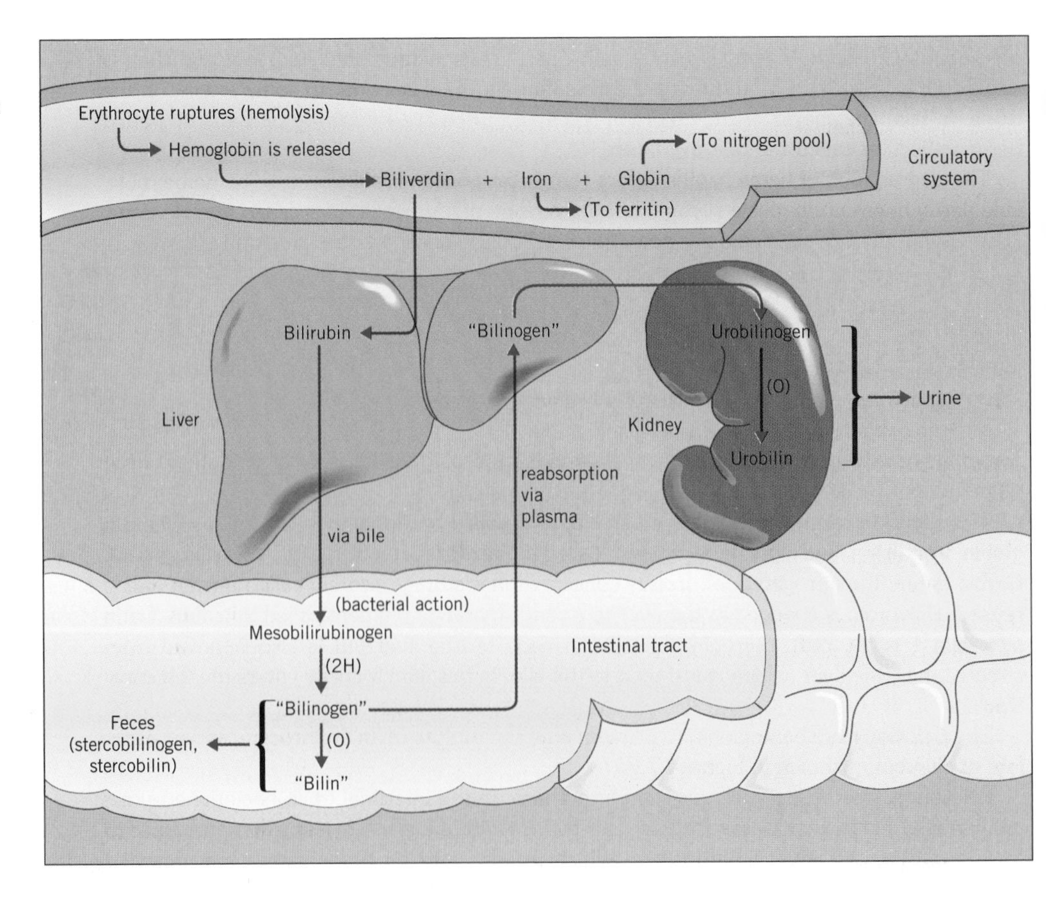

FIGURE 17.7 The formation and the elimination of the products of the catabolism of hemoglobin.

SUMMARY

Amino acid distribution The nitrogen pool receives amino acids from the diet, from the breakdown of proteins in body fluids or tissues, and from any synthesis of nonessential amino acids that occurs. Amino acids are used to build and repair tissue, replace proteins of body fluids, make nonprotein nitrogen compounds, provide chemical energy if needed, and supply molecular parts for gluconeogenesis or fatty acid synthesis.

Amino acid metabolism By reactions of transamination, oxidative deamination, direct deamination, and decarboxylation, the α-amino acids shuttle amino groups between themselves and intermediates of the citric acid cycle, or the synthesis of urea and nonprotein nitrogen compounds.

Deaminated amino acids eventually become acetyl CoA, acetoacetyl CoA, pyruvate, or an intermediate in the citric acid cycle. The skeletons of most amino acids can be used to make glucose, or fatty acids, or the ketone bodies. Their nitrogen atoms become part of urea.

Metabolism of other nitrogen compounds The nitrogen atoms in some of the sidechain bases of nucleic acids end up in urea and those of the others are excreted as sodium urate. Urea is made by a complex cycle of reactions—the urea cycle.

Heme is catabolized to bile pigments and its iron is reused. The pigments—first, biliverdin (green), then bilirubin (red), then mesobilirubinogen (colorless), and finally bilinogen and bilin (brown)—become stercobilin or stercobilinogen, urobilin or urobilinogen, depending on the route of elimination.

REVIEW EXERCISES

The answers to Review Exercises whose numbers are in color are found in Appendix A. The answers to the other Review Exercises are found in the Study Guide that accompanies this book. The more challenging questions are marked with asterisks.

Nitrogen Pool

17.1 What is the nitrogen pool?

17.2 What are four ways in which amino acids are used in the body?

17.3 When the body retains more nitrogen than it excretes in all forms, the system is said to be on a *positive nitrogen balance*. Would this state characterize infancy or old age?

17.4 What happens to amino acids that are obtained in the diet but aren't needed to make any nitrogeneous compounds?

Biosynthesis of Amino Acids

17.5 Which amino acid is the major supplier of amino groups for the synthesis of the nonessential amino acids?

17.6 Write the equation for the reductive amination that produces glutamate. (Use NADPH as the reducing agent.)

*17.7 Write the structure of the keto acid that forms when phenylalanine undergoes transamination with α-ketoglutarate.

***17.8** When valine and α-ketoglutarate undergo transamination, what new keto acid forms? Write its structure.

The Catabolism of Amino Acids

17.9 Cysteine is a glucogenic amino acid. What does this mean?

17.10 Lysine is exclusively a ketogenic amino acid. What does this mean?

17.11 An amino acid that can generate acetyl CoA without going through pyruvate is glucogenic or ketogenic?

***17.12** By means of two successive equations, one a transamination and the other an oxidative deamination, write the reactions that illustrate how the amino group of alanine can be removed as NH_4^+.

*17.13 Arrange the following compounds in the order in which they would be produced if the carbon skeleton of alanine were to appear in one of the ketone bodies.

Pyruvate	Acetoacetyl CoA	Acetoacetate
1	**2**	**3**
Alanine	Acetyl CoA	
4	**5**	

***17.14** In what order would the following compounds appear if some of the carbon atoms in glutamate were to become part of glycogen?

Oxaloacetate	α-Ketoglutarate	Glucose
1	**2**	**3**
Glycogen	Glutamate	
4	**5**	

17.15 Can any of the carbon atoms of glucose become part of alanine? If so, explain (in general terms).

17.16 Write the structure of the keto acid that forms by the direct deamination of threonine.

*17.17 When tyrosine undergoes decarboxylation, what forms? Write its structure.

***17.18** Write the structure of the product of the decarboxylation of tryptophan.

17.19 In the conditions of starvation or diabetes, what can the amino acids be used for?

The Formation of Urea

17.20 In the biosynthesis of urea, what are the sources of (a) the two NH_2 groups and (b) the C=O group?

17.21 What is hyperammonemia and, in general terms, how does it arise and how can it be handled in infants?

17.22 What is the overall equation for the synthesis of urea?

17.23 Would a deficiency in ornithine transcarbamylase (the enzyme for step 2 of the urea cycle, Figure 17.5) cause hypoammonemia or hyperammonemia? Explain.

The Catabolism of Nonprotein Nitrogen Compounds

17.24 What compounds are catabolized to make uric acid?

17.25 What product of catabolism accumulates in the joints in gout?

*17.26 Arrange the names of the following substances in the order in which they appear during the catabolism of heme.

Biliverdin	Heme	Hemoglobin	Bilirubin
1	**2**	**3**	**4**

Mesobilirubinogen	Bilin	Bilinogen
5	**6**	**7**

Tetrapyrrole Pigments and Jaundice (Special Topic 17.1)

17.27 Briefly describe what jaundice does to the body.

17.28 Describe how the following kinds of jaundice arise.
(a) hemolytic jaundice
(b) the jaundice of hepatic diseases

17.29 Why should an obstruction of the bile ducts cause jaundice?

Additional Exercises

***17.30** From a study of the figures in this chapter, can the carbon atoms of serine become a part of a molecule of palmitic acid? If so, write the names of the compounds, beginning with serine, in the sequence to the start of fatty acid synthesis.

*17.31 The carbon atoms of a ketogenic amino acid eventually are present among intermediates in the citric acid cycle, including oxaloacetate, a starting material for gluconeogenesis. Why, then, aren't all ketogenic amino acids also glucogenic?

NUTRITION

18

Specialists in nutrition counsel us to have a variety of foods in our diets, and from both land and sea a bountiful nature obliges. This chapter describes the kinds and amounts of nutrients we must have for maximum health.

18.1
GENERAL NUTRITIONAL REQUIREMENTS
18.2
PROTEIN REQUIREMENTS
18.3
VITAMINS
18.4
MINERALS AND TRACE ELEMENTS IN NUTRITION

18.1

GENERAL NUTRITIONAL REQUIREMENTS

The intent of the recommended dietary allowances (RDAs) is to help meal planners provide for the nutritional needs of practically all healthy people. **Nutrition** can be defined in both technical and personal terms. Technically, nutrition is a field of science that investigates the identities, the quantities, and the sources of substances, called *nutrients,* that are needed for health.

■ Two of the great scientific triumphs of the nineteenth century were the germ theory of disease and the birth of the science of nutrition.

In personal terms, we speak of our own nutrition, the sum of the foods taken in the proper proportions at the best moments as we try to maintain a state of well-being and avoid diet-related diseases and infirmities.

The science of **dietetics** is the application of the findings of the science of nutrition to feeding individual humans, whether they are ill or well.

This chapter is mostly about some of the major findings of the science of nutrition, a field that is so huge and complex that we can do little more than introduce some of its most important features and terms.

TABLE 18.1
Recommended Daily Dietary Allowances[a] of the Food and Nutrition Board, National Academy of Sciences National Research Council, Revised 1989

							Fat-Soluble Vitamins			
		Weight		Height						
Persons	Age (years)	(kg)	(lb)	(cm)	(in.)	Protein (g)	A (μg)[b]	D (μg)[c]	E (mg)[d]	K (μg)
Infants	0.0–0.5	6	13	60	24	13	375	7.5	3	5
	0.5–1.0	9	20	71	28	14	375	10	4	10
Children	1–3	13	29	90	35	16	400	10	6	15
	4–6	20	44	112	44	24	500	10	7	20
	7–10	28	62	132	52	28	700	10	7	30
Males	11–14	45	99	157	62	45	1000	10	10	45
	15–18	66	145	176	69	59	1000	10	10	65
	19–24	72	160	177	70	58	1000	10	10	70
	25–50	79	174	176	70	63	1000	5	10	80
	51+	77	170	173	68	63	1000	5	10	80
Females	11–14	46	101	157	62	46	800	10	8	45
	15–18	55	120	163	64	44	800	10	8	55
	19–24	58	128	164	65	46	800	10	8	65
	25–50	63	138	163	64	50	800	5	8	65
	51+	65	143	160	63	50	800	5	8	65
Pregnant						60	800	10	10	65
Lactating	First 6 months					65	1300	10	12	65
	Second 6 months					62	1200	10	11	65

[a] The allowances are intended to provide for individual variations among most normal persons as they live in the United States under usual environmental stresses. Diets should be based on a variety of common foods in order to provide other nutrients for which human requirements have been less well defined.

[b] Retinol equivalents. 1 retinol equivalent = 1 μg retinol or 6 μg β-carotene.

***Nutrients* Are Chemicals Required by Healthy Metabolism, and *Foods* Are Substances That Supply Them** When James Lind discovered in 1852 that oranges, lemons, and limes could cure scurvy, and K. Takadi in the 1880s found that a proper diet could cure beriberi, the science of nutrition was on its way. Good personal nutrition prevents a number of specific diseases and it cures several. It provides the correct nutrients for whatever happens to be the metabolic or health status of the body.

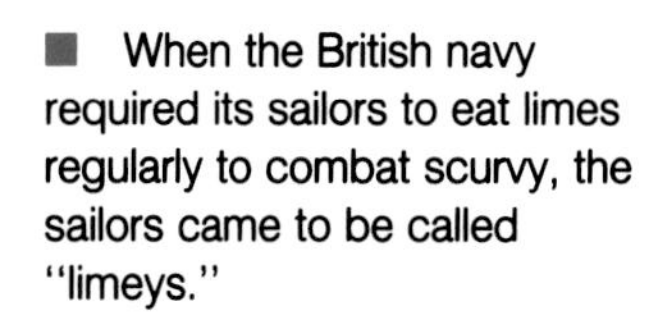
■ When the British navy required its sailors to eat limes regularly to combat scurvy, the sailors came to be called "limeys."

The **nutrients** are any chemical substances that take part in any nourishing, health-supporting, or health-promoting metabolic activity. Carbohydrates, lipids, proteins, vitamins, minerals, and trace elements are nutrients. Oxygen and water are usually not called nutrients, although they formally qualify under the definition. Materials that supply one or more nutrients are called **foods.**

The *Recommended Dietary Allowances* Describe What Nutrients Are Adequate for Healthy People For several decades, the Food and Nutrition Board of the National Research Council of the National Academy of Sciences has published what are known as the **recommended dietary allowances,** or the **RDAs.** In the judgment of this Board, these allowances are the intake levels of essential nutrients that are "adequate to meet the known nutritional needs of practically all healthy persons." The values of the RDAs, as revised in 1989, are given in Table 18.1.

A number of points and qualifications about the RDAs must be emphasized.

1. *The RDAs are not the same as the U.S. Recommended Daily Allowances (USRDAs).* The USRDAs are set by the U. S. Food and Drug Administration, based on the RDAs, as standards for nutritional information on food labels.
2. *The RDAs are not the same as the Minimum Daily Requirements (the MDRs) for any one individual.*

Water-Soluble Vitamins							Minerals						
Ascorbic acid (mg)	Folate (μg)	Niacin[e] (mg)	Riboflavin (mg)	Thiamine (mg)	Vitamin B_6 (mg)	Vitamin B_{12} (mg)	Calcium (mg)	Phosphorus (mg)	Iodine (μg)	Iron (mg)	Magnesium (mg)	Selenium (μg)	Zinc (mg)
30	25	5	0.4	0.3	0.3	0.3	400	300	40	6	40	10	5
35	35	6	0.5	0.4	0.6	0.5	600	600	50	10	60	15	5
40	50	9	0.7	0.7	1.0	0.7	800	800	70	10	80	20	10
45	75	12	0.9	0.9	1.1	1.0	800	800	90	10	120	20	10
45	100	13	1.0	1.0	1.4	1.4	800	800	120	10	170	30	10
50	150	17	1.5	1.3	1.7	2.0	1200	1200	150	12	270	40	15
60	200	20	1.8	1.5	2.0	2.0	1200	1200	150	12	400	55	15
60	200	19	1.7	1.5	2.0	2.0	1200	1200	150	10	350	70	15
60	200	19	1.7	1.5	2.0	2.0	800	800	150	10	350	70	15
60	200	15	1.4	1.2	2.0	2.0	800	800	150	10	350	70	15
50	150	15	1.3	1.1	1.4	2.0	1200	1200	150	15	280	45	12
60	180	15	1.3	1.1	1.5	2.0	1200	1200	150	15	300	50	12
60	180	15	1.3	1.1	1.6	2.0	1200	1200	150	15	280	55	12
60	180	15	1.3	1.0	1.6	2.0	800	800	150	15	280	55	12
60	180	13	1.2	1.1	1.6	2.0	800	800	150	10	280	55	12
70	400	17	1.6	1.5	2.2	2.2	1200	1200	175	30	320	65	15
95	280	20	1.9	1.6	2.1	2.6	1200	1200	200	15	355	75	19
90	280	20	1.7	1.6	2.1	2.6	1200	1200	200	15	340	75	16

[c] As cholecalciferol. 10 μg cholecalciferol = 400 IU of vitamin D.

[d] α-Tocopherol equivalents. 1 mg D-α-tocopherol = 1 α-tocopherol equivalent.

[e] 1 niacin equivalent (NE) equals 1 mg of niacin or 60 mg of dietary tryptophan.

The MDRs are just that, minimums. They are set very close to the levels at which actual signs of deficiencies occur. The RDAs are two to six times the MDRs. Just as individuals differ greatly in height, weight, and appearance, they also differ greatly in specific biochemical needs. Therefore an effort has been made to set the RDAs far enough above average requirements so that "practically all healthy people" will thrive. Most will receive more than they need; a few will not receive enough.

■ It's the job of hospital dieticians to devise the large variety of diets needed by the patient population.

One of the controversies over the RDAs concerns the validity of the statistical research and analysis by which "average requirements" were determined. Moreover, some nutritionists insist that among healthy people the range of daily need for a particular nutrient can vary far more widely than the Food and Nutrition Board has determined.

3. *The RDAs do not define therapeutic nutritional needs.*
People with chronic diseases, such as prolonged infections or metabolic disorders; people who take certain medications on a continuing basis; and prematurely born infants all require special diets. Recently the importance of certain vitamins to pregnant women, particularly in the first few weeks of pregnancy, has been established by controlled studies.

■ Many scientists urge women planning pregnancies to begin supplementing their diets with certain vitamins, like folate, *before* pregnancies.

The RDAs do cover people according to age, sex, and size, and they indicate special needs for pregnant and lactating women, but they do not include any other special needs. Therapeutic needs for water and salt increase during strenuous physical activity and prolonged exposure to high temperatures.

In some areas of the United States and in many parts of the world, intestinal parasites are common. These organisms rob the affected people of some of their food intake each day, and such people also need special diets.

4. *The RDAs can (and ought to) be provided in the diet from a number of combinations and patterns of food.*
No single food contains all nutrients. People take dangerous risks with their health when they go on fad diets limited to one particular food, like brown rice, gelatin, yogurt, or liquid protein. The ancient wisdom of a varied diet that includes meat, fruit, vegetables, grains, nuts, pulses (e.g., beans), and dairy products may seem to be supported solely by cultural and aesthetic factors. A varied diet, however, also assures us of getting any trace and needed nutrients that might not yet have been discovered. The U.S. Department of Agriculture uses a "food pyramid" for educating the public about how to organize personal nutrition to obtain all needed nutrients in healthy proportions through a varied diet (see Figure 18.1). Some objections to the food pyramid have been voiced by those who see it as reducing the value of meat in the diet.

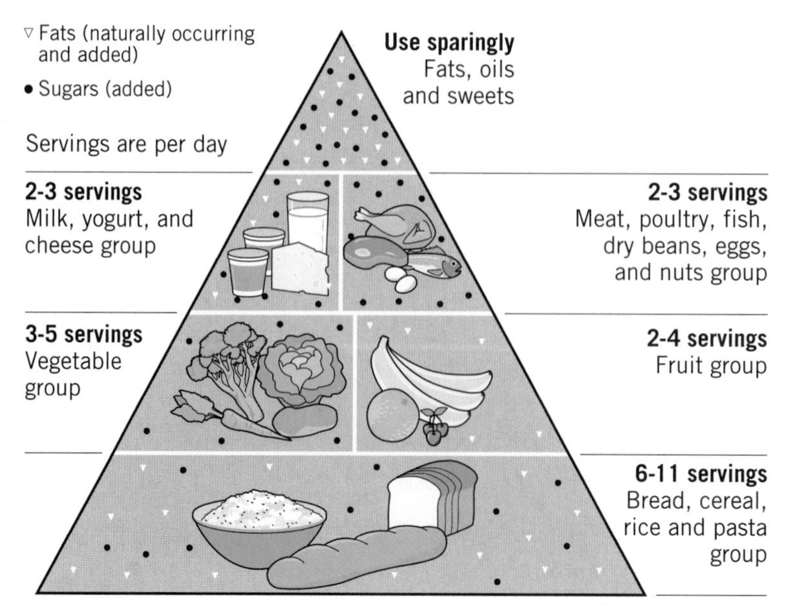

FIGURE 18.1
The Food Group Pyramid—a device for educating the public about nutrition. (*Source:* U.S. Department of Agriculture.)

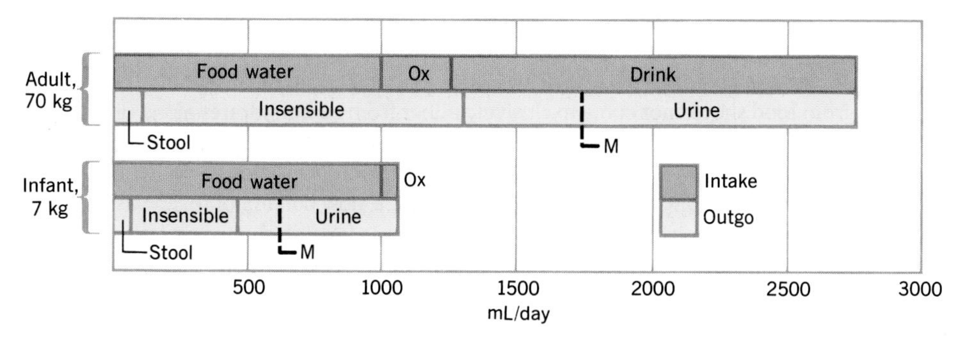

FIGURE 18.2
Water intake and outgo by the principal routes (excluding sensible perspiration or sweating). The dashed lines at **M** are the minimal volumes of urine at the maximal concentrations of solutes. "Ox" refers to water formed by the oxidation of foods. (*Source: Recommended Dietary Allowances,* 9th ed., 1980. National Academy of Sciences, Washington, D.C.)

Adults Have a Water Budget of 2.5 to 3 L per Day Our water intake comes from the fluids we drink, the water in the foods we eat, and the water made by the oxidations of nutrients in cells. Most comes in response to the thirst mechanism.

Water leaves the body via the urine, the feces, exhaled air, and perspiration. Figure 18.2 gives the relative quantities by each route for adults and infants.

Not shown in Figure 18.2 is the water that can be lost by sensible perspiration (sweat). Any activity in the hot sun of an arid desert can cause the loss of as much as 10 L of water per day along with body salts. If these losses aren't made up at a reasonable rate, the results can be heat exhaustion, heat stroke, heat cramps, and even death.

■ No amount of physical training or willpower can condition anyone to go without water.

Energy Needs Determine Oxygen Requirements Our daily oxygen needs vary with how much chemical energy we have to import each day to maintain metabolism under different activities. At rest, the largest user of oxygen in the body is the brain. Its mass is only about 2% of the body mass, but it consumes 20% of the oxygen used by the resting body. Much of the energy used in the brain goes to keep up concentration gradients of ions across the membranes of the billions of brain cells. In general, energy demands in the range of 2.0×10^3 to 3.0×10^3 kcal/day means a need for roughly 18 to 25 moles of oxygen per day. Table 18.2 gives data on energy demands for several activities.

TABLE 18.2
Daily Energy Expenditures of Adult Men and Women in Light Occupations

		Man (70 kg)		Woman (58 kg)	
		Rate		Rate	
Activity Category	Time (h)	kcal/min	Total (kcal)	kcal/min	Total (kcal)
Sleeping, reclining	8	1.0–1.2	540	0.9–1.1	440
Very light[a]	12	up to 2.5	1300	up to 2.0	900
Light[b]	3	2.5–4.9	600	2.0–3.9	450
Moderate[c]	1	5.0–7.4	300	4.0–5.9	240
Heavy[d]	0	7.5–12.0		6.0–10.0	
Total	24		2740		2030

Source: J. V. G. A. Durin and R. Passmore, 1967. In *Recommended Dietary Allowances,* 9th ed., 1980. National Academy of Sciences, Washington, D.C.

[a] Seated and standing activities; driving cars and trucks; secretarial work; laboratory work; sewing, ironing; playing musical instruments.

[b] Walking (on the level, 2.5 to 3 mph); tailoring and pressing; carpentry; electrical trades; restaurant work; washing clothes; light recreation such as golf, table tennis, volleyball, sailing.

[c] Walking at 3.5 to 4 mph; garden work; scrubbing floors; shopping with heavy load; moderate sports such as skiing, tennis, dancing, bicycling.

[d] Uphill walking with a load; pick-and-shovel work; lumbering; heavy sports, such as swimming, climbing, football, basketball.

Fad Diets Meant to Control Energy Intake Can Cause Grave Harm The energy obtained from food should not come exclusively either from carbohydrates alone or from fats alone. If you went on a zero-carbohydrate diet, your body would make glucose to meet its needs. (The brain derives nearly all its energy from the breakdown of glucose.) If the body has to make glucose over a long period of time, say weeks, a slow buildup of toxic wastes would cause harm. On the other hand, if you were to go on a zero-fat diet, then some fatty acids that are important to the body would be unavailable, and your system would not absorb the fat-soluble vitamins very well. A daily diet that includes at least 15 to 25 g of food fat (the equivalent of 2 to 4 pats of margarine or butter) and 50 to 100 g of digestible carbohydrate prevents these problems.

■ High-protein diets, if eaten for an extended period, can make the kidneys overwork and become damaged.

Linoleic Acid Must Be in the Diet Linoleic acid is called an **essential fatty acid,** and it is particularly needed for the health and growth of infants.[1] This acid is so poorly supplied by the animal fats in the diet, including milk fat, that we rely almost totally on vegetable oils for it. Our bodies use arachidonic acid to make prostaglandins (Special Topic 9.1, page 226), and this acid is made in the body from linoleic acid. Animals on a diet free of linoleic acid show very poor growth, impaired healing of wounds, and skin problems, and they have shortened lives. Thanks to the widespread presence of linoleic acid in all edible vegetable oils, it is virtually impossible for an adult not to obtain enough. Dietary linoleic acid lowers the serum level of the *low-density lipoprotein complex* or LDL, the "bad cholesterol," which gives another reason for adults to have vegetable oils in the diet. (See Special Topic 16.1, page 408.)

■ *Hydrogenated* vegetable oils are, like animal fats, deficient in the important fatty acids because hydrogenation removes the carbon–carbon double bonds of linoleic, linolenic, and arachidonic acids.

18.2

PROTEIN REQUIREMENTS

A protein's biological value is determined largely by its digestibility and its ability to supply the essential amino acids.
The total protein requirement is 56 g/day for adult men and 46 g/day for adult women. Small as these numbers are, they are still more than the bare minimum. Their intent is to allow about 30% extra to cover a wide range of variations in the protein needs of individuals. Moreover, they assume that we digest and absorb into the bloodstream only 75% of the protein in the diet. (These allowances are considered too narrow by some nutritionists.)

■ This is only about 2 ounces per day.

Our Protein Requirement Actually Means a Requirement for Essential Amino Acids Proteins supply α-amino acids, and these are what protein requirements are all about. We particularly need daily intakes of a small number of specific amino acids whose presence in the diet is so critical that they are called the **essential amino acids.**

Of the 20 amino acids used to make proteins, adults can synthesize 12 from parts and pieces of other molecules, including other amino acids (as discussed in Section 17.1). This leaves eight essential amino acids, those that the body cannot synthesize, at least not at rates rapid enough to make much difference. They are listed in the margin.

Our total daily needs for the essential amino acids add up only to about 13 g, less than half an ounce. When a cell is actually making a protein, however, all of the amino acids needed for it must be available *at the same time,* even if only a trace of any one is required. The absence of lysine, for example, when a protein that includes it has to be made, would prevent the synthesis of the protein altogether.

■

Essential Amino Acids[a]	Daily Needs (g/day)
Isoleucine	1.4
Leucine	2.2
Lysine	1.6
Methionine	2.2
Phenylalanine	2.2
Threonine	1.0
Tryptophan	0.5
Valine	1.6

[a] Histidine is believed to be essential to infants. Data from F. E. Deatherage, *Food for Life,* Plenum Press, New York, 1975.

[1] All sources agree that linoleic acid is an *essential* fatty acid. Some add arachidonic acid to the list, but the body is able to make arachidonic acid from linoleic acid. Other lists include linolenic acid as an essential fatty acid, but the body can also make this acid from linoleic acid. No doubt in feeding studies, marked improvements in essential fatty acid deficiency occurs with giving any of the three—linoleic, linolenic, and arachidonic acids. It is arachidonic acid that is finally crucial internally, because many prostaglandins are made from it, and it is possible that infants (but not adults) lack the ability to make arachidonic acid from linoleic acid.

The Availability of the Amino Acids in Food Proteins Varies with the Digestibility of the Protein Not all proteins are easily and completely digested. The variations are expressed by a **coefficient of digestibility,** defined by Equation 18.1.

$$\text{Coefficient of digestibility} = \frac{(\text{N in food eaten}) - (\text{N in feces})}{(\text{N in food eaten})} \tag{18.1}$$

The difference of terms in the numerator:

$$(\text{N in food eaten}) - (\text{N in feces})$$

is *the nitrogen that is actually absorbed by the bloodstream.*

Animal proteins have higher coefficients of digestibility than those of plants or fruits. For the average animal protein, this coefficient is 0.97, which means 97% digestible. The coefficients for the proteins in fruits and fruit juices average about 0.85; for whole wheat flour, 0.79; and for whole rye flour, 0.67. Milling improves the digestibility coefficients, raising them to 0.83 for all-purpose bread flour and 0.89 for cake flour.

Milling does the same to rice. The digestibility coefficient for brown rice is 0.75 but 0.84 for polished white rice. Milling, unfortunately, also reduces the quantities of vitamins, minerals, and fiber in the grain product. But these are usually put back, as in enriched flours, for example.

The average digestibility coefficient of the proteins in legumes and nuts is 0.78, and in vegetables the range is 0.65 to 0.74.

Proteins Low in Essential Amino Acids Have Low Nutritional Value The digestibility of a protein is just one factor in the quality of a protein source. If a protein is lacking or low in one or more of the essential amino acids, it is a low-quality protein in nutrition.

When the body gets amino acids from a low-quality protein, it simply will not be able to use the amino acids efficiently *to make protein.* When a low-quality protein diet fails to supply one essential amino acid, like lysine, the body still gets a load of other amino acids it can't use to make lysine-containing proteins. As we learned in Chapter 17, there is no way to store these excess amino acids until lysine arrives, and they cannot be excreted as amino acids. To get rid of them, the body has to break them down so their nitrogen can be excreted as urea. This leaves other compounds.

The remaining, nitrogen-free products can be used by the body for energy, as we'll study later. But if the body isn't demanding this energy by its level of activity, then it converts these breakdown products into fat. You can get fat on a high-protein, low-activity diet, and you strain your kidneys besides.

■ On a very-high-protein diet, the formation and removal of urea can place a strain on the liver and the kidneys.

One Measure of Protein Quality Is Called the Protein's *Biological Value* Measured when the body operates under the stress of receiving not quite enough protein overall, the percent of the nitrogen retained out of the total nitrogen eaten is called the **biological value** of the protein. It is a measure of the efficiency with which the body uses the nitrogen of *the actually absorbed amino acids* furnished by the protein.

The largest single factor in a protein's biological value is its amino acid composition. What most limits biological value is the extent to which the protein supplies the essential amino acids in sufficient quantities for human use. Human milk protein is the best of all proteins in terms of digestibility and biological value. Whole-egg protein is very close, and it is often taken as the reference for experimental work.

When the diet is well balanced in absorbable amino acids, the amount of nitrogen ingested equals the amount excreted. This condition is called **nitrogen balance.** Most 70-kg men could be in nitrogen balance by ingesting 35 g/day of the proteins in human milk, the value used as a reference standard in rating other proteins.

■ Infants and schoolchildren must have a positive nitrogen balance, meaning that they must ingest more nitrogen than they excrete, in order to grow.

The Essential Amino Acids Most Poorly Supplied Put an Upper Limit on a Protein's Biological Value Table 18.3 summarizes some information about most of the proteins, named in the first column, that are prominent in various diets of the world's peoples. We now study what the data in the other columns mean.

TABLE 18.3
Comparisons of Food Proteins with Human Milk Protein

Food	Limiting Amino Acid	Food's Protein Equivalent to 35 g Human Milk Protein (g)	Digestibility Coefficient of Food's Protein[a]	Amount of Food's Protein[b] (g)	Percentage of Protein in the Food	Amount of Food Needed[c]	Kilocalories of Food Received
Wheat	Lysine	80.6	79.0	102	13.3%	767 g (1.7 lb)	2560
Corn	Tryptophan and lysine	72.4	60.0	120	7.8%	1540 g (3.4 lb)	5660
Rice	Lysine	51.7	75.0	68.9	7.5%	919 g (2.0 lb)	3310
Beans	Valine	50.5	78.0	64.8	24.0%	270 g (0.59 lb)	913
Soybeans	Methionine and cysteine	43.8	78.0	56.2	34.0%	165 g (0.36 lb)	665
Potatoes	Leucine	71.6	74.0	96.7	2.1%	4600 g (10.1 lb)	3500
Cassava	Methionine and cysteine	82.4	60.0	137	1.1%	12,500 g (27.5 lb)	16,400
Eggs	Leucine	36.6	97.0	37.8	12.8%	295 g (0.65 lb)	477
Meat	Tryptophan	43.1	97.0	44.4	21.5%	206 g (0.45 lb)	295
Cow's milk	Methionine and cysteine	43.8	97.0	45.2	3.2%	1410 g (3.1 lb)	903

Source: Data from F. E. Deatherage, *Food for Life,* Plenum Press, New York, 1975. Used by permission.

[a] Expressed as percentages.

[b] The grams of protein that have to be obtained from each food source in order that the protein be nutritionally equivalent (with respect to essential amino acids) to 35 g of human milk protein, allowing for the poorer digestibility of that food's protein (i.e., its digestibility coefficient).

[c] The grams of each food that are equivalent in nutritional value (with respect to essential amino acids) to 35 g human milk protein allowing for the digestibility coefficient and the percent protein in the food.

Column 2 The essential amino acid that is most poorly supplied by a given protein is called its **limiting amino acid.** These are named in this column of Table 18.3.

Column 3 This column gives the number of grams of each food that a 70-kg man would have to digest *and absorb* per day to get the same amount of its limiting amino acid that is available from 35 g of human milk protein. Such an intake, of course, would also supply all the other needed but nonlimiting amino acids. For example, 80.6 g of wheat protein—not wheat, but wheat protein—would have to be digested and absorbed to obtain the lysine in 35 g of human milk protein. But this figure, 80.6 g, assumes 100% digestion, so we have to adjust for a lower percentage digestion.

Column 4 This column gives the digestibility coefficients of all the proteins listed. Wheat protein has a digestibility coefficient of 0.790 so, by a factor of 100/79.0, we need more wheat protein than 80.6 g in order to get the necessary lysine.

Column 5 Here is the result of multiplying 80.6 by (100/79.0). We need 102 g of wheat protein to get the necessary limiting amino acid, lysine. But remember, we're talking about wheat *protein,* not actual wheat, so we move on to the next column.

Column 6 Wheat, on the average, is only 13.3% protein, so we have to multiply our 102 g of wheat protein by the factor 100/13.3. The grim result, 767 g, is in the next column.

Column 7 A 70-kg man would have to eat 767 g (1.69 lb) of wheat if his daily needs for all amino acids are to be met by wheat alone. But all this wheat also has calories. The implication of a daily diet of 767 g of wheat is in the last column.

Column 8 Anyone eating 767 g of wheat also gets 2560 kcal of food energy. This is nearly as much total energy per day as a 70-kg adult should have, so that isn't prohibitive, but can you imagine a diet this boring? Just imagine his problems if he had to get all his amino acids from potatoes, 10.1 lb/day, or cassava, 27.5 lb/day!

The data in Table 18.3 are important to those concerned about both the protein and total calorie needs of the world's burgeoning population. Neither children nor adults can eat enough corn, rice, potatoes, or cassava per day to meet both their protein and energy needs.

Proteins That Provide a Balanced Supply of Essential Amino Acids Are Called Adequate Proteins Proteins of eggs and meat, as you can see in Table 18.3, are particularly good. They are said to be **adequate proteins,** because they include all the essential amino acids in suitable proportions to make it possible to satisfy amino acid and total nitrogen needs without excess intake of calories.

Soybeans are the best of the nonanimal sources of proteins. Cassava, a root, is an especially inadequate protein, and corn (or maize) is also poor. Unhappily, huge numbers of the world's peoples, especially those in Africa, Central and South America, India, and the countries of the Middle and Far East, rely heavily on these two foods. Although they adequately provide energy needs, not enough of them can be eaten per day to give the essential amino acids, so there are widespread dietary deficiency diseases in these regions.

Variety in the Diet Offers Many Nutritional Advantages The data in Table 18.3 also provide scientific support for the long-standing practices in all major cultures of including a wide variety of foods in the daily diet. Meat and eggs can ensure adequate protein while they leave room for an attractive variety that is important to good eating habits. Milk, soybeans, and other beans also leave room for variety.

With Planning and Good Timing, Vegetarians Obtain All Essential Amino Acids When both rice (low in lysine) and beans (low in valine) are included in equal proportions, about 43 g of a rice–bean combination is equal in protein value to 35 g of human milk protein. Less of this combined diet is needed than rice alone or beans alone, which means that room is left for other foods that our taste buds crave.

The rice and beans should, of course, be eaten fairly closely together to ensure that all essential amino acids are simultaneously available during protein synthesis. Eating just the rice early in the morning and the beans in the evening works against efficient protein synthesis. All of us, including vegetarians, must also include vitamins, studied next.

18.3

VITAMINS

The vitamins essential to health must be in the diet because the body cannot make them.

The term **vitamin** applies to any compound or a closely related group of compounds satisfying the following criteria.

■ Vitamin = "vital amine," from an early belief that vitamins might all be amines.

1. It is organic rather than inorganic or an element.
2. It cannot be synthesized at all (or at least in sufficient amounts) by the body and it must be in the diet.
3. Its absence causes a specific **vitamin deficiency disease.**
4. Its presence is essential to normal growth and health.
5. It is present in foods in *small* concentrations, and it is not a carbohydrate, a saponifiable lipid, an amino acid, or a protein.

Vitamins function in the body either as precursors for coenzymes or as coenzymes themselves. Most of the nutritionally required minerals function as cofactors to enzymes.

Sometimes Any Member of a Set of Compounds Prevents a Vitamin Deficiency Disease The members of a set of related compounds that prevent a specific vitamin deficiency disease are called *vitamers*. The body can convert any one of them into the active forms needed. Our needs for vitamin A, for example, are satisfied by several structurally related compounds, including a family of plant pigments called the *carotenoids*.

Some Individuals Require Greater Daily Vitamin Intakes because of Genetic Defects In addition to vitamin deficiency diseases, there are at least 25 disorders classified as *vitamin-responsive inborn errors of metabolism*. These arise from genetic faults that can be partly and sometimes entirely overcome by the daily ingestion of 10 to 1000 times as much of a particular vitamin as normally should be present in the diet.

Genetic faults can compromise the body's use of a vitamin at several places along the vitamin's metabolic path. There might be impaired absorption of the vitamin, or impaired transport in the blood. Many vitamins are needed to provide a prosthetic group for an enzyme, so the genetic error could be a defect in building the vitamin molecule into the enzyme.

■ Dietary surpluses of *water-soluble* vitamins leave the body in the urine.

Vitamins Are Classified as Fat-Soluble or Water-Soluble As we should now expect, fat-soluble vitamins are largely hydrocarbonlike, and water-soluble vitamins are polar or ionic. In this section we study what the vitamins are, their related deficiency diseases, and their sources.

The Fat-Soluble Vitamins Are A, D, E, and K The fat-soluble vitamins occur in the fat fractions of living systems. As a group, the fat-soluble vitamins pose dangers when taken in excess. We all have at least some fatty tissue, and "like dissolves like." So when we ingest surpluses of the fat-soluble vitamins, fatty tissue absorbs some of the excess. Such accumulations can be dangerous, as we will see.

Because molecules of the fat-soluble vitamins have alkene double bonds or phenol rings, oxidizing agents readily attack them. These vitamins are destroyed, therefore, by prolonged exposures to air or to organic peroxides found in fats and oils that are turning rancid.

■ Notice the alkene groups, which are easily oxidized and so can trap undesirable oxidizing agents, like peroxides.

Vitamin A activity is given by several vitamers. In the body, the active forms—retinol, retinal, and retinoic acid—all have five alkene groups. β-Carotene is a source for all three because the liver has enzymes that can cleave its molecules to give the retinol skeleton. This is why β-carotene, the yellow pigment in many plants and available in liver and egg yolk, is an important source of vitamin A activity. It is an example of a *provitamin*, a compound that the body can change into an active vitamin.

CH_3 CH_3 G CH_3

Retinol, G = CH_2OH
Retinal, G = CH=O
Retinoic acid, G = CO_2H

β-Carotene

The deficiency disease for vitamin A is nyctalopia, or night blindness—impaired vision in dim light. Special Topic 2.2 (page 46) discussed the primary chemical event in vision and how it directly involves retinal. We also need vitamin A for healthy mucous membranes.

Several studies have shown that a high intake of β-carotene in a diet rich in vegetables and fruit is associated with a reduced risk of cancer of the epithelial cells in several locations. (Epithelial cell cancers, like colon cancer and lung cancer, account for over 90% of cancer deaths in the United States.) The ease of oxidation of the multiply unsaturated skeleton of β-carotene might account for this. Two kinds of oxidizing agents, excited (activated) oxygen and free radicals, have been implicated in the onset of cancer, and β-carotene might serve to trap and destroy them.

■ Epithelial cells make up the tissue that lines tubes and cavities.

As Special Topic 1.2 (page 30) described, free radicals are species having at least one unpaired electron. The hydroxy free radical is only one example and is represented as HO·, where the dot signifies one unpaired electron. Free radicals lack outer octets and so are reactive and attack almost any organic substance with double bonds, including such crucial materials as genes, hormones, enzymes, and vitamins. Free radicals also attack the cholesterol present in low-density lipoprotein complexes, LDL. Partially oxidized cholesterol attracts white blood cells, which can lead to a build up of a plaque in a blood capillary more rapidly than usual. Thus anything that suppresses free radicals helps to provide protection against heart disease.

■ The cholesterol molecule has one double bond (page 234).

Carefully controlled studies in both Africa and India have shown that vitamin A supplements among peoples with chronically low availability of vitamin A significantly reduce infant mortality and morbidity. When given to children with severe measles, the mortality rate dropped appreciably.

■ *Morbidity* is the ratio of diseased to healthy individuals in a population.

Excess doses of vitamin A must not be taken because it is toxic to adults. The livers of polar bears and seals are particularly rich in vitamin A, and Eskimos are careful about eating these otherwise desirable foods. Siberian huskies also have very high levels of vitamin A in their livers, and early explorers, when driven to eating their sled dogs to survive, risked serious harm and death by eating too much husky liver. In early 1913 the Antarctic explorer X. Mertz almost certainly died in just this way, and his companion, Douglas Mawson, barely survived.

■ Those suffering from excess vitamin A recover quite quickly when they stop taking vitamin A.

Some and maybe all forms of vitamin A, when taken in excess, are also known to be teratogenic (they cause birth defects), so it is not recommended that vitamin supplements with extra vitamin A be taken by pregnant women.

Vitamin D also exists in a number of forms. Cholecalciferol (D_3) and ergocalciferol (D_2) are two that occur naturally.

HO

Ergocalciferol (D_2)

HO

Cholecalciferol (D_3)

These two are equally useful in humans, and either can be changed in the body to the slightly oxidized forms that the body uses as hormones to stimulate the absorption and uses of calcium ions and phosphate ions. Vitamin D is especially important during the years of early growth when bones and teeth, which need calcium and phosphate ions, are developing. Lack of vitamin D causes the deficiency disease known as rickets, a bone disorder.

■ Poor management of Ca^{2+} metabolism contributes to a disfiguring bone condition of old age called *osteoporosis*.

Eggs, butter, liver, fatty fish, and fish oils such as cod liver oil are good natural sources of vitamin D precursors. The most common source today is vitamin D fortified milk. We are able to make some vitamin D ourselves from certain steroids that we can manufacture, provided we get enough direct sunlight on the skin. Energy absorbed from sunlight converts these steroids to the vitamin. Youngsters who worked from dawn to dusk in dingy factories or mines during the early years of the industrial revolution were particularly prone to rickets simply because they saw little if any sun.

No advantage is gained by taking large doses of vitamin D, and in sufficient excess it is dangerous. Excesses promote an increase in the calcium ion level of the blood, and this damages the kidneys and causes soft tissue to calcify and harden.

■ The benzene ring with the phenolic OH is easily oxidized, which makes the tocopherols effective in trapping unwanted oxidizing agents.

Vitamin E needs are satisfied by any member of the tocopherol family. The most active member is α-tocopherol.

α-Tocopherol

Vitamin E occurs so widely in vegetable oils that it is almost impossible not to obtain enough of it. Vitamin E detoxifies peroxides, compounds which have the general formula R—O—O—H. These can form when oxygen attacks CH_2 groups that are adjacent to alkene double bonds, as in the many —CH_2—CH=CH— units in molecules of the unsaturated fatty acids. Because the acyl units of these acids are present in the phospholipids of which cell membranes are made, vitamin E is needed to protect all membranes. β-Carotene also protects cell membranes. Because β-carotene and vitamin E are both fat-soluble, they are actually able to enter the lipid-rich environment of the membrane.

In the absence of vitamin E, the activities of certain enzymes are reduced, and red blood cells hemolyze more readily. Anemia and edema are reported in infants whose feeding formulas are low in vitamin E.

Because vitamin E detoxifies peroxides it is called an *antioxidant.* Because vitamin E is a fat-soluble vitamin, some of it tends to be carried in the bloodstream within the lipid-rich environment of lipoprotein complexes, particularly the low-density lipoprotein complexes (LDL) described on page 407. Vitamin E molecules with their antioxidant properties are thus strategically positioned to inhibit what has lately come to be recognized as an important part of atherosclerosis: the oxidation of LDL. Vitamin E inhibits LDL oxidation and thus might act to thwart the onset of coronary disease. In two huge studies reported in 1993, both men and women who regularly took vitamin E supplements for more than two years experienced a significantly reduced risk of coronary disease. Whether the long-term intake (many years, up to a lifetime) of vitamin E supplements will cause toxic results is not known. Until this question is answered, specialists in coronary disease counsel caution in taking large doses of vitamin E.

■ Much of the vitamin K that we need is made for us by our own intestinal bacteria.

Vitamin K is the antihemorrhagic vitamin. It is present in green, leafy vegetables, and deficiencies are rare. It works as a cofactor in the blood-clotting mechanism. Sometimes women about to give birth and their newborn infants are given vitamin K to provide an extra measure of protection against possible hemorrhaging. Vitamin K also aids in the absorption of calcium ion into bone, so it works against the development of osteoporosis.

The Water-Soluble Vitamins Are Vitamin C, Choline, Thiamine, Riboflavin, Nicotinamide, Folate, Vitamin B_6, Vitamin B_{12}, Pantothenic Acid, and Biotin These are the water-soluble vitamins recognized by the Food and Nutrition Board. It is widely believed that they are among the safest substances known, but this generalization is too broad, as we will see. Most water-soluble vitamins are now known to be essential components of cofactors for enzymes (page 273).

■ Collagen is the protein in bone that holds the minerals together, much as steel rods work in reinforced concrete.

Vitamin C, or ascorbic acid, prevents scurvy, a sometimes fatal disease in which collagen is not well made (see page 254). Whether it prevents other diseases is the subject of enormous controversy, speculation, and research. It is an antioxidant, much as is vitamin E. A variety of studies have found that vitamin C is involved in the metabolism of amino acids, in the synthesis of some adrenal hormones, and in the healing of wounds. There are probably millions of people who believe that vitamin C in sufficient dosages—up to several grams per day—acts to prevent or to reduce the severity of the common cold. The vitamin appears to be nontoxic at these high levels.

Vitamin C is also a destroyer of free radicals. Because it is water-soluble, it is able to do this work within a cell's cytosol. Where low levels of vitamin C exist in seminal fluid, high levels of oxidative damage to sperm have been observed, which may be a factor either in reduced fertility or in birth defects.

Vitamin C is present in citrus fruits, potatoes, leafy vegetables, and tomatoes. It is destroyed by extended cooking, heating over steam tables, or prolonged exposure to air or to ions of iron or copper. Even when vitamin C is kept in a refrigerator in well-capped bottles, it slowly deteriorates.

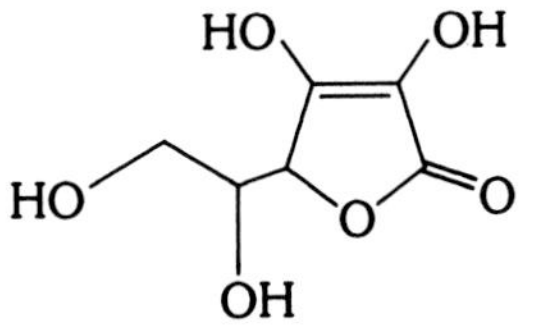

HO OH HO O O OH

Ascorbic acid (vitamin C)

$HOCH_2CH_2\overset{+}{N}(CH_3)$

Choline

Choline is needed to make complex lipids, as we discussed in Section 9.3. Acetylcholine, which is made from choline, is one of several substances that carry nerve signals from one nerve cell to another. The body can make choline, but the Food and Nutrition Board calls it a vitamin because there are ten species of higher animals that have dietary requirements for it. We make it too slowly to meet all our daily needs, so having it in the diet provides protection. It occurs widely in meats, egg yolk, cereals, and legumes. No choline deficiency disease has been demonstrated in humans, but in animals the lack of dietary choline leads to fatty livers and to hemorrhagic kidney disease.

■ Acetylcholine is one of several neurotransmitters (page 342).

Thiamine is needed for the breakdown of carbohydrates. Its deficiency disease is beriberi, a disorder of the nervous system. Good sources are lean meats, legumes, and whole (or enriched) grains. It is stable when dry but is destroyed by alkaline conditions or prolonged cooking. Thiamine is not stored, and excesses are excreted in the urine. One's daily thiamine requirement is proportional to the number of calories that are represented by the diet.

■ In rice, most of the thiamine is in the husk, which is lost when raw rice is milled.

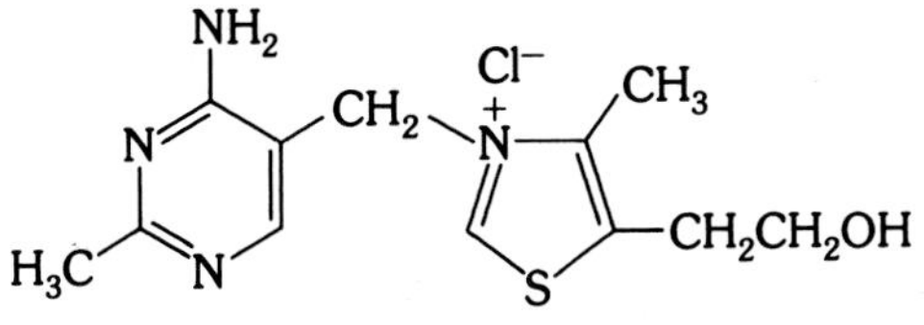

NH_2 N H_3C N CH_2 Cl^- $\overset{+}{N}$ CH_3 S CH_2CH_2OH

Thiamine

CH_2OH H—C—OH H—C—OH H—C—OH CH_2 H_3C H_3C N N N O NH O

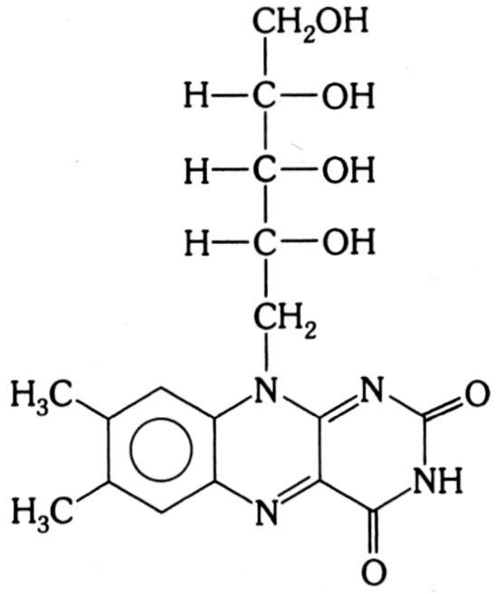

Riboflavin

Riboflavin is required by a number of oxidative processes in metabolism. How riboflavin works was discussed in Special Topic 11.1 (page 275). Deficiencies lead to the inflammation and breakdown of tissue around the mouth and nose, as well as the tongue, a scaliness of the skin, and burning, itching eyes. Wound healing is impaired. The best source is milk, but certain meats (e.g., liver, kidney, and heart) also supply it. Cereals are poor sources unless they have been enriched. Little if any riboflavin is stored, and excesses in the diet are excreted. Alkaline substances, prolonged cooking, and irradiation by light destroy this vitamin.

Niacin, meaning both nicotinic acid and nicotinamide, is essential for nearly all biological oxidations. (Special Topic 11.1, page 275, described how niacin is built into NAD^+, $NADP^+$, and their reduced forms.) It's needed by every cell of the body every day. Its deficiency disease

■ Severe niacin deficiency can cause delerium and dementia.

is pellagra, a deterioration of the nervous system and the skin. Pellagra is particularly a problem where corn (or maize) is the major item of the diet. Corn (maize) is low in niacin and the essential amino acid tryptophan, from which we are able to make some of our own niacin. Where the diet is low in tryptophan, niacin must be provided in other foods, such as enriched grains. Prolonged cooking destroys niacin.

■ We can make about 1 mg of niacin for every 60 mg of dietary tryptophan—nearly all we need.

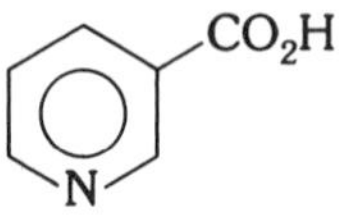

CO_2H $CONH_2$

Nicotinic acid (niacin)

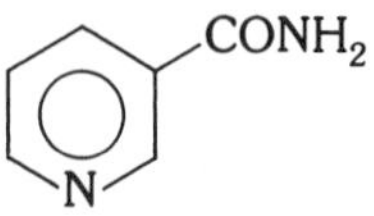

Nicotinamide (niacinamide)

■ Chronic alcoholism causes folacin deficiency.

Folate is the name used by the Food and Nutrition Board for folic acid and related compounds. Its deficiency disease is megaloblastic anemia. Several drugs, including alcohol, promote folic acid deficiency. The enzymes that use folic acid as a cofactor are largely involved in reactions that transfer one-carbon units, like $—CH_3$ and $—CH{=}O$. These reactions include the synthesis of nucleic acids and heme. Good sources are fresh, leafy green vegetables, asparagus, liver, and kidney. Folacin is relatively unstable to heat, air, and ultraviolet light, and its activity is often lost in both cooking and food storage.

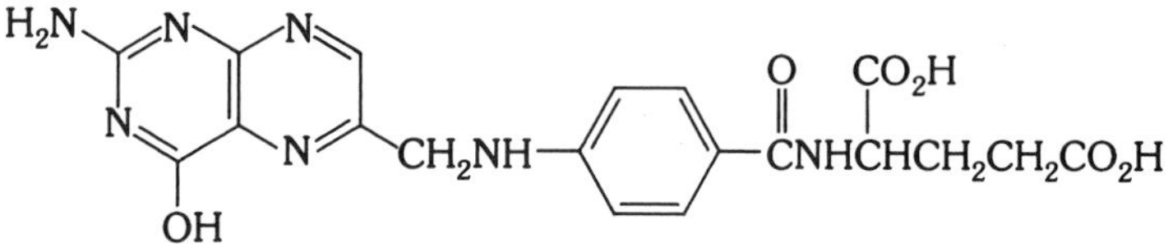

H_2N N N N N OH CH_2NH O CO_2H $CNHCHCH_2CH_2CO_2H$

Folic acid

In a careful study sponsored by the British Medical Research Council, it was found that supplementary folic acid (4 mg/day) prevented the recurrence of neural-tube defects (e.g., spina bifida) in 72% of the women who had a history of delivering babies with such defects. Its use during the first four weeks of pregnancy was particularly important. A large controlled study in Hungary showed that vitamin–mineral supplements, including folic acid, resulted in 50% fewer congenital malformations of all kinds, when taken prior to pregnancy.

Vitamin B_6 activity is supplied by pyridoxine, pyridoxal, or pyridoxamine. All can be changed in the body to the active form, pyridoxal phosphate. The activities of at least 60 enzymes involved in the metabolism of various amino acids depend on pyridoxal phosphate. One deficiency disease is hypochromic microcytic anemia, and disturbances in the central nervous system also occur. The vitamin is present in meat, wheat, yeast, and corn. It is relatively stable to heat, light, and alkali.

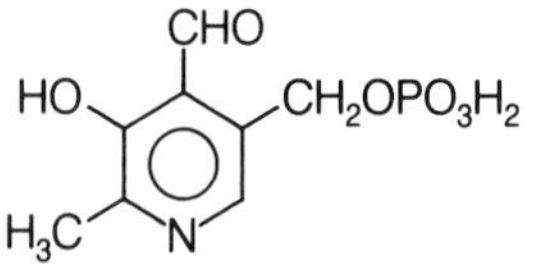

■ CHO HO $CH_2OPO_3H_2$ H_3C N

Pyridoxal phosphate

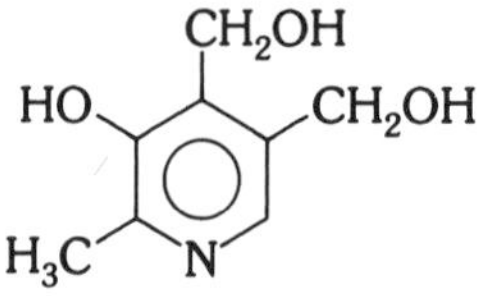

CH_2OH HO CH_2OH H_3C N

Pyridoxine

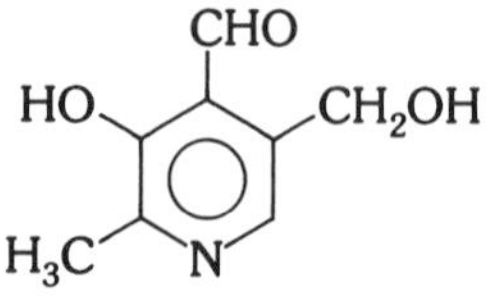

CHO HO CH_2OH H_3C N

Pyridoxal

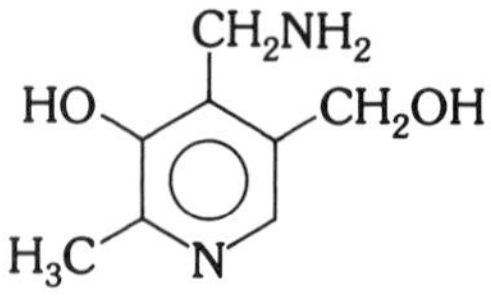

CH_2NH_2 HO CH_2OH H_3C N

Pyridoxamine

Pyridoxine is widely used as a component of body-building diets and in the treatment of premenstrual syndrome. *Massive doses ("megavitamin doses") at levels of 500 mg/day and higher, however, can severely disable parts of the nervous system and should be avoided.* Pyridoxine is clearly not "among the safest substances known."

Vitamin B_{12}, or cobalamin, is a controlling factor for pernicious anemia. This deficiency disease, however, is very rare, because it is very difficult to design a diet that lacks this vitamin. Enzymes that use vitamin B_{12} are, like folate, involved in the transfers of one-carbon units. The synthesis of DNA, for example, depends on vitamin B_{12}.

TABLE 18.4
Estimated Safe and Adequate Daily Dietary Intakes of Biotin and Pantothenic Acid

Category	Age (years)	Biotin (μg)	Pantothenic acid (mg)
Infants	0–0.5	10	2
	0.5–1	15	3
Children and adolescents	1–3	20	3
	4–6	25	3–4
	7–10	30	4–5
	11+	30–100	4–7
Adults		30–100	4–7

[a] Food and Nutrition Board, National Academy of Sciences National Research Council, Revised 1989. Because less information is available for basing allowances, these are not in Table 18.1.

Animal products such as liver, kidney, and lean meats as well as milk products and eggs are good sources—and virtually the only sources. Thus most people who develop B_{12} deficiency are true vegetarians (or are infants born of true vegetarian mothers). The onset of any symptoms of B_{12} deficiency occurs slowly because the body stores it fairly well and because such minute traces are needed.

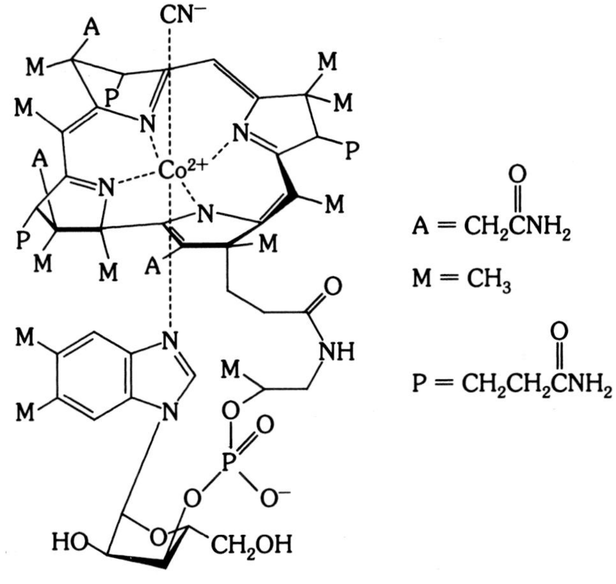

■ The replacement of the CN group by other groups (e.g., OH or CH_3 and a few others) gives the members of a small family of compounds that function as cofactors for enzymes.

$A = CH_2C(=O)NH_2$

$M = CH_3$

$P = CH_2CH_2C(=O)NH_2$

Cyanocobalamin

■ Dorothy Crowfoot Hodgkins, a British chemist, won the 1964 Nobel prize in chemistry for determining the structure of cyanocobalamin.

Pantothenic acid is used to make a coenzyme, coenzyme A (symbol: CoASH). As we learned in Chapter 16, the body needs coenzyme A to metabolize fatty acids. Signs of a deficiency disease for this vitamin have not been observed clinically in humans, but the deliberate administration of compounds that work to lower the availability of pantothenic acid in the body causes symptoms of cellular damage in vital organs. This vitamin is supplied by many foods, and especially by liver, kidney, egg yolks, and skim milk.

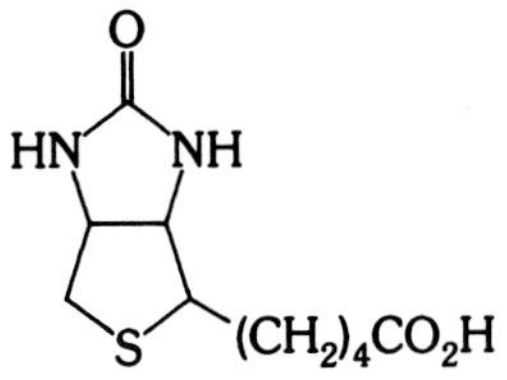

Biotin

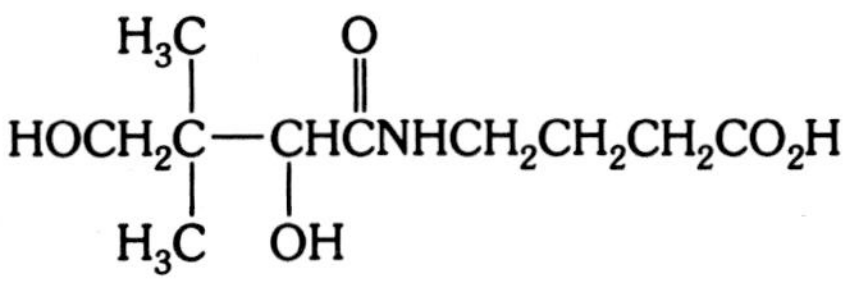

$HOCH_2C(CH_3)_2-CH(OH)C(=O)NHCH_2CH_2CH_2CO_2H$

Pantothenic acid

Table 18.4 describes safe and adequate daily dietary intakes of biotin and pantothenic acid.

Biotin is required for all pathways in which carbon dioxide is temporarily used as a reactant, as in the synthesis of fatty acids (Section 16.4). Signs of biotin deficiency are hard to find, but when such a deficiency is deliberately induced, the individual experiences nausea, pallor, dermatitis, anorexia, and depression. When biotin is given again, the symptoms disappear. Our own intestinal microorganisms probably make biotin for us. Egg yolks, liver, tomatoes, and yeast are good sources.

18.4

MINERALS AND TRACE ELEMENTS IN NUTRITION

Many metal ions are necessary for the activities of enzymes.
Most of the minerals and trace elements needed in the diet are metal ions, but some anions are also required. The distinction between a *mineral* and a *trace element* is a matter of quantity.

Minerals Are Needed at Levels of 100 mg/Day or More The dietary **minerals** are calcium (Ca^{2+}), phosphorus (phosphate ion, P_i), magnesium (Mg^{2+}), sodium (Na^+), potassium (K^+), and chlorine (Cl^-). As we learned in Section 12.2, these are the chief electrolytes in the body. Besides its involvement in bones, the calcium ion is an essential part of the pathways that operate the nervous system and that activate enzymes (Sections 11.6 and 11.7).

Trace Elements Are Needed at Levels of 20 mg/Day or Less The Food and Nutrition Board recognizes 17 **trace elements,** all of which have been found to have various biological functions in animals, and which therefore are quite likely used in the human body. Ten of them are *known* to be needed by humans. All are toxic in excessive amounts. They are fluorine (as F^-) and iodine (as I^-), and the ions of chromium, manganese, iron, cobalt, copper, zinc, and molybdenum; selenium occurs in selenocysteine. All occur widely in food and drink, but some are removed by food refining and processing. For trace elements not listed in Table 18.1, see Table 18.5 for estimated safe and adequate daily dietary intakes.

Selenocysteine is an amino acid with a CH_2SeH sidechain in place of the CH_2SH sidechain of normal cysteine.

1 ppm F^- = 1 mg/L

Fluorine (as fluoride ion) is essential to the growth and development of sound teeth, and the Food and Nutrition Board recommends that public water supplies be fluoridated at a level of about 1 ppm wherever natural fluoride levels are too low. Both the medical and dental associations in the United States strongly support this. Those in some other countries do not, France, West Germany, Denmark, for example. A controversy over the appropriateness of fluoridating drinking water supplies has gone on for nearly half a century.

TABLE 18.5
Estimated Safe and Adequate Daily Dietary Intakes of Selected Minerals[a]

Category	Age (years)	Copper (mg)	Manganese (mg)	Fluoride (mg)	Chromium (μg)	Molybdenum (μg)
Infants	0–0.5	0.4–0.6	0.3–0.6	0.1–0.5	10–40	15–30
	0.5–1	0.6–0.7	0.6–1.0	0.2–1.0	20–60	20–40
Children and adolescents	1–3	0.7–1.0	1.0–1.5	0.5–1.5	20–80	25–50
	4–6	1.0–1.5	1.5–2.0	1.0–2.5	30–120	30–75
	7–10	1.0–2.0	2.0–3.0	1.5–2.5	50–200	50–150
	11+	1.5–2.5	2.0–5.0	1.5–2.5	50–200	75–250
Adults		1.5–3.0	2.0–5.0	1.5–4.0	50–200	75–250

[a] The upper levels in this table should not be habitually exceeded because the toxic levels for many trace elements may be only several times the usual intakes. Data from Food and Nutrition Board, National Academy of Sciences National Research Council, Revised 1989. Because less information is available for basing allowances, these are not in Table 18.1.

Iodine (as iodide ion) is essential in the synthesis of certain hormones made by the thyroid gland. Diets deficient in iodide ion cause an enlargement of the thyroid gland known as a goiter. It takes only about 1 μg (1×10^{-6} g) of I^- per kilogram of body weight each day to prevent a goiter.

Seafood is an excellent source of iodide ion, but iodized salt with 75 to 80 μg of iodide ion equivalent per gram of salt is the surest way to obtain the iodine that is needed. Before the days of iodized salt, goiters were quite common, especially in regions where little if any fish or other seafood was in the diet.

Chromium (Cr^{3+}) is required for the work of insulin and normal glucose metabolism. Chromium that occurs naturally in foods is absorbed significantly more easily than chromium given as the simple salts added to vitamin–mineral supplements. Most animal proteins and whole grains supply this trace element. The daily intake should be 0.05 to 0.2 mg.

■ High-fiber diets can work against the absorption of some of the trace elements.

Manganese (Mn^{2+}) is required for normal nerve function, for the development of sound bones, and for reproduction. It is essential to the activities of certain enzymes in the metabolism of carbohydrates. Nuts, whole grains, fruits, and vegetables supply this element, but a recommended daily allowance has yet to be set by the Food and Nutrition Board. Some nutritionists recommend a daily intake of 2.5 to 5.0 mg/day.

Iron (Fe^{2+}) is an essential cofactor for heme, certain cytochromes, and the iron–sulfur proteins (Section 25.3). The intestines regulate how much dietary iron is absorbed, so a proper level of iron in circulation is maintained in this way. A high serum iron level apparently renders an individual more susceptible to infections. Pregnant women must have larger than usual amounts of iron to keep pace with the needs of fetal blood.

Cobalt (Co^{2+}) is part of the vitamin B_{12} molecule cyanocobalamin, page 453). Apparently there is no other use for it in humans. But without it there can be anemia and growth retardation.

Copper (Cu^{2+}) occurs in a number of proteins and enzymes, including certain cytochromes (Section 14.3). If deficient in copper, the individual synthesizes lower-strength collagen and elastin and will tend to suffer anemia, skeletal defects, and degeneration of the myelin sheaths of nerve cells. Ruptures and aneurysms of the aorta become more likely. The structure of hair is affected, and reproduction tends to fail.

Fortunately, copper occurs widely in foods, particularly in nuts, raisins, liver, kidney, certain shellfish, and legumes. An intake of just 2 mg/day assures a copper balance for nearly all people. Unfortunately, however, the copper contents of the foods in a typical American diet have declined over the last four decades. Some scientists believe that the increase in the incidence of heart disease in the United States during this period has been caused partly by diets low in copper. In one study that involved women, the lower the copper level in the blood the higher were their blood cholesterol levels. (We discussed cholesterol and heart disease in Chapter 16.)

Zinc (Zn^{2+}) is required for the activities of several enzymes. Without sufficient zinc in the diet, an individual will experience loss of appetite and poor wound healing. Insufficient zinc in an infant's diet, which is a chronic problem in the Middle East, causes dwarfism and poor development of the gonads. If too much zinc is in the diet, and too little copper is present, the extra zinc acts to inhibit the absorption of copper. The ratio of zinc to copper probably doesn't matter as long as sufficient copper is available. When enough zinc is present, it acts to inhibit the absorption of a rather toxic pollutant, cadmium (as Cd^{2+}), which is just below zinc in the periodic table.

Selenium is known to be essential in many animals, including humans. It is needed for thyroid hormone action. How much we need is not known, but probably we need 0.05 to 0.2 mg/day. *Too much is very toxic.*

A strong statistical correlation exists between high levels of selenium in livestock crops and low incidences of human deaths by heart disease. In the United States, those who live where selenium levels are high—the Great Plains between the Mississippi River and the eastern Rocky Mountains—have one-third the chance of dying from heart attack and strokes as those who live where levels are very low—the northeastern quarter of the United States, Florida, and the Pacific Northwest. Rats, lambs, and piglets on selenium-poor diets develop damage to heart tissue and abnormal electrocardiograms.

Selenium reduces the occurrence of certain cancers in animals, and it may possibly provide a similar benefit in humans; at excessive levels it causes cancer in experimental animals.

Molybdenum is in an enzyme required for the metabolism of nucleic acids as well as in some enzymes that catalyze oxidations. Deficiencies in humans are unknown, meaning that almost any reasonable diet furnishes enough.

Nickel, silicon, tin, vanadium, and boron are possibly trace elements for humans, because deficiency diseases for these elements have been induced in experimental animals.

SUMMARY

Nutrition Good nutrition entails the ingestion of all the substances needed for health—water, oxygen, food energy, essential amino acids and fatty acids, vitamins, minerals, and trace elements. Foods that best supply various nutrients have been identified. The Food and Nutrition Board of the National Academy of Sciences regularly publishes *recommended daily allowances* that are intended to be amounts that will meet the nutritional needs of practically all healthy people. Some people need more, most need less. But neither more nor less of anything is necessarily better. The RDAs should be obtained by a varied diet because it promotes good eating habits and because such a diet might supply a nutrient no one yet knows is essential.

Water needs The thirst mechanism leads to our chief source of water. Our principal routes of exporting water are the urine, perspiration (both sensible and insensible), and exhaled air. When excessive water losses occur during vigorous exercise, the body also loses electrolytes.

Energy Both carbohydrates and lipids should be in the diet as sources of energy. Without carbohydrates for several days, certain poisons build up in the blood as the body works to make glucose internally from amino acids. A zero-lipid diet means zero ingestion of essential fatty acids. Our oxygen requirements adjust to the caloric demands of our activities.

Protein in the diet What we need most from proteins are certain essential amino acids, and we need some extra nitrogen if we have to make the nonessential amino acids. When a diet excretes as much nitrogen as it takes in, the individual is in nitrogen balance. The most superior, balanced proteins—those that are highly digestible and that supply the essential amino acids in the right proportions—are proteins associated with animals such as the proteins in milk and whole eggs. (Human milk protein is the standard of excellence, and whole-egg protein is very close to it.)

The most important factor in the biological value of a protein is its limiting amino acid—the essential amino acid that it supplies in the lowest quantity. Another factor is the digestibility of the protein. For several reasons—poor source of an essential amino acid, low digestibility coefficients, low concentration of the protein—several foods cannot be used as exclusive or even major components of a healthy diet. Foods that have inadequate proteins include corn (maize), rice, potatoes, and cassava.

Vitamins Organic compounds, called vitamins, or sets of closely related compounds that satisfy the same need, must be in the diet in at least trace amounts or an individual will suffer from a vitamin deficiency disease. Vitamins can't be sufficiently made by the body. (Essential amino acids and essential fatty acids are generally not classified as vitamins, nor are carbohydrates, proteins, or triacylglycerols and other saponifiable lipids.) Excessive amounts of the fat-soluble vitamins (A, D, E, and K)—especially A and D—must be avoided. These vitamins accumulate in fatty tissue, and in excess they can cause serious trouble. Any excesses of the water-soluble vitamins are eliminated.

Minerals and trace elements The minerals are inorganic cations and anions that are needed in the diet in amounts in excess of 100 mg/day. The trace elements are needed at levels of 20 mg/day or less. The whole body quantities of the minerals are large relative to the trace elements. (More about minerals is given in other chapters.) The trace elements that are metal ions are essential to several enzyme systems. Two anions are trace elements, F^- and I^-. Fluoride ion is needed to make strong teeth, and iodide ion is needed to make thyroid hormones and to prevent goiter.

REVIEW EXERCISES

The answers to Review Exercises whose numbers are in color are found in Appendix A. The answers to the other Review Exercises are found in the Study Guide that accompanies this book. The more challenging questions are marked with asterisks.

Nutrition

18.1 What does the science of nutrition study?

18.2 What is meant by the term *nutrient?*

18.3 What is the relationship of nutrients to *foods?*

18.4 What is the relationship of the science of dietetics to nutrition?

18.5 Why are the recommended dietary allowances higher than minimum daily requirements?

18.6 What are seven situations that require special therapeutic diets?

18.7 Why should our diets be drawn from a variety of foods?

18.8 What is potentially dangerous about the following diets?
(a) a carbohydrate-free diet
(b) a lipid-free diet

18.9 All agree that one fatty acid is *essential*. What is its name and what does "essential" mean in this context?

18.10 What are two other fatty acids often cited as "essential" besides the one of Review Exercise 18.9?

18.11 Which fatty acid is needed to make prostaglandins and stands closest to prostaglandins in the synthetic pathway?

Protein and Amino Acid Requirements

18.12 Because the full complement of 20 amino acids is required to make all the body's proteins, why are fewer than half of this number considered essential amino acids?

18.13 If we are able to make all the nonessential amino acids, why should our daily protein intake include more than what is represented solely by the essential amino acids?

18.14 What does the body do with the amino acids that it absorbs from the bloodstream but doesn't use?

18.15 What is the equation that defines the coefficient of digestibility of a protein? What does the numerator in this equation stand for?

18.16 Which kind of protein is generally more fully digested by the body, the protein from an animal or a plant source?

18.17 What can be done to whole grains to improve the digestibility of their proteins? What also happens in this process that reduces the food value of the grains?

18.18 What is the most important factor in determining the biological value of a given protein?

18.19 Which specific protein has the highest coefficient of digestibility and the highest biological value for humans? Which protein comes so close to this on both counts that it is possible to use it as a substitute for research purposes?

18.20 What is meant by the *limiting* amino acid of a protein?

18.21 Why does the protein in corn have a lower biological value than the protein in whole eggs?

***18.22** In one variety of hybrid corn, the limiting amino acids are lysine and tryptophan. It takes 75 g of the protein of this corn to be equivalent to 35 g of human milk protein. The coefficient of digestibility of the protein in this corn is 0.59.
(a) In order to match the nutritional value of human milk protein, how many grams of the protein in this corn must be ingested?
(b) To get this much protein from this variety of corn, how many grams of the corn must be eaten? The corn is only 7.6% protein.
(c) How many kilocalories are also consumed with the quantity of corn calculated in part (b) if the corn has 360 kcal/100 g?
(d) Could a child eat enough corn per day to satisfy all its protein needs and still have room for any other food?

Vitamins

18.23 Make a table that lists each vitamin, at least one good source of each, and a serious consequence of a deficiency of each. (Use only the information available in this book.) Set up the table with the following column heads.

Vitamin	Source(s)	Problem(s) if Deficient

18.24 Why aren't the essential amino acids listed as vitamins?

18.25 On a strict vegetarian diet—no meat, eggs, and dairy products of any sort—which one vitamin is hardest to obtain?

18.26 Why should strict vegetarians use two or more different sources of proteins?

18.27 Which are the fat-soluble vitamins?

18.28 Which vitamin is activated when the skin is exposed to sunlight?

18.29 Which vitamin acts as a hormone in its active forms?

18.30 Which vitamin has been shown to be teratogenic when used in excess?

18.31 The yellow-orange pigment in carrots, β-carotene, can serve as a source of the activity of which vitamin?

18.32 The vitamin needed to participate in the blood-clotting mechanism is which one?

18.33 What is a free radical and why is this species dangerous in cells and their membranes?

18.34 What vitamins provide protection against free radicals within membranes of cells?

18.35 How might the attack of free radicals on the cholesterol in low-density lipoprotein complexes (LDL) accelerate the formation of a plaque in a capillary?

18.36 Name a vitamin that helps to destroy free radicals within the cytosol.

18.37 The Food and Nutrition Board of the National Academy of Sciences recognizes which substances as the water-soluble vitamins?

18.38 Scurvy is prevented by which vitamin?

18.39 Name at least two vitamins that tend to be destroyed by prolonged cooking.

18.40 Which vitamin prevents beriberi?

18.41 When corn (maize) is the chief food in the diet, which vitamin is likely to be in short supply because a raw material for making it is in short supply?

18.42 Which vitamin, when taken during (or before) pregnancies offers protection against neural tube defects?

Minerals and Trace Elements

18.43 What criterion makes the distinction between *minerals* and *trace elements?*

18.44 Name and give the chemical forms of the six minerals.

18.45 Make a table of the ten trace elements and at least one particular function of each.

18.46 What can be the body's response to an iodine-deficient diet?

Additional Exercises

*18.47 If an individual must take in 20.0 mol O_2 in one day to satisfy all needs, how many liters of *air* must be inhaled to supply this? (Air is 21.0% oxygen, on a volume/volume basis, and assume that the measurement is at STP.)

18.48 The limiting amino acid in peanuts is lysine, and 62 g protein obtained from peanuts is equivalent to 35 g human milk protein. The coefficient of digestibility of pe nut protein is 0.78. In order to match human milk protein nutritional value,

(a) How many grams of peanut protein must be eaten?
(b) How many grams of peanuts must be eaten if peanu are 26.2% protein?
(c) How many kilocalories are also ingested with this ma grams of peanuts if peanuts have 282 kcal/100 g?
(d) Could a child eat enough peanuts per day to satisfy protein needs and still have room for other foods?

ANSWERS TO PRACTICE EXERCISES AND SELECTED REVIEW EXERCISES

If you have computed an answer that differs in a small way with the answer given, the difference most likely is caused by the timing of the rounding off of a calculated result after one or more intermediate steps. The answers given here were in almost every instance obtained by chain calculations. When working with atomic masses, remember the rule to round an atomic mass to the first decimal place *before* using it (except round hydrogen's atomic mass to 1.01).

CHAPTER 1

Practice Exercises, Chapter 1

1. (a) $CH_3—CH_2—CH_3$ (b)
```
CH3—CH—CH3
    |
    CH3
```

(c)
```
         CH3  CH3  CH3
         |    |    |
    CH3—C————CH———CH———CH3
         |
         CH3
```

2. (a) $CH_3CH_2CH_3$ (b)
```
CH3CHCH3
    |
    CH3
```

(c)
```
      CH3  CH3  CH3
      |    |    |
   CH3C————CH———CHCH3
      |
      CH3
```

3. (a)
```
     H  H
     |  |
   H—C——C—H
     |  |
     H  H
```

(b)
```
              H
              |
     H     H—C—H   H          H
     |        |    |          |
   H—C————————C————C——————————C—H
     |        |    |          |
     H        H  H—C—H        H
                   |
                   H
```

(c)
```
                        H
                        |
        H  H         H—C—H     H  H  H
        |  |            |      |  |  |
    H—C——C——————————C——————C——C——C—H
        |  |            |      |  |  |
        H  H         H—C—H     H  H  H
                        |
                        H
```

4. Structures (b) and (c) violate the tetravalences of carbon at one point.

5. (a)
```
        H   H H   H
         \ / H \ /
    H     C  |  C     H
     \  /  \ | /  \  /
  H—C       C       C—H
  H—|       |       |—H
  H—C       C       C—H
     \  \  / | \  /  /
  H     C    |   C     H
      H |    H   | H
      H—C————————C—H
        |        |
        H        H
```

(b)
```
     H HH H H HH H
      \/ \/   \/ \/
      C—C     C—C
  H  /     \ /     \  H
   —C       C       C—
  H  \     / \     /  H
      C—C     C—C
      /\ /\   /\ /\
     H HH H H HH H
```

6.
```
CH3CH2—⬡—⬠
```

7. (a) identical (b) isomers
(c) identical (d) isomers
(e) different in another way

8. 2-methyl-1-butanol (less polar)

9. (a) 3-methylhexane
(b) 4-ethyl-2,3-dimethylheptane
(c) 5-ethyl-2,4,6-trimethyloctane

10.

```
(a) BrCH2CHCH2CH2CH3
            |
            NO2

         CH3   CH3   CH3   CH(CH3)2
          |     |     |     |
(b) CH3C------C-----C-----CHCH2CH2CH3
          |     |     |
         CH3   CH3   CH3

        I CH3       CH3CHCH2CH3
        | |            |
(c) CH3CCHCH----------CHCHCH2CH2CH3
        |  |             |
        I  CH(CH3)2      C(CH3)3

(d) BrCHCHCH3
       |  |
       Cl CH3

                    CH3CHCH2CH3
                       |
(e) CH3CH2CH2CH2CCH2CH2CH2CH2CH3
                       |
                    CH3CHCH2CH3
```

11.

```
                      ↓
               CH3(C)HCH2CH3
  ↓    ↓    ↓    |↓   ↓    ↓    ↓
CH3CH2CH2CH2CCH2CH2CH2CH2CH3
                 |
               CH3(C)HCH2CH3
                      ↑
```

12. (a) ethyl chloride (b) butyl bromide
(c) isobutyl chloride (d) *t*-butyl bromide

13. (a) butyl chloride, 1-chlorobutane
sec-butyl chloride, 2-chlorobutane
(b) isobutyl chloride, 2-methyl-1-chloropropane
t-butyl chloride, 2-methyl-2-chloropropane

Review Exercises, Chapter 1

1.6 Compounds b, d, and e are considered to be inorganic.

1.9 (a) Molecular; low melting and flammable
(b) Ionic; water soluble (and likely a carbonate or a bicarbonate)
(c) Molecular; no ionic compound is a gas (or a liquid) at room temperature.
(d) Ionic; high melting and nonflammable
(e) Molecular; most liquid organic compounds are insoluble in water but will burn.
(f) Molecular; no ionic compound is a liquid at room temperature.

1.11 Compounds a, d, and e are possible.

1.12

```
        H                      Cl
        |                      |
(a) H—C—O—H           (b) H—C—H
        |                      |
        H                      Cl

     H  H                   H  H
     |  |                   |  |
(c) H—N—N—H            (d) H—C—C—H
                            |  |
                            H  H

      O                       O
      ‖                       ‖
(e) H—C—H              (f) H—C—O—H

      H                     H  H
      |                     |  |
(g) H—N—O—H            (h) H—C=C—H

      Cl
      |
(i) H—C—Cl             (j) H—C≡N
      |
      Cl

      H                     H  H
      |                     |  |
(k) H—C=C=N—H          (l) H—C—N—H
                            |
        H                   H
        |
   or H—C—C≡N
        |
        H
```

1.19 Unsaturated compounds are (b), (c), and (d).

1.21 Identical compounds are a, b, e, f, and m. Isomers are d, g, i, j, k, and l. Unrelated structures occur at c and n.

1.22 (a) alkane (b) alcohol
(c) thioalcohol (d) alkyne
(e) aldehyde (f) ester
(g) ester (h) ketone
(i) amine (j) ether

1.23 (a) alcohol
(b) alcohol
(c) both thioalcohols
(d) first, alkene; second, cycloalkane
(e) ketone
(f) alkane
(g) both amines
(h) first, carboxylic acid; second, alcohol + ketone
(i) first, ester; second, carboxylic acid (Why is an ester [cut off] first and not an aldehyde, will be explained later.)
(j) first, ester + alcohol; second, alcohol + carboxylic a[cut off]
(k) first, ether + ketone; second, ester
(l) both alkanes
(m) ketone

1.25 B. Its molecules have water-like OH groups.

1.27 Add a drop of the liquid to water. If it dissolves, it is met[cut off] alcohol.

1.31 $CH_3CH_2CH_2CH_2CH_2CH_2CH_3$ heptane

```
   CH3
   |
CH3CHCH2CH2CH2CH3        2-methylhexane

       CH3
       |
CH3CH2CHCH2CH2CH3        3-methylhexane

       CH2CH3
       |
CH3CH2CHCH2CH3           3-ethylpentane
```

CH_3
$CH_3CCH_2CH_2CH_3$ — 2,2-dimethylpentane
CH_3

CH_3
$CH_3CH_2CCH_2CH_3$ — 3,3-dimethylpentane
CH_3

$H_3C\ CH_3$
$CH_3CHCHCH_2CH_3$ — 2,3-dimethylpentane

$CH_3\ \ CH_3$
$CH_3CHCH_2CHCH_3$ — 2,4-dimethylpentane

$H_3C\ \ CH_3$
$CH_3C{-}CHCH_3$ — 2,2,3-trimethylbutane
CH_3

32 (a) $CH_3CH_2CH_2CH_2CH_2CH_2CH_3$
(b) $CH_3CHCH_2CH_2CH_2CH_3$ (with CH_3 on C-2)

33 (a) 5-*sec*-butyl-5-ethyl-2,3,3,9-tetramethyldecane
(b) 7-*t*-butyl-5-isobutyl-2-methyl-6-propyldecane

35 (a) $CH_3CH_2CH_2Cl$ (b) CH_3CHCH_2I (with CH_3 on C-2)
(c) CH_3CBr (with two CH_3 on C) (d) CH_3CH_2Br

37 (a) CH_3CHCH_2Cl (with CH_3 on C-2)
1-chloro-2-methylpropane
(b) 1,3-dichlorocyclopentane
(c) $CH_3CHCH_2CH_3$ (with CH_2CH_3 on C-2)
3-methylpentane
(d) 1,2,4-trimethylcyclohexane

39 $CH_3CH_2OH + 3O_2 \rightarrow 2CO_2 + 3H_2O$

41 CH_3CHCl_2, 1,1-dichloroethane
$ClCH_2CH_2Cl$, 1,2-dichloroethane

44 Petroleum is a mixture of liquid (crude oil) and gas.

51 (a) $H_2 + Cl_2 \rightarrow 2HCl$
(b) $Cl_2 + \text{UV or heat} \rightarrow 2Cl\cdot$
$Cl\cdot + H_2 \rightarrow HCl + H\cdot$
$H\cdot + Cl_2 \rightarrow HCl + Cl\cdot$

54 No reaction with any of the given reactants.

56 11.1 g chlorocyclopentane

CHAPTER 2

Practice Exercises, Chapter 2

1. (a) 2-methylpropene (or 2-methyl-1-propene)
(b) 4-isobutyl-3,6-dimethyl-3-heptene
(c) 1-chloropropene (or 1-chloro-1-propene)
(d) 3-bromopropene (or 3-bromo-1-propene)
(e) 4-methyl-1-hexene
(f) 4-methylcyclohexene

2. (a) $CH_3CH{=}CHCHCH_3$ (with CH_3 on C-4)
(b) $CH_2{=}CHCHCH_2CH_2CH_2CH_3$ (with $CH_2CH_2CH_3$ on C-3)
(c) $CH_2{=}CHCCH_2Cl$ (with two CH_3 on C-3)
(d) $CH_3C{=}CCH_3$ (with H_3C and CH_3 on the double-bond carbons)

3. Cis–trans isomerism is possible for (a) and (b). For part (a), cis versus trans is based on the way the main chain passes through the double bond.

(a) cis isomer: CH_3CH_2 and CH_3 (upper), CH_3 and H (lower) on C=C; trans isomer: CH_3CH_2 and H (upper), CH_3 and CH_3 (lower) on C=C

(b) cis isomer: Cl and Cl (upper), H and H (lower) on C=C; trans isomer: Cl and H (upper), H and Cl (lower) on C=C

4. (a) $CH_3CH_2CH_3$
(b) no reaction
(c) methylcyclohexane
(d) $CH_3(CH_2)_{16}CO_2H$

5. (a) CH_3CCH_2Br (with CH_3 and Br on C-2)
(b) no reaction (in the absence of heat or UV radiation)
(c) $ClCH_2CHCH_2CH_3$ (with Cl on C-2)
(d) $CH_3CH_2CH_2CH_3$

6. (a) $CH_3CHCH_2CH_3$ (with Cl on C-2)
(b) CH_3CCH_3 (with CH_3 and Br on C-2)

(c) $CH_3CH_2C(CH_3)(OH)$–cyclohexyl

(d) 3-methylcyclohexanol + 4-methylcyclohexanol (CH_3, OH on cyclohexane rings)

(e) no reaction

7. (a) $CH_3CH_2CH_2\overset{+}{C}H_2$ and ($CH_3CH_2\overset{+}{C}HCH_3$) circled

$CH_3CH_2CHClCH_3$

(b) $CH_3CH(CH_3)\overset{+}{C}H_2$ and ($CH_3\overset{+}{C}(CH_3)CH_3$) circled; $CH_3CCl(CH_3)CH_3$

(c) cyclohexyl–$\overset{+}{C}H_2$ and (1-methylcyclohexyl cation, $^+$–CH_3) circled

1-chloro-1-methylcyclohexane (Cl, CH_3)

(d) Only one carbocation is possible:

($CH_3CH_2\overset{+}{C}HCH_3$) circled; $CH_3CH_2CHClCH_3$

(e) $CH_3\overset{+}{C}HCH_2CH_2CH_3$ $CH_3CHClCH_2CH_2CH_3$

and

$CH_3CH_2\overset{+}{C}HCH_2CH_3$ $CH_3CH_2CHClCH_2CH_3$

(Both carbocations are 2°, so both are equally possible and both form. Thus two isomeric chloropentanes form.)

(f) $CH_3CH(OH)CH_2CH_2CH_3$ and $CH_3CH_2CH(OH)CH_2CH_3$

Two 2° carbocations can form, so two isomeric pentanols form. Only one carbocation can form from propene, namely, the more stable isopropyl carbocation.

Review Exercises, Chapter 2

2.1
(a) A, B, E
(b) C
(c) E (Structure C has "-pent-" in "cyclopentane.")
(d) All are insoluble in water.
(e) B, and C. No, C is not an alkene.
(f) A, D, and E. Only D is an alkyne.
(g) Not well; the presence of rings or multiple bonds can reduce the ratio of carbon atoms to hydrogen atoms of a hydrocarbon.

2.4
(a) 1-octene
(b) 1-bromo-3-methyl-1-butene
(c) 4-methyl-2-propyl-1-hexene
(d) 4,4-dimethyl-2-hexene

2.5
$CH_2{=}CHCH_2CH_2CH_3$ 1-pentene
$CH_3(H)C{=}C(H)CH_2CH_3$ (same side) *cis*-2-pentene
$CH_3(H)C{=}C(H)CH_2CH_3$ (opposite sides) *trans*-2-pentene
$CH_2{=}C(CH_3)CH_2CH_3$ 2-methyl-1-butene
$CH_2{=}CHCH(CH_3)CH_3$ 3-methyl-1-butene
$CH_3C(CH_3){=}CHCH_3$ 2-methyl-2-butene

2.7
$HC{\equiv}CCH_2CH_3$ 1-butyne
$CH_3C{\equiv}CCH_3$ 2-butyne

2.9
$CH_2{=}C{=}CHCH_2CH_3$ 1,2-pentadiene
$CH_2{=}CHCH{=}CHCH_3$ 1,3-pentadiene
$CH_2{=}CHCH_2CH{=}CH_2$ 1,4-pentadiene
$CH_3CH{=}C{=}CHCH_3$ 2,3-pentadiene
$CH_2{=}C(CH_3)CH{=}CH_2$ 2-methyl-1,3-butadiene

2.13
(a) identical
(b) identical
(c) isomers
(d) identical
(e) identical

2.15
(a) F(Br)C=C(H)Cl and F(Br)C=C(Cl)H
(b) No geometric isomers.
(c) $H_3C(H)C{=}C(H)CH{=}CH_2$ (H$_3$C and H on top) and $H_3C(H)C{=}C(H)CH{=}CH_2$ (H$_3$C and CH=CH$_2$ on top)

2.17 Compound A could be either of the following:

1,2-dimethylcyclopentene (H_3C, CH_3) and 1,5-dimethylcyclopentene (H_3C, CH_3)

2.19 (a) $CH_3C(CH_3){=}CHCH_3 + H{-}O{-}SO_2{-}OH \rightarrow CH_3C(CH_3)(OSO_2OH)CH_2CH_3$

(b) $CH_3C(CH_3){=}CHCH_3 + H_2 \xrightarrow{\text{catalyst}} CH_3CH(CH_3)CH_2CH_3$

(c) $CH_3C(CH_3){=}CHCH_3 + H_2O \xrightarrow{H^+} CH_3C(CH_3)(OH)CH_2CH_3$

(d) $CH_3C(CH_3){=}CHCH_3 + HCl \rightarrow CH_3C(CH_3)(Cl)CH_2CH_3$

(e) $CH_3C(CH_3){=}CHCH_3 + HBr \rightarrow CH_3C(CH_3)(Br)CH_2CH_3$

(f) $CH_3C(CH_3){=}CHCH_3 + Br_2 \rightarrow CH_3C(CH_3)(Br){-}CH(Br)CH_3$

2.22 Cycloalkanes with the formula C_5H_{10} have no alkene groups and so cannot react with the given reactants. Possible structures are

[cyclopentane; methylcyclobutane (CH_3); 1,1-dimethylcyclopropane (H_3C, CH_3); 1,2-dimethylcyclopropane (CH_3, CH_3)]

2.24 11.1 g $K_2C_6H_8O_4$

2.27 (a) etc.$-CH_2C(CH_3)_2-CH_2C(CH_3)_2-CH_2C(CH_3)_2-CH_2C(CH_3)_2-$etc.

(b) $\left(CH_2C(CH_3)_2\right)_n$

2.31 Dipentene has no benzene ring.

2.33 The products of the reactions are the following, where C_6H_5 is the phenyl group.

(a) $C_6H_5SO_2OH$ (or $C_6H_5SO_3H$, benzenesulfonic acid) + H_2O
(b) $C_6H_5NO_2 + H_2O$
(c) no reaction
(d) no reaction
(e) no reaction (No iron or iron salt catalyst is specified.)
(f) no reaction
(g) $C_6H_5Br + HBr$

2.35 [benzene ring with two adjacent CH_3 groups (o-xylene)]

2.52 (a) $CH_3CH_2CH_2CH_3$ (b) $CH_3CCl_2{-}CCl_2CH_3$

(c) $CH_3CBr_2{-}CBr_2CH_3$

2.56 (a) $CH_3CH_2CH_2CH(OH)CH_2CH_3$

(b) $CH_3CH(CH_3)CH_2CH_3$
(c) $C_6H_5Br + HBr$
(d) no reaction
(e) [cyclopentane ring with OH] (cyclopentanol)
(f) $CH_3CH_2CH_2CH_2CH_3$
(g) no reaction
(h) $C_6H_5CH_2CH(OH)C_6H_5$
(i) no reaction
(j) $C_5H_{12} + 8O_2 \rightarrow 5CO_2 + 6H_2O$
(k) no reaction

CHAPTER 3

Practice Exercises, Chapter 3

1. (a) monohydric, secondary
(b) monohydric, secondary
(c) dihydric, unstable (two OH groups on the same carbon)
(d) dihydric, both are secondary
(e) monohydric, primary
(f) monohydric, primary
(g) monohydric, tertiary
(h) monohydric, secondary
(i) trihydric, unstable (three OH groups on the same carbon)

2. (a) alcohol
(b) phenol
(c) carboxylic acid
(d) alcohol (but unstable; an enol)
(e) alcohol
(f) alcohol

3. (a) 4-methyl-1-pentanol
(b) 2-methyl-2-propanol
(c) 2-ethyl-2-methyl-1-pentanol
(d) 2-methyl-1,3-propanediol

4. In 1,2-propanediol. Its boiling point is 189 °C, much higher than that of 1-butanol (bp 117 °C). 1,2-Propanediol is more soluble in water.

5. (a) $CH_3CH{=}CH_2$ (b) $CH_3CH{=}CH_2$

(c) $CH_2{=}\overset{CH_3}{\overset{|}{C}}CH_3$ (d) 

6. (a) $CH_3\overset{CH_3}{\overset{|}{C}}HCH{=}O$ then $CH_3\overset{CH_3}{\overset{|}{C}}HCO_2H$

(b) $C_6H_5CH{=}O$ then $C_6H_5CO_2H$

7. (a) $CH_3\overset{O}{\overset{\|}{C}}CH_2CH_3$ (b) $C_6H_5\overset{O}{\overset{\|}{C}}CH_3$ (c) 

8. (a) $CH_3\overset{O}{\overset{\|}{C}}CH_2CH{=}O$ and $CH_3\overset{O}{\overset{\|}{C}}CH_2CO_2H$

(b) no reaction

(c) $CH_3\underset{CH_3}{\underset{|}{\overset{CH_3}{\overset{|}{C}}}}CH{=}O$ and $CH_3\underset{CH_3}{\underset{|}{\overset{CH_3}{\overset{|}{C}}}}CO_2H$

(d) $CH_3\underset{CH_3}{\underset{|}{C}}H\overset{O}{\overset{\|}{C}}CH_3$

9. (a) CH_3OCH_3 (b) $CH_3CH_2CH_2OCH_2CH_2CH_3$

(c) 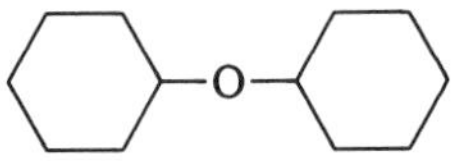

10. (a) $2CH_3SH$ (b) $CH_3\overset{CH_3}{\overset{|}{C}}HSS\overset{CH_3}{\overset{|}{C}}HCH_3$

(c)  (d) 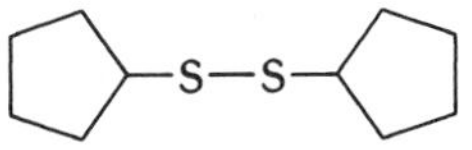

Review Exercises, Chapter 3

3.1 1, 1°alcohol; 2, ketone; 3, 2°alcohol; 4, ketone; 5, alkene

3.3 (a) $CH_3\overset{CH_3}{\overset{|}{C}}HCH_2OH$ (b) $CH_3\overset{OH}{\overset{|}{C}}HCH_3$

(c) $CH_3CH_2CH_2OH$ (d) $HOCH_2\underset{OH}{\underset{|}{C}}HCH_2OH$

3.8 (a) 2-methyl-1-propanol (b) 2-propanol
(c) 1-propanol (d) 1,2,3-propanetriol

3.12 Hydrogen bonds of the following type form.

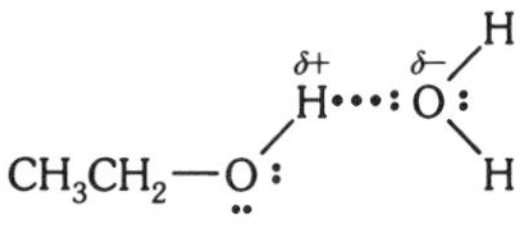

3.16 (a) $HOCH_2CH_2CH_3$ or $CH_3\underset{OH}{\underset{|}{C}}HCH_3$

(b) 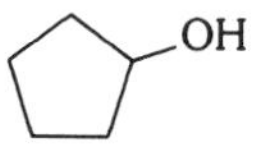

(c) $CH_3\overset{CH_2OH}{\overset{|}{C}}HCH_3$ or $CH_3\underset{OH}{\underset{|}{\overset{CH_3}{\overset{|}{C}}}}CH_3$

(d) 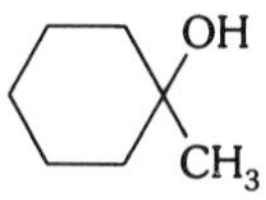 or 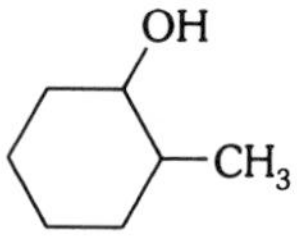

3.17 (a) $HOCH_2CH_2CH_2CH_3$

(b) $CH_3CH_2CH_2\overset{OH}{\overset{|}{C}}HCH(CH_3)_2$

(c) 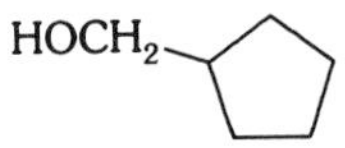

(d) $C_6H_5CH_2OH$

3.19 **A** reacts with aqueous NaOH; **B** does not. **B** can be dehydrated to an alkene; **A** cannot. (Both react with oxidizing agents, but in different ways.)

3.22 (a) $CH_3CH_2CH_2OH$ (b) $CH_3\underset{CH_3}{\underset{|}{C}}HCH_2OH$

(c) 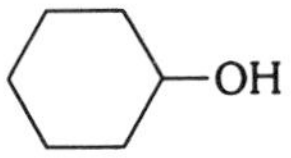 (d) 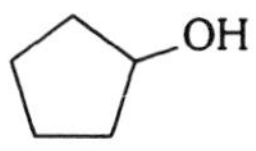

3.24 No reaction occurs. Ethers are stable in base.

3.26 (a) $CH_3CH_2CH_2SSCH_2CH_2CH_3$
(b) $(CH_3)_2CHCH_2SH$

(c) 

(d) $(CH_3)_2CHSSCH(CH_3)_2$

3.27 Hydrogen bonding in the thioalcohol family does exist, but the hydrogen bonds are not as strong as those in the alcohol family.

3.43 (a)  (b) $CH_3CH_2OCH_2CH_3$

(c) $CH_3\overset{O}{\overset{\|}{C}}CH_2CH_3$ (d) 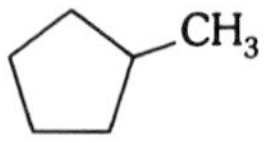

(e) no reaction (f) no reaction

(g) no reaction (h) $CH_3CH_2\underset{CH_3}{\underset{|}{C}}HCH_3$

(i) 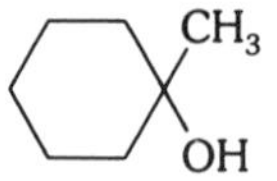 (j) 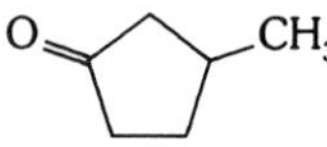

3.46 (a) $C_3H_8O + 2Cr_2O_7^{2-} + 16H^+ \rightarrow 3C_3H_6O_2 + 4Cr^{3+} + 11H_2O$
(b) $3C_3H_8O + 2K_2Cr_2O_7 + 16HCl \rightarrow 3C_3H_6O_2 + 4CrCl_3 + 4KCl + 11H_2O$
(c) 2/3 mol $K_2Cr_2O_7$
(d) $K_2Cr_2O_7 \cdot 2H_2O$; 2/3 mol
(e) 15.3 g propanoic acid
(f) 53.1 g $K_2Cr_2O_7 \cdot 2H_2O$

Chapter 4

Practice Exercises, Chapter 4

1. (a) 2-methylpropanal (b) 3-bromobutanal
(c) 4-ethyl-2,4,6-trimethylheptanal

2. 2-Isopropylpropanal would have the structure:

$$CH_3\underset{\underset{CH_3CHCH_3}{|}}{C}H\overset{O}{\overset{\|}{C}}H$$

and it should be named 2,3-dimethylbutanal.

3. (a) 2-butanone (b) 6-methyl-2-heptanone
(c) 2-methylcyclohexanone

4. (a) $CH_3CH_2\overset{O}{\overset{\|}{C}}\underset{\underset{CH_3}{|}}{C}HCH_3$

(b) $CH_3\overset{O}{\overset{\|}{C}}C_6H_5$

(c) $CH_3CH_2CH_2\overset{O}{\overset{\|}{C}}CH_2CH_2CH_3$

(d) $(CH_3)_3C\overset{O}{\overset{\|}{C}}C(CH_3)_3$

5. (a) $CH_3CH_2\overset{OH}{\overset{|}{C}}HCH_3$ (b) $CH_3\underset{\underset{CH_3}{|}}{C}HCH_2CH_2OH$

(c) 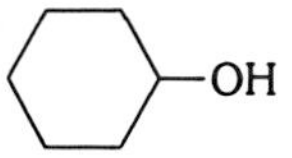 cyclohexane ring—OH

6. (a) not a hemiacetal (b) not a hemiacetal
(c) $HO\underset{\uparrow}{C}H_2OCH_2CH_3$ (hemiacetal position) (d) 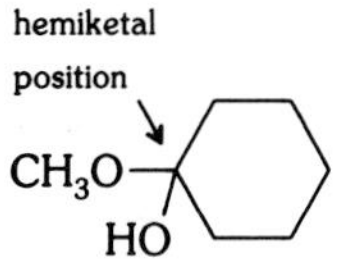 CH_3O—cyclohexane ring—HO (hemiketal position at the ring carbon bearing CH_3O and HO)

7. (a) $CH_3\overset{OH}{\overset{|}{C}}HOCH_3$

(b) $CH_3CH_2CH_2\overset{OH}{\overset{|}{C}}HOCH_2CH_3$

(c) $C_6H_5\overset{OH}{\overset{|}{C}}HOCH_2CH_2CH_3$

(d) $HOCH_2OCH_3$

8. (a) $CH_3CH_2\overset{O}{\overset{\|}{C}}H + HOCH_3$

(b) $CH_3CH_2OH + H\overset{O}{\overset{\|}{C}}CH_2CH_3$

9. (a) neither an acetal nor a ketal
(b) A ketal. The carbon atom that holds two oxygen atoms is a keto group carbon. The breakdown products are:

$$2CH_3CH_2OH + CH_3\overset{O}{\overset{\|}{C}}CH_3$$

10. (a) $2CH_3OH + H\overset{O}{\overset{\|}{C}}H$ (b) no reaction

(c) $2CH_3OH + CH_3\overset{H_3C}{\overset{|}{C}}H\overset{O}{\overset{\|}{C}}CH_3$

Review Exercises, Chapter 4

4.2 $CH_3CH_2\overset{O}{\overset{\|}{C}}H$ (aldehyde) $CH_3\overset{O}{\overset{\|}{C}}CH_3$ (ketone) $CH_3CH_2\overset{O}{\overset{\|}{C}}OH$ (carboxylic acid) $CH_3\overset{O}{\overset{\|}{C}}OCH_3$ (ester)

4.3 (a) $CH_3CH_2\underset{\underset{CH_3}{|}}{C}HCHO$

(b) 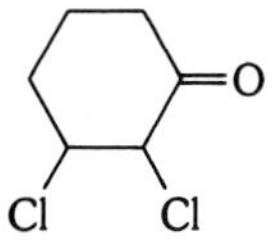 cyclohexanone ring (=O) with Cl on each of the two adjacent ring carbons (Cl, Cl)

(c) $C_6H_5\overset{O}{\overset{\|}{C}}CH_3$

(d) $(CH_3)_2CHCH_2\overset{O}{\overset{\|}{C}}CH_2CH(CH_3)_2$

(e) $CH_3\overset{O}{\overset{\|}{C}}CH_2CH_2\overset{O}{\overset{\|}{C}}CH_3$

4.5 (a) 2-methylcyclohexanone (b) propionic acid
(c) 2-pentanone (d) propanal
(e) 2-methylpentanal

4.7 5-ketohexanal

4.9 (a) 3-methylbutanal
(b) 3-methylpentanal
(c) 3-ethyl-5-methylhexanal
(d) 4-methyl-2-pentanone
(e) 3-propylcyclopentanone

4.11 valeraldehyde

4.13 **B < A < D < C**

4.15 **B < A < D < C**

4.17 CH_3CH_2 and H on C: $\text{C}\overset{\delta-}{=}\text{O}\cdots\overset{\delta+}{\text{H}}—\text{O}—\text{H}$

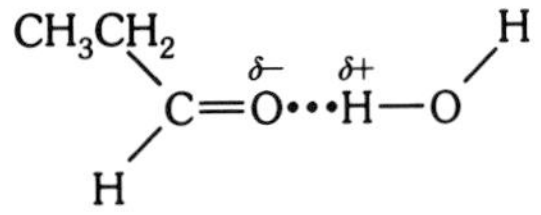

4.19 (a) $CH_3CH_2\overset{O}{\overset{\|}{C}}CH_3$
2-butanone

(b) $CH_3\overset{CH_3}{\overset{|}{C}}HCH_2CH_2CHO$
4-methylpentanal

(c) 3-methylcyclopentanone (structure: H_3C–cyclopentane ring =O)

(d) $C_6H_5CH_2CH(CH_3)CHO$

2-methyl-3-phenylpropanal

4.21 C_3H_6O is $CH_3CH_2CH{=}O$; $C_3H_6O_2$ is $CH_3CH_2CO_2H$.

4.23 Positive Benedict's tests are given by (a) and (b).

4.25 (a) The cations of transition metals
(b) Electron-rich particles, whether electrically neutral or negatively charged.
(c) F^-, Cl^-, Br^-, and I^- are four examples in the same family.
(d) H_2O, which forms $Cu(H_2O)_4^{2+}$. NH_3, which forms $Cu(NH_3)_4^{2+}$

4.30 $CH_3CH(OH)CO_2^-$

4.34 (a) $CH_3CH_2O^-$
(b) $CH_3CH_2O^- + H_2O \rightarrow CH_3CH_2OH + OH^-$
(c) ethanol

4.36 $H_3\overset{+}{N}CH(CH_2CH_2O^-)CO_2^-$ **A** $\quad H_3\overset{+}{N}CH(CH_2CH_2OH)CO_2^-$ **B**

4.38 (a) $CH_3C(=O)CH_2CH_3$ (b) $HC(=O)CH(CH_3)CH_2CH_3$
(c) cyclohexanone (ring =O) (d) CH_3–C_6H_4–$CH{=}O$

4.40 (a) hemiacetal (b) acetal
(c) something else (a 1,2-diether) (d) ketal

4.42 (a) $CH_3CH_2CH(OH)OCH_3$ $\quad CH_3CH_2CH(OCH_3)OCH_3$
(b) $CH_3CH_2CH(OH)OCH_2CH_3$ $\quad CH_3CH_2CH(OCH_2CH_3)OCH_2CH_3$

4.44 CH_3–$CH(OH)$–CH_2–CH_2–CH_2–$CH{=}O$ (drawn in ring-like arrangement)

4.46 $HOCH_2$–CH_2–CH_2–$C(=O)$–CH_3 (drawn in ring-like arrangement)

4.48 (a) $CH_3CH_2CHO + 2CH_3OH$
(b) no reaction
(c) $CH_3C(=O)CH_3 + 2CH_3CH_2OH$
(d) cyclopentanone (ring =O) $+ 2CH_3OH$

4.53 (a) $CH_3CH(CH_3)CH_2OH$ (b) $(CH_3)_2CHC(=O)CH_3$
(c) no reaction (d) $CH_3CH_2CH_2CH_2CH_3$
(e) $CH_3CH_2CH(OH)OCH_3$ (f) $CH_3CH_2OH + Mtb^+$
(g) $CH_3CH(OCH_2CH_3)OCH_2CH_3$ (h) $CH_3CHO + 2CH_3OH$
(i) cyclopentanecarboxylic acid (ring–CO_2H) (j) no reaction

4.55 $CH_3CH_2C(=O)H$ **A** $\quad CH_3CH_2CH_2OH$ **B** $\quad CH_3CH{=}CH_2$ **C** $\quad CH_3CH(OH)CH_3$ **D** $\quad CH_3C(=O)CH_3$ **E**

4.57 (a) 0.173 mol butanal
(b) 1.23 mol CH_3OH
(c) Yes, 11.1 g CH_3OH needed but 39.4 g taken
(d) 3.11 g H_2O obtained
(e) To ensure that the equilibria involved in the reaction are all shifted as much as possible to the right, in favor of the products.

CHAPTER 5

Practice Exercises, Chapter 5

1. (a) 2,2-dimethylpropanoic acid
(b) 5-ethyl-5-isopropyl-3-methyloctanoic acid
(c) sodium ethanoate
(d) 5-chloro-3-methylheptanoic acid
2. pentanedioic acid
3. 9-octadecenoic acid
4. (a) $CH_3CH_2CO_2^-$ (b) CH_3O–C_6H_4–CO_2^-
(c) $CH_3CH{=}CHCO_2^-$
5. (a) CH_3O–C_6H_4–CO_2H (b) $CH_3CH_2CO_2H$
(c) $CH_3CH{=}CHCO_2H$
6. (a) $CH_3C(=O)OCH_3$ (b) $CH_3C(=O)OCH_2CH_2CH_3$
(c) $CH_3C(=O)OCH(CH_3)CH_3$
7. (a) $HC(=O)OCH_2CH_3$ (b) $CH_3CH_2C(=O)OCH_2CH_3$

(c) $C_6H_5\overset{O}{\overset{\|}{C}}OCH_2CH_3$

8. (a) methyl propanoate
(b) propyl 3-methylpentanoate

9. (a) *t*-butyl acetate (b) ethyl butyrate

10. (a) $CH_3OH + CH_3CO_2H$
(b) $(CH_3)_2CHOH + CH_3CH_2CO_2H$
(c) $CH_3CH_2CH_2OH + (CH_3)_2CHCO_2H$

11. (a) $C_6H_5OH + CH_3CO_2^-$
(b) 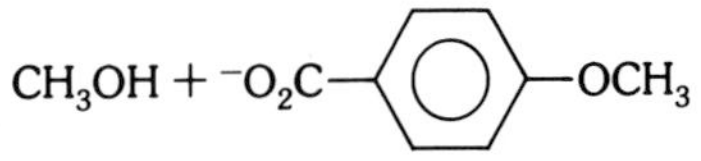 $CH_3OH + {}^-O_2C-C_6H_4-OCH_3$ (para)

Review Exercises, Chapter 5

5.1 (a) **B** (b) **A** (c) **B** (d) **C**

5.6 (a) 2-methylpropanoic acid
(b) 2,2-dimethylbutanoic acid
(c) sodium 3-chloropentanoate
(d) potassium benzoate

5.8 (a) sodium 3-hydroxybutanoate
(b) sodium β-hydroxybutyrate

5.11 etc.•••HOC(H)=O•••HOC(H)=O•••HOC(H)=O•••etc. ($\delta-$ on O, $\delta+$ on H)

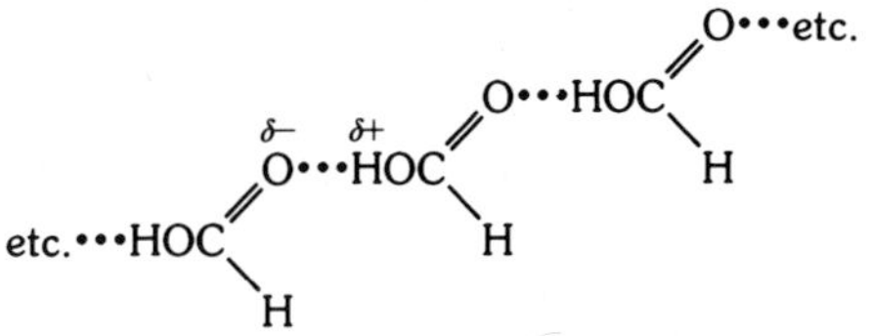

5.12 **C < A < B**

5.13 **B < C < D < A**

5.15 (a) $HCO_2H + H_2O \rightleftharpoons HCO_2^- + H_3O^+$
(b) Toward the formate ion. The added OH^- (from NaOH) neutralizes H_3O^+ and so reduces the concentration of H_3O^+ in the equilibrium. The equilibrium thus must shift to the right to replace the lost H_3O^+.
(c) $K_a = \dfrac{[HCO_2^-][H^+]}{[HCO_2H]}$
(d) stronger

5.17 **B < A < C < D**

5.19 (a) $HO_2CCH_2CH_2CO_2H + 2OH^- \rightarrow {}^-O_2CCH_2CH_2CO_2^- + 2H_2O$
(b) $HOCH_2CH_2CH_2CO_2H + OH^- \rightarrow HOCH_2CH_2CH_2CO_2^- + H_2O$
(c) $H\overset{O}{\overset{\|}{C}}CH_2CH_2CH_2CO_2H + OH^- \rightarrow H\overset{O}{\overset{\|}{C}}CH_2CH_2CH_2CO_2^- + H_2O$
(d) 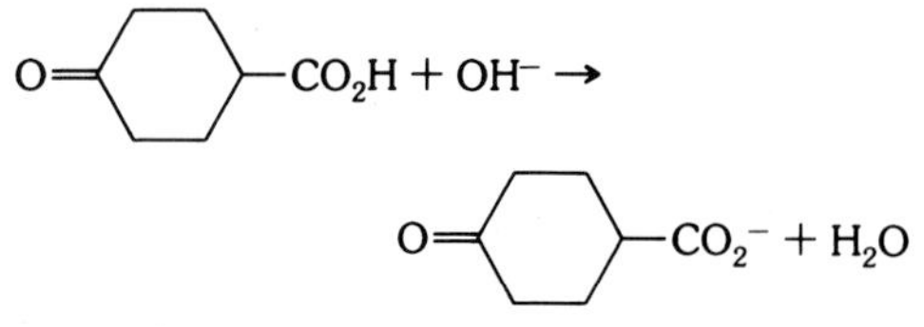 O=(cyclohexane)$-CO_2H + OH^- \rightarrow$ O=(cyclohexane)$-CO_2^- + H_2O$

5.21 **A,** an ionic compound, is much more soluble in water than the molecular compound, **B**.

5.23 (a) $HOCH_2CH_2CO_2^- + H^+ \rightarrow HOCH_2CH_2CO_2H$
(b) no reaction
(c) $C_6H_5O^- + H^+ \rightarrow C_6H_5OH$

5.24 (a) $CH_3CH_2\overset{O}{\overset{\|}{C}}Cl$ (b) $CH_3CH_2\overset{O}{\overset{\|}{C}}O\overset{O}{\overset{\|}{C}}CH_2CH_3$
(c) $CH_3CH_2\overset{O}{\overset{\|}{C}}OH$

5.26 (a) $CH_3\overset{O}{\overset{\|}{C}}OCH_2CH_3$
(b) The electronegativities of O and Cl place a relatively larg $\delta+$ charge on the carbon atom of the carbonyl group in acetyl chloride. This charge is able quite strongly to attract the $\delta-$ charge on the O atom of the alcohol molecule. In addition, the Cl^- ion is a very stable leaving group and so quite readily leaves the carbonyl carbon atom of the acetyl chloride molecule when the alcohol molecule attacks.

5.28 (a) $CH_3CH_2\overset{O}{\overset{\|}{C}}OCH_2CH_3 + H_2O$
(b) $(CH_3)_2CH\overset{O}{\overset{\|}{C}}OCH_2CH_3 + H_2O$
(c) 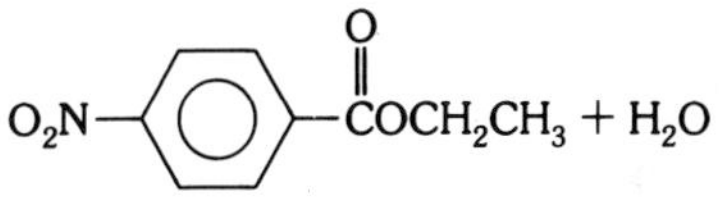 $O_2N-C_6H_4-\overset{O}{\overset{\|}{C}}OCH_2CH_3 + H_2O$
(d) 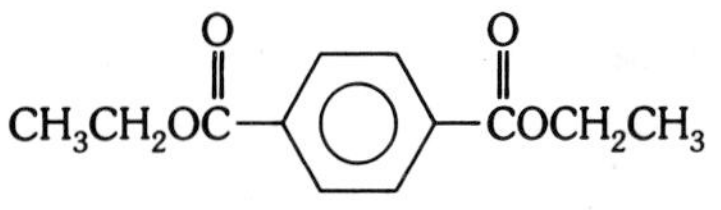 $CH_3CH_2O\overset{O}{\overset{\|}{C}}-C_6H_4-\overset{O}{\overset{\|}{C}}OCH_2CH_3$

5.31 A much stronger attraction can exist between the OH^- ion and the $\delta+$ charge on the carbonyl carbon atom of the ester, the specific site attacked in both hydrolysis and saponification, than between this $\delta+$ site and the $\delta-$ charge on a water molecule.

5.33 The acid chloride has a more stable leaving group (the weakly basic Cl^- ion) than the ester, for which the leaving group is a very strongly basic anion of an alcohol.

5.34 (a) $H\overset{O}{\overset{\|}{C}}OCH_2CH_3$ (b) 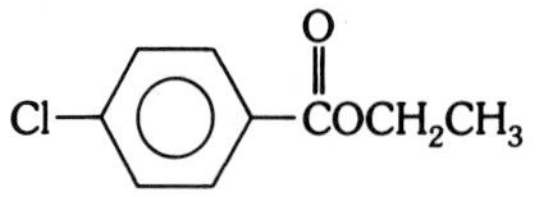 $Cl-C_6H_4-\overset{O}{\overset{\|}{C}}OCH_2CH_3$

5.36 **C < A < B < D**

5.38 (a) $CH_3\overset{O}{\overset{\|}{C}}OCH_2CH(CH_3)CH_3 + H_2O \xrightarrow{H^+} CH_3\overset{O}{\overset{\|}{C}}OH + HOCH_2CH(CH_3)CH_3$
(b) 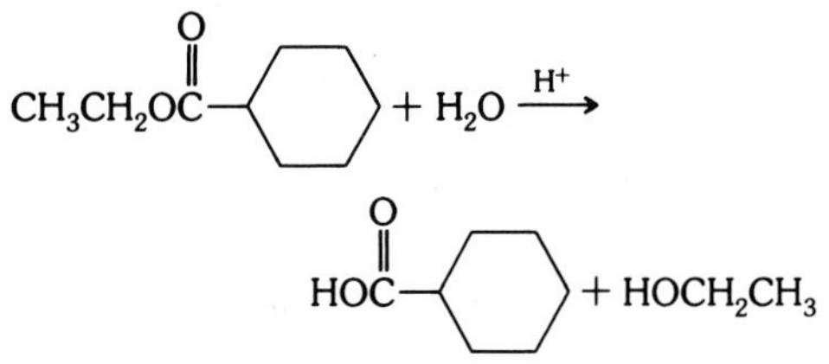 $CH_3CH_2O\overset{O}{\overset{\|}{C}}-C_6H_{11} + H_2O \xrightarrow{H^+} HO\overset{O}{\overset{\|}{C}}-C_6H_{11} + HOCH_2CH_3$ (cyclohexyl)

(c) no reaction (d) no reaction

5.40 $HOCH_2CHCH_2OH + CH_3(CH_2)_{12}CO_2H +$
(OH on the central carbon)

$CH_3(CH_2)_{14}CO_2H + CH_3(CH_2)_{10}CO_2H$

5.42 (a) $CH_3\overset{O}{\overset{\|}{C}}O^-Na^+ + HOCH_2\overset{CH_3}{\overset{|}{C}}HCH_3$

(b) $Na^{+-}O\overset{O}{\overset{\|}{C}}$–(cyclohexyl) $+ HOCH_2CH_3$

(c) no reaction (d) no reaction

5.44 $HOCH_2CHCH_2OH + CH_3(CH_2)_{12}CO_2^-Na^+$
(OH on the central carbon)

$+ CH_3(CH_2)_{14}CO_2^-Na^+ + CH_3(CH_2)_{10}CO_2^-Na^+$

5.46 If a large molar excess of ethyl alcohol is used, the following equilibrium will lie so much on the side of the products that essentially all of the expensive acid will be converted to the ester.

$$RCO_2H + CH_3CH_2OH \rightleftharpoons RCO_2CH_2CH_3 + H_2O$$

5.48 (a) $CH_3O—\overset{O}{\overset{\|}{P}}—OH$, with OH on P

(b) $CH_3CH_2O—\overset{O}{\overset{\|}{P}}—O—\overset{O}{\overset{\|}{P}}—OH$, with OH on each P

(c) $CH_3CH_2CH_2O—\overset{O}{\overset{\|}{P}}—O—\overset{O}{\overset{\|}{P}}—O—\overset{O}{\overset{\|}{P}}—OH$, with OH on each P

5.60 (a) 6.29 g methyl benzoate
(b) 1.48 g CH_3OH; 1.88 mL CH_3OH
(c) The reaction involves an equilibrium. By using a large excess of methyl alcohol, the equilibrium shifts in accordance with Le Châtelier's principle so that essentially all of the benzoic acid is converted to the ester.

5.61 (a) $CH_3CHCO_2^-Na^+$
(with CH_3O on the CH carbon)

(b) $CH_3CH_2CO_2H + CH_3OH$
(c) CH_3CO_2H
(d) (cyclohexene)
(e) $(CH_3)_2CHCH_2CO_2^-Na^+ + CH_3OH$
(f) $(CH_3)_2CH\overset{O}{\overset{\|}{C}}CH_3$
(g) $CH_3CH_2\overset{O}{\overset{\|}{C}}OCH_2CH_3$
(h) $CH_3CH_2CHO + 2CH_3OH$
(i) no reaction
(j) $C_6H_5\overset{O}{\overset{\|}{C}}OCH_2CH_3$
(k) $CH_3\underset{Cl}{\underset{|}{C}}HCH_2CH_3$
(l) no reaction

CHAPTER 6

Practice Exercises, Chapter 6

1. (a) isopropyldimethylamine
(b) cyclohexylamine
(c) *t*-butylisobutylamine

2. (a) $(CH_3)_3CNH\underset{CH_3}{\underset{|}{C}}HCH_2CH_3$
(b) NO_2–(benzene ring)–NH_2
(c) NH_2–(benzene ring)–CO_2H

3. (a) $C_6H_5NH_3^+$ (b) $(CH_3)_3NH^+$
(c) $^+NH_3CH_2CH_2NH_3^+$

4. (a) HO–(benzene ring, with a second HO)–$\overset{OH}{\overset{|}{\underset{H}{\underset{|}{C}}}}—CH_2—NHCH_3$
(b) CH_3O–(benzene ring, with OCH_3 above and OCH_3 below)–$CH_2CH_2NH_2$

5. (a) 4-methylhexanamide
(b) 2-ethylbutanamide

6. (a) $(CH_3)_2CH\overset{O}{\overset{\|}{C}}NHCH_3$ (b) $CH_3\overset{O}{\overset{\|}{C}}NHC_6H_5$
(c) No amide forms. (d) No amide forms.

7. (a) $C_6H_5CO_2H + NH_2CH_3$
(b) No hydrolysis occurs.
(c) $C_6H_5NH_2 + HO_2CCH_3$
(d) $NH_2CH_2CH_2NH_2 + 2CH_3CO_2H$

8. $NH_2CH_2CO_2H + NH_2\underset{CH_3}{\underset{|}{C}}HCO_2H +$

$NH_2CHCO_2H + NH_2CHCO_2H$
(first: CH_3CH with CH_3 on the alpha carbon, i.e. isopropyl; second: CH_2SH on the alpha carbon)

Review Exercises, Chapter 6

6.1 (a) aliphatic amide + ether group
(b) aliphatic amine + ester group
(c) aliphatic, heterocyclic amide
(d) Aromatic *compound* overall because of the benzene ring, but the amine is an *aliphatic* amine. (The amino group is not attached directly to the ring.)

6.3 (a) amine; heterocyclic
(b) 1, amine; 2, ester; 3, amine (aromatic)
(c) 1, heterocyclic amine; 2, amine (heterocyclic) amine
(d) 1, alcohol; 2, amine

6.5 (a) isopropylpropylamine
(b) ethylmethylpropylamine
(c) *p*-bromoaniline
(d) dipropylamine

6.7 (a) $CH_3CH_2CH_2NH_3^+$ (b) $CH_3CH_2CH_2NH_2$
(c) no reaction (d) no reaction

6.9 **A** is the stronger base; it is an amine (plus a ketone), and **B** is an amide.

6.11 (a) butanamide (b) 3-methylbutanamide

6.13 $C_6H_5CON(CH_3)_2$

6.17 $CH_3\overset{O}{\overset{\|}{C}}Cl + 2NH_3 \rightarrow CH_3\overset{O}{\overset{\|}{C}}NH_2 + NH_4^+Cl^-$

$CH_3\overset{O}{\overset{\|}{C}}O\overset{O}{\overset{\|}{C}}CH_3 + 2NH_3 \rightarrow CH_3\overset{O}{\overset{\|}{C}}NH_2 + NH_4^+\ {}^-O\overset{O}{\overset{\|}{C}}CH_3$

6.19 (a) $NH_2CH_2\overset{O}{\overset{\|}{C}}NH\underset{CH_3}{\underset{|}{C}H}\overset{O}{\overset{\|}{C}}—$ (b) two

6.21 (a) $CH_3CH_2NH_2 + CH_3CO_2H$
(b) $(CH_3)_2CHNH_2 + CH_3CH_2CO_2H$
(c) $CH_3NH_2 + (CH_3)_2CHCO_2H$
(d) Does not hydrolyze

6.33 (a) Water reacts quantitatively with alkenes, acetals or ketals, esters, and amides. (Not shown are the hydrolyses of acid chlorides, acid anhydrides, and esters of phosphoric acid.) The R groups can be alike or different. (H)R means that the group can be H or R.

$$\text{>C=C<} + H_2O \xrightarrow{H^+} -\underset{H}{\underset{|}{\overset{|}{C}}}-\underset{OH}{\underset{|}{\overset{|}{C}}}-$$
Alkene → Alcohol

$$\text{(H)R}\overset{\text{(H)R}}{\overset{|}{C}}(OR)_2 + H_2O \xrightarrow{H^+} \text{(H)R}\overset{O}{\overset{\|}{C}}\text{R(H)} + 2HOR$$
Acetal or ketal → Aldehyde or ketone + Alcohol

$$\text{(H)R}\overset{O}{\overset{\|}{C}}OR + H_2O \xrightarrow{H^+} \text{(H)R}\overset{O}{\overset{\|}{C}}OH + HOR$$
Ester → Carboxylic acid + Alcohol

$$\text{(H)R}\overset{O}{\overset{\|}{C}}NH_2 + H_2O \xrightarrow{H^+} \text{(H)R}\overset{O}{\overset{\|}{C}}OH + NH_3$$
Amide → Carboxylic acid + Ammonia

(The H's on N of the amide can be replaced by one or two alkyl groups.)

(b) The groups that can be hydrogenated are alkenes, aldehydes and ketones, and disulfides.

$$\text{>C=C<} + H_2 \xrightarrow{\text{catalyst}} -\underset{H}{\underset{|}{\overset{|}{C}}}-\underset{H}{\underset{|}{\overset{|}{C}}}-$$
Alkene → Alkane

$$\text{(H)R}\overset{O}{\overset{\|}{C}}\text{R(H)} + H_2 \xrightarrow{\text{catalyst}} \text{(H)R}\overset{OH}{\overset{|}{C}}\text{HR(H)}$$
Aldehyde or ketone → Alcohol

$$RSSR + H_2 \xrightarrow{\text{catalyst}} 2RSH$$
Disulfide → Thioalcohol

(c) Oxidizable groups in our study are 1° and 2° alcohols, aldehydes, and thioalcohols.

$$RCH_2OH + (O) \longrightarrow RCH{=}O$$
1° Alcohol → Aldehyde

$$R\overset{OH}{\overset{|}{C}}HR + (O) \longrightarrow R\overset{O}{\overset{\|}{C}}R$$
2° Alcohol → Ketone

$$RCHO + (O) \longrightarrow RCO_2H$$
Aldehyde → Carboxylic acid

$$2RSH + (O) \longrightarrow RSSR$$
Thioalcohol → Disulfide

6.35 (a) 1.66 g benzoic acid (b) 28.3 mL 0.482 *M* HCl

6.36 (a) $CH_3CO_2H + CH_3OH$
(b) $CH_3\overset{OH}{\overset{|}{C}}HCH_2CH_3$
(c) no reaction
(d) $CH_3CH_2\overset{O}{\overset{\|}{C}}NH_2$
(e) no reaction
(f) no reaction
(g) $CH_3CHO + 2HOCH_2CH_3$
(h) $CH_3CH_2CO_2H$
(i) no reaction
(j) $CH_3CH_2\overset{O}{\overset{\|}{C}}OCH_3$
(k) $CH_3CH_2CO_2^-Na^+ + NH_3$

(l) 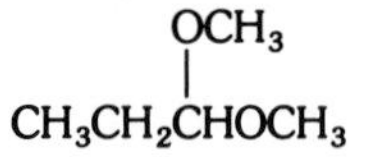 $CH_3CH_2CH(OCH_3)OCH_3$ [drawn with OCH_3 above the CH carbon: $CH_3CH_2\overset{OCH_3}{\overset{|}{C}}HOCH_3$]

(m) $CH_3CH_2SSCH_2CH_3$

(n) $C_6H_5CO_2^-Na^+ + HOCH_2CH(CH_3)_2$

(o) $Cl^-\ {}^+NH_3CH_2CH_2CH(CH_3)_2$

(p) $CH_3CH_2CH_2CH_2OCH_3$

6.39 (a) sodium benzoate (b) propylamine (c) aniline (d) propionic acid (e) butyraldehyde (f) isobutyl alcohol (g) phenol (h) diethyl ether (i) ethyl butyrate (j) acetamide (k) acetone

6.40 **B.** It is a carboxylic acid that will become an anion at the basic pH and so more soluble in water. (**A** is an ester and **C** is an amine.)

CHAPTER 7

Practice Exercises, Chapter 7

1. (a) 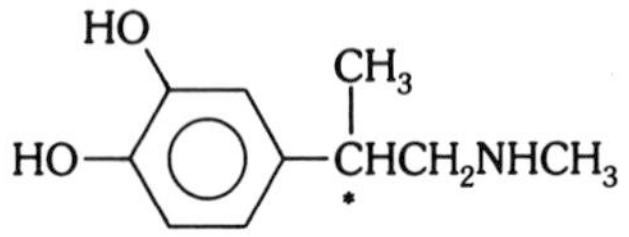 3,4-(HO)$_2C_6H_3$—$\overset{\bullet}{C}H(CH_3)CH_2NHCH_3$ [HO groups on the benzene ring; CH_3 above CH; stereocenter marked with •]

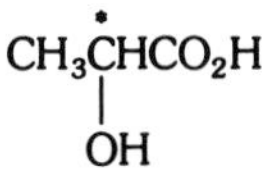

(b) $CH_3\overset{\bullet}{C}HCO_2H$ with OH on the marked carbon

(c) $CH_3\overset{\bullet}{C}H\overset{\bullet}{C}HCO_2^-$ with HO on the first and NH_3^+ on the second marked carbon

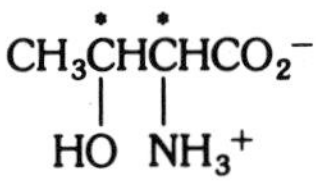

(d) 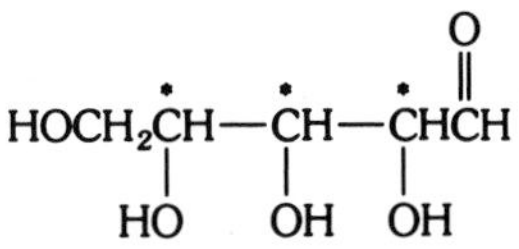 $HOCH_2\overset{\bullet}{C}H—\overset{\bullet}{C}H—\overset{\bullet}{C}HCH{=}O$ with HO, OH, OH on the marked carbons

2. (a) 3 (b) 8 (c) 4

3. 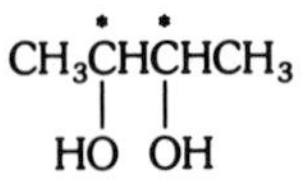 $CH_3\overset{\bullet}{C}H\overset{\bullet}{C}HCH_3$ with HO and OH on the marked carbons

The two tetrahedral stereocenters are identical; they hold identical sets of four different groups, CH_3, OH, H, and $CH_3CH(OH)$.

Review Exercises, Chapter 7

7.2 $CH_3CH_2CH_2OH$ and $(CH_3)_2CHOH$

7.5 (a) stereoisomers (b) constitutional

7.7 (a) 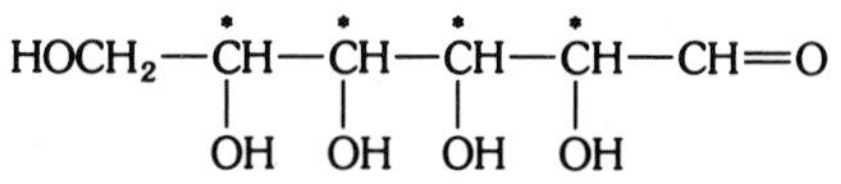 $HOCH_2—\overset{\bullet}{C}H—\overset{\bullet}{C}H—\overset{\bullet}{C}H—\overset{\bullet}{C}H—CH{=}O$ with OH on each marked carbon

(b) All are different (c) 16 (d) 8

7.8 No; glycine has no tetrahedral stereocenter.

7.10 148.5 °C. The designations (+) and (−) placed before otherwise identical names tell us that the two compounds are enantiomers, and enantiomers have identical physical properties.

7.14 −0.375°

7.18 3.01 g/100 mL

7.20 Strychnine. The calculated specific rotation for the sample is −139°, which corresponds to the value for strychnine, not for brucine.

7.22 (a) They are not related as object to mirror image.
(b) Because they are stereoisomers of each other
(c) What makes them different is not a lack of free rotation.
(d) diastereomers

7.23 Methane lacks a tetrahedral stereocenter and is not a member of a *set* of stereoisomers.

7.26 (a) 2-butanol (b) 3-methylhexane

CHAPTER 8

Practice Exercises, Chapter 8

1. (a) a = b
c = e
(b) Compounds a and d are enantiomers.
(c) Compound c (or e) is a meso compound.

2. 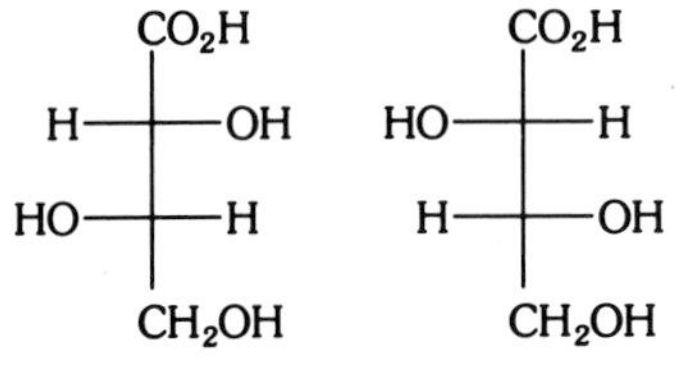 Fischer projections (CO_2H top, CH_2OH bottom):

First: H—C—OH; HO—C—H and HO—C—H; H—C—OH

A pair of enantiomers

Second: H—C—OH; H—C—OH and HO—C—H; HO—C—H

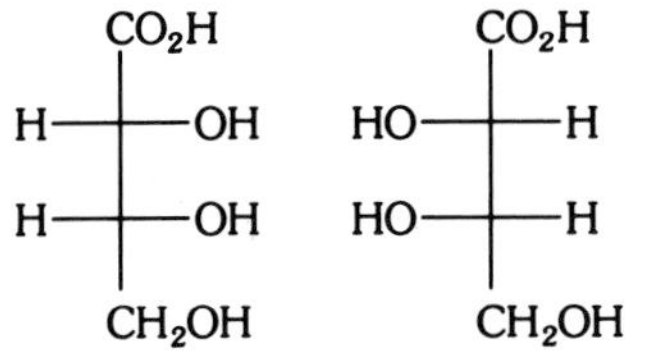

A pair of enantiomers

Review Exercises, Chapter 8

8.5 (a) C (b) D (c) A (d) B and D

8.7 $HOCH_2CH(OH)CH{=}O$

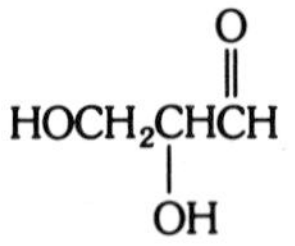

glyceraldehyde

8.9 polysaccharide

8.11 (a) 2 (b) trisaccharide

8.13 $HOCH_2CH(OH)C({=}O)CH_2OH$

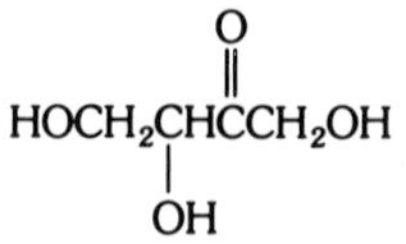

8.16 (a) 2

(b) 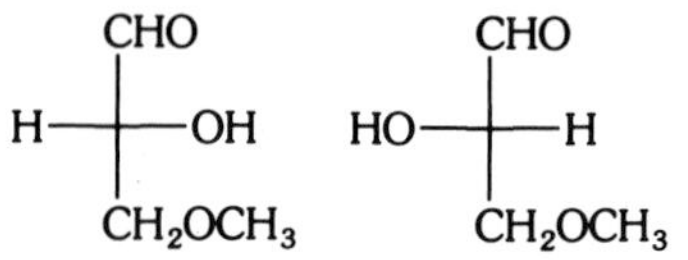 Fischer projections (CHO top, CH_2OCH_3 bottom): H—C—OH and HO—C—H

(c) D L

8.17 It has no tetrahedral stereocenter.

8.19

```
      CH=O
      |
      CH2
      |
  H—C—OH
      |
  H—C—OH
      |
      CH2OH
```

8.21 (a) L-family

(b)

```
        CH2OH
        |
        C=O
        |
   H————OH
        |
  HO————H
        |
   H————OH
        |
        CH2OH
```

(c)

```
        CH2OH
        |
        C=O
        |
        CH2
        |
  HO————H
        |
   H————OH
        |
        CH2OH
```

8.23

```
       CO2H
       |
   H—C—OH
       |
  HO—C—H
       |
   H—C—OH
       |
   H—C—OH
       |
       CH2OH
```

8.24 (a)

CH₂OH, C—OH, H, H, C, H, H, CH=O, HO, C—C, OH OH

(b) at carbon 3

(c) D-family. The relative positions of the CH_2OH group and the O atom of the ring tell us that the compound is in the D-family.

(d) D-allose

(e) an epimer

8.26

H, CH_2OH, O, HO, OH, H, HO, H, H, H, OH

α-mannose

H, CH_2OH, O—H, HO, OH, O, H, HO, C, H, H, H

open form of mannose

H, CH_2OH, O, HO, OH, H, OH, HO, H, H, H

β-mannose

8.29 As the beta form is used, molecules of the other forms continuously change into it as the equilibria shift.

8.31 Something else: β-2-deoxyfructose

8.33 No, an OH group is required at position 4 to make possible the formation of the five-membered ring.

8.36

HO, CH_2OH, O, H, H, H, H, HO, OH, H, OCH_3

methyl α-galactoside

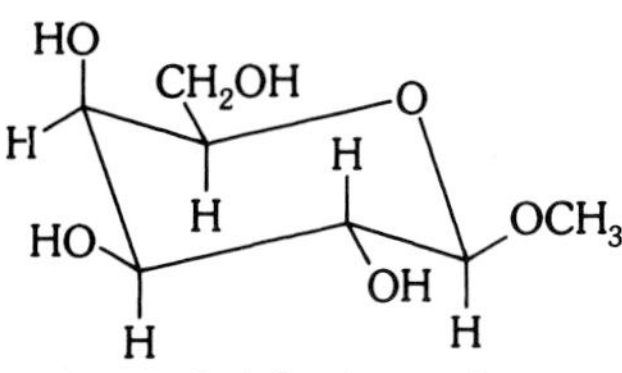

methyl β-galactoside

These are cis–trans isomers in the sense that we have used this concept, because the OCH_3 group is on opposite sides of the rings. However, the term is just not used in connection with glycosides.

8.40 (a) yes, see arrow

(b) yes, see enclosure

acetal system, CH_2OH, CH_2OH, O, O, O, OH, OH, HO, OH, OH, OH, hemiacetal system

(c) a β(1 → 4) bridge

(d) Yes, it has the hemiacetal system, so the open form of the corresponding ring (on the right) has an aldehyde group.

(e) Maltose has an α(1 → 4) bridge between the two rings.

(f) two glucose molecules

8.42

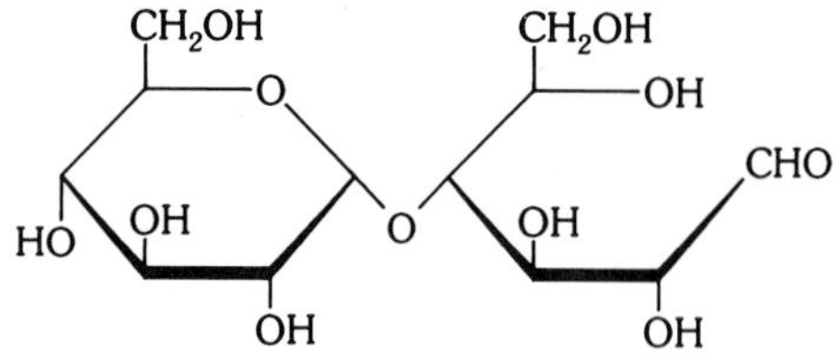

CH_2OH CH_2OH
O OH
CHO
HO OH O OH
OH OH

8.59 (a)

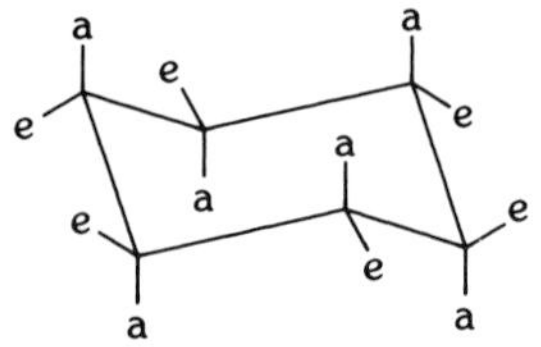

a a e e e a a e e e a a

(b) 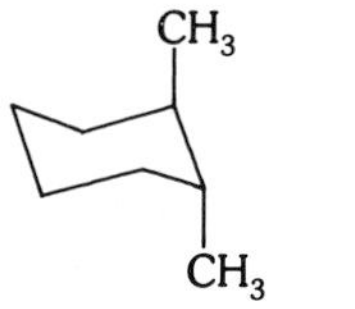 CH_3 CH_3

(c) 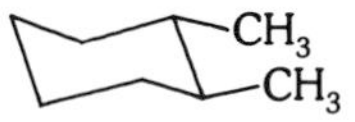 CH_3 CH_3

(d)

H CH_2OH O
HO H
H OH
HO
OH
H H

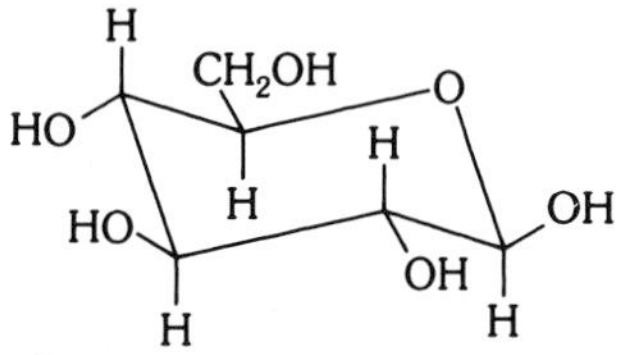

β-glucose
(all substituents are equatorial)

(d)

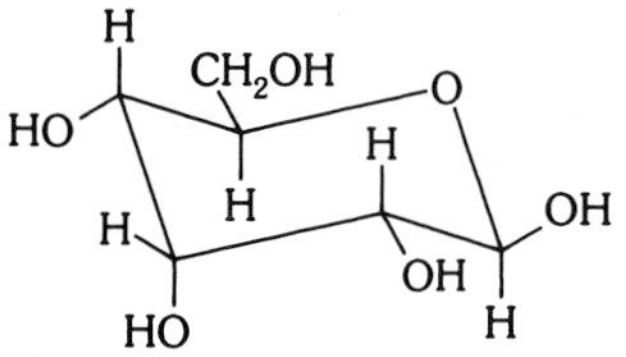

H CH_2OH O
HO H
H OH
H
OH
HO H

β-allose
(one substituent axial)

8.61 An enzyme (amylase) in the saliva catalyzes the hydrolysis of enough of the starch so that the resulting solution fails to give the iodine test.

8.63 To form a cyclic hemiacetal, the ring would be limited to four atoms. Although four-membered rings are known, they are difficult to form because of the unfavorable bond angle (90° as compared to the normal angle of 109.5°).

CHAPTER 9

Practice Exercises, Chapter 9

1. 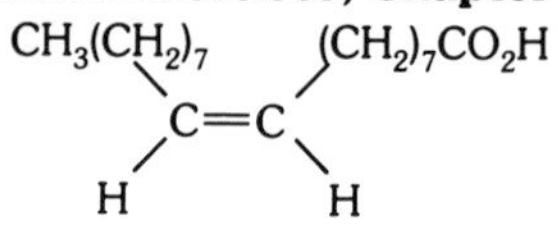 $CH_3(CH_2)_7$ and $(CH_2)_7CO_2H$ on C=C, with H and H

$$\mathrm{CH_3(CH_2)_7CH{=}CH(CH_2)_7CO_2H}$$

2. $CH_3(CH_2)_{26}CO_2(CH_2)_{25}CH_3$

3. 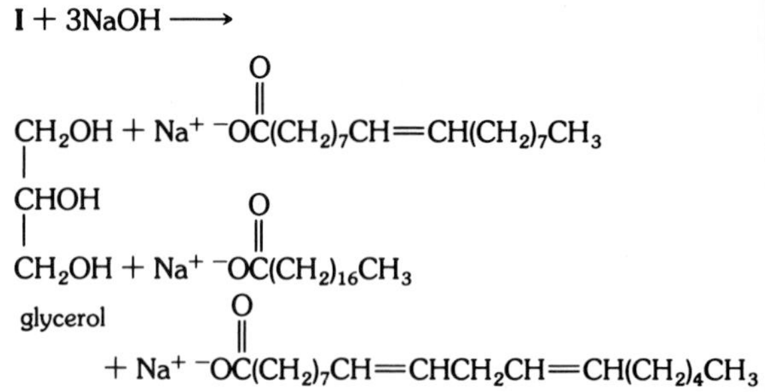 $\mathrm{I} + 3NaOH \longrightarrow$

$$\mathrm{CH_2OH + Na^+\ ^-OC(O)(CH_2)_7CH{=}CH(CH_2)_7CH_3}$$
$$\mathrm{CHOH}$$
$$\mathrm{CH_2OH + Na^+\ ^-OC(O)(CH_2)_{16}CH_3}$$

glycerol

$$\mathrm{+\ Na^+\ ^-OC(O)(CH_2)_7CH{=}CHCH_2CH{=}CH(CH_2)_4CH_3}$$

4.

$$\mathrm{I} + 3H_2 \xrightarrow{\text{catalyst}} \mathrm{CH_2{-}O{-}C(O)(CH_2)_{16}CH_3}$$
$$\mathrm{CH{-}O{-}C(O)(CH_2)_{16}CH_3}$$
$$\mathrm{CH_2{-}O{-}C(O)(CH_2)_{16}CH_3}$$

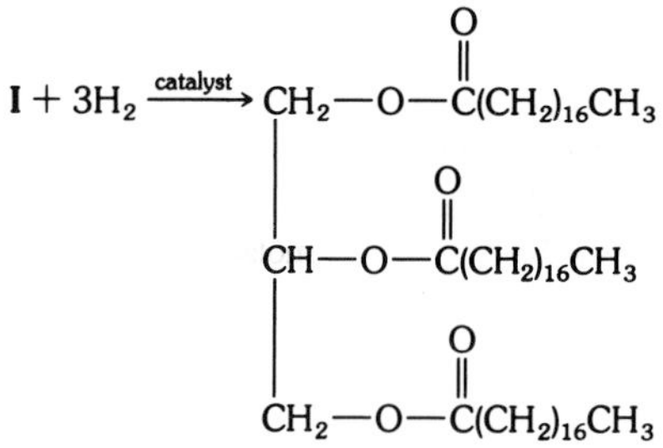

Review Exercises, Chapter 9

9.2 It is extractable from animal and plant sources by relatively nonpolar solvents.

9.4 It is present in undecomposed plant or animal materials and is extractable by relatively nonpolar solvents.

9.8 The organic products of the reactions are the following.

(a) 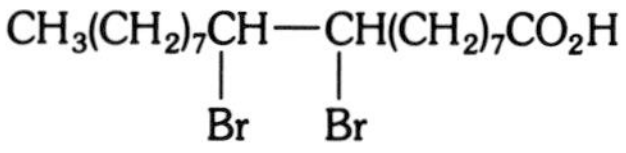 $\mathrm{CH_3(CH_2)_7CH(Br){-}CH(Br)(CH_2)_7CO_2H}$

(b) $\mathrm{CH_3(CH_2)_7CH{=}CH(CH_2)_7CO_2^-K^+}$

(c) $\mathrm{CH_3(CH_2)_{16}CO_2H}$

(d) $\mathrm{CH_3(CH_2)_7CH{=}CH(CH_2)_7CO_2CH_2CH_3}$

9.11

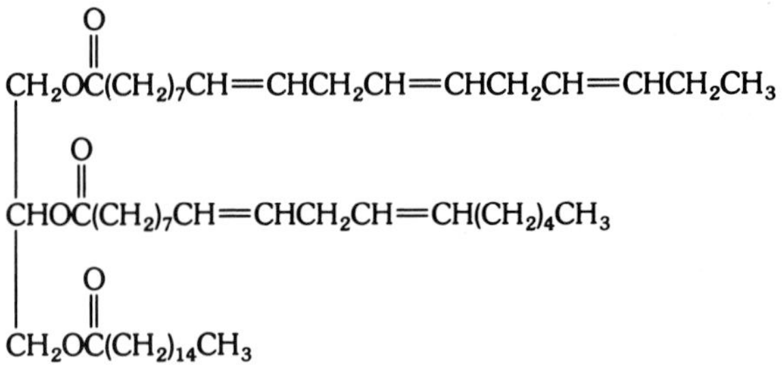

$$\mathrm{CH_2OC(O)(CH_2)_7CH{=}CHCH_2CH{=}CHCH_2CH{=}CHCH_2CH_3}$$
$$\mathrm{CHOC(O)(CH_2)_7CH{=}CHCH_2CH{=}CH(CH_2)_4CH_3}$$
$$\mathrm{CH_2OC(O)(CH_2)_{14}CH_3}$$

9.13 $\mathrm{HOCH_2CH(OH)CH_2OH}$

$$\mathrm{+\ HO_2C(CH_2)_7CH{=}CHCH_2CH{=}CH(CH_2)_4CH_3}$$
$$\mathrm{+\ HO_2C(CH_2)_{12}CH_3}$$
$$\mathrm{+\ HO_2C(CH_2)_7CH{=}CH(CH_2)_7CH_3}$$

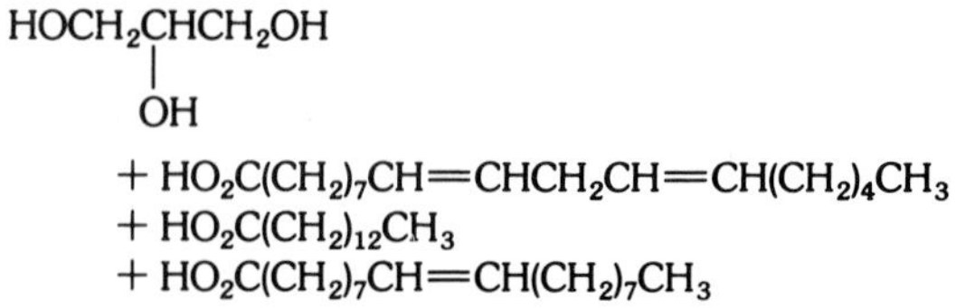

9.15 More than one structure is possible because the three different acyl groups can be joined in different orders to the glycerol unit. One possible structure is

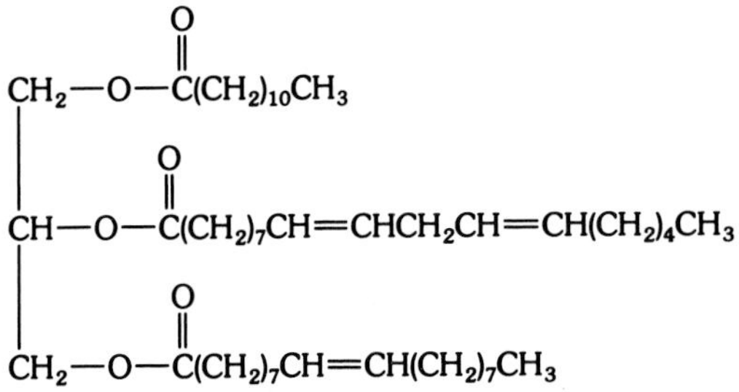

$$\begin{array}{l} CH_2{-}O{-}\overset{O}{\overset{\|}{C}}(CH_2)_{10}CH_3 \\ | \\ CH{-}O{-}\overset{O}{\overset{\|}{C}}(CH_2)_7CH{=}CHCH_2CH{=}CH(CH_2)_4CH_3 \\ | \\ CH_2{-}O{-}\overset{O}{\overset{\|}{C}}(CH_2)_7CH{=}CH(CH_2)_7CH_3 \end{array}$$

9.22 **C. A** is ruled out because *both* the acid and alcohol portions of the wax molecule are usually long chain. **B** is ruled out because *both* of these portions are likely to have an even number of carbons.

9.23 The molecules of both types give glycerol and phosphoric acid plus a fatty acid when fully hydrolyzed.

9.25 Molecules of both types are derivatives of sphingosine (rather than glycerol). Sphingomyelin molecules have a phosphate ester unit, but those of the cerebrosides have a monosaccharide unit instead.

9.27 Their molecules bear electrical charges at different locations.

9.29 sphingomyelins and cerebrosides

9.31

(a) 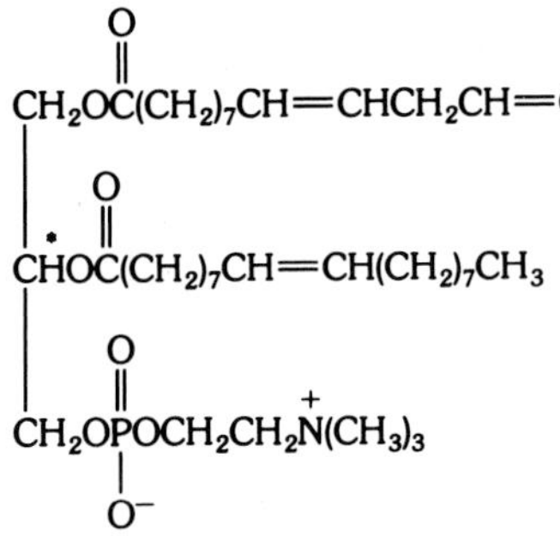
$$\begin{array}{l} CH_2O\overset{O}{\overset{\|}{C}}(CH_2)_7CH{=}CHCH_2CH{=}CHCH_2CH{=}CHCH_2CH_3 \\ | \\ \overset{\bullet}{C}HO\overset{O}{\overset{\|}{C}}(CH_2)_7CH{=}CH(CH_2)_7CH_3 \\ | \\ CH_2O\overset{O}{\overset{\|}{P}}OCH_2CH_2\overset{+}{N}(CH_3)_3 \\ \quad\;\; | \\ \quad\;\; O^- \end{array}$$

(b) A glycerophospholipid, because it is based on glycerol, not sphingosine

(c) Yes, the asterisk in the structure of part (a) marks the tetrahedral stereocenter.

(d) A lecithin, because its hydrolysis would give 2-(trimethylamino)ethanol

9.40 The hydrophobic tails intermesh with each other between the two layers of the bilayer.

9.42 The water-avoiding properties of the hydrophobic units and the water-attracting properties of the hydrophilic units

9.54 (a) yes (b) yes (c) ester groups

(d) 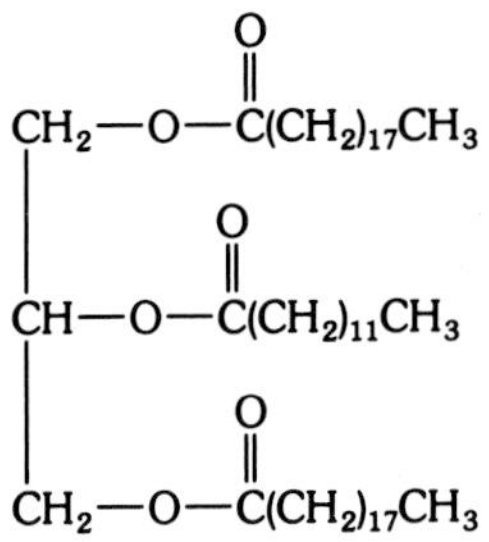
$$\begin{array}{l} CH_2{-}O{-}\overset{O}{\overset{\|}{C}}(CH_2)_{17}CH_3 \\ | \\ CH{-}O{-}\overset{O}{\overset{\|}{C}}(CH_2)_{11}CH_3 \\ | \\ CH_2{-}O{-}\overset{O}{\overset{\|}{C}}(CH_2)_{17}CH_3 \end{array}$$

(e) No. Its fatty acid units have odd numbers of carbon atoms.

CHAPTER 10

Practice Exercises, Chapter 10

1. glycine $^+NH_3CH_2CO_2^-$

alanine $\begin{array}{l} ^+NH_3CHCO_2^- \\ \qquad\;\; | \\ \qquad\; CH_3 \end{array}$

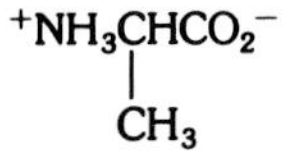

lysine $\begin{array}{l} ^+NH_3CHCO_2^- \\ \qquad\;\; | \\ \qquad\; CH_2CH_2CH_2CH_2NH_2 \end{array}$

glutamic acid $\begin{array}{l} ^+NH_3CHCO_2^- \\ \qquad\;\; | \\ \qquad\; CH_2CH_2CO_2H \end{array}$

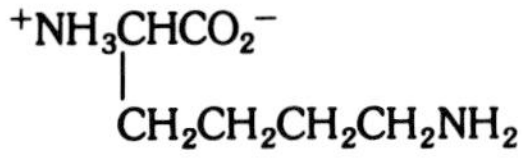

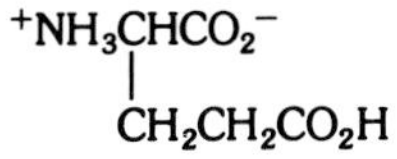

2. (a) 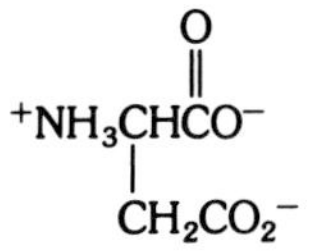 $\begin{array}{l} ^+NH_3CH\overset{O}{\overset{\|}{C}}O^- \\ \qquad\;\; | \\ \qquad\; CH_2CO_2^- \end{array}$ (b) 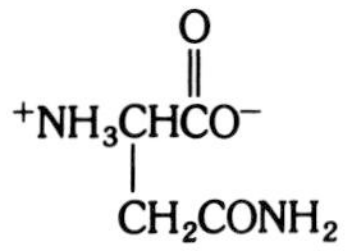 $\begin{array}{l} ^+NH_3CH\overset{O}{\overset{\|}{C}}O^- \\ \qquad\;\; | \\ \qquad\; CH_2CONH_2 \end{array}$

3. 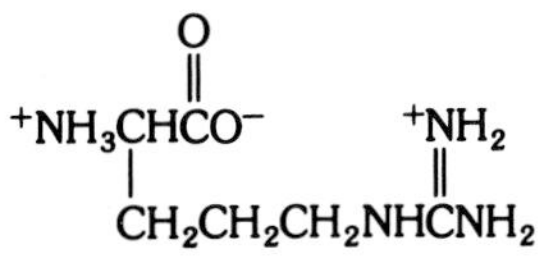 $\begin{array}{l} ^+NH_3CH\overset{O}{\overset{\|}{C}}O^- \qquad\quad {}^+NH_2 \\ \qquad\;\; | \qquad\qquad\qquad\;\; \| \\ \qquad\; CH_2CH_2CH_2NHCNH_2 \end{array}$

4. Hydrophilic; neutral (The side chain has an amide group, not an amino group.)

5. 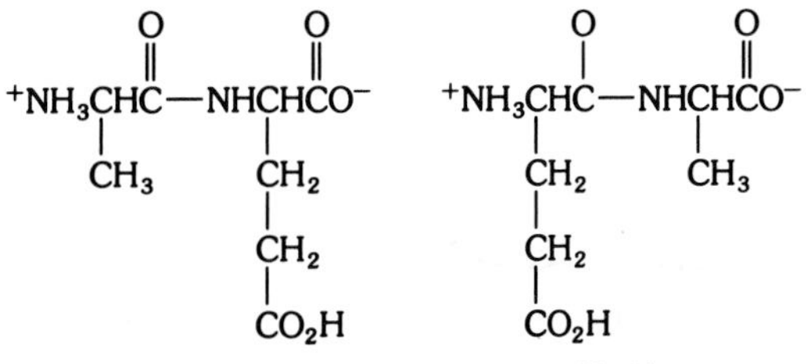 $\begin{array}{l} ^+NH_3CH\overset{O}{\overset{\|}{C}}{-}NHCH\overset{O}{\overset{\|}{C}}O^- \\ \qquad\;\; | \qquad\qquad\; | \\ \qquad\; CH_3 \qquad\;\; CH_2 \\ \qquad\qquad\qquad\quad\; | \\ \qquad\qquad\qquad\;\; CH_2 \\ \qquad\qquad\qquad\quad\; | \\ \qquad\qquad\qquad\;\; CO_2H \end{array}$ $\begin{array}{l} ^+NH_3CH\overset{O}{\overset{|}{C}}{-}NHCH\overset{O}{\overset{\|}{C}}O^- \\ \qquad\;\; | \qquad\qquad\; | \\ \qquad\; CH_2 \qquad\;\; CH_3 \\ \qquad\;\; | \\ \qquad\; CH_2 \\ \qquad\;\; | \\ \qquad\; CO_2H \end{array}$

Ala-Glu Glu-Ala

Review Exercises, Chapter 10

10.1 **B.** Its NH_3^+ group is not on the same carbon that holds the CO_2^- group.

10.3 $^+NH_3CH_2CO_2H$

10.5 $\begin{array}{l} NH_2CHCO_2CH_2CH_3 \\ \qquad\; | \\ \qquad CH_3 \end{array}$

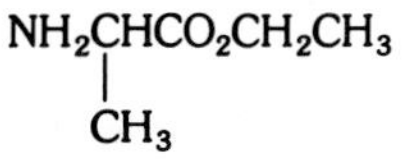

The polarity of this molecule is much, much less than the polarity of the dipolar ionic form of alanine; thus, the ester molecules stick together with forces weaker than those present between alanine molecules, and the ester has a lower melting point than alanine.

10.7 **A.** It has amine-like groups that can both donate hydrogen bonds to water molecules and accept them. (**B** has an alkyl group side chain, which is hydrophobic.)

10.11 oxidizing agent

10.12 In the presence of additional acid, the following equilibrium shifts to the right in accordance with Le Châtelier's principle. This neutralizes the extra acid.

$$^+NH_3CH_2CO_2^- + H^+ \rightleftharpoons {}^+NH_3CH_2CO_2H$$

In the presence of additional base, the following equilibrium shifts to the right in accordance with Le Châtelier's principle. This neutralizes the extra base.

$$^+NH_3CH_2CO_2^- + OH^- \rightleftharpoons NH_2CH_2CO_2^- + H_2O$$

10.14 **A** has an amide bond not to the amino group of the α-position of an amino acid unit but to an amino group of a side chain (that of lysine). **B** has a proper peptide bond.

10.17 Lys-Glu-Cys Glu-Cys-Lys Cys-Lys-Glu
Lys-Cys-Glu Glu-Lys-Cys Cys-Glu-Lys

10.20

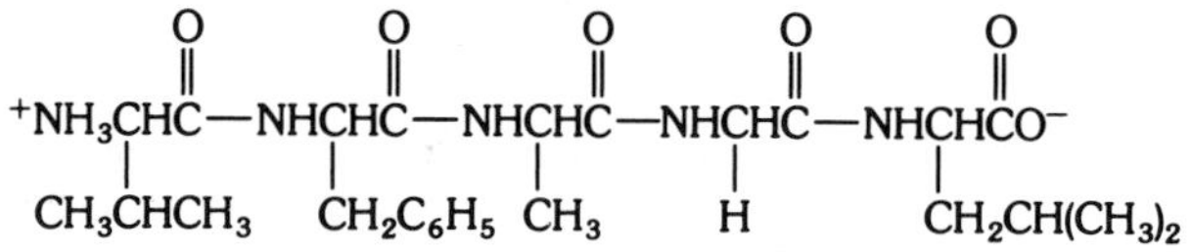

$$^+NH_3\underset{\displaystyle CH_3CHCH_3}{CH}\overset{\displaystyle O}{\overset{\|}{C}}-NH\underset{\displaystyle CH_2C_6H_5}{CH}\overset{\displaystyle O}{\overset{\|}{C}}-NH\underset{\displaystyle CH_3}{CH}\overset{\displaystyle O}{\overset{\|}{C}}-NH\underset{\displaystyle H}{CH}\overset{\displaystyle O}{\overset{\|}{C}}-NH\underset{\displaystyle CH_2CH(CH_3)_2}{CH}\overset{\displaystyle O}{\overset{\|}{C}}O^-$$

10.22 (a) **A**
(b) **B**; it has only hydrophilic side chains. All those in **A** are hydrophobic.

10.24 Gly-Cys-Ala
|
Gly-Cys-Ala

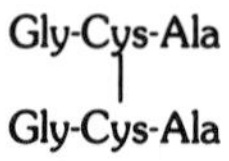

10.30 It forms *after* a polypeptide with a cysteine side chain has been put together, so it forms after the primary structure has become set.

10.32 reduce

10.34 A left-handed helix structure

10.36 It consists of three left-handed collagen helices wound together as a right-handed, cablelike triple helix. Between the strands occur molecular "bridges." A fibril forms when individual triple helices overlap lengthwise.

10.37 Covalent linkages fashioned from lysine side chains

10.39 No, they represent portions of the secondary structure of a polypeptide and often both features are present.

10.44 (a) Myoglobin is single stranded; hemoglobin has four subunits.
(b) Myoglobin is in muscle tissue; hemoglobin is in red cells.
(c) Both have the heme unit.
(d) Myoglobin accepts and stores O_2 molecules carried into tissue by hemoglobin molecules.

10.45

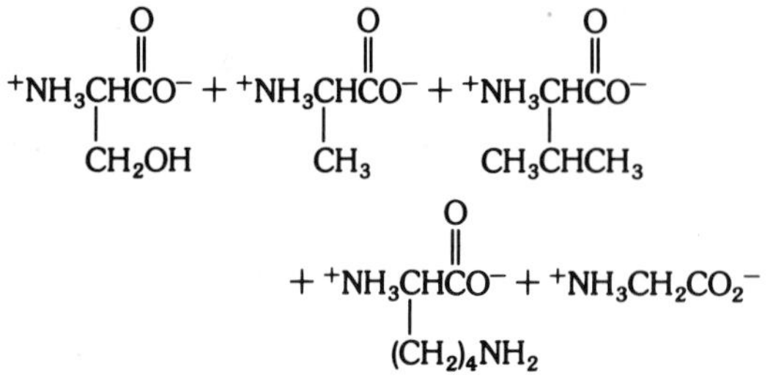

$$^+NH_3\underset{\displaystyle CH_2OH}{CH}\overset{\displaystyle O}{\overset{\|}{C}}O^- + {}^+NH_3\underset{\displaystyle CH_3}{CH}\overset{\displaystyle O}{\overset{\|}{C}}O^- + {}^+NH_3\underset{\displaystyle CH_3CHCH_3}{CH}\overset{\displaystyle O}{\overset{\|}{C}}O^-$$

$$+ {}^+NH_3\underset{\displaystyle (CH_2)_4NH_2}{CH}\overset{\displaystyle O}{\overset{\|}{C}}O^- + {}^+NH_3CH_2CO_2^-$$

10.52 cell fluid

10.58 An oligosaccharide is made from more than two monosaccharide units, and it never has the thousands of such units commonly present in polysaccharide molecules.

10.60 D-glucosamine

10.63 The resiliency of ground substance depends on the hydrogen bonds increasing the "stickiness" of the molecules of ground substance and their abilities to hold large amounts of water as water of hydration.

CHAPTER 11

Practice Exercises, Chapter 11

1. (a) sucrose (b) glucose
(c) protein (d) an ester

2. feedback inhibition

Review Exercises, Chapter 11

11.5 It catalyzes the rapid reestablishment of the equilibrium after it has been disturbed.

11.9 $H^+ + NADH + FAD \rightarrow NAD^+ + FADH_2$

11.11 (a) an oxidation
(b) the transfer of a methyl group
(c) a reaction with water
(d) an oxidation–reduction equilibrium

11.13 Hydrolysis is a kind of reaction catalyzed by a hydrolase enzyme.

11.20 (a) $V \propto [E_0]$
(b) $V \propto [S]$

11.22 At the another site. *Allosteric* describes an action induced at a site on an enzyme molecule at some distance from the active site.

11.24 As the substrate binds to one active site it induces changes in the shape of the enzyme that enable all active sites to become active, so the rate of the reaction suddenly increases rapidly.

11.26 Calmodulin and troponin, the latter being in muscle cells

11.28 At higher Ca^{2+} concentration, $Ca_3(PO_4)_2$ would precipitate.

11.30 Ca^{2+} converts them to activated effectors.

11.32 When a zymogen is cleaved properly, an active enzyme emerges. Trypsinogen is the zymogen for trypsin.

11.43 The levels of these enzymes increase in blood as the result of a disease or injury to particular tissues, which causes tissue cells to release their enzymes.

11.45 CK(*MM*).

11.54 They are primary chemical messengers.

11.56 cyclic AMP and inositol phosphate

11.58 When activated, adenylate cyclase catalyzes the formation of cyclic AMP (from ATP), which then activates an enzyme inside the target cell.

11.60 The cyclic AMP is hydrolyzed to AMP.

11.62 One helps keep the cellular glucose level high and the other helps to bring the Ca^{2+} level of the cytosol up.

11.64 They are hydrocarbon-like and so slip through a hydrocarbon-like lipid bilayer.

11.67 It is hydrolyzed back to acetic acid and choline. The enzyme is cholinesterase. Nerve poisons inactivate this enzyme.

11.69 It blocks the receptor protein for acetylcholine.

11.71 They catalyze the deactivation of neurotransmitters such as norepinephrine and thus reduce the level of signal-sending activity that depends on such neurotransmitters.

11.73 They inhibit the reabsorption of norepinephrine by the presynaptic neuron and thus reduce the rate of its deactivation by the monoamine oxidases.

11.75 dopamine

11.77 They accelerate the release of dopamine from the presynaptic neuron.

11.83 By reducing the flow of Ca^{2+} into cells of heart muscles, the heart beats with reduced vigor.

11.85 They consist of very tiny, gaseous molecules that easily slip through cell membranes.

11.102 (a) The lock-and-key theory, perhaps as modified by induced fit.
(b) Add water to the alkene group and then oxidize the resulting 2° alcohol to a ketone. (An enzyme would have to guide the addition of the water molecule to give the specific 2° alcohol needed.)

11.104 150.0 mg

CHAPTER 12

Review Exercises, Chapter 12

12.2 Saliva, gastric juice, pancreatic juice, and intestinal juice

12.4 Molecules of a competitive inhibitor occupy the active sites of an enzyme. Cimetidine shuts down the K^+-H^+ pump and prevents the secretion of gastric juice, which gives the ulcer time to heal in a relatively acid-free environment.

12.6 (a) α-amylase
(b) pepsinogen and gastric lipase
(c) α-amylase, lipase, nuclease, trypsinogen, chymotrypsinogen, procarboxypeptidase, and proelastase
(d) no enzymes
(e) amylase, aminopeptidase, sucrase, lactase, maltase, lipase, nucleases, enteropeptidase

12.8 (a) amino acids
(b) glucose, fructose, and galactose
(c) fatty acids and monoacylglycerols (plus some diacylglycerols)

12.10 It catalyzes the conversion of trypsinogen to trypsin. Then trypsin catalyzes the conversion of other zymogens to chymotrypsin, carboxypeptidase, and elastin. Thus enteropeptidase turns on enzyme activity for three major protein-digesting enzymes.

12.12 They are surface-active agents that help to break up lipid globules, wash lipids from the particles of food, and aid in the absorption of fat-soluble vitamins.

12.14 (a) HCl (b) enteropeptidase
(c) trypsin (d) trypsin
(e) trypsin

12.16 This enzyme is inactive at the high acidity of the digesting mixture in the adult stomach, but the mixture's acidity in the infant's stomach is less.

12.20 The flow of bile normally delivers colored breakdown products from hemoglobin in the blood, and these products give the normal color to feces. When no bile flows, no colored products are available to the feces.

12.21 Cholesterol has no hydrolyzable groups.

12.23 serum-soluble proteins (albumins, mostly)

12.36 hypercalcemia

12.38 chloride ion, Cl^-

12.41 Blood pressure and osmotic pressure. The natural return of fluids to the blood on the venous side from the interstitial compartment is not balanced by the now reduced blood pressure on the venous side, so fluids return to the blood from which they left on the arterial side.

12.43 (a) Blood proteins leak out, which allows water to leave the blood and enter interstitial spaces throughout various tissues.
(b) Blood proteins are lost to the blood by being consumed, which also leads to the loss of water from the blood and its appearance in interstitial compartments.
(c) Capillaries are blocked at the injured site, reducing the return of blood in the veins, so fluids accumulate at the site.

12.46 The first oxygen molecule to bind changes the shapes of other parts of the hemoglobin molecule and makes it much easier for the remaining three oxygen molecules to bind. This ensures that all four oxygen-binding sites of each hemoglobin molecule will leave the lungs fully loaded with oxygen.

12.47 $HHb + O_2 \rightleftharpoons HbO_2^- + H^+$
(a) to the left (b) to the left
(c) to the right (d) to the left
(e) to the left (f) to the right

12.49 It generates H^+ needed to convert HCO_3^- to CO_2 and H_2O and to convert $HbCO_2^-$ to HHb and CO_2.

12.51 It helps to shift the following equilibrium to the left:

$$HHb + O_2 \rightleftharpoons HbO_2^- + H^+$$

12.56 For oxygenation:

$$HHb{-}BPG + O_2 + HCO_3^- \longrightarrow HbO_2^- + BPG + CO_2 + H_2O$$

For deoxygenation:

$$HbO_2^- + BPG + CO_2 + H_2O \longrightarrow HHb{-}BPG + O_2 + HCO_3^-$$

12.58 Oxygen affinity is lowered. Where the partial pressure of CO_2 is relatively high (as in actively metabolizing tissue) there is a need for oxygen, so the lowering effect of CO_2 on oxygen affinity helps to release O_2 precisely where O_2 is most needed.

12.63 The pH of the blood decreases in both, but both pCO_2 and $[HCO_3^-]$ increase in respiratory acidosis and both decrease in metabolic acidosis.

12.65 Hypoventilation is observed in metabolic alkalosis, and isotonic ammonium chloride can be given to neutralize the excess base. Involuntary hypoventilation is observed in respiratory acidosis, and isotonic sodium bicarbonate might be given to neutralize excess acid.

12.67 In respiratory alkalosis. The involuntary loss of CO_2 reduces the level of H_2CO_3 in the blood and thereby reduces the level of H^+.

12.69 In respiratory acidosis

12.71 (a) respiratory alkalosis (b) metabolic alkalosis (c) respiratory acidosis (d) respiratory acidosis (e) metabolic acidosis (f) respiratory alkalosis (g) metabolic acidosis (h) metabolic alkalosis (i) respiratory acidosis (j) respiratory acidosis

12.72 (a) hyperventilation (b) hypoventilation (c) hypoventilation (d) hypoventilation (e) hyperventilation (f) hyperventilation (g) hyperventilation (h) hypoventilation (i) hypoventilation (j) hypoventilation

12.74 Hypoventilation in emphysema lets the blood retain carbonic acid, and the pH decreases.

12.76 The loss of alkaline fluids from the duodenum and lower intestinal tract leads to a loss of base from the bloodstream, too. The result is a decrease in the blood's pH and thus acidosis.

12.78 hypercapnia

12.80 It acts to prevent the loss of water via the urine by letting the hypophysis secrete vasopressin, whose target cells are in the kidneys. The retention of water helps to keep the blood's osmotic pressure from increasing further. The thirst mechanism is also activated to bring in more water to dilute the blood.

12.82 The kidneys secrete a trace of renin into the blood. This catalyzes the conversion of angiotensinogen to angiotensin I, which catalyzes the formation of angiotensin II, a neurotransmitter and powerful vasoconstrictor. With constricting of the capillaries, the blood pressure has to increase to keep the delivery of blood going.

12.84 They transfer hydrogen ions into the urine and put bicarbonate ions into the bloodstream.

12.86 Without the bile salts in the bile, normally obtained from the gallbladder, there is less emulsifying action to aid in the digestion of triacylglycerols with the usual long fatty acyl side chains. The shorter chain molecules are a little more soluble in the digestive medium.

12.88 These animals can store more oxygen in heart muscle, which helps them to go longer without breathing.

CHAPTER 13

Practice Exercises, Chapter 13

1. (a) proline (b) arginine (c) glutamic acid (d) lysine

2. (a) serine (b) CT(chain termination) (c) glutamic acid (d) isoleucine

Review Exercises, Chapter 13

13.1 (a) cytosol (b) protoplasm (c) cytoplasm (d) ribosome (e) mitochondrion (f) deoxyribonucleic acid (g) chromatin (h) histone (i) gene

13.6 nucleotides

13.9 The main chains all have the same phosphate–deoxyribose–phosphate–deoxyribose repeating system.

13.10 In the sequence of bases attached to the deoxyribose units of the main chain

13.13 A and T pair to each other, so they must be in a 1 : 1 ratio regardless of the species. Similarly, G and C pair to each other and must be in a 1 : 1 ratio.

13.15 Hydrophobic interactions stabilize the helices. (Hydrogen bonds hold two helices together.)

13.19 The introns are b, d, and f, because they are the longer segments.

13.26 ATA. A codon is RNA material, and T does not occur in RNA.

13.28 Writing them in the 5′ to 3′ direction:
(a) AAA (b) GGA
(c) UGU (d) AUC

13.29 (a) $5' \rightarrow 3'$ UUUCUUAUAGAGUCCCCAACAGAU
(b) $5' \rightarrow 3'$ UUUUCCACAGAU
(c) Phe-Ser-Thr-Asp

13.32 (a) A large number of sequences are possible because three of the specific amino acid residues are coded by more than one codon. The possibilities are indicated by

Met—Ala—Trp—Ser—Tyr
AUG GCU UGG UCU UAU ($5' \rightarrow 3'$)
GCC UCC UAC
GCA UCA
GCG UCG

(b) CAU ($5' \rightarrow 3'$) or, if ($3' \rightarrow 5'$), then UAC

13.34 How polypeptide synthesis can be controlled by the use of repressors

13.40 The codons specify the same amino acids in all organisms. No.

13.47 RNA replicase is able to direct the synthesis of RNA from the "directions" encoded on RNA. A normal host cell does not contain RNA replicase, so the virus either must bring it along or it must direct its synthesis inside the host cell.

13.49 (+)mRNA

13.51 RNA replicase

13.53 A silent gene is a unit provided by a virus particle to a host cell but which does not take over the genetic apparatus of the cell until activated.

13.57 The theory is that AZT molecules will bind to reverse transcriptase and inhibit the work of this enzyme in HIV.

13.59 restriction enzyme

13.63 a defect in a gene

13.64 The synthesis of a transmembrane protein that lets chloride ion pass through the membranes of mucous cells in the lungs and the digestive tract. The reduced movement of

Cl^- out of the cell means that less water is outside of the cell, and the mucous is thereby thickened and made more viscous. Breathing is thereby impaired.

13.66 Retroviruses can bring about the insertion of DNA into a chromosome of a host cell.

13.70 It is the synthesis of clones of DNA. DNA samples at crime scenes generally involve tiny amounts of DNA, and DNA typing depends on making enough cloned DNA for the next steps of the procedure.

13.76 (a) by one methyl group
(b) no
(c) Both uracil and thymine can form a base pair with adenine.

CHAPTER 14

Practice Exercise, Chapter 14

1. (a) yes (b) yes (c) no

Review Exercises, Chapter 14

14.1 All of them, but chiefly fatty acids and carbohydrates

14.3 Combustion produces just heat. The catabolism of glucose uses about half of the energy to make ATP, and the remainder appears as heat.

14.5 Adenosine—O—P(=O)(O⁻)—O—P(=O)(O⁻)—O—P(=O)(O⁻)—O^-

14.7 The singly and doubly ionized forms of phosphoric acid, $H_2PO_4^- + HPO_4^{2-}$.

14.9 The first two have phosphate group transfer potentials equal to or higher than that of ATP, whereas this potential is lower than that of ATP for glycerol-3-phosphate.

14.11 (a) yes (b) yes (c) no (d) no

14.13 It stores phosphate group energy and transfers phosphate to ADP to remake the ATP consumed by muscular work.

14.15 (a) The aerobic synthesis of ATP
(b) The synthesis of ATP when a tissue operates anaerobically
(c) The supply of metabolites for the respiratory chain
(d) The supply of metabolites for the respiratory chain and for the citric acid cycle

14.17 An increase in its supply of ADP. The need to convert ADP back to ATP is met by metabolism, which requires oxygen.

14.19 (a) E < B < C < A < D (b) B < C < A < D

14.23 When a cell is temporarily low on oxygen for running the respiratory chain (as a source of ATP), the cell can continue (for a while) to make ATP anyway.

14.25 An acetyl unit:

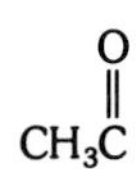

14.27 two

14.30 9

14.32 An enzyme like isocitrate dehydrogenase, because the reaction is a dehydrogenation (not an addition of water to a double bond, catalyzed by an enzyme like aconitase)

14.35 $CH_3CH(OH)CO_2^- + NAD^+ \rightarrow CH_3C(=O)CO_2^- + NAD{:}H + H^+$

(a) $CH_3CH(OH)CO_2^-$ (b) NAD^+

14.36 $^-O_2CCH_2CH_2CO_2^-$ + FAD → $^-O_2CCH{=}CHCO_2^-$ + $FADH_2$

14.40 Across the inner membrane of the mitochondrion. The value of $[H^+]$ is higher on the outer side of the inner membrane than in the mitochondrial matrix.

14.42 If the membrane is broken, then the simple process of diffusion defeats any mitochondrial effort to set up a gradient of H^+ ions across the membrane, but the chain itself can still operate.

14.44 Oxidative phosphorylation is the kind made possible by the energy released from the operation of the respiratory chain. Substrate phosphorylation arises from the direct transfer of a phosphate unit from a higher to a lower energy phosphate.

14.46 An enzyme catalyzes the formation of ATP from ADP and P_i, but the new ATP sticks tightly to the enzyme. But this enzyme is at the end of the channel for protons in the inner mitochondrial membrane, and as protons flow through they change the enzyme so that it expels the ATP.

14.48 $CH_3CH_2C(=O)CH(CH_3)CO_2CH_2CH_3$

14.49 $CH_3CH{=}O + NAD^+ + H_2O \rightarrow C_2H_3O_2^- + NADH + 2H^+$; $E'^{\circ}_{cell} = +0.28$ V. The spontaneous reaction is the oxidation of ethanal.

14.50 (a) $CH_3CH{=}O$
(b) acetic acid, CH_3CO_2H (or acetate ion, $CH_3CO_2^-$)
(c) α-ketoglutarate

14.51 (a) 4
(b) The nucleus of a hydrogen atom, because $H{:}^-$ transfers
(c) dehydrogenase

CHAPTER 15

Review Exercises, Chapter 15

15.5 99 mg/dL

15.8 The lack of glucose means the lack of the one nutrient most needed by the brain.

15.12 D < C < E < A < B

15.14 Glucose might be changed back to glycogen as rapidly as it is released from glycogen, and no glucose would be made available to the cell.

15.16 Glucose-1-phosphate is the end product, and phosphoglucomutase catalyzes its change to glucose-6-phosphate.

15.19 Glucagon, because it works better at the liver than epinephrine in initiating glycogenolysis, and when glycogenolysis occurs at the liver there is a mechanism for releasing glucose into circulation.

15.23 Too much insulin leads to a sharp decrease in the blood sugar level and therefore a decrease in the supply of glucose, the chief nutrient for the brain.

15.26 By its conversion to glucose-6-phosphate, which either enters a pathway that makes ATP or is converted to glycogen for storage

15.31 An overrelease of epinephrine (as in a stressful situation) that induces an overrelease of glucose

15.32 $C_6H_{12}O_6 + 2ADP + 2P_i \rightarrow 2C_3H_5O_3^- + 2H^+ + 2ATP$

15.34 Glyceraldehyde-3-phosphate is in the direct pathway of glycolysis, so changing dihydroxyacetone phosphate into it ensures that all parts of the original glucose molecule are used in glycolysis.

15.36 It is reoxidized to pyruvate, which then undergoes oxidative decarboxylation to the acetyl group in acetyl CoA. This enters the citric acid cycle.

15.38 NADPH forms, and the body uses it as a reducing agent to make fatty acids.

15.40 All are catabolized, and parts of some of their molecules are used to make fatty acids and, thence, fat.

15.42 α-Ketoglutarate is an intermediate in the citric acid cycle (Fig. 14.2) and so is changed eventually to oxaloacetate (normally the acceptor of acetyl units at the start of the citric acid cycle). Gluconeogenesis also can use oxaloacetate to make "new" glucose.

15.43 Succinyl units are in the citric acid cycle, which ends with the formation of oxaloacetate, and the latter can be used in gluconeogenesis.

15.57 (a) $CH_3CH(OH)CH_2CH(=O)$

(b) $CH_3CH_2CH(OH)CH(CH_3)CH(=O)$

(c) $CH_3C(OH)(CH_3)CH_2C(=O)CH_3$

15.58 $C_6H_5CH_2CH(=O)$

15.59 (a) Yes, either pyruvate or lactate containing carbon-13 may reenter circulation.

(b) Yes, either pyruvate or lactate containing carbon-13 might be absorbed by the liver from the bloodstream.

(c) Yes, glucose with carbon-13 atoms might be made via gluconeogenesis from either pyruvate or lactate containing carbon-13.

15.60 (a) 15 ATP

(b) 3 ATP

(c) 3 glucose (because 6 ATP are needed to make each molecule of glucose by gluconeogenesis)

CHAPTER 16

Review Exercises, Chapter 16

16.2 They become reconstituted into triacylglycerols.

16.4 They unload some of their triacylglycerol.

16.6 Some cholesterol has originated in the diet and some has been synthesized in the liver.

16.8 triacylglycerol

16.10 the liver

16.12 The synthesis of steroids and the fabrication of cell membranes

16.14 When the receptor proteins are reduced in number, the liver cannot remove cholesterol from the blood, so the blood cholesterol level increases.

16.19 fasting and diabetes

16.21 $E < D < A < C < B$

16.23 The citric acid cycle processes the acetyl units manufactured by the β-oxidation pathway and so fuels the respiratory chain.

16.25 An increase in the blood sugar level triggers the release of insulin which inhibits the release of fatty acids from adipose fat.

16.28 $CH_3CH_2CH_2C(=O)SCoA + FAD \rightarrow CH_3CH{=}CHC(=O)SCoA + FADH_2$

$CH_3CH{=}CHC(=O)SCoA + H_2O \rightarrow CH_3CH(OH)CH_2C(=O)SCoA$

$CH_3CH(OH)CH_2C(=O)SCoA + NAD^+ \rightarrow CH_3C(=O)CH_2C(=O)SCoA + NAD{:}H + H^+$

$CH_3C(=O)CH_2C(=O)SCoA + CoASH \rightarrow 2CH_3C(=O)SCoA$

No more turns of the β-oxidation pathway are possible.

16.30 NADH passes its hydrogen into the respiratory chain and is changed back to NAD^+.

16.32 (a) inside mitochondria (b) cytosol

16.34 The pentose phosphate pathway of glucose catabolism

16.36 Cholesterol inhibits the synthesis of HMG–CoA reductase.

16.40 A proton or hydrogen ion, H^+. If the level of hydrogen ion increases, the problem is acidosis.

16.47 acetoacetic acid

16.49 Their *over*production leads to acidosis.

16.51 Each Na^+ ion that leaves corresponds to the loss of one HCO_3^- ion, the true base, because HCO_3^- neutralizes acid generated as the ketone bodies are made. And for every negative ion that leaves with the urine a positive ion, mostly Na^+, has to leave to ensure electrical neutrality.

16.63 (a) 7 (b) 6 (c) 6 (d) 112

CHAPTER 17

Review Exercises, Chapter 17

17.3 infancy

17.5 glutamic acid (glutamate)

17.7 $C_6H_5CH_2CCO_2H$ (with C=O on the second C)

17.9 The body can use it to make glucose by gluconeogenesis.

17.11 ketogenic

17.13 $4 < 1 < 5 < 2 < 3$

17.15 Yes: Glucose $\xrightarrow[\text{glycolysis}]{\text{aerobic}}$ pyruvate $\xrightarrow{\text{transamination}}$ alanine

17.17

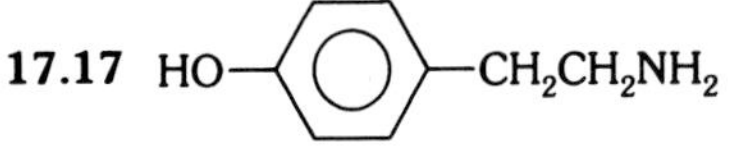

tyramine

17.19 To synthesize glucose by means of gluconeogenesis

17.20 (a) originally, the amino groups of amino acids
(b) carbon dioxide

17.23 Hyperammonemia. Step 2 consumes carbamoyl phosphate, which is made using ammonia. If carbamoyl phosphate levels rise, a backup occurs to cause ammonia levels to increase.

17.26 $3 < 2 < 1 < 4 < 5 < 7 < 6$

17.31 The exclusively ketogenic amino acids, those not also glucogenic, cannot produce net extra oxaloacetate.

CHAPTER 18

Review Exercises, Chapter 18

18.3 Foods are complex mixtures of nutrients.

18.5 To allow for individual differences among people and to ensure that practically all people can thrive

18.7 No food contains all of the essential nutrients, and there might still be nutrients yet to be discovered but which are routinely provided by a varied diet.

18.9 Linoleic acid. "Essential" means that it must be provided in the diet. If linoleic acid is absent in the diet, the prostaglandins are not made at a sufficient rate.

18.12 The body can make several amino acids itself.

18.14 It breaks them down and eliminates the products.

18.16 From an animal source

18.18 The proportions of essential amino acids available from it

18.20 The essential amino acid most poorly supplied by the protein

18.22 (a) 1.3×10^2 g
(b) 1.7×10^3 g
(c) 6.1×10^3 kcal
(d) Very likely not, since 1.7×10^3 g is nearly 3 lb.

18.24 They are needed in much more than trace amounts, and they come from proteins.

18.26 No single vegetable source has a balanced supply of essential amino acids.

18.28 vitamin D

18.30 vitamin A

18.32 vitamin K

18.34 vitamins C and E

18.36 vitamin C

18.38 vitamin C

18.40 thiamin

18.42 folate

18.43 The quantity needed per day. Minerals are needed in the amount of more than 100 mg/day and trace elements in the amount of less than 20 mg/day.

18.47 2.13×10^3 L air

GLOSSARY[1]

Absolute Configuration The actual arrangement in space about each tetrahedral stereocenter in a molecule. (8.3)

Acetal Any organic compound in which two ether-like linkages extend from one CH unit. (14.5)

Acetyl Coenzyme A The molecule from which acetyl groups are transferred into the citric acid cycle or into the synthesis of fatty acids. (14.1)

Achiral Not possessing chirality; that quality of a molecule (or other object) that allows it to be superimposed on its mirror image. (7.2)

Acid Anhydride In organic chemistry, a compound formed by splitting water out between two OH groups of the acid functions of two organic acids. (5.3)

Acid Chloride A derivative of a carboxylic acid in which the OH group of the acid has been replaced by Cl. (5.3)

Acid Derivative Any organic compound that can be made from an organic acid or that can be changed back to the acid by hydrolysis. (Examples are acid chlorides, acid anhydrides, esters, and amides.) (5.3)

Acidosis A condition in which the pH of the blood is below normal. *Metabolic acidosis* is brought on by a defect in some metabolic pathway. *Respiratory acidosis* is caused by a defect in the respiratory centers or in the mechanisms of breathing. (12.4, 16.6)

Active Transport The movement of a substance through a biological membrane against a concentration gradient and caused by energy-consuming chemical changes that involve parts of the membrane. (10.8)

Acyl Group

$$R{-}\overset{\overset{\displaystyle O}{\|}}{C}{-} \quad (5.3)$$

Acyl group

[1] The entries in this Glossary include the terms that appear in boldface within the chapters, including the margin comments, as well as several additional entries. The numbers in parentheses following the definitions are the section numbers (or Special Topics) where the entry was introduced or discussed.

Acyl Group Transfer Reaction Any reaction in which an acyl group transfers from a donor to an acceptor. (5.3)

Addition Reaction Any reaction in which two parts of a reactant molecule add to a double or a triple bond. (2.4)

Adenosine Diphosphate (ADP) A high-energy diphosphate ester obtained from adenosine triphosphate (ATP) when part of the chemical energy in ATP is tapped for some purpose in a cell. (14.1)

Adenosine Monophosphate (AMP) A low-energy phosphate ester that can be obtained by the hydrolysis of ATP or ADP; a monomer for the biosynthesis of nucleic acids. (14.1)

Adenosine Triphosphate (ATP) A high-energy triphosphate ester used in living systems to provide chemical energy for metabolic needs. (14.1)

Adequate Protein A protein that, when digested, makes available all of the essential amino acids in suitable proportions to satisfy both the amino acid and total nitrogen requirements of good nutrition without providing excessive calories. (18.2)

ADP (See *Adenosine Diphosphate*)

Aerobic Sequence An oxygen consuming sequence of catabolism that starts with glucose and proceeds through glycolysis, the citric acid cycle, and the respiratory chain. (14.1)

Agonist A compound whose molecules can bind to a receptor on a cell membrane and cause a response by the cell. (11.7)

Albumin One of a family of globular proteins that tend to dissolve in water, and that in blood contribute to the blood's colloidal osmotic pressure and aid in the transport of metal ions, fatty acids, cholesterol, triacylglycerols, and other water insoluble substances. (10.9, 12.2)

Alcohol Any organic compound whose molecules have the OH group attached to a saturated carbon; ROH. (3.1)

Alcohol Group The OH group when it is joined to a saturated carbon. (3.1)

Aldehyde An organic compound that has a carbonyl group joined to H on one side and C on the other. (4.1)

Aldehyde Group —CH=O (4.1)

Aldohexose A monosaccharide whose molecules have six carbon atoms and an aldehyde group. (8.2)

Aldose A monosaccharide whose molecules have an aldehyde group. (8.2)

Aldosterone A steroid hormone, made in the adrenal cortex, secreted into the bloodstream when the sodium ion level is low, and that signals the kidneys to leave sodium ions in the bloodstream. (12.5)

Aliphatic Compound Any organic compound whose molecules lack a benzene ring or a similar structural feature. (1.5)

Alkaloid A physiologically active, heterocyclic amine isolated from plants. (6.2)

Alkalosis A condition in which the pH of the blood is above normal. *Metabolic alkalosis* is caused by a defect in metabolism. *Respiratory alkalosis* is caused by a defect in the respiratory centers of the brain or in the apparatus of breathing. (12.4)

Alkane A saturated hydrocarbon, one that has only single bonds. A *normal* alkane is any whose molecules have straight chains. (1.5, 2.1)

Alkene A hydrocarbon whose molecules have one or more double bonds. (1.5, 2.1)

Alkyl Group A substituent group that is an alkane minus one H atom. (1.6)

Alkyne A hydrocarbon whose molecules have triple bonds. (1.5, 2.1)

Allosteric Activation The activation of an enzyme's catalytic site by the binding of some molecule at a position elsewhere on the enzyme. (11.4)

Allosteric Inhibition The inhibition of the activity of an enzyme caused by the binding of an inhibitor molecule at some site other than the enzyme's catalytic site. (11.4)

Amide An organic compound whose molecules have a carbonyl–nitrogen single bond. (5.3, 6.1)

Amide Bond The single bond that holds the carbonyl group to the nitrogen atom in an amide. (6.1)

Amine An organic compound whose molecules have a trivalent nitrogen atom, as in R—NH_2, R—NH—R, or R_3N. (6.1)

Amine Salt An organic compound whose molecules have a positively charged, tetravalent, protonated nitrogen atom, as in RNH_3^+, $R_2NH_2^+$, or R_3NH^+. (6.2)

Amino Acid Any organic compound whose molecules have both an amino group and a carboxyl group. (10.1)

Amino Acid Residue A structural unit in a polypeptide,

$$-NH-\underset{\displaystyle R}{\underset{|}{CH}}-\overset{\displaystyle O}{\overset{\|}{C}}-$$

furnished by an α-amino acid, where R is the sidechain group of a particular amino acid. (10.1)

Aminoacyl Group

$$NH_2-\underset{\displaystyle R}{\underset{|}{CH}}-\overset{\displaystyle O}{\overset{\|}{C}}-$$

where R is one of the amino acid sidechains. (10.1)

AMP (See *Adenosine Monophosphate*)

Amphipathic Compound A substance whose molecules have both hydrophilic and hydrophobic groups. (9.5)

Anaerobic Sequence The oxygen-independent catabolism of glucose to lactate ion. (14.1, 15.3)

Anoxia A condition of a tissue in which it receives no oxygen. (12.4)

Antagonist A compound that can bind to a membrane receptor but not cause any response by the cell. (11.7)

Antibiotics Antimetabolites made by bacteria and fungi. (11.4)

Anticodon A sequence of three adjacent sidechain bases on a molecule of tRNA that is complementary to a codon and that fits to its codon on an mRNA chain during polypeptide synthesis. (13.3)

Antimetabolite A substance that inhibits the growth of bacteria. (11.4)

Apoenzyme The wholly polypeptide part of an enzyme. (11.1)

Aromatic Compound Any organic compound whose molecules have a benzene ring (for a feature very similar to this). (1.5, 2.7)

ATP (See *Adenosine Triphosphate*)

Base Heterocyclic A heterocyclic amine obtained from the hydrolysis of nucleic acids: adenine, thymine, guanine, cytosine, or uracil. (13.2)

Base Ionization Constant (K_b) For the equilibrium (where *B* is some base)

$$B + H_2O \rightleftharpoons BH^+ + OH^-$$

$$K_b = \frac{[BH^+][OH^-]}{[B]} \qquad (6.2)$$

Base Pairing In nucleic acid chemistry, the association by means of hydrogen bonds of two heterocyclic, sidechain bases—adenine with thymine (or uracil) and guanine with cytosine. (13.2)

Benedict's Reagent A solution of copper(II) sulfate, sodium citrate, and sodium carbonate that is used in the Benedict's test. (4.3)

Benedict's Test The use of Benedict's reagent to detect the presence of any compound whose molecules have easily oxidized functional groups—α-hydroxyaldehydes and α-hydroxyketones—such as those present in monosaccharides. In a positive test the intensely blue color of the reagent disappears and a reddish precipitate of copper(I) oxide separates. (4.3)

Beta Oxidation The catabolism of a fatty acid by a series of repeating steps that produce acetyl units (in acetyl CoA); the fatty acid cycle of catabolism. (16.3)

Bile A secretion of the gallbladder that empties into the upper intestine and furnishes bile salts; a route of excretion for cholesterol and bile pigments. (12.1)

Bile Pigment Colored products of the partial catabolism of heme that are transferred from the liver to the gallbladder for secretion via the bile. (17.4)

Bile Salts Steroid-based detergents in bile that emulsify fats and oils during digestion. (12.1)

Bilin The brownish pigment that is the end product of the catabolism of heme and that contributes to the characteristic colors of feces and urine. (17.4)

Bilinogen A product of the catabolism of heme that contributes to the characteristic colors of feces and urine and some of which is oxidized to bilin. (17.4)

Bilirubin An reddish-orange substance that forms from biliverdin during the catabolism of heme and which enters the intestinal tract via the bile and is eventually changed into bilinogen and bilin. (17.4)

Biliverdin A greenish pigment that forms when partly catabolized hemoglobin (as verdohemoglobin) is further broken down, and which is changed in the liver to bilirubin. (17.4)

Biochemistry The study of the structures and properties of substances found in living systems. (8.1)

Biological Value In nutrition, the percentage of the nitrogen of ingested protein that is absorbed from the digestive tract and retained by the body when the total protein intake is less than normally required. (18.2)

Biotin A water-soluble vitamin needed to make enzymes used in fatty acid synthesis. (18.3)

2,3-Bisphosphoglycerate (BPG) An organic ion that nestles within the hemoglobin molecule in deoxygenated blood but is expelled from the hemoglobin molecule during oxygenation. (12.3)

Blood Sugar The carbohydrates—mostly glucose—that are present in blood. (8.2)

Blood Sugar Level The concentration of carbohydrate—mostly glucose—in the blood; usually stated in units of mg/dL. (15.1)

Boat Form A conformation of a six-membered ring that resembles a boat. (Special Topic 8.2)

Bohr Effect The stimulation of hemoglobin to bind oxygen caused by the removal (neutralization) of the hydrogen ion released by oxygen binding. (12.3)

BPG (See *2,3-Bisphosphoglycerate*)

Branched Chain A sequence of atoms to which additional atoms are attached at points other than the ends. (1.2)

Butyl Group $CH_3CH_2CH_2CH_2—$ (1.6)

sec-Butyl-Group $CH_3CH_2CH(CH_3)—$ (1.6)

t-Butyl Group $(CH_3)_3C—$ (1.6)

Carbaminohemoglobin Hemoglobin that carries chemically bound carbon dioxide. (12.3)

Carbocation Any cation in which a carbon atom has just six outer level electrons. (2.5)

Carbohydrate Any naturally occurring substance whose molecules are polyhydroxyaldehydes or polyhydroxyketones or can be hydrolyzed to such compounds. (8.2)

Carbonyl Group The atoms carbon and oxygen joined by a double bond, C=O. (4.1)

Carboxylic Acid A compound whose molecules have the carboxyl group, CO_2H. (5.1)

Catabolism The reaction of metabolism that break molecules down. (14.1)

Chair Form A conformation of a six-membered ring that resembles a chair. (8.4, Special Topic 8.2)

Chemiosmotic Theory An explanation of how oxidative phosphorylation is related to a flow of protons in a proton gradient established by the respiratory chain, a gradient that extends across the inner membrane of a mitochondrion. (14.3)

Chiral Having handedness in a molecular structure. (See also *Chirality*) (7.2)

Chiral Carbon (See *Tetrahedral Stereocenter*)

Chirality The quality of handedness that a molecular structure has that prevents this structure from being superimposable on its mirror image. (7.2)

Chloride Shift An interchange of chloride ions and bicarbonate ions between a red blood cell and the surrounding blood serum. (12.3)

Choline A compound needed to make complex lipids and acetylcholine; classified as a vitamin. (18.3)

Chromosome Small threadlike bodies in a cell nucleus that carry genes in a linear array and that are microscopically visible during cell division. (13.1)

Citric Acid Cycle A series of reactions that dismantle acetyl units and send electrons (and protons) into the respiratory chain; a major source of metabolites for the respiratory chain. (14.1, 14.2)

Codon A sequence of three adjacent sidechain bases in a molecule of mRNA that codes for a specific amino acid residue when the mRNA participates in polypeptide synthesis. (13.3)

Coefficient of Digestibility The proportion of an ingested protein's nitrogen that enters circulation rather than elimination (in feces); the difference between the nitrogen ingested and the nitrogen in the feces divided by the nitrogen ingested. (18.2)

Coenzyme An organic compound needed to make a complete enzyme from an apoenzyme. (11.1)

Cofactor A nonprotein compound or ion that is an essential part of an enzyme. (11.1)

Collagen The fibrous protein of connective tissue that changes to gelatin in boiling water. (10.9)

Competitive Inhibition The inhibition of an enzyme by the binding of a molecule that can compete with the substrate for the occupation of the catalytic site. (11.4)

Complex Ion A combination of a metal ion with one or more ligands—negatively charged or neutral electron-rich species. (4.3)

Conformation One of the infinite number of contortions of a molecule that are permitted by free rotations around single bonds. (1.2)

Constitutional Isomerism The existence of two or more compounds with identical molecular formulas but different atom-to-atom sequences. (7.1)

Constitutional Isomers Compounds with identical molecular formulas but different atom-to-atom sequences. (1.3, 7.1)

Cori Cycle The sequence of chemical events and transfers of substances in the body that describes the distribution, storage, and mobilization of blood sugar, including the reconversion of lactate to glycogen. (15.2)

Deamination The removal of an amino group from an amino acid. (17.2)

Decarboxylation The removal of a carboxyl group. (17.2)

Denatured Protein A protein whose molecules have suffered the loss of their native shape and form as well as their ability to function biologically. (10.2)

Deoxyribonucleic Acid (DNA) The chemical of a gene; one of a large number of polymers of deoxyribonucleotides and whose sequences of sidechain bases constitute the genetic messages of genes. (13.2)

Detergent A surface-active agent; a soap. (Special Topic 9.3)

Dextrorotatory That property of an optically active substance by which it can cause the plane of plane-polarized light to rotate clockwise. (7.3)

D-Family; L-Family The names of the two optically active families to which substances can belong when they are considered solely according to one kind of molecular chirality (molecular handedness) or the other. (8.3)

Diabetes Mellitus A disease in which there is an insufficiency of effective insulin and an impairment of glucose tolerance. (15.2)

Diastereomers Stereoisomers whose molecules are not related as an object is to its mirror image. (7.2)

Dietetics The application of the findings of the science of nutrition to the feeding of individual humans, whether well or ill. (18.1)

Digestive Juice A secretion into the digestive tract that consists of a dilute aqueous solution of digestive enzymes (or their zymogens) and inorganic ions. (12.1)

Dihydric Alcohol An alcohol with two OH groups; a glycol. (3.1)

Dipeptide A compound whose molecules have two α-amino acid residues joined by a peptide (amide) bond. (10.3)

Dipolar Ion A molecule that carries one plus charge and one minus charge, such as an α-amino acid. (10.1)

Disaccharide A carbohydrate that can be hydrolyzed into two monosaccharides. (8.2)

Disulfide Link The sulfur–sulfur covalent bond in polypeptides. (10.1)

Disulfide System S—S as in R—S—S—R. (3.6)

DNA (See *Deoxyribonucleic Acid; see Double Helix DNA Model*)

Double Bond A covalent bond in which two pairs of electrons are shared. (13.1, 13.3)

Double Helix DNA Model A spiral arrangement of two intertwining DNA molecules held together by hydrogen bonds between sidechain bases. (13.2)

Duplex DNA DNA in its double-stranded form. (13.2)

Edema The swelling of tissue caused by the retention of water. (12.2)

Effector A chemical other than a substrate that can allosterically activate an enzyme. (11.4)

Elastin The fibrous protein of tendons and arteries. (10.9)

Electrolytes, Blood The ionic substances dissolved in the blood. (12.2)

Enantiomers Stereoisomers whose molecules are related as an object is related to its mirror image but that cannot be superimposed. (7.2)

Enzyme A catalyst in a living system. (11.1)

Enzyme Induction The chemical process whereby the synthesis of an enzyme is prompted. (13.4)

Enzyme–Substrate Complex The temporary combination that an enzyme must form with its substrate before catalysis can occur. (11.2)

Epinephrine A hormone of the adrenal medulla that activates the enzymes needed to release glucose from glycogen. (15.1)

Erythrocyte A red blood cell. (12.3)

Essential Amino Acid An α-amino acid that the body cannot make from other amino acids and that must be supplied by the diet. (18.2)

Essential Fatty Acid A fatty acid that must be supplied by the diet. (18.1)

Ester A derivative of an acid and an alcohol that can be hydrolyzed to these parent compounds. Esters of carboxylic acids and phosphoric acid occur in living systems. (15.3)

Esterification The formation of an ester. (5.3)

Ether An organic compound whose molecules have an oxygen attached by single bonds to separate carbon atoms neither of which is a carbonyl carbon atom: R—O—R′. (3.5)

Ethyl Group CH_3CH_2— (1.6)

Exon A segment of a DNA strand that eventually becomes expressed as a corresponding sequence of aminoacyl residues in a polypeptide. (13.2)

Extracellular fluids Body fluids that are outside of cells. (12.1)

Fatty Acid Any carboxylic acid that can be obtained by the hydrolysis of animal fats or vegetable oils. (5.1, 9.1)

Fatty Acid Cycle (See *Beta Oxidation*)

Feedback Inhibition The competitive inhibition of an enzyme by a product of its own action. (11.4)

Fibrin The fibrous protein of a blood clot that forms from fibrinogen during clotting. (10.9, 12.2)

Fibrinogen A protein in blood that is changed to fibrin during clotting. (10.9, 12.2)

Fibrous Proteins Water-insoluble proteins found in fibrous tissues. (10.9)

Fischer Projection Structure A two-dimensional representation, prepared according to rules, of the configuration at a tetrahedral stereocenter. (8.3)

Folate A vitamin supplied by folic acid or pteroylglutamic acid and that is needed to prevent megaloblastic anemia. (18.3)

Food A material that supplies one or more nutrients without contributing materials that, either in kind or quantity, would be harmful to most healthy people. (18.1)

Free Rotation The absence of a barrier to the rotation of two groups with respect to each other when they are joined by a single, covalent bond. (1.2)

Functional Group An atom or a group of atoms in a molecule that is responsible for the particular set of reactions that all compounds with this group have. (1.4)

Gap Junctions Tubules made of membrane-bound proteins that interconnect one cell to neighboring cells and through which materials can pass directly. (10.8)

Gastric Juice The digestive juice secreted into the stomach that contains pepsinogen, hydrochloric acid, and gastric lipase. (12.1)

Gene A unit of heredity carried on a cell's chromosomes and consisting of DNA. (13.1, 13.2)

Genetic Code The set of correlations that specify which codons on mRNA chains are responsible for which amino acyl residues when the latter are steered into place during the mRNA-directed synthesis of polypeptides. (13.3)

Genetic Engineering The use of recombinant DNA to manufacture substances or to repair genetic defects. (13.6)

Genome The entire complement of genetic information of a species; all the genes of an individual. (13.7)

Geometric Isomerism Stereoisomerism caused by restricted rotation that gives different geometries to the same structural organization; cis–trans isomerism. (2.3)

Geometric Isomers Stereoisomers whose molecules have identical atomic organizations but different geometries; cis–trans isomers. (2.3)

Globular Proteins Proteins that are soluble in water or in water that contains certain dissolved salts. (10.9)

Globulins Globular proteins in the blood that include γ-globulin, an agent in the body's defense against infectious diseases. (10.9, 12.2)

Glucagon A hormone, secreted by the α-cells of the pancreas, in response to a decrease in the blood sugar level, that stimulates the liver to release glucose from its glycogen stores. (15.1)

Gluconeogenesis The synthesis of glucose from compounds with smaller molecules or ions. (15.4)

Glucose Tolerance The ability of the body to manage the intake of dietary glucose while keeping the blood sugar level from fluctuating widely. (15.2)

Glucose Tolerance Test A series of measurements of the blood sugar level after the ingestion of a considerable amount of glucose; used to obtain information about an individual's glucose tolerance. (15.2)

Glucoside An acetal formed from glucose (in its cyclic, hemiacetal form) and an alcohol. (8.5)

Glucosuria The presence of glucose in urine. (15.1)

Glycerophospholipid A hydrolyzable lipid that has an ester linkage between glycerol and one phosphoric acid unit (this, in turn, forming another ester link to a small molecule.) In *phosphatides,* the remaining two OH units of glycerol are esterified with fatty acids. In *plasmalogens,* one OH is esterified with a fatty acid and the other is joined by an ether link to a long-chain unsaturated alcohol. (9.3)

Glycogenesis The synthesis of glycogen. (15.1)

Glycogenolysis The breakdown of glycogen to glucose. (15.1)

Glycol A dihydric alcohol. (3.1)

Glycolipid A lipid whose molecules include a glucose unit, a galactose unit, or some other carbohydrate unit. (9.3, 10.8)

Glycolysis A series of chemical reactions that break down glucose or glucose units in glycogen until pyruvate remains (when the series is operated aerobically) or lactate forms (when the conditions are anaerobic). (14.1, 15.3)

Glycoprotein A protein, often membrane-bound, that is joined to a carbohydrate unit. (10.8)

Glycoside An acetal or a ketal formed from the cyclic form of a monosaccharide and an alcohol. (8.5)

Gradient The presence of a change in value of some physical quantity with distance, as in *concentration* gradient in which the concentration of a solute is different in different parts of the system. (10.8)

Ground Substance A gel-like material present in cartilage and other extracellular spaces that gives flexibility to collagen and other fibrous proteins. (10.8)

α-Helix One kind of secondary structure of a polypeptide in which its molecules are coiled. (10.4)

Heme The deep-red, iron-containing prosthetic group in hemoglobin and myoglobin. (10.5, 12.3)

Hemiacetal Any compound whose molecules have both an OH and an OR group coming to a CH unit. (4.5)

Hemiketal Any compound whose molecules have both an OH and an OR group coming to a carbon that otherwise bears no H atoms. (4.5)

Hemoglobin The oxygen-carrying protein in red blood cells. (10.5, 12.3)

Heterocyclic Compound An organic compound with a ring in which an atom other than carbon takes up at least one position in the ring. (1.2)

Heterogeneous Nuclear RNA (hnRNA) RNA made directly at the guidance of DNA and from which messenger RNA (mRNA) is made. (Formerly called primary transcript RNA, ptRNA.) (13.3)

High-Energy Phosphate An organophosphate with a phosphate group transfer potential equal to or higher than that of ADP or ATP. (14.1)

Homeostasis The response of an organism to a stimulus such that the organism is restored to its prestimulated state. (11.4)

Hormone A primary chemical messenger made by an endocrine gland and carried by the bloodstream to a target organ where a particular chemical response is initiated. (11.6)

Human Growth Hormone One of the hormones that affects the blood sugar level; a stimulator of the release of the hormone glucagon. (15.1)

Hydrocarbon An organic compound that consists entirely of carbon and hydrogen. (1.5)

Hydrolase An enzyme that catalyzes a hydrolysis reaction. (11.1)

Hydrolyzable Lipid A lipid that can be hydrolyzed or saponified. (Formerly called a saponifiable lipid.) (9.1)

Hydrophilic Group Any part of a molecular structure that attracts water molecules; a polar or ionic group such as OH, CO_2^-, NH_3^+, or NH_2. (9.5)

Hydrophobic Group Any part of a molecular structure that has no attraction for water molecules; a nonpolar group such as an alkyl group. (9.5)

Hydrophobic Interaction The water avoidance by nonpolar groups or sidechains that is partly responsible for the shape adopted by a polypeptide or nucleic acid molecule in an aqueous environment. (10.1, 13.2)

Hyperammonemia An elevated level of ammonium ion in the blood. (17.3)

Hypercapnia An elevated level of carbon dioxide in the blood as indicated by a partial pressure of CO_2 in venous blood above 50 mm Hg. (12.4)

Hyperglycemia An elevated level of glucose in the blood—above 110 mg/dL in whole blood. (15.1)

Hyperkalemia An elevated level of potassium ion in blood—above 5.0 meq/L. (12.2)

Hypernatremia An elevated level of sodium ion in blood—above 145 meq/L. (12.2)

Hypocapnia A condition of a below-normal concentration of carbon dioxide in the blood as indicated by a partial pressure of CO_2 in venous blood of less than 35 mm Hg. (12.4)

Hypoglycemia A low level of glucose in blood—below 65 mg/dL of whole blood. (15.1)

Hypokalemia A low level of potassium ion in blood—below 3.5 meq/L. (12.2)

Hyponatremia A low level of sodium in blood—below 135 meq/L. (12.2)

Hypoxia A condition of a low supply of oxygen. (12.4)

Induced Fit Model Many enzymes are induced by their substrate molecules to modify their shapes to accommodate the substrate. (11.2)

Inducer A substance whose molecules remove repressor molecules from operator genes and so open the way for structural genes to direct the overall syntheses of particular polypeptides. (13.4)

Inhibitor A substance that interacts with an enzyme to prevent its acting as a catalyst. (11.4)

Inorganic Compound Any compound that is not an organic compound. (1.1)

Insulin A protein hormone made by the pancreas, released in response to an increase in the blood sugar level, and used by certain tissues to help them take up glucose from circulation. (15.1)

Internal Environment Everything enclosed within an organism. (15.1)

International Union of Pure and Applied Chemistry System (IUPAC System) A set of systematic rules for naming compounds and designed to give each compound one unique name and for which only one structure can be drawn. (1.6)

Interstitial Fluids Fluids in tissues but not inside cells or in the blood. (12.1)

Intestinal Juice The digestive juice that empties into the duodenum from the intestinal mucosa and whose enzymes also work within the intestinal mucosa as molecules migrate through. (12.1)

Intracellular Fluids Fluids inside cells. (12.1)

Intron A segment of a DNA strand that separates exons and that does not become expressed as a segment of a polypeptide. (13.2)

Invert Sugar A 1 : 1 mixture of glucose and fructose. (8.5)

In Vitro Occurring in laboratory vessels. (3.3)

In Vivo Occurring within a living system. (3.3)

Iodine Test A test for starch by which a drop of iodine reagent produces an intensely purple color if starch is present. (8.6)

Isobutyl Group $(CH_3)_2CHCH_2$— (1.6)

Isoelectric Molecule A molecule which has an equal number of positive and negative sites. (10.1)

Isoelectric Point (pI) The pH of a solution in which a specified amino acid or a protein is in an isoelectric condition; the pH at which there is no net migration of the amino acid or protein in an electric field. (10.1)

Isoenzymes Enzymes that have identical catalytic functions but which are made of slightly different polypeptides. (11.1)

Isohydric Shift In actively metabolizing tissue, the use of a hydrogen ion released from newly formed carbonic acid to react with and liberate oxygen from oxyhemoglobin; in the lungs, the use of hydrogen ion, released when hemoglobin oxygenates, to combine with bicarbonate ion and liberate carbon dioxide for exhaling. (12.3)

Isomerase An enzyme that catalyzes the conversion of a compound into one of its isomers. (11.1)

Isomerism The phenomenon of the existence of two or more compounds with identical molecular formulas but different structures. (1.3)

Isomers Compounds with identical molecular formulas but different structures. (1.3)

Isopropyl Group $(CH_3)_2CH$— (1.6)

Isozyme (See *Isoenzymes*)

IUPAC System (See *International Union of Pure and Applied Chemistry System*)

Keratin The fibrous protein of hair, fur, fingernails, and hooves. (10.9)

Ketal A substance whose molecules have two OR groups joined to a carbon that also holds two hydrocarbon groups. (4.5)

Ketoacidosis The acidosis caused by untreated ketonemia. (16.6)

Keto Group The carbonyl group when it is joined on each side to carbon atoms. (4.1)

Ketohexose A monosaccharide whose molecules contain six carbon atoms and have a keto group. (8.2)

Ketone Any compound with a carbonyl group attached to two carbon atoms, as in $R_2C{=}O$. (4.1)

Ketone Bodies Acetoacetate, β-hydroxybutyrate—or their parent acids—and acetone. (16.6)

Ketonemia An elevated concentration of ketone bodies in the blood. (16.6)

Ketonuria An elevated concentration of ketone bodies in the urine. (16.6)

Ketose A monosaccharide whose molecules have a ketone group. (8.2)

Ketosis The combination of ketonemia, ketonuria, and acetone breath. (16.6)

Kinase An enzyme that catalyzes the transfer of a phosphate group. (11.1)

Kreb's Cycle (See *Citric Acid Cycle*)

Levorotatory The property of an optically active substance that causes a counterclockwise rotation of the plane of plane-polarized light. (7.3)

Ligand An electron-rich species, either negatively charged or electrically neutral, that binds with a metal ion to form a complex ion. (4.3)

Ligase An enzyme that catalyzes the formation of bonds at the expense of triphosphate energy. (11.1)

Like-Dissolves-Like Rule Polar solvents dissolve polar or ionic solutes and nonpolar solvents dissolve nonpolar or weakly polar solutes. (1.5)

Limiting Amino Acid The essential amino acid most poorly provided by a dietary protein. (18.2)

Lipid A plant or animal product that tends to dissolve in such nonpolar solvents as ether, carbon tetrachloride, and benzene. (9.1)

Lipid Bilayer The sheet-like array of two layers of lipid molecules, interspersed with molecules of cholesterol and proteins, that make up the membranes of cells in animals. (9.5)

Lipoprotein Complex A combination of lipid and protein molecules that serves as the vehicle for carrying the lipid in the bloodstream. (16.1)

Lock-And-Key Theory The specificity of an enzyme for its substrate is caused by the need for the substrate molecule to fit to the enzyme's surface much as a key fits to and turns only one tumbler lock. (11.2)

Lyase An enzyme that catalyzes an elimination reaction to form a double bond. (11.1)

Markovnikov's Rule In the addition of an unsymmetrical reactant to an unsymmetrical double bond of a simple alkene, the positive part of the reactant molecule (usually H^+) goes to the carbon that has the greater number of hydrogen atoms and the negative part goes to the other carbon of the double bond. (2.4)

Mercaptan A thioalcohol; R—S—H. (3.6)

Meso Compound One of a set of optical isomers whose own molecules are not chiral and which, therefore, is optically inactive. (Special Topic 7.1)

Messenger RNA (mRNA) RNA that carries the genetic code, in the form of a specific series of codons for a specific polypeptide, from the cell's nucleus to the cytoplasm. (13.3)

Methyl Group CH_3— (1.6)

Micelle A globular arrangement of the molecules of an amphipathic compound in water in which their hydrophobic parts intermingle inside the globule and their hydrophilic parts are exposed to the water. (9.5)

Minerals, Dietary Ions that must be provided in the diet at levels of 100 mg/day or more; Ca^{2+}, Mg^{2+}, Na^+, K^+, Cl^-, and phosphate. (18.4)

Mitochondrion A unit inside a plant or animal cell in which the machinery for making high-energy phosphates by oxidative phosphorylation is located. (13.1)

Monoamine Oxidase An enzyme that catalyzes the inactivation of neurotransmitters or other amino compounds of the nervous system. (11.7)

Monohydric Alcohol An alcohol whose molecules have one OH group. (3.1)

Monomer A compound that can be used to make a polymer. (2.6)

Monosaccharide A carbohydrate that cannot be hydrolyzed. (8.2)

Mucin A viscous glycoprotein released in the mouth and the stomach that coats and lubricates food particles and protects the stomach from the acid and pepsin of gastric juice. (12.1)

Mutarotation The gradual change in the specific rotation of a substance in solution but without a permanent, irreversible chemical change occurring. (8.4)

Myosins Proteins in contractile muscle. (10.9)

Native Protein A protein whose molecules are in the configuration and shape they normally have within a living system. (10.2)

Neurotransmitter A substance released by one nerve cell to carry a signal to the next nerve cell. (11.6)

Niacin A water-soluble vitamin needed to prevent pellagra and essential to the coenzymes in NAD^+ and $NADP^+$; nicotinic acid or nicotinamide. (18.3)

Nitrogen Balance A condition of the body in which it excretes as much nitrogen as it receives in the diet. (18.2)

Nitrogen Pool The sum total of all nitrogen compounds in the body. (17.1)

Nomenclature The system of names and the rules for devising such names, given structures, or for writing structures, given names. (1.6)

Nonfunctional Group A section of an organic molecule that remains unchanged during a chemical reaction at a functional group. (1.4)

Nonhydrolyzable Lipid Any lipid, such as the steroids, that cannot be hydrolyzed or similarly broken down by aqueous alkali. (9.1)

Normal Fasting Level The normal concentration of something in the blood, such as sugar, after about four hours without food. (15.1)

Nucleic Acid A polymer of nucleotides in which the repeating units are pentose phosphate diesters, each pentose unit bearing a sidechain base (one of five heterocyclic amines); polymeric compounds that are involved in the storage, transmission, and expression of genetic messages. (13.2)

Nucleotide A monomer of a nucleic acid that consists of a pentose phosphate ester in which the pentose unit carries one of five heterocyclic amines as a sidechain base. (13.2)

Nucleus In biology, the organelle in a cell that houses DNA. (13.1)

Nutrient Any one of a large number of substances in food and drink that is needed to sustain growth and health. (18.1)

Nutrition The science of the susbtances of the diet that are necessary for growth, operation, energy, and repair of body tissues. (18.1)

Olefin An alkene. (2.6)

One-Substance–One-Structure Rule If two samples of matter have identical physical and chemical properties, they have identical molecules. (7.1)

Optical Isomer One of a set of compounds whose molecules differ only in their chiralities. (7.3)

Optically Active The ability of a substance to rotate the plane of polarization of plane-polarized light. (7.3)

Optical Rotation The degrees of rotation of the plane of plane-polarized light caused by an optically active solution; the observed rotation of such a solution. (7.3)

Organic Chemistry The chemistry of carbon compounds. (1.1)

Organic Compounds Compounds of carbon other than those related to carbonic acid and its salts, or to the oxides of carbon, or to the cyanides. (1.1)

Oxidase (See *Oxidoreductase*)

Oxidative Deamination The change of an amino group to a keto group with loss of nitrogen. (17.2)

Oxidative Phosphorylation The synthesis of high-energy phosphates such as ATP from lower energy phosphates and inorganic phosphate by the reactions that involve the respiratory chain. (14.1, 14.3)

Oxidoreductase An enzyme that catalyzes the formation of an oxidation–reduction equilibrium. (11.1)

Oxygen Affinity The percentage to which all of the hemoglobin molecules in the blood are saturated with oxygen molecules. (12.3)

Oxygen Debt The condition in a tissue when anaerobic glycolysis has operated and lactate has been excessively produced. (15.3)

Oxyhemoglobin Hemoglobin carrying its capacity of oxygen. (12.3)

P_i Inorganic phosphate ion(s) of whatever mix of PO_4^{3-}, HPO_4^{2-}, $H_2PO_4^-$, and even traces of H_3PO_4 that is possible at the particular pH of the system, but almost entirely HPO_4^{2-} + $H_2PO_4^-$. (5.6)

Pancreatic Juice The digestive juice that empties into the duodenum from the pancreas. (12.1)

Pantothenic Acid A water-soluble vitamin needed to make coenzyme A. (18.3)

Pentose Phosphate Pathway The synthesis of NADPH that uses chemical energy in glucose-6-phosphate and that involves pentoses as intermediates. (15.3)

Peptide Bond The amide linkage in a protein; a carbonyl–nitrogen bond. (10.1, 10.3)

Phenol Any organic compound whose molecules have an OH group attached to a benzene ring. (3.4)

Phenyl Group The benzene ring minus one H atom; C_6H_5. (2.7)

Phosphate Group Transfer Potential The relative ability of an organophosphate to transfer a phosphate group to some acceptor. (14.1)

Phosphatide A glycerophospholipid whose molecules are esters between glycerol, two fatty acids, phosphoric acid, and a small alcohol. (9.3)

Phosphoglyceride (See *Glycerophospholipid*)

Phospholipid Lipids such as the glycerophospholipids (phosphatides and plasminogens) and the sphingomyelins whose molecules include phosphate ester units. (9.3)

Photosynthesis The synthesis in plants of complex compounds from carbon dioxide, water, and minerals with the aid of sunlight captured by the plant's green pigment, chlorophyll. (8.2)

pI (See *Isoelectric Point*)

Plane-Polarized Light Light whose electric field vibrations are all in the same plane. (7.3)

Plasmalogens Glycerophospholipids whose molecules include an unsaturated fatty alcohol unit. (9.3)

Plasmid A circular molecule of supercoiled DNA in a bacterial cell. (9.6)

β-Pleated Sheet A secondary structure for a polypeptide in which the molecules are aligned side by side in a sheet-like array with the sheet partially pleated. (10.4)

Poison A substance that reacts in some way in the body to cause changes in metabolism that threaten health or life. (11.4)

Polarimeter An instrument for detecting and measuring optical activity. (7.3)

Polymer Any substance with a very high formula mass whose molecules have a repeating structural unit. (2.6)

Polymerization A chemical reaction that makes a polymer from a monomer. (2.6)

Polypeptide A polymer with repeating α-aminoacyl units joined by peptide (amide) bonds. (10.1)

Polysaccharide A carbohydrate whose molecules are polymers of monosaccharides. (8.2)

PP_i Inorganic diphosphate ion(s). (14.1)

Primary Alcohol An alcohol in whose molecules an OH group is attached to a primary carbon, as in RCH_2OH. (3.1)

Primary Carbon In a molecule, a carbon atom that is joined directly to just one other carbon, such as the end carbons in $CH_3CH_2CH_3$. (1.6)

Primary Structure The sequence of amino acyl residues held together by peptide bonds in a polypeptide. (10.2)

Primary Transcript RNA (ptRNA) [See *Heterogeneous Nuclear RNA (hnRNA)*]

Proenzyme (See *Zymogen*)

Propyl Group $CH_3CH_2CH_2$— (1.6)

Prosthetic Group A nonprotein molecule joined to a polypeptide to make a biologically active protein. (10.5)

Protein A naturally occurring polymeric substance made up wholly or mostly of polypeptide molecules. (10.1)

Proton-Pumping ATPase The enzyme on the matrix side of the inner mitochondrial membrane that catalyzes the formation of ATP from ADP and P_i under the influence of a flow of protons across this membrane. (14.3)

Quaternary Structure An aggregation of two or more polypeptide strands each with its own primary, secondary, and tertiary structure. (10.2, 10.6)

Racemic Mixture A 1:1 mixture of enantiomers and therefore optically inactive. (7.3)

Radiomimetic Substance A substance whose chemical effect in a cell mimics the effect of ionizing radiation. (13.4)

Receptor Molecule A molecule of a protein built into a cell membrane that can accept a molecule of a hormone or a neurotransmitter. (10.8, 11.6)

Recombinant DNA DNA made by combining the natural DNA of plasmids in bacteria or the natural DNA in yeasts with DNA from external sources, such as the DNA for human insulin, and made as a step in a process that uses altered bacteria or yeasts to make specific proteins (e.g., interferons, human growth hormone, and insulin). (13.6)

Recommended Dietary Allowance (RDA) The level of intake of a particular nutrient as determined by the Food and Nutrition Board of the National Research Council of the National Academy of Sciences to meet the known nutritional needs of most healthy individuals. (18.1)

Reducing Carbohydrate A carbohydrate that gives a positive Benedict's test. (8.2)

Reductase An enzyme that catalyzes a reduction. (See *Oxidoreductase*) (11.1)

Reductive Amination The conversion of a keto group to an amino group by the action of ammonia and a reducing agent. (17.1)

Renal Threshold That concentration of a substance in blood above which it appears in the urine. (15.1)

Replication The reproductive duplication of a DNA double helix. (13.1)

Repressor A substance whose molecules can bind to a gene and prevent the gene from directing the synthesis of a polypeptide. (13.4)

Respiration The intake and chemical use of oxygen by the body and the release of carbon dioxide. (12.3)

Respiratory Chain The reactions that transfer electrons from the intermediates made by other pathways to oxygen; the mechanism that creates a proton gradient across the inner membrane of a mitochondrion and that leads to ATP synthesis; the enzymes that handle these reactions. (14.1, 14.2)

Respiratory Enzymes The enzymes of the respiratory chain. (14.3)

Respiratory Gases Oxygen and carbon dioxide. (12.3)

Riboflavin A B vitamin needed to give protection against the breakdown of tissue around the mouth, the nose, and the tongue, as well as to aid in wound healing. (18.3)

Ribonucleic Acids (RNA) Polymers of nucleotides made using ribose that participate in the transcription and the translation of the genetic messages into polypeptides. (See also *Heterogeneous Nuclear RNA, Messenger RNA, Ribosomal RNA, and Transfer RNA*) (13.2)

Ribosomal RNA (rRNA) RNA that is incorporated into cytoplasmic bodies called ribosomes. (13.3)

Ribosome A granular complex of rRNA that becomes attached to a mRNA strand and that supplies some of the enzymes for mRNA-directed polypeptide synthesis. (13.1)

Ribozyme An enzyme whose molecules consist of ribonucleic acid, not polypeptide. (13.1)

Ring Compound A compound whose molecules consist three or more atoms joined in a ring. (1.2)

RNA (See *Ribonucleic Acid*)

Saliva The digestive juice secreted in the mouth whose enzyme, amylase, catalyzes the partial digestion of starch. (12.1)

Salt Bridge A force of attraction between (+) and (−) sites on polypeptide molecules. (10.5)

Saponifiable Lipid (See *Hydrolyzable Lipid*)

Saponification The reaction of an ester with base to give an alcohol and the salt of an acid. (5.5)

Saturated Compound A compound whose molecules have only single bonds. (1.2)

Secondary Alcohol An alcohol in whose molecules an OH group is attached to a secondary carbon atom; R_2CHOH. (3.1)

Secondary Carbon Any carbon atom in an organic molecule that has two and only two bonds to other carbon atoms, such as the middle carbon atom in $CH_3CH_2CH_3$. (1.6)

Secondary Structure A shape, such as the α-helix or a unit in a β-pleated sheet, that all or a large part of a polypeptide molecule adopts under the influence of hydrogen bonds, salt bridges, and hydrophobic interactions after its peptide bonds have been made. (10.2)

Shock, Traumatic A medical emergency in which relatively large volumes of blood fluid leave the vascular compartment and enter the interstitial spaces. (12.2)

Simple Lipid (See *Triacylglycerol*)

Simple Sugar Any monosaccharide. (8.2)

Soap A detergent that consists of salts of long-chain fatty acids. (Special Topic 9.3)

Somatostatin A hormone of the hypothalamus that inhibits or slows the release of glucagon and insulin from the pancreas. (15.1)

Specific Rotation $[\alpha]$ The optical rotation of a solution per unit of concentration per unit of path length in decimeters:

$$[\alpha] = \frac{\alpha}{cl}$$

where α = observed rotation; c = concentration in g/mL, and l = path length (in dm). (7.3)

Sphingolipid A lipid that, when hydrolyzed, gives sphingosine instead of glycerol, plus fatty acids, phosphoric acid, and a small alcohol or a monosaccharide; sphingomyelins and cerebrosides. (9.3)

Stereoisomers Isomers whose molecules have the same atom-to-atom sequences but different geometric arrangements; a geometric (cis–trans) or optical isomer. (7.1)

Stereoisomerism The existence of stereoisomers. (7.1)

Steroids Nonhydrolyzable lipids such as cholesterol and several sex hormones whose molecules have the four fused rings of the steroid nucleus. (9.4)

Straight Chain A continuous, open sequence of covalently bound carbon atoms from which no additional carbon atoms are attached at interior locations of the sequence. (1.2)

Structural Formula A formula that uses lines representing covalent bonds to connect the atomic symbols in the pattern that occurs in one molecule of a compound. (1.2)

Structural Isomer One of a set of isomers whose molecules differ in their atom-to-atom sequence. (1.3)

Structure Synonym for structural formula. (See *Structural Formula*) (1.2)

Substitution Reaction A reaction in which one atom or group replaces another atom or group in a molecule. (1.7)

Substrate The substance on which at enzyme performs its catalytic work. (7.2)

Substrate Phosphorylation The direct transfer of a phosphate unit from an organophosphate to a receptor molecule. (14.1)

Superimposition A chirality testing operation to see if one molecular model can be made to blend simultaneously at exactly every point with another model. (7.2)

Synapse The fluid-filled gap between the end of the axon of one nerve cell and the next nerve cell. (11.7)

Target Cell A cell at which a hormone molecule finds a site where it can become attached and then cause some action that is associated with the hormone. (11.6)

Target Tissue The organ whose cells are recognizable by the molecules of a particular hormone. (11.6)

Tertiary Alcohol An alcohol in whose molecules an OH group is held by a carbon from which three bonds extend to other carbon atoms; R_3COH. (3.1)

Tertiary Carbon A carbon in an organic molecule that has three and only three bonds to adjacent *carbon* atoms. (1.6)

Tertiary Structure The shape of a polypeptide molecule that arises from further folding or coiling of secondary structures. (10.2, 10.5)

Tetrahedral Descriptive of the geometry of bonds at a central atom in which the bonds project to the corners of a regular tetrahedron. (12.2)

Tetrahedral Stereocenter An atom in a molecule with four single bonds arranged tetrahedrally and holding four different atoms or groups. (7.2)

Thiamine A B vitamin needed to prevent beriberi. (18.3)

Thioalcohol A compound whose molecules have the SH group attached to a saturated carbon atom; a mercaptan. (3.6)

Tollens' Reagent A slightly alkaline solution of the diammine complex of the silver ion, $Ag(NH_3)_2^+$, in water. (4.3)

Tollens' Test The use of Tollens' reagent to detect an easily oxidized group such as the aldehyde group. (4.3)

Trace Element, Dietary Any element that the body needs each day in an amount of no more than 20 mg. (18.4)

Transamination The transfer of an amino group from an amino acid to a receiver with a keto group such that the keto group changes to an amino group. (17.1)

Transcription The synthesis of messenger RNA under the direction of DNA. (13.3)

Transferase An enzyme that catalyzes the transfer of some group. (11.1)

Transfer RNA (tRNA) RNA that serves to carry an amino acyl group to a specific acceptor site of an mRNA molecule at a ribosome where the amino acyl group is placed into a growing polypeptide chain. (13.3)

Translation The synthesis of a polypeptide under the direction of messenger RNA. (13.4)

Triacylglycerol A lipid that can be hydrolyzed to glycerol and fatty acids; a triglyceride; sometimes, simply called a glyceride or a simple lipid. (9.1)

Tricarboxylic Acid Cycle (See *Citric Acid Cycle*)

Triglyceride (See *Triacylglycerol*)

Trihydric Alcohol An alcohol with three OH groups per molecule. (3.1)

Triple Helix The quaternary structure of tropocollagen in which three polypeptide chains are coiled together. (10.6)

Unsaturated Compound Any compound whose molecules have a double or a triple bond. (1.2)

Urea Cycle The reactions by which urea is made from amino acids. (17.3)

Vascular Compartment The entire network of blood vessels and their contents. (12.2)

Vasopressin A hypophysis hormone that acts at the kidneys to help regulate the concentrations of solutes in the blood by instructing the kidneys to retain water (if the blood is too concentrated) or to excrete water (if the blood is too dilute). (12.5)

Virus One of a large number of substances that consist of nucleic acid surrounded by a protein overcoat and that can enter host cells, multiply, and destroy the host. (13.5)

Vital Force Theory A discarded theory that organic compounds could be made in the laboratory only if the chemicals possessed a vital force contributed by some living thing. (1.1)

Vitamin An organic substance that must be in the diet, whose absence causes a deficiency disease, which is present in foods in trace concentrations, and that isn't a carbohydrate, lipid, protein, or amino acid. (18.3)

Vitamin A Retinol; a fat-soluble vitamin in yellow-colored foods and needed to prevent night blindness and certain conditions of the mucous membranes. (18.3)

Vitamin B_6 Pyridoxine, pyridoxal, or pyridoxamine; a vitamin needed to prevent hypochromic microcytic anemia and used in enzymes of amino acid catabolism. (18.3)

Vitamin B_{12} Cobalamin; a vitamin needed to prevent pernicious anemia. (18.3)

Vitamin C Ascorbic acid; a vitamin needed to prevent scurvy. (18.3)

Vitamin D Cholecalciferol (D_3) or ergocalciferol (D_2); a fat-soluble vitamin needed to prevent rickets and to ensure the formation of healthy bones and teeth. (18.3)

Vitamin Deficiency Diseases Diseases caused not by bacteria or viruses but by the absence of specific vitamins, such as pernicious anemia (B_{12}), hypochromic microcytic anemia (B_6), pellagra (niacin), the breakdown of certain tissues (riboflavin), megaloblastic anemia (folate), beriberi (thiamin), scurvy (C), hemorrhagic disease (K), rickets (D), and night blindness (A). (18.3)

Vitamin E A mixture of tocopherols; a fat-soluble vitamin apparently needed for protection against edema and anemia (in infants) and possibly against dystrophy, paralysis, and heart attacks. (18.3)

Vitamin K The antihemorrhagic vitamin that serves as a cofactor in the formation of a blood clot. (18.3)

Wax A lipid whose molecules are esters of long-chain monohydric alcohols and long-chain fatty acids. (9.1)

Zwitterion (See *Dipolar Ion*)

Zymogen A polypeptide that is changed into an enzyme by the loss of a few amino acid residues or by some other change in its structure; a proenzyme. (11.4)

PHOTO CREDITS

Chapter 1
Opener: Brian Stablyk/AllStock, Inc. *Pages 4, 8, and 10:* Tripos Associates. *Special Topic 1.1, Figure 1:* Kent & Donna Dannen. *Special Topic 1.1, Figure 2:* Bob Anderson/Masterfile. *Pages 23, 24, and 28:* Tripos Associates. *Special Topic 1.2, Figure 1:* A. Bradshaw/Sipa Press.

Chapter 2
Opener: C. Falkenstein/Mountain Stock Photography & Film, Inc. *Pages 43 and 44:* Tripos Associates. *Figures 2.3, 2.4, and 2.5:* Andy Washnik. *Special Topic 2.3:* Jim Mendenhall. *Figure 2.6:* Russ Schliepman/Medichrome. Figure 2.7: Photo Researchers.

Chapter 3
Opener: Tom Benoit/AllStock, Inc.

Chapter 4
Opener: Trevor Mein/Tony Stone Images. *Figure 4.1:* Andy Washnik. *Figure 4.2:* Michael Watson. *Page 112* (left): Michael Watson.

Chapter 5
Opener: Johnny Johnson/AllStock, Inc.*Special Topic 5.1, Figure 1:* Andy Washnik. *Special Topic 5.2, Figure 2:* Harry J. Przekop, Jr./Medichrome.

Chapter 6
Opener: Dan Smith/Tony Stone Images. *Special Topic 6.2:* Lafoto/AllStock, Inc.

Chapter 7
Opener: Richard J. Green/Photo Researchers. *Page 183:* Robert J. Capece. *Page 184:* Michael Watson. *Figure 7.7:* Courtesy of Polaroid Corporate Archives.

Chapter 8
Opener: Kathleen Hanzel/AllStock, Inc. *Page 197:* Paul Barton/The Stock Market. *Page 216:* Andy Washnik.

Chapter 9
Opener: Comstock, Inc. *Pages 224, 225, and 228:* Tripos Associates, *Pages 232, 233, and 234:* Courtesy of Richard Pastor, FDA.

Chapter 10
Opener: Irving Geiss/Peter Arnold. *Special Topic 10.1:* Bill Longcore/Photo Researchers.

Chapter 11
Opener: Darrell Jones/AllStock, Inc. *Page 288:* Courtesy Boehringer Mannheim Diagnostics.

Chapter 12
Opener: Galen Rowell/Peter Arnold.

Chapter 13
Opener: Werner H. Muller/Peter Arnold. *Special Topic 13.1, Figure 1:* Courtesy Lifecodes Corporation.

Chapter 14
Opener: Duomo. *Figure 14.3(a):* Courtesy of Dr. Keith R. Porter. *Figure 14.6(a):* From Parsons, D.F., *Science* 140, 985 (1973).

Chapter 15
Opener: Bruce Wilson/AllStock, Inc.

Chapter 16
Opener: Tim Davis/AllStock, Inc. *Special Topic 16.1, Figure 1:* W. Ober/Visuals Unlimited.

Chapter 17
Opener: FPG International.

Chapter 18
Opener: Barry L. Runk/Grant Heilman Photography.

INDEX

age numbers in italics refer to tables.

RELATIONSHIP OF UNITS
(Values in boldface are exact.)

Length

1 in. = **2.54** cm
1 ft. = **30.48** cm
1 yd. = **91.44** cm

Volume

1 liq oz = **29.57353** mL
1 liq qt = **946.352946** mL
1 gallon = **3.785411784** L

Mass

1 oz = **28.349523125** g
1 lb = **453.59237** g

Pressure

$$1 \text{ mm Hg} = 1 \text{ torr} = \frac{\mathbf{1}}{\mathbf{760}} \text{ atm}$$

Energy

1 cal = **4.184** joule

STRONG AQUEOUS ACIDS

Hydrochloric acid, HCl
Hydrobromic acid, HBr
Hydriodic acid, HI
Nitric acid, HNO_3
Sulfuric acid, H_2SO_4

STRONG, HIGHLY SOLUBLE AQUEOUS BASES

Sodium hydroxide, NaOH
Potassium hydroxide, KOH

STRONG, SLIGHTLY SOLUBLE AQUEOUS BASES

Calcium hydroxide, $Ca(OH)_2$
Magnesium hydroxide, $Mg(OH)_2$

PHYSICAL CONSTANTS

Atomic mass unit (u) = $1.66056520 \times 10^{-24}$ g
Avogadro's number = 6.0221367×10^{23}

Gas constant, R $= 6.24 \times 10^{4} \dfrac{\text{mm Hg mL}}{\text{mol K}}$

$= 0.0821 \dfrac{\text{L atm}}{\text{mol K}}$

Molar volume = 22.41383 L/mol (ideal gas, at 273.15 K and 760 mm Hg)

ABBREVIATIONS OF UNITS

atm	atmosphere
°C	degree Cels
cal	calories
cc	cubic centir
cm	centimeter
dL	deciliter
eq	equivalent
°F	degree Fah
ft	foot
g	gram
gal	gallon
in.	inch
J	joule
K	kelvin
kcal	kilocalorie
kg	kilogram
km	kilometer
L	liter
lb	pound
m	meter
M	mol/L (mola
meq	milliequivale
mg	milligram
μg	microgram
mi	mile
mL	milliliter
μL	microliter
mm	millimeter
μm	micrometer
mm Hg	millimeter of mercury
mmol	millimole
mol	mole
mOs	milliosmole
Os	osmole
oz	ounce
ppb	parts per billion
ppm	parts per million
pt	pint
qt	quart
s	second
T	Kelvin temperature
t_c	Celsius temperature
t_f	Fahrenheit temperature
u	atomic mass unit
yd	yard